STUDENT WOR
AND
SELECTED SOLUTIONS

Saundra Yancy McGuire
Louisiana State University

Elzbieta Cook
Louisiana State University

INTRODUCTORY

Chemistry

ATOMS FIRST

Fifth Edition

Steve Russo • Mike Silver

PEARSON

Boston   Columbus   Indianapolis   New York   San Francisco   Upper Saddle River
Amsterdam   Cape Town   Dubai   London   Madrid   Milan   Munich   Paris   Montréal   Toronto
Delhi   Mexico City   São Paulo   Sydney   Hong Kong   Seoul   Singapore   Taipei   Tokyo

Editor in Chief: Adam Jaworski
Acquisitions Editor: Chris Hess
Executive Marketing Manager: Jonathan Cottrell
Project Manager: Lisa R. Pierce
Associate Team Lead, Program Management, Chemistry and Geosciences: Jessica Moro
Team Lead, Project Management, Chemistry and Geosciences: Gina M. Cheselka
Full-Service Project Management/Composition: PreMediaGlobal
Operations Specialist: Christy Hall
Supplement Cover Designer: Seventeenth Street Studios
Cover Image Credit: Vera Kuttelvaserova/Fotolia

PEARSON

www.pearsonhighered.com

2 3 4 5 6 7 8 V069 19 18 17 16 15

ISBN-10: 0-321-95693-1; ISBN-13: 978-0-321-95693-4

Contents

To the Student

Your decision to purchase this *Student Workbook and Selected Solutions* manual is an indication of your desire and intent to succeed in your study of chemistry. Hundreds of thousands of students before you have achieved success in chemistry, and you can too. We have combined the *Student Workbook and Selected Solutions* into one book for your convenience. However, you should treat them as if they were two separate books. You should refer to the workbook often, using it to preview material before lecture, to get an overview of the chapter, to practice problem-solving techniques, and to test your understanding of the concepts presented in each chapter. The selected solutions section, however, should be used very sparingly. You should *always* attempt a problem yourself before looking at the solution; this is the only way to develop problem-solving skills.

Each component of the *Student Workbook and Selected Solutions* manual has a specific objective.

- The *Learning Objectives* list what you should be able to do after you have mastered the content of the chapter. These objectives can be reviewed before you begin your study of a new chapter to give you a preview of the chapter content. In addition, they can be used to check your mastery of the chapter material when you are studying for examinations.

- The *Chapter Outline* shows the topics included in each chapter. It can serve as a general guide for taking notes during the course lectures.

- The *Review of Key Concepts from the Text* provides a summary of material included in each section of the textbook chapter, and includes examples and practice problems. The review is intended to help you gain additional insight into the chapter content and provide you with examples and problems that allow you to use the problem-solving techniques presented in the chapter.

- The *Exercises for Self-Testing* are designed to help you test your mastery of the content of the chapter. You should complete these exercises after you have studied the entire chapter in depth. Completion and True-False items test your knowledge of the fundamental concepts. The questions and problems test your ability to explain phenomena described in the chapter and your ability to apply the problem-solving techniques presented. Answers to Exercises are provided at the end of the workbook so that you will know which areas you have mastered and which areas need additional work.

- The *Selected Solutions* section provides comprehensive answers and detailed solutions to selected problems in the textbook. Although most problems can be solved by more than one method, the solutions given generally use the same method used in the textbook. You should not look at the solution to a problem until you have spent time and effort attempting to work the problem in its entirety by yourself. Problem-solving skills can only be developed by earnestly attempting to solve problems that you initially thought you could not solve. This skill will come in very handy on tests. The solutions section should always be used to check your solution to a problem; never to provide you with a solution to problems you have not attempted yourself.

Study Strategies for Mastering Chemistry

In order to achieve a high level of success, you will need to implement a variety of learning strategies in your study of chemistry. Some of these are listed below.

- Preread the material that will be covered in lecture before the lecture. Use the syllabus provided by your instructor to determine the appropriate material to preread. This brief preview of the material will help tremendously in your understanding of the lecture material.

- Be an active participant in lecture. Sit near the front of your class where you will be able to focus on the lecturer, avoid distractions, and see everything clearly. Try to form mental pictures of the phenomena being discussed because visualizing the material will help you to retain more of the concepts.

- Take good notes, and review and rework them as soon after the lecture as possible. Develop a note-taking system that works for you. (Several different types are discussed in books on developing study strategies.) Research studies have shown that reviewing lecture notes within a few hours after a lecture considerably increases retention and understanding. Note areas where you are unsure and seek assistance from your instructor.

- Form a study group with classmates. Actively discussing the material with fellow students will help you to understand concepts better, and will help you remember the material.

- Visit your instructor's office hours regularly. Discussing concepts with your instructor will help you clarify concepts you find confusing, and will help you get to know a faculty member better. This may come in quite handy when you are seeking letters of recommendation.

- Rework all missed items on homework, quizzes, and previous tests as soon as possible after they have been returned to you. This will help you to clear up any misconceptions you may have developed and will help you to master material that you had not mastered before.

- Use tutorial and other resources available through the Learning Center on your campus. Workshops on time management, listening and note taking, test preparation, test taking, and other topics will be quite helpful as you develop the learning strategies that will enable you to achieve academic excellence.

If you use these strategies consistently you will find that your studying will be more efficient, and you will achieve success in your study of chemistry.

The *Student Workbook and Selected Solutions* manual will aid you in your study of chemistry if you use it wisely, and not as a crutch or as a substitute for working through the material yourself. If used correctly, this book will help you to master chemical concepts and will help you to enjoy your study of chemistry. We welcome comments or suggestions for improvement, and we wish you the best of success in your study of chemistry. Below are a few questions (and answers) to guide you in your study of chemistry.

What does acing chemistry require? Nothing that you can't do!

- *Mastering the concepts* (not just memorizing facts, definitions, and formulas).
- *Spending at least 30 minutes studying chemistry* (outside of class) 3–5 times per week.
- *Effectively using resources* (your instructor's office hours, tutoring, the *Student Workbook and Selected Solutions*, etc.).
- *Developing a systematic plan for problem solving* and using the right mathematical procedures to arrive at the right answer.

How will this book *guarantee* your success in chemistry? Glad you asked!

- It provides a chart with *step-by-step procedures* for solving all of the types of problems in each chapter.
- It provides additional examples, *practice problems, practice quizzes,* and cumulative quizzes after every three chapters.
- It provides *answers to the practice problems and quizzes* so that you will know immediately whether you have done the problem correctly.

What do you need to do? Get started right now by doing the following things!

- *Commit to yourself (and recommit each week) to use this book* to learn how to do every type of problem your instructor covers in class.
- *Study each chart carefully*, making sure you understand why each step is done.
- *Develop your own chart* if you want to do the problem differently (there are lots of different ways to do chemistry problems; this book presents just one of them).
- *Resist the temptation to look at the answers* before you have really, really tried to solve the problem yourself! (This is the most important thing you can do to ace your chemistry course).

Are there other keys to acing chemistry? You bet!

- *Preview the material* (for about 10 minutes) that will be covered in class, before you go to class.
- Consistently attend class.
- *Review your notes* and use this problem-solving guide to review problems as soon after class as possible.
- *Don't peek at the solutions* while you're doing your homework!

   How can we guarantee this will work? We've seen it work for thousands of other students! So get started right now, and you will probably find that you are enjoying studying chemistry. Imagine that!

Acknowledgments

We would like to express our heartfelt gratitude to Steve Russo and Mike Silver, authors of the textbook, for including us in this project. We are indebted to the many individuals who have worked tirelessly to ensure the utility, accuracy, and clarity of the information presented in the fifth edition of this workbook. They include Lisa Pierce, Project Manager at Pearson, accuracy checkers Christopher G. Hamaker and Vanessa Castleberry, and Project Managers at PreMediaGlobal Revathi Viswanathan and Jenna Gray. Specific thanks go to Editor-in-Chief Adam Jaworski, Senior Chemistry Acquisitions Editor Terry Haugen, and Acquisitions Editor Chris Hess, for their continuous support.

With regard to the previous editions, our most sincere thanks go to Melissa Bailey Crawford, who carefully checked most of the solutions in the fourth edition and to the editorial team, in particular Associate Team Lead, Jessica Moro. Last, but certainly not least, we want to thank our husbands, Dr. Steve McGuire and Dr. Robert L. Cook, for their patience, love, and support during the many long hours we spent working on this project.

Saundra Y. McGuire, PhD
Professor of Chemistry
Director Emerita, Center for Academic Success
Louisiana State University
Baton Rouge, LA 70803

Elzbieta (Elizabeth) Cook, PhD
Instructor
Department of Chemistry
Louisiana State University
Baton Rouge, LA 70803

Additional resources available for instructors:

Instructor Teaching Guide and Complete Solutions (Download Only) for *Introductory Chemistry: Atoms First*, 5th edition (0321-97643-6)
By Saundra Yancy McGuire and Elzbieta Cook of Louisiana State University. Includes chapter summaries, complete descriptions of appropriate chemical demonstrations for lecture, suggestions for addressing common student misconceptions, and examples of everyday applications of selected topics for lecture use, as well as the solutions for all the problems in the text.

Instructor's Resource Materials (Download Only) for *Introductory Chemistry: Atoms First*, 5th edition (0-321-95734-2)
The online instructor resources include all the art, photos, and tables from the book in high-resolution format for use in classroom projection or for creating study materials and tests. In addition, the Instructor can access modifiable PowerPoint® lecture outlines to highlight key points in his or her lecture. Also available are downloadable files of the *Instructor Teaching Guide*, the Test Bank with more than 1000 questions, and a set of "clicker questions" suitable for use with classroom-response systems.

Test Bank (Download Only) for *Introductory Chemistry: Atoms First*, 5th edition (0-321-95691-5)
By Christine Hermann of Radford University
This printed test bank includes over 1700 questions that correspond to the major topics in the text.

Instructor Manual for the *Laboratory Manual for Introductory Chemistry* (0-321-73026-7)
By Wendy Gloffke and coauthored by Doris Kimbrough of the University of Colorado at Denver with contributions from Chris Bahn of Montana State University. This manual includes lists of equipment and chemicals needed to perform each lab.

Additional resources available for students:

Mastering Chemistry with Pearson eText for *Introductory Chemistry: Atoms First*, 5th edition (www.masteringchemistry.com) (0-321-92456-8)
Mastering Chemistry is the leading online homework and tutorial system for the sciences. Instructors can create online assignments for their students by choosing from a wide range of items, including end-of- chapter problems and research-enhanced tutorials. Assignments are automatically graded with up-to-date diagnostic information, helping instructors pinpoint where students struggle either individually or as a whole class.

Laboratory Manual for Introductory Chemistry, 4th Edition (0-321-73025-9)
By Wendy Gloffke and coauthored by Doris Kimbrough of the University of Colorado at Denver with contributions from Chris Bahn of Montana State University. Helps students develop data acquisition, organization, and analysis skills while teaching basic techniques. Written to accompany the text, this manual offers 25 experiments. This lab manual is available via Catalyst: The Pearson Custom Laboratory Program for Chemistry organization. This program allows you to custom build a chemistry lab manual that matches your content needs and course organization. You can either write your own labs using the Lab Authoring Kit tool or select from the hundreds of labs available at http://www.pearsonlearningsolutions.com/custom-library/catalyst. This program also allows you to add your own course notes, syllabi, or other materials.

CHAPTER

1

What Is Chemistry?

Learning Outcomes

1. Describe the difference between science and technology.
2. Define the types of mixtures in terms of the types of matter.
3. Describe how an elemental substance differs from a compound.
4. Explain what a chemical formula is (what it tells you).
5. Name the states of matter and the types of physical transformation.
6. Explain what happens to a substance or substances after they undergo a chemical change.
7. Describe the parts of the scientific method and how they are related to each other.

Chapter Outline

1.1 Science and Technology
   A. Science—The experimental investigation and explanation of natural phenomena
   B. Technology—The application of scientific knowledge
   C. Chemistry—The study of matter and the changes that matter undergoes
1.2 Matter
   A. Pure substances
      1. Elements—the simplest form of matter; basic building blocks of matter
      2. Compounds—formed by the chemical combination of elements
      3. Chemical formula—indicates the number of atoms of each element in a compound
   B. Mixtures
      1. Homogeneous mixture (solution)—identical composition throughout
      2. Heterogeneous mixture—composition not identical throughout

---

1

1.3 Matter and Its Physical Transformations

    A. Physical changes—Changes producing the same substance in a different state

    B. States of matter—Solid, liquid, gas (or vapor)

    C. Changes in states of matter—Freezing, melting, evaporation, condensation, sublimation, and deposition

    D. Physical properties—Characterize the physical state and physical behavior of a substance (e.g., melting point, color, odor, taste)

1.4 Matter and Its Chemical Transformations

    A. Chemical properties—Ways in which a pure substance behaves when it is combined with other pure substances

    B. Chemical transformation (chemical reaction)—New substances (products) are formed from starting substances (reactants)

1.5 How Science Is Done: The Scientific Method

    A. Components of the scientific method

        1. Experiments—collecting data to study a phenomenon

        2. Law—a summary of the data collected in experiments

        3. Theory—an explanation that attempts to explain why a law is true

    B. Steps in the scientific method

        1. Making observations and collecting data from experiments

        2. Using the data to propose a law

        3. Postulating a theory or model to account for the law

        4. Repeating Steps 1–3 to postulate new laws and test new theories

    C. Bias—A strong preference or inclination that inhibits impartial judgment

1.6 Learning Chemistry with This Textbook

    A. Understanding fundamental concepts

    B. Testing understanding of concepts with WorkPatches

    C. Summarizing chapter content

    D. Delving deeper into the material using the "One More Thing" section

Review of Key Concepts from the Textbook

1.1 Science and Technology

Science involves the pursuit of knowledge for its sake, whereas technology is the application of scientific knowledge. Science without technology would do little to move the human race forward in our capability to control our environment. Technology without science is impossible; without science, there *is* no technology. The two sides of the coin have operated throughout history to improve living conditions for humankind.

    There are many branches of science, each focusing on investigating a different set of questions. For example, biology is the study of living things; geology is the study of Earth; astronomy is the study of celestial bodies. Chemistry, called the central science, is the study of matter and the transformations (or changes) that matter undergoes.

1.2 Matter

Matter is defined as anything that has mass and occupies space (or has volume). Matter is divided into categories as shown in the diagram below.

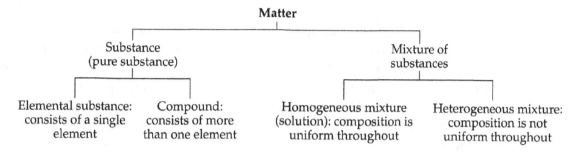

As you can see, all matter can first be classified as a pure substance or a mixture of pure substances. If the matter is a pure substance, it must be either an element or a compound. If it is not an element or a compound, it is not a pure substance but a mixture of pure substances.

The 113 currently known elements are the building blocks of all matter. They are given names and represented by symbols. Examples are hydrogen (H), oxygen (O), aluminum (Al), and chlorine (Cl). (Note that when more than one letter appears in the symbol, only the first letter is capitalized.) The elements are organized in the periodic table, which is shown below.

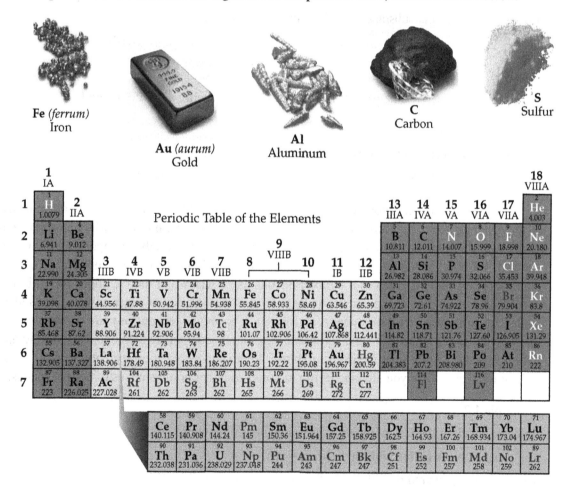

You should memorize the names and symbols of the first 20 elements as well as these other common elements: barium (Ba), bromine (Br), copper (Cu), gold (Au), iodine (I), iron (Fe), mercury (Hg), silver (Ag), tin (Sn), and zinc (Zn). You will learn the symbols of other elements as they arise in your study of chemistry. Knowing the names and symbols of elements will be a tremendous advantage to you.

Elements (also called elemental substances) are made up of only one type of atom. Compounds are pure substances that are made from atoms of more than one element. For example, the compounds water (H_2O) and carbon dioxide (CO_2) are each made up of atoms from two different elements. A compound that has interesting electrical properties, $YBa_2Cu_3O_6$, is made from atoms of four different elements.

Although elements contain only one kind of atom, there may be more than one of these atoms in the basic building blocks of the element. For example, oxygen and hydrogen both contain two atoms in the basic unit of the element, and are written as O_2 and H_2, respectively, when describing the elemental form. Elements that contain two atoms in the basic unit are called diatomic elements.

The diatomic elements are hydrogen (H_2), nitrogen (N_2), oxygen (O_2), fluorine (F_2), chlorine (Cl_2), bromine (Br_2), and iodine (I_2). You should memorize the diatomic elements because knowing them will be very important throughout your study of chemistry.

Example 1.1

Identify the following substances as elements or compounds:

(a) NO_2 (b) N_2 (c) He   (d) S_8 (e) NH_4NO_3

Solution

(a) compound   (b) element   (c) element   (d) element   (e) compound

Practice Exercise 1.1

Identify the following substances as elements or compounds:

(a) P_4 (b) SO_3 (c) $C_6H_{12}O_6$ (d) NaCl   (e) H_2

Any matter that is not a pure substance is a mixture. Mixtures are either homogeneous or heterogeneous. Homogeneous mixtures (also called solutions) are mixtures whose composition is constant throughout, whereas heterogeneous mixtures have a variable composition throughout the mixture. It is generally true that if a mixture looks like it has only one component, it is a homogeneous mixture, whereas if a mixture looks like it has two or more components, it is a heterogeneous mixture. For example, a glass containing a solution of sugar in water looks like a glass of pure water (if all the sugar is dissolved). It is a homogeneous mixture. However, a glass containing a mixture of sand and water looks like it contains two components, a solid and a liquid. It is therefore a heterogeneous mixture. Although there are some exceptions to this general rule, it is helpful when you are classifying mixtures as either homogeneous or heterogeneous.

Example 1.2

Indicate which of the following are *not* homogeneous mixtures:

(a) a cup of black coffee with sugar (dissolved)

(b) a mixture prepared by grinding salt and sugar together

(c) a glass of pure water

(d) a glass of sweetened iced tea

(e) the air in your chemistry classroom

Solution

(a) It is a homogeneous mixture (a solution of coffee and sugar).

(b) It is not a homogeneous mixture because solids ground together cannot be uniform at the submicroscopic level; it is a heterogeneous mixture.

(c) It is not a mixture of any kind; pure water is a compound.

(d) It is a homogeneous mixture (a solution of tea and water).

(e) It is a homogeneous mixture composed of nitrogen, oxygen, and some other gases in very small amounts.

Practice Exercise 1.2

Indicate whether each of the following is a homogeneous mixture, a heterogeneous mixture, or neither:

(a) the Atlantic Ocean

(b) a cup of salt water

(c) tincture of iodine (made by completely dissolving iodine in alcohol)

(d) a bowl of cereal and milk

(e) a cloud in the sky

1.3 Matter and Its Physical Transformations

The transformations (changes) that matter undergoes are classified as either physical or chemical changes. A physical transformation is a change that produces no new substances; it results in a change of state or shape for an existing substance. The three most common states of matter are solid, liquid, and gas. Physical transformations involve converting a substance from one of these states to another. Examples of physical transformations are water freezing, ice melting, or water condensing as dew on the grass. Generally, we think of physical transformations as easily reversible changes. For example, we can easily refreeze water into ice that melted to produce it. Or, the energy in sunlight will quickly evaporate the water that has condensed on the leaves of grass. The terms used for different changes of state are listed below.

Change of State	Phase Transformation
Melting (or fusion)	Solid to liquid
Freezing	Liquid to solid
Sublimation	Solid to gas
Deposition	Gas to solid
Boiling (or evaporation)	Liquid to gas
Condensation	Gas to liquid

In addition to the state of matter of a substance, some other physical properties often useful to chemists are mass, volume, temperature, density, melting point, boiling point, general appearance, including color, and solubility. Knowing these properties often helps chemists identify an unknown substance or make predictions about the structure or behavior of the substance.

1.4   Matter and Its Chemical Transformations

A chemical transformation results in the formation of new substances that have new physical and chemical properties. Examples of chemical transformations are the formation of rust from iron, the burning of gasoline in an automobile, and the decomposition of dead organisms. Such changes produce different substances from the ones that were present before the chemical change occurred, and they are not easily reversible.

Example 1.3

Indicate whether each of the following is a physical or a chemical change:

(a) corrosion of steel   (b) evaporation of alcohol   (c) burning of coal

(d) frying of an egg   (e) sublimation of dry ice

Solution

(a) The corrosion of steel is a chemical change. Elemental iron is converted to a compound of iron and oxygen (iron oxide).

(b) The evaporation of alcohol is a physical change. Alcohol changes from the liquid to the gaseous state.

(c) The burning of coal is a chemical change. Elemental carbon (coal) is converted to a compound of carbon and oxygen (carbon dioxide).

(d) The frying of an egg is a chemical change. The proteins in the egg are converted to new compounds that are solid at room temperature.

(e) The sublimation of dry ice is a physical change. Solid carbon dioxide (CO_2) changes to the gaseous state.

Practice Exercise 1.3

Indicate which of the following are *not* physical changes:

(a) deposition of silver vapor to form a layer of solid silver

(b) explosion of a firecracker

(c) sweetening a cup of coffee by adding sugar

(d) burning a piece of paper

(e) evaporating all the water from a saltwater solution

Matter is described in terms of its properties, the characteristics of the material that makes up the matter. Properties are described as physical or chemical, depending on what is necessary to observe the property. Physical properties can be observed without changing the identity of the material (without subjecting the material to a chemical change). However, chemical properties can only be observed by causing the material to undergo a chemical change. Examples of physical properties are boiling point, melting point, and density. Examples of chemical properties are rusting (e.g., of iron), tarnishing (e.g., of silver), and combustion. All these properties can be observed only as the result of a chemical change. Because new substances

---

STUDENT WORKBOOK AND SELECTED SOLUTIONS

are formed during chemical transformations, we use different terms to describe the substance(s) present before and after the change. The starting materials are called the reactants, and the new substances present after the change are called the products. The chemical change itself is called a chemical reaction. In writing chemical reactions, we use an arrow to separate the reactants from the products.

Example 1.4

Identify the following properties as physical or chemical:

(a) color   (b) boiling point   (c) combustion point (or flash point)   (d) density

Solution

(a) physical   (b) physical   (c) chemical   (d) physical

Practice Exercise 1.4

Identify the property (or properties) below that are chemical:

(a) reactivity with oxygen   (b) mass   (c) freezing point   (d) solubility

1.5   How Science Is Done: The Scientific Method

We defined science as the pursuit of knowledge. The scientific method describes the steps involved in that pursuit. The first step in the scientific method involves observation and then stating the question or hypothesis to be tested. After the question is stated, experiments are conducted to collect data to answer the question. Next, scientists analyze the data collected and postulate a law based on the data. The next step involves the formulation of a theory (or model) to account for the law. Finally, the theory is tested by performing additional experiments. If the new experiments do not produce results that the theory would predict, this theory is rejected and a new one is formulated. The new one is then tested by performing additional experiments. This process is repeated until a theory is put forth that is consistent with the experimental data. It is important to understand that a theory can be *disproved* if the experimental evidence contradicts it, but a theory can never be *proved*, even if all experimental evidence supports it. The best we can say is that the theory is possibly correct, but there may be other theories consistent with the experimental data that may also be correct.

The scientific method appears to be very objective; therefore, the results would seem to be independent of who is executing the method to solve a problem. In real life, however, sometimes scientists (who are human) have a bias (a strong preference) for a particular theory, and this bias inhibits their ability to design appropriate experiments to test the theory or impartially interpret the results of their experiments. It is for this reason that a very important part of the scientific method involves having many other scientists repeat the experiments that have been done, in order to verify that the results reported are reliable and accurate and not the result of any bias that the original scientist(s) may have had. This repetition to confirm results is a very good method of checks and balances used by scientists. Even this system is not perfect, however, because sometimes the other scientists who are repeating the experiments have the same biases as the original investigator(s) (or they may have their biases). Fortunately for science, it is rare that most of the scientists performing an experiment will have the same biases. Therefore, it is usually possible to get unbiased results when a large number of scientists repeat a set of experiments.

1.6 Learning Chemistry with This Textbook

This text is a valuable tool in your study of introductory chemistry. It is written in a manner that emphasizes the fundamental concepts and processes of chemistry. However, because chemistry is a quantitative science, there are also problems throughout this text that involve calculations.

Read each chapter slowly, with a pad of paper and pencil handy. You should take notes on the reading, work the problems in the WorkPatches, and write down questions about concepts that you do not fully understand.

The WorkPatches, Concept Questions, Practice Problems, and the section entitled "Have You Learned This?" are all designed to help you *actively* learn the material in each chapter, instead of merely passively reading each chapter. The more active you are when you are reading this text, the more thorough your understanding of chemistry will be.

Finally, near the end of each chapter, there will be a short section called "One More Thing." Here, you will have an opportunity to dig deeper into the concepts just covered and formulate your opinion about issues brought up in the chapter.

Strategies for Working Problems in Chapter 1

Overview: What You Should Be Able to Do

Chapter 1 provides an overview of matter and its transformations and describes the steps in the scientific method. The types of problems that you should be able to solve after mastering Chapter 1 are as follows:

1. Identify a sample of matter as an elemental substance, a compound, a homogeneous mixture, or a heterogeneous mixture.

2. Indicate whether a given transformation of matter is a chemical or a physical transformation.

Flowcharts

Flowchart 1.1 **Identifying a sample of matter as an elemental substance, a compound, a homogeneous mixture, or a heterogeneous mixture**

Flowchart 1.1

Summary

Method for determining whether a sample of matter is an elemental substance, a compound, a homogeneous mixture, or a heterogeneous mixture

Step 1

- **Determine if the sample appears to consist of more than one substance. If more than one substance is visible, the sample is a heterogeneous mixture. If only one substance is visible, proceed to Step 2.**

$\downarrow$

Step 2

- **Determine the composition of the sample (provided by the information in the problem). If only one element is present, the sample is an element; if only one compound is present, the sample is a compound. If the sample contains two or more substances, it is a homogeneous mixture.**

Example 1

Classify a sample containing a tablespoon of salt dissolved in a cup of water as an elemental substance, a compound, a homogeneous mixture, or a heterogeneous mixture.

Solution

Perform Step 1

Only one substance is visible in a tablespoon of salt dissolved in a cup of water. We therefore proceed to Step 2.

Perform Step 2

Because the composition of the sample is two substances (salt and water), the sample is a homogeneous mixture.

Example 2

Would a scoop of vanilla ice cream with chocolate chips be classified as an elemental substance, a compound, a homogeneous mixture, or a heterogeneous mixture?

Solution

Perform Step 1

Because it is visibly obvious that a scoop of vanilla ice cream with chocolate chips contains at least two substances, the sample is a heterogeneous mixture.

Example 3

Would a gold bar be classified as an elemental substance, a compound, a homogeneous mixture, or a heterogeneous mixture?

Solution

Perform Step 1

Only one substance is visible in a gold bar. We therefore proceed to Step 2.

Perform Step 2

Because the composition of the sample is gold, an element, the sample is an elemental substance.

Practice Problems

Indicate whether each of the following is an elemental substance, a compound, a homogeneous mixture, or a heterogeneous mixture.

1.1 a bowl of peaches and cream
1.2 a bar of zinc metal
1.3 a cup of distilled water
1.4 a tablespoon of maple syrup

Flowchart 1.2 **Indicating whether a given transformation of matter is a chemical or a physical transformation**

Flowchart 1.2

Summary

Method for determining whether a given transformation of matter is a chemical or a physical transformation

Step 1
• **Determine if the transformation results in a change in the physical state of the substance or in the production of a new substance.**

↓

Step 2
• **Classify the transformation as physical if there is simply a change in the physical state, or classify the transformation as chemical if there is a new substance produced (evidence of a new substance may include an odor change, a color change, or fizzing).**

Example 1

Classify the melting of a bar of platinum to form molten platinum as a physical or a chemical transformation.

Solution

Perform Step 1

The transformation results in a change in the physical state of platinum.

Perform Step 2

Because the transformation results in only a change in physical state, a physical transformation has occurred.

Example 2

Classify the burning of a block of wood as a physical or a chemical transformation.

Solution

Perform Step 1

The transformation results in the production of a new substance, as evidenced by the formation of smoke.

Perform Step 2

Because a new substance is produced, a chemical transformation has occurred.

Practice Problems

Indicate whether each of the following is a physical or a chemical transformation.

1.5 the combination of H_2 and O_2 to form water
1.6 vaporizing a sample of liquid mercury
1.7 dissolving a tablespoon of sugar in water
1.8 the spoiling of a carton of orange juice

Quiz for Chapter 1 Problems

NOTE: The chapter quiz items will involve problem solving and may require additional information provided in the textbook.

1. Classify the following as elemental substances, compounds, homogeneous mixtures, or heterogeneous mixtures.
   (a) a glass of iced tea with lemon slices
   (b) a tablespoon of instant ice tea dissolved in water
   (c) a cup of homogenized milk
   (d) a beaker of liquid mercury

2. Indicate whether the change described below results in a physical or a chemical transformation.
   (a) A block of iron metal is heated until it becomes a liquid.
   (b) An iron car rusts in an open field.
   (c) A sample of liquid bromine is converted to vapor.

3. Complete the following statement: In a chemical transformation, a(n) _____ _____ is produced.

4. Identify the following as elemental substances or compounds:
   (a) sulfur dioxide   (b) helium gas   (c) ammonia   (d) sulfur

5. Classify each of the following transformations as physical or chemical:
   (a) sublimation   (b) burning   (c) fermentation   (d) condensation

6. Explain the difference between a heterogeneous and a homogeneous mixture.

7. When a carton of milk spoils, a chemical transformation has occurred. Explain what indicates that this is a chemical and not a physical transformation.

8. Give an example of each of the following:
   (a) an elemental substance   (b) a compound   (c) a homogeneous mixture
   (d) a heterogeneous mixture

9. Indicate whether the following statement is true or false. If false, explain why it is false. A glass of fizzing seltzer water is a homogeneous mixture.

10. Complete the following statement: When water is electrolyzed to produce hydrogen and oxygen, a(n) _____ transformation has occurred.

Exercises for Self-Testing

A. Completion. *Write the correct word(s) to complete each statement below.*

1. Another name for a solution is a(n) _____ _____.
2. Elements and compounds are classified as _____ _____.
3. Chemistry is the study of _____ and the _____ that matter undergoes.
4. A(n) _____ is the smallest possible piece of an element.
5. A(n) _____ is a substance that is made up of only one type of atom, whereas a(n) _____ is made up of atoms of two or more different substances.
6. _____ is the name for the change from a gas to a liquid; _____ is the name for the change from a liquid to a solid.
7. When a pure substance goes directly from the solid to the gaseous state, without passing through the liquid state, the phase change is called _____.
8. When a(n) _____ transformation occurs, no new substances are produced.
9. In a chemical reaction, the new substances produced are called the _____ of the reaction, and the starting substances are called the _____.
10. A strong preference is also called a(n) _____.

B. True/False. *Indicate whether each statement below is true or false, and explain why the false statements are false.*

_____ 1.   All solutions are homogeneous mixtures, and all homogeneous mixtures are solutions.

_____ 2.   A fizzing soft drink is a homogeneous mixture.

_____ 3.   Theories can be disproved by experimental data, but they can never be proved by experimental data.

_____ 4. P_4 is a compound because there is more than one atom in the formula.

_____ 5.   A physical property can be determined without changing the composition of the substance.

_____ 6.   S is the symbol for the element sodium.

_____ 7.   In the scientific method, theories are postulated based on experimental data.

	8.	Compounds are made up of two or more elements.
_____	9.	The scientific method always yields the same results and is never affected by the scientist using the method.
_____	10.	Elements are the basic building blocks of matter.

C. Questions

1. What is the relationship between compounds and elemental substances? Give an example of each.
2. Explain why a mixture of copper and zinc is a homogeneous mixture if the two metals are melted, mixed together, and then allowed to solidify, but a heterogeneous mixture if the two metals are ground into powders and mixed together as solids.
3. Define sublimation. Give an example of a substance that sublimes under normal atmospheric conditions.
4. Would you expect the existence of a reverse change of state to sublimation?
5. Explain the role of experiments in the scientific method.
6. Explain how biases might interfere with scientific development. Give an example, complete with a hypothetical problem being investigated, of how bias can lead to incorrect conclusions.

2

The Numerical Side of Chemistry

Learning Outcomes

1. Explain what an exact number is and what a measured number is.
2. Describe the difference between precision and accuracy.
3. Explain why measured numbers always have uncertainty associated with them.
4. Identify leading zeros and trailing zeros in a measured number.
5. Identify the amount of uncertainty in a written measured quantity.
6. Convert a number from standard to scientific notation and vice versa.
7. Identify the degree of uncertainty in a number written in scientific notation.
8. Correctly add or subtract numbers according to the rules of significant figures.
9. Correctly multiply or divide numbers according to the rules of significant figures.
10. Identify the common SI units used by science.
11. Define density and use it to do calculations.
12. Convert the units of a measured quantity to some other units using unit analysis.
13. Algebraically rearrange a mathematical equation to solve it for a desired variable.
14. Calculate the amount of heat energy using specific heat.

Chapter Outline

2.1 Numbers in Chemistry—Precision and Accuracy

    A. Precision—The closeness of a series of measurements to one another

    B. Accuracy—The closeness of a measured result to the true value

2.2 Numbers in Chemistry—Uncertainty and Significant Figures

    A. Uncertainty—Indicated by the number of digits in a measurement

    B. Significant figures—The numbers that are within the ability of the measuring device to accurately measure

2.3 Zeros and Significant Figures
  A. Rules for determining number of significant figures
    1. All nonzero digits are significant
    2. Zeros between nonzero digits are significant
    3. Leading zeros are not significant
    4. Trailing zeros after a decimal point are significant
    5. Trailing zeros before a decimal point may or may not be significant
  B. Determining uncertainty in numbers with trailing zeros before a decimal point
2.4 Scientific Notation
  A. Converting numbers to scientific notation
  B. Converting from scientific to standard notation
  C. Assigning uncertainty to numbers written in scientific notation
2.5 How to Handle Significant Figures and Scientific Notation When Doing Math
  A. Multiplying and dividing numbers written in scientific notation
  B. Adding and subtracting numbers written in scientific notation
  C. Significant figures and exact numbers
2.6 Numbers with a Name—Units of Measure
  A. Common SI units used in chemistry
  B. Derived SI units used in chemistry
  C. Greek prefixes used with SI units
  D. Non-SI units for pressure and temperature
  E. Temperature scales and conversions
  F. Unit conversions
2.7 Density: A Useful Physical Property of Matter
  A. Definition of density—Mass per unit volume
  B. Using density to identify and characterize an unknown substance
  C. Intensive property—Does not depend on the amount of material
  D. Extensive property—Depends on the amount of material
2.8 Doing Calculations in Chemistry—Unit Analysis
  A. Steps in the unit analysis method
    1. Write down the given measurement
    2. Multiply the measurement by the appropriate conversion factor to get the desired unit
  B. Conversion factor—A ratio of two equivalent quantities (e.g., 60 seconds/1 minute)
  C. Using unit analysis in problems
2.9 Rearranging Equations—Algebraic Manipulations with Density
  A. Rules for algebraic manipulations
  B. Problems involving algebraic manipulations
2.10 Quantifying Energy
  A. Energy units
    1. Calorie—amount of heat required to raise the temperature of 1 g of water from 25 °C to 26 °C

2. Joule—the SI unit of energy often used in science and engineering
   (1 cal = 4.184 J)

B. Specific heat—Energy required to raise the temperature of 1 g of substance by 1 °C

C. Energy calculation for temperature change
   Heat = Specific heat × Mass × Change in temperature

D. Bomb calorimetry calculations
   Heat lost by substance = Heat gained by water

Review of Key Concepts from the Textbook

The numbers that are generated as data during the experimental process are a crucial component in the scientific method. The correct interpretation of the numbers and the units that accompany them is of utmost importance. This chapter teaches us how to decide whether our data are accurate, precise, both, or neither; determine the uncertainty associated with our measurements; express numbers in a kind of scientific shorthand; and do chemical calculations. In addition, we are introduced to several Système International d'Unités (SI units) and shown how to convert units to others using conversion factors. A thorough understanding of this chapter is vital to your success in chemistry. Be sure to carefully work the problems in the textbook and in this *Study Guide* in order to become proficient with these operations.

2.1 Numbers in Chemistry—Precision and Accuracy

Precision refers to the agreement among repeated measurements of the same quantity. A set of data with a smaller spread (highest minus lowest value in the data set) will have more precision than a data set with a higher spread.

Accuracy refers to the agreement of a particular value (or the average of a set of values) with the true value. Systematic errors (e.g., errors due to the measuring device) will lead to low accuracy even though precision may be high. Random errors (e.g., errors due to misreading a value) will lead to measurements that are neither accurate nor precise.

If you have trouble keeping these terms straight, just remember that *accuracy* is related to the *absolutely correct value*, whereas *precision* is related to the *proximity of a set of values to each other*. The following example illustrates the difference between precision and accuracy.

Example 2.1

Three students, A, B, and C, took three measurements of a length of pipe whose true length is 53.2 cm. The values for their measurements are shown in the table below. The units on all values are centimeters.

| | **Measurement** | | | | |
Student	1	2	3	Average	Spread
A	51.3	54.2	53.1	52.9	2.9
B	53.3	53.2	53.4	53.3	0.2
C	58.9	58.8	58.9	58.9	0.1

(a) Which student is most accurate?

(b) Which student is most precise?

(c) Suggest a possible cause for student C to be very precise but very inaccurate.

Solution

(a) Student B is the most accurate of the three students because the average of her data (53.3 cm) is closest to the absolutely correct value of 53.2 cm.

(b) Student C is most precise. His data have a very small spread of 0.1.

(c) Student C may have had a meterstick whose values were mislabeled. He could have done a good job of measuring precisely but was unaware that his meterstick was giving him bad values. Another possibility is that student C had an accurate meterstick but had very poor technique, which he employed consistently.

Practice Exercise 2.1

A student took three measurements of the mass of a sample whose actual mass was 3.24 g. The three masses obtained by the student were 3.25, 3.22, and 3.23 g.

(a) What is the average of the student's measurements?

(b) What is the range of the student's measurements?

(c) Were the student's measurements accurate, precise, both, or neither?

(d) List a set of three measurements that would be precise but not accurate.

(e) List a set of three measurements that would be accurate but not precise.

2.2 Numbers in Chemistry—Uncertainty and Significant Figures

Whenever a number is obtained from a measurement, there will be some uncertainty. For example, if a length is measured using a meterstick marked in centimeters, it will be impossible to read the length to a better accuracy than 0.1 cm. (See Section 2.2 of the textbook for a full discussion of this concept.) Therefore, in a length reported as, say, 30.6 cm, the 0.6 is estimated based on where the end of the object falls between the centimeter markings.

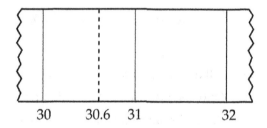

The conventions governing uncertainty in any numerical measurement are as follows:

1. The last place in the number is where the uncertainty lies.
2. The uncertainty is ±1 of these units.

Therefore, in the number 30.6, the uncertainty is ±0.1 (because the last place is the tenths place).

Example 2.2

Express the uncertainty in the following numbers:

(a) 58.32   (b) 1002   (c) 0.479   (d) 592.8

Solution

(a) ± 0.01 (b) ± 1 (c) ± 0.001 (d) ± 0.1

Practice Exercise 2.2

Express the uncertainty in the following numbers:

(a) 198.003   (b) 23.56   (c) 1279.2   (d) 363

The numbers that have meaning (those within the ability of the measuring device to measure directly) are called significant figures.

2.3   Zeros and Significant Figures

Expressing values to the correct number of significant figures is important for communicating where the uncertainty in a measured or calculated value lies. (See Section 2.3 of the textbook for a detailed discussion of significant figures.) The rules for determining the number of significant figures in a value are as follows:

1. All nonzero digits are *always* significant.
2. Zeros between nonzero digits are *always* significant.
3. Leading zeros (zeros before the first nonzero number) are *never* significant. (They are merely placeholders.)
4. Trailing zeros (zeros after the last nonzero digit) after a decimal point are *always* significant.
5. Trailing zeros before an implied decimal point may or may not be significant, depending on where the uncertainty lies.

Example 2.3

Indicate the number of significant figures in the following values. Explain your answers.

(a) 3052   (b) 0.0036   (c) 0.3500   (d) 35,000

Solution

(a) Four significant figures. All digits are significant, per rules 1 and 2.

(b) Two significant figures. Only the last two digits are significant, per rules 1 and 3.

(c) Four significant figures. All digits are significant, per rules 1, 3, and 4.

(d) This is an ambiguous case. The number of significant figures depends on where the uncertainty lies. Two, three, four, or five digits may be significant, per rule 5. If there is no information about *how* this number was arrived at, it may be the safest to assume that it only has two significant figures.

Practice Exercise 2.3

Indicate the number of significant figures in the following values. Explain your answers.

(a) 0.000 67   (b) 360,001   (c) 12,500   (d) 3.456 00

STUDENT WORKBOOK AND SELECTED SOLUTIONS

2.4 Scientific Notation

Scientists often use very large and/or very small numbers that are cumbersome to write in the standard way. For example, the speed of light is 300,000,000 m/s. And the federal limit on the amount of lead allowed in tap water is 0.000 015 g/L of water. There is a less cumbersome way to express these types of numbers, called scientific notation. Scientific notation is the expression of a given number as a power of 10 times a number (usually between 1 and 10). (See Section 2.4 of the textbook for a full discussion.) In scientific notation, 300,000,000 can be written as 3×10^8. This notation indicates that 3 is to be multiplied by 10 eight times, or that 3 is to be multiplied by 100,000,000, which is 10^8.

Any number can be expressed using a power of 10 by writing the number as the product of 0.01, 0.1, 10, 100, 1000, and so on times another number. Consider the following examples:

1. Expressing 583 as positive powers of ten:

$$583 = 58.3 \times 10 = 5.83 \times 100 = 0.583 \times 1000$$

or

$$583 = 58.3 \times 10^1 = 5.83 \times 10^2 = 0.583 \times 10^3$$

2. Expressing 583 as negative powers of ten:

$$583 = 5830 \times 0.1 = 58,300 \times 0.01 = 583,000 \times 0.001$$

or

$$583 = 5830 \times 10^{-1} = 58,300 \times 10^{-2} = 583,000 \times 10^{-3}$$

Because $10^0 = 1$ (any number raised to the zero power $= 1$), we can also express 583 as 583×10^0.

Example 2.4

Convert the following numbers to scientific notation:

(a) 45,895   (b) 0.0356   (c) 98,000,000,000   (d) 0.000 56

Solution

(a) 4.5895×10^4 (b) 3.56×10^{-2} (c) 9.8×10^{10} (d) 5.6×10^{-4}

You should notice that whenever the number gets smaller when going from the original number to the number expressed in scientific notation (e.g., from 45,895 to 4.5895), the exponent on the power of 10 is *positive*. Conversely, whenever the number gets larger when going from the original number to the number expressed in scientific notation (e.g., from 0.0356 to 3.56), the exponent on the power of 10 is *negative*. Demonstrate this pattern as you do Practice Exercise 2.4 below.

Practice Exercise 2.4

Convert the following numbers to scientific notation:

(a) 0.678   (b) 5700   (c) 79,345,000   (d) 0.000 009 2

Numbers written in scientific notation can be expressed as standard numbers by moving the decimal point to the right (if the power of 10 is positive) or to the left (if the power of 10 is negative). Zeros can be added if there are not enough numbers to move the decimal point the required number of places. For example, $4.89 \times 10^3 = 4890$.

Example 2.5

Convert the following numbers, written in scientific notation, to standard notation:

(a) 8.345×10^{-3} (b) 6.02×10^5 (c) 9.34×10^1 (d) 5.6×10^2

Solution

(a) 0.008 345   (b) 602,000   (c) 93.4   (d) 560

Notice that positive powers of 10 result in standard numbers that are greater than the number part of the quantity expressed in scientific notation, and negative powers of 10 result in standard numbers that are smaller than the number part of the quantity expressed in scientific notation.

Practice Exercise 2.5

Convert the following numbers, written in scientific notation, to standard notation:

(a) 6.94×10^{-1} (b) 4.61×10^{-7} (c) 5.78×10^4 (d) 6.542×10^8

2.5 How to Handle Significant Figures and Scientific Notation When Doing Math

When you are performing arithmetic operations, the following rules for significant figures hold true:

A. **Multiplying and Dividing.** The answer has no more significant digits than the number with the fewest significant digits (the least precise figure). Round off after calculations have been performed. The rules for rounding are straightforward but very important. They are listed below:

1. If the last digit to be kept is followed by a number less than 5, drop all the digits after the last one to be kept. For example, 2.854 rounded to three significant digits is 2.85.
2. If the last digit to be kept is followed by a number greater than 5, add 1 to the last digit to be kept and drop the rest. For example, 0.038 756 rounded to two significant digits is 0.039.
3. If the last digit to be kept is followed by a 5 and any other nonzero digits, add 1 to the last digit to be kept and drop the rest. For example, 23.6523 rounded to three significant digits is 23.7.
4. If the last digit to be kept is followed by a 5 and no other nonzero digits, add 1 if the last digit to be kept is odd, but do not add 1 if the last digit to be kept is even. (This rule may seem overly complicated, but it diminishes the errors associated with rounding numbers.) For example, 2.3500 rounded to two significant digits is 2.4, and 2.4500 rounded to two significant digits is also 2.4. (You won't encounter this situation often, but the rule comes in handy when you do.)

B. **Adding and Subtracting.** The answer has no more decimal places than the addend, minuend, or subtrahend with the fewest number of decimal places. Note that the *number* of significant figures is irrelevant in adding and subtracting; it is the number of decimal places that counts.

Example 2.6

Perform the following calculations, expressing the answers to the correct number of significant figures:

(a) 2.37×0.0052 (b) $98.321 + 0.038\,45$ (c) $0.0004 \div 56$

Solution

(a) 0.012 (The answer can have two significant digits.)

(b) 98.359 (The answer can have three significant digits after the decimal point.)

(c) 0.000 007 (The answer can have one significant digit.)

Practice Exercise 2.6

Perform the following calculations, expressing the answers to the correct number of significant figures:

(a) $6.3321 - 2.46$ (b) 0.083×4 (c) $874 \div 25,867$

Section 2.5 of the textbook has additional examples and practice problems dealing with performing arithmetic operations on numbers written in scientific notation. You should go over this section carefully.

2.6 Numbers with a Name—Units of Measure

Units must always be specified when discussing a measurement of a quantity. There are a variety of different units that can be used to express quantities, but there is one standard set of units, SI units, that is designated for general use. These units are shown in the table below.

Physical Quantity	Name of Unit	Symbol
Length	meter	m
Mass	kilogram	kg
Time	second	s
Temperature	kelvin	K
Amount of substance	mole	mol

You should note that we do use other units to describe these quantities, but the units in the table are what are used when SI units are called for. Some additional SI units are presented in Table 2.2 of the textbook. These units are called derived units because they come from a combination of the basic SI units. You should memorize the units in the table above and be very familiar with the ones in Table 2.2 of the textbook.

The SI units are called *metric* units because the prefixes indicating the relative sizes of the units are all powers of ten. Some of the prefixes used with SI units and their meanings are shown in the table below. (See also Table 2.3 in the textbook.) You should memorize the meanings of these prefixes because you will encounter them often in your study of chemistry. You will also have to use them to convert from one unit to another.

Greek Prefix	Meaning
giga-	10^9
mega-	10^6
kilo-	10^3
deci-	10^{-1}
centi-	10^{-2}
milli-	10^{-3}
micro-	10^{-6}
nano-	10^{-9}
pico-	10^{-12}

Example 2.7

A field is 125 m in length.

(a) How many picometers is this?

(b) How many kilometers?

Solution

(a) 125 m is 125×10^{12} pm. Because pico $= 10^{-12}$, a picometer is 10^{-12} the size of a meter, and there are 10^{12} pm in 1 m. In scientific notation, the length of the field is 1.25×10^{14} pm.

(b) 125 m is 0.125 km. There are 1000 m in 1 km.

Whenever you are converting between units, you can check to make sure that your answer makes sense. If the unit you are converting *to* is *smaller* than the one you are converting *from*, the new number will be *larger* than the old number. For example, 4 ft = 48 in.; 1.3 L = 1300 mL. Conversely, if the unit you are converting *to* is *larger* than the one you are converting *from*, the new number will be *smaller* than the old number. For example, 12 ft = 4 yd; 5.67 m = 0.0057 km.

Practice Exercise 2.7

A 12-oz bottle of soft drink contains 473 mL.

(a) How many kiloliters are there in 473 mL? (*Hint*: Convert the mL to L first, and then convert the L to kL.)

(b) Will the number of kiloliters be smaller or larger than the number of milliliters?

Temperature is an important property in the study of matter. It serves as a measure of the amount of heat in a substance. Temperature and heat are not the same thing, however. Heat is a form of energy and is typically expressed in units of calories or joules. Temperature, on the other hand, is a measure of the *amount* of energy contained in a substance.

The temperature scales in common use today are the Fahrenheit scale and the Celsius scale (sometimes referred to as the centigrade scale). The two scales differ in the temperature each assigns to the freezing and boiling points of water and in the size of each degree. On the

Fahrenheit scale, water freezes at 32 °F and boils at 212 °F. On the Celsius scale, water freezes at 0 °C and boils at 100 °C. There are 180 degrees between the freezing and boiling points of water on the Fahrenheit scale (212 − 32) and 100 degrees between these two points on the Celsius scale (100 − 0). Therefore, the size of a Fahrenheit degree is smaller than that of a Celsius degree by a factor of $\frac{100}{180}$, or $\frac{5}{9}$. Alternatively, we can say that the size of a Celsius degree is larger than the size of a Fahrenheit degree by a factor of $\frac{180}{100}$, or $\frac{9}{5}$.

The formulas for converting temperatures from one scale to the other are as follows:

$$°C = \frac{5}{9}(°F − 32) \text{ and } °F = 32 + \frac{9}{5}°C$$

Typically, the temperature in °C will be a lower number than the temperature in °F.

Two common mistakes in solving problems involving temperature conversions are (1) adding 32 when it should be subtracted and (2) using $\frac{5}{9}$ when $\frac{9}{5}$ should be used. The general rules listed below will help to guard against these errors.

1. Remember that the value of the temperature in °F is generally larger than the value of the same temperature in °C. (See the figure below.)

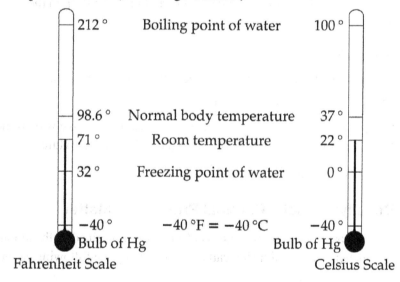

212 °	Boiling point of water	100 °
98.6 °	Normal body temperature	37 °
71 °	Room temperature	22 °
32 °	Freezing point of water	0 °
−40 °	−40 °F = −40 °C	−40 °
Bulb of Hg		Bulb of Hg

Fahrenheit Scale   Celsius Scale

2. When converting to °F (the larger number), always multiply by $\frac{9}{5}$ (the ratio greater than 1), and then *add* 32.

3. When converting to °C (the smaller number), always *subtract* 32, and then multiply by $\frac{5}{9}$ (the ratio less than 1).

4. Remember that the subtraction must be done *before* the multiplication by $\frac{5}{9}$, but the addition is done *after* the multiplication by $\frac{9}{5}$.

Example 2.8

Perform the following temperature conversions:

(a) 27 °C to °F   (b) 250 °F to °C

Solution

(a) We use the formula for converting °C to °F and substitute in the values given:

$$°F = 32 + \frac{9}{5}°C$$

$$°F = 32 + \frac{9}{5}(27) = 81 \, °F$$

(b) We use the formula for converting °F to °C and substitute in the values given:

$$°C = \frac{5}{9}(°F - 32)$$

$$°C = \frac{5}{9}(250 - 32) = 121 \, °C$$

As a check of accuracy, you can convert your result back to the original temperature to verify that you converted the original value properly.

Practice Exercise 2.8

Use the appropriate formulas to perform the following conversions:

(a) 81 °F to °C   (b) 121 °C to °F

A third temperature scale, the Kelvin scale, is divided in the same way as the Celsius scale. To convert Celsius to Kelvin, you add 273.15 to the Celsius temperature.

$$K = °C + 273.15$$

2.7 Density: A Useful Physical Property of Matter

Density is a measure of the compactness of matter; it is defined as the amount of mass in a given volume of matter. This relationship can be expressed by the following equation:

$$\text{Density} = \frac{\text{Mass}}{\text{Volume}}$$

Density can also be thought of as a conversion factor between mass and volume, allowing you to find the mass from the volume or the volume from the mass as long as you have the value of the density. Alternative expressions of this relationship are the following:

$$\text{Mass} = \text{Density} \times \text{Volume} \qquad \text{Volume} = \frac{\text{Mass}}{\text{Density}}$$

The mass is generally expressed in grams, and the volume in milliliters, so the units used to describe density are typically grams per milliliters. One exception to this generalization is gaseous matter. Because there is very little matter in a milliliter of gas, the gas volume is generally expressed in liters, and the density of gases is then expressed in grams/liter.

Example 2.9

The density of osmium is 22.6 g/mL. What is the mass of a sample of osmium whose volume is 2.38 L?

Solution

Because the problem asks us to calculate the mass, we will use

$$Mass = Density \times Volume$$

Note, however, that we must convert the volume of the osmium to milliliters because that is the volume unit for the density.

$$2.38 \text{ L} \times \frac{1000 \text{ mL}}{1 \text{ L}} = 2380 \text{ mL}$$

$$Mass = 22.6 \text{ g/mL} \times 2380 \text{ mL} = 53.8 \times 10^3 \text{ g (or 53.8 kg)}$$

A second way to do this problem is to treat density as a conversion factor.

$$Mass = 2380 \text{ mL} \times \frac{22.6 \text{ g}}{1 \text{ mL}} = 53.8 \times 10^3 \text{ g}$$

Practice Exercise 2.9

Diamond has a density of 3.51 g/mL. What volume will a 10.0 g sample of diamond occupy?

Unlike mass and volume, the density of a substance does not depend on how much of the substance is present. The density of a 3.0 cm^3 block of gold is exactly the same as that of a 100 ft^3 block of gold. This is certainly not the case with the mass. The 100 ft^3 block of gold would weigh much more than the 3.0 cm^3 block of gold. Properties that depend on the amount of substance present (e.g., mass and volume) are called *extensive properties*. Properties that do not depend on the amount of substance present (e.g., density and temperature) are called *intensive properties*.

2.8   Doing Calculations in Chemistry—Unit Analysis

A very effective method of solving problems in chemistry (and many other subjects) is referred to as unit analysis (or the factor units method). This method relies on multiplying a given quantity by conversion factors in such a way as to cancel all units except the unit in which the answer will be expressed. (See Section 2.8 of the textbook for a detailed discussion of this technique.) Consider the following examples.

Example 2.10

Given that there are 12 eggs in 1 dozen, and 12 dozen eggs in 1 gross, how many gross are there in 452 eggs?

Solution

In the unit analysis method, we always first write the given quantity with its unit:

452 eggs

We next write the conversion factor that involves the given unit and some other unit. The conversion factor is written such that the given unit is in the denominator so that it will cancel:

$$452 \text{ eggs} \times \frac{1 \text{ dozen}}{12 \text{ eggs}}$$

We continue this process until the unit that remains is the unit of what is asked for in the problem. In this case, the remaining unit will be gross, because that is what the problem asks us to find:

$$452 \text{ eggs} \times \frac{1 \text{ dozen}}{12 \text{ eggs}} \times \frac{1 \text{ gross}}{12 \text{ dozen}} = 3.14 \text{ gross}$$

You should note that in the example above we could have used the conversion factor involving the units gross and eggs. There are 12 dozen objects in 1 gross, which means there are 144 objects in 1 gross:

$$452 \text{ eggs} \times \frac{1 \text{ gross}}{144 \text{ eggs}} = 3.14 \text{ gross}$$

Notice that we get the same answer no matter which set of conversion factors we use. This will always be the case if we have set up the conversion factors correctly.

One of the common mistakes that students make in performing conversions is writing the conversion factor incorrectly—that is, placing the incorrect number with the unit. For example, a student might write the conversion factor involving eggs and dozen as

$$\frac{12 \text{ dozen}}{1 \text{ egg}}$$

Another common mistake is writing the conversion factor in the wrong direction (inverting the factor) and not recognizing that the units don't cancel. For example, in the conversion of 452 eggs to gross, writing the conversion as

$$452 \text{ eggs} \times \frac{144 \text{ eggs}}{1 \text{ gross}}$$

and not noticing that the conversion factor is upside down will lead to an incorrect result.

Although these mistakes may look ridiculous to you now, when you are doing conversion problems, it is easy to make these types of errors. Therefore, you should always think carefully and check your numbers every time you write a conversion factor to make sure you've done it correctly.

Important Note About Using the Factor Units Method

You should *always* think through a problem to get a ballpark answer before you do the calculations. You will then have a very rough idea of what the answer should be. That way, if you make a mistake (e.g., putting the wrong number with a unit in a conversion factor), it will be obvious to you when you get an answer that is not anywhere near the value that it should be.

Let's go back to Example 2.10 and try getting a ballpark answer. We would estimate that 452 eggs is a little less than 40 dozen (because 480 eggs = 40 dozen). A little less than 40 dozen would be around 3 gross (because 3 gross = 36 dozen). Therefore, our answer of 3.14 is reasonable. If we had mistakenly written one of the conversion factors incorrectly, we could have had something like:

$$452 \text{ eggs} \times \frac{1 \text{ dozen}}{12 \text{ eggs}} \times \frac{12 \text{ gross}}{1 \text{ dozen}} = 452 \text{ gross}$$

In this case, we would have known to check each step to see where the mistake was, and we would then have recognized that we had written the conversion factor involving gross and dozen incorrectly. You should form the habit of always getting a ballpark answer before you start a problem and then checking to see if your numerical answer is consistent with that ballpark estimate. If it's not, you have probably misplaced a number in your unit analysis calculations, and you should check each one carefully.

Practice Exercise 2.10

Perform the following conversions:

(a) 376 in. to cm (1.00 in. = 2.54 cm)   (b) 0.087 km to m   (c) 3000 L to mL

2.9 Rearranging Equations—Algebraic Manipulations with Density

An equation is a mathematical sentence that states that two expressions are equal. Formulas, such as $PV = nRT$, are equations that often require us to solve for one of the variables—say, T. To solve for a variable, we must isolate it on one side of the equation by algebraically moving everything else that is on that side of the equation. We must do this while keeping both sides of the equation equal. (Because an equation expresses an equality, whatever operation is done to one side of the equation must be done to the other to maintain the equality.) We are allowed to perform the following operations on any equation:

1. Add the same quantity to both sides of the equation.
2. Subtract the same quantity from both sides of the equation.
3. Multiply both sides of the equation by the same quantity.
4. Divide both sides of the equation by the same quantity.

(See Section 2.9 of the textbook for a more detailed discussion of this.)

Example 2.11

Solve for T in the equation $PV = nRT$.

Solution

We need to isolate T. In the equation, T is being multiplied by n and R. Therefore, we must divide both sides of the equation by n and R to isolate T. (Remember, whatever you do to one side of the equation must be done to both sides.)

$$\frac{PV}{nR} = \frac{nRT}{nR} \quad \text{or} \quad T = \frac{PV}{nR}$$

Example 2.12

In the equation

$$\frac{1}{X} = kt + \frac{1}{Y}$$

solve for k.

Solution

We must isolate k, first by subtracting $1/Y$ from both sides to get:

$$\frac{1}{X} - \frac{1}{Y} = kt$$

We then divide both sides by t:

$$\frac{\frac{1}{X} - \frac{1}{Y}}{t} = \frac{kt}{t}$$

Recall that dividing a fraction by a quantity (t in this case) is the same as multiplying the fraction by the inverse of t ($1/t$). The above expression then becomes:

$$\frac{1}{tX} - \frac{1}{tY} = \frac{kt}{t}$$

which simplifies to

$$\frac{1}{tX} - \frac{1}{tY} = k$$

Practice Exercise 2.11

Solve for T in the formula $G = H - TS$.

Practice Exercise 2.12

Solve for k in the formula $t = 1/kA$.

2.10 Quantifying Energy

Quantifying energy is an important part of the study of chemistry because chemical changes involve changes in energy. One of the most common units for measuring energy is the calorie (cal). One calorie is defined as the amount of energy required to raise the temperature of 1 g of water from 25 °C to 26 °C (1 Cal = 1000 cal). Another unit commonly used to measure energy is the joule (J). A joule is approximately one-fourth the size of a calorie (1 cal = 4.184 J). The conversion factor is therefore:

$$\frac{4.184 \text{ J}}{1 \text{ cal}} \quad \text{or alternatively} \quad \frac{1 \text{ cal}}{4.184 \text{ J}}$$

The conversion factor involving kilojoules and kilocalories is simply:

$$\frac{4.184 \text{ kJ}}{1 \text{ kcal}} \quad \text{or alternatively} \quad \frac{1 \text{ kcal}}{4.184 \text{ kJ}}$$

Although one calorie will raise the temperature of 1 g of water by 1 °C, a different amount of heat would be required to raise the temperature of another substance by 1 °C. In fact, the amount of heat required to raise the temperature of 1 g of a substance by 1 °C is a characteristic of the substance and is given the name *specific heat*. The units of specific heat are cal/g · °C, or J/g · °C.

Table 2.5 of the textbook provides the values of the specific heat of four common substances. If you have the specific heat, the mass, and the temperature change that a substance undergoes, you can calculate the heat absorbed (or released) during the change by using the formula:

Heat = Specific heat × Mass of substance × Change in temperature of substance

Notice that this equation can be algebraically rearranged to solve for specific heat, mass, or change in temperature.

Example 2.13

Calculate the joules required to raise the temperature of a 387.0 g bar of iron metal from 25.0 °C to 40.0 °C. The specific heat of iron is 0.449 J/g · °C.

Solution

Heat = Specific heat × Mass of substance × Change in temperature of substance

Heat = $0.449 \, \text{J/g} \cdot {}^{\circ}\text{C} \times 387.0 \, \text{g} \times 15 \, {}^{\circ}\text{C} = 2.61 \times 10^3 \, \text{J} = 2.61 \, \text{kJ}$

Example 2.14

Calculate the change in temperature resulting from adding 2000.0 cal of heat to 150.0 g of ethanol. The specific heat of ethanol is 0.581 cal/g · °C.

Solution

Algebraically solving for change in temperature, we get

$$\text{Change in temperature} = \frac{\text{Heat}}{\text{Specific heat} \times \text{Mass}}$$

$$\text{Change in temperature} = \frac{2000.0 \, \text{cal}}{0.581 \, \text{cal/g} \cdot {}^{\circ}\text{C} \times 150.0 \, \text{g}} = 22.9 \, {}^{\circ}\text{C}$$

Practice Exercise 2.13

Calculate the calories required to raise the temperature of a 2.00 kg bar of aluminum metal from 10.0 °C to 27.0 °C. The specific heat of aluminum is 0.215 cal/g · °C.

Practice Exercise 2.14

250.0 g of a bar made of an unknown metal experiences a rise in temperature of 8.30 °C when 370.0 J of heat is added. What is the specific heat of the metal?

The amount of heat released upon burning a substance can be determined by using a *bomb calorimeter*. The bomb calorimeter contains a mass of water whose temperature increases upon burning the substance. Because all the heat released by the substance is absorbed by the water in the calorimeter, the following relationship is true:

Heat lost by the substance = Heat gained by the water

Because we know the specific heat of water and the mass of water in the calorimeter, and we have measured the change in temperature of the water, we can calculate the heat gained by the water by using the equation presented earlier for calculating heat. This procedure can be used to determine the heat content of any substance that can be burned in a calorimeter. (See Section 2.10 of the textbook for a complete discussion of the use of bomb calorimeters.)

Strategies for Working Problems in Chapter 2

Overview: What You Should Be Able to Do

Chapter 2 provides an introduction to the quantitative aspects of chemistry, including topics such as interpreting data, expressing numbers in different types of notation, converting quantities from one unit to another, and quantifying energy. The types of problems that you should be able to solve after mastering Chapter 2 are as follows:

1. Determine the number of significant figures in a numerical value.
2. Convert a number from scientific to standard notation.
3. Convert a number from standard to scientific notation.
4. Express the answer to the correct number of significant figures when multiplying or dividing measured values.
5. Express the answer to the correct number of significant figures when adding or subtracting measured values.
6. Express the answer to the correct number of significant figures when performing calculations involving both multiplication and/or division *and* addition and/or subtraction.
7. Use unit analysis to convert a measurement to another set of units.
8. Rearrange algebraic equations to solve for any variable that appears in the equation.

Flowcharts

Flowchart 2.1   Determining the number of significant figures in a numerical value

Flowchart 2.1

Summary

Method for determining the number of significant figures in a numerical value

Step 1
- **Count the number of digits beginning with the first nonzero digit and ending with the last nonzero digit. (Zeros *before* the first nonzero digit are never significant, but all zeros between the first and last nonzero digit *are* significant.)**

$\downarrow$

Step 2
- **Determine if the zeros after the last nonzero digit (if there are any) are significant by using the following rule:**
  - **Trailing zeros after a decimal point are significant.**
  - **Trailing zeros before a decimal point are significant only if the decimal point is written at the end of the number. (If the decimal point is not written at the end of the number, the trailing zeros are not significant.)**

$\downarrow$

Step 3
- **Determine the number of significant figures in the value by adding the significant digits from Steps 1 and 2.**

Example 1

Determine the number of significant figures in 0.002030.

Solution

Perform Step 1

There are three digits beginning with the 2 (the first nonzero digit) and ending with the 3 (the last nonzero digit).

Perform Step 2

The one trailing zero is significant because it occurs after the decimal point.

Perform Step 3

The number of significant figures in 0.002030 is four (the sum of the three digits from Step 1 and the one trailing zero from Step 2).

Example 2

Determine the number of significant figures in 1.8000×10^5.

Solution

Perform Step 1

There are two digits beginning with the 1 (the first nonzero digit) and ending with the 8 (the last nonzero digit).

Perform Step 2

The three trailing zeros are significant because they occur after the decimal point.

Perform Step 3

The number of significant figures in 1.8000×10^5 is five (the sum of the three digits from Step 1 and the three trailing zeros from Step 2).

Example 3

The number 265,000,000 has how many significant figures?

Solution

Perform Step 1

There are three digits beginning with the 2 (the first nonzero digit) and ending with the 5 (the last nonzero digit).

Perform Step 2

The six trailing zeros are not significant because a decimal point is not written at the end of the number.

Perform Step 3

The number of significant figures in 265,000,000 is three (the sum of the three digits from Step 1 and no significant trailing zeros from Step 2).

Example 4

The number 265,000,000 has how many significant figures?

Solution

Perform Step 1

There are three digits beginning with the 2 (the first nonzero digit) and ending with the 5 (the last nonzero digit).

Perform Step 2

The six trailing zeros are significant because the decimal point has been written in at the end of the number.

Perform Step 3

The number of significant figures in 265,000,000 is nine (the sum of the three digits from Step 1 and the six significant trailing zeros from Step 2).

Example 5

Determine the number of significant figures in 856.612.

Solution

Perform Step 1

There are six digits beginning with the 8 (the first nonzero digit) and ending with the 2 (the last nonzero digit).

Perform Step 2

There are no trailing zeros.

Perform Step 3

The number of significant figures in 856.612 is six (the sum of the six digits from Step 1 and no trailing zeros).

Practice Problems

Determine the number of significant figures in each numerical value below. (Assume all values are measurements.)

2.1	0.0035	2.2	8.3021500	2.3	1205.0
2.4	9365	2.5	8.150×10^{-3}	2.6	560,000
2.7	0.005600	2.8	500	2.9	0.500

Flowchart 2.2   Converting a number from scientific to standard notation

Flowchart 2.2

Summary

Method for converting a number from scientific to standard notation

---

Step 1
- **Determine if the exponent on the power of 10 is positive or negative.**

$\downarrow$

Step 2
- **If the exponent is positive, move the decimal point to the right to produce the number in standard notation. If the exponent is negative, move the decimal point to the left to produce the number in standard notation. (Add zeros if there are not enough places to move the decimal point the appropriate number of places.)**

Example 1

Convert 8.79×10^4 to standard notation.

Solution

Perform Step 1

The exponent on the power of 10 is positive (4).

Perform Step 2

Move the decimal point four places to the right to produce the number in standard notation. (Two zeros must be added to allow the decimal to be moved four places.)
Therefore, $8.79 \times 10^4 = 87,900$.

Example 2

Convert 4.3×10^{-2} to standard notation.

Solution

Perform Step 1

The exponent on the power of 10 is negative (-2).

Perform Step 2

Move the decimal point two places to the left to produce the number in standard notation. (One zero must be added to allow the decimal to be moved two places.)
Therefore, $4.3 \times 10^{-2} = 0.043$.

Practice Problems

Convert the following numbers that are expressed in scientific notation to numbers expressed in standard notation.

2.10 6.74×10^{-3} **2.11** 6.74×10^3

2.12 8.3421×10^9 **2.13** 3.58×10^{-5}

Flowchart 2.3   Converting a number from standard to scientific notation

Flowchart 2.3

Summary

Method for converting a number from standard to scientific notation

Step 1
- Move the decimal point to the right or left until the decimal is located just after the first nonzero digit.

↓

Step 2
- Determine the value of the exponent (x) on the power of 10. The value of x reflects the number of places the decimal point was moved. If the decimal point was moved to the left, the x is positive; if the decimal point was moved to the right, the x is negative.

Example 1

Convert 0.00295 to scientific notation.

Solution

Perform Step 1

The decimal must be moved three places to the right so that it will be located just after the first nonzero digit.

$$0.00295 \text{ becomes } 2.95 \times 10^x$$

Perform Step 2

The x is equal to -3 because the decimal point was moved three places to the right. Therefore, $0.00295 = 2.95 \times 10^{-3}$.

Example 2

Convert 3,592,000 to scientific notation.

Solution

Perform Step 1

The decimal must be moved six places to the left so that it will be located just after the first nonzero digit.

$$3{,}592{,}000 \text{ becomes } 3.592000 \times 10^x$$

Perform Step 2

The x is equal to 6 because the decimal point was moved six places to the left. Therefore, $3{,}592{,}000 = 3.592000 \times 10^6$.

Example 3

Convert 0.0000000527 to scientific notation.

Solution

Perform Step 1

The decimal must be moved eight places to the right so that it will be located just after the first nonzero digit.

$$0.0000000527 \text{ becomes } 5.27 \times 10^x$$

Perform Step 2

The x is equal to -8 because the decimal point was moved eight places to the right. Therefore, $0.0000000527 = 5.27 \times 10^{-8}$.

Practice Problems

Convert the following numbers, which are expressed in standard notation, to numbers expressed in scientific notation.

2.14 83,000,000   **2.15** 0.00078200   **2.16** 503

2.17 0.467   **2.18** 29   **2.19** 0.0006

Flowchart 2.4 **Determining the number of significant figures in a numerical value obtained by multiplying and/or dividing**

Flowchart 2.4

Summary

Method for expressing the answer to the correct number of significant figures when multiplying and/or dividing numerical values

Step 1
• **Determine the number of significant figures in each number being multiplied and/or divided. The smallest number of significant figures in these values is the number of significant figures allowed in the answer.**

$$\downarrow$$

Step 2
• **Perform the arithmetic operation(s).**

$$\downarrow$$

Step 3
• **Round the answer to the correct number of significant figures.**

Example 1

Perform the following calculation, expressing the answer to the correct number of significant figures: 0.034×2.30.

Solution

Perform Step 1

There are two significant figures in 0.034, and there are three significant figures in 2.30. Therefore, the answer can have only two significant figures.

Perform Step 2

$$0.036 \times 2.30 = 0.0828$$

Perform Step 3

There must be two significant figures in the answer, so 0.0828 must be rounded to 0.083. The answer can also be expressed as 8.3×10^{-2}.

Example 2

Perform the following calculation, expressing the answer to the correct number of significant figures: $\dfrac{5.87650 \times 10^5}{0.00241 \times 10^8}$.

Solution

Perform Step 1

There are six significant figures in 5.87650×10^5, and there are three significant figures in 0.00241×10^8. Therefore, the answer can have only three significant figures.

Perform Step 2

$$\frac{5.87650 \times 10^5}{0.00241 \times 10^8} = 2.4384 \times 10^1$$

Perform Step 3

There must be three significant figures in the answer, so 2.4384×10^1 must be rounded to 2.44×10^1. The answer can also be expressed as 24.4.

Example 3

Perform the following calculation, expressing the answer to the correct number of significant figures: $\dfrac{0.9842 \times 0.389}{0.0812 \times 298.15}$.

Solution

Perform Step 1

There are four significant figures in 0.9842, three significant figures in 0.389, three significant figures in 0.0812, and five significant figures in 298.15. Therefore, the answer can have only three significant figures.

Perform Step 2

$$\frac{0.9842 \times 0.389}{0.0812 \times 298.15} = 0.015814$$

Perform Step 3

There must be three significant figures in the answer, so 0.015814 must be rounded to 0.0158. The answer can also be expressed as 1.58×10^{-2}.

Practice Problems

Perform the following calculations and express the answers to the correct number of significant figures.

2.20 $\dfrac{2.35}{40.08}$

2.21 $0.065 \times 6.023 \times 10^{23}$

2.22 $\dfrac{2.500 \times 453.6}{1.21}$

2.23 $\dfrac{5.000 \times 0.001300}{0.1256}$

2.24 $\dfrac{2.0 \times 20.0}{0.080 \times 8.0}$

Flowchart 2.5 **Determining the number of significant figures in a numerical value obtained by adding and/or subtracting**

Flowchart 2.5

Summary

Method for expressing the answer to the correct number of significant figures when adding and/or subtracting numerical values

Step 1
- **Determine the number of decimal places (digits behind the decimal point) in each number being added and/or subtracted. The smallest number of decimal places in these values is the number of decimal places allowed in the answer.**

↓

Step 2
- **Perform the arithmetic operation(s).**

↓

Step 3
- **Round the answer to the correct number of decimal places.**

Example 1

Perform the following calculation, expressing the answer to the correct number of significant figures: $7.32 + 0.0056$.

Solution

Perform Step 1

There are two decimal places in 7.32, and there are four decimal places in 0.0056. Therefore, the answer can have only two decimal places.

Perform Step 2

7.32 + 0.0056 = 7.3256

Perform Step 3

There must be two decimal places in the answer, so 7.3256 must be rounded to 7.33.

Example 2

Perform the following calculation, expressing the answer to the correct number of significant figures: 92.75 − 16.

Solution

Perform Step 1

There are two decimal places in 92.75, and no decimal places in 16. Therefore the answer can have no decimal places.

Perform Step 2

92.75 − 16 = 76.75

Perform Step 3

There must be no decimal places in the answer, so 76.75 must be rounded to 77.

Example 3

Perform the following calculation, expressing the answer to the correct number of significant figures: 23.113 + 76.2 − 5.31.

Solution

Perform Step 1

There are three decimal places in 23.113, one decimal place in 76.2, and two decimal places in 5.31. Therefore, the answer can have only one decimal place.

Perform Step 2

23.113 + 76.2 − 5.31 = 94.003

Perform Step 3

There must be one decimal place in the answer, so 94.003 must be rounded to 94.0.

Practice Problems

Perform the following calculations and express the answers to the correct number of significant figures.

2.25 1.351 + 12.0236   **2.26** 0.653 + 1.2   **2.27** 35 − 1.356

2.28 36.3 + 0.159 − 0.12   **2.29** 359 − 2.21 − 0.167

Flowchart 2.6 Expressing the answer to the correct number of significant figures when performing calculations involving both multiplication and/or division *and* addition and/or subtraction

Flowchart 2.6

Summary

Method for expressing the answer to the correct number of significant figures when performing calculations involving both multiplication and/or division *and* addition and/or subtraction

Step 1
* Perform the addition/subtraction steps and note the correct number of significant figures that the answer will have (after rounding to the correct number of decimal places), but do not round.

↓

Step 2
* Perform the multiplication/division step(s) and note the correct number of significant figures allowed in the answer.

↓

Step 3
* Round the answer to the correct number of significant figures.

Example 1

Perform the following calculation, expressing the answer to the correct number of significant figures: $\dfrac{18.6 + 0.071}{12.02}$.

Solution

Perform Step 1

$18.6 + 0.071 = 18.671$, which will become 18.7 when rounded to one decimal place. 18.7 has three significant figures.

Perform Step 2

$$\frac{18.671}{12.02} = 1.5533278$$

There must be three significant figures in the answer because 18.7 (from Step 1) has three significant figures and 12.02 has four significant figures.

Perform Step 3

1.5533278 must be rounded to three significant figures.
The correct answer is 1.55.

Example 2

Perform the following calculation, expressing the answer to the correct number of significant figures: $(1.800)(99.6 - 32)$.

Solution

Perform Step 1

$99.6 - 32 = 67.6$, which will become 68 when rounded to no decimal places. 68 has two significant figures.

Perform Step 2

$(1.800)(67.6) = 121.68$

There must be two significant figures in the answer because 68 (from Step 1) has two significant figures and 1.800 has four significant figures.

Perform Step 3

121.68 must be rounded to two significant figures.
The correct answer is 120, or 1.2×10^1.

Example 3

Perform the following calculation, expressing the answer to the correct number of significant figures: $\dfrac{(0.7577 \times 34.969) + (0.2423 \times 36.966)}{6.0 \times 10^{23}}$.

Solution

Perform Step 1

In this case, we must multiply to determine which two numbers will be added. After doing the multiplication steps (and rounding to the correct number of significant figures), we obtain

$$\frac{26.50 + 8.957}{6.0 \times 10^{23}}$$

Now, we add the two numbers.
There are two decimal places in 26.50, and three decimal places in 8.957. Therefore, the sum of these two numbers can have only two decimal places, but we will not round yet. We will use 35.457 in the next step.

Perform Step 2

$$\frac{35.457}{6.0 \times 10^{23}} = 5.8899 \times 10^{-23}$$

There must be two significant figures in the answer because 35.46 (from Step 1) has four significant figures and 6.0×10^{23} has two significant figures.

Perform Step 3

5.8899×10^{-23} must be rounded to 5.9×10^{-23}.

Practice Problems

Perform the following calculations and express the answers to the correct number of significant figures.

2.30 $\dfrac{1.89 + 13.957}{5.1}$

2.31 $\dfrac{(0.632)(2.33 - 0.867)}{12}$

2.32 $\dfrac{(0.667)(109.57) + (0.333)(83.65)}{3.22}$

Flowchart 2.7   **Using unit analysis to convert a measurement to another set of units**

Flowchart 2.7

Summary

Method for using unit analysis to convert a measurement to another set of units

Step 1
- **Write the measurement with its units.**
- **Identify the units to which you need to convert.**

↓

Step 2
- **Multiply the measurement by a conversion factor that will let you cancel the units you don't want.**
- **Cancel units. You can cancel identical units on the top and bottom of the expression.**

↓

- **If one conversion factor won't yield the units you need, multiply by additional conversion factors.**

↓

Step 3
- **Perform the calculation.**

Example 1

Convert 253 miles to centimeters.

Solution

Perform Step 1

253 miles must be converted to centimeters. However, the conversion factor we have available will only allow us to convert miles to kilometers.

Perform Step 2

We can convert miles to kilometers, using the conversion factor:
    1 km = 0.62137 mi
Therefore,

$$253 \text{ mi} \times \frac{1 \text{ km}}{0.62137 \text{ mi}} = 407 \text{ km}$$

Note that we could have used the conversion factor 1 mi = 1.6093 km in the following way:

$$253 \text{ mi} \times \frac{1.6093 \text{ km}}{1 \text{ mi}} = 407 \text{ km}$$

Notice that both conversion factors yield identical results.
We must now multiply by the additional conversion factors to convert kilometers to centimeters.

Perform Step 3

$$407 \text{ km} \times \frac{1000 \text{ m}}{1 \text{ km}} \times \frac{100 \text{ cm}}{1 \text{ m}} = 40{,}700{,}000 \text{ cm} = 4.07 \times 10^7 \text{cm}$$

Example 2

How many grams of silver are there in 750.0 mL? The density of silver is 10.5 g/mL.

Solution

Perform Step 1

750.0 mL Ag

This must be converted to grams, using the density as the conversion factor.

Perform Step 2

$$750.0 \text{ mL Ag} \times \frac{10.5 \text{ g Ag}}{1 \text{ mL Ag}}$$

Perform Step 3

$$750.0 \text{ mL Ag} \times \frac{10.5 \text{ g Ag}}{1 \text{ mL Ag}} = 7875 \text{ g Ag}$$

7875 g Ag = 7880, or 7.88×10^3 g Ag to three significant figures.

Example 3

How many liters of oxygen gas are occupied by 170.2 g of oxygen at standard temperature and pressure? At standard temperature and pressure, 1 g of oxygen occupies 0.700 L.

Solution

Perform Step 1

170.2 g oxygen

This must be converted to liters, using the conversion factor provided in the problem.

Perform Step 2

$$170.2 \text{ g oxygen} \times \frac{0.700 \text{ L oxygen}}{1 \text{ g oxygen}}$$

Perform Step 3

$$170.2 \text{ g oxygen} \times \frac{0.700 \text{ L oxygen}}{1 \text{ g oxygen}} = 119 \text{ L oxygen}$$

Example 4

Convert 70.0 miles per hour to kilometers per minute.

Solution

Perform Step 1

70.0 miles per hour must be converted to kilometers per minute, using the appropriate conversion factors.

Perform Step 2

$$\frac{70.0 \text{ mi}}{1 \text{ hr}} \times \frac{1.6093 \text{ km}}{1 \text{ min}} \times \frac{1 \text{ hr}}{60 \text{ min}}$$

Perform Step 3

$$\frac{70.0 \text{ mi}}{1 \text{ hr}} \times \frac{1.6093 \text{ km}}{1 \text{ min}} \times \frac{1 \text{ hr}}{60 \text{ min}} = 1.88 \text{ km/ min}$$

Example 5

Convert 15 square feet to square centimeters.

Solution

Perform Step 1

15 square feet must be converted to square centimeters. The conversion factors we have available will allow us to convert feet to inches and inches to centimeters. Since the problem involves square feet and square centimeters, we will have to square each of our conversion factors.

Perform Step 2

Convert square feet to square centimeters using the conversion factors: 1 ft = 12 in. and 1 in. = 2.54 cm.
Therefore,

$$15 \text{ ft}^2 \times \frac{(12 \text{ in.})^2}{(1 \text{ ft})^2} \times \frac{(2.54 \text{ cm})^2}{(1 \text{ in.})^2} = 1.4 \times 10^4 \text{ cm}^2$$

Practice Problems

Perform the following conversions, using unit analysis, and express the answers to the correct number of significant figures.

2.33 100.0 grams of iron to liters of iron. The density of iron is 7.87 g/mL.

2.34 120 kilometers per hour to miles per minute

2.35 536 milligrams to kilograms

2.36 25 square meters to square centimeters

2.37 25 dollars per square feet to dollars per square inch

Flowchart 2.8   Rearranging algebraic equations to solve for any variable that appears in the equation

Flowchart 2.8

Summary

Method for rearranging algebraic equations to solve for any variable that appears in the equation, when given the values for all other variables in the equation

Step 1
- Write the equation.
- Identify the variable that must be solved for in the problem.

$\downarrow$

Step 2
- Isolate this variable on one side of the equation by algebraically solving the equation for this variable. (Consult the mathematics review in the appendix if you are not sure how to do this.)
- If the variable is already isolated on one side of the equation, proceed to Step 3.

$\downarrow$

Step 3
- Substitute the values given in the problem for the variables other than the one for which you are solving.
- Make sure that the values being substituted are in the correct units. If not, perform the appropriate conversions before substituting.

$\downarrow$

Step 4
- Perform the arithmetic operation(s) to arrive at the answer.

Example 1

Calculate the amount of heat (in joules) needed to raise the temperature of 550.0 g of iron by 25.0 °C. The specific heat of iron is 0.449 J/g · °C. The algebraic equation relating the variables is shown below:

$$\text{Heat} = \text{Specific heat} \times \text{Mass of iron} \times \text{Change in temperature}$$

Solution

Perform Step 1

The variable to be solved for is heat, which is already isolated on the left side of the equation.

Perform Step 2

$$\text{Heat} = 0.449 \, \text{J/g} \cdot {}^\circ\text{C} \times 550.0 \, \text{g} \times 25.0 \, {}^\circ\text{C}$$

Perform Step 3

$$\text{Heat} = 0.449 \, \text{J/g} \cdot {}^\circ\text{C} \times 550.0 \, \text{g} \times 25.0 \, {}^\circ\text{C} = 6.17 \times 10^3 \, \text{J}$$

Example 2

Calculate the change in temperature resulting from adding 3.134×10^5 J of heat to a 5.00 kg block of aluminum. The specific heat of aluminum is 0.901 J/g · °C.

The algebraic equation relating the variables is shown below:

Heat = Specific heat × Mass of aluminum × Change in temperature

Solution

Perform Step 1

The variable to be solved for is change in temperature, which must be isolated on one side of the equation.

Perform Step 2

Change in temperature can be isolated by dividing both sides of the equation by specific heat and by mass of aluminum. The resulting equation is:

$$\text{Change in temperature} = \frac{\text{Heat}}{\text{Specific heat} \times \text{Mass of aluminum}}$$

Perform Step 3

Note that before substituting the values, we must convert the mass from kilograms to grams, so that the mass unit is consistent with the units of specific heat.

$$\text{Change in temperature} = \frac{3.134 \times 10^5 \text{J}}{0.901 \text{ J/g} \,°\text{C} \times 5.00 \times 10^3 \text{g}}$$

Perform Step 4

Change in temperature = 69.6 °C.

Practice Problems

2.38 How many joules are required to raise the temperature of 972.0 g of aluminum from 0.0 °C to 50.0 °C? The specific heat of aluminum is 0.901 J/g · °C.

2.39 25.0 kilojoules of heat are added to a 500. g bar of iron metal at 25 °C. What is the final temperature of the iron bar? The specific heat of iron is 0.449 J/g · °C.

Quiz for Chapter 2 Problems

1. Indicate the number of significant figures in each measured number below.
   (a) 357 mL   (b) 1.0600 L   (c) 0.000501   (d) 23,000   (e) 0.08720100

2. Indicate the number of zeros that are significant in each measured number below.
   (a) 0.0038100   (b) 500.00   (c) 0.030010   (d) 5200.   (e) 5200

3. Convert the following numbers from scientific to standard notation.
   (a) 8.59×10^{-3} (b) 2.76×10^2 (c) 2.76×10^{-2} (d) 7.2×10^9

4. Convert the following numbers from standard to scientific notation.
   (a) 0.0000000008304   (b) 9,500,000   (c) 0.013   (d) 58.3   (e) 0.583

5. Perform the following arithmetic operations and express the answer to the correct number of significant figures.

   (a) 0.392 + 51.4   (b) 0.001 + 5.32   (c) 273.15 − 28.3   (d) 1582 + 0.59

6. Perform the following arithmetic operations and express the answer to the correct number of significant figures.

   (a) 8.63 × 0.58   (b) $\dfrac{6.02}{3.0}$ (c) $\dfrac{5260 \times 12.0}{2.1}$ (d) $\dfrac{22.4 \times 3.1}{0.3214}$

7. Perform the following arithmetic operations and express the answer to the correct number of significant figures.

   (a) (8.39 × 4.7) + 6.23   (b) $\dfrac{6.8 \times 10^5}{7.31 - 4.2}$ (c) $\dfrac{3.90 \times 8.631}{6.0}$ (d) $\dfrac{3.9 \times 8.631}{6}$

8. Use unit analysis to perform each of the following conversions.

   (a) 58.3 mL to L   (b) 0.00463 cm to m   (c) 839 kg to g   (d) 92.0 in.^{2} to cm^2

9. Copper has a density of 8.96 g/mL. Calculate the volume occupied by 125.0 g of copper.

10. Mercury has a density of 13.6 g/mL. What is the weight of 75.3 mL of mercury?

11. Calculate the number of cubic inches in 5000 cm^3.

12. A carpet is advertised at a price of $30.00/ft^2. What is the cost of the carpet per square inch?

13. Calculate the heat (in joules) required to raise the temperature of 1000.0 g of aluminum from 25.0 °C to 50.0 °C. (The specific heat of aluminum is 0.901 J/g · °C.)

14. A sample of iron experiences a temperature increase of 35.0 °C when 5892 J of heat are added to it. What is the mass of the iron? (The specific heat of iron is 0.449 J/g · °C.)

15. The maximum speed of some U.S. Interstate highways is 75 mi/hr. What is this speed in kilometers per hour?

16. How are the measured values 156,000 and 156,000 different? Which one assumes a more accurate measuring device?

17. The formula for converting temperature from °C to °F is given below.
    °F = 32 + 9/5°C
    Convert 37 °C to °F and express the answer to the correct number of significant figures.

18. Fill in the blanks with the words greater than, less than, or equal to.
    (a) 5.32 is _____ _____ 53.2 × 10^1.
    (b) 5.32 is _____ _____ 53.2 × 10^{-1}.
    (c) 5.32 is _____ _____ 53.2 × 10^0.
    (d) 5.32 × 10^{-1} is _____ _____ 0.532.
    (e) 5.32 × 10^{-1} is _____ _____ 53.2.

19. Calculate the density of a piece of metal if the metal weighs 386 g and occupies a volume of 20.0 mL.

20. The formula for converting °C to K is shown below.
    K = °C + 273.15
    Convert 83.1 °C to K, and express the answer to the correct number of significant figures.

Exercises for Self-Testing

A. Completion. *Write the correct word(s) to complete each statement below.*

1. _____ is the term used to describe the closeness of a measured result to the true value of the number; _____ is the term used to describe the proximity of a set of values to each other.
2. The _____ _____ in a number is where the uncertainty is presumed to be.
3. A(n) _____ property depends on the amount of material present; a(n) _____ property does not depend on the amount of material present.
4. The unit that represents 1/100 of a meter is the _____.
5. When you are determining the number of significant digits, all _____ _____ are always significant.
6. The Greek prefix meaning 10^6 is _____.
7. _____ is a measure of the compactness of matter; it is defined as the mass per unit volume.
8. The boiling point of water is _____ on the Celsius scale, _____ on the Fahrenheit scale, and _____ on the Kelvin scale.
9. 0.0854 is equal to _____ $\times 10^{-4}$.
10. If $a = bx + y$, then $x =$ _____.
11. The amount of heat required to raise the temperature of 1 g of a substance by 1 °C is called the _____ _____.

B. True/False. *Indicate whether each statement below is true or false, and explain why the false statements are false.*

_____ 1.   A millimeter is smaller than a meter.
_____ 2.   A length of pipe that is actually 31.21 cm is measured by a student. The student obtains three measurements: 34.78, 34.80, and 34.79 cm. These measurements can be considered precise but not accurate.
_____ 3.   If a liquid has a density of 1.5 g/mL, 6 mL of the liquid weighs 9 g.
_____ 4.   There are four significant digits in the number 0.0450.
_____ 5.   The SI unit of mass is the kilogram.
_____ 6.   There are 5.6×10^{23} mm in 5.6 m.
_____ 7.   An extensive property depends on the amount of material present.
_____ 8.   The size of a kelvin is equal to the size of a Fahrenheit degree.
_____ 9.   If only one measurement is taken of a length, the measurement can be properly described as precise if the measured value is close to the true value.
_____ 10.   −40 °C is equal to −40 °F
_____ 11.   The amount of heat gained by the water in a bomb calorimeter is equal to the amount of heat lost by the substance burned in the calorimeter.

C. Problems and Questions

1. Perform the following conversions:
   (a) 543 liters to milliliters   (b) 3.6 inches to feet   (c) 43.0 years to months
2. Convert 70 °F
   (a) to °C   (b) to K

3. An object is determined to weigh 96.3 g and occupy a volume of 0.082 L. What is the density of the object in grams per milliliter?

4. A child weighs 52 lb. What is her mass in milligrams? (*Hint*: 1 kg = 2.2046 lb.)

5. Explain the difference between an intensive and an extensive property. Give an example of each.

6. Write the conversion factor involving liters and kiloliters in two different ways.

7. Solve the following equation for n: $PV = nRT$.

8. The density of mercury is 13.6 g/mL. Calculate the volume that a 123.2 g sample of mercury will occupy.

9. Perform the following arithmetic operations. Express the answer to the correct number of significant digits.

    (a) $59.3 \times 10^2 + 8.13 \times 10^1$ (b) $0.0032 \times 5.63 \times 10^5$

10. Explain the difference between an exact value (e.g., the number 5) and a measured value in terms of the number of significant digits in each.

11. The specific heat of copper is 0.385 J/g · °C. Calculate the heat required to raise the temperature of a 1.50 kg block of copper from 28.5 °C to 75.0 °C.

3

The Evolution of Atomic Theory

Learning Outcomes

1. Calculate the percent composition of each element in a compound.
2. Describe how Dalton explained formulas of compounds.
3. Name and describe the subatomic particles in an atom.
4. Describe Rutherford's alpha-particle experiment and what it proved.
5. Explain what the atomic number and the mass number tell you about an atom.
6. Explain the meaning of an atom's atomic mass and mass number.
7. Define *isotopes.*
8. Explain Mendeleev's discovery.
9. Identify groups, periods, metals, and nonmetals in the periodic table.
10. Explain what cations and anions are and how to form them.
11. Understand what a mass spectrometer is and how it works.

Chapter Outline

3.1 Dalton's Atomic Theory

    A. Laws leading to the formulation of the theory

        1. Law of conservation of matter

        2. Law of definite proportions and calculation of percent by mass

        3. Law of multiple proportions

    B. Five tenets of Dalton's atomic theory

Review of Key Concepts from the Textbook

3.1 Dalton's Atomic Theory

The existence of atoms was proposed by the philosopher Democritus as early as 400 B.C. However, it was not until the early 1800s that John Dalton formulated an atomic theory based on scientific evidence. The evidence that Dalton used was compiled from the experiments that were used to formulate the law of conservation of matter and the law of constant composition (or definite proportions). Each of these laws is discussed briefly below.

1. *The law of conservation of matter* states that when a chemical reaction takes place, matter is neither created nor destroyed. In other words, the mass of the substances reacting equals the mass of the substances produced.

2. *The law of constant composition (or definite proportions)* states that a compound always consists of the same percentages by mass of all its elements. This is illustrated by the percentage composition of water:

$$\% \text{ H} = \frac{1.0079 \times 2 \text{ g H}}{18.015 \text{ g H}_2\text{O}} \times 100\% = 11.189\% \text{ H}$$

$$\% \text{ O} = \frac{15.9994 \text{ g H}}{18.015 \text{ g H}_2\text{O}} \times 100\% = 88.811\% \text{ O}$$

In other words, water always consists of 11.189% hydrogen by mass and 88.811% oxygen by mass, no matter whether the sample of water is found on Mars or Earth.

Based on the law of conservation of matter and the law of constant composition (or definite proportions), Dalton formulated the five tenets of his atomic theory:

1. All matter is composed of discrete, indestructible atoms.
2. Atoms can be neither created nor destroyed.
3. All atoms of a particular element are exactly alike.
4. The atoms of different elements are different.
5. Chemical reactions involve the combination, separation, or rearrangement of individual atoms.

3.2   Development of a Model for Atomic Structure

Once it was demonstrated that atoms exist, the next problem scientists addressed was the composition and structure of the atom. The presence of negatively charged electrons and positively charged protons had been established by 1907. The third subatomic particle, the neutron, was discovered in 1932. The relative masses and charges of the three subatomic particles are shown in the following table:

Particle	Mass	Charge
Electron (e)	1/1836	−1
Proton (p)	1	+1
Neutron	1	0

After the discovery of the electron in 1897, J. J. Thomson proposed a "plum pudding" model of the atom, in which the electrons were dispersed in a spherical cloud of positive charge, as shown in the figure below. The model was called the plum pudding model because the electrons are like plums dispersed in a pudding (the spherical cloud of positive charge).

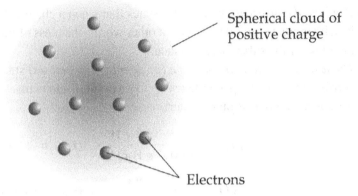

Spherical cloud of positive charge

Electrons

3.3 The Nucleus

After the discovery of the proton in 1907, Ernest Rutherford conducted a clever experiment involving alpha particles. (See Section 3.3 of the textbook for a complete discussion of the experiment.) Rutherford was able to establish that the atom consists of mostly empty space but contains an extremely small, extremely dense positively charged mass at its center. This small, dense center is the nucleus. Rutherford proposed a model in which the small, dense nucleus is at the center of an atom of mostly empty space, as shown in the figure below.

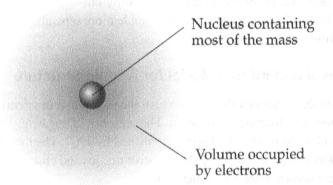

Nucleus containing most of the mass

Volume occupied by electrons

3.4 The Structure of the Atom

We have established that the atom contains positively charged protons, neutrons with no charge, and negatively charged electrons. The number of protons is called the *atomic number*, which is abbreviated by the letter Z. The atomic number is a defining characteristic of an element because the number of protons determines the identity of the element.

The number of electrons in a neutral atom is always equal to the number of protons. So, once you know how many protons an atom contains (its atomic number, Z), you also know how many electrons it has.

Determining the number of neutrons inside the nucleus is a bit more involved because there is no simple relationship between the number of neutrons and the number of protons in an atom. However, chemists have defined a quantity called the *mass number*, which is abbreviated by the letter A. The mass number is the sum of the number of protons plus the number of neutrons (abbreviated by the letter N).

$$\text{Protons} + \text{Neutrons} = \text{Mass number}$$
$$\text{Atomic number} + \text{Neutrons} = \text{Mass number}$$
$$Z + N = A$$

It should now be apparent that we can determine the number of neutrons in an atom if we know the number of protons and the mass number. We simply subtract the atomic number (number of protons) from the mass number (protons + neutrons) to get the number of neutrons:

$$\text{Neutrons} = \text{Mass number} - \text{Atomic number}$$
$$N = A - Z$$

Example 3.1

An atom of iron has an atomic number of 26 and a mass number of 59. How many protons, electrons, and neutrons does the atom contain?

Solution

Protons = Atomic number = 26 protons

Electrons = Protons = 26 electrons

Neutrons = Mass number − Atomic number = 59 − 26 = 33 neutrons

Practice Exercise 3.1

An atom has 18 protons and 22 neutrons. What is the atomic number of the element? What is the mass number of the element?

We stated earlier that the number of protons defines the identity of an element. This means that all atoms of a given element have exactly the same number of protons. However, this is not the case with neutrons. Two atoms of the same element can contain different numbers of neutrons. Whenever this occurs, the two versions of the element are called *isotopes*. Isotopes are atoms of the same element containing different numbers of neutrons. Because the number of neutrons helps to determine the mass number, isotopes of an element have different mass numbers and, therefore, different masses.

Example 3.2

The atomic number of iron is 26. The mass number of one isotope of iron (isotope A) is 56, and the mass number of another isotope (isotope B) is 58. How many protons, electrons, and neutrons are in each isotope?

Solution

Isotope A:	26 protons	26 electrons	30 neutrons
Isotope B:	26 protons	26 electrons	32 neutrons

Notice that the atomic number (number of protons) is always the same for different isotopes of the same element.

Practice Exercise 3.2

Indicate the number of protons, electrons, and neutrons in the two isotopes of an element with the following atomic and mass numbers:

Isotope A:	Atomic number	= 27	Mass number = 59
Isotope B:	Atomic number	= ?	Mass number = 60

Note that you should be able to determine the atomic number of isotope B from the information given in the problem.

Chemists symbolize isotopes using full atomic symbols, which include the symbol for the element, the atomic number, and the mass number, as shown here:

$$^A_Z X$$

where X is the elemental symbol, Z is the atomic number (number of protons), and A is the mass number (protons + neutrons); the number of neutrons $= A - Z$. *Note*: All atoms of an element have the same number of protons; only the number of neutrons (for different isotopes) or the number of electrons (in the case of an ion of the element) can change.

Isotopes are often abbreviated by writing the symbol of the element and the mass number, such as ^{14}C or ^{238}U. This format provides as much information as the full atomic symbol, because once we know the element, we know the atomic number.

Example 3.3

How many protons, electrons, and neutrons are there in an atom of ^{14}C?

Solution

We know that ^{14}C represents the isotope of carbon with a mass number of 14. We also know that the atomic number of carbon is 6 (see Section 3.4 of the textbook). Therefore, ^{14}C has six protons, six electrons, and eight neutrons.

Practice Exercise 3.3

How many protons, electrons, and neutrons are there in an atom of ^{238}U?

We will now consider the *atomic mass* of an element. Although the atomic mass of an element is usually close to the mass number of an isotope of the element, it is different from the mass number (unless the element exists in only one isotopic form). The atomic mass of an element is the *weighted average* of all the isotopes of an element. We get the atomic mass by adding the mass of each isotope multiplied by its percent abundance in decimal form (the percent abundance divided by 100%). In other words, we use the following mathematical equation to calculate the atomic mass from the masses of individual isotopes:

$$\text{Atomic mass} = \left(\text{Mass of isotope 1} \times \frac{\% \text{ of isotope 1}}{100\%} \right) + \left(\text{Mass of isotope 2} \times \frac{\% \text{ of isotope 2}}{100\%} \right)$$

Example 3.4

Calculate the atomic mass of naturally occurring copper (Cu), which is composed of two isotopes. The atomic number of Cu is 29, and the percent abundances and masses of the two Cu isotopes are as follows:

^{63}Cu	69.17%	Atomic mass	= 62.94 amu
^{65}Cu	30.83%	Atomic mass	= 64.93 amu

Solution

$$\text{Atomic mass} = \left(\text{Mass of isotope 1} \times \frac{\% \text{ of isotope 1}}{100\%} \right) + \left(\text{Mass of isotope 2} \times \frac{\% \text{ of isotope 2}}{100\%} \right)$$

$$= (62.94 \text{ amu} \times 0.6917) + (64.93 \text{ amu} \times 0.3083) = 63.55 \text{ amu}$$

Practice Exercise 3.4

Calculate the atomic mass of naturally occurring silver (Ag), which is composed of two isotopes. The atomic number of Ag is 47, and the percent abundances and masses of the two Ag isotopes are as follows:

^{107}Ag	51.84%	Atomic mass	= 106.9051 amu
^{109}Ag	48.16%	Atomic mass	= 108.9048 amu

3.5 The Law of Mendeleev—Chemical Periodicity

When the properties of the elements were examined, Mendeleev discovered that their chemical properties repeated in a regular way when they were arranged according to increasing atomic mass. The repeating behavior of the chemical properties of the elements is called chemical periodicity or periodic behavior. Mendeleev had discovered an important relationship among the elements. However, his arrangement of the elements in order of increasing atomic masses was later found to be close to but not exactly consistent with an arrangement according to chemical properties.

3.6 The Modern Periodic Table

After more elements were discovered and their properties studied, it became clear that the correct arrangement of the elements is in order of increasing atomic *number*, not increasing atomic *mass*. The modern periodic table is arranged according to increasing atomic number and consists of vertical groups (or families) and horizontal periods (see the figure below). The elements in the table are separated into three main classes: representative or main group elements, transition metals, and rare earths. Some of the groups have special names.

Group IA	Alkali metals
Group IIA	Alkaline earth metals
Group VIA	Chalcogens
Group VIIA	Halogens
Group VIIIA	Noble gases

The metals are separated from the nonmetals by a stair-step division line. Those elements touching the line are the metalloids, except for aluminum, which is a metal.

3.7 An Introduction to Ions and the First Ionization Energy

Two other properties of elements that vary according to position in the periodic table are atomic radius (or size) and ionization energy (the minimum energy needed to remove the outermost electron from a neutral atom). The atomic radius gradually decreases from left to right across a period and gradually increases from top to bottom down a group. The ionization energy gradually increases from left to right across a period and gradually decreases from top to bottom down a group.

Note that when an electron is removed from a neutral atom, there is no longer an equal number of protons and electrons. When the number of protons differs from the number of electrons, the species is called an *ion*. If an electron is *removed* from a neutral atom, there are more protons, and the resulting positive ion is called a *cation*. If an electron is *added* to a neutral atom, there are more electrons, and the resulting negative ion is called an *anion*.

The complete ionic symbol, as shown below, contains the elemental symbol (O), the atomic number (8), the mass number (16), and the charge on the ion (2−).

$$^{16}_{8}O^{2-}$$

Example 3.5

Write the complete ionic symbol for an ion of sodium-23 (^{23}Na) that has one less electron than the number of protons.

Solution

$$^{23}_{11}Na^{+}$$

Practice Exercise 3.5

Write the complete ionic symbol for an ion of sulfur-32 (^{32}S) that has two more electrons than the number of protons.

In Chapter 4, you will learn why atoms of certain elements, such as sodium and sulfur, form ions with specific charges: 1+ and 2−, respectively. You will also find out why some charges are not possible for ions of certain atoms. For instance, you will never encounter a 2+ cation of sodium or a 1− anion of sulfur. The main key to deciding on the actual charges ions lies in the location of the elements in the periodic table.

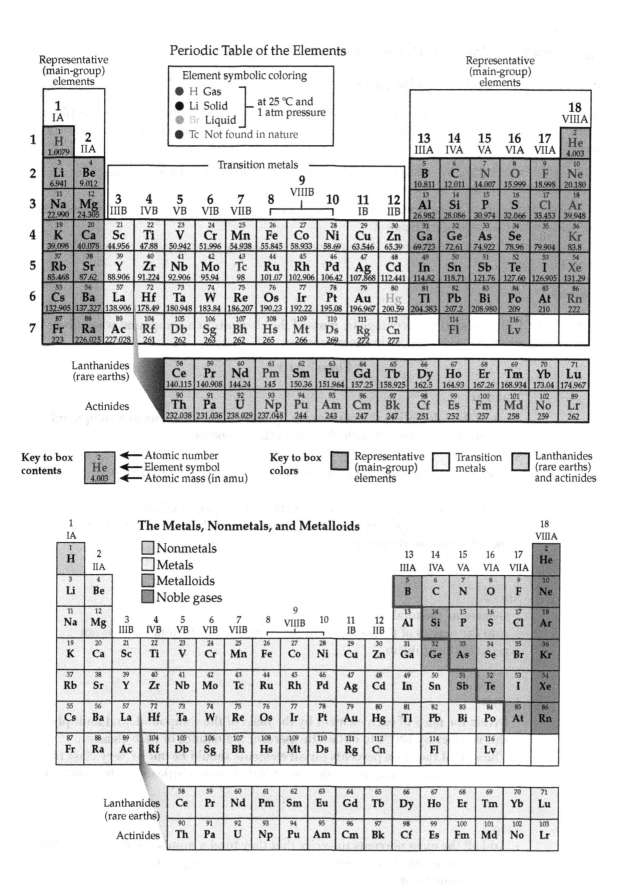

Periodic Table of the Elements

Strategies for Working Problems in Chapter 3

Overview: What You Should Be Able to Do

Chapter 3 provides an introduction to the structure of the atom, the periodic table, and the properties of the elements. The types of problems that you should be able to solve after mastering Chapter 3 are as follows:

1. Calculate the percentage by mass of the elements in a compound when given the grams of each element in the compound.
2. Calculate the weighted average atomic mass of any element when given the percent abundance and the mass of each isotope of the element.
3. Arrange a group of elements in order of increasing (or decreasing) atomic radius (atomic size) or ionization energy.

Flowcharts

Flowchart 3.1   **Calculating the percentage by mass of the elements in a compound when given the grams of each element in the compound**

Flowchart 3.1

Summary

Method for calculating the percentage by mass of the elements in a compound when given the grams of each element in the compound

Step 1
- **Determine the mass of each element in the compound from the information given.**
- **If all but one of the masses are given, the missing mass value can be obtained by subtracting the sum of the masses of the other elements from the total mass of the compound.**

$\downarrow$

Step 2
- **Determine the total mass of the compound.**
- **Add the masses of the elements in the compound if the total mass is not given in the problem.**

$\downarrow$

Step 3
- **Divide the mass of each element by the mass of the compound, and multiply this number by 100% to obtain the percent abundance of each element.**
- **The sum of the mass percents for all elements should be one or very close to 100%. (Rounding may make the sum a little less or a little more than 100%.)**

$\downarrow$

Step 4
- **Add all the mass percents to verify that they total a number very close to 100%.**
- **If the total is not close to 100%, redo the problem, beginning at Step 1, to correct your error.**

Example 1

A 25.00 g sample of PCl_3 contains 5.64 g of phosphorus. Calculate the mass percent of P and Cl in PCl_3.

Solution

Perform Step 1

The mass of phosphorus $= 5.64$ g P

The mass of chlorine $= 25.00 - 5.64 = 19.36$ g Cl

Perform Step 2

The total mass of the compound is 25.00 g.

Perform Step 3

$$\% \text{ P} = \frac{5.64 \text{ g P}}{25.00 \text{ g cmpd}} \times 100\% = 22.6\% \text{ P}$$

$$\% \text{ Cl} = \frac{19.36 \text{ g Cl}}{25.00 \text{ g cmpd}} \times 100\% = 77.44\% \text{ Cl}$$

Perform Step 4

Sum of mass percents:

22.6% P $+ 77.44\%$ Cl $= 100.04\%$ (i.e., close to 100%)

Example 2

Calculate the mass percent of each element in a compound containing aluminum and oxygen if 203.92 g of compound contain 105.8 g of aluminum.

Solution

Perform Step 1

The mass of aluminum $= 105.8$ g Al

The mass of oxygen $= 203.92 - 105.8 = 98.1$ g O

Perform Step 2

The total mass of the compound is 203.92 g.

Perform Step 3

$$\% \text{ Al} = \frac{105.8 \text{ g Al}}{203.92 \text{ g cmpd}} \times 100\% = 51.88\% \text{ Al}$$

$$\% \text{ O} = \frac{98.1 \text{ g O}}{203.92 \text{ g cmpd}} \times 100\% = 48.1\% \text{ O}$$

Perform Step 4

Sum of mass percents:

51.88% Al $+ 48.1\%$ O $= 100.0\%$

Example 3

Calculate the mass percent of each element in a compound containing 15.00 g sulfur and 22.44 g oxygen.

Solution

Perform Step 1

The mass of sulfur $= 15.00$ g S
The mass of oxygen $= 22.44$ g O

Perform Step 2

The total mass of the compound is 15.00 g S $+ 22.44$ g O $= 37.44$ g.

Perform Step 3

$$\% \text{ S} = \frac{15.00 \text{ g S}}{37.44 \text{ g cmpd}} \times 100\% = 40.06\% \text{ S}$$

$$\% \text{ O} = \frac{22.44 \text{ g O}}{37.44 \text{ g cmpd}} \times 100\% = 59.94\% \text{ O}$$

Perform Step 4

Sum of mass percents:
40.06% S $+ 59.94\%$ O $= 100\%$

Practice Problems

3.1 Calculate the mass percent of each element in a compound containing carbon and sulfur if the compound contains 24.0 g of carbon and 128.0 g of sulfur.

3.2 Calculate the mass percent of each element in a compound containing sodium and oxygen if 39.0 g of the compound contains 16.0 g of oxygen.

3.3 Calculate the mass percent of each element in a compound that contains 5.40 g of aluminum and 4.80 g of oxygen.

Flowchart 3.2 **Calculating the weighted average atomic mass of any element when given the percent abundance and the mass of each isotope of the element**

Flowchart 3.2

Summary

Method for calculating the weighted average atomic mass of any element when given the percent abundance and the mass of each isotope of the element

Step 1
- Determine the mass of each isotope in the compound from the information given.

↓

Step 2
- Determine the percent abundance of each isotope in the compound from the information given.

- **If all but one of the percent abundances are given, the missing percent abundance value can be obtained by subtracting the sum of the percent abundances of the other isotopes from 100%.**

$$\downarrow$$

Step 3
- **For each isotope, divide the percent abundance by 100%, and multiply the result by the mass of the isotope.**
- **This result represents the portion of the weight of the element contributed by each isotope.**

$$\downarrow$$

Step 4
- **Add all the portions obtained in Step 3.**
- **This sum is the weighted average atomic mass of the element.**

Example 1

Calculate the weighted average atomic mass for bromine. Bromine consists of two isotopes, ^{79}Br (atomic mass 78.9183 amu, abundance 50.69%) and ^{81}Br (atomic mass 80.9163 amu, abundance 49.31%).

Solution

Perform Step 1

Atomic masses:

$^{79}Br = 78.9183$ amu

$^{81}Br = 80.9163$ amu

Perform Step 2

Percent abundance of each isotope:

$^{79}Br = 50.69\%$

$^{81}Br = 49.31\%$

Perform Step 3

Portion of the mass contributed by each isotope:

$$^{79}Br = \frac{50.69\%}{100\%} \times 78.9183 \text{ amu} = 40.00 \text{ amu from } ^{79}Br$$

$$^{81}Br = \frac{49.31\%}{100\%} \times 80.9163 \text{ amu} = 39.90 \text{ amu from } ^{81}Br$$

Perform Step 4

Add all portions from Step 3:

Weighted average atomic mass for Br = 40.00 amu + 39.90 amu = 79.90 amu

Example 2

Calculate the weighted average atomic mass for magnesium. Magnesium consists of three isotopes, ^{24}Mg (atomic mass 23.9850 amu, abundance 78.99%), ^{25}Mg (atomic mass 24.9858 amu, abundance 10.00%), and ^{26}Mg (atomic mass 25.9826 amu).

Solution

Perform Step 1

Atomic masses:

^{24}Mg = 23.9850 amu

^{25}Mg = 24.9858 amu

^{26}Mg = 25.9826 amu

Perform Step 2

Percent abundance of each isotope:

^{24}Mg = 78.99%

^{25}Mg = 10.00%

^{26}Mg = 100% − (78.99% + 10.00%) = 11.01%

Perform Step 3

Portion of the mass contributed by each isotope:

$$^{24}\text{Mg} = \frac{78.99\%}{100\%} \times 23.9850 \text{ amu} = 18.95 \text{ amu from } ^{24}\text{Mg}$$

$$^{25}\text{Mg} = \frac{10.00\%}{100\%} \times 24.9858 \text{ amu} = 2.499 \text{ amu from } ^{25}\text{Mg}$$

$$^{26}\text{Mg} = \frac{11.01\%}{100\%} \times 25.9826 \text{ amu} = 2.861 \text{ amu from } ^{26}\text{Mg}$$

Perform Step 4

Add all portions from Step 3:

Weighted average atomic mass for Mg = 18.95 amu + 2.499 amu + 2.861 amu = 24.31 amu

Practice Problems

3.4 Boron consists of two naturally occurring isotopes, ^{10}B and ^{11}B. If the abundance of ^{10}B is 19.9%, what is the abundance of ^{11}B?

3.5 Argon consists of three naturally occurring isotopes, ^{36}Ar (atomic mass 35.9675 amu, abundance 0.3365%), ^{38}Ar (atomic mass 37.9627 amu, abundance 0.0632%), and ^{40}Ar (atomic mass 39.9634 amu, abundance 99.6003%). Calculate the weighted average atomic mass for Ar.

Flowchart 3.3 **Arranging a group of elements in order of increasing (or decreasing) atomic radius (atomic size) or ionization energy**

Flowchart 3.3

Summary

Method for arranging a group of elements in order of increasing (or decreasing) atomic radius (atomic size) or ionization energy

Step 1

- Determine the periodic trend for the specified property (e.g., atomic radius increases from top to bottom and decreases from left to right on the periodic table).

↓

Step 2

- Note whether the problem specifies an arrangement in increasing or decreasing order.
- If increasing order is specified, the smallest will be first; if decreasing order is specified, the largest will be first.

↓

Step 3

- Arrange the elements in the appropriate order.

Example 1

Arrange the following elements in order of increasing atomic radius:

Mg, S, Si, P, Na

Solution

Perform Step 1

The atomic radius decreases from left to right.

Perform Step 2

The problem asks for arranging in order of increasing atomic radius; therefore, the smallest atom (the one farthest to the right in the periodic table) will be first. This is S.

Perform Step 3

S < P < Si < Mg < Na

Example 2

Arrange the following elements in order of decreasing ionization energy:

Se, O, Te, Po, S

Solution

Perform Step 1

The ionization energy decreases from top to bottom.

Perform Step 2

The problem asks for arranging in order of decreasing ionization energy; therefore, the element with the largest ionization energy (the one closest to the top in the periodic table) will be first. This is O.

Perform Step 3

O > S > Se > Te > Po

Example 3

Arrange the following elements in order of decreasing atomic radius:

Sn, Rb, Xe, Sr, Te

Solution

Perform Step 1

The atomic radius decreases from left to right.

Perform Step 2

The problem asks for arranging in order of decreasing atomic radius; therefore, the element with the largest atomic radius (the one closest to the left in the periodic table) will be first. This is Rb.

Perform Step 3

Rb > Sr > Sn > Te > Xe

Practice Problems

3.6 Arrange the following elements in order of increasing ionization energy:

Br, F, Cl, I

3.7 Arrange the following elements in order of decreasing atomic radius:

Br, F, Cl, I

3.8 Which element below has the smallest ionization energy?

Ge, Sn, Si, C

Quiz for Chapter 3 Problems

1. Carbonic acid has the formula H_2CO_3. Calculate the mass percents of the elements in carbonic acid.

2. Calculate the weighted average atomic mass for silver. Silver consists of two isotopes, ^{107}Ag (atomic mass 106.9051 amu, abundance 51.840%) and ^{109}Ag (atomic mass 108.9048, abundance 48.162%).

3. Arrange the following elements in order of increasing atomic radius:

Pb, Po, Tl, Cs

4. Potassium consists of three naturally occurring isotopes, ^{39}K (atomic mass 38.9637 amu), ^{40}K (atomic mass 39.9640 amu), and ^{41}K (atomic mass 40.9618). If the weighted average atomic mass of potassium is 39.098, which of the three isotopes is present in the greatest amount in naturally occurring potassium? Explain your choice.

5. Arrange the following elements in order of increasing ionization energy:

O, C, Ne, F, Be

6. Circle the element that has the greatest atomic radius:

Al, In, Tl, B, Ga

7. Indium consists of two naturally occurring isotopes, ^{113}In (atomic mass 112.9041 amu) and ^{115}In (atomic mass 114.9039 amu). The percent abundance of ^{113}In is 4.29%. What is the percent abundance of ^{115}In?

8. Arrange the following elements in order of decreasing atomic radius:

   Na, Rb, K, Cs, Li

9. Circle the element with the smallest ionization energy:

   Ar, Ne, He, Xe, Kr

10. Arrange the following elements in order of decreasing ionization energy:

    O, C, Ne, F, Be

Exercises for Self-Testing

A. Completion. *Write the correct word(s) to complete each statement below.*

1. All matter is composed of discrete, indestructible _____.
2. The Thomson model of the atom is known as the _____ _____ model.
3. Rutherford used _____ _____ to bombard gold foil in his famous experiment.
4. The three major subatomic particles are _____, _____, and _____.
5. The atomic number is the number of _____ in an atom's nucleus.
6. The small, dense center of the atom is called the _____, and it contains the _____ and _____.
7. _____ are atoms with the same atomic number but different mass numbers.
8. An atom of ^{13}C contains _____ protons, _____ electrons, and _____ neutrons.
9. To determine the number of neutrons in an atom, the _____ number is subtracted from the _____ number.
10. The _____ _____ of an element is the weighted average of all the isotopes of that element.
11. The vertical columns in the periodic table are called _____, and the horizontal rows are called _____.
12. The elements in group IIA of the periodic table are known as the _____ _____ metals.
13. The atomic radii of the elements _____ going from left to right and _____ going from top to bottom in the periodic table.
14. The name of the element with the symbol K is _____.
15. The _____ _____ is the minimum energy required to remove the outermost electron from a neutral atom.
16. An atom that has lost three electrons has a charge of _____.
17. Positively charged ions are called _____, and negatively charged ions are called _____.

B. True/False. *Indicate whether each statement below is true or false, and explain why the false statements are false.*

_____ 1.   All atoms of a particular element are exactly alike.

_____ 2.   All atoms of carbon contain six protons.

_____ 3.   The mass number of an element is the sum of its protons and neutrons.

_____ 4.   All atoms of nitrogen contain seven neutrons.

_____ 5.   Isotopes are atoms that have the same number of protons but different numbers of electrons.

_____ 6.   In the modern periodic table, the elements are arranged according to increasing atomic mass.

_____ 7.   Anions are ions that contain more protons than electrons.

_____ 8.   A 2+ cation of calcium-40 isotope has 40 protons and 38 electrons.

_____ 9.   A 1− anion of chlorine-35 isotope has 17 protons and 18 electrons.

_____ 10.   A 2+ cation of calcium and a 1− anion of chlorine contain the same number of electrons.

C. Questions and Problems

1. Explain why Dalton's statement that "all atoms of a particular element are exactly alike" is false.

2. Write the symbols for the following elements:
   (a) helium   (b) potassium   (c) sodium   (d) iron   (e) silver   (f) gold

3. Indicate the number of protons, neutrons, and electrons in a neutral atom of each of the following isotopes:
   (a) $^{200}_{85}At$ (b) $^{108}_{47}Ag$ (c) ^{15}N (d) ^{23}Na

4. Write the full atomic symbol and the abbreviated symbol for each of the following:

   (a) a carbon atom that contains eight neutrons

   (b) a hydrogen atom with a mass number of 3

5. Identify each of the subatomic particles described below:
   (a) −1 charge, mass = 1/1836   (b) 0 charge, mass = 1   (c) +1 charge, mass = 1

6. Calculate the atomic mass of bromine (Br). Br is composed of two isotopes, ^{79}Br and ^{81}Br. The percent abundances and masses of these isotopes are as follows:
   ^{79}Br 50.5%   Atomic mass = 78.9 amu
   ^{81}Br 49.5%   Atomic mass = 80.92 amu

7. Arrange the following elements in order of increasing atomic radius (from smallest to largest). Explain your order.
   N   As   O   F   P

8. Arrange the elements in Problem 7 in order of decreasing ionization energy.

9. Indicate the number of protons, neutrons, and electrons in each of the following cations:
   (a) $^{56}_{26}Fe^{2+}$ (b) $^{56}_{26}Fe^{3+}$ (c) $^{107}_{47}Ag^{+}$ (d) $^{109}_{47}Ag^{+}$

10. Indicate the number of protons, neutrons, and electrons in each of the following anions:
   (a) $^{14}_{7}N^{3-}$ (b) $^{80}_{34}Se^{2-}$ (c) $^{127}_{53}I^{-}$ (d) $^{80}_{35}Br^{-}$

4

The Modern Model of the Atom

Learning Outcomes

1. Describe electromagnetic radiation in terms of its wavelength, frequency, and energy.
2. Calculate the energy of light.
3. Explain what is meant by "quantized energy."
4. Describe Bohr's shells (orbits) and how they relate to quantized energy and n.
5. Explain the difference between a ground-state and an excited-state electron configuration.
6. Explain how the Bohr model accounts for atomic line spectra.
7. Explain how the Bohr model accounts for periodicity of the elements.
8. Describe Bohr's subshells and why they are needed.
9. Write the ground-state electron configuration for an atom and an ion.
10. Identify the period, group, and valence electrons given an electron configuration.
11. Describe the relationship between the periodic table and an atom's electron configuration.
12. Describe the periodic trends in atomic size and ionization energy.
13. Describe the octet rule and how it applies to metals and nonmetals.
14. Predict the formula of a compound formed from a metal and a nonmetal using the octet rule.
15. Describe what an orbital is and the types of orbitals that exist in an atom.

Chapter Outline

4.1 Seeing the Light—A New Model for the Atom
    A. Properties of light—Speed, wavelength, and energy
    B. Mathematical relationship between speed, wavelength, and energy
    C. Electromagnetic spectrum
    D. Line spectra of the elements

Review of Key Concepts from the Textbook

4.1 Seeing the Light—A New Model for the Atom

The current model of the atom is a vast improvement over the Rutherford model because it is more consistent with the way atoms behave. This model was developed after scientists began studying the properties of light and discovered that the light given off by electrified gases produced *line spectra* (spectra containing discrete lines of color) instead of continuous spectra (a rainbow of colors).

Light can be viewed as a wave and characterized by its wavelength (the distance between similar adjacent points on the wave, such as between two crests or two troughs) and by the energy of the wave. The relationship between the energy of light and its wavelength is given by the equation:

$$E = \frac{hc}{\lambda}$$

where $E =$ energy of light

$c = 3.00 \times 10^8$ m/s (speed of light)

$h = 6.63 \times 10^{-34}$ J $\cdot$ s (Planck's constant)

$\lambda =$ wavelength of light expressed in meters

The visible spectrum is one small section of the electromagnetic spectrum. The colors in the visible region of the spectrum, going from lowest to highest energy (longest to shortest wavelength) are red, orange, yellow, green, blue, indigo, and violet. You can remember this order by the word ROYGBIV (pronounced Roy G Biv) formed from the first letter of each of the colors.

Knowing the relationship of the energies associated with the colors in the visible spectrum (e.g., blue light has more energy than red light) is useful in solving problems involving the energy and wavelength of visible light.

Example 4.1

What is the energy of yellow light that has a wavelength of 580.3 nm?

Solution

$$E = \frac{hc}{\lambda} = \frac{(6.626 \times 10^{-34}\text{ J} \cdot \text{s})(3.00 \times 10^8\text{ m/s})}{5.803 \times 10^{-7}\text{ m}} = 3.42 \times 10^{-19}\text{ J}$$

(See Practice Problem 4.2 of the textbook for a similar problem.)

Practice Exercise 4.1

What is the wavelength (in nanometers) of light that has an energy of 4.42×10^{-19} J?

4.2   A New Kind of Physics—Energy Is Quantized

The Rutherford model of the atom was a great improvement over Thomson's plum pudding model, but there were some experimental observations for which the Rutherford model could not account. It was Niels Bohr who developed a new model for the atom. Bohr added two new components to the Rutherford model. He stated that (1) the electrons travel in fixed orbits around the nucleus and (2) the energies of the electrons are quantized (i.e., the energies can have only particular, allowable values).

The idea that the energies of electrons could take on only particular values was hard for physicists to accept at the time. Each of the objects with which they were familiar could have any energy, depending on how much energy was put into it. There were no energy values that were not allowed. Bohr maintained, however, that electrons were different and could have only certain energy values and nothing in between. This was a revolutionary idea at the time, and many physicists disagreed vehemently with it.

We now know that both Bohr and the physicists who disagreed with him were correct, because there are two kinds of physics—classical and quantum physics. In classical physics, objects can have any energy. In quantum physics, objects can have only certain, particular energies. The vast majority of objects obey the laws of classical physics; objects the size of an atom (or smaller) obey the laws of quantum physics. (This is actually an oversimplification because the laws of quantum physics do apply to large objects; we just don't observe them. See Section 4.2 of the textbook for a more complete discussion of this concept.)

The concept of quantization was (and still is) very important in explaining the behavior of atoms. To help you better understand the concept, we will consider an example of quantization with which you are familiar.

Example 4.2

Explain how the concept of quantization applies to the volume of soft drink obtainable from a vending machine.

Solution

The volume of soft drink that can be obtained from a vending machine is determined by the size of the containers in the machine. If the machine sells 12-oz cans, the only volumes of drink you can get are 12, 24, 36 oz, and so on. You cannot get 18, 30, or 42 oz because these volumes are "not allowed" from the machine. You can only obtain the volume in one can, two cans, three cans, and so on.

Practice Exercise 4.2

Contrast the volume of water obtainable from a store that sells only 16-oz bottles of water to the volume of water obtainable from a water fountain. In which case is the volume quantized?

4.3 The Bohr Theory of Atomic Structure

Bohr stated that electrons travel in fixed orbits around the nucleus. The energy values available to the electrons are the energies associated with these orbits. Because the electrons cannot be "between" orbits, there are only certain allowable energies. The orbits that are closer to the nucleus are lower in energy than those that are farther away because the stabilizing attraction of the positively charged nucleus for the negatively charged electrons is strongest for those electrons close to the nucleus.

The electron orbits in the Bohr model are called *shells*. Each shell is assigned a primary quantum number (n), beginning with $n = 1$ for the orbit closest to the nucleus. As n gets larger for orbits farther from the nucleus, the energy associated with the orbits increases, and the stability decreases. (Remember that higher energy means less stability.)

The electrons in an atom occupy the lowest energy shells available to them. Therefore, electrons go first into the $n = 1$ shell until it is full. Then electrons enter the $n = 2$ shell, then the $n = 3$ shell, and so on. The maximum number of electrons that each shell can hold is $2n^2$, where n is the shell number; see the following table.

Shell (n)	Maximum Number of Electrons ($2n^2$)
1	2
2	8
3	18
4	32
5	50
6	72

The farther we get from the nucleus, the higher the energy of the shell. The shells also get larger as we move out from the nucleus, and they can hold more electrons.

The Bohr model was quite successful in explaining experimental observations that the Rutherford model could not explain. However, new facts emerged later that could not be explained by the Bohr model, and a still newer model was developed. In spite of this, the Bohr model is still useful, and chemists still rely heavily on it.

Example 4.3

Draw a Bohr model for an atom with 24 electrons.

Solution

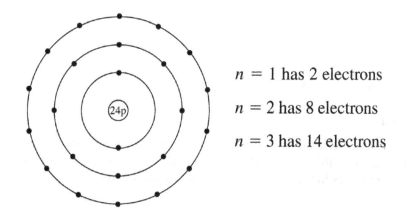

$n = 1$ has 2 electrons

$n = 2$ has 8 electrons

$n = 3$ has 14 electrons

Practice Exercise 4.3

Suggest a reason why the smaller inner shells can hold fewer electrons than the larger outer shells.

4.4 Periodicity and Line Spectra Explained

Three of the observations that the Bohr model successfully explained are the following:

1. *That electrons do not lose energy while "circling" in their orbits.* The Bohr model states that electrons are quantized and cannot lose energy to collapse into the nucleus.
2. *That chemical properties of atoms recur in a periodic fashion.* The Bohr model states that valence (outermost) electrons determine the properties of an element. Because the same

number of valence electrons recurs periodically, this would result in properties recurring periodically (see Section 4.4 of the textbook). Elements in the same group have the same number of valence electrons. The number of valence electrons is equal to the group number (for the A group elements).

3. *That electrically excited gaseous atoms produce line spectra (spectra containing discrete lines of color).* According to the Bohr model, the electrons in an atom occupy the lowest energy shells available. This configuration, with the electrons in the lowest energy shells, is called the *ground state* of an atom. When energy is added to the atom, an electron can "jump" to a higher energy level. This configuration, with one or more electrons in a higher energy shell than in the ground state, is called an *excited state* of an atom.

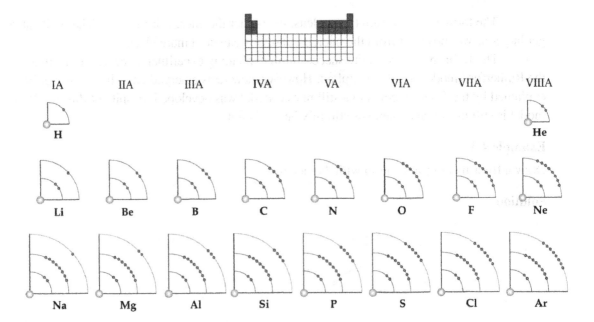

Example 4.4

What is the energy of the light absorbed when a hydrogen atom goes from an excited state where the electron occupies the $n = 3$ shell to an excited state where the electron occupies the $n = 4$ shell? (Use the values in the figure in Section 4.4 of the textbook.)

Solution

The $n = 3$ level has an energy of 13.1 eV. The $n = 4$ level has an energy of 13.8 eV. The energy of the light is $13.8 - 13.1 = 0.7$ eV.

Practice Exercise 4.4

How much energy is released for a hydrogen atom to go from an excited state, where the electron occupies the $n = 4$ shell, to the ground state, where the electron occupies the $n = 1$ shell?

The energy released when an electron falls from an excited state to the ground state determines the region of the spectrum the emitted light will be associated with and where the lines in the line spectrum will appear. (See Table 4.1 of the textbook.) If the energy is in the 1.69 eV to 3.34 eV range, the lines will be in the visible region of the spectrum. If the energy is greater than 3.34 eV, the lines will be in the ultraviolet region or beyond. See the following table.

Energy Given Off	Region of Spectrum
1.69 to 3.34 eV	Visible
<1.69 eV	Infrared
>3.34 eV	Ultraviolet

Example 4.5

What will be the color of the light emitted if an electron in a hydrogen atom emits 2.15 eV when falling from an excited state to the ground state? (Use Table 4.1 of the textbook.)

Solution

2.15 eV is in the region of *yellow* light.

Practice Exercise 4.5

What will be the color of the light emitted if an electron in a hydrogen atom emits 1.80 eV when falling from an excited state to the ground state?

The Bohr model correctly reproduced the experimental data for the hydrogen atom, but it didn't work for other atoms. So yet another model of atomic structure had to be developed. Although not entirely correct, the Bohr model was and still is quite useful in predicting the chemical composition of many compounds.

4.5 Subshells and Electron Configuration

A refinement to the Bohr model allowed the model to be even more useful in predicting the chemical behavior of the elements. This refinement was the introduction of subshells within the principal shells. Without the introduction of subshells, the Bohr model predicts a potassium atom with nine valence electrons in its $n = 3$ valence shell. However, its position in the periodic table indicates that it should have one valence electron. The introduction of subshells within shells solved this problem and correctly predicted the chemistry of potassium.

The subshells are given the single letter designations s, p, d, f, and so on (s is the first subshell within a level, p is the next, d is the next, and so on). The number of subshells within a given shell (n) is equal to the value of n, as shown in the following table:

n Value	Number of Subshells	Letter(s) of Subshell(s)
1	1	s
2	2	s, p
3	3	s, p, d
4	4	s, p, d, f
5	5	s, p, d, f, g

We can figure out how many electrons each subshell holds if we use the fact that each level has $2n^2$ electrons. See the following table.

n Value	Electron Capacity of Shells ($2n^2$)	Subshells	Electrons Capacity of Subshells
1	2	s	$s = 2$
2	8	s, p	$s = 2$
			$p = 6$
3	18	s, p, d	$s = 2$
			$p = 6$
			$d = 10$
4	32	s, p, d, f	$s = 2$
			$p = 6$
			$d = 10$
			$f = 14$

We stopped the table above at four subshells, because the elements currently existing use only the s, p, d, and f subshells. However, it should be apparent that there are five subshells in the $n = 5$ shell, six subshells in the $n = 6$ shell, and so on. The letters for the subshells that first appear in shells with $n = 5, 6, 7$, and so on are g, h, i, and so on, respectively.

Example 4.6

How many electrons can the $n = 5$ shell hold? How many electrons fit into the g sublevel?

Solution

When $n = 5, 2n^2 = 50$. The $n = 5$ shell will hold 50 electrons.

The s sublevel holds 2 electrons, p holds 6, d holds 10, and f holds 14, for a total of 32 electrons that will be accommodated by the first four subshells. The g subshell must therefore hold 18 electrons ($50 - 32 = 18$).

Practice Exercise 4.6

How many more electrons does the p subshell hold than the s? How many more does the f subshell hold than the d? Based on this pattern and the answers in Example 4.6, how many electrons would an h subshell hold?

When electrons enter the energy shells and subshells in atoms, they do so in a definite pattern. There are two ways in which to determine the order of electron filling for an atom: by using an electron-filling diagram (shown below) or by using the periodic table (which we will see shortly).

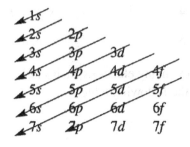

To use the electron-filling diagram, follow each arrow from tail to head (beginning with the top arrow) to get the order of filling for the electrons. The periodic table provides the same order of electron filling. (See Section 4.5 of the textbook for a thorough discussion of this method.) It is important to know that the order of electron filling predicted by this diagram and the periodic table is not correct for all atoms. There are some exceptions, mostly occurring in the larger atoms. One notable exception is the occurrence of s^1d^5 and s^1d^{10} instead of s^2d^4 and s^2d^9 among the d-block elements. In the larger elements, the energy levels are close together, and there is considerable overlap. Following the diagram or the periodic table will yield the correct electron configuration most of the time, however.

The electron configuration obtained by using the diagram or the periodic table is called the *ground-state electron configuration* because it is the configuration in which the electrons occupy the lowest energy levels available to them. If any electrons are in higher energy levels than the lowest available, the electron configuration is called an *excited-state electron configuration.*

Writing the electron configuration of an atom, instead of drawing the Bohr diagram, is called the shorthand notation for the electron configuration. The electrons in the outermost energy shell (the shell with the largest n value) are the *valence electrons.*

Example 4.7

Write the electron configuration for bromine (Br). How many valence electrons does a Br atom have?

Solution

Br has an atomic number of 35, so it has 35 electrons. Using the electron-filling diagram to get the order of electron filling, we find the configuration to be

$$1s^2 2s^2 2p^6 3s^2 3p^6 4s^2 3d^{10} 4p^5$$

The valence shell is $n = 4$ (the largest n value). There are seven electrons in the $n = 4$ shell (the $4s^2$ and the $4p^5$), so the Br atom has seven valence electrons. (Note that the valence electrons may be separated in the electron configuration, so be careful to count them all. You should also remember that the number of valence electrons is the same as the group number for the A main group elements. Because Br is in group VIIA, we know that it has seven valence electrons.)

Practice Exercise 4.7

How many valence electrons does tellurium (Te) have? (The atomic number is 52.) What is tellurium's group number in the periodic table?

When you are writing the electron configuration for large atoms such as lead (Pb), which has an atomic number of 82, it is useful to introduce one additional abbreviation. In this shorthand form, the noble gas atom that directly precedes the element is used to represent the electrons contained in the noble gas atom; the other electrons are written in the usual manner. This abbreviated form is illustrated in the following example.

Example 4.8

Write the electron configuration for lead (Pb), giving both (a) the full notation and (b) the noble gas abbreviated notation.

Solution

Pb has 82 electrons. The noble gas preceding Pb is xenon (Xe). Xe has 54 electrons.

(a) The full notation for the electron configuration for Pb is

$$1s^2 2s^2 2p^6 3s^2 3p^6 4s^2 3d^{10} 4p^6 5s^2 4d^{10} 5p^6 6s^2 4f^{14} 5d^{10} 6p^2$$

(b) The 54 electrons from Xe can be abbreviated [Xe]. The abbreviated notation is therefore

$$[Xe]6s^2 4f^{14} 5d^{10} 6p^2$$

Practice Exercise 4.8

Write the electron configuration for iodine (I), giving both (a) the full notation and (b) the noble gas abbreviated notation.

4.6 Regular Variations in the Properties of Elements

We have already seen that the atomic radii of the elements decrease going from left to right and increase going from top to bottom in the periodic table. The orbital diagrams in the Bohr model allow us to explain this trend.

When moving across a period, we are adding protons and electrons (into the same shell) to the atom. Because the number of positively charged protons in the nucleus is increasing, the nuclear attraction for the electrons increases, thereby decreasing the size of the atom.

When moving down a group, we are adding more protons to the nucleus and therefore increasing the nuclear charge. However, the outer electrons are situated in the next larger shell than the electrons in the previous period. The result is that the atom gets larger.

When moving across the third period from sodium to argon, we see that the increased nuclear charge pulls the shells closer to the nucleus, making the atom smaller. However, when moving down group IA from sodium to potassium, we see that the addition of a fourth shell more than compensates for the decreased distance from the nucleus of the first three shells, and the potassium atom is larger than the sodium atom. (Although this is an oversimplified view, it does lead to correct predictions of the relative sizes of atoms.) These trends are shown in the figure below.

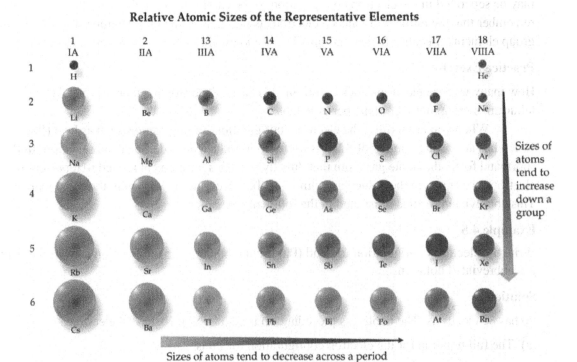

Relative Atomic Sizes of the Representative Elements

Sizes of atoms tend to increase down a group

Sizes of atoms tend to decrease across a period

Note that the trend in atomic size is also responsible for the trend in ionization energy. The smaller the atom, the larger the ionization energy; the larger the atom, the smaller the ionization energy.

4.7 Compound Formation and the Octet Rule

We have already seen that the formula of a compound is intrinsically fixed by the nature of the elements combining and that the formula is always the same. (See the law of definite composition.) Water (H_2O) always contains two atoms of hydrogen and one atom of oxygen, no matter where the water is found. If we change the subscripts in the formula, we change the compound. Where do the subscripts come from? The Bohr model answered this question.

When we look at the arrangement of electrons in the unreactive noble gases, it is apparent that elements having eight valence (outermost) electrons are unreactive, for the most part. In fact, elements without eight valence electrons will participate in chemical reactions in order to obtain eight valence electrons. This is such an important concept that it is known as the *octet rule*. Simply stated, *elements will react to form compounds that will allow them to have eight valence electrons in their valence shells*. (There are several exceptions to this rule: Hydrogen and helium can hold a maximum of two valence electrons, beryllium forms compounds in which it has four valence electrons, and boron forms compounds in which it has six valence electrons. There are also other elements, such as phosphorus and sulfur, that form some compounds in which they have more than eight valence electrons.)

In general, an element with *fewer* than four valence electrons will *lose* its valence electrons to obtain eight valence electrons (from the next inner shell), whereas an element with *more* than four valence electrons will *gain* electrons until it has reached a total of eight valence electrons. Thus, groups IA through IIIA will lose their valence electrons to form positive ions (cations), whereas groups VA through VIIA will gain enough electrons to obtain eight valence electrons, thus forming negative ions (anions). An element with four valence electrons will generally share electrons to obtain an octet. This is true for C, Si, and Ge; Sn and Pb, however, form ions.

Example 4.9

Indicate the ion that will be formed from each element below:

(a) Ca   (b) I   (c) N   (d) Rb

Solution

Fix ion charges

(a) Ca^{2+} (b) I^{2-} (c) N^{3-} (d) Rb^{+}

Practice Exercise 4.9

Indicate the ion that will be formed from each element below:

(a) K   (b) P   (c) O   (d) Al

Ionic compounds are formed when an atom that gains electrons to form an anion with an octet reacts with an atom that loses electrons to form a cation with an octet. Both atoms can have their needs satisfied by a transfer of electrons between atoms, thereby forming an ionic compound. We can determine the formula of the ionic compound that will be formed from the combination of two elements by considering the numbers of electrons that will be lost by one element and gained by the other.

For example, if we look at the case of the compound formed from barium (which will lose two electrons) and chlorine (which will gain one electron), we see that one barium atom will combine with two chlorine atoms for a transfer of two electrons between them. The formula of the ionic compound will be $BaCl_2$. Next, consider the ionic compound that will be formed from the reaction of aluminum and sulfur. Two atoms of aluminum (which need to give up three electrons each) will combine with three atoms of sulfur (which need to gain two electrons each) in order for a total of six electrons to be involved in the exchange. The formula of the ionic compound formed will be Al_2S_3. Let's consider one final example, the combination of calcium and oxygen. The calcium needs to lose two electrons, and the oxygen needs to gain two electrons. One oxygen atom can satisfy the needs of one calcium atom, so the formula of the compound formed will be CaO.

A shorthand method for determining the formula of ionic compounds is known as the *crisscross method*. The steps in the crisscross method are as follows:

1. Write the elemental symbols, with the cation written first.
2. Write the charge on each ion.
3. Transpose or "crisscross" the values of the charges (if the values are not the same), leaving off the signs. (Note that when the charge is $1+$ or $1-$, the number is left out of both the superscript and the subscript.) If the values are the same, there will be one ion of each element needed to make the compound.

There are some limitations to the crisscross method, but it generally works well for ionic compounds. The method is illustrated in the following example.

Example 4.10

Predict the formula of the ionic compound formed from the following pairs of elements:

(a) potassium and nitrogen   (b) gallium and arsenic

Solution

(a) Potassium (K) has one valence electron and will become K^+. Nitrogen (N) has five valence electrons and will become N^{3-}. Using the crisscross method, we first write K^+N^{3-}. Then, we transpose the values of the charges to get the formula of the ionic compound:

$$K^+_{\diagdown}N^{3-}_{\diagdown} = K_3N$$

(b) Gallium (Ga) has three valence electrons and will become Ga^{3+}. Arsenic (As) has five valence electrons and will become As^{3-}. Because the value of the charges is the same (three in each case), one Ga will be required for each As. The formula of the ionic compound is GaAs.

Practice Exercise 4.10

Predict the formula of the ionic compound formed from the following pairs of elements:

(a) calcium and phosphorus   (b) magnesium and sulfur

4.8   The Modern Quantum Mechanical Model of the Atom

The problem with the Bohr model of the atom is that it only worked completely for hydrogen. The model was incorrect in its assumption that electrons circle the nucleus in specific, fixed orbits. This is not really the case. Instead, electrons occupy a *region* of space but cannot be precisely placed at any given point in that region of space.

The Heisenberg uncertainty principle states that it is impossible to know precisely (1) where an electron is (its current position) *and* (2) how fast and in what direction it is going. The more precisely we know one of these things, the less precisely we know the other. To illustrate this concept, we can consider a hypothetical bug in a dark room. This bug uses light energy to scamper about the room. As long as the room is totally dark, the bug is still but we cannot see it. We know its speed and direction precisely (its speed is zero and it is moving in no direction). If we shine a flashlight (which has a very narrow beam of intense light) on the bug, we see it for an instant, but it quickly scampers out of sight. We may have seen in what direction it took off, but we do not know how fast it is moving or its precise position in the room. (At this point, we do not know its position or its velocity.) For the instant that we saw it in the light, we knew its exact position, but because it took off that same instant, we do not know how fast it was moving or in what direction it took off.

Because we do not know exactly where an electron in an atom is, we can only speak about the probability of finding an electron in a region of space. These regions of space are called *orbitals*, and each subshell contains a specific number of orbitals, as shown in the following table.

Subshell	Number of Orbitals
s	1
p	3
d	5
f	7

The three shapes for orbitals that you should know about are shown in the figure below.

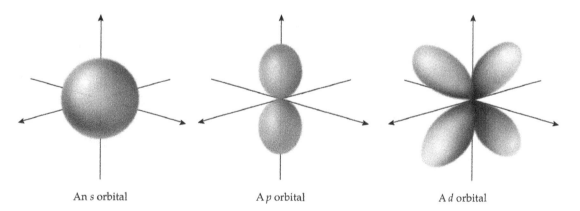

An *s* orbital   A *p* orbital   A *d* orbital

Example 4.11

Given the total number of electrons in a p subshell and the number of p orbitals, how many electrons does a p orbital hold?

Solution

There are a total of six electrons in the p subshell, and there are three p orbitals. Therefore, a p orbital holds two electrons.

Practice Exercise 4.11

Given the total number of electrons in a d subshell and the number of d orbitals, how many electrons does a d orbital hold? How many electrons does an f orbital hold?

Strategies for Working Problems in Chapter 4

Overview: What You Should Be Able to Do

Chapter 4 provides a model of atomic structure that describes the wave properties of electrons. The arrangement of electrons in atoms is presented, and a distinction is made between ground-state and excited-state arrangement of electrons in atoms. The types of problems that you should be able to solve after mastering Chapter 4 are as follows:

1. Calculate the energy when given the wavelength (or the wavelength when given the energy) of electromagnetic radiation.
2. Calculate the energy (scaled) absorbed or released by a hydrogen atom when an electron "jumps" to a higher energy shell or "falls" to a lower energy shell.
3. Write the electron configuration showing the electron distribution in an atom.
4. Write the shorthand electron configuration for an atom.

Flowcharts

Flowchart 4.1 **Calculating the energy when given the wavelength (or the wavelength when given the energy) of electromagnetic radiation**

Flowchart 4.1

Summary

Method for calculating the energy, E, when given the wavelength, λ (or the wavelength when given the energy), of electromagnetic radiation

Step 1
• **Determine the property that is to be calculated (i.e., energy or wavelength).**

Step 2

• **Use the relationship $E = \dfrac{hc}{\lambda}$ to find the value sought.**

• **If solving for energy, simply substitute the values h, c, and λ into the equation. (λ must be expressed in meters. Convert any other length unit to meters before substituting into the equation.)**

• **If solving for wavelength, algebraically manipulate the equation to isolate λ on the left side of the equation.**

• **The new equation becomes $\lambda = \dfrac{hc}{E}$.**

• **$h = 6.626 \times 10^{-34}\,\text{J} \cdot \text{s}$ (Planck's constant).**
• **$c = 3.00 \times 10^{8}\,\text{m/s}$ (speed of light).**

Step 3
• **Carry out the calculation.**

Example 1

Calculate the energy of light that has a wavelength of 600 nm.

Solution

Perform Step 1

The quantity to be calculated is energy.

Perform Step 2

$$E = \frac{hc}{\lambda}$$

$$E = \frac{(6.626 \times 10^{-34}\,\text{J} \cdot \text{s})\,(3.00 \times 10^8\,\text{m/s})}{6.00 \times 10^{-7}\,\text{m}}$$

Perform Step 3

$$E = \frac{(6.626 \times 10^{-34}\,\text{J} \cdot \text{s})(3.00 \times 10^8\,\text{m/s})}{6.00 \times 10^{-7}\,\text{m}} = 3.31 \times 10^{-19}\,\text{J}$$

Example 2

What is the wavelength of electromagnetic radiation with energy of 4.42×10^{-19} J?

Solution

Perform Step 1

The quantity to be calculated is wavelength.

Perform Step 2

$$E = \frac{hc}{\lambda}$$

Algebraically manipulating the equation to solve for λ, we obtain

$$\lambda = \frac{hc}{E}$$

$$\lambda = \frac{(6.626 \times 10^{-34}\,\text{J} \cdot \text{s})(3.00 \times 10^8\,\text{m/s})}{4.42 \times 10^{-9}\,\text{J}}$$

Perform Step 3

$$\lambda = \frac{(6.626 \times 10^{-34}\,\text{J} \cdot \text{s})(3.00 \times 10^8\,\text{m/s})}{4.42 \times 10^{-9}\,\text{J}} = 4.50 \times 10^{-7}\,\text{m} = 450\,\text{nm}$$

Example 3

Calculate the energy of electromagnetic radiation that has a wavelength of 0.0100 m.

Solution

Perform Step 1

The quantity to be calculated is energy.

Perform Step 2

$$E = \frac{hc}{\lambda}$$

$$E = \frac{(6.626 \times 10^{-34}\,\text{J} \cdot \text{s})(3.00 \times 10^8\,\text{m/s})}{0.0100\,\text{m}}$$

Perform Step 3

$$E = \frac{(6.626 \times 10^{-34}\,\text{J} \cdot \text{s})(3.00 \times 10^8\,\text{m/s})}{0.0100\,\text{m}} = 1.99 \times 10^{-23}\,\text{J}$$

Practice Problems

4.1 Calculate the energy of electromagnetic radiation that has a wavelength of 350 nm.

4.2 What is the wavelength of electromagnetic radiation that has 4.09×10^{-19} J of energy?

4.3 Green light has a wavelength of 486 nm. Calculate the energy of green light.

Flowchart 4.2   Calculating the energy (scaled) absorbed or released by a hydrogen atom when an electron "jumps" to a higher energy shell or "falls" to a lower energy shell

Flowchart 4.2

Summary

Method for calculating the energy (scaled) absorbed or released by a hydrogen atom when an electron "jumps" to a higher energy shell or "falls" to a lower energy shell

Step 1
• **Determine the energy of the two shells involved in the electron transition. (Use the energy-level diagram in Section 4.4 of the textbook to determine the energy values.)**

Step 2
• **Subtract the smaller energy value from the larger energy value.**
• **If the electron "jumps" from a lower energy shell to a higher energy shell, energy will be absorbed.**
• **If the electron "falls" from a higher energy shell to a lower energy shell, energy will be released.**

Example 1

Calculate the amount of energy (scaled) absorbed when an electron in the $n = 1$ shell of the hydrogen atom jumps to the $n = 4$ shell.

Solution

Perform Step 1

Scaled energy of the $n = 1$ shell $= 1.0$ eV

Scaled energy of the $n = 4$ shell $= 13.8$ eV

Perform Step 2

The difference in energy $= 13.8\,\text{eV} - 1.0\,\text{eV} = 12.8\,\text{eV}$

Therefore, 12.8 eV of energy (scaled) will be absorbed.

Example 2

Calculate the amount of energy (scaled) released when an electron in the $n = 5$ shell of the hydrogen atom falls to the $n = 3$ shell.

Solution

Perform Step 1

Scaled energy of the $n = 5$ shell $= 14.1\,\text{eV}$

Scaled energy of the $n = 3$ shell $= 13.1\,\text{eV}$

Perform Step 2

The difference in energy $= 14.1\,\text{eV} - 13.1\,\text{eV} = 1.0\,\text{eV}$

Therefore, 1.0 eV of energy (scaled) will be released.

Example 3

Calculate the amount of energy (scaled) released when an electron in the $n = 3$ shell of the hydrogen atom falls to the $n = 1$ shell.

Solution

Perform Step 1

Scaled energy of the $n = 3$ shell $= 13.1\,\text{eV}$

Scaled energy of the $n = 1$ shell $= 1.0\,\text{eV}$

Perform Step 2

The difference in energy $= 13.1\,\text{eV} - 1.0\,\text{eV} = 12.1\,\text{eV}$

Therefore, 12.1 eV of energy (scaled) will be released.

Practice Problems

4.4 Calculate the amount of energy (scaled) released when an electron in the $n = 6$ shell of the hydrogen atom falls to the $n = 2$ shell.

4.5 Calculate the amount of energy (scaled) released when an electron in the $n = 3$ shell of the hydrogen atom jumps to the $n = 5$ shell.

4.6 Calculate the amount of energy (scaled) released when an electron in the $n = 1$ shell of the hydrogen atom jumps to the $n = 6$ shell.

Flowchart 4.3   Writing the electron configuration showing the electron distribution is an atom

Flowchart 4.3

Summary

Method for writing the electron configuration showing the electron distribution is an atom

STUDENT WORKBOOK AND SELECTED SOLUTIONS

Step 1
- **Determine the number of electrons in the atom.**
- **The number of electrons is equal to the atomic number for a neutral atom.**

$\downarrow$

Step 2
- **Use the periodic table and place the electrons in subshells according to the periodic table shown in Section 4.5 of the textbook.**

Example 1

Write the electron configuration for silicon.

Solution

Perform Step 1

The atomic number of silicon is 14. There are 14 electrons in the silicon atom.

Perform Step 2

Following the order of placing electrons as demonstrated by the figure in Section 4.5 of the textbook, we arrive at the following electron configuration for silicon.

$1s^2 2s^2 2p^6 3s^2 3p^2$

Example 2

Write the electron configuration for polonium.

Solution

Perform Step 1

The atomic number of polonium is 84. There are 84 electrons in the polonium atom.

Perform Step 2

Following the order of placing electrons as demonstrated by the figure in Section 4.5 of the textbook, we arrive at the following electron configuration for polonium.

$1s^2 2s^2 2p^6 3s^2 3p^6 4s^2 3d^{10} 4p^6 5s^2 4d^{10} 5p^6 6s^2 4f^{14} 5d^{10} 6p^4$

Example 3

Write the electron configuration for zinc.

Solution

Perform Step 1

The atomic number of zinc is 30. There are 30 electrons in the zinc atom.

Perform Step 2

Following the order of placing electrons as demonstrated by the figure in Section 4.5 of the textbook, we arrive at the following electron configuration for zinc.

$1s^2 2s^2 2p^6 3s^2 3p^6 4s^2 3d^{10}$

Practice Problems

4.7 Write the electron configuration for calcium.

4.8 Write the electron configuration for selenium.

4.9 Write the electron configuration for lead.

Flowchart 4.4   **Writing the shorthand electron configuration for an atom**

Flowchart 4.4

Summary

Procedure for writing the shorthand electron configuration for an atom

Step 1
• **Determine the number of electrons in the atom.**
• **The number of electrons is equal to the atomic number for a neutral atom.**

$\downarrow$

Step 2
• **Identify the noble gas directly preceding the element whose configuration is to be written.**
• **Place the symbol of the noble gas in brackets.**
• **This symbol represents the electron configuration of the noble gas.**

$\downarrow$

Step 3
• **Write in the electrons following those in the noble gas whose symbol appears in the brackets.**

Example 1

Using the noble gas abbreviated notation, write the electron configuration for germanium.

Solution

Perform Step 1

The atomic number of germanium is 32. There are 32 electrons in the germanium atom.

Perform Step 2

The noble gas immediately preceding germanium is argon (atomic number 18). Therefore, the notation [Ar] represents the electron through $3p^6$.

Perform Step 3

Writing the bracketed noble gas, and then following the order of placing electrons as demonstrated by the figure in Section 4.5 of the textbook, we arrive at the following electron configuration for germanium.

$[Ar]\, 4s^2 3d^{10} 4p^2$

STUDENT WORKBOOK AND SELECTED SOLUTIONS

Example 2

Using the noble gas abbreviated notation, write the electron configuration for barium.

Solution

Perform Step 1

The atomic number of barium is 56. There are 56 electrons in the barium atom.

Perform Step 2

The noble gas immediately preceding barium is xenon (atomic number 54). Therefore, the notation [Xe] represents the electron through $5p^6$.

Perform Step 3

Writing the bracketed noble gas, and then following the order of placing electrons as demonstrated by the figure in Section 4.5 of the textbook, we arrive at the following electron configuration for barium.

$$[\text{Xe}]\,6s^2$$

Example 3

Using the noble gas abbreviated notation, write the electron configuration for tellurium.

Solution

Perform Step 1

The atomic number of tellurium is 52. There are 52 electrons in the tellurium atom.

Perform Step 2

The noble gas immediately preceding tellurium is krypton (atomic number 36). Therefore, the notation [Kr] represents the electron through $4p^6$.

Perform Step 3

Writing the bracketed noble gas, and then following the order of placing electrons as demonstrated by the figure in Section 4.5 of the textbook, we arrive at the following electron configuration for tellurium.

$$[\text{Kr}]\,5s^2 4d^{10} 5p^4$$

Practice Problems

4.10 Write the electron configuration for magnesium, using the noble gas abbreviated notation.

4.11 Write the electron configuration for bismuth, using the noble gas abbreviated configuration.

4.12 Write the electron configuration for vanadium, using the noble gas abbreviated configuration.

Quiz for Chapter 4 Problems

1. How much energy (scaled) is absorbed when an electron in a hydrogen atom jumps from the $n = 3$ shell to the $n = 6$ shell?

2. Calculate the energy of electromagnetic radiation that has a wavelength of 400 nm.

3. How many electrons are in the s subshell in an atom of potassium?

4. What noble gas will appear in brackets in the noble gas abbreviated electron configuration for iodine?

5. How much energy (scaled) is released when an electron in a hydrogen atom falls from the $n = 4$ shell to the $n = 1$ shell?

6. Red light has a wavelength of 750 nm. What is the energy of red light?

7. Write the electron configuration for the phosphorus atom.

8. How does the energy released when an electron falls from the $n = 6$ shell to the $n = 5$ shell compare to the amount of energy absorbed when an electron jumps from the $n = 5$ shell to the $n = 6$ shell in the hydrogen atom?

9. Write the electron configuration for cadmium, using the noble gas abbreviated configuration.

10. What is the wavelength of electromagnetic radiation that has an energy of 3.50×10^{-18} J?

Exercises for Self-Testing

A. Completion. *Write the correct word(s) to complete each statement below.*

1. The _____ is the distance between similar adjacent points on the wave.
2. The greater the wavelength of light, the _____ the energy.
3. _____ spectra are spectra containing discrete lines of color, whereas the familiar rainbow of colors is a _____ spectrum.
4. _____ is a term that is used to describe objects that can have only particular, allowable energy values.
5. The _____ atomic theory states that electrons travel in orbits around the nucleus.
6. The _____ _____ _____ is designated by the letter n.
7. The maximum number of electrons that a principal energy shell can hold is given by the formula _____.
8. The outermost shell of an atom is called the _____ _____.
9. For the representative elements, the number of valence electrons is equal to the _____ _____.
10. The _____ _____ states that elements react to form compounds in such a way as to put eight electrons in their valence shell.
11. The _____ _____ electron configuration of an atom is the arrangement of electrons in which the electrons are in the lowest energy levels available.
12. The elements in the alkaline earth group have _____ valence electrons.
13. The _____ _____ _____ states that it is impossible to know precisely where an electron is and how fast and in what direction it is going.
14. The atomic size _____ as you move across a row in the periodic table and _____ as you move down a group.
15. Each subshell in an atom consists of _____, each of which will hold two electrons.
16. The Bohr model of the atom was developed after the _____ model to explain some experimental data that the former model could not explain.
17. An atom of cesium has _____ valence electron(s), which is (are) found in the $n =$ _____ shell.
18. The number of subshells in the $n = 3$ shell is _____.

B. True/False. *Indicate whether each statement below is true or false, and explain why the false statements are false.*

_____ 1.   The energy of light increases as the wavelength decreases.

_____ 2.   The wavelength of red light is longer than that of blue light.

_____ 3.   In the Bohr model of the atom, electrons can start to fill the $n = 4$ shell before the $n = 3$ shell is completely filled.

_____ 4.   The number of valence-shell electrons for all elements can be determined from the group number in the periodic table.

_____ 5.   Energy is emitted when an electron jumps from a lower energy level to a higher energy level.

_____ 6.   Red light has a shorter wavelength than blue light.

_____ 7.   The light associated with an $n = 4$ to $n = 3$ transition has a longer wavelength than the light associated with an $n = 3$ to $n = 2$ transition.

_____ 8.   The atoms that gain electrons to satisfy the octet rule become positively charged cations.

_____ 9.   Atoms get larger when moving from left to right on the periodic table because the atomic number is increasing.

_____ 10.   The $n = 2$ shell contains only s and p subshells.

_____ 11.   Gallium (Ga) has three valence electrons in the $n = 3$ shell.

_____ 12.   The transition metals are found in the d block on the periodic table.

C. Questions and Problems

1. Calculate the energy of orange light that has a wavelength of 650 nm. Would blue light have a shorter or longer wavelength than this?
2. Sketch the electromagnetic spectrum, showing the colors in the visible region of the spectrum and the position of ultraviolet light, infrared light, X-rays, gamma rays, and radio waves.
3. Illustrate by the use of a diagram what is meant by a line spectrum.
4. Draw the Bohr diagram for an atom of silicon (Si).
5. Is energy emitted or absorbed when an electron in a hydrogen atom falls from the $n = 4$ to the $n = 1$ shell?
6. Use a Bohr diagram to explain the difference between the ground state and an excited state of sulfur.
7. Use only the periodic table to determine the number of valence electrons in:
    (a) Rb   (b) Ba   (c) Al   (d) Ge   (e) Sb   (f) Te   (g) I
8. Use only the periodic table to write the electron configuration for polonium (Po). How many valence electrons does Po have?
9. Indicate the formula of the compound formed from the combination of the following pairs of elements:
    (a) Ba and Cl   (b) Al and O   (c) Li and O
10. Write the predicted electron configuration for the following elements, using both the full and noble gas shorthand notation:
    (a) Sr   (b) Ag   (c) V   (d) U
11. Sketch the periodic table, and indicate the positions of the s-, p-, d-, and f-block elements in the table.
12. Explain the difference between metals and nonmetals in terms of valence electrons.

13. For each element whose electron configuration is shown below, indicate the group number and the period number:
   (a) $1s^2 2s^2 2p^6 3s^2 3p^6 4s^2 3d^{10} 4p^6 5s^2 4d^{10} 5p^6 6s^2 4f^{14} 5d^{10} 6p^4$
   (b) $1s^2 2s^2 2p^6 3s^2 3p^6 4s^2$
   (c) $[Ar]4s^2 3d^{10} 4p^5$

14. Explain the major difference between the Bohr model and the quantum mechanical model of the atom.

5

Chemical Bonding and Nomenclature

Learning Outcomes

1. Define an ionic bond and predict when it is likely to occur.
2. Predict the formula of an ionic compound.
3. Describe what an ionic lattice and lattice energy is.
4. Describe what a molecule is and give some examples.
5. Describe what a covalent bond is and what gives rise to it in terms of attractions.
6. Draw a covalent bond and count the valence electrons owned by each atom.
7. Draw dot diagrams for atoms.
8. Draw dot diagrams for molecules.
9. Draw dot diagrams for molecules that have multiple bonds.
10. Draw resonance forms.
11. Define *electronegativity* and describe its periodic trends.
12. Describe bond polarity using electronegativity.
13. Predict if a bond is covalent, polar covalent, or ionic.
14. Name and give formulas for binary ionic compounds.
15. Name and give formulas for compounds possessing transition metals.
16. Name and give formulas for binary covalent compounds.
17. Name and give formulas for compounds possessing polyatomic ions.
18. Name and give formulas for acids.
19. Draw dot diagrams for molecules that possess electron deficient atoms.
20. Draw dot diagrams for molecules that possess expanded octet atoms.

Chapter Outline

5.1 Ionic Bonding—Bring on the Metals

   A. Formation of ions and ionic compounds

   B. Coulomb's law, ionic lattices, and lattice energy

5.2 Molecules: What Are They? Why Are They?

   A. Definition of molecule

   B. Differences in properties of compounds and elements

5.3 Holding Molecules Together—The Covalent Bond

   A. Increased positive/negative electron-nuclei attraction as atoms approach each other

   B. Definition of covalent bond

   C. Valence electron role in bond formation

   D. Energy changes in forming and breaking covalent bonds

   E. The octet rule

5.4 Molecules, Dot Structures, and the Octet Rule

   A. Determination of valence electrons in neutral atoms

   B. Acquisition of eight electrons (an octet) by electron sharing

   C. Predicting the number of bonds an atom will form

   D. Predicting the formula of covalent compounds

   E. Lewis dot diagrams for elements

      1. Drawing dot diagrams of elements

      2. Paired electrons and unpaired electrons

   F. Lewis dot diagrams for compounds

      1. Drawing dot diagrams of compounds

      2. Bonding pairs and lone pairs

5.5 Multiple Bonds

   A. Reason for multiple-bond formation

   B. Single, double, and triple bonds

   C. Drawing dot diagrams of compounds containing multiple bonds

   D. Rules for Lewis dot diagrams for compounds

   E. Resonance forms

   F. Dot diagrams for polyatomic ions

5.6 Equal Versus Unequal Sharing of Electrons—Electronegativity and the Polar Covalent Bond

   A. Definition of electronegativity

   B. Periodic trend in electronegativity values

   C. Four most electronegative elements—N, O, F, Cl

   D. Polar covalent bonds

   E. Predicting bond type from difference in electronegativity

STUDENT WORKBOOK AND SELECTED SOLUTIONS

5.7 Nomenclature—Naming Chemical Compounds

    A. Rules for naming binary ionic compounds

        1. Compounds containing cation with only one possible charge state

        2. Compounds containing cation with more than one charge state

          a. Roman numeral system

          b. System using *-ous* or *-ic* to indicate charge on cation

    B. Naming compounds that contain polyatomic ions

        1. Names and formulas of common polyatomic ions

        2. Determining formulas of compounds containing polyatomic ions

    C. Naming binary covalent compounds

        1. Using Greek prefixes to indicate numbers of atoms

        2. Common names of binary covalent compounds

5.8 One More Thing: Exceptions to the Octet Rule

    A. Electron deficient atoms

    B. Expanded octet atoms

Review of Key Concepts from the Textbook

5.1 Ionic Bonding—Bring on the Metals

Ionic bonding involves a transfer of electrons, as contrasted with the sharing of electrons that occurs in covalent compounds. (When electrons are transferred, positive and negative ions are formed, and their attraction to each other is what holds ionic compounds together.) Another difference between covalent and ionic compounds is that ionic compounds exist as a three-dimensional network of ions (an ionic lattice), whereas covalent compounds exist as individual molecules.

    Ionic compounds are made from a metal and a nonmetal, a metal and a negative polyatomic ion (e.g., SO_4^{2-}), a positive polyatomic ion (e.g., NH_4^+) and a nonmetal, or two polyatomic ions. Covalent compounds, on the other hand, are made from nonmetals and exist as individual molecules.

Example 5.1

Indicate whether each compound below exists as individual molecules or as an ionic lattice:

(a) NH_3 (b) KF   (c) $C_6H_{12}O_6$ (d) Na_3PO_4 (e) NH_4NO_3

Solution

(a) individual molecules   (b) ionic lattice   (c) individual molecules

(d) ionic lattice   (e) ionic lattice

Practice Exercise 5.1

Indicate whether each compound below exists as individual molecules or as an ionic lattice:

(a) CO_2 (b) PH_3 (c) NH_4Cl (d) $NaNO_3$ (e) $Mg_3(PO_4)_2$

5.2 Molecules: What Are They? Why Are They?

Molecules are stable collections of atoms that are held together by covalent and ionic bonds. Molecules can be as small as two atoms (as in H_2, O_2, F_2, and HCl) or as large as thousands of atoms (as in many biologically important molecules). Molecules have very different properties than the atoms from which they are formed. For example, the chemical and physical properties of carbon monoxide, CO, a toxic gas, are entirely different from the properties of pure carbon and pure oxygen.

5.3 Holding Molecules Together—The Covalent Bond

When we studied individual atoms in Chapter 4, we discovered that most atoms prefer to have eight valence electrons, according to the octet rule. (Hydrogen is the exception, preferring to have two valence electrons, which completely fill its $n = 1$ valence shell.) Forming covalent bonds with other atoms is one way that atoms satisfy the octet rule. The covalent bonds are stable because the shared electrons are attracted to both nuclei more than the electrons repel each other. The increased attractive forces present in the molecule (as compared to individual atoms) result in the molecules being more stable than the separate atoms, favoring the bond-making process.

We generally show the covalent bonding in molecules by the use of a diagram that gives the symbols for the atoms plus the electrons, as in the H_2 molecule shown below:

$$H:H$$

When we draw a bond diagram in this way, each pair of shared electrons counts as part of the valence electrons for *each* atom sharing the bond. We call this approach *double counting* the shared electrons. In the diagram for H_2 above, the hydrogen on the right is considered to have two valence electrons, and the hydrogen on the left is also considered to have two valence electrons (i.e., the double counting), even though there are only two electrons in the entire dot diagram.

Example 5.2

The SiH_4 molecule has the dot diagram shown below.

$$
\begin{array}{c}
H \\
\overset{\cdot\cdot}{H:Si:H} \\
\overset{\cdot\cdot}{H}
\end{array}
$$

How many valence electrons does the Si atom have? How many does each H atom have?

Solution

The shared electrons are counted as valence electrons for both the Si and the H atoms. Therefore, the Si atom has eight valence electrons, and each H atom has two valence electrons.

Practice Exercise 5.2

How many electrons did the Si atom gain in making the SiH_4 molecule?

5.4 Molecules, Dot Structures, and the Octet Rule

The number of bonds that an atom will want to form depends on how many valence electrons it has. (Recall that for the main group elements, the group number is equal to the number of valence electrons.) Table 5.1 of the textbook provides information on the number of covalent bonds an atom will form. You should study this table carefully, as this information will be important when we draw Lewis dot diagrams of molecules. We can use the information in Table 5.1 to predict the formula of the compound that forms between two elements, as shown in Practice Problem 5.1 of the textbook.

The crisscross method we used for ionic compounds sometimes works for covalent compounds if we substitute the charge on the ions with the number of bonds the atom wants to form. (Note that there are many exceptions to this generalization, such as N_2O_5, NO, and NO_3. The method is useful for predicting the formulas of *some* covalent compounds but is not as reliable with covalent compounds as it is with ionic compounds.)

When using the crisscross method to determine the formula of a covalent compound, we do the following:

1. Write the element symbol, with the number of bonds the atom wants to form (from Table 5.1 of the textbook) as the superscript.
2. Crisscross the superscripts to get the formula of the compound (which will now include subscripts for any number of atoms greater than 1). Notice that the number of bonds the element wants to form is equal to the number of electrons the element needs to gain to obtain an octet (or a duet in the case of hydrogen). The number of electrons needed to obtain an octet is simply 8 minus the group number for the main group elements.

Example 5.3

Use the crisscross method to determine the formula of the compound formed between phosphorus (P) and fluorine (F).

Solution

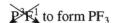 to form PF_3

(PF_5 is also formed by phosphorus and fluorine, but this molecule requires the phosphorus to have more than an octet of electrons around it.)

Practice Exercise 5.3

Determine the formula of the compound formed from the combination of the following pairs of elements:

(a) hydrogen and sulfur   (b) arsenic and oxygen

We can use dot diagrams to show an atom's valence electrons. When drawing dot diagrams, we put the electrons in each of the four positions around the atomic symbol one at a time before we begin to pair them. (See Section 5.3 of the textbook for a thorough discussion of this.) The single electrons are called unpaired electrons, and the paired electrons are simply called electron pairs. The number of unpaired electrons is equal to the number of covalent bonds the atom will form.

Example 5.4

Draw the dot diagram for an atom of oxygen. How many unpaired electrons does oxygen have? How many paired electrons? How many bonds will the oxygen atom form?

Solution

$\cdot \overset{\displaystyle \cdot \cdot}{\underset{\displaystyle \cdot}{O}} :$

There are two unpaired electrons and four paired electrons. The oxygen atom will form two bonds.

Practice Exercise 5.4

Draw the dot diagram for arsenic. How many unpaired electrons does arsenic have? How many paired electrons? How many bonds will the arsenic atom form?

   Atoms share unpaired electrons to form covalent bonds. When $H \cdot$ and $\cdot \overset{\cdots}{\underset{\cdot}{O}} :$ combine, they make $H : \overset{\cdots}{\underset{\cdots}{O}} :$; that is, water. Note that because all four positions around the $\overset{\cdots}{O}$ are equivalent,
$\quad\quad\quad\quad\quad$ H

we could have written H_2O as $H : \overset{\cdots}{\underset{\cdots}{O}} : H$.

5.5 Multiple Bonds

All the dot structures that we have drawn so far have placed only one pair of electrons between two atoms. (One pair of electrons between two atoms is called a single bond.) However, molecules exist in which there are *two* pairs of electrons between atoms (double bonds) or even *three* pairs (triple bonds). These double and triple bonds are also referred to as multiple covalent bonds (or multiple bonds).

Example 5.5

Write the dot diagram for the AsF_3 molecule.

Solution

We combine $\cdot \overset{\cdots}{As} \cdot$ with 3 $: \overset{\cdots}{\underset{\cdots}{F}} \cdot$ atoms to obtain $: \overset{\cdots}{\underset{\cdots}{F}} : \overset{\cdots}{As} : \overset{\cdots}{\underset{\cdots}{F}} :$
$\quad\quad\quad\quad\quad\quad\quad\quad\quad\quad\quad\quad\quad\quad\quad\quad : \overset{\cdots}{\underset{\cdots}{F}} :$

Practice Exercise 5.5

Write the dot diagram for the SBr_2 molecule.

   When you are drawing dot diagrams for more complicated molecules, it is helpful to recall the information given in Table 5.1 of the textbook concerning the number of bonds each atom forms in covalent molecules. The information is presented again in the table below, along with the number of lone pairs (pairs of electrons not shared with other atoms) that an atom of each element usually has. There are some exceptions to these general guidelines, particularly when ions are involved, but they generally apply when you are drawing dot structures of neutral compounds.

Atom	Number of Covalent Bonds the Atom Will Form	Number of Lone Pairs on the Atom
H	1	0
C, Si, Ge	4	0
N, P, As, Sb	3	1
O, S, Se, Te	2	2
F, Cl, Br,	1	3

Example 5.6

Draw the dot diagram for CH_2Cl_2.

Solution

Because the carbon forms four bonds and the hydrogen and chlorine form only one bond each, we know that the carbon has to be at the center of the molecule, with the other four atoms around it. The correct structure is as follows:

$$\begin{array}{c} \ddot{H} \quad \ddot{} \\ H\!:\!\ddot{C}\!:\!\ddot{\ddot{C}l}\!: \\ :\!\ddot{\ddot{C}l}\!: \end{array}$$

Practice Exercise 5.6

Draw the dot diagram for NHF_2.

Example 5.7

Draw a dot diagram for C_3H_6O. The three carbons are all bonded to each other, with an $O\!-\!H$ group at the end. (Note that this compound is different from acetone, whose structure you were asked to draw in Practice Problem 5.13 of the textbook.)

Solution

$$\begin{array}{c} H \quad H\;H \quad \ddot{} \\ H\!:\!\ddot{C}\!::\!\ddot{C}\!:\!\ddot{C}\!:\!\ddot{O}\!:\!H \\ \quad\quad\quad H \end{array}$$

Each shared pair can be represented by a line instead of dots, making drawing dot structures a little less tedious.

Practice Exercise 5.7

Draw the dot diagram for C_2H_4.

There is a set of rules that we can use to draw dot diagrams of molecules without first drawing the dot diagrams of the individual atoms. These rules are as follows:

1. Determine the total number of valence electrons (dots) to be shown in the diagram of the molecule or polyatomic ion. (Don't forget to subtract or add electrons based on the charge of the polyatomic ion.)
2. Connect the atoms using single covalent bonds. (Use the guidelines presented in the table on page 96 of this book to help you decide the correct way to connect the atoms.)
3. Count up all the electrons used for the single bonds in Step 2. Subtract this number from the total determined in Step 1 to determine how many more electrons you can put in the diagram. Distribute the remaining electrons as lone pairs, satisfying the octet rule for each element as you proceed. (Remember that hydrogen is satisfied with two electrons.)
4. If there are atoms that do not have an octet after you have put in all the valence electrons available, move lone pairs into positions between the atom they are on and an adjacent atom, thereby forming multiple bonds.

Example 5.8

Draw a dot diagram for $COCl_2$. (The carbon is the central atom.)

Solution

We'll follow the steps given above.

1. There are a total of 24 valence electrons: 4 from C, 6 from O, and 2(7) from Cl.

2. The atoms are connected as follows:

$$\begin{array}{c} O \\ | \\ Cl-C-Cl \end{array}$$

(This is the only arrangement that allows us to satisfy all the rules listed in the table on page 96 of this book.)

3. Because the single bonds require 6 electrons, we have a total of 18 remaining electrons that can be put in as lone pairs. (We will put them in starting with the Cl on the left, moving across to C, then across to the Cl on the right, and finally up to the O.)

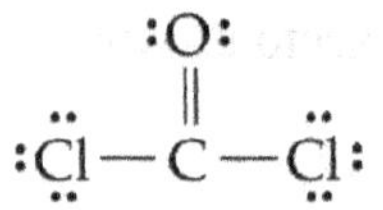

$$\begin{array}{c} :O: \\ \| \\ :\ddot{C}l-C-\ddot{C}l: \end{array}$$

4. We have run out of electrons after putting just two lone pairs on the oxygen. Oxygen does not yet have an octet of electrons. We therefore move the lone pair of electrons on the carbon to a position between the carbon and the oxygen, creating a double bond between the carbon and the oxygen. The completed dot structure is shown below:

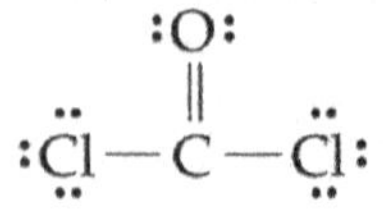

$$\begin{array}{c} :O: \\ \| \\ :\ddot{C}l-C-\ddot{C}l: \end{array}$$

Note that if you had started filling in the lone pairs in a different order, you might have gotten a double bond between the carbon and the chlorine. However, this would have resulted in an arrangement in which the Cl would have had two bonds and two lone pairs and the O would have had one bond and three lone pairs. This would not have been the best arrangement because it violates the general guidelines described in the table on page 96 of this book. Whenever there is more than one possible arrangement for placing the electrons, use the general guidelines in this table to help you to choose the best one.

Practice Exercise 5.8

Draw the dot diagram for N_2H_2.

When there is more than one possible arrangement of electrons for a molecule or ion, the different diagrams are called *resonance structures* (or resonance forms). Resonance structures differ *only* in the position of electrons in the dot diagram. In order for structures to qualify as resonance structures, the positions of all atoms must be the same, as demonstrated in Example 5.9.

Example 5.9

Indicate whether the dot diagrams below are resonance structures. Explain your reasoning.

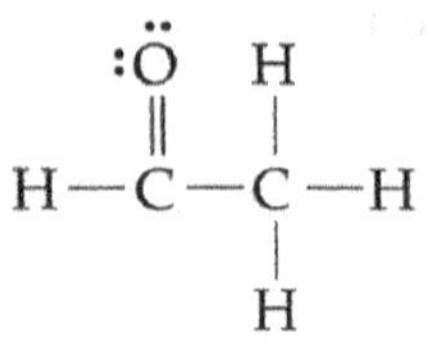

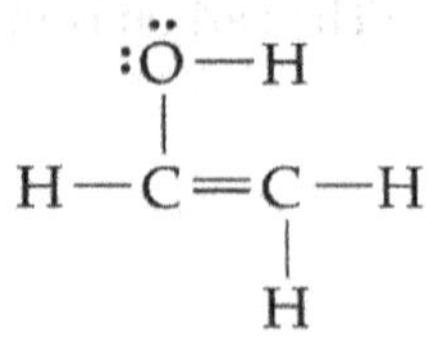

$$\begin{array}{ccc} :\ddot{O} \quad H & & :\ddot{O}-H \\ \| \quad | & & | \\ H-C-C-H & \text{and} & H-C=C-H \\ | & & | \\ H & & H \end{array}$$

Solution

These are not resonance structures because the position of one of the H atoms is different in the two structures.

Practice Exercise 5.9

Indicate whether the dot diagrams below are resonance structures. Explain your reasoning.

$$\left[\begin{array}{c} :O: \\ \parallel \\ H-C-\ddot{O}: \end{array} \right]^{-} \quad \text{and} \quad \left[\begin{array}{c} :\ddot{O}: \\ \mid \\ H-C=\ddot{O}: \end{array} \right]^{-}$$

We draw dot diagrams for polyatomic ions (ions made from many atoms, covalently bonded to each other) much as we have done for molecules, with the exception that we subtract one electron for each positive charge on the ion and add one electron for each negative charge on the ion. Note that the general guidelines listed in the table on page 96 in this book are not reliable when we are drawing the dot structures of ionic compounds. Use the rules stated earlier to develop the correct structure.

Example 5.10

Draw the dot diagram for the NO_2^- ion.

Solution

$$\left[:\ddot{O}:\ddot{N}::\ddot{O}: \right]^{-} \quad \text{or} \quad \left[:\ddot{O}::\ddot{N}:\ddot{O}: \right]^{-}$$

(These are two resonance forms.)

Practice Exercise 5.10

Draw the dot diagram for SO_3^{2-}.

5.6   Equal Versus Unequal Sharing of Electrons — Electronegativity and the Polar Covalent Bond

We have discussed ionic bonds (formed from the transfer of electrons) and covalent bonds (formed from the sharing of electrons) as the two types of bonds that exist between atoms (or ions). The property of atoms that determines whether electrons will be shared or transferred is called *electronegativity*. Electronegativity is defined as "the ability of an atom to attract shared electrons." Electronegativity values for the main group elements are shown in the chart on the next page.

There are three important aspects of this chart that you should commit to memory:

1. The four most electronegative elements (from highest to lowest) are fluorine, oxygen, nitrogen, and chlorine.
2. The electronegativity increases from left to right across a period, and generally decreases from top to bottom down a group. (The elements in groups IIIA and IVA behave a little differently.)
3. The noble gases have no electronegativity values assigned to them. This is because electronegativity is a measure of an atom's ability to attract *shared* electrons. Since the noble gases generally do not form compounds (they have a stable octet of electrons), they do not share electrons.

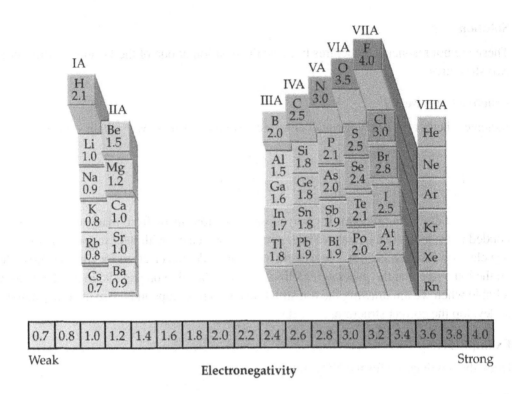

IA

H
2.1

IIA

Li 1.0   Be 1.5
Na 0.9   Mg 1.2
K 0.8   Ca 1.0
Rb 0.8   Sr 1.0
Cs 0.7   Ba 0.9

IIIA   IVA   VA   VIA   VIIA
B 2.0   C 2.5   N 3.0   O 3.5   F 4.0
Al 1.5   Si 1.8   P 2.5   S 3.0   Cl 3.0
Ga 1.6   Ge 1.8   As 2.4   Se 2.8   Br 2.8
In 1.7   Sn 1.8   Sb 1.9   Te 2.1   I 2.5
Tl 1.8   Pb 1.9   Bi 1.9   Po 2.0   At 2.1

VIIIA
He
Ne
Ar
Kr
Xe
Rn

| 0.7 | 0.8 | 1.0 | 1.2 | 1.4 | 1.6 | 1.8 | 2.0 | 2.2 | 2.4 | 2.6 | 2.8 | 3.0 | 3.2 | 3.4 | 3.6 | 3.8 | 4.0 |

Weak   Strong

Electronegativity

It is the *difference* in electronegativity, ΔEN, that determines whether two atoms will form an ionic bond, a covalent bond (nonpolar covalent bond), or a polar covalent bond.

If there is a large difference in the ability to attract shared electrons ($\Delta EN > 1.8$), the more electronegative atom will completely take the electrons from the less electronegative atom, resulting in a transfer of electrons, thus forming an ionic bond. (1.8 is not a hard-and-fast cutoff. See Section 5.6 of the textbook for a complete discussion of this.)

If the electronegativities of the atoms are equal or nearly equal ($\Delta EN < 0.4$), the electrons will be equally shared between the two atoms, resulting in a covalent (sometimes called nonpolar covalent) bond.

There is a third possibility, however. If the electronegativities of the atoms are between 0.4 and 1.8, the electrons will be unequally shared between the two atoms, resulting in what is called a polar covalent bond.

This information is summarized in the table below:

ΔEN	Type of Bond
<0.4	Covalent (or nonpolar covalent)
Between 0.4 and 1.8	Polar covalent
>1.8 (with some exceptions)	Ionic

The partial charges in polar covalent bonds are indicated by placing a $\delta+$ sign over the partially positive atom and $\delta-$ sign over the partially negative atom. Alternatively, a $\longleftrightarrow$ can be used, in which the arrow is placed over the partially negative atom. Both conventions are shown in the HCl molecule below:

$$\overset{\delta+}{H}-\overset{\delta-}{Cl} \qquad \overset{\longleftrightarrow}{H-Cl}$$

Example 5.11

Use the electronegativity values and the ΔEN line to determine whether the bond between each of the following pairs of atoms would be (1) covalent, (2) polar covalent, or (3) ionic. If the bond is polar covalent, use δ+ and δ− to indicate the partial charges.

(a) C and O   (b) Li and Br   (c) C and H   (d) Mg and Cl   (e) P and Cl

Solution

(a) polar covalent   (b) ionic   (c) covalent   (d) ionic   (e) polar covalent

For the polar covalent bonds:

(a) $\overset{\delta+}{C}—\overset{\delta-}{O}$ (e) $\overset{\delta+}{P}—\overset{\delta-}{Cl}$

Practice Exercise 5.11

Use the electronegativity values and the ΔEN line to determine whether the bond between each of the following pairs of atoms would be (1) covalent, (2) polar covalent, or (3) ionic. If the bond is polar covalent, use δ+ and δ− to indicate the partial charges.

(a) K and F   (b) H and Br   (c) C and Si   (d) Na and N   (e) B and I

5.7   Nomenclature—Naming Chemical Compounds

Naming Binary Ionic Compounds

Binary ionic compounds (those containing two elements) are named according to the following rules:

1. Name the cation (positive ion) first.
2. Name the anion (negative ion), giving it an ending of -ide.

You should note that there will be no prefixes (*mono-*, *di-*, *tri-*, etc.) in the names of binary ionic compounds. The number of ions of each type is determined by the charge on each ion, so there is no need to specify this information in the name.

Example 5.12

Name the following binary ionic compounds:

(a) $BaCl_2$ (b) Na_3N (c) K_2S (d) LiI

Solution

(a) barium chloride   (b) sodium nitride   (c) potassium sulfide   (d) lithium iodide

Practice Exercise 5.12

Name the following binary ionic compounds:

(a) BaO   (b) Ca_3N_2 (c) K_2O (d) MgF_2

Naming binary compounds that contain transition elements (also Pb and Sn) is a little tricky because these elements can form more than one kind of ion. The charge on the metal ion must be specified in the name of the compound. The charge is placed in parentheses after the metal is named. Generally, we determine the charge on the metal ion from the charges on the ions to which it is attached. For example, the charge on $CuCl_2$ is determined from the charge on the chloride ions. Because each chloride ion has a −1 charge, the charge on the Cu must be

---

+2 in order for the compound to be electrically neutral. Hence, the name of $CuCl_2$ is copper(II) chloride.

Example 5.13

Name the following compounds:

(a) $FeCl_2$ (b) Cr_2O_3 (c) CoS (d) SnO_2

Solution

(a) iron(II) chloride   (b) chromium(III) oxide   (c) cobalt(II) sulfide   (d) tin(IV) oxide

Practice Exercise 5.13

Name the following compounds:

(a) $CrCl_3$ (b) CuS (c) $FeBr_3$ (d) MnF_2

An older naming system involving the suffixes *-ous* and *-ic* is discussed in the textbook. You should be familiar with this system of naming transition metal ionic compounds as well as the one presented above.

Naming Compounds That Contain Polyatomic Ions

Polyatomic ions are ions that contain covalently bonded atoms. Some examples that we have already encountered are NO_3^-, SO_4^{2-}, and NH_4^+. Table 5.5 of the textbook lists the names and formulas of some common polyatomic ions. Take the time to memorize them. You will find this information very helpful as you continue your study of chemistry.

Naming compounds that contain polyatomic ions is analogous to naming binary ionic compounds; the name of the polyatomic ion is the name of the cation or the anion, depending on the polyatomic ion present. The following example illustrates this concept.

Example 5.14

Name the following compounds:

(a) $CaSO_4$ (b) Li_2CO_3 (c) $(NH_4)_3PO_4$ (d) $Ba(ClO)_2$

Solution

(a) calcium sulfate   (b) lithium carbonate   (c) ammonium phosphate

(d) barium hypochlorite

Practice Exercise 5.14

Name the following compounds:

(a) $Mg(NO_3)_2$ (b) $Al(NO_2)_3$ (c) $NaHCO_3$ (d) $Fe(OH)_2$

Naming Binary Covalent Compounds

Binary covalent compounds are named almost like binary ionic compounds, with the exception that the number of atoms of each element is specified with a Greek prefix (*mono-*, *di-*, *tri-*, etc.). The following rules apply:

1. The less electronegative element is named first.
2. A Greek prefix is used to indicate how many atoms of each element are present. The Greek prefix is omitted if there is only one atom of the less electronegative element.

The following example illustrates the use of these rules in naming binary covalent compounds.

Example 5.15

Name the following compounds:

(a) CCl_4 (b) N_2O_5 (c) CO (d) SO_3

Solution

(a) carbon tetrachloride   (b) dinitrogen pentoxide   (c) carbon monoxide
(d) sulfur trioxide

Practice Exercise 5.15

Name the following compounds:

(a) SiO_2 (b) PI_3 (c) CBr_4 (d) BCl_3

Strategies for Working Problems in Chapter 5

Overview: What You Should Be Able to Do

Chapter 5 provides a complete discussion of molecular and ionic bonding and presents the system for naming compounds. The types of problems that you should be able to solve after mastering Chapter 5 are as follows:

1. Draw dot diagrams for molecules and polyatomic ions.
2. Name binary ionic compounds when the metal has the same ionic charge in all compounds.
3. Name binary ionic compounds when the metal has different ionic charges in different compounds (most transition metals and some heavy representative metals).
4. Name binary covalent compounds.
5. Name binary ionic compounds that contain a polyatomic ion.
6. Name acids that are dissolved in water. (Acids that are not dissolved in water are named using the rules for naming binary covalent compounds or compounds containing polyatomic ions.)

Flowcharts

Flowchart 5.1 Drawing dot diagrams for molecules and polyatomic ions

Flowchart 5.1

Summary

Method for drawing dot diagrams for molecules and polyatomic ions

Step 1
- **Determine the total number of valence electrons (dots) that will be in the diagram. Recall that the number of valence electrons is equal to the group number of the representative elements. (For example, N is in group 5A and therefore has five valence electrons.)**
- **For a polyatomic ion, adjust the electron count to reflect the charge (for cations, subtract the number of electrons equal to the charge; for anions, add the number of electrons equal to the absolute value of the charge).**

↓

Step 2
- **Connect the atoms using single covalent bonds.**
- **If the arrangement is not known, use the first non-hydrogen atom in the formula as the central atom.**
- **Add up the total number of electrons used in the single bonds, and subtract this number from the total number of electrons (calculated in Step 1). This tells you the remaining electrons that must be placed as lone pairs in the dot diagram.**

↓

Step 3
- **Add the remaining electrons as lone pairs, making sure that you satisfy the octet rule for each atom. Place the lone pairs on the central atom last.**

↓

Step 4
- **Check to see if there are any atoms lacking an octet. If not, the dot structure is completed.**

- **If there are atoms lacking an octet, move lone pairs from adjacent atoms into bonding positions on the octet-deficient atom(s).**

$$\downarrow$$

Step 5 (for polyatomic ions only)
- **Place the entire dot structure in brackets, with the charge in the upper right-hand corner.**

Example 1

Draw the electron dot diagram for H_2S.

Solution

Perform Step 1

The total number of valence electrons in H_2S is 8, as determined below.

H: 2×1 valence electron per H $= 2$ electrons

S: 1×6 valence electrons per S $= 6$ electrons

Total valence electrons $= 2 + 6 = 8$

Perform Step 2

Connecting the atoms (with S as the central atom) using single bonds, we obtain:

   H:S:H

There are now four electrons in the structure and four electrons remaining to be put in.

Perform Step 3

Placing the remaining electrons as lone pairs on the sulfur (remember that H never has lone pairs), we get the following dot structure for H_2S:

   H:S̈:H

Example 2

Draw the electron dot diagram for SO_3^{2-}.

Solution

Perform Step 1

The total number of valence electrons in SO_3^{2-} is 26, as determined below.

S: 1×6 valence electrons per S $= 6$ electrons

O: 3×6 valence electrons per O $= 18$ electrons

$2-$ charge: add 2 electrons $= 2$ electrons

Total electrons $= 6 + 18 + 2 = 26$ electrons

Perform Step 2

Connecting the atoms (with S as the central atom) using single bonds, we obtain:

   O:S:O
   O

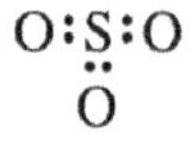

There are now 6 electrons in the structure, leaving 20 electrons to be placed as lone pairs.

Perform Step 3

Placing the remaining 20 electrons as lone pairs on the atoms (leaving the S for last) and satisfying the octet rule, we get the following dot diagram for SO_3^{2-}:

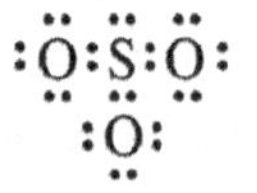

$$:\ddot{\text{O}}:\ddot{\text{S}}:\ddot{\text{O}}:$$
$$:\ddot{\text{O}}:$$

Perform Step 4

There are no atoms lacking an octet, so the dot structure is complete.

Perform Step 5

Because this is a polyatomic ion, the entire structure must be placed in brackets, with the 2− charge in the upper right-hand corner, as shown here:

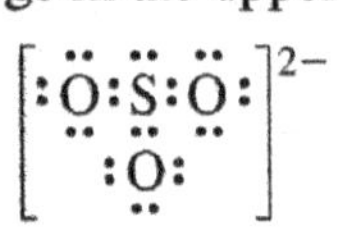

$$\left[:\ddot{\text{O}}:\ddot{\text{S}}:\ddot{\text{O}}:\atop :\ddot{\text{O}}:\right]^{2-}$$

Example 3

Draw the electron dot diagram for N_2H_2.

Solution

Perform Step 1

The total number of valence electrons in N_2H_2 is 12, as determined below.

N: 2 × 5 valence electrons per N = 10 electrons
H: 2 × 1 valence electrons per H = 2 electrons
Total electrons = 10 + 2 = 12 electrons

Perform Step 2

Connecting the atoms (with the N's as the central atoms) using single bonds, we obtain:

H:N:N:H

There are now six electrons in the structure, leaving six electrons to be placed as lone pairs.

Perform Step 3

Placing the remaining six electrons as lone pairs on the atoms (until we run out of electrons) and satisfying the octet rule, we get the following dot diagram:

$$\text{H}:\ddot{\text{N}}:\ddot{\text{N}}:\text{H}$$

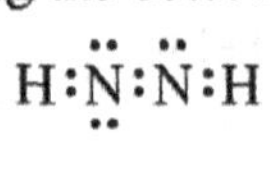

Perform Step 4

The nitrogen atom on the right is lacking an octet, so we move one of the lone pairs on the nitrogen on the left into a bonding position between the two nitrogen atoms to obtain the dot diagram for N_2H_2:

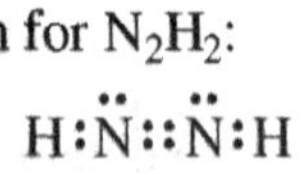

$$\text{H}:\ddot{\text{N}}::\ddot{\text{N}}:\text{H}$$

Practice Problems

5.1 Draw the electron dot diagram for H_2O_2.
5.2 Draw the electron dot diagram for C_2H_4.
5.3 Draw the electron dot diagram for ClO_3^-.

Flowchart 5.2 Naming binary ionic compounds when the metal has the same ionic charge in all compounds

Flowchart 5.2

Summary

Method for naming binary ionic compounds when the metal has the same ionic charge in all compounds

Step 1
- **Name the metal ion (the positively charged cation) using the name of the metal.**

↓

Step 2
- **Name the nonmetal ion (the negatively charged anion) by giving it an *-ide* ending.**

↓

Step 3
- **Combine the name of the metal ion with that of the nonmetal ion to get the name of the compound.**

Example 1

Name the compound Na_2O.

Solution

Perform Step 1

The name of the metal ion is sodium.

Perform Step 2

The name of the nonmetal ion is oxide.

Perform Step 3

The name of the compound is sodium oxide.

Example 2

Name the compound CaS.

Solution

Perform Step 1

The name of the metal ion is calcium.

Perform Step 2

The name of the nonmetal ion is sulfide.

Perform Step 3

The name of the compound is calcium sulfide.

Example 3

Name the compound Mg_3P_2.

Solution

Perform Step 1

The name of the metal ion is magnesium.

Perform Step 2

The name of the nonmetal ion is phosphide

Perform Step 3

The name of the compound is magnesium phosphide.

Practice Problems

5.4 Name the compound $BaCl_2$.
5.5 Name the compound Ce_3N.
5.6 Name the compound KI.

Flowchart 5.3 **Naming binary ionic compounds when the metal has different ionic charges in different compounds (most transition metals and some heavy representative metals)**

Flowchart 5.3

Summary

Method for naming binary ionic compounds when the metal has different ionic charges in different compounds (most transition metals and some heavy representative metals)

Step 1
• **Name the metal ion (the positively charged cation) using the name of the metal.**

$\downarrow$

Step 2
• **Specify the charge on the metal ion by placing a Roman numeral indicating the charge in parentheses after the metal name, if giving the systematic name.**
• **Indicate the charge on the metal ion by using the appropriate suffix, if giving the old name.**

$\downarrow$

Step 3
• **Name the nonmetal ion (the negatively charged anion) by giving it an *-ide* ending.**

$\downarrow$

Step 4
• **Combine the name of the metal ion, the charge (if giving the systematic name), and the nonmetal ion to get the name of the compound.**

Example 1

Name the compound $CuBr_2$, giving the systematic name.

Solution

Perform Step 1

The name of the metal ion is copper.

Perform Step 2

The charge on the copper ion is 2+, which must be indicated by the Roman numeral II placed in parentheses. The name of the cation thus becomes copper(II).

Perform Step 3

The name of the nonmetal ion is bromide.

Perform Step 4

The name of the compound is copper(II) bromide.

Example 2

Name the compound $CuBr_2$, giving the old name.

Solution

Perform Step 1

The name of the metal ion is copper.

Perform Step 2

The charge on the copper ion is 2+, which must be indicated by the using the suffix *-ic*. The name of the cation thus becomes cupr*ic*.

Perform Step 3

The name of the nonmetal ion is bromide.

Perform Step 4

The name of the compound is cupric bromide.

Example 3

Name the compound FeO, giving the systematic name.

Solution

Perform Step 1

The name of the metal ion is iron.

Perform Step 2

The charge on the iron ion is 2+, which must be indicated by the Roman numeral II placed in parentheses. The name of the cation thus becomes iron(II).

Perform Step 3

The name of the ion is oxide.

Perform Step 4

The name of the compound is iron(II) oxide.

Practice Problems

    5.7 Name the compound CoI_2, giving the systematic name.

    5.8 Name the compound CuF, giving the old name.

    5.9 Name the compound $HgBr_2$, giving the systematic name.

Flowchart 5.4   Naming binary covalent compounds

Flowchart 5.4

Summary

Method for naming binary covalent compounds

Step 1
- **Name the less electronegative element first (this is usually the first element written in the formula).**
- **The element name is preceded by a Greek prefix (from Table 5.4 of the textbook) indicating the number of atoms of the element.**
- **If there is only one atom of the first element, the prefix *mono-* is not used.**

↓

Step 2
- **Name the more electronegative element second, giving it the suffix *-ide*.**
- **The name is preceded by a Greek prefix indicating the number of atoms of the element.**

↓

Step 3
- **Combine the names of both elements to get the name of the compound.**

Example 1

Name the compound P_2O_5.

Solution

Perform Step 1

The name of the less electronegative element is phosphorus. There are two phosphorus atoms, so the prefix *di-* is used, giving the name *di*phosphorus.

Perform Step 2

The name of the more electronegative element, with the suffix *-ide*, is oxide. There are five oxygen atoms present, so the prefix *penta-* is used, giving the name *pent*oxide. (Note that the "a" is dropped from penta to avoid having two vowels together.)

Perform Step 3

The name of the compound is diphosphorus pentoxide.

Example 2

Name the compound B_2H_6.

Solution

Perform Step 1

The name of the less electronegative element is boron. There are two boron atoms, so the prefix *di-* is used. The name is *di*boron.

Perform Step 2

The name of the more electronegative element, with the suffix *-ide*, is hydr*ide*. There are six hydrogen atoms present, so the prefix *hexa-* is used, giving the name *hexa*hydride.

Perform Step 3

The name of the compound is diboron hexahydride.

Example 3

Name the compound OF_2.

Solution

Perform Step 1

The name of the less electronegative element is oxygen. There is only one oxygen atom, so the prefix *mono-* is not used.

Perform Step 2

The name of the more electronegative element, with the suffix *-ide*, is fluor*ide*. There are two fluoride atoms present, so the prefix *di-* is used, giving the name *di*fluoride.

Perform Step 3

The name of the compound is oxygen difluoride.

Practice Problems

5.10 Name the compound B_2O_3.
5.11 Name the compound NF_3.
5.12 Name the compound XeF_4.

Flowchart 5.5   Naming binary ionic compounds that contain a polyatomic ion

Flowchart 5.5

Summary

Method for naming binary ionic compounds that contain a polyatomic ion

Step 1
- **Name the metal ion (the positively charged cation) using the name of the metal.**
- **Specify the charge on the metal ion if the metal has different ionic charges in different compounds.**
- **The positive ion NH_4^+ is named ammonium.**

Step 2
- **Name the polyatomic anion, using the names given in Table 5.5 of the textbook.**

Step 3
- **Combine the name of the metal ion, the charge (if giving the systematic name), and the polyatomic ion to get the name of the compound.**

Example 1

Name the compound Li_3PO_4.

Solution

Perform Step 1

The name of the metal ion is lithium.

Perform Step 2

The name of the polyatomic ion is phosphate.

Perform Step 3

The name of the compound is lithium phosphate.

Example 2

Name the compound $Ca(NO_3)_2$.

Solution

Perform Step 1

The name of the metal ion is calcium.

Perform Step 2

The name of the polyatomic ion is nitrate.

Perform Step 3

The name of the compound is calcium nitrate.

Example 3

Name the compound $Fe(NO_2)_2$, giving the systematic name.

Solution

Perform Step 1

The name of the metal ion is iron.

Perform Step 2

The charge on the iron ion is 2+, which must be indicated by the Roman numeral II placed in parentheses. The name of the cation thus becomes iron(II).

Perform Step 3

The name of the polyatomic ion is nitrite.

Perform Step 4

The name of the compound is iron(II) nitrite.

Practice Problems

5.13 Name the compound TiF_3, giving the systematic name.
5.14 Name the compound NiO, giving the old name.
5.15 Name the compound $SnCl_4$, giving the systematic name.

Flowchart 5.6   **Naming acids that are dissolved in water**

Flowchart 5.6

Summary

Method for naming acids that are dissolved in water

Step 1
- **Determine if the acid contains oxygen.**
- **If no oxygen is present, proceed to Step 2.**
- **If oxygen is present, proceed to Step 3.**

Step 2
- **If the formula does not contain oxygen, add the prefix *hydro-* and the suffix *-ic* acid to the name of the non-hydrogen element or the polyatomic ion to get the name of the acid.**

Step 3
- **If the formula contains oxygen, name the acid based on the name of the polyatomic ion the acid contains.**
- **If the polyatomic ion name ends in *-ite*, replace the suffix *-ite* with *-ous* acid to name the acid.**
- **If the polyatomic ion name ends in *-ate*, replace the suffix *-ate* with *-ic* acid to name the acid.**

Example 1

Name the acid HI.

Solution

Perform Step 1

The acid contains no oxygen; proceed to Step 2.

Perform Step 2

Adding the prefix *hydro-* and the suffix *-ic* acid to the iodine gives the name hydroiodic acid.

Example 2

Name the acid $HClO_4$.

Solution

Perform Step 1

The acid contains oxygen; proceed to Step 3.

Perform Step 3

The polyatomic ion ends in *-ate* (perchlor*ate*), so we replace the suffix *-ate* with *-ic* acid, giving us the name perchlor*ic* acid.

Example 3

Name the acid HNO_3.

Solution

Perform Step 1

The acid contains oxygen; proceed to Step 3.

Perform Step 3

The polyatomic ion ends in *-ate* (nitr*ate*), so we replace the suffix *-ate* with *-ic* acid, giving us the name nitr*ic* acid.

Practice Problems

5.16   Name the acid $HClO_2$.
5.17   Name the acid $HBrO_4$.
5.18   Name the acid HF.

Quiz for Chapter 5 Problems

1. Draw electron dot diagrams for the following molecules:
   (a) PCl_3 (b) SO_2

2. Draw electron dot diagrams for the following polyatomic ions:
   (a) NH_4^+ (b) NO_2^-

3. Name the following compounds:
   (a) Ca_3P_2 (b) FeI_3

4. Name the following acids:
   (a) H_2SO_4 (b) HNO_2

5. The charge on the copper in the compound CuO is _____, and the charge on the copper in the compound Cu_2O is _____.

6. Change each of the following incorrect names to correct names.
   (a) trimagnesium dinitride for Mg_3N_2
   (b) nitric acid for HNO_2
   (c) chlorous acid for $HClO$

7. Indicate the number of valence electrons that will appear in the dot structure for each species below:
   (a) $HClO_4$ (b) SO_4^{2-} (c) NH_4^+ (d) O_3

8. Which of the following atoms will never satisfy the octet rule in electron dot structures?
   (a) oxygen   (b) boron   (c) chlorine   (d) hydrogen   (e) sulfur

9. In which of the compounds below is it necessary to specify the charge on the metal ion when naming the compound? More than one answer is possible.
   (a) $BaCl_2$ (b) $CoCl_2$ (c) $BeCl_2$ (d) $HgCl_2$ (e) $CrCl_2$

10. Explain why it would be impossible to write the formula of a compound whose name is given as nitrogen oxide.

STUDENT WORKBOOK AND SELECTED SOLUTIONS

Exercises for Self-Testing

A. Completion. *Write the correct word(s) to complete each statement below.*

1. A stable collection of bonded atoms is a(n) _____.

2. The number of bonds that a main group element will form is determined by the number of _____ _____.

3. In the molecule formed from the combination of aluminum and sulfur, there will be _____ aluminum atoms and _____ sulfur atoms.

4. A bond that involves the sharing of two pairs of electrons is a(n) _____ _____.

5. Different Lewis dot structures that differ only in the position of electrons are called _____ _____.

6. In neutral, covalent compounds, oxygen typically forms _____ bonds and has _____ lone pairs.

7. In the periodic table, the electronegativity generally _____ across a period and _____ down a group.

8. _____ is the most electronegative element.

9. If element A has an electronegativity of 2.2 and element B has an electronegativity of 1.5, the partially positive end of the molecule AB will be in the direction of element _____, and the partially negative end will be in the direction of element _____.

10. If the ΔEN for two elements is 1.9, the bond between atoms of these elements will be _____.

11. The correct formula (with the charge) for the carbonate ion is _____.

12. Covalent compounds exist as discrete _____, whereas ionic compounds exist as a(n) _____ _____.

13. A covalent bond in which the electrons are shared unequally is called a(n) _____ _____ bond.

14. The correct name of FeO is _____ _____.

15. Greek prefixes are used when naming _____ _____ compounds but not when naming _____ compounds.

B. True/False. *Indicate whether each statement below is true or false, and explain why the false statements are false.*

_____ 1. The physical properties of molecules are generally similar to those of the atoms from which they are made.

_____ 2. Covalent bonds always involve equal sharing of electrons.

_____ 3. A compound formed between an element in group IA and an element in group VIA will probably be ionic.

_____ 4. The Lewis dot diagram for the N atom has two paired electrons and one unpaired electron.

_____ 5. The Lewis dot diagram for the H_2S molecule has two lone pairs of electrons.

_____ 6. Resonance forms differ only in the position of the electrons in the dot diagram.

_____ 7. All atoms react chemically in order to obtain eight valence electrons.

_____ 8. $MgCl_2$ exists as discrete molecules.

_____ 9. The correct name for $CaCl_2$ is calcium dichloride.

_____ 10. In polar covalent bonds, the difference in electronegativity will be greater than 0.4 and less than 1.8.

C. Questions and Problems

1. Classify bonds in the following compounds as covalent, polar covalent, or ionic. Explain your reasoning.
   - (a) CaO
   - (b) $SiBr_4$
   - (c) CH_3Cl
   - (d) H_2O

2. Name the following compounds:
   - (a) Mg_3N_2
   - (b) P_2O_5
   - (c) NH_4Br
   - (d) Fe_2O_3

3. Arrange the following molecules in order of increasing polarity:
   - (a) HCl
   - (b) HI
   - (c) HBr
   - (d) HF
   - (e) H_2

4. Explain why it is not necessary to use Greek prefixes when naming binary ionic compounds.

5. Draw the electron dot diagram for the hydronium ion, H_3O^+.

6. Draw the electron dot diagrams for BCl_3 and PCl_3. Would you expect these two compounds to have similar chemical properties? Why or why not?

7. Write the formulas of the following compounds:
   - (a) lithium carbonate
   - (b) iron(III) iodide
   - (c) ammonium sulfate
   - (d) sodium acetate
   - (e) boron trichloride

8. Draw the Lewis dot structure for CH_3CN. (The two carbons are bonded together.)

9. Write the names for the following compounds:
   - (a) P_2S_3
   - (b) K_3PO_4
   - (c) $Co_2(SO_4)_3$
   - (d) XeF_4

10. Explain why it is not correct to refer to *molecules* of sodium chloride.

6

The Shape of Molecules

Learning Outcomes

1. Explain how molecular shape is important with regard to disease and drugs that might cure it.
2. Explain the principles behind VSEPR theory.
3. Draw and name a molecule's shape starting from a dot diagram.
4. Draw the molecular shape for molecules possessing multiple bonds.
5. Explain polarity in terms of individual bonds and an entire molecule.
6. Predict the relative polarity of a molecule.
7. Explain how molecular polarity leads to intermolecular attraction.
8. Deduce and draw a dipole moment for an entire molecule.
9. Describe how intermolecular attractions arise between dipolar molecules.
10. Draw dot diagrams and describe the shapes of molecules possessing central atoms with an expanded octet of electrons.

Chapter Outline

6.1 The Importance of Molecular Shape
  A. Need for antibacterial agent prior to World War II
  B. Sulfanilamide mimicking of PABA based on shape
6.2 Valence Shell Electron Pair Repulsion Theory
  A. Electron pair repulsion in methane
  B. Tetrahedral geometry of methane
  C. Steric number
  D. Geometry of molecules containing multiple bonds
  E. Geometry of molecules containing lone pairs
  F. VSEPR shape table

6.3 Polarity of Molecules, or When Does 2 + 2 Not Equal 4?
   A. Intermolecular attractive forces
   B. Behavior of polar and nonpolar molecules in an electric field
   C. Conventions for indicating bond polarity
   D. Dipole moments
      1. Bond dipole moment vectors
      2. Resultant dipole moment vector for entire molecule

6.4 Intermolecular Forces—Dipolar Interactions
   A. Dipole–dipole forces due to polarity
   B. Determination of polarity from Lewis dot diagram

6.5 One More Thing: VSEPR Theory for Molecules Possessing Expanded Octet Atoms
   A. Hypervalent molecules
   B. Placement of lone pairs in hypervalent molecules

Review of Key Concepts from the Textbook

6.1 The Importance of Molecular Shape

Up to this point, we have represented molecules as two-dimensional structures, when in fact most are three-dimensional structures. The shape of a molecule can determine its reactivity, its effectiveness as an enzyme or metabolite, or any number of other properties. We will learn how to predict the shapes of molecules in this chapter.

6.2 Valence Shell Electron Pair Repulsion Theory

Valence shell electron pair repulsion, or VSEPR (pronounced "vesper"), theory is based on the concept that the valence electrons in a molecule will repel each other. The shape that a molecule (or ion) assumes is the one that allows the valence electrons to be as far apart as possible.

To correctly determine molecular shape, we must always begin with a correct Lewis dot structure, which will tell us how many valence electrons are in the molecule. (If we start with an *incorrect* dot diagram, we will most probably assign the molecule an incorrect shape. Always exercise extreme care when drawing dot diagrams to make sure they are correct.)

When we have the correct dot structure, we can determine the steric number—the number of electron groups around the central atoms of the molecule. The molecular shape associated with each number of groups of electrons is shown in the table on the next page.

When we consider molecules that have lone pairs of electrons, we determine the shape of the molecule by ignoring the position of the lone pairs. The possible molecular shapes (ignoring the position of lone pairs) are shown in the table at the top of the next page. Notice the slight change in bond angles when lone pairs are present. For example, for a molecule in which the central atom has four electron groups around it, the bond angle changes from 109.5° (when no lone pairs are present) to approximately 107° (when one lone pair is present) to approximately 105° (when two lone pairs are present).

The steps involved in determining the shape of a molecule are as follows:

1. Draw the correct electron dot structure.
2. Count the number of electron groups (bonding groups plus lone pairs) around the central atom. This is the steric number.

Steric Number (Number of Electron Groups)	Geometry of Electron Groups	Corresponding Molecular Shape	Example
Two	180° Bond angle 180°	Linear	$C{=}O{=}C$ Carbon dioxide, CO_2
Three	120° Bond angles all 120°	Trigonal planar	Sulfur trioxide, SO_3
Four	109.5° Bond angles all 109.5°	Tetrahedral	Carbon tetrachloride, CCl_4

3. Determine the electron arrangement around the central atom. (This is the shape of the molecule if there are no lone pairs.)
4. If lone pairs are present, describe the shape of the molecule, ignoring the position(s) occupied by any lone pairs.

Total Number of Electron Groups	Structure	Electron Group Arrangement	Bond Angles (idealized)*	Molecular Shape
Four				
4 bonding groups no lone pairs		Tetrahedral	109.5°	Tetrahedral
3 bonding groups 1 lone pair		Tetrahedral	~107° (109.5°)*	Pyramidal
2 bonding groups 2 lone pairs		Tetrahedral	~105° (109.5°)*	V-shaped or bent
Three				
3 bonding groups no lone pairs		Trigonal planar	120°	Trigonal planar
2 bonding groups 1 lone pair		Trigonal planar	~118° (120°)*	V-shaped or bent
Two				
2 bonding groups no lone pairs		Linear	180°	Linear

*If central atom is in second period, −2° for each lone pair.

We will use this information to determine the shape of the molecule in the following example.

Example 6.1

Predict the shape of the NO_3^- ion. What are the bond angles?

Solution

We must first draw a correct Lewis dot diagram, using the rules presented in Chapter 5 of the textbook and Chapter 5 of this *Study Guide*.

$$\left[\begin{array}{c} :\ddot{O}: \\ \| \\ :\ddot{O}:N:\ddot{O}: \end{array} \right]^-$$

Nitrogen is the central atom. There are three total groups of electrons around the N. The nitrogen atom has a steric number of 3. (Remember to count double and triple bonds as one group because all the electrons are in the same region.) All three of the groups are bonding pairs. (There are no lone pairs on the N.) The molecular shape associated with three groups of electrons around a central atom is trigonal planar. The bond angles associated with a trigonal planar shape are 120°.

Practice Exercise 6.1

Predict the shape of the NO_2^- ion. What is the bond angle?

6.3   Polarity of Molecules, or When Does 2 + 2 Not Equal 4?

We discussed the polarity of bonds in Chapter 5. We will now see how the polarity of molecules is determined, and how molecular polarity affects the properties of compounds.

Many properties of compounds, such as boiling point and solubility, are determined by the intermolecular attractive forces between the molecules. One of the most common types of intermolecular attractive forces (or intermolecular forces) is the dipole–dipole interaction. Dipole–dipole interactions occur when a compound is composed of polar molecules.

In polar molecules, there is a partially positive end and a partially negative end of the molecule. Because opposite charges attract, when polar molecules are placed between two oppositely charged metal plates, the molecules line up such that the partially positive ends are near the negative plate and the partially negative ends are near the positive plate.

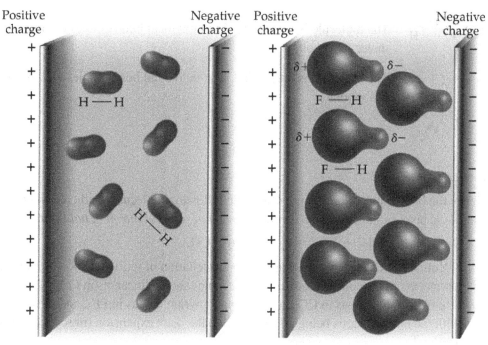

No lining up. H_2 must be nonpolar.   They line up. HF must be polar.

This behavior is the reason that these molecules are called polar molecules. Molecules that have no partially charged regions do not line up when placed between oppositely charged metal plates and are said to be nonpolar. (See the figure above.)

Two factors must be considered when determining whether a molecule is polar or nonpolar:

1. Whether any polar bonds are present in the molecule, due to differences in electronegativity between bonded atoms. If there are no polar bonds in the molecule, the molecule will be nonpolar.

2. Whether the individual bond dipoles cancel each other, based on the shape of the molecule. (We use vectors to determine whether the individual bond dipoles cancel, as described in Section 6.3 of the textbook.) If all the bond dipoles cancel, the molecule will be nonpolar.

Example 6.2

Indicate the bond dipoles (if any) present in the following molecules. Indicate whether the molecule is polar or nonpolar. *The central atom (if present) is shown in **bold font**.*

(a) **Si**SO   (b) HI   (c) **N**H$_3$ (d) **C**Cl$_4$

Solution

(a) S⇌**Si**⇌O   The molecule is linear. However, the individual dipoles do *not* cancel, because oxygen is more electronegative than sulfur. Consequently, the Si—O bond dipole moment vector is a little longer than that for the Si—S bond and the molecule is slightly polar.

(b) H⇌I   The molecule is polar.

(c) The molecule is pyramidal. The individual bond dipoles do not cancel. The molecule is polar.

(d) The molecule is tetrahedral. The individual bond dipoles cancel. The molecule is nonpolar.

Practice Exercise 6.2

Indicate the bond dipoles (if any) present in the following molecules. Indicate whether the molecule is polar or nonpolar. *The central atom (if present) is shown in **bold font**.*

(a) CO_2 (b) PH_3 (c) F_2 (d) CH_3I

Notice that the geometries that lead to cancellation of individual bond dipoles (if the atoms connected to the central atom are all the same) are (1) linear, as in CO_2, (2) trigonal planar, as in BF_3, (3) tetrahedral, as in CCl_4, (4) trigonal bipyramidal, as in PF_5, and (5) octahedral, as in SeF_6, with the last two cases being possible due to the octet expansion (introduced in Section 6.5 of the textbook).

6.4 Intermolecular Forces—Dipolar Interactions

Polar molecules attract each other because of the attraction of the unlike partial charges. This attraction is what causes the molecules to aggregate with one another to form a liquid or a solid under normal conditions. The intermolecular attractive forces between polar molecules are called dipole–dipole forces (or polar forces). Dipole–dipole forces are some of the strongest intermolecular forces known, but they are considerably weaker than covalent and ionic bonds (referred to as *intra*molecular forces). It is therefore much easier (requires a lot less energy) to evaporate or boil water than it is to separate water into its component elements of hydrogen and oxygen. Evaporation requires only the breaking of intermolecular forces to separate water molecules from the liquid as they pass into the gaseous phase. On the other hand, separating water into its component elements requires breaking intramolecular forces (covalent bonds).

In order to determine whether a molecule is polar—and thus has dipole–dipole intermolecular attractive forces—you must do the following:

1. Draw the correct Lewis dot diagram for the molecule.
2. Apply the VSEPR theory to determine the shape of the molecule.
3. Draw the three-dimensional shape of the molecule, showing all bond dipole moments.
4. Determine whether the individual bond dipoles cancel (making the molecule nonpolar) or whether there is a net dipole moment in the molecule (making the molecule polar).

Example 6.3

Indicate which of the following molecules experience dipole–dipole intermolecular attractive forces. *The central atom (if present) is shown in **bold font**.*

(a) CH_2Br_2 (b) H_2Se (c) SO_2 (d) $SiCl_4$

Solution

(a) CH_2Br_2 is a polar molecule. (The molecule is tetrahedral, and the individual bond dipoles do not cancel each other.) CH_2Br_2 experiences dipole–dipole intermolecular attractive forces.

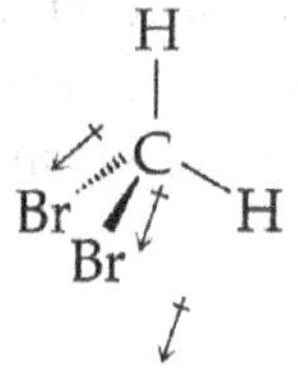

Net dipole

(b) H_2Se is a polar molecule. (The molecule is bent, and the individual bond dipoles do not cancel each other.) H_2Se experiences dipole–dipole intermolecular attractive forces.

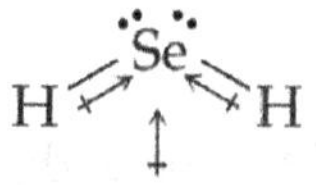

Net dipole

(c) SO_2 is a polar molecule. (The molecule is bent, and the individual bond dipoles do not cancel each other.) SO_2 experiences dipole–dipole intermolecular attractive forces.

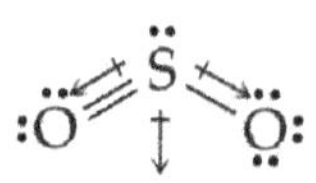

Net dipole

(d) $SiCl_4$ is nonpolar. (The molecule is tetrahedral, and the individual bond dipoles cancel each other.) $SiCl_4$ does not experience dipole–dipole intermolecular attractive forces. Note that lone pairs on Cl atoms are not shown for clarity.

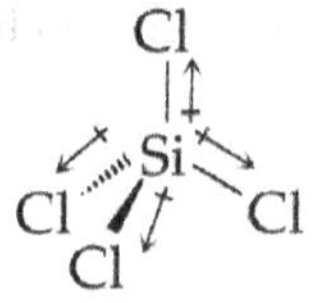

No net dipole

Practice Exercise 6.3

Indicate which of the following molecules experience dipole–dipole intermolecular attractive forces. *The central atom (if present) is shown in **bold font**.*

(a) **Si**HCl$_3$ (b) **C**S$_2$ (c) **C**F$_4$ (d) **P**F$_3$

6.5 VSEPR Theory for Molecules Possessing Expanded Octet Atoms

For some atoms in periods three, four, and beyond, there is a possibility of expansion of the octet. In these cases, there will be more than eight valence electrons in the vicinity of these atoms. It is important to know that the exceptions to the octet rule, especially the octet expansion, as almost as common as the examples of where the rule is followed. The list of atoms that commonly expand their octets includes, but is not limited to, sulfur, chlorine, phosphorus, bromine, and antimony. Interestingly, if not for the octet expansion, we would not have molecules and polyatomic ions involving noble gases, such as XeF_2, XeF_4, and XeF_6^{2-}.

The procedures for determining molecular shapes and polarities of molecules possessing expanded octet atoms is the same as those outlined in Sections 6.2 and 6.3. Whether these molecules will or will not experience intermolecular attractions can be determined in a manner outlined in Section 6.4.

Example 6.4

Indicate which of the following molecules experience dipole–dipole intermolecular attractive forces. *The central is shown in **bold font** and has an expanded octet.*

(a) **S**F$_4$ (b) **Xe**F$_2$ (c) **P**F$_5$ (d) **Xe**F$_4$

Solution

(a) SF_4 is polar. (SN = 5; the molecule has a seesaw shape. The individual bond dipoles in the linear positions cancel each other, but these in the triangular positions do not.) SF_4 experiences dipole–dipole intermolecular attractive forces. Note that the lone pairs on the F atoms are not shown for clarity.

(b) XeF_2 in nonpolar. (SN = 5; the molecule has a linear shape due to the presence of three lone pairs in triangular positions of the trigonal bipyramid, and two Xe—F bonds being in the linear positions. The individual bond dipoles cancel each other.) XeF_2 does not experience dipole–dipole intermolecular attractive forces. Note that the three lone pairs on the central Xe atom and the F atoms are not shown for clarity.

(c) PF_5 is nonpolar. (SN = 5; the molecule has trigonal bipyramidal shape, and all individual bond dipoles cancel each other.) PF_5 does not experience dipole–dipole intermolecular attractive forces. Note that the lone pairs on the F atoms are not shown for clarity.

(d) XeF_4 is nonpolar. (SN = 6; the molecule has a square planar shape due to the presence of two lone pairs on the opposite sides of the molecule. The four Xe—F individual bond dipoles in the square planar positions cancel each other.) XeF_4 does not experience dipole–dipole intermolecular attractive forces. Note that the lone pairs on the F atoms are not shown for clarity.

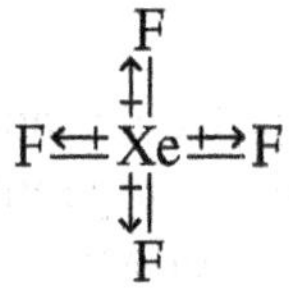

$$
\begin{array}{c}
\text{F} \\
\updownarrow| \\
\text{F} \leftrightharpoons \text{Xe} \rightleftharpoons \text{F} \\
\updownarrow| \\
\text{F}
\end{array}
$$

Practice Exercise 6.4

Indicate which of the following molecules experience dipole–dipole intermolecular attractive forces. *The central is shown in **bold font** and has an expanded octet.*

(a) $\textbf{Cl}F_3$ (b) $\textbf{Se}F_6$ (c) $\textbf{Se}Cl_4$ (d) $\textbf{Sb}F_5$

Strategies for Working Problems in Chapter 6

Overview: What You Should Be Able to Do

Chapter 6 provides the procedures for determining the geometry and polarity of molecules and discusses how the intermolecular attractive forces between molecules are dependent on molecular geometries. The types of problems that you should be able to solve after mastering Chapter 6 are as follows:

1. Use VSEPR theory to determine the shape and bond angles of molecules and polyatomic ions.
2. Use molecular shape to determine the polarity of molecules.

Flowcharts

Flowchart 6.1 **Using VSEPR theory to determine the shape and bond angles of molecules and polyatomic ions**

Flowchart 6.1

Summary

Method for using VSEPR theory to determine the shape and bond angles of molecules and polyatomic ions

Step 1
- **Draw the correct electron dot structure for the molecule.**

$$\downarrow$$

Step 2
- **Count the number of electron groups (bonding groups plus lone pairs) around the central atom. This is the steric number.**
- **Count each multiple bond as a single electron group.**

$$\downarrow$$

Step 3
- **Determine the electron-group geometry around the central atom, using the number of electron groups from Step 2 and (VSEPR Shape Table) in Section 6.2 of the textbook.**

$$\downarrow$$

Step 4
- **Pretend the lone pairs are invisible (if there are any lone pairs), and describe the resulting shape of the molecule.**

$$\downarrow$$

Step 5
- **Use the VSEPR Shape Table in page 118 to determine the bond angles in the molecule.**

Example 1

Determine the molecular shape and bond angles for $SiBr_4$.

Solution

Perform Step 1

The correct electron dot structure is

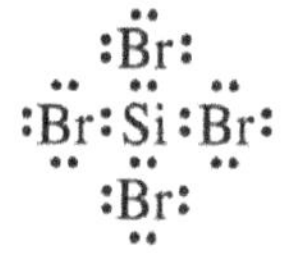

$$\begin{array}{c} :\overset{\displaystyle ..}{Br}: \\ :\overset{\displaystyle ..}{Br}:\overset{\displaystyle ..}{Si}:\overset{\displaystyle ..}{Br}: \\ :\overset{\displaystyle ..}{Br}: \end{array}$$

Perform Step 2

There are four groups of electrons around the central atom Si. The steric number is 4.

Perform Step 3

According to the VSEPR Shape Table, a molecule with four groups of electrons around the central atom has a tetrahedral electron-group geometry.

Perform Step 4

There are no lone pairs on the central atom, so the molecular shape is also tetrahedral.

Perform Step 5

The bond angles for a tetrahedral molecule are 109.5°.

Example 2

Determine the molecular shape and bond angles for NF_3.

Solution

Perform Step 1

The correct electron dot structure is

$$\begin{array}{c} :\overset{\displaystyle ..}{F}:\overset{\displaystyle ..}{N}:\overset{\displaystyle ..}{F}: \\ :\overset{\displaystyle ..}{F}: \end{array}$$

Perform Step 2

There are four groups of electrons around the central N atom. The steric number is 4.

Perform Step 3

According to the VSEPR Shape Table, a molecule with four groups of electrons around the central atom has a tetrahedral electron-group geometry.

Perform Step 4

There is a lone pair on the central atom, so the molecular shape is pyramidal.

Perform Step 5

The bond angles for a pyramidal molecule are 107°.

Example 3

Determine the molecular shape and bond angles for SO_3.

Solution

Perform Step 1

The correct electron dot structure is

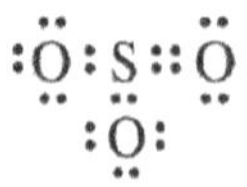

$$\ddot{\underset{..}{O}} : \underset{..}{S} :: \ddot{\underset{..}{O}}$$
$$: \ddot{\underset{..}{O}} :$$

Perform Step 2

There are three groups of electrons around the central S atom. The steric number is 3. (The double bond counts as one electron group.)

Perform Step 3

According to the VSEPR Shape Table, a molecule with three groups of electrons around the central atom has a trigonal planar electron-group geometry.

Perform Step 4

There are no lone pairs on the central atom, so the molecular shape is trigonal planar.

Perform Step 5

The bond angles for a trigonal planar molecule are 120°.

Example 4

Determine the molecular shape and bond angles for XeF_6^{2-}.

Solution

Perform Step 1

The correct electron dot structure is

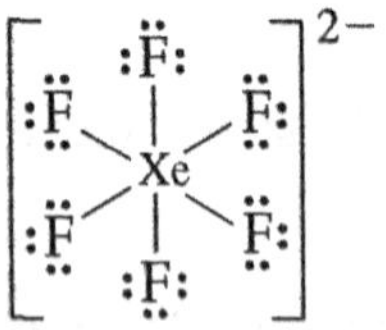

Perform Step 2

There are six groups of electrons around the central atom Xe. The steric number is 6.

Perform Step 3

According to the VSEPR Shape Table, a molecule with six groups of electrons around the central atom has an octahedral electron-group geometry.

Perform Step 4

There are no lone pairs on the central atom, so the molecular shape is also octahedral.

Perform Step 5

The bond angles for an octahedral molecule are 90°. Note that 180° angles exist between Xe — F bonds positioned exactly opposite of each other.

Practice Problems

6.1 Determine the molecular shape and bond angles for Cl_2O.
6.2 Determine the molecular shape and bond angles for CO_3^{2-}.
6.3 Determine the molecular shape and bond angles for $CHCl_3$.
6.4 Determine the molecular shape and bond angles for CS_2.
6.5 Determine the molecular shape and bond angles for ICl_2^-.

Flowchart 6.2   Using molecular shape to determine the polarity of molecules

Flowchart 6.2

Summary

Method for using molecular shape to determine the polarity of molecules

Step 1
- **Determine the molecular shape from the dot structure by applying VSEPR theory.**

Step 2
- **Draw vectors representing all bond dipole moments (using electronegativities).**
- **Determine the molecular dipole moment, if any.**

Example 1

Determine whether OCS is a polar or nonpolar molecule. Indicate the direction of the molecular dipole moment if one is present.

Solution

Perform Step 1

The correct electron dot structure is

$$:\ddot{O}::C::\ddot{S}:$$

VSEPR theory indicates that the molecule is linear because there are two electron groups around the central atom.

Perform Step 2

The individual bond dipoles are in the direction of the O and the S because both O and S are more electronegative than carbon. The C — O bond dipole moment is larger because oxygen is more electronegative than sulfur.

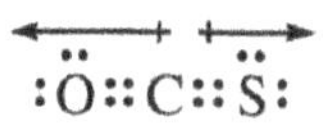

$$:\ddot{O}::C::\ddot{S}:$$

The molecular dipole is in the direction of the oxygen because the C — O bond dipole moment is larger than the C — S bond dipole moment.

Example 2

Determine whether CF_4 is a polar or nonpolar molecule. Indicate the direction of the molecular dipole moment if one is present.

Solution

Perform Step 1

The correct electron dot structure is

$$\ddot{:}\underset{\cdot\cdot}{\ddot{F}}\ddot{:}$$
$$:\ddot{F}:\underset{\cdot\cdot}{\ddot{C}}:\ddot{F}:$$
$$:\underset{\cdot\cdot}{\ddot{F}}:$$

VSEPR theory indicates that the molecule is tetrahedral because there are four electron groups around the central atom.

Perform Step 2

The bond dipoles are

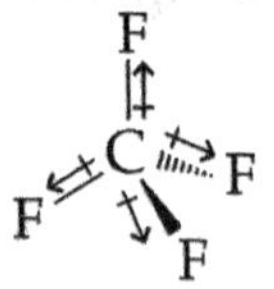

The individual bond dipoles are in the direction of the F because F is more electronegative than C. There is no net dipole because the individual dipole moments cancel one another. Therefore, the CF_4 molecule is nonpolar. Note that, for clarity, the lone pairs of electrons on the F atoms are not shown.

Example 3

Determine whether $BFCl_2$ is a polar or nonpolar molecule. Indicate the direction of the molecular dipole moment if one is present.

Solution

Perform Step 1

The correct electron dot structure is

$$:\ddot{F}:B:\ddot{Cl}:$$
$$:\ddot{Cl}:$$

VSEPR theory indicates that the molecule is trigonal planar because there are three electron groups around the central atom.

Perform Step 2

The bond dipoles are

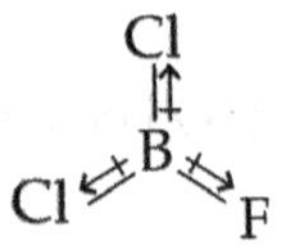

The individual bond dipoles are in the direction of the fluorine and the chlorines because fluorine and chlorine are more electronegative than boron. The net dipole is in the direction of the fluorine because fluorine is more electronegative than chlorine. Therefore, the molecule is polar. Note that, for clarity, the lone pairs of electrons on the F atoms are not shown.

Practice Problems

6.6 Determine whether PI_3 is a polar or nonpolar molecule. Indicate the direction of the molecular dipole moment if one is present.

6.7 Determine whether ONCl is a polar or nonpolar molecule. Indicate the shape of the molecule and the direction of the molecular dipole moment if one is present.

6.8 Determine whether BCl_3 is a polar or nonpolar molecule. Indicate the shape of the molecule and the direction of the molecular dipole moment if one is present.

Quiz for Chapter 6 Problems

1. Write the electron dot structure and indicate the molecular shape and bond angles for the HCOCl molecule. (The C is the central atom; all other atoms are connected to it.)

2. Indicate the electron-group geometry for a molecule with the following number of electron groups around the central atom.
   (a) two electron groups   (b) three electron groups   (c) four electron groups

3. Predict the molecular shape and the bond angles for the ClO_3^- ion.

4. Use VSEPR theory to determine whether the following molecules are polar or nonpolar. If the molecular is polar, indicate the direction of the net dipole moment.
   (a) CH_4 (b) CH_3F (c) CH_2F_2 (d) CHF_3 (e) CF_4

5. Arrange the following bonds in order of increasing bond dipole moment.
   C—O   C—F   C—Br   C—H   C—Cl

6. Indicate whether the following statement is true or false. If false, explain why "The molecular shape is tetrahedral for all molecules in which the central atom is surrounded by four electron groups."

7. The molecular shape of a molecule containing two electron groups and no lone pairs around the central atom is _____.

8. Circle the nonpolar molecular molecule(s) below.
   NF_3 BF_3 SO_3 PF_3 SF_4

9. The electron-group geometry for Cl_2O is _____, and the molecular shape is

   _____.

10. There are two different bent molecular shapes, one with bond angles of approximately 105° and the other with bond angles of approximately 118°. Indicate (a) the number of total electron groups, (b) the number of bonding pairs, and (c) the number of lone pairs that are associated with each of these two molecular shapes.

Exercises for Self-Testing

A. Completion. *Write the correct word(s) to complete each statement below.*

   1. The acronym VSEPR stands for _____ _____ _____

      _____ _____.

2. The molecular geometry that results from a total of four groups of electrons, including two lone pairs around the central atom, is _____.

3. When two atoms of different electronegativities share electrons, the bond formed is a(n) _____ _____ bond.

4. When the ΔEN between two atoms is greater than 1.8, the type of bond formed is _____.

5. In the covalent compound formed between nitrogen and chlorine, one molecule of the compound will contain _____ chlorine atoms.

6. The intermolecular attractive forces that operate in polar molecules are called _____–_____ forces.

7. The molecular shape of a molecule in which there are three electron groups around the central atom and no lone pairs is _____ _____.

8. In Lewis electron dot diagrams, nitrogen will generally form _____ bonds and have _____ lone pair(s).

9. When determining the molecular shape, we ignore the positions of _____ _____.

10. A bond in which one end is partially negative and one end is partially positive is called a(n) _____ bond.

B. True/False. *Indicate whether each statement below is true or false, and explain why the false statements are false.*

_____ 1. If a molecule has polar bonds, it must be polar.

_____ 2. A pyramidal molecule will have two lone pairs of electrons.

_____ 3. A tetrahedral molecule in which all the atoms attached to the central atom are equivalent will always be nonpolar.

_____ 4. The intermolecular attractive forces involved in HCl are dipole–dipole forces.

_____ 5. A hydrogen atom can form a double bond to an adjacent atom if this is required to satisfy the octet rule.

_____ 6. The CH_2Cl_2 molecule has a square shape.

_____ 7. The presence of lone pairs in a molecule makes the bond angles a little smaller.

_____ 8. One can identify linear molecules in which the steric number of the central atom is 2, 5, or 6.

_____ 9. In determining molecular geometry, we treat multiple bonds as single bonds.

_____ 10. Intermolecular attractive forces are usually stronger than intramolecular forces.

C. Questions and Problems

1. Indicate the shape of the following molecules:
   (a) CF_4 (b) H_2S (c) CO_2 (d) O_3

2. Classify each of the following compounds as either polar or nonpolar:
   (a) CH_3F (b) NBr_3 (c) $SiCl_4$ (d) CS_2

3. Draw the Lewis dot diagrams for the following molecules or polyatomic ions:
   (a) NO_2^- (b) CO_2 (c) AsF_3

4. Would you expect a compound that is a liquid at room temperature to have larger or smaller intermolecular attractive forces than a substance that is a gas at room temperature? Explain.

5. Does the OCS molecule have a larger or smaller dipole moment than CO_2? Draw a Lewis dot diagram for both molecules, showing the correct geometry, to explain your answer.

6. Which electron-group geometries can give rise to linear molecules or polyatomic ions? Explain.

CHAPTER

7

Intermolecular Forces and the Phases of Matter

Learning Outcomes

1. Describe the three phases of matter from a molecular viewpoint.
2. Explain what is necessary for one phase to transform into another.
3. Describe London forces and their magnitude.
4. Describe hydrogen bonding and predict when it can occur.
5. Predict the most important intermolecular force present and its effect on physical properties.
6. Describe what a nonmolecular substance is on an atomic level.
7. Explain why nonmolecular compounds typically have very high melting points.
8. Describe some interesting and important effects due to hydrogen bonding.

Chapter Outline

7.1 Why Does Matter Exist in Different Phases?
    A. Viewing phase changes from the macroscopic level
    B. Viewing phase changes from the molecular level
    C. Relationship between temperature and speed of gas molecules
    D. Dependence of phase on the kinetic energy of the molecules
    E. Dependence of phase on intermolecular attractive forces
7.2 Intermolecular Forces
    A. London (or dispersion) forces (induced dipoles) in nonpolar molecules
    B. Dipolar forces in polar molecules
7.3 A Closer Look at Dipolar Forces—Hydrogen Bonding
    A. Comparison of boiling points of acetone and water
    B. Description and representation of hydrogen bonds

C. Elements that give rise to hydrogen bonding (O, N, F)

D. Role of hydrogen bonding in DNA

7.4 Nonmolecular Substances

A. Ionic compounds

B. Network solids (network covalent substances)

C. Pure metals

7.5 One More Thing: Vancomycin—The Antibiotic of Last Resort and Its Five Live-Saving Hydrogen Bonds

Review of Key Concepts from the Textbook

7.1 Why Does Matter Exist in Different Phases?

The three common phases (or states) of matter are liquid, solid, and gas (or vapor). (A fourth state of matter, plasma, exists only at extremely high temperatures such as those present in the sun. The plasma state is not discussed in the textbook.) A substance can change phases if the temperature and/or pressure changes. All phase changes are physical changes, not chemical changes, because the chemical identity of the substance undergoing the phase change is not altered. A substance in the gas phase can be converted to the liquid phase by removing heat from the gas. This process (converting a gas to a liquid) is called *condensation*. The liquid can be converted to a solid (*freezing*) by removing heat from the liquid. The reverse set of phase changes (solid → liquid → gas) can be accomplished by adding heat. These relationships are shown in the diagram below.

When we consider the phases from a molecular standpoint, the physical properties of each phase can be easily understood. The molecules in the gas phase are very energetic, traveling at velocities of over 1500 miles per hour. The molecules are widely separated from each other and come into contact only when two molecules collide. When the gas is cooled, the molecules lose energy and move much more slowly. (The speed of gas molecules is directly proportional to the temperature of the gas. As the temperature increases, the speed of the molecules increases, and as the temperature decreases, the speed of the molecules decreases.) When a gas is cooled, the speed eventually slows down enough that the attraction between the molecules causes them to condense into a liquid.

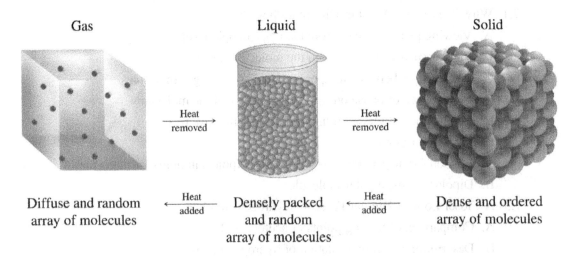

Gas	Liquid	Solid

Diffuse and random array of molecules   Densely packed and random array of molecules   Dense and ordered array of molecules

STUDENT WORKBOOK AND SELECTED SOLUTIONS

The molecules in a liquid are in close proximity to each other but are mobile enough that they can easily move past one another and travel throughout the volume of the liquid. When energy is removed from the liquid, however, the molecules slow down even further and are eventually moving so slowly that the attractive forces holding them together keep them fixed in one position. It is at this point that the liquid has become a solid; freezing has occurred.

There are specific names for each of the phase changes that substances undergo. You should memorize these names, which are shown below.

Phase Change	Name
Gas $\rightarrow$ liquid	Condensation
Liquid $\rightarrow$ solid	Freezing
Gas $\rightarrow$ solid	Deposition
Solid $\rightarrow$ liquid	Melting (or fusion)
Liquid $\rightarrow$ gas	Boiling (or vaporization)
Solid $\rightarrow$ gas	Sublimation

The liquid and solid phases are sometimes called *condensed phases* because the molecules are very close to each other. The properties of liquids and solids differ greatly from the properties of gases. Solids and liquids are only slightly compressible, whereas gases are very compressible. In addition, liquids and solids do not fill the volume of their container. Liquids take the shape of their containers but do not fill them, whereas solids maintain their rigid shape. Gas molecules feel almost no attraction for each other and expand to fill the volume of their container, leaving lots of empty space between the molecules. Molecules of liquids and solids, however, are in relatively close proximity to each other. Liquids and solids cannot be compressed the way a gas can.

The combination of temperature and intermolecular attractive forces determines the phase of a substance. We will look at intermolecular forces in more detail in the next section.

Example 7.1

Indicate whether heat must be added or removed to produce the changes listed below. Explain your answers.

(a) Dew formation on grass

(b) Evaporation of liquid hydrogen

(c) Sublimation of dry ice (solid carbon dioxide)

Solution

(a) Gaseous water vapor is condensing into a liquid when dew forms. Heat must be removed in order for this to occur.

(b) Evaporation is the conversion of a liquid to a gas. Heat must be added to convert liquid hydrogen to hydrogen gas.

(c) Sublimation is the conversion of a solid to a vapor. Heat must be added to convert dry ice to the gas phase.

Practice Exercise 7.1

Indicate whether heat must be added or removed to produce the changes listed below. Explain your answers.

(a) Clearing of early morning fog as the day progresses

(b) Freezing of rain on roadways

(c) Vaporization of liquid mercury

7.2   Intermolecular Forces

Intermolecular forces are the attractive forces that hold molecules together. The types of intermolecular forces acting in a substance depend on whether the substance is ionic, covalent, polar, or nonpolar. The different types of intermolecular forces are discussed briefly below. (A full discussion appears in Section 7.2 of the textbook.)

London Forces (or Dispersion Forces)

London forces (which are also sometimes referred to as van der Waals attractions) are the sole intermolecular attractive forces operating in nonpolar substances, such as He, CH_4, and BF_3. (Recall that we can use the Lewis dot structure and the electronegativity to determine whether a molecule is polar or nonpolar.) It may seem odd that nonpolar molecules have attractive forces operating in them because attractive forces arise from unlike full or partial charges that attract each other. Nonpolar molecules have no charged regions; the electron distribution is evenly spread throughout the molecule. However, even though nonpolar molecules have no permanently partially charged regions, nonpolar molecules do experience momentary, instantaneous partial charges. Let's see why this is true.

When we think of molecules as a collection of mobile electrons, it is easy to envision that at one instant the electrons in a molecule might "slosh" to one side of the molecule, as shown below.

Momentary excess of electrons in one part of molecule

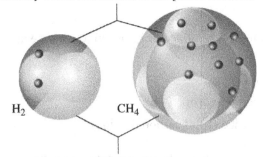

Momentary deficit in the other part

When this happens, one side of the molecule is partially negative (the side to which the electrons have moved) in relation to the other side of the molecule, which is partially positive. This imbalance sets up a temporary dipole that causes the electrons in adjacent molecules to shift due to electron–electron repulsions.

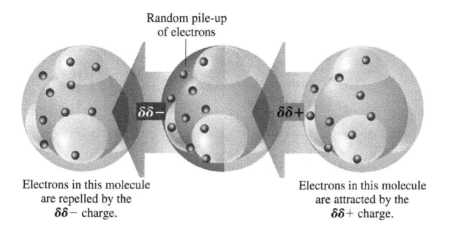

Random pile-up
of electrons

$\delta\delta-$ $\delta\delta+$

Electrons in this molecule
are repelled by the
$\delta\delta-$ charge.

Electrons in this molecule
are attracted by the
$\delta\delta+$ charge.

The net effect is that the instantaneous partially positive end of one molecule is attracted to the instantaneous partially negative end of another. In the next instant, the electrons "slosh" to a different region of the molecule or atom, and the process begins again.

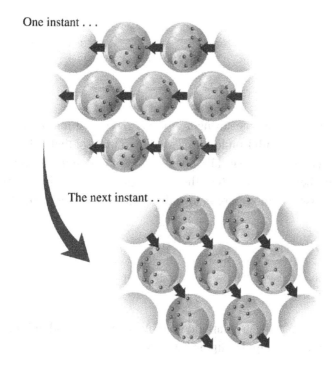

One instant . . .

The next instant . . .

The net effect is that these instantaneous dipoles result in attractive forces that hold nonpolar molecules together. London forces are the only forces holding nonpolar molecules together. Small molecules with few electrons have weaker London forces; large molecules with many electrons have stronger London forces. One additional factor that determines the strength of London forces is the shape of the molecule. Long molecules have stronger London forces than more spherically shaped molecules because the long molecules have greater contact with each other. More contact allows the attractive forces to be felt more strongly, as shown in the comparison of *n*-pentane and neopentane. Difference in boiling points (309.4 K for *n*-pentane versus 282.7 K for neopentane) illustrates this point nicely.

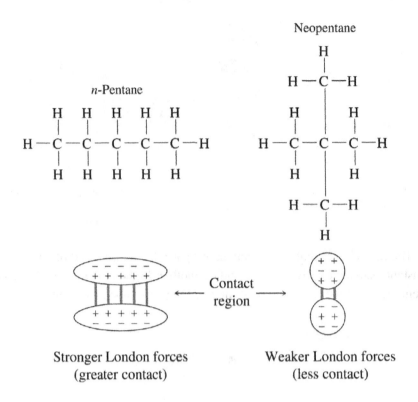

Neopentane

n-Pentane

Contact region

Stronger London forces
(greater contact)

Weaker London forces
(less contact)

Dipole–Dipole Interactions

Dipole–dipole interactions are the stronger of the two types of intermolecular forces that operate in polar molecules (along with London forces), due to the fact that polar molecules have a permanent dipole moment. The partially positive end of one molecule attracts the partially negative end of another. The strength of the dipole–dipole interaction increases as the polarity of the molecules increases. In the figure below, the dotted lines denote dipole–dipole interactions.

$$\overset{\delta+}{H}-\overset{\delta-}{Cl}\cdots\overset{\delta+}{H}-\overset{\delta-}{Cl}$$
$$\overset{\delta+}{H}-\overset{\delta-}{Cl}$$

Note that all molecules (polar and nonpolar) contain London forces; polar molecules contain London forces *and* dipole–dipole interactions.

Example 7.2

Indicate which of the following substances have only London forces holding the molecules together, and which have London forces *and* dipole–dipole interactions:

(a) CF_4 (b) Br_2 (c) HCl   (d) BF_3 (e) Cl_2O

Solution

Nonpolar molecules will have only London forces, and polar molecules will have London forces and dipole–dipole interactions. The electronegativities of the atoms in the molecule and the Lewis dot structure of the molecule will allow us to determine if the molecule is polar or nonpolar.

(a) CF_4 is nonpolar. Only London forces are involved.

(b) Br_2 is nonpolar. Only London forces are involved.

(c) HCl is polar. London forces *and* dipole–dipole interactions are involved.

(d) BF_3 is nonpolar. Only London forces are involved.

(e) Cl_2O is polar. London forces *and* dipole–dipole interactions are involved.

Practice Exercise 7.2

Indicate which of the following substances have only London forces holding the molecules together, and which have London forces *and* dipole–dipole interactions:

(a) CH_3Cl (b) NF_3 (c) H_2 (d) CO_2 (e) H_2S

7.3 A Closer Look at Dipolar Forces—Hydrogen Bonding

Some polar molecules have a boiling point that is much higher than would be predicted based on the size of the molecule. The common feature of all these molecules is that they contain a hydrogen atom bonded to an oxygen, a nitrogen, or a fluorine atom. When hydrogen is bonded to one of these three very electronegative atoms, the bond is so strongly polar that there are very large partial charges that give rise to superstrong dipole–dipole attractive forces. These superstrong dipole–dipole forces are known as *hydrogen bonds* and are indicated by dotted lines between the atoms involved.

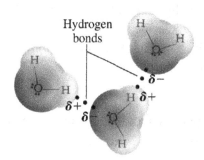

Hydrogen bonding forms between a hydrogen atom covalently bonded to a very electronegative atom (specifically, O, N, or F) and a lone pair of electrons of another electronegative atom. Hydrogen bonds usually occur intermolecularly (between two atoms on two different molecules) but sometimes intramolecularly (between two atoms on the same molecule). Hydrogen bonds should not be confused with covalent bonds within the same molecule. Note that hydrogen bonding occurs only in compounds that contain an O—H, N—H, or F—H bond. There must also be a lone pair of electrons on the O, N, or F atom.

Example 7.3

Indicate which of the following molecules form hydrogen bonds:

(a) CH_3F (b) CH_3CH_2OH (c) CH_3OCH_3 (d) CH_3NH_2

Solution

Molecules containing an O—H, N—H, or F—H bond will form hydrogen bonds. Their Lewis dot diagrams will allow us to determine whether the molecule has one of these bonds.

(a) CH_3F molecules will not form hydrogen bonds. The H atoms are bonded to the C atom.

(b) CH_3CH_2OH molecules will form hydrogen bonds due to the presence of an O—H bond. In addition, there are two lone pairs of electrons on the O atom.

(c) CH_3OCH_3 molecules will not form hydrogen bonds. The oxygen is connected to two carbons in the Lewis dot structure.

(d) CH_3NH_2 molecules will form hydrogen bonds due to presence of the N—H bonds. In addition, there is a lone pair of electrons on the N atom.

Practice Exercise 7.3

Indicate which of the following molecules form hydrogen bonds:

(a) CH_3NHCH_3 (b) CH_3OH (c) CHF_3 (d) HF

The strength of the intermolecular attractive forces in substances has a significant impact on physical properties. For example, the boiling points of substances with strong intermolecular forces are higher than those of substances with weak intermolecular forces. The stronger the attractive forces between molecules, the more energy must be added to separate the molecules of liquid into the gas phase. Thus, the strength of the intermolecular attractive forces and the boiling point are directly proportional to each other. The stronger the attractive forces, the higher the boiling point.

When we try to compare the relative strengths of the different types of intermolecular forces, we run into difficulty. Although it is generally true that London forces are weaker than dipole–dipole forces, and that dipole–dipole forces are weaker than hydrogen bonding, this is not always the case. A very large molecule may have strong London forces and a higher boiling point than a small molecule in which hydrogen bonding is present. (See Section 7.3 of the textbook for a complete discussion of this concept.)

Example 7.4

Arrange the following compounds in order of increasing boiling points. Explain the order.

(a) CH_3CH_2OH (b) CH_3OCH_3 (c) H_2 (d) CH_4 (e) CH_3OH

Solution

H_2 and CH_4 will have the lowest boiling points because they are nonpolar and contain only London forces. CH_4 will boil at a higher temperature than H_2 because CH_4 is the larger molecule. CH_3OH and CH_3CH_2OH will have the highest boiling points due to hydrogen bonding. CH_3CH_2OH will have the higher boiling point of the two due to its larger size. CH_3OCH_3, having London forces and dipole–dipole interactions, will boil at a temperature higher than the nonpolar molecules but lower than those molecules that contain hydrogen bonding. The order is therefore as follows:

$$H_2 < CH_4 < CH_3OCH_3 < CH_3OH < CH_3CH_2OH$$

Practice Exercise 7.4

Arrange the following compounds in order of increasing boiling points.

(a) H_2O (b) CH_3F (c) $CH_3CH_2CH_3$ (d) F_2 (e) C_2H_5OH

7.4 Nonmolecular Substances

Up to this point we have been discussing intermolecular forces between molecules of substances that exist as discrete molecules. However, there are many substances that do not consist of discrete molecules. These substances, called *nonmolecular solids*, include ionic compounds, network covalent substances (network solids), and metals. Instead of discrete molecules held together by intermolecular attractive forces, these substances are made of a lattice of ions or a network of atoms bound together by ionic or covalent bonds. The ionic or covalent bonds holding these substances together are much stronger than the intermolecular attractive forces holding the molecules of molecular substances together.

We can use this information to predict the relative melting and boiling points of some of these types of compounds. The stronger the forces holding the molecules together, the higher the melting point. Therefore, molecular solids (held together by intermolecular attractive forces) have much lower melting points than ionic solids (held together by ionic bonds) or network covalent solids (held together by covalent bonds). Many metals (held together by metallic bonds) have high melting points, reflecting strong metallic bonds, but there are some metals (e.g., mercury and sodium) in which the metallic bonding is weak, and the melting point is low. (See Section 7.4 of the textbook for a complete discussion of these concepts.)

To distinguish between the three different types of solids, we use the following generalizations:

1. *Ionic solids* are composed of cations of elements in groups IA and IIA (and also transition metal cations) with anions of elements in groups VIA and VIIA. In addition, compounds made up of polyatomic ions are also ionic solids.
2. *Network solids* are composed of elements linked together by a network of covalent bonds that extend throughout the solid. Network solids often contain carbon and/or silicon.
3. *Metallic solids* are composed of a large array of atoms of metallic elements, such as iron and sodium.

Example 7.5

Classify each of the following compounds as a molecular solid, a network solid, a metal, or an ionic solid:

(a) LiBr   (b) $C_{22}H_{46}$ (c) Fe   (d) Diamond

Solutions

(a) Ionic solid   (b) Molecular solid   (c) Metal   (d) Network solid

Practice Exercise 7.5

Classify each of the following compounds as a molecular solid, a network solid, a metal, or an ionic solid:

(a) Solid CO_2 (dry ice)   (b) Cr   (c) SiO_2 (d) NaI

Strategies for Working Problems in Chapter 7

Overview: What You Should Be Able to Do

Chapter 7 provides a discussion of the various forces that hold molecules together to form liquids and solids. The types of problems that you should be able to solve after mastering Chapter 7 are as follows:

1. Determine the type(s) of intermolecular attractive forces operating in a compound.
2. Determine the relative boiling points of substances based on their structures.

Flowcharts

Flowchart 7.1 **Determining the type(s) of intermolecular attractive forces operating in a compound**

Flowchart 7.1

Summary

Method for determining the type(s) of intermolecular attractive forces operating in a compound

Step 1
- **Determine whether the molecule is polar or nonpolar.**
- **If the compound is nonpolar, London dispersion forces are the only intermolecular attractive forces in the compound.**
- **If the compound is polar, proceed to Step 2.**

↓

Step 2
- **Determine if the compound contains N — H, O — H, or F — H bonds.**
- **If there are no N — H, O — H, or F — H bonds, the compound has dipolar forces in addition to London dispersion forces.**
- **If the compound contains N — H, O — H, or F — H bonds, the intermolecular forces present are hydrogen bonding, dipolar interactions, and London dispersion forces.**
- **Generally speaking, hydrogen bonding is stronger than dipole–dipole interactions, which are stronger that London dispersion forces.**

Example 1

What type(s) of intermolecular attractive force(s) is(are) present in CH_3OCH_3?

Solution

Perform Step 1

The molecule is polar.

Perform Step 2

There are no O — H, N — H, or F — H bonds. Therefore, the compound has dipolar and London dispersion forces present, but no hydrogen bonds.

Example 2

What type(s) of intermolecular attractive forces is(are) present in CBr_4?

Solution

Perform Step 1

> The molecule is nonpolar.

Perform Step 2

> Therefore, London dispersion forces are the only forces present in the compound.

Example 3

What type(s) of intermolecular attractive forces is(are) present in CH_3NH_2?

Solution

Perform Step 1

> The molecule is polar.

Perform Step 2

> There are two N—H bonds and a lone pair of electrons on the N atom. Therefore, the compound has hydrogen bonds, dipolar forces, and London dispersion forces.

Practice Problems

7.1 What type(s) of intermolecular attractive force(s) is(are) present in CH_2F_2?

7.2 What type(s) of intermolecular attractive force(s) is(are) present in CH_3OH?

7.3 What type(s) of intermolecular attractive force(s) is(are) present in C_2H_6?

Flowchart 7.2   Determining the relative boiling points of substances based on their structures

Flowchart 7.2

Summary

Method for determining the relative boiling points of substances based on their structures

Step 1
- **Determine the types of intermolecular attractive forces present in the compounds being ranked.**
- **If the compounds have different intermolecular attractive forces present, use the general rule that hydrogen bonding is stronger than dipolar forces, which are stronger than London dispersion forces.**
- **The stronger the intermolecular attractive forces, the higher the boiling point.**
- **If all compounds have the same types of intermolecular attractive forces, proceed to Step 2.**

↓

Step 2
- **Consider the relative sizes of the molecules of the compounds being compared.**

- **Generally, the larger the molecule, the higher the boiling point if the compounds all have the same type of intermolecular attractive forces.**

Example 1

Arrange the following compounds in order of increasing boiling points:
CH_3OCH_3, $CH_3CH_2CH_3$, CH_3CH_2OH

Solution

Perform Step 1

CH_3OCH_3 has dipolar and London dispersion intermolecular attractive forces.
$CH_3CH_2CH_3$ has only London dispersion forces.
CH_3CH_2OH has hydrogen bonding, dipolar forces, and London dispersion forces.

Perform Step 2

Therefore, $CH_3CH_2CH_3$ has the lowest and CH_3CH_2OH has the highest boiling points. The arrangement of compounds in order of increasing boiling point is $CH_3CH_2CH_3 < CH_3OCH_3 < CH_3CH_2OH$.

Example 2

Arrange the following compounds in order of increasing boiling points:
$CH_3CH_2CH_3$, CH_3CH_3, CH_4

Solution

Perform Step 1

All three compounds are nonpolar and have only London dispersion forces.

Perform Step 2

The larger the molecule, the more electrons and the stronger the intermolecular attractive forces.

The order, is therefore, $CH_4 < CH_3CH_3 < CH_3CH_2CH_3$.

Practice Problems

7.4 Which has a higher boiling point, CH_3Cl or CH_3OH?
7.5 Arrange in order of increasing boiling point:
$CH_3CH_2CH_2CH_3$, $CH_3CH_2CH_2NH_2$, $CH_3CH_2OCH_3$
7.6 Which has a higher boiling point, HF or HCl?

Quiz for Chapter 7 Problems

1. Indicate the type of intermolecular attractive forces in CH_3Cl.

2. Two compounds, A and B, have boiling points of 78 °C and 30 °C, respectively. If the two compounds are known to be CH_3OCH_3 and CH_3CH_2OH, which compound had which boiling point? Explain.

3. Which compound in each pair had the lower boiling point?
   (a) CH_4 or CH_3Cl (b) CH_3CH_3OH or CH_3CH_2Cl

STUDENT WORKBOOK AND SELECTED SOLUTIONS

4. Indicate the type(s) of intermolecular attractive forces operating in each type of compound:
   (a) nonpolar
   (b) polar but without O—H, N—H, or F—H bonds
   (c) polar and with O—H, N—H, or F—H bonds

5. For compounds that have the same type(s) of intermolecular attractive forces present, the larger the molecule, the _____ the boiling point.

6. Which compound in each pair has stronger intermolecular attractive forces?
   (a) CH_3CH_3 or $CH_3CH_2CH_2CH_3$ (b) CH_3OH or CH_3CH_2OH

7. Which molecule(s) below can participate in hydrogen bonding?

$$CH_3OCH_2CH_3 \qquad \overset{\displaystyle CH_3NCH_3}{\underset{\displaystyle CH_3}{|}} \qquad CH_3NH_2 \qquad H_2S$$

8. Which molecules have dipolar intermolecular attractive forces?
   (a) CF_4 (b) CH_2Cl_2 (c) CH_3OH (d) BF_3

9. Arrange the following compounds in order of increasing boiling point.
   (a) CH_3CH_3 (b) CH_3CH_2Cl (c) CH_3OH (d) CH_3F (e) CH_4

10. Which compound has the higher boiling point, CH_3OH or CH_3NH_2? Explain.

Exercises for Self-Testing

A. Completion. *Write the correct word(s) to complete each statement below.*

1. The condensed states of matter are the _____ and _____ states.
2. _____ _____ are due to the instantaneous partial charges that arise from temporary changes in the distributions of electrons in a molecule.
3. Hydrogen bonding occurs between molecules that have a hydrogen atom bonded to an atom of _____, _____, or _____.
4. _____ is the phase change in which a solid is converted directly to a gas, without going through the liquid phase.
5. In general, the strongest of the intermolecular attractive forces is _____.
6. The larger the molecule, the _____ the London forces.
7. The three atoms that, when bonded to hydrogen, are involved in hydrogen bonding are _____, _____, and _____.
8. The _____ the intermolecular forces between the molecules of a substance, the higher the boiling point of the substance.
9. Ionic and network compounds are both considered _____ solids.
10. The type of bonding that occurs between iron atoms is called _____ bonding.

B. True/False. *Indicate whether each statement below is true or false, and explain why the false statements are false.*

_____ 1.   If the molecules are approximately the same size, compounds in which hydrogen bonding is present have a higher boiling point than compounds in which hydrogen bonding is not present.

_____ 2.   Network solids generally have a higher boiling point than molecular solids.

_____ 3.   London forces are present only in nonpolar molecules.

_____ 4.   All molecules with polar bonds will experience dipole–dipole intermolecular attractive forces.

_____ 5.   Some nonpolar molecules have a higher boiling point than some polar molecules.

_____ 6.   All compounds in which hydrogen bonding is present will have a higher boiling point than compounds that involve only London forces.

_____ 7.   Intramolecular attractive forces are stronger than intermolecular attractive forces.

_____ 8.   The melting points of molecular solids are generally higher than those of network solids.

_____ 9.   The molecules of substances that exist as liquids and solids at room temperature are held together by stronger intermolecular attractive forces than the molecules of substances that exist as gases at room temperature.

_____ 10. CH_3CBr_3 is a polar molecule with dipole–dipole interactions but no London forces.

C. Questions and Problems

1. Both dimethyl ether, CH_3OCH_3, and ethyl alcohol, CH_3CH_2OH, have the same molecular formula, C_2H_6O. Which one would you expect to have a higher boiling point? Why?

2. Use the molecular description of solids, liquids, and gases to explain the following phenomena:
   (a) Liquids take the shape of their container, but solids are rigid.
   (b) Gases are compressible, but liquids and solids are not.
   (c) Gases fill the volume of their container, but liquids and solids do not.

3. Molecules of _n_-pentane, C_5H_{12}, are linear molecules, whereas molecules of isopentane (with the same molecular formula) are spherically shaped. Which compound would you expect to have the higher boiling point? Why?

4. Indicate all the intermolecular attractive forces operating in the following substances:
   (a) HI   (b) CF_4 (c) CH_3NH_2 (d) He   (e) H_2S

5. Explain why CCl_4 is a nonpolar molecule even though it has polar bonds.

6. Explain why H_2O has a much higher boiling point than H_2S even though H_2S is the larger molecule.

7. Explain why evaporation of perspiration from the skin causes the skin to feel cooler.

8. Which intermolecular attraction is responsible for the liquefaction of helium?

9. Explain why iodine, which forms diatomic molecules of I_2, is solid at room temperature, while the other elements in group VIIA are either gaseous (F_2 and Cl_2) or liquid (Br_2).

10. Explain why, at normal conditions, solid carbon dioxide undergoes sublimation (a direct transformation of solid into gas) instead of first undergoing melting and then boiling.

8

Chemical Reactions

Learning Outcomes

1. Describe what a chemical reaction is in general terms and in terms of chemical bonds.
2. Explain how collisions play a role in chemical reactions.
3. Write and balance chemical equations.
4. Describe and identify the different reaction types.
5. Predict the precipitate and write a chemical equation for precipitation reactions.
6. Write chemical reactions that involve acids and bases.
7. Plan a simple chemical synthesis.

Chapter Outline

8.1 What Is a Chemical Reaction?
    A. Example and definition of a chemical reaction
    B. Components of a chemical reaction

8.2 How Are Reactants Transformed into Products?
    A. Collisions of reactant molecules in gas phase reactions
    B. Energetics of bond-breaking and bond-making processes

8.3 Balancing Chemical Equations
    A. Balanced equations and the law of conservation of matter
    B. Procedure for balancing chemical equations

8.4 Types of Reactions
    A. Single-replacement reaction: $A + BC \rightarrow AC + B$
    B. Double-replacement reaction: $AB + CD \rightarrow AD + CB$
    C. Decomposition reaction: $AB \rightarrow A + B$
    D. Combination (or synthesis) reaction: $A + B \rightarrow C$
    E. Combustion reaction: Substance $+ O_2 \rightarrow$ Product

Review of Key Concepts from the Textbook

8.1 What Is a Chemical Reaction?

A chemical reaction is a chemical change in which new substances (products) are produced from starting substances (reactants). Chemical reactions are represented by chemical equations, in which the reactants and products are separated by an arrow.

Reactants $\rightarrow$ Products

It is important to remember that chemical reactions *always* involve the formation of one or more new substances. A change that does not involve forming new substances is a physical change, and no chemical reaction is involved.

8.2 How Are Reactants Transformed into Products?

When chemical reactions occur, bonds in the reactant molecules break, and new bonds are formed to create the products. The bond-breaking process always requires an input of energy, and the bond-making process always gives off energy. The energy for the bond-breaking process comes from collisions between the reactant molecules. We will learn more about this later.

8.3 Balancing Chemical Equations

Chemical equations tell us what changes occur during a chemical reaction. A balanced chemical equation shows the correct chemical formula of all the reactants and products. In addition, a balanced chemical equation will have the same number of each kind of atom on the reactant side as on the product side, because no atoms are created or destroyed during the chemical reaction.

It is crucial to have the correct formulas of the substances involved in a chemical reaction. If the formulas are incorrect, the equation cannot possibly be a correct representation of the change that is occurring, and the equation will be impossible to balance. Chemical reactions will always involve elements and compounds, and we must know how to represent them correctly.

Elements

Whenever an element is part of a chemical reaction, we must know whether the element is diatomic or monatomic. Diatomic elements are those, such as H_2 and O_2, that occur naturally as two atoms of the element bonded together. All the diatomic elements must be written as such in chemical equations. They are listed below, and you should memorize all of them. The diatomic elements are hydrogen (H_2), nitrogen (N_2), oxygen (O_2), fluorine (F_2), chlorine (Cl_2), bromine (Br_2), and iodine (I_2). Some other elements come in molecular form (e.g., C_{12}, P_4, and S_8), but you will not have to memorize these. Their formulas will be provided for you.

Compounds

Chemical formulas tell us how many of each kind of atom there are in one molecule of a substance. Recall that the number of valence electrons and the octet rule determine the correct formula of compounds. To write an accurate chemical equation, you must correctly assign formulas to all the reactants and products.

Example 8.1

Aluminum metal reacts with oxygen gas to form aluminum oxide. Write the unbalanced chemical equation for the reaction.

Solution

Aluminum metal is monatomic, and oxygen is diatomic. The formula for aluminum oxide can be determined by the crisscross method, developed in Chapter 4. (We determine the charge the atom will have if it obeys the octet rule, and then crisscross the charges, leaving off the sign.) Using this method, we get the formula Al_2O_3 for aluminum oxide. The unbalanced equation is

$$Al + O_2 \rightarrow Al_2O_3$$

Practice Exercise 8.1

Magnesium metal reacts with iodine gas to form magnesium iodide. Write the unbalanced chemical equation for the reaction.

Balancing Equations

The equations that we have written thus far are not correct because they do not obey the law of conservation of matter. Chemical equations must be balanced so as to have the same number of each kind of atom on both sides of the equation. The procedure involved in balancing equations is presented in Section 8.3 of the textbook. Remember that balancing a chemical equation involves changing *only the coefficients* (the numbers in front) of the reactants and products. It is *never* permissible to change the subscripts in a molecular formula to balance an equation. The subscripts are fixed and cannot be changed. (Remember the law of definite proportions.)

Example 8.2

Balance the following equation: $C_6H_{12}N_4 + H_2O \rightarrow H_2CO + NH_3$.

Solution

We will use the method described in the textbook. First, we will balance the carbons, placing a 6 in front of the H_2CO.

$$C_6H_{12}N_4 + H_2O \rightarrow 6H_2CO + NH_3$$

Next, we will balance the nitrogens, placing a 4 in front of NH_3.

$$C_6H_{12}N_4 + H_2O \rightarrow 6H_2CO + 4NH_3$$

Next, we will balance the oxygens, placing a 6 in front of H_2O.

$$C_6H_{12}N_4 + 6H_2O \rightarrow 6H_2CO + 4NH_3$$

Next, we will balance the hydrogens. When we count them on each side, we see that they are already balanced, with 24 on each side. The final step is to count each type of atom on both sides of the equation to make sure the equation is balanced:

C: 6 atoms

H: 24 atoms

N: 4 atoms

O: 6 atoms

The equation is balanced!

Practice Exercise 8.2

Balance the following equation: $PH_3 + O_2 \rightarrow P_4O_{12} + H_2O$.

8.4 Types of Reactions

There are hundreds of thousands of different chemical reactions, but many of these reactions can be classified as one of four different types: single-replacement, double-replacement, decomposition, or combination (or synthesis) reactions. An example of each type is shown below.

1. Single-replacement reactions are those in which an element reacts with a compound to form a new element and a new compound. The general representation of a single-replacement reaction is $A + BC \rightarrow AC + B$. An example is shown below.

$$Na + AgCl \rightarrow NaCl + Ag$$

2. Double-replacement reactions are those in which two compounds simply exchange partners to form two new compounds. The general representation of a double-replacement reaction is $AB + CD \rightarrow AD + CB$. An example is shown below.

$$CaCl_2 + 2KF \rightarrow CaF_2 + 2KCl$$

3. Decomposition reactions are those in which a compound breaks down into two or more simpler substances. The general representation of a decomposition reaction is $AB \rightarrow A + B$. An example is shown below.

$$CaCO_3 \rightarrow CaO + CO_2$$

4. Combination (or synthesis) reactions are those in which two or more substances combine to form a larger substance. The general representation of a combination reaction is $A + B \rightarrow C$. An example is shown below.

$$Ni + 4CO \rightarrow Ni(CO)_4$$

5. A combustion reaction is the reaction of any substance with oxygen, O_2.

Example 8.3

Classify each of the following reactions as single replacement, double replacement, decomposition, or combination.

(a) $2Al + 3H_2SO_2 \rightarrow Al_2(SO_4)_3 + 6H_2$

(b) $2AsI_3 \rightarrow 2As + I_2$

(c) $FeCO_3 + H_2CO_3 \rightarrow Fe(HCO)_2$

(d) $Mg(NO_3)_2 + HCl \rightarrow HNO_3 + MgCl_2$

Solution

(a) single replacement

(b) decomposition

(c) combination

(d) double replacement

Practice Exercise 8.3

Classify each of the following reactions as single replacement, double replacement, decomposition, or combination.

(a) $3CaBr_2 + 2H_3PO_4 \rightarrow Ca_3(PO_4)_2 + 6HBr$

(b) $Cu_2S \rightarrow 2Cu + S$

(c) $3MgO + 2Al \rightarrow Al_2O_3 + 3Mg$

(d) $2H_2 + CO \rightarrow CH_3OH$

8.5 Solubility and Precipitation Reactions

Precipitation reactions are those in which a precipitate (insoluble solid) is formed when two or more aqueous solutions are mixed. An example of a precipitation reaction is the reaction that occurs when a KCl solution is combined with $AgNO_3$, as shown below. (The solid AgCl comes out of solution as a white precipitate.)

$$KCl(aq) + AgNO_3(aq) \rightarrow KNO_3(aq) + AgCl(s)$$

In order to determine whether a reaction is a precipitation reaction, we must know which compounds are insoluble in water. If no insoluble solid is formed when the two reactants are mixed together, the reaction is not a precipitation reaction. The solubility rules presented in Chapter 8 of the textbook allow us to determine whether a compound is insoluble and therefore a precipitate. For example, according to Table 8.1 of the textbook, AgCl is an insoluble solid (precipitate) but LiCl is not. Therefore, if we combine KCl with $LiNO_3$, the reaction will not produce a precipitate and is not a precipitation reaction. The rules in Table 8.1 are important, and you should memorize them. You can do this by looking at them often and reciting them until they are firmly planted in your memory. Also, you should try to do all the problems related to solubility and precipitation reactions without referring to the table. Use the table to check your answers.

Example 8.4

Indicate whether each of the following reactions is a precipitation reaction, and indicate which product is the precipitate if one is produced.

(a) $Ba(NO_3)_2 + Na_2CO_3 \rightarrow BaCO_3 + 2NaNO_3$

(b) $NaNO_3 + NH_4Br \rightarrow NaBr + NH_4NO_3$

(c) $CaCl_2 + MgSO_4 \rightarrow CaSO_4 + MgCl_2$

Solution

(a) It is a precipitation reaction. $BaCO_3$ is the precipitate.

(b) It is not a precipitation reaction.

(c) It is a precipitation reaction. $CaSO_4$ is the precipitate.

Practice Exercise 8.4

Indicate whether each of the following reactions is a precipitation reaction, and indicate which product is the precipitate if one is produced.

(a) $CaBr_2 + 2NaCl \rightarrow CaCl_2 + 2NaBr$

(b) $MgI_2 + KOH \rightarrow Mg(OH)_2 + KI$

(c) $Na_2CO_3 + CaF_2 \rightarrow CaCO_3 + 2NaF$

When precipitation reactions occur, some ions combine to form the precipitate and others remain in solution. We can write a form of the equation for the reaction that shows all the ions that are present before and after the reaction. For example, the ionic form of the equation for the reaction $AgNO_3 + KCl \rightarrow AgCl + KNO_3$ is

$$Ag^+ + NO_3^- + K^+ + Cl^- \rightarrow AgCl(s) + K^+ + NO_3^-$$

We notice that some ions are present on both the reactant and the product side. We call these ions *spectator ions* because they do not take part in the reaction. In a sense, they are simply on the sidelines while the actual change occurs. We cross these ions out of the equation to obtain the net ionic equation.

Ionic equation: $Ag^+ + \cancel{NO_3^-} + \cancel{K^+} + Cl^- \rightarrow AgCl(s) + \cancel{K^+} + \cancel{NO_3^-}$
Net ionic equation: $Ag^+ + Cl^- \rightarrow AgCl(s)$

You may wonder how we know which substances to break into ions and which to leave combined. The answer to this question is simple. All ionic compounds that are soluble in water can be written as ions. (Review the section on ionic compounds in Chapter 5 if you've forgotten how to determine whether a compound is ionic or not.)

Example 8.5

Write the balanced molecular equation, the ionic equation, and the net ionic equation for the precipitation reaction below.

$$(NH_4)_2SO_4 + MgF_2 + MgSO_4 + NH_4F$$

Solution

Balanced molecular equation:

$$(NH_4)_2SO_4 + MgF_2 \rightarrow MgSO_4 + 2NH_4F$$

Ionic equation:

$$2\cancel{NH_4^+} + SO_4^{2-} + Mg^{2+} + 2\cancel{F^-} \rightarrow MgSO_4(s) + 2\cancel{NH_4^+} + 2\cancel{F^-}$$

Net ionic equation:

$$SO_4^{2-} + Mg^{2+} \rightarrow MgSO_4(s)$$

Practice Exercise 8.5

Write the balanced molecular equation, the ionic equation, and the net ionic equation for the precipitation reaction below.

$$Pb(NO_3)_2 + NaCl \rightarrow PbCl_2 + NaNO_3$$

8.6 Introduction to Acid–Base Reactions

Acids are compounds that produce H^+ ions when added to water, and bases are compounds that produce OH^- ions when added to water. Some of the common acids are HCl, HNO_3, H_2SO_4, and $HC_2H_3O_2$. Some common bases are $NaOH$, KOH, $Ca(OH)_2$, and NH_3.

When strong acids and strong bases are combined, they neutralize each other to produce water. This type of reaction is called a neutralization reaction. The net ionic equation for neutralization reactions is $H^+ + OH^- \rightarrow H_2O$.

Example 8.6

Write the balanced molecular equation, the ionic equation, and the net ionic equation for the reaction below.

$$HNO_3 + Ca(OH)_2 \rightarrow Ca(NO_3)_2 + H_2O$$

Solution

Molecular equation: $2HNO_3 + Ca(OH)_2 \rightarrow Ca(NO_3)_2 + 2H_2O$
Ionic equation: $2H^+ + 2\cancel{NO_3^-} + \cancel{Ca^{2+}} + 2OH^- \rightarrow \cancel{Ca^{2+}} + 2\cancel{NO_3^-} + 2H_2O$
Net ionic equation: $2H^+ + 2OH^- \rightarrow 2H_2O$, which reduces to $H^+ + OH^- \rightarrow H_2O$

Practice Exercise 8.6

Write the balanced molecular equation, the ionic equation, and the net ionic equation for the reaction below.

$$HCl + Mg(OH)_2 \rightarrow MgCl_2 + H_2O$$

Strategies for Working Problems in Chapter 8

Overview: What You Should Be Able to Do

Chapter 8 presents the various types of chemical reactions and provides a method for balancing chemical equations. The types of problems that you should be able to solve after mastering Chapter 8 are as follows:

1. Balance chemical equations by the inspection method.

2. Identify the precipitate that will form in a potential precipitation reaction and write the net ionic equation for the reaction.

3. Write the equation for the neutralization reaction that will occur when an acid reacts with a base and write the net ionic equation for the reaction.

Flowcharts

Flowchart 8.1   **Balancing chemical equations by the inspection method**

Flowchart 8.1

Summary

Method for balancing chemical equations by the inspection method

Step 1
- **Scan the equation from left to right, placing appropriate coefficients to make the number of atoms of each element the same on both sides of the equation.**
- **Never change the subscripts in a formula to balance an equation; this will change the identity of the substance.**
- **It is often helpful to save for last substances that appear pure elements (e.g., H_2, N_2, O_2, Al).**
- **Fractional coefficients (e.g., $1/2$, $3/2$) can be used to balance the equation.**

$$\downarrow$$

Step 2
- **Convert all fractions to whole number by multiplying each coefficient by the denominator of the fraction.**
- **Check to make sure that each side of the equation contains the same number of each element.**

Example 1

Balance the following equation.

$$C_5H_{12} + O_2 \rightarrow CO_2 + H_2O$$

Solution

Perform Step 1

Scanning the reaction from left to right (leaving the oxygens for last), and adding coefficients to balance each element, we obtain:

$$C_5H_{12} + O_2 \rightarrow 5CO_2 + 6H_2O$$

Balancing the oxygens, we obtain:

$$C_5H_{12} + 8O_2 \rightarrow 5CO_2 + 6H_2O$$

Perform Step 2

Checking the number of atoms on each side of the equation:

C: 5 atoms on each side

H: 12 atoms on each side

O: 16 atoms on each side

Therefore, the equation is balanced.

Example 2

Balance the following equation.

$$C_5H_{10}O_2 + O_2 \rightarrow CO_2 + H_2O$$

Solution

Perform Step 1

Scanning the reaction from left to right (leaving the oxygens for last), and adding coefficients to balance each element, we obtain:

$$C_5H_{10}O_2 + O_2 \rightarrow 5CO_2 + 5H_2O$$

To balance the oxygens, we need an additional 13 Os on the left. We must use the coefficient $^{13}/_2$ to obtain 13 Os from O_2.

The equation then becomes:

$$C_5H_{10}O_2 + \frac{13}{2} O_2 \rightarrow 5CO_2 + H_2O$$

Perform Step 2

The coefficients can be converted to whole numbers by multiplying all of them by 2 (to convert the $^{13}/_2$ to 13). The equation then becomes:

$$2C_5H_{10}O_2 + 13O_2 \rightarrow 10CO_2 + 10H_2O$$

Checking the number of atoms on each side of the equation:

C: 10 atoms on each side

H: 20 atoms on each side

O: 30 atoms on each side

Therefore, the equation is balanced.

Example 3

Balance the following equation.

$$NH_3 + O_2 \rightarrow NO_2 + H_2O$$

Solution

Perform Step 1

Scanning the reaction from left to right, and adding coefficients to balance each element, we obtain:

$$2NH_3 + \frac{7}{2} O_2 \rightarrow 2NO_2 + 3H_2O$$

Perform Step 2

The coefficients can be converted to whole numbers by multiplying all of them by 2 (to convert the $^7/_2$ to 7). The equation then becomes:

$4NH_3 + 7O_2 \rightarrow 4NO_2 + 6H_2O$

Checking the number of atoms on each side of the equation:

N: 4 atoms on each side

H: 12 atoms on each side

O: 14 atoms on each side

Therefore, the equation is balanced.

Practice Problems

8.1 Balance the following equation:

$NCl_3 + H_2O \rightarrow NH_3 + HOCl$

8.2 Balance the following equation:

$Al_2O_3 \rightarrow Al + O_2$

8.3 Balance the following equation:

$PCl_5 + H_2O \rightarrow H_3PO_4 + HCl$

Flowchart 8.2 **Identifying the precipitate that will form in a potential precipitation reaction and writing the net ionic equation for the reaction**

Flowchart 8.2

Summary

Method for identifying the precipitate that will form in a potential precipitation reaction and writing the net ionic equation for the reaction

Step 1
- **Identify all ions present when the starting solutions are mixed. (These are the ions produced when the starting materials dissolved.)**

Step 2
- **Identify all possible cation–anion combinations that are different from the ones in the starting materials. These combinations represent possible precipitates.**

Step 3
- **Determine any precipitates (insoluble salts) that will form from any of the cation–anion combinations.**

Step 4
- **If a precipitation reaction occurs, write the balanced equation, the ionic equation, and the net ionic equation for it, making sure that the molecular formula for the precipitate is electrically neutral.**

Example 1

A solution of sodium chloride is combined with a solution of silver nitrate. Write the balanced equation for the precipitation reaction that occurs, the ionic equation, and the net ionic equation.

Solution

Perform Step 1

The ions present in the solution will be Na^+, Cl^-, Ag^+, and NO_3^-.

Perform Step 2

The possible ion combinations to form a precipitate are Na^+ with NO_3^- and Ag^+ with Cl^-.

Perform Step 3

According to the solubility rules, only Ag^+ and Cl^- will form a precipitate.

Perform Step 4

The balanced equation for the reaction that occurs is:

$$NaCl(aq) + AgNO_3(aq) \rightarrow NaNO_3(aq) + AgCl(s)$$

The ionic equation for the reaction that occurs is:

$$Na^+(aq) + Cl^-(aq) + Ag^+(aq) + NO_3^-(aq) \rightarrow Na^+(aq) + NO_3^-(aq) + AgCl(s)$$

The net ionic equation is:

$$Ag^+(aq) + Cl^-(aq) \rightarrow AgCl(s)$$

Example 2

A solution of $Ca(NO_3)_2$ is combined with a solution of Na_2SO_4. Write the balanced equation for the precipitation reaction that occurs, the ionic equation, and the net ionic equation.

Solution

Perform Step 1

The ions present in the solution will be Ca^{2+}, NO_3^-, Na^+, and SO_4^{2-}.

Perform Step 2

The possible ion combinations to form a precipitate are Ca^{2+} with SO_4^{2-} to form $CaSO_4$, and Na^+ with NO_3^- to form $NaNO_3$.

Perform Step 3

According to the solubility rules, only $CaSO_4$ will form a precipitate.

Perform Step 4

The balanced equation for the precipitation reaction that occurs is:

$$Ca(NO_3)_2(aq) + Na_2SO_4(aq) \rightarrow CaSO_4(s) + 2NaNO_3(aq)$$

The ionic equation is:

$$Ca^{2+}(aq) + 2NO_3^-(aq) + 2Na^+(aq) + SO_4^{2-}(aq) \rightarrow CaSO_4(s) + 2Na^+(aq) + 2NO_3^-(aq)$$

The net ionic equation is:

$$Ca^{2+}(aq) + SO_4^{2-}(aq) \rightarrow CaSO_4(s)$$

Practice Problems

8.4 A solution of NaOH is mixed with a solution of $MgCl_2$. Write the balanced equation for the precipitation reaction that occurs, the ionic equation, and the net ionic equation.

8.5 A solution of K_3PO_4 is mixed with a solution of $Al(NO_3)_3$. Write the balanced equation for the precipitation reaction that occurs, the ionic equation, and the net ionic equation.

Flowchart 8.3 **Writing the equation for the neutralization reaction that will occur when an acid reacts with a base and writing the net ionic equation for the reaction**

Flowchart 8.3

Summary

Method for writing the equation for the neutralization reaction that will occur when an acid reacts with a base and writing the net ionic equation for the reaction

Step 1
• Write the balanced intact molecule equation (the molecular equation) for the reaction.

Step 2
• Write the reactants for the reaction (remember the coefficients).
• Break the acid into its H^+ ions and anion, and break the base into its OH^- ions and cation.

Step 3
• Write the products of the reaction.
• Combine the cation from the base with the anion from the acid to make a salt. (The number of cations and anions in the formula of the salt must be such that the salt is electrically neutral.)
• Combine the H^+ and the OH^- to form water.

Step 4
• Write the ionic neutralization reaction equation.
• Write the net ionic equation for the reaction $H^+ + OH^- \rightarrow H_2O$.

Example 1

Write the neutralization reaction equation and the net ionic equation for the reaction that occurs between HI and KOH.

Solution

Perform Step 1

$HI(aq) + KOH(aq) \rightarrow KI(aq) + H_2O(l)$

Perform Step 2

The reactants are $H^+(aq) + I^-(aq) + K^+(aq) + OH^-(aq)$.

Perform Step 3

The products are $K^+(aq) + I^-(aq)$ and $H_2O(l)$.

Perform Step 4

The ionic equation for the neutralization reaction is:

$$H^+(aq) + I^-(aq) + K^+(aq) + OH^-(aq) \rightarrow K^+(aq) + I^-(aq) + H_2O(l)$$

The net ionic equation is $H^+(aq) + OH^-(aq) \rightarrow H_2O(l)$.

Example 2

Write the neutralization reaction equation and the net ionic equation for the reaction that occurs between HNO_3 and $Ca(OH)_2$.

Solution

Perform Step 1

$$2HNO_3(aq) + Ca(OH)_2(aq) \rightarrow Ca(NO_3)_2(aq) + 2H_2O(l)$$

Perform Step 2

The reactants are $2H^+(aq) + 2NO_3^-(aq) + Ca^{2+}(aq) + 2OH^-(aq)$.

Perform Step 3

The products of the reaction are $Ca^{2+}(aq) + 2NO_3^-(aq)$ and $2H_2O(l)$.

Perform Step 4

The ionic equation for the neutralization reaction is:

$$2H^+(aq) + 2NO_3^-(aq) + Ca^{2+}(aq) + 2OH^-(aq) \rightarrow Ca^{2+}(aq) + 2NO_3^-(aq) + 2H_2O(l)$$

The net ionic equation is $2H^+(aq) + 2OH^-(aq) \rightarrow 2H_2O(l)$, which reduces to $H^+(aq) + OH^-(aq) \rightarrow H_2O(l)$.

Practice Problems

8.7 Write the neutralization reaction equation and the net ionic equation for the reaction that occurs between HBr and KOH.

8.8 Write the neutralization reaction equation and the net ionic equation for the reaction that occurs between HNO_3 and $Ba(OH)_2$.

8.9 Write the neutralization reaction equation and the net ionic equation for the reaction that occurs between H_2SO_4 and NaOH.

Quiz for Chapter 8 Problems

1. Balance the equation for the reaction of $C_{10}H_{20} + O_2 \rightarrow CO_2 + H_2O$.

2. The coefficient in front of the PH_3 when the equation below is balanced is _____.

   $$PH_3 + O_2 \rightarrow P_4O_{12} + H_2O$$

3. Which of the following ionic salts are insoluble in water?
   (a) $Ba(NO)_3$ (b) $MgCO_3$ (c) $CaSO_4$ (d) NH_4Cl

4. Write the formula of the precipitate that forms when a solution of K_2CO_3 is mixed with a solution of $CaBr_2$.

5. Write all the ions that are present in solution when a solution of LiCl is mixed with a solution of $Ba(NO_3)_2$.

6. Write the molecular equation, the ionic equation, and the net ionic equation for the reaction that occurs when Na_2SO_4 reacts with $Pb(NO_3)_2$.

7. When an acid is mixed with a base, the products of the reaction are a(n) _____ and
_____.

8. Write the molecular equation, the ionic equation, and the net ionic equation for the precipitation
reaction that occurs when a solution of NH_4I is mixed with a solution of $Pb(NO_3)_2$.

9. Which of the following ionic salts are insoluble in water?
   (a) Na_2CO_3 (b) $BaCl_2$ (c) $CaSO_4$ (d) NH_4NO_3

10. The reaction between an acid and a base is called a(n) _____ reaction.

11. When balancing chemical equations, only the _____ of the reactants and products can be
changed because changing the _____ changes the chemical identity of a substance.

Exercises for Self-Testing

A. Completion. *Write the correct word(s) to complete each statement below.*

1. _____ _____ occur when new substances are produced due to bond
breaking in reactants and bond formation in products.
2. It always takes _____ to break a chemical bond, and _____ is released
when a new bond is formed.
3. The diatomic elements are _____, _____, _____,
_____, _____, _____, and _____.
4. Chemical reactions always involve the formation of one or more _____
_____.
5. The net ionic equation for a neutralization reaction between a strong acid and a strong base is
_____.
6. When an element reacts with a compound to form a new element and a new compound, the
reaction is called a _____-_____ reaction.
7. The insoluble solid that is formed during some chemical reactions is called a(n)
_____.
8. Ions that appear on both sides of a chemical equation are called
_____.
9. In the balanced equation for the reaction of pentane, C_5H_{12}, with oxygen gas to produce carbon
dioxide and water, the coefficient in front of oxygen is _____.
10. Of the elements hydrogen, nitrogen, phosphorous, chlorine, and iodine, all are diatomic
elements except _____.

B. True/False. *Indicate whether each statement below is true or false, and explain why the false
statements are false.*

_____ 1. $CaCl_2$ is insoluble in water.
_____ 2.   In the balanced equation for the reaction of HCl with oxygen gas to produce
water and chlorine gas, the coefficient in front of HCl is 4.
_____ 3.   A decomposition reaction is one in which a substance breaks down to form two
or more simpler substances.
_____ 4.   If necessary, the subscripts can be changed to balance a chemical equation.
_____ 5.   In a balanced chemical equation, the number of atoms of each element must be
equal on both sides of the equation.
_____ 6.   In a balanced chemical equation, the sum of the coefficients on the reactant side
must equal the sum of the coefficients on the product side.

_____ 7.   In a balanced chemical equation, the coefficient in front of any element will equal the number of atoms of the element.

_____ 8.   The products of a neutralization reaction are a salt and water.

_____ 9.   All chemical reactions can be classified as single-replacement, double-replacement, decomposition, or combination reactions.

_____ 10.   The spectator ions in a chemical reaction never appear in the net ionic equation.

C. Questions and Problems

1. Write the formula of the precipitate that forms when a solution of Li_2CO_3 is mixed with a solution of $MgBr_2$.

2. Trichloroamine, NCl_3, an antibacterial agent, is produced according to the following reaction: $Cl_2 + NH_3 \rightarrow NCl_3 + HCl$. Write the balanced chemical equation.

3. Which of the following ionic salts are insoluble in water?
   (a) $Mg(NO)_3$ (b) $CaCO_3$ (c) $SrSO_4$ (d) NH_4I

4. Write the balanced molecular equation, the ionic equation, and the net ionic equation for the reaction of a solution of NH_4I with a solution of $AgNO_3$.

5. Write the balanced molecular equation, the ionic equation, and the net ionic equation for the reaction of HBr with LiOH.

6. Which of the following compounds of lead(II) are insoluble in water?
   (a) $Pb(NO_3)_2$ (b) $Pb(OH)_2$ (c) PbS (d) $Pb_3(PO_4)_2$

7. Which ions are typically spectator ions in double-replacement reactions?
   (a) K^+ (b) Na^+ (c) NO_3^- (d) CO_3^{2-}

8. Briefly describe what will happen when a solution of $Pb(NO_3)_2$ is mixed with a solution of NaI. Write the balanced net ionic equation for the reaction that takes place.

9

Stoichiometry and the Mole

Learning Outcomes

1. Calculate amounts of ingredients using a recipe.
2. Describe what a mole is and how it relates to formulas of substances, molar mass, and chemical equations.
3. Balance chemical reactions and calculate the theoretical yield and percent yield for a reaction.
4. Identify a limiting reactant if present and calculate the theoretical and percent yield when a limiting reactant is present.
5. Calculate a compound's empirical and actual formula from percent by mass elemental compositions.
6. Calculate percent by mass elemental composition from a molecular formula.
7. Describe how it is possible for a formula to have fractional subscripts.

Chapter Outline

9.1 Stoichiometry: What Is It?

    A. Meaning of stoichiometry—Study of quantitative aspects of reactions

    B. Stoichiometry of baking cheesecakes

    C. Limiting and excess reactants

    D. Balanced equation as source of conversion factors

9.2 The Mole

    A. Definition of the mole

        1. Avogadro's number—6.022×10^{23} particles

        2. Relative atomic mass in grams

        3. Molar mass

    B. Conversion between number of atoms, moles, and grams

9.3 Reaction Stoichiometry

    A. Solving stoichiometry problems

    B. Theoretical yield, actual yield, and percent yield

    C. Calculations involving reaction stoichiometry

9.4 Dealing with a Limiting Reactant

    A. Limiting reactant—The reactant present in the smallest (relative) amount limiting the amount of product

    B. Excess reactant—Reactants present in excess relative to the limiting reactant

    C. Procedure for determining the limiting reactant

    D. Determining theoretical yield from the limiting reactant

    E. Solving limiting-reactant problems

9.5 Combustion Analysis

    A. Combustion analysis reaction

    B. Determination of empirical and molecular formula from combustion analysis data

       1. Determination of empirical formula

       2. Determination of actual formula from empirical formula and molar mass

9.6 Going Back and Forth Between Formulas and Percent Composition

    A. Elemental mass percent composition from formula

    B. Formula from elemental mass percent composition

9.7 One More Thing: Nonstoichiometric Compounds

    A. Superconductors

    B. Semiconductors

Review of Key Concepts from the Text

9.1 Stoichiometry: What Is It?

Stoichiometry involves the quantitative relationships between the components of a chemical reaction, and can be thought of as *chemical arithmetic*. Stoichiometry allows us to determine things such as how many grams of product can be made from a certain number of grams of reactants, or how many grams of one reactant are needed to react with a certain number of grams of another. Without stoichiometry, we would not be able to translate the language of chemical equations from the submicroscopic atoms and molecules in which they are written into quantities we can observe and measure in the laboratory. The quantity that allows us to do the translation is the mole.

9.2 The Mole

A practical way to become familiar with the term *mole* is to think of it in the following way: A mole is the quantity of a substance whose mass in grams is equal to the sum of the atomic masses of the elements (molar mass) that make up the substance. (The atomic mass of each element must be multiplied by the number of atoms of that element in the substance.) For the elements, the molar mass is simply equal to the mass of the element (from the periodic table) in grams. For example, the molar mass of aluminum is 26.98 g, the molar mass of uranium is 238.03 g, and the molar mass of chlorine is 70.91 g.

To get the molar mass of a compound, we use the compound's formula and add up the molar masses of all the elements in the compound. The molar mass of each element is multiplied by the number of atoms of that element according to the compound's formula.

Example 9.1

Calculate the molar mass (grams in 1 mole) of $(NH_4)_2CO_3$.

Solution

We determine the mass contributed by each element in the compound:

	Number of atoms	$\times$	Atomic mass	$=$	Mass of element
Mass of N	$=$ 2	$\times$	14.0067	$=$	28.0134
Mass of H	$=$ 8	$\times$	1.00794	$=$	8.06352
Mass of C	$=$ 1	$\times$	12.011	$=$	12.011
Mass of O	$=$ 3	$\times$	15.9994	$=$	47.9982
		Molar mass of compound		$=$	96.086 g

Practice Exercise 9.1

Calculate the molar mass of $Al_2(SO_4)_3$.

We now know that the mass of 1 mole of a compound equals the compound's molar mass in grams. We can determine the mass of 1 mole of any substance, but there is one more aspect of the mole that we need to consider. We need to know how many particles (atoms or molecules) there are in 1 mole. It turns out that if you count the number of atoms of Al in 26.98 g of aluminum, or the number of molecules of H_2O in 18.015 g of water, or the number of molecules of H_2 in 2.016 g of hydrogen gas, the number is the same. That number, 6.022×10^{23}, is called *Avogadro's number*.

One mole of any substance contains 6.022×10^{23} particles.

We now have a quantity—the mole—that will allow us to translate chemical equations from molecules to quantities we can measure and handle.

Given the two definitions of a mole, we can develop two different mathematical conversion factors for calculating moles. We can calculate moles from grams of substance, and vice versa. We can also calculate moles from particles of substance. Note that the *mass* of 1 mole depends on the substance involved, but the number of particles in 1 mole is always the same, 6.022×10^{23}. The conversion factors are as follows:

$$\frac{\text{Molar mass of substance}}{1 \text{ mole of substance}} \quad \text{and} \quad \frac{6.022 \times 10^{23} \text{ particles}}{1 \text{ mole}}$$

We can also write the conversion factors as:

$$\frac{1 \text{ mole of substance}}{\text{Molar mass of substance}} \quad \text{and} \quad \frac{1 \text{ mole}}{6.022 \times 10^{23} \text{ particles}}$$

The mathematical relationship for the conversions are shown below:

$$\text{Mole} \underset{\div \text{ molar mass}}{\overset{\times \text{ mole mass}}{\rightleftarrows}} \text{Grams}$$

$$\text{Mole} \underset{\div \ 6.022 \times 10^{23}}{\overset{\times \ 6.022 \times 10^{23}}{\rightleftarrows}} \text{Atoms or molecules}$$

Combining the mathematical relationships into one, we have:

$$\text{Grams} \underset{\div \text{ molar mass}}{\overset{\times \text{ molar mass}}{\rightleftarrows}} \text{Moles} \underset{\div \ 6.022 \times 10^{23}}{\overset{\times \ 6.022 \times 10^{23}}{\rightleftarrows}} \text{Atoms or molecules}$$

You should notice from the conversion factors and the above diagram that it is not possible to interconvert directly between grams and atoms or molecules. You must go through moles first.

Example 9.2

Calculate the number of molecules in 3.5 moles of CO_2.

Solution

To convert from moles to molecules, we use the conversion factor involving Avogadro's number.

$$3.5 \ \text{mol} \ CO_2 \times \frac{6.022 \times 10^{23} \text{ molecules}}{1 \ \text{mol} \ CO_2} = 2.1 \times 10^{24} \text{ molecules } CO_2$$

Practice Exercise 9.2

Calculate the number of moles in 1.8×10^{24} molecules of H_2O.

Example 9.3

Calculate the number of grams in 3.78 moles of glucose ($C_6H_{12}O_6$).

Solution

To convert from moles to grams, we need the molar mass of glucose. The molar mass of glucose = $12.011 \times 6 + 1.0079 \times 12 + 15.999 \times 6 = 180.155$ g/mol.

$$\text{Grams of glucose} = 3.78 \ \text{mol glucose} \times \frac{180.16 \text{ g glucose}}{1 \ \text{mol glucose}} = 681 \text{ g glucose}$$

Practice Exercise 9.3

Calculate the number of moles in 150 g of NH_4Cl.

Example 9.4

Calculate the number of molecules of C_2H_4 in 450.0 g of C_2H_4.

Solution

To convert from grams to molecules, we must go through moles first. The molar mass of $C_2H_4 = 12.011 \times 2 + 1.007 \ 94 \times 4 = 28.054$ g/mol.

$$\text{Moles of } C_2H_4 = 450.0 \text{ g } C_2H_4 \times \frac{1 \text{ mol } C_2H_4}{28.054 \text{ g } C_2H_4} = 16.04 \text{ mol } C_2H_4$$

$$\text{Molecules of } C_2H_4 = 16.04 \text{ mol } C_2H_4 \times \frac{6.022 \times 10^{23} \text{ } C_2H_4 \text{ molecules}}{1 \text{ mol } C_2H_4}$$

$$= 9.66 \times 10^{24} \text{ } C_2H_4 \text{ molecules}$$

Practice Exercise 9.4

Calculate the number of grams in 2.5×10^{27} molecules of H_2O.

9.3 Reaction Stoichiometry

The coefficients in a balanced chemical equation represent the number of moles of each substance in the reaction. We can use the coefficients along with the molar masses to determine the number of grams of each substance in the reaction. Therefore,

> – **before you do anything else, write and balance the reaction equation,**
> – **check for correctness.**

This is your starting point. If your reaction is not properly balanced, everything you do from this point on will be wrong.

Example 9.5

In the reaction of hydrogen and oxygen to produce water, how many grams of each substance take part in the reaction as it is written in the balanced equation?

Solution

We first write the balanced equation.

$$2 H_2 + O_2 \rightarrow 2 H_2O$$

Then, we determine the grams of each substance.

$$\text{Grams of } H_2 = 2 \text{ mol } H_2 \times \frac{2.0158 \text{ g } H_2}{1 \text{ mol } H_2} = 4.0318 \text{ g } H_2$$

$$\text{Grams of } O_2 = 1 \text{ mol } O_2 \times \frac{31.998 \text{ g } O_2}{1 \text{ mol } O_2} = 31.998 \text{ g } O_2$$

$$\text{Grams of } H_2O = 2 \text{ mol } H_2O \times \frac{18.015 \text{ g } H_2O}{1 \text{ mol } H_2O} = 36.030 \text{ g } H_2O$$

Note that the sum of the masses of the reactants (4.03 g + 32.00 g) equals the mass of the product (36.03 g). This must always be the case due to the law of conservation of matter.

Practice Exercise 9.5

In the reaction of pentane, C_5H_{12}, with oxygen to produce CO_2 and H_2O, how many grams of each substance take part in the reaction as it is written in the balanced equation?

Reactions are not usually run in the exact molar amounts that appear in the balanced equation. Using stoichiometry, however, we can determine exactly how much of any reactant will be used, or any product formed, when given the mass of any of the reactants or products. The maximum amount of product that can be formed from a given amount of reactants in a chemical reaction is called the *theoretical yield*. The following relationships will be useful in calculating the theoretical yield, or in determining the grams of reactants needed, for any chemical reaction:

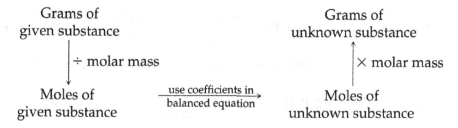

The above diagram outlines the three possible steps to complete when doing problems involving grams and moles in stoichiometry calculations.

1. Converting grams of the given substance to moles of the given substance.
2. Converting moles of the given substance to moles of the unknown substance.
3. Converting moles of the unknown substance to grams of the unknown substance.

We will consider the following examples to see how these calculations work.

Example 9.6

Tin (Sn) reacts with hydrofluoric acid to produce tin(II) fluoride and hydrogen gas.

(a) How many moles of tin(II) fluoride will be produced from the reaction of 100.0 g of HF? (Assume that there is enough Sn to completely react with the HF.)

(b) How many grams of tin(II) fluoride will be produced?

Solution

(a) We must always start with the correctly balanced equation for any stoichiometry problem. The balanced equation for this reaction is:

$$Sn + 2\,HF \rightarrow SnF_2 + H_2$$

Recall the diagram

Grams   Grams

Moles $\longrightarrow$ Moles

First, we must calculate the moles of HF.

$$\text{Moles of HF} = 100.0\ \text{g HF} \times \frac{1\ \text{mol HF}}{20.01\ \text{g HF}} = 4.998\ \text{mol HF}$$

We next calculate the moles of SnF_2 from the moles of HF, using the coefficients in the balanced equation.

$$\text{Moles of } SnF_2 = 4.998 \text{ mol HF} \times \frac{1 \text{ mol } SnF_2}{2 \text{ mol HF}} = 2.499 \text{ mol } SnF_2$$

Combining the two steps, we would have:

$$\text{Moles of } SnF_2 = 100.0 \text{ g HF} \times \frac{1 \text{ mol HF}}{20.0 \text{ g HF}} \times \frac{1 \text{ mol } SnF_2}{2 \text{ mol HF}} = 2.499 \text{ mol } SnF_2$$

(b) We can get the grams of SnF_2 from the moles, using the final step in the diagram.

$$\text{Grams of } SnF_2 = 2.499 \text{ mol } SnF_2 \times \frac{15.671 \text{ g } SnF_2}{1 \text{ mol } SnF_2} = 391.6 \text{ g } SnF_2$$

Combining all three steps, we would have:

$$100.0 \text{ g HF} \times \frac{1 \text{ mol HF}}{20.0 \text{ g HF}} \times \frac{1 \text{ mol } SnF_2}{2 \text{ mol HF}} \times \frac{156.71 \text{ g } SnF_2}{1 \text{ mol } SnF_2} = 391.6 \text{ g } SnF_2$$

The 392 g of SnF_2 is the maximum amount that could be produced from 100.0 g of HF with excess Sn, and is called the *theoretical yield*.

Practice Exercise 9.6

How many grams of NH_3 can be produced from the reaction of 425 g of nitrogen gas with hydrogen gas? (Assume that there is enough hydrogen gas to completely react with the nitrogen gas.)

The theoretical yield is the maximum amount of product that can be made from a chemical reaction if all the reactants are converted to products. However, it is generally not the case that all the reactants will be converted to products, and the *actual yield* is less than the theoretical yield. We can calculate the percent yield from the actual yield and the theoretical yield.

$$\% \text{ yield} = \frac{\text{Actual yield}}{\text{Theoretical yield}} \times 100\%$$

Example 9.7

If the actual yield in the SnF_2 reaction in Example 9.6 is 311 g, what is the percent yield?

Solution

$$\% \text{ yield} = \frac{\text{Actual yield}}{\text{Theoretical yield}} \times 100\%$$

$$\% \text{ yield} = \frac{311 \text{ g}}{392 \text{ g}} \times 100\% = 79.3\%$$

Practice Exercise 9.7

If the actual yield in the SnF_2 reaction in Example 9.6 is 278 g, what is the percent yield?

You should carefully study Section 9.3 of the textbook for additional information and examples of stoichiometry problems. Note that when using the stoichiometry diagram,

Grams   Grams

↓   ↑

Moles ⟶ Moles ,

you may be given grams or moles, and you may be asked to calculate grams or moles. If you are given moles, you begin with Step 2 in the process. If you are asked for moles, you end with Step 2.

9.4 Dealing with a Limiting Reactant

Sometimes we are given the amounts of more than one reactant in a chemical reaction and asked to determine the theoretical yield. If the reactants are not there in exactly the right amounts, one of them will run out before the other(s). The one that runs out first is called the limiting reactant.

The procedure for determining limiting reactants is provided in Section 9.4 of the textbook. The steps are as follows:

1. Calculate the molar mass of each reactant.
2. Determine the number of moles of each reactant.
3. Divide each of the molar amounts by its respective coefficient in the balanced equation.
4. The reactant that yields the smallest number in Step 3 will be limiting.
5. Calculate the theoretical yield based on the amount of the limiting reactant.

Example 9.8

How many grams of NH_3 can be produced from the reaction of 25.0 g of H_2 and 50.0 g of N_2?

Solution

We must first write the balanced equation for the reaction.

$$N_2 + 3H_2 \rightarrow 2NH_3$$

Then, we follow the steps described above.

1. Calculate the molar mass of each reactant:

   $N_2 = 2 \times 14.0067 = 28.0134$ g/mol
   $H_2 = 2 \times 1.007\,94 = 2.015\,88$ g/mol

2. Determine the number of moles of each reactant.

   $$\text{Moles of } H_2 = 25.0 \text{ g } H_2 \times \frac{1 \text{ mol } H_2}{2.015\,88 \text{ g } H_2} = 12.4 \text{ mol } H_2$$

   $$\text{Moles of } N_2 = 50.0 \text{ g } N_2 \times \frac{1 \text{ mol } N_2}{28.0134 \text{ g } N_2} = 1.78 \text{ mol } N_2$$

3. Divide each of the molar amounts by its coefficient in the balanced equation.

$$H_2: \frac{12.4 \text{ mol}}{3} = 4.13$$

$$N_2: \frac{1.78 \text{ mol}}{1} = 1.78$$

4. The reactant that yields the smallest number in Step 3 will be limiting. Because 1.78 is smaller than 4.13, N_2 is the limiting reactant, and the theoretical yield is based on it.

5. Calculate the theoretical yield based on the amount of the limiting reactant.

$$\text{Moles of NH}_3 = 1.78 \text{ mol N}_2 \times \frac{2 \text{ mol NH}_3}{1 \text{ mol N}_2} = 3.56 \text{ mol NH}_3$$

$$\text{Grams of NH}_3 = 3.56 \text{ mol NH}_3 \times \frac{17.031 \text{ g NH}_3}{1 \text{ mol NH}_3} = 60.6 \text{ g NH}_3$$

Practice Exercise 9.8

How many grams of PBr_3 can be produced from the reaction of 100.0 g of P and 200.0 g of Br_2? The unbalanced equation for the reaction is $P + Br_2 \rightarrow PBr_3$.

9.5   Combustion Analysis

The formula of a compound gives us the number of each type of atom in a molecule of the compound. When we have a compound whose formula we do not know, we may be able to use combustion analysis to determine the formula. Combustion analysis is useful for compounds known to contain carbon and hydrogen. The process involves reacting the compound with oxygen to produce CO_2 from the carbon in the compound and H_2O from the hydrogen in the compound. Combustion analysis is used to determine what is known as the empirical formula of the compound. The empirical formula is the simplest formula that shows the ratio of moles of each element in the compound. Empirical formulas can be calculated from percent composition or from combustion data.

To calculate the empirical formula from combustion data, do the following:

Step 1: Check that the percent by mass elemental compositions add up to 100%. If they do not, calculate the percent by mass of the additional element(s) by subtracting the total of the given percentages from 100%.

Step 2: Assume a 100 g sample of the compound.

Step 3: Convert grams of each element to moles of each element.

Step 4: Divide each number in Step 3 by the smallest number in order to get the subscripts in the empirical formula. If one or more of the subscripts is not a whole number, multiply all the numbers by a factor that will make them all whole numbers.

Example 9.9

When 10.01 g of a compound containing C, H, and O are subjected to combustion analysis, the compound was found to consist of 39.46% C and 8.84% H. What is the empirical formula of the compound?

Solution

We will follow the steps outlined above.

1. 39.46% C + 8.84% H = 48.30%, which is less than 100%. Therefore, the percent O = 100% − 48.30% = 51.70% O

2. Assume 100 g of compound (percentages then become grams).

   39.46 g of carbon, 8.84 g of hydrogen, 48.30 g of oxygen

3. Calculate the moles of each element (moles = grams/atomic mass).

$$\text{Moles of C} = 39.46 \text{ g C} \times \frac{1 \text{ mol C}}{12.011 \text{ g C}} = 3.288 \text{ mol C}$$

$$\text{Moles of H} = 8.84 \text{ g H} \times \frac{1 \text{ mol H}}{1.0079 \text{ g H}} = 8.77 \text{ mol H}$$

$$\text{Moles of O} = 48.30 \text{ g O} \times \frac{1 \text{ mol O}}{15.999 \text{ g O}} = 3.019 \text{ mol O}$$

4. Divide by the smallest number to obtain the subscripts in the empirical formula. The smallest number is 2.262.

$$\text{Subscript for C: } \frac{3.288}{3.019} = 1.09 = 1$$

$$\text{Subscript for H: } \frac{8.771}{3.019} = 2.91 = 3$$

$$\text{Subscript for O: } \frac{3.019}{3.019} = 1 = 1$$

The empirical formula is therefore CH_3O.

Practice Exercise 9.9

When 3.70 g of a compound containing C, H, and O were subjected to combustion analysis, the compound was found to consist of 48.6% C and 8.11% H. What is the empirical formula of the compound?

The empirical formula is the simplest formula. The actual formula may be the empirical formula, or it may be some multiple of the empirical formula. To find the actual formula from the empirical formula, use the following equation:

$$\frac{\text{Molar mass}}{\text{Empirical formula molar mass}} = \text{Multiplier of each subscript in empirical formula}$$

Example 9.10

The molar mass of a compound whose empirical formula is CH_2 is 56 g/mol. What is the actual formula of the compound?

Solution

$$\frac{\text{Molar mass}}{\text{Empirical formula molar mass}} = \text{Multiplier of each subscript in empirical formula}$$

$$\frac{56 \text{ g/mol}}{14 \text{ g/mol}} = 4$$

The molecular formula of the compound is therefore C_4H_8.

Practice Exercise 9.10

The molar mass of a compound whose empirical formula is $C_4H_5N_2O$ is 194 g/mol. What is the actual formula of the compound?

9.6 Going Back and Forth Between Formulas and Percent Composition

The formula of a compound gives the number of each type of element in the compound. We can determine the percent composition of a compound by determining the mass of each element in the compound.

$$\text{Percent of each element in a compound} = \frac{\text{Grams of element in 1 mole of the compound}}{\text{Molar mass of the compound}} \times 100\%$$

Example 9.11

Determine the percent composition of $CaSO_4$.

Solution

$$\% \text{ Ca} = \frac{\text{Grams of Ca in CaSO}_4}{\text{Molar mass of CaSO}_4} \times 100\% = \frac{40.078}{136.14} \times 100\% = 29.439\%$$

$$\% \text{ S} = \frac{\text{Grams of S in CaSO}_4}{\text{Molar mass of CaSO}_4} \times 100\% = \frac{32.066}{136.14} \times 100\% = 23.554\%$$

$$\% \text{ O} = \frac{\text{Grams of O in CaSO}_4}{\text{Molar mass of CaSO}_4} \times 100\% = \frac{4 \times 15.999}{136.14} \times 100\% = 47.007\%$$

Practice Exercise 9.11

Calculate the percent composition of N_2H_4CO.

An additional example of calculating the empirical formula from the percent composition is given below.

Example 9.12

Calculate the empirical formula of a compound that has the following percent composition: 24.39% carbon, 4.08% hydrogen, and 71.53% chlorine.

Solution

1. Check to see if the percents add to 100%.

   24.39% C + 4.08% H + 71.53% Cl = 100%. All elements are accounted for.

2. Assume 100 g of compound (percentages then become grams).

   24.39 g of carbon, 4.08 g of hydrogen, 71.53 g of chlorine

3. Calculate the moles of each element (moles = grams/atomic mass).

$$\text{Moles of C} = 24.39 \text{ g C} \times \frac{1 \text{ mol C}}{12.011 \text{ g C}} = 2.031 \text{ mol C}$$

$$\text{Moles of H} = 4.08 \text{ g H} \times \frac{1 \text{ mol H}}{1.0079 \text{ g H}} = 4.05 \text{ mol H}$$

$$\text{Moles of Cl} = 71.53 \text{ g Cl} \times \frac{1 \text{ mol Cl}}{35.453 \text{ g Cl}} = 2.018 \text{ mol Cl}$$

4. Divide by the smallest number to obtain the subscripts in the empirical formula. The smallest number is 2.018.

$$\text{Subscript for C: } \frac{2.031}{2.018} = 1.006$$

$$\text{Subscript for H: } \frac{4.048}{2.018} = 2.006$$

$$\text{Subscript for Cl: } \frac{2.018}{2.018} = 1.000$$

The empirical formula is therefore CH_2Cl.

Practice Exercise 9.12

Calculate the empirical formula of a compound that has the following percentage composition: 54.55% carbon, 9.09% hydrogen, and 36.36% oxygen.

Note: Empirical formulas can be calculated from any data that will allow you to determine the number of grams of each element in the compound. Simply calculate the number of moles of each element and divide by the smallest number, as demonstrated above.

Strategies for Working Problems in Chapter 9

Overview: What You Should Be Able to Do

Chapter 9 introduces the quantitative aspects of chemical reactions and the procedures for performing stoichiometric calculations. The types of problems that you should be able to solve after mastering Chapter 9 are as follows:

1. Calculate theoretical yield of products from grams (or moles) of one or more reactants.

2. Solve limiting reactant problems to determine the theoretical yield when amounts of two reactants are given.

3. Determine the empirical formula from the percent composition.

4. Convert an empirical formula to a molecular formula.

5. Determine the percent composition from the molecular formula.

Flowcharts

Flowchart 9.1 **Calculating theoretical yield of products from grams (or moles) of one or more reactants**

Flowchart 9.1

Summary

Method for calculating theoretical yield of products from grams (or moles) of one or more reactants

Step 1
• **Write the balanced chemical equation for the reaction.**

↓

Step 2
• **Convert the given information (grams, atoms, or molecules) for all reactants to moles.**
• **If moles are given, proceed to Step 3.**
• **Use the molar mass to convert grams to moles.**
• **Use Avogadro's number to convert atoms or molecules to moles.**

↓

Step 3
• **Use the coefficients from the balanced equation to find the relationship between the number of moles of the given substance and the number of moles of the unknown substance.**
• **The ratio of the coefficients is used as a conversion factor, with the coefficient in front of the unknown as the numerator and the coefficient in front of the given as the denominator.**
• **The conversion factor is multiplied by the number of moles of the given substance.**

↓

Step 4
- **Convert the moles of the unknown, calculated in Step 3, to grams (or atoms or molecules).**

Example 1

How many grams of H_2 are required to produce 82.00 g of ammonia, NH_3, from N_2 and H_2.

Solution

Perform Step 1
The balanced equation is $N_2 + 3H_2 \rightarrow 2NH_3$.

Perform Step 2

$$\text{Moles of } NH_3 = 82.00 \text{ g} \times \frac{1 \text{ mol } NH_3}{17.03 \text{ g } NH_3} = 4.184 \text{ mol } NH_3$$

Perform Step 3

$$\text{Moles of } H_2 = 4.184 \text{ mol } NH_3 \times \frac{3 \text{ mol } H_2}{2 \text{ mol } NH_3} = 6.276 \text{ mol } H_2$$

Perform Step 4

$$\text{Grams of } H_2 = 6.276 \text{ mol } H_2 \times \frac{2.0159 \text{ g } H_2}{1 \text{ mol } H_2} = 12.65 \text{ g } H_2$$

Example 2

In the reaction of O_2 with C_2H_4 to produce CO_2 and H_2O, how many moles of O_2 are needed to completely react with 25.0 g of C_2H_4? How many molecules of O_2 are needed?

Solution

Perform Step 1
The balanced equation is $C_2H_4 + 3O_2 \rightarrow 2CO_2 + 2H_2O$.

Perform Step 2

$$\text{Moles of } C_2H_4 = 25.0 \text{ g} \times \frac{1 \text{ mol } C_2H_4}{28.05 \text{ g } C_2H_4} = 0.891 \text{ mol } C_2H_4$$

Perform Step 3

$$\text{Moles of } O_2 = 0.891 \text{ mol } C_2H_4 \times \frac{3 \text{ mol } O_2}{1 \text{ mol } C_2H_4} = 0.267 \text{ mol } O_2$$

Perform Step 4 to determine number of atoms:

$$\text{Atoms of } O_2 = 0.267 \text{ mol } O_2 \times \frac{6.022 \times 10^{23} \text{ molecules } O_2}{1 \text{ mol } O_2} = 1.61 \times 20^{23} \text{ molecules } O_2$$

Practice Problems

9.1 Calculate the number of grams of NH_3 that can be produced from 500.0 g of NCl_3 according to the following (unbalanced) equation.

$$NCl_3 + H_2O \rightarrow NH_3 + HOCl$$

9.2 Calculate the number of moles of NaOH that are needed to react with 500.0 g of H_2SO_4 according to the following (unbalanced) equation.

$$H_2SO_4 + NaOH \rightarrow Na_2SO_4 + H_2O$$

Flowchart 9.2   **Solving limiting reactant problems to determine the theoretical yield when amounts of two reactants are given**

Flowchart 9.2

Summary

Method for solving limiting reactant problems to determine the theoretical yield when amounts of two reactants are given

Step 1
- **Write the balanced chemical equation for the reaction.**

↓

Step 2
- **Convert the given information (grams, atoms, or molecules)** *for all reactants* **to moles.**
- **If moles for all reactants are given, proceed to Step 3.**
- **Use the molar mass to convert grams to moles.**
- **Use Avogadro's number to convert atoms or molecules to moles.**

↓

Step 3
- **Determine the mole-to-coefficient ratio for each reactant, using the appropriate coefficients from the balanced equation.**
- **The reactant that has the smallest mole-to-coefficient ratio is the limiting reactant.**

↓

Step 4
- **Use the coefficients from the balanced equation and the molar amount of the limiting reactant to calculate the theoretical yield of the product in moles.**

↓

Step 5
- **Convert the moles of the unknown, calculated in Step 3, to grams (or atoms or molecules).**

Example 1

In the reaction of HCl with Zn to produce H_2 gas and $ZnCl_2$, how many grams of H_2 gas can be made from 90.0 g of HCl with 50.0 g of Zn?

Solution

Perform Step 1

The balanced equation for the reaction is:

$2HCl + Zn \rightarrow H_2 + ZnCl_2$

Perform Step 2

Moles of HCl $= 90.0 \text{ g HCl} \times \dfrac{1 \text{ mol HCl}}{36.46 \text{ g HCl}} = 2.46 \text{ mol HCl}$

Moles of Zn $= 50.0 \text{ g Zn} \times \dfrac{1 \text{ mol Zn}}{65.39 \text{ g Zn}} = 0.765 \text{ mol Zn}$

Perform Step 3

Mole-to-coefficient ratios:

HCl: $\dfrac{2.46}{2} = 1.23$ Zn: $\dfrac{0.765}{1} = 0.765$

Zn has the smaller mole-to-coefficient ratio and is therefore the limiting reactant.

Perform Step 4

Moles of $H_2 = 0.765 \text{ mol Zn} \times \dfrac{1 \text{ mol } H_2}{2 \text{ mol HCl}} = 0.383 \text{ mol } H_2$

Perform Step 5

Grams of $H_2 = 0.383 \text{ mol } H_2 \times \dfrac{2.016 \text{ g } H_2}{1 \text{ mol } H_2} = 0.772 \text{ g } H_2$

Example 2

In the production of NBr_3 from N_2 and Br_2, according to the unbalanced equation below, how many grams of NBr_3 can be made from 20.0 g of N_2 with 300.0 g of Br_2?

$N_2 + Br_2 \rightarrow NBr_3$

Solution

Perform Step 1

The balanced equation for the reaction is:

$N_2 + 3Br_2 \rightarrow 2NBr_3$

Perform Step 2

Moles of $N_2 = 20.0 \text{ g } N_2 \times \dfrac{1 \text{ mol } N_2}{28.014 \text{ g } N_2} = 0.714 \text{ mol } N_2$

Moles of $Br_2 = 300.0 \text{ g } Br_2 \times \dfrac{1 \text{ mol } Br_2}{159.8 \text{ g } Br_2} = 1.877 \text{ mol } Br_2$

Perform Step 3

Mole-to-coefficient ratios:

$N_2: \dfrac{0.714}{1} = 0.714 \qquad Br_2: \dfrac{1.877}{3} = 0.6258$

Br_2 is the limiting reactant because it has the smaller mole-to-coefficient ratio. Note that although there are many more grams of Br_2 than grams of N_2, when the moles of each substance and the coefficients in the balance equation are considered, Br_2 is limiting.

Perform Step 4

Moles of NBr_3 = 1.877 mol $Br_2 \times \dfrac{2 \text{ mol } NBr_3}{3 \text{ mol } Br_2} = 1.251$ mol NBr_3

Perform Step 5

Grams of NBr_3 = 1.251 mol $NBr_3 \times \dfrac{253.7 \text{ g } NBr_3}{1 \text{ mol } NBr_3} = 317.1$ g NBr_3

Practice Problems

9.3 In the reaction of H_2 with O_2 to produce H_2O, according to the unbalanced equation below, how many grams of water can be produced from the reaction of 150.0 g of H_2 with 200.0 g of O_2?

$H_2 + O_2 \rightarrow H_2O$

9.4 In the reaction of C_3H_8 with O_2 to produce CO_2 and H_2O, according to the unbalanced equation below, how many grams of CO_2 can be produced from the reaction of 25.0 g of C_3H_8 with 75.00 g of O_2?

$C_3H_8 + O_2 \rightarrow CO_2 + H_2O$

9.5 In the reaction of $Ca(OH)_2$ with HCl to produce $CaCl_2$ and H_2O, according to the unbalanced equation below, how many grams of HCl will be have to be used up to produce 12.4 g of $CaCl_2$?

$Ca(OH)_2 + HCl \rightarrow CaCl_2 + H_2O$

Flowchart 9.3   Determining the empirical formula from the percent composition

Flowchart 9.3

Summary

Procedure for determining the empirical formula from the percent composition

Step 1
- **Assume you have exactly 100 g (unlimited significant digits) of compound.**

- **In 100 g of compound, all percents are equivalent to the number of grams of each element. (If they don't add up to 100, the amount of the last element can be obtained by subtracting the sum of the other elements from 100—see Practice Problem 9.28 of the textbook.)**

↓

Step 2
- Convert the grams of each element to moles by dividing by the atomic mass of the element.

↓

Step 3
- Use the calculated number of moles of each element as the subscripts in the empirical formula.
- Divide the values of all subscripts by the smallest one, and if the resulting values are very close to whole numbers, round them to whole numbers.
- If one or more subscripts are far from a whole number, find some multiplier that makes them whole.
- Multiply all the subscripts by the multiplier that makes whole numbers.

Example 1

Determine the empirical formula of a compound that has the following percent composition: 24.39% carbon, 4.08% hydrogen, and 71.53% chlorine.

Solution

Perform Step 1

100 g of compound contain 24.39 g carbon, 4.08 g hydrogen, and 71.53 g chlorine.

Perform Step 2

$$\text{Moles of C} = 24.39 \text{ g C} \times \frac{1 \text{ mol C}}{12.011 \text{ g C}} = 2.031 \text{ mol C}$$

$$\text{Moles of H} = 4.08 \text{ g H} \times \frac{1 \text{ mol H}}{1.0079 \text{ g H}} = 4.048 \text{ mol H}$$

$$\text{Moles of Cl} = 71.53 \text{ g Cl} \times \frac{1 \text{ mol Cl}}{35.453 \text{ g Cl}} = 2.018 \text{ mol Cl}$$

Perform Step 3

Using the moles from Step 3 and dividing by the smallest number, we get

$$C_{\frac{2.031}{2.018}} H_{\frac{4.048}{2.018}} Cl_{\frac{2.018}{2.018}} \cdot$$

$$C_{\frac{2.031}{2.018}} H_{\frac{4.048}{2.018}} Cl_{\frac{2.018}{2.018}} \rightarrow CH_2Cl$$

The empirical formula is therefore CH_2Cl.

Example 2

Determine the empirical formula of a compound that has the following percent composition: 85.71% carbon and 14.29% hydrogen.

Solution

Perform Step 1

100 g of compound contain 85.71 g carbon and 14.29 g hydrogen.

Perform Step 2

$$\text{Moles of C} = 85.71 \text{ g C} \times \frac{1 \text{ mol C}}{12.011 \text{ g C}} = 7.136 \text{ mol C}$$

$$\text{Moles of H} = 14.29 \text{ g H} \times \frac{1 \text{ mol H}}{1.0079 \text{ g H}} = 14.18 \text{ mol H}$$

Perform Step 3

Using the moles from Step 3 and dividing by the smallest number, we get $C_{\frac{7.136}{7.136}}H_{\frac{14.18}{7.136}}$.

$$C_{\frac{7.136}{7.136}}H_{\frac{14.18}{7.136}} \rightarrow CH_2$$

The empirical formula is therefore CH_2.

Practice Problems

9.6 Determine the empirical formula of a compound that has the following percent composition: 29.71% carbon, 6.23% hydrogen, and 64.06% lead.

9.7 Determine the empirical formula of a compound that has the following percent composition: 92.30% carbon and 7.70% hydrogen.

Flowchart 9.4   Converting an empirical formula to a molecular formula

Flowchart 9.4

Summary

Method for converting an empirical formula to a molecular formula

Step 1
- **Determine the molar mass of the empirical formula.**

↓

Step 2
- **Divide the molar mass of the compound by the molar mass of the empirical formula.**
- **This will yield a number very close to a whole number, and should be rounded to a whole number.**

↓

Step 3
- **Multiply all subscripts in the empirical formula by the number calculated in Step 1.**

Example 1

A compound whose empirical formula is CH_2 has a molar mass of 98.00 g/mol. What is the molecular formula of the compound?

Solution

Perform Step 1

The molar mass of the empirical formula is the sum of the masses of one carbon atom and two hydrogen atoms.

C: $12.011 \times 1 = 12.011$

H: $1.0079 \times 2 = 2.0158$

Molar mass of empirical formula is $12.011 + 2.0158 = 14.027$.

Perform Step 2

$$\frac{\text{Molar mass of compound}}{\text{Molar mass of empirical formula}} = \frac{98.00}{14.027} = 6.99 = 7$$

Perform Step 3

Multiplying the subscripts in the empirical formula by 7, we get the molecular formula, C_7H_{14}.

Example 2

A compound whose empirical formula is CH_3O has a molar mass of 62.0 g/mol. What is the molecular formula of the compound?

Solution

Perform Step 1

The molar mass of the empirical formula is the sum of the masses of one carbon atom and two hydrogen atoms.

C: $12.011 \times 1 = 12.011$

H: $1.0079 \times 3 = 3.0237$

O: $15.999 \times 1 = 15.999$

Molar mass of empirical formula is $12.011 + 3.0237 + 15.999 = 31.034$.

Perform Step 2

$$\frac{\text{Molar mass of compound}}{\text{Molar mass of empirical formula}} = \frac{62.0}{31.034} = 1.998 = 2$$

Perform Step 3

Multiplying the subscripts in the empirical formula by 2, we get the molecular formula, $C_2H_6O_2$.

Practice Problems

9.8   A compound whose empirical formula is C_3H_3O has a molar mass of 110.0 g/mol. What is the molecular formula of the compound?

9.9   A compound whose empirical formula is CH_2 has a molar mass of 322 g/mol. What is the molecular formula of the compound?

9.10 Another compound whose empirical formula is CH_2 has a molar mass of 84 g/mol. What is the molecular formula of the compound?

Flowchart 9.5   Determining the percent composition from the molecular formula

Flowchart 9.5

Summary

Method for determining the percent composition from the molecular formula

Step 1
- Assume exactly 1 mole (unlimited number of significant digits) of compound.
- The molecular formula gives the number of moles of each element in the compound.

$$\downarrow$$

Step 2
- Determine the number of grams of each element in the compound by multiplying the moles of the element by its atomic mass.
- Determine the molar mass of the compound by summing the masses of grams of each element.

$$\downarrow$$

Step 3
- Divide the grams of each element by the grams of 1 mole of compound (the molar mass of the compound).
- Convert this number to a percent by multiplying by 100%.

$$\downarrow$$

Step 4
- Check the answer by adding all the percents from Step 3 to ensure that the sum is 100% (or a number very close to 100%).

Example 1

What is the mass percent of each element in $C_3H_4O_3$?

Solution

Perform Step 1

One mole of compound contains 3 moles of C, 4 moles of H, and 3 moles of O.

Perform Step 2

Grams of C: $12.011 \times 3 = 36.033$ g C
Grams of H: $1.0079 \times 4 = 4.0316$ g H
Grams of O: $15.999 \times 3 = 47.997$ g O
Molar mass of compound $= \overline{88.062 \text{ g/mol}}$

Perform Step 3

$$\% \text{ C} = \frac{36.033 \text{ g C}}{88.062 \text{ g cmpd}} \times 100\% = 40.918\% \text{ C}$$

$$\% \text{ H} = \frac{4.0316 \text{ g H}}{88.062 \text{ g cmpd}} \times 100\% = 4.5781\% \text{ H}$$

$$\% \text{ O} = \frac{47.982 \text{ g O}}{88.062 \text{ g cmpd}} \times 100\% = 54.491\% \text{ O}$$

Perform Step 4

Sum of the percents = 40.92% + 4.58% + 54.50% = 100%

Example 2

What is the mass percent of each element in $Al(SO_4)_3$?

Solution

Perform Step 1

One mole of compound contains 1 mole of Al, 3 moles of S, and 12 moles of O.

Perform Step 2

Grams of Al: 26.982 × 1 = 26.982 g Al
Grams of S: 32.066 × 3 = 96.198 g S
Grams of O: 15.999 × 12 = 191.99 g O
  Molar mass of compound = 315.17 g/mol

Perform Step 3

$$\% \text{ Al} = \frac{26.982 \text{ g Al}}{315.17 \text{ g cmpd}} \times 100\% = 8.5611\% \text{ Al}$$

$$\% \text{ S} = \frac{96.198 \text{ g S}}{315.17 \text{ g cmpd}} \times 100\% = 30.523\% \text{ S}$$

$$\% \text{ O} = \frac{191.93 \text{ g O}}{315.17 \text{ g cmpd}} \times 100\% = 60.916\% \text{ O}$$

Perform Step 4

Sum of the percents = 8.5611% + 30.523% + 60.916% = 100%

Practice Problems

9.11 What is the mass percent of each element in SnF_4?

9.12 What is the mass percent of each element in $Mg(NO_3)_2$?

Quiz for Chapter 9 Problems

1. In the reaction of Ca with O_2 to produce CaO, how many grams of CaO can be produced from 640.0 g of O_2 according to the unbalanced equation:

$$Ca + O_2 \rightarrow CaO$$

2. If 350.0 g of H_2 are allowed to react with 350.0 g of O_2, how much H_2O will be produced? The unbalanced equation for the reaction is shown below.

$$H_2 + O_2 \rightarrow H_2O$$

3. When 4.50 g of a compound were subjected to combustion analysis, the compound was found to consist of 62.0% C, 10.3% H, and possibly oxygen. What is the empirical formula of the compound?

4. A compound is found to contain 32.13% Al and 67.87% F. Determine the empirical formula of the compound.

5. A compound has an empirical formula of CH_2O. What is the molecular formula of the compound if the molar mass is 180 g/mol.

6. Nitrogen gas can be prepared by a reaction that occurs by the following unbalanced equation $NaN_3 \rightarrow Na + N_2$. How many grams of N_2 can be prepared from the reaction of 1500 g of NaN_3?

7. In the reaction of CH_4 with S to produce CS_2 and H_2S, how many grams of S are needed to exactly react with 450.0 g of CH_4 according to the unbalanced equation:

$$CH_4 + S \rightarrow CS_2 + H_2S$$

8. To determine the molecular formula of a compound, the _____ mass must be divided by the mass of the _____ formula.

9. What is the mass percent of each element in $C_7H_{16}O$?

10. A fellow student tells you that it is easy to determine the limiting reactant in a reaction because it is the reactant that is present in the smallest number of grams. Is she correct? Explain.

11. Calculate the number of grams of $SiCl_4$ needed to exactly react with 500.0 g of water according to the following (unbalanced) equation:

$$SiCl_4 + H_2O \rightarrow SiO_2 + HCl$$

12. How many grams of SiO_2 will be produced by the reaction described in Problem 11?

13. What is the empirical formula of a compound that contains C, H, and, O, in the following percentages: 65.45% C, 5.492% H, 29.06% O.

14. By what number must each subscript in the empirical formula be multiplied to obtain the molecular formula for a compound that has a molar mass of 582 g/mol, and an empirical formula weight of 97 g/mol?

15. If 200.00 g of Cl_2 are allowed to react with 10.00 g of H_2 to produce HCl, which reactant is limiting? The reaction occurs by the following (unbalanced) equation: $H_2 + Cl_2 \rightarrow HCl$

16. How many moles of HCl will be produced in Problem 15? How many grams of HCl will be produced?

Exercises for Self-Testing

A. Completion. *Write the correct word(s) to complete each statement below.*

1. _____ involves the quantitative relationships between the components of a chemical reaction, or simply chemical arithmetic.

2. The number of molecules in 1 mole of any substance is called _____ _____, and its value is _____.

3. The molar mass of $(NH_4)_3PO_4$ is _____.

4. The number of grams of $CaCl_2$ in 7.77 moles is _____.

5. The number of moles of HBr containing 6.58×10^{27} molecules is _____.

6. The maximum amount of product that can be produced from a chemical reaction if all the reactants are converted to products is called the _____ _____.

7. The reactant that runs out first during a chemical reaction and determines the amount of product that will be formed is called the _____ _____.

8. In the reaction of hydrogen gas with chlorine gas to produce hydrogen chloride, _____ moles of hydrogen chloride can be produced from the reaction between 8 moles of hydrogen gas and 12 moles of chlorine gas.

9. The percent yield for a chemical reaction is determined by dividing the _____ _____ by the _____ _____ and multiplying by _____.

10. _____ _____ is the technique that involves burning a compound (reacting it with oxygen) to determine the amount of carbon and hydrogen in the compound.

11. The _____ _____ of a compound is the simplest ratio of the moles of each element in the compound.

12. The percent composition of $CaCl_2$ is _____% calcium and _____ % chlorine.

13. In the balanced equation for the reaction of pentane, C_5H_{12}, with oxygen gas to produce carbon dioxide and water, the coefficient in front of oxygen is _____.

14. The number of grams of carbon in 10.67 g of carbon dioxide is _____.

15. The actual formula of a compound can be determined from the empirical formula and the _____ _____ of the compound.

B. True/False. *Indicate whether each statement below is true or false, and explain why the false statements are false.*

_____ 1. There are 1.20×10^{24} molecules of CO_2 in 88 g of CO_2.

_____ 2. CH_2 can be the actual formula of a known compound.

_____ 3. When water is produced from the reaction of hydrogen and oxygen, a maximum of 36 g of water can be produced from 32 g of oxygen. (Assume that there is enough hydrogen to react with all the oxygen.)

_____ 4. The number of molecules contained in 1 mole of a substance depends on the identity of the substance.

_____ 5. The theoretical yield of a chemical reaction is generally greater than the actual yield produced by the reaction.

_____ 6. The limiting reactant is the reactant that is present in the least number of grams in the reaction mixture.

_____ 7. There are more molecules in 10 g of O_2 than in 2 g of H_2.

_____ 8. All the carbon is converted to CO_2 in combustion analysis.

_____ 9. Combustion analysis data is sufficient to determine the actual formula of a compound.

_____ 10. The molar mass of a compound is always greater than the mass of the empirical formula.

_____ 11. According to the following (unbalanced) reaction: $H_2 + N_2 \rightarrow NH_3$, it is possible to produce 12.2 g of NH_3 from 1.88 g of H_2.

_____ 12. It is possible to have yields greater than 100% if you are an experienced scientist.

C. Questions and Problems

1. Calculate the number of grams of CH_4 in each of the following:
   (a) 5.78 moles of CH_4 (b) 1.8×10^{25} molecules of CH_4

2. Trichloroamine, NCl_3, an antibacterial agent, is produced according to the following reaction: $Cl_2 + NH_3 \rightarrow NCl_3 + HCl$.
   (a) Write the balanced chemical equation.
   (b) How many grams of NCl_3 can be made from the reaction of 426 g of Cl_2 and 51 g of NH_3?
   (c) How many grams of the excess reactant will be left over after the reaction is complete?

3. Upon chemical analysis, a compound was found to contain 42.4% nitrogen, 9.09% hydrogen, and 48.5% oxygen.
   (a) What is the empirical formula of the compound?
   (b) If the molar mass of the compound is 33 g/mol, what is the actual formula of the compound?

4. 1.65 g of an unknown compound containing C, H, and possibly O produces 3.96 g of carbon dioxide and 0.81 g of water when subjected to a combustion analysis.
   (a) Is there any oxygen in the compound? If so, how many grams?
   (b) What is the empirical formula of the compound?

5. Indicate the number of molecules in each of the samples below:
   (a) 300 g of H_2 (b) 300 g of N_2 (c) 300 g of I_2

6. Indicate the number of molecules in each of the samples below:
   (a) 3.0 moles of H_2 (b) 3.0 moles of N_2 (c) 3.0 moles of I_2

7. Calculate the percent yield of a reaction whose theoretical yield is 36.34 g but whose actual yield is 20.32 g.

8. Explain why the theoretical yield of a chemical reaction is usually larger than the actual yield.

9. Explain the difference between the empirical formula and the actual formula of a compound. Are they always different?

10. Calculate the percent composition of chloropentane, $C_5H_{11}Cl$.

11. Which of the following formulas cannot be empirical formulas?
    (a) H_3BO_3 (b) C_6H_{12} (c) $C_2H_4O_2$

12. What is the empirical formula of benzene, C_6H_6? Are there other compounds with the same empirical formula?

10

Electron Transfer in Chemical Reactions

Learning Outcomes

1. Describe what electricity is.

2. Assign oxidation states to atoms in compounds and molecules.

3. Use oxidation states to decide if electron transfer is occurring during a chemical reaction.

4. Identify the oxidizing agent and reducing agent in an electron transfer reaction.

5. Construct (diagram) a battery and label its parts and direction of electron flow given a balanced redox reaction.

6. Use the EMF series to construct (diagram) a battery and label its parts and direction of electron flow.

7. Describe the corrosion of metals in terms of electron-transfer reactions.

8. Describe the reactivity that might result from the presence of a less common oxidation state.

Chapter Outline

Review of Key Concepts from the Text

10.1 What Is Electricity?

Electricity is the flow of electrons through a wire. The rate at which the electrons flow (the current) depends partly on the push on the electrons traveling through the wire (voltage). Electricity can be generated either by an electric turbine in a power plant or by chemical reactions that involve a transfer of electrons.

10.2 Electron Bookkeeping—Oxidation States

Chemical reactions that involve a transfer of electrons are called electron-transfer reactions. They are also called oxidation-reduction reactions, or simply redox reactions. Not all chemical reactions are redox reactions, but the vast majority of them are. We can tell whether a reaction involves a transfer of electrons by determining the oxidation state of each of the elements involved in the reaction, both before and after the reaction. (The terms *oxidation state* and *oxidation number* are equivalent and can be used interchangeably.) The oxidation state is the apparent charge on an atom based on a set of predetermined rules. The oxidation state of an atom depends on how many electrons the atom "owns" in the Lewis dot diagram of the compound (or ion) it is in, as compared to the number of valence electrons on a free atom of the element (obtained from its position in the periodic table). For the purposes of assigning oxidation states, an atom in a Lewis dot structure owns all of its lone-pair electrons plus all of the electrons it is sharing with a less electronegative element. This concept is illustrated in the example below. Following this logic, the less electronegative atom in a bond does not own any of the shared electrons. Note that for nonzero oxidation numbers, the sign ($+$ or $-$) comes *before* the number, i.e., -2, while for ionic charges, the sign comes second, i.e., $2-$. For example, the oxidation number of sulfur in S^{2-} anion is -2.

Example 10.1

Determine how many electrons each atom owns in CS_2, NH_3, and HNCS. Also indicate the number of valence electrons on the free atom.

Solution

Molecule	Dot diagram	Valence electrons on free atom	Electrons owned by each atom
CS_2	:S̈::C::S̈:	C: 4 electrons	C: 0 electrons
		S: 6 electrons	S: 8 electrons
NH_3	H:N̈:H Ḧ	N: 5 electrons	N: 8 electrons
		H: 1 electron	H: 0 electrons
HNCS	H—N̈=C=S̈	N: 5 electrons	N: 8 electrons
		C: 4 electrons	C: 0 electrons
		S: 6 electrons	S: 8 electrons
		H: 1 electron	H: 0 electrons

Practice Exercise 10.1

Determine how many electrons each atom owns in NO_3^- and PF_3.

The oxidation state of an atom is equal to the number of valence electrons on a free atom of the element *minus* the number of electrons owned by the atom in the Lewis dot structure (when electrons are assigned to the atom by oxidation state bookkeeping).

Example 10.2

Determine the oxidation state of each element in CBr_3F.

Solution

Molecule	Dot diagram	Valence electrons on free atom	Electrons owned by each atom	Oxidation state
CBr_3F	:B̈r: :B̈r:C:B̈r: :F̈:	C: 4 electrons	C: 0 electrons	$4 - 0 = +4$
		Br: 7 electrons	Br: 8 electrons	$7 - 8 = -1$
		F: 7 electrons	F: 8 electrons	$7 - 8 = -1$

Practice Exercise 10.2

Determine the oxidation state of each element in $PHCl_2$.

Shortcut Method for Assigning Oxidation States

The electron bookkeeping method of determining oxidation states can be very time-consuming because we must first draw a correct Lewis dot structure. Fortunately, there is another method we can use that does not require the use of Lewis dot structures. This shortcut method has advantages and disadvantages, however. The major advantage is that it saves time and effort. The major disadvantage is that it does not produce a correct answer 100% of the time. However, it is accurate for the vast majority of the substances chemists work with and is therefore used almost exclusively to determine oxidation states. The shortcut method is based on a set of rules, listed below. You should memorize these rules thoroughly.

1. The oxidation state of a neutral atom in its most abundant naturally occurring elemental form is 0.
2. The oxidation state of an oxygen atom in a compound is almost always -2. There are two exceptions to this rule. One exception is when oxygen is in a peroxide, such as hydrogen peroxide (H_2O_2). The oxidation state of oxygen in a peroxide is -1. The other exception is when oxygen is bonded to fluorine (which is more electronegative than oxygen). The oxidation state of fluorine is always -1. The oxidation state of oxygen depends on the formula of the compound.
3. The oxidation state of a hydrogen atom in a compound is $+1$, except when hydrogen is combined with a less electronegative metal atom in compounds called hydrides. The oxidation state of hydrogen in these compounds is -1.
4. The oxidation state of any group I atom (alkali metal) is always $+1$.
5. The oxidation state of any group II atom (alkaline earth metal) is always $+2$.
6. The oxidation state of a monatomic ion is equal to the charge on the ion.
7. The sum of the oxidation states for every atom in a compound (or ion) must be equal to the overall charge of the compound (or ion). This rule is crucial when assigning oxidation states to atoms for which there is no other rule. (See Section 9.2 of the textbook for a complete discussion of this.) This rule is also crucial when checking to make sure that correct oxidation states have been assigned to all of the atoms in a compound.

The text also mentions a final shortcut rule called the halide rule:

The oxidation state of a halogen atom is most likely -1. Exceptions occur when a halogen *other than fluorine* is bonded to oxygen (which is more electronegative than the halogen) and when a less electronegative halogen is bonded to a more electronegative halogen (such as Br–Cl). Of course, if a halogen occurs in a diatomic molecule— the most abundant form of occurrence of halogens, e.g., Cl_2—rule 1 kicks in and the oxidation number of both atoms is 0.

Note: When assigning oxidation states using the shortcut rules, it is useful to have two rows of numbers above each element. The bottom row of numbers will represent the oxidation state of the atom. The top row will represent the total charge contributed by the element, taking into account both the oxidation number and the number of atoms of the element. This method is demonstrated in the next examples.

Example 10.3

Assign oxidation states to the atoms in the following compounds or ions:

(a) In_2O_3 (b) $C_2H_4O_2$ (c) PO_4^{3-} (d) $K_2Cr_2O_7$ (e) Na_2O_2

Solutions

(a) total charge: $+6 - 6 = 0$

oxidation number: $+3 - 2$

$$In_2O_3$$

There is no rule for the assignment of In, so the oxidation number is assigned first for O (oxidation state $= -2$). Because there are three O atoms the total charge from the oxygen is -6. In order for the total charge on the compound to be 0, the total charge from the In must be $+6$. Because there are two In atoms, each one must have a $+3$ oxidation state.

(b) total charge: $0 + 4 - 4 = 0$

oxidation number: $0 + 1 - 2$

$$C_2H_4O_2$$

There is no rule for the assignment of C, so the oxidation numbers are assigned first for O (oxidation state $= -2$) and H (oxidation state $= +1$). Because there are two O atoms, the total charge from the oxygen is -4. The total charge from the four H atoms is $+4$. In order for the total charge on the compound to be 0, the total charge from the C must be 0, because the sum of the charges from the O and the H already equals 0. (Carbon is unusual because it can have an oxidation number ranging from -4 to $+4$, depending on the compound in which it appears.)

(c) total charge: $+5 - 8 = 3-$

oxidation number: $+5 - 2$

$$PO_4{}^{3-}$$

There is no rule for the assignment of P, so the oxidation number is assigned first for O (oxidation state $= -2$). Because there are four Os, the total charge from the oxygen is -8. In order for the total charge on the ion to be 3^- (from the formula), the total charge from the P must be $+5$. Because there is only one P atom, it must have a $+5$ oxidation state.

(d) total charge: $+2 + 12 - 14 = 0$

oxidation number: $+1 + 6 - 2$

$$Na_2\,Cr_2\,O_7$$

There is no rule for the assignment of Cr, so the oxidation numbers are assigned first for O (oxidation state $= -2$) and Na (oxidation state $= +1$). Because there are seven Os, the total charge from the oxygen is -14. The total charge from the two Na atoms is $+2$. In order for the total charge on the compound to be 0, the total charge from the Cr must be $+12$. Because there are two Cr atoms, each has an oxidation state of $+6$.

(e) total charge: $+2 - 2 = 0$

oxidation number: $+1 - 1$

$$Na_2O_2 \text{ (Na}_2\text{O}_2 \text{ is a peroxide)}$$

Because Na_2O_2 is a peroxide, the oxidation state of oxygen is -1. Because there are two Os, the total charge from the oxygen is -2. In order for the total charge in the compound to be 0, the total charge for the Na must be $+2$. Because there are two Na atoms, each has an

---

oxidation state of +1, which is consistent with rule 4. (Notice that we would have gotten the same result if we had begun with rule 4, which indicates that the oxidation state of all alkali metals is +1.)

Practice Exercise 10.3

Assign oxidation states to the atoms in the following compounds or ions:

(a) C_2H_6 (b) NO_3^- (c) $KMnO_4$ (d) H_2O_2 (e) Ga_2O_3

10.3 Recognizing Electron-Transfer Reactions

Once we assign oxidation states to the atoms in a chemical reaction, we can determine whether an electron-transfer reaction has occurred. Because oxidation states represent the apparent charge on an atom, a change in oxidation states indicates that a redox reaction has occurred.

Example 10.4

Indicate whether each of the following reactions is a redox reaction:

(a) $HBr + KOH \rightarrow KBr + H_2O$ (b) $2H_2 + O_2 \rightarrow 2H_2O$

Solutions

We must assign oxidation states to determine whether any element has undergone a change in state from reactants to products. For reaction (a),

total charge:	$+1 - 1 = 0$	$+1 - 2 + 1 = 0$	$+1 - 1 = 0$	$+2 - 2 = 0$
oxidation number:	$+1 - 1$	$+1 - 2 + 1$	$+1 - 1$	$+1 - 2$

$$HBr \quad + \quad KOH \quad \rightarrow \quad KBr \quad + \quad H_2O$$

Because none of the atoms in this reaction has undergone a change in oxidation number, electron transfer has not occurred. Therefore, this reaction is *not* an electron-transfer (redox) reaction. For reaction (b),

total charge:	0	0	$+2 - 2 = 0$
oxidation number:	0	0	$+1 - 2$

$$2H_2 \quad + \quad O_2 \quad \rightarrow \quad 2H_2O$$

(Notice that the coefficients in the balanced equation have no effect on the oxidation states. You can ignore them when assigning oxidation states.) Because the oxidation numbers of hydrogen and oxygen *have* changed in going from reactants to products, this reaction *is* an electron-transfer (redox) reaction.

Practice Exercise 10.4

Indicate whether each of the following reactions is a redox reaction:

(a) $2HCl + Br_2 \rightarrow 2HBr + Cl_2$ (b) $2Fe_2O_3 \rightarrow 4Fe + 3O_2$

STUDENT WORKBOOK AND SELECTED SOLUTIONS

In all redox reactions, electrons have been transferred *from* one element *to* another. The element that has lost electrons has been oxidized, and the element that has gained electrons has been reduced. Because electrons have a negative charge, this means that the oxidation number of the element that has lost electrons will become larger (as in more positive or less negative), and the oxidation number of the element that has gained electrons will become smaller (as in more negative or less positive). This should make sense intuitively. When something is reduced, it gets smaller; when an element gets reduced, its oxidation number gets smaller. In reaction (b) in Example 10.4, the hydrogen has been oxidized and the oxygen has been reduced.

As stated above, oxidation is the loss of electrons and reduction is the gain of electrons. A phrase that may help you to remember this distinction is OIL RIG, which stands for "Oxidation Is Loss, Reduction Is Gain." Another mnemonic is "LEO the lion goes GER." LEO stands for "a Loss of Electrons is Oxidation" and GER stands for "a Gain of Electrons is Reduction." If you can remember OIL RIG, or that LEO goes GER, you should have no trouble recalling the definitions of oxidation and reduction.

Example 10.5

Identify the element oxidized and the element reduced in the following redox reaction:
$2K + 2H_2O \rightarrow 2KOH + H_2$.

Solution

To determine the element being oxidized and the element being reduced, we must first assign the oxidation numbers.

total charge:	0	$+2 - 2 = 0$	$+1 - 2 + 1 = 0$	0
oxidation number:	0	$+1 - 2$	$+1 - 2 + 1$	0
	$2K$ +	$2H_2O \rightarrow$	$2KOH$ +	H_2

The element oxidized will be the one that has lost electrons and has a higher oxidation state as a product than as a reactant. This is the case for the K. The H has decreased in oxidation number (from $+1$ to 0) and is therefore the element that is reduced.

Practice Exercise 10.5

Identify the element oxidized and the element reduced in the following redox reaction:
$Si + 2Cl_2 \rightarrow SiCl_4$.

In all redox reactions, oxidation and reduction must be occurring simultaneously. Electrons are transferred *from* the element that is oxidized *to* the element that is reduced. The substance containing the element that is oxidized (loses electrons) makes it possible for the other substance to be reduced (gain electrons). In addition, the substance that is reduced (gains electrons) makes it possible for the other substance to be oxidized (lose electrons). For this reason, the substance containing the element that is oxidized is called the *reducing agent*, and the substance containing the element that is reduced is called the *oxidizing agent*. This relationship can be seen in the following example.

Example 10.6

In the reaction, $2NaCl + Br_2 \rightarrow 2NaBr + Cl_2$, identify what is oxidized, what is reduced, the oxidizing agent, and the reducing agent.

Solution

We must first assign the oxidation numbers.

total charge:	$+1 - 1 = 0$	0	$+1 - 1 = 0$	0
oxidation number:	$+1 - 1$	0	$+1 - 1$	0

$$2NaCl \quad + \quad Br_2 \rightarrow 2NaBr \quad + \quad Cl_2$$

We can see that chlorine has been oxidized (its oxidation number has *increased*, from -1 to 0) and that bromine has been reduced (its oxidation number has *decreased*, from 0 to -1). Therefore, the oxidizing agent is Br_2 (substance containing the atom that is reduced), and the reducing agent is NaCl (substance containing the atom that is oxidized). Note that when we identify what has been oxidized and what has been reduced, we specify the element, and when we identify the oxidizing and reducing *agents*, we specify the entire compound.

Practice Exercise 10.6

In the reaction of sodium and chlorine to produce sodium chloride, identify what is oxidized, what is reduced, the oxidizing agent, and the reducing agent.

10.4 Electricity from Redox Reactions

We indicated at the beginning of the chapter that electricity is the flow of electrons, and we have seen that redox reactions involve a transfer of electrons from one substance to another. The electron transfer can occur directly as the two substances touch each other, or the electrons can be made to flow through a wire if the two substances are not in direct contact. This arrangement is shown in the figure below:

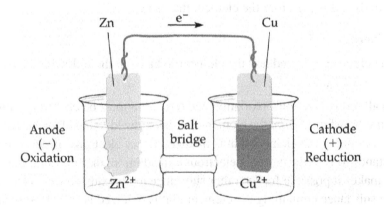

The metal strips are called electrodes, and the apparatus is called a galvanic cell, or battery. The reaction occurring in the galvanic cell above is one in which the redox reaction (and the electron flow) occurs spontaneously. The salt bridge shown is necessary to maintain the electrical neutrality of the cell. The salt bridge contains positive and negative ions (any ions that do not participate in the redox reaction are suitable) that migrate to keep charges from building up in the separate compartments of the cell. In the absence of the salt bridge, the cell could not remain neutral, and the spontaneous flow of electrons would stop. (See Section 10.4 of the textbook for a complete discussion of the components of a galvanic cell.)

The electrons flow through the wire from the substance that is losing electrons (being oxidized) to the substance that is gaining electrons (being reduced). The electrode at which oxidation occurs is the *anode*, and the electrode at which reduction occurs is the *cathode*. (You will find it easy to remember which reaction occurs at which electrode if you remember that the two vowel terms go together and the two consonant terms go together—*o*xidation at the *a*node, *r*eduction at the *c*athode).

The electrons in a galvanic cell always travel from the anode to the cathode. (You can think of them as traveling in alphabetical order—from *a* to *c*.) We also assign charges to the electrodes. The anode is the negative electrode, and the cathode is the positive electrode. (This is easy to remember when you recall that anions are negative and cations are positive.)

In a galvanic cell, the electrons flowing through the wire produce usable electricity. The "pressure" pushing the electrons through the wire is the *voltage*. The voltage produced by a specific galvanic cell reaction is determined by which substances are being oxidized and reduced in the reaction.

There are many types of batteries that can be produced from redox reactions. One of the more familiar ones is the dry-cell battery, described in Section 10.4 of the textbook.

You should practice drawing the diagram for a galvanic cell, including the metal electrodes (anode and cathode), the ions in the beakers, the salt bridge, and the wire (indicating the direction of electron flow through it). You should study the figure thoroughly.

Example 10.7

Construct a galvanic cell (battery) from the following redox reaction: $Zn + 2Ag^+ \rightarrow Zn^{2+} + 2Ag$. Label the cathode, the anode, and the salt bridge, and indicate the direction of electron flow. Also indicate which electrode is positive and which is negative.

Solution

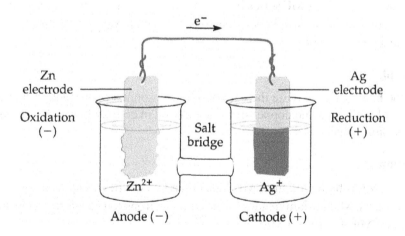

Practice Exercise 10.7

Construct a galvanic cell (battery) from the following redox reaction: $Cu^{2+} + Mg \rightarrow Mg^{2+} + Cu$. Label the cathode, the anode, and the salt bridge, and indicate the direction of electron flow. Also indicate which electrode is positive and which is negative.

10.5 Which Way Do Electrons Flow? The EMF Series

In the galvanic cells we have discussed so far, we knew what was being oxidized and what was being reduced because we had the redox equation presented to us. For example, in Example 10.7 we were given the reaction $Zn + 2Ag^+ \rightarrow Zn^{2+} + 2Ag$. We can tell from the reaction that Zn loses electrons to Ag^+. The Zn is oxidized (loss of electrons is oxidation), and the Ag^+ is reduced (gain of electrons is reduction). The reaction occurs spontaneously in the direction written (electrons transferred from Zn to Ag^+) but would not occur spontaneously in the opposite direction (electrons transferred from Ag to Zn^{2+}).

What determines the direction of electron flow in redox reactions? It is the tendency of the metal to lose electrons. Some metals lose valence electrons more easily than others, and will tend to become oxidized (form positive ions) when reacting with the positive ions of the less active metal. (Recall that the alkali metals are very reactive and have a strong tendency to lose electrons.) In the example of Zn and Ag^+, we can see that Zn has a greater tendency to lose electrons than Ag because Zn ends up as the positive ion in the products. If Ag had a greater tendency to lose electrons than Zn, it would have ended up as the positive ion in the products, and there would be no reaction. A metal that easily loses valence electrons is called an *active metal*. The greater the tendency to lose electrons, the more active the metal. Because zinc can lose electrons more easily than silver, we say that zinc is a more active metal than silver. In other words, in the battle to lose electrons (become positively charged), zinc can beat silver and end up as the positively charged ion.

To determine the order of activity of metals, we could do a series of "electron losing" battles to see exactly where metals fall in their ability to lose electrons. When this is done (by carrying out a series of chemical reactions), we obtain the electromotive force (EMF) series, which is the relative ordering of metal activities. The EMF series is shown here. We can use this series to predict which redox reactions will be spontaneous and which way the electrons will flow in a galvanic cell.

EMF series

Most active
(loses electrons
most easily)

Li
K
Ba
Sr
Ca
Na
Mg
Al
Mn
Zn
Cr
Fe
Cd
Co   Activity
Ni
Sn
Pb
H
Sb
As
Bi
Cu
Ag
Pd
Hg
Pt
Au

Least active
(loses electrons
least easily)

Example 10.8

Use the EMF series to write the equation for the spontaneous reaction that will occur when a beaker of Cu^{2+} ions and copper metal are placed in contact (through a wire) with Mg^{2+} ions and magnesium metal.

Solution

Because Mg is higher on the EMF series than Cu, it will end up as the positively charged ion in the products. Magnesium metal will give up electrons to Cu^{2+} ions, according to the following equation: $Mg + Cu^{2+} \rightarrow Mg^{2+} + Cu$.

Practice Exercise 10.8

Use the EMF series to write the equation for the spontaneous reaction that will occur when a beaker of Cd^{2+} ions and cadmium metal are placed in contact (through a wire) with Zn^{2+} ions and zinc metal.

10.6 Another Look at Oxidation: The Corrosion of Metals

The corrosion of metals is simply the reaction of metals with oxygen to form an oxide of the metal. For example, rust is formed when iron metal (Fe) reacts with oxygen gas (O_2), in the presence of water, to form iron oxide (Fe_2O_3), or rust. Copper also has a tendency to form an oxide, but less so than iron. (Iron is a more active metal than copper.) A complete discussion of the mechanism of rust formation is presented in Section 10.6 of the textbook.

There are numerous ways to protect against corrosion. Painting or applying sealant to the metal so the metal surface does not come in contact with oxygen is one way. Another way is to place the metal needing protection from corrosion in contact with a more active metal. When the less active metal begins to corrode, the more active metal simply gives up its electrons to the less active metal, reducing the less active metal back to its elemental form. The more active metal is called the sacrificial metal, because it becomes oxidized (corrodes) instead of the less active metal. In other words, the more active metal is sacrificed (becomes oxidized) so that the less active metal can remain in its elemental form.

Example 10.9

Indicate which metal is the sacrificial metal when the following pairs of elements are placed in contact:

(a) Fe and Zn   (b) Cr and Mg   (c) Ag and Au

Solutions

The more active metal in each pair will be the sacrificial metal. We use the EMF series to identify the more active metal.

(a) Zn   (b) Mg   (c) Ag

Practice Exercise 10.9

Indicate which metal is the sacrificial metal when the following pairs of elements are placed in contact:

(a) Pb and Cu   (b) Co and Zn   (c) Pt and Cu

Strategies for Working Problems in Chapter 10

Overview of Problem Types: What You Should Be Able to Do

Chapter 10 provides a discussion of what happens to electrons during chemical reactions and a method for determining whether a reaction will occur spontaneously in the direction written or in the opposite direction. The types of problems that you should be able to solve after mastering Chapter 10 are:

1. Assign oxidation states by the electron-bookkeeping method.
2. Assign oxidation states by the shortcut method.
3. Determine whether a reaction is an oxidation-reduction (redox) reaction or not.
4. Use the electromotive force series (EMF series) to write the spontaneous redox reaction that will occur between two metals and the cations of the metals.

Flowcharts

Flowchart 10.1   **Assigning oxidation states by the electron bookkeeping method**

Flowchart 10.1

Summary
Method for assigning oxidation states by the electron-bookkeeping method

Step 1
- **Draw the correct electron dot diagram for the molecule.**

↓

Step 2
- **Assign every valence electron to an atom.**
- **Assign each lone pair to the atom on which it is drawn in the dot diagram.**
- **Assign all electrons in each bond to the atom with the higher electronegativity.**
- **If bonding electrons are shared by atoms of equal electronegativity, divide the electrons evenly between the two atoms.**

↓

Step 3
- **Determine the oxidation state of each atom by subtracting the number of electrons assigned to the atom (as determined in Step 2) from the number of valence electrons in a free atom of that element (as determined from its group in the periodic table).**

↓

Step 4
- **Check your answer.**
- **The sum of the oxidation states for all the atoms should equal the charge on the molecule or ion. (The charge on a molecule is always 0 because molecules are neutral. The charge on a polyatomic ion is the charge written in the formula of the ion.)**

STUDENT WORKBOOK AND SELECTED SOLUTIONS

Example 1

Use the electron bookkeeping method to determine the oxidation state of each atom in HCl.

Solution

Perform Step 1

The correct electron dot diagram for HCl is

H:Cl:

Perform Step 2

Cl will be assigned six electrons from its lone pairs.
Because Cl is more electronegative than hydrogen, chlorine will be assigned the two electrons in the H—Cl bond.

Perform Step 3

The oxidation state of each atom = number of valence electrons in free atom–number of electrons assigned to the atom from Step 2.

Oxidation state of H $= 1 - 0 = +1$
Oxidation state of Cl $= 7 - 8 = -1$

Perform Step 4

Sum of oxidation states =
H: $+1$
Cl: -1

$+1 + (-1) = 0$, which is the charge on the HCl molecule.

Example 2

Use the electron bookkeeping method to determine the oxidation state of each atom in SO_4^{2-}.

Solution

Perform Step 1

The correct electron dot diagram for SO_4^{2-} is:

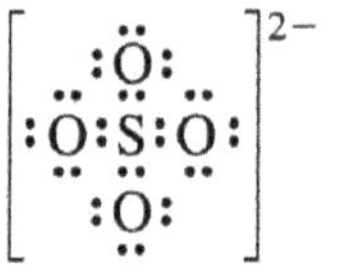

$$\left[\begin{array}{c} :\ddot{O}: \\ :\ddot{O}:\ddot{S}:\ddot{O}: \\ :\ddot{O}: \end{array} \right]^{2-}$$

Perform Step 2

Each O will be assigned six electrons from the lone pairs.
Because oxygen is more electronegative than sulfur, the oxygen atom will be assigned the two electrons in each S—O bond. Therefore, each oxygen atom is assigned eight electrons, and the sulfur atom is assigned no electrons.

Perform Step 3

The oxidation state of each atom = number of valence electrons in free atom − number of electrons assigned to the atom from Step 2.

Oxidation state of each O $= 6 - 8 = -2$
Oxidation state of S $= 6 - 0 = +6$

Perform Step 4

Sum of oxidation states =
O: -2×4 (for each of four O atoms) $= -8$
S: $+6 \times 1 = +6$

Sum $= -8 + 6 = -2$, which is the charge on the SO_4^{2-} polyatomic ion.

Example 3

Use the electron bookkeeping method to determine the oxidation state of each atom in $COCl_2$. The carbon is the central atom; the other three atoms are attached to carbon.

Solution

Perform Step 1

The correct electron dot diagram for $COCl_2$ is:

$$:\overset{\displaystyle ..}{\underset{}{O}}:$$
$$:\overset{..}{\underset{..}{Cl}}:\overset{..}{\underset{}{C}}:\overset{..}{\underset{..}{Cl}}:$$

Perform Step 2

Each Cl will be assigned six electrons from the lone pairs.
Because chlorine is more electronegative than carbon, each chlorine atom will be assigned the two electrons from the C—Cl bond. Therefore, each chlorine atom is assigned eight electrons.

The C is assigned zero electrons because it has no lone pairs and will be assigned no electrons from the bonds because carbon is less electronegative than chlorine and oxygen.

The O atom will be assigned the four electrons from the two lone pairs.
Because oxygen is more electronegative than carbon, oxygen atom will be assigned the four (double bond) electrons in the C=O bond as well. Therefore, the oxygen atom is assigned a total of eight electrons.

Perform Step 3

The oxidation state of each atom = number of valence electrons in free atom − number of electrons assigned to the atom from Step 2.

Oxidation state of each Cl $= 7 - 8 = -1$
Oxidation state of C $= 4 - 0 = +4$
Oxidation state of O $= 6 - 8 = -2$

Perform Step 4

Sum of oxidation states =
Cl: -1×2 (for each of two Cl atoms) $= -2$
C: $+ 4 \times 1 = +4$
O: $-2 \times 1 = -2$

Sum $= -2 + 4 - 2 = 0$, which is the charge on the $COCl_2$ molecule.

Example 4

Use the electron bookkeeping method to determine the oxidation state of each atom in HCN. The carbon is the central atom.

Solution

Perform Step 1

The correct electron dot diagram for HCN is

H:C:::N:

Perform Step 2

H will be assigned zero electrons from the bonding electrons (H is less electronegative than C). C will be assigned two electrons from the H—C bond, and zero electrons from the C≡N bond (C is less electronegative than N). Therefore, the carbon is assigned a total of two electrons.

The N will be assigned two electrons from the lone pairs.
Because nitrogen is more electronegative than carbon, nitrogen will be assigned the six electrons from the C≡N bond. Therefore, nitrogen is assigned a total of eight electrons.

Perform Step 3

The oxidation state of each atom = number of valence electrons in free atom − number of electrons assigned to the atom from Step 2.

Oxidation state of H $= 1 - 0 = +1$
Oxidation state of C $= 4 - 2 = +2$
Oxidation state of N $= 5 - 8 = -3$

Perform Step 4
Sum of oxidation states =
H: $+1$
C: $+2 \times 1 = +2$
N: $-3 \times 1 = -3$

Sum $= +1 + 2 + (-3) = 0$, which is the charge on the HCN molecule.

Practice Problems

10.1 Use the electron bookkeeping method to determine the oxidation state of each atom in CH_3F.

10.2 Use the electron bookkeeping method to determine the oxidation state of each atom in H_2O.

10.3 Use the electron bookkeeping method to determine the oxidation state of each atom in ClO_4^-.

10.4 Use the electron bookkeeping method to determine the oxidation state of each atom in CO.

Flowchart 10.2 Assigning oxidation states by the shortcut method

Flowchart 10.2

Summary

Method for assigning oxidation states by the shortcut method

Step 1
- **Use the rules below for assigning oxidation numbers for each element in the molecule or polyatomic ion:**
  1. **The oxidation state of a neutral atom in its most abundant naturally occurring elemental form is typically zero.**
  2. **The oxidation state of an oxygen atom in a compound is almost always −2.**

3. The oxidation state of a hydrogen atom in a compound is +1.
4. The oxidation state of any group I atom (1) in a compound is +1.
5. The oxidation state of any group II (2) atom in a compound is +2.
6. The oxidation state of a monatomic ion is equal to the charge on the ion.
7. The sum of the oxidation states of the atoms in a chemical formula must add up to the overall charge on the molecule or ion represented by the formula. This rule is used to determine the oxidation number on elements for which where is no rule specified.

Halide rule:
A halogen atom in a compound usually has an oxidation state of −1.

Step 2
• Recheck your answer.
• The sum of the oxidation states for all the atoms should equal the charge on the molecule or ion. (The charge on a molecule is always zero because molecules are neutral. The charge on a polyatomic ion is the charge written in the formula of the ion.)

Important Note: When assigning oxidation states using the shortcut rules, it is very useful to have two rows of numbers above each element. The bottom row of numbers will represent the oxidation state of the atom. The top row will represent the total charge contributed by the element, taking into account both the oxidation number and the number of atoms of the element. This method is presented in the examples below.

Example 1

Use the shortcut method to determine the oxidation state of each atom in HCl.

Solution

Perform Step 1

Assigning oxidation numbers by the rules, we get the following oxidation numbers.
H: +1
Cl: −1

Viewing the oxidation numbers and total charges in two rows, we get:

total charge: $+1 - 1 = 0$
oxidation number: $+1 - 1$
                       HCl

Perform Step 2

Sum of oxidation states =
H: +1
Cl: −1

$+1 + (-1) = 0$, which is the charge on the HCl molecule.

Example 2

Use the shortcut method to determine the oxidation state of each atom in SO_4^{2-}.

Solution

Perform Step 1

Assigning oxidation numbers by the rules, we get the following oxidation numbers.
O: -2
S: must be determined by using rule 7. There is no rule for the assignment of an oxidation number to S, so the oxidation number is assigned first for O (oxidation state $= -2$). Since there are four O atoms, the total charge from the oxygen atoms will be -8. In order for the total charge on the ion to be $2-$ (from the formula), the total charge from the S atom must be $+6$. Since there is only one S atom, it must have a $+6$ oxidation state.

Viewing the oxidation numbers and total charges in two rows, we get:

total charge: $\quad +6 - 8 = -2$
oxidation number: $\quad +6 - 2$
$$SO_4^{2-}$$

Perform Step 2

Sum of oxidation states $=$
S: $+6 \times 1 = +6$
O: $-2 \times 4 = -8$

$+6 + (-8) = -2$, which is the charge on the SO_4^{2-} polyatomic ion.

Example 3

Use the shortcut method to determine the oxidation state of each atom in $COCl_2$.

Solution

Perform Step 1

Assigning oxidation numbers by the rules, we get the following oxidation numbers.
O: -2
Cl: -1
C: must be determined by using rule 7. There is no rule for the assignment of an oxidation number to C, so the oxidation number is assigned first for O (oxidation state $= -2$) and Cl (oxidation state $= -1$). Because there is one O atom, the total charge from the oxygen will be -2. Because there are two Cl atoms, the total charge from the chlorines will be -2. In order for the total charge on the molecule to be zero, the total charge from the C must be $+4$. Since there is only one C atom, it must have a $+4$ oxidation state.

Viewing the oxidation numbers and total charges in two rows, we get:

total charge: $\quad +4 - 2 - 2 = 0$
oxidation number: $\quad +4 - 2 - 1$
$$C \ O \ Cl_2$$

Perform Step 2

Sum of oxidation states $=$
C: $+4 \times 1 = +4$
O: $-2 \times 1 = -2$
Cl: $-1 \times 2 = -2$

$+4 + (-2) + (-2) = 0$, which is the charge on the $COCl_2$ molecule.

Example 4

Use the shortcut method to determine the oxidation state of each atom in $C_4H_8O_2$.

Solution

Perform Step 1

Assigning oxidation numbers by the rules, we get the following oxidation numbers.
H: $+1$
O: -2
C: must be determined by using rule 7. There is no rule for the assignment of an oxidation number to C, so the oxidation number is assigned first for H (oxidation state $= +1$) and O (oxidation state $= -2$). Because there are two O atoms, the total charge from the oxygens will be -4. Because there are eight H atoms, the total charge from the hydrogen atoms will be $+8$. In order for the total charge on the molecule to be 0, the total charge from the C must be -4. Since there are four C atoms, each one must have a -1 oxidation state.

Viewing the oxidation numbers and total charges in two rows, we get:

total charge: $-4 + 8 - 4 = 0$
oxidation number: $-1 + 1 - 2$
              $C_4H_8O_2$

Perform Step 2

Sum of oxidation states $=$
C: $-1 \times 4 = -4$
H: $+1 \times 8 = +8$
O: $-2 \times 2 = -4$

$-4 + 8 + (-4) = 0$, which is the charge on $C_4H_8O_2$ molecule.

Practice Problems

10.5 Use the shortcut method to determine the oxidation state of each atom in CH_3F.

10.6 Use the shortcut method to determine the oxidation state of each atom in H_2S.

10.7 Use the shortcut method to determine the oxidation state of each atom in ClO_4^-.

10.8 Use the shortcut method to determine the oxidation state of each atom in CO.

10.9 Use the shortcut method to determine the oxidation state of each atom in $C_2O_4^{2-}$.

Flowchart 10.3   Determining whether a reaction is an oxidation-reduction (redox) reaction or not

Flowchart 10.3

Summary

Method for determining whether a reaction is an oxidation-reduction (redox) reaction or not

Step 1

- Assign oxidation numbers to each atom on the reactants side and each atom on the products side. (Be careful to correctly assign oxidation numbers. If numbers are incorrect, Step 2 will not yield correct information.)

$\downarrow$

Step 2

- Compare oxidation numbers of each element on the reactants side with the oxidation number of the same element on the products side.
- If the oxidation number of an element on the products side is different from its oxidation number on the reactants side, a redox reaction has occurred.
- If there is no change in oxidation number for any elements, the reaction is not a redox reaction.

Example 1

Indicate whether the following reaction is a redox reaction.

$H_2 + Cl_2 \rightarrow 2HCl$

Perform Step 1

Assigning oxidation numbers we get:

total charge	0	0	$+1 - 1 = 0$
oxidation no:	0	0	$+1 - 1$
	H_2 +	Cl_2 $\rightarrow$	$2HCl$

(Notice that the coefficients in the balanced equation have no effect on the oxidation states. You can ignore them when assigning oxidation sates.)

Perform Step 2

Because there are changes in the oxidation numbers of hydrogen (0 to +1) and chlorine (0 to −1) in going from reactants to products, this reaction is a redox reaction.

Example 2

Indicate whether the following reaction is a redox reaction.

$HCl + LiOH \rightarrow LiCl + H_2O$

Perform Step 1

Assigning oxidation numbers we get:

total charge:	$+1 - 1 = 0$	$+1 - 2 + 1 = 0$	$+1 - 1 = 0$	$+2 - 2 = 0$
oxidation no:	$+1 - 1$	$+1 - 2 + 1$	$+1 - 1$	$+1 - 2$
	HCl +	LiOH $\rightarrow$	LiCl +	H_2O

Perform Step 2

Because none of the atoms in this reaction has undergone a change in oxidation number in going from reactants to products, a redox reaction has not occurred.

Practice Problems

10.10 Indicate whether the following reaction is a redox reaction.

$$CaCl_2 + Na_2O \rightarrow 2NaCl + CaO$$

10.11 Indicate whether the following reaction is a redox reaction.

$$2Al_2O_3 \rightarrow 4Al + 3O_2$$

Flowchart 10.4   Use the electromotive force series (EMF series) to write the spontaneous redox reaction that will occur between two metals and their cations.

Flowchart 10.4

Summary

Method for using the electromotive force series (EMF series) to write the spontaneous redox reaction that will occur between two metals and the cations of the metals

Step 1
- **Use the EMF series in Section 10.5 of the textbook to find out which metal is more active (higher) in the series.**

$\downarrow$

Step 2
- **Write the spontaneous redox reaction by writing the more active metal on the reactants side and the less active metal on the products side.**
- **The cation of the less active metal will be on the reactants side and the cation of the more active metal will be on the products side.**

Example 1

Write the spontaneous redox reaction that will occur when a beaker of Cd^{2+} ions and Cd metal are placed in contact (through a wire + a salt bridge) with Mg^{2+} ions and Mg metal.

Solution

Perform Step 1

Magnesium is the more active metal, so Mg will be a reactant and Cd will be a product in the spontaneous reaction.

Perform Step 2

The spontaneous reaction will be:

$$Mg + Cd^{2+} \rightarrow Mg^{2+} + Cd$$

Example 2

Write the spontaneous redox reaction that will occur when a beaker of Ni^{2+} ions and Ni metal are placed in contact (through a wire + a salt bridge) with Pb^{2+} ions and Pb metal.

Solution

Perform Step 1

Nickel is the more active metal, so Ni will be a reactant and Pb will be a product in the spontaneous reaction.

Perform Step 2

The spontaneous reaction will be: $Ni + Pb^{2+} \rightarrow Ni^{2+} + Pb$

Practice Problems

10.12 Write the spontaneous redox reaction that will occur when a beaker of Co^{2+} ions and Co metal are placed in contact (through a wire + a salt bridge) with Cu^{2+} ions and Cu metal.

10.13 Write the spontaneous redox reaction that will occur when a beaker of Zn^{2+} ions and Zn metal are placed in contact (through a wire + a salt bridge) with Sn^{2+} ions and Sn metal.

Quiz for Chapter 10 Problems

1. Use the electron bookkeeping method to assign oxidation numbers for each atom in the following species.
   (a) HI   (b) CH_2Cl_2 (c) SO_3 (d) C_2H_6

2. Use the shortcut method to assign oxidation numbers for each atom in the following species.
   (a) HNO_3 (b) $S_2O_3^{2-}$ (c) $CaSO_4$

3. Indicate whether each of the following reactions is a redox reaction.
   (a) $B_2O_3 + 3Mg \rightarrow 2B + 3MgO$
   (b) $HF + KOH \rightarrow KF + H_2O$
   (c) $Zn + 2HCl \rightarrow ZnCl_2 + H_2$

4. Indicate whether each of the reactions below is spontaneous in the direction written or in the opposite direction.
   (a) $Zn + Cu^{2+} \rightarrow Cu + Zn^{2+}$ (b) $Li + K^+ \rightarrow K + Li^+$
   (c) $Fe + Sn^{2+} \rightarrow Sn + Fe^{2+}$ (d) $Ni + Fe^{2+} \rightarrow Fe + Ni^{2+}$

5. Write the spontaneous reaction that will occur when a beaker of Ni^{2+} ions and Ni metal are placed in contact (through a wire) with Mn^{2+} ions and Mn metal.

6. Complete the following sentence: Metals higher in the electromotive series tend to lose electrons _____ easily than metals lower in the EMF series.

7. Indicate the oxidation number of carbon in each of the compounds below.
   (a) CO_2 (b) C_2H_6 (c) C_2H_4 (d) C_2H_2

8. Indicate the change in oxidation number for each element in the following reaction. Which element(s) undergo no change in oxidation state?

   $Mg + H_2SO_4 \rightarrow MgSO_4 + H_2$

9. Some mercury is accidentally dropped onto a ring that is known to be either gold or silver. If there is no reaction when the mercury is in contact with the ring, which metal is the ring made of? Explain.

10. A strip of copper metal is placed in a solution containing nickel ions. Will a reaction occur or not? Why or why not?

11. Show that the electron bookkeeping method and the shortcut method yield the same result when assigning oxidation numbers to each element in the polyatomic ion PO_4^{3-}.

12. Which of the following metals could serve as a sacrificial anode to protect an iron pipe?
(a) Mg   (b) Zn   (c) Cu   (d) Cr

Exercises for Self-Testing

A. Completion. *Write the correct word(s) to complete each statement below.*

1. In the oxidation state method of electron bookkeeping, electrons in a bond are completely owned by the more _____ atom.
2. For the representative elements, the number of valence electrons on a free atom is equal to the _____ _____ of the element on the periodic table.
3. The oxidation state of an atom in its most abundant elemental state is always _____.
4. The halide rule, which states that a halogen has a -1 oxidation number, is not true when the Br, Cl, or I is bonded to _____ or _____.
5. A loss of electrons is _____; a gain of electrons is _____.
6. In a galvanic cell constructed with magnesium (Mg) and silver (Ag) electrodes, the _____ electrode will get smaller as the redox reaction proceeds, and the _____ electrode will get larger.
7. In a battery, the _____ _____ is necessary to maintain the electrical neutrality of the cell.
8. In a redox reaction, electrons are _____ (lost, gained) by the oxidizing agent and _____ (lost, gained) by the reducing agent.
9. In a spontaneous redox reaction, a _____ active metal will always force electrons onto ions of a _____ active metal.
10. The term used to describe the slow oxidation of metals by oxygen is _____.

B. True-False. *Indicate whether each statement below is true or false, and explain why the false statements are false.*

_____ 1.   The reaction $LiNO_3 + KCl \rightarrow LiCl + KNO_3$ is a redox reaction.
_____ 2.   In a galvanic cell in which the two electrodes are silver and nickel, the nickel electrode is the anode.
_____ 3.   The oxidation number of Mn in MnO_4^- is $+5$.
_____ 4.   A sacrificial metal is more easily oxidized than the metal it is protecting.
_____ 5.   In a galvanic cell, the oxidation occurs at the positive electrode.
_____ 6.   The reaction of gold ions (Au^{3+}) with silver (Ag) to produce silver ions (Ag^+) and gold (Au) is nonspontaneous.
_____ 7.   In the electromotive force (EMF) series, a metal will reduce the cations of the metals above it in the series.
_____ 8.   An oxidizing agent gains electrons and a reducing agent loses electrons during the course of a redox reaction.
_____ 9.   Corrosion involves the reaction of a metal with oxygen gas.
_____ 10.   Lewis dot diagrams can be used to assign oxidation states to the elements in a compound.
_____ 11.   When SO_4^{2-} becomes SO_3^{2-} in an electron transfer reaction, S is being reduced.
_____ 12.   In the reaction: $Fe + 2H^+ \rightarrow Fe^{2+} + H_2$, iron is the oxidizing agent.

STUDENT WORKBOOK AND SELECTED SOLUTIONS

C. Questions and Problems

1. Using Lewis dot structures, assign oxidation states to all of the atoms in the following molecules or ions:
   (a) NH_4^+ (b) O_3 (c) NO_2^-
   (d) $COCl_2$ (the O and two Cl atoms are bonded to the carbon atom)

2. Determine the oxidation number of carbon in each of the following compounds:
   (a) CO_2 (b) C_2H_6 (c) C_2H_4 (d) C_2H_2

3. Iron metal can be produced from its ore by the following reaction:
   $2Fe_2O_3 + 3C \rightarrow 4Fe + 3CO_2$. What is the oxidizing agent and what is the reducing agent in the reaction?

4. Magnesium metal does not occur naturally in the Earth's crust, whereas gold does. Explain this in terms of the relative positions of the two elements on the electromotive force (EMF) series.

5. Draw a galvanic cell that could be made from iron (Fe) and manganese (Mn) electrodes and solutions of the cations of these metals. Label all components, give the signs of the electrodes, and indicate the direction of electron flow.

6. Use the EMF series to write the equation for the reaction that will occur in the galvanic cell constructed from the following metals and solutions of their ions:
   (a) Ca and Co   (b) Ag and Ni   (c) Mn and Mg

7. Explain why copper can be used as a sacrificial metal to protect silver but cannot be used to protect iron.

8. What must occur on the anode in order for the reduction to take place on a cathode in a galvanic cell?

9. Explain the role of a salt bridge in a galvanic cell.

10. A double-replacement acid-base reaction is not a redox reaction. Give an example and explain why this is so.

11

What If There Were No Intermolecular Forces? The Ideal Gas

Learning Outcomes

1. Describe the units of pressure and how pressure is created by a gas.
2. Describe the relationship between any two variables that characterize a gas when the remaining variables are held constant.
3. Perform calculations using the ideal gas law.
4. Calculate the molar mass of a gas using the ideal gas law.
5. Calculate the density of a gas using the ideal gas law.
6. Calculate a result for an "initial condition" to "final condition" gas problem.
7. Calculate a result for a reaction stoichiometry problem involving gases.
8. Calculate the pressure of a gas and the deviation from ideal behavior using the van der Waals equation.

Chapter Outline

B. The ideal gas equation

    1. Statement of the equation: $P = \dfrac{nRT}{V}$

    2. Problems involving the equation

11.3 Getting the Most from the Ideal Gas Law

  A. Determining molar mass from the ideal gas law

  B. Calculating the density of gas from the ideal gas law

  C. Solving initial-condition, final-condition gas problems

  D. Molar volume of a gas

  E. Solving reaction stoichiometry problems

11.4 One More Thing: Deviations from Ideality

  A. Van der Waals equation

Review of Key Concepts from the Text

11.1 Describing the Gas Phase—*P*, *V*, *n*, and *T*

Gases have many properties that are quite different from those of liquids and solids. For example, gases expand to fill their containers and can be compressed. These properties can be explained using a model that describes a gas as a collection of molecules that are spaced far apart and move around rapidly and randomly, occasionally colliding with one another or the walls of the container. Intermolecular forces are essentially nonexistent because the molecules are too far apart and moving too rapidly to attract or repel each other. This model of an *ideal gas* includes the assumption that there are no intermolecular attractive forces at all. Therefore, the molecules in an ideal gas travel in straight lines.

    Gas samples are typically described in terms of four variables: pressure, volume, temperature, and number of moles present. There is a mathematical relationship between these variables, the *ideal gas law*, which we will describe shortly. First, however, we will consider the meaning of each of the variables.

Pressure (P)

The pressure of a gas in a container is due to the molecules of the gas hitting the walls of the container. The atmospheric pressure is due to the weight of the nitrogen and oxygen (and a few other gases) in the atmosphere. Atmospheric pressure is measured by a barometer and is described in units of atmospheres (atm) or millimeters of mercury (mm Hg). Normal atmospheric pressure is defined to be 1 atmosphere, which is equivalent to 760 mm Hg. According to the SI metric units (review Chapter 2), the unit of pressure is Pascal (Pa). Clearly, there are several different units of pressures still in use, and each of them calls for somewhat different units of the remaining variables and that of the ideal gas constant (read on).

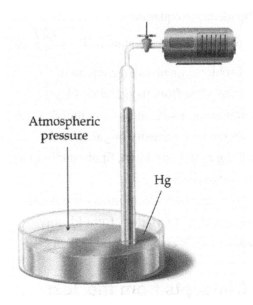

Atmospheric pressure

Hg

Volume (*V*)

The volume of any gas sample is equal to the volume of its container. Gas volumes are most often described in units of liters (L) or milliliters (mL).

Temperature (*T*)

The temperature of a gas sample is a measure of how much kinetic energy the gas contains, which is demonstrated by how fast the molecules are traveling. The higher the temperature, the more kinetic energy in the molecules, and the faster the molecules are moving. Gas temperatures are often described in terms of the Kelvin (or absolute) temperature scale. Temperatures reported in °C or °F are easily converted to K using the relationships defined below.

$$K = °C + 273.15$$

$$°C = \frac{5}{9}(°F - 32)$$

You can practice temperature unit conversions using examples from Chapter 2 in this *Study Guide*.

Number of Moles Present (*n*)

The amount of gas present can be specified in terms of the number of moles. Recall that the mass of 1 mole of any substance is equal to the molecular mass of the substance in grams. The relationships between the four variables are summarized in the next section.

11.2 Describing a Gas Mathematically—The Ideal Gas Law

The state or condition of a gas can be characterized by the values of the quantities defined in the previous section (*P*, *V*, *T*, and *n*). Because these variables are not completely independent of each other, changing one will cause changes in the others. The equation that illustrates the relationship between all of these variables is called the *ideal gas equation*. We will consider the relationship between pairs of variables as we develop the general ideal gas equation that includes all of them.

Pressure and Volume

The pressure of a gas *increases* when the volume it occupies *decreases*. In other words, the pressure of a gas is inversely proportional to its volume ($P \propto 1/V$). This means that when the volume available to a gas is decreased, the gas molecules hit the walls of the container more often, resulting in an increase in pressure.

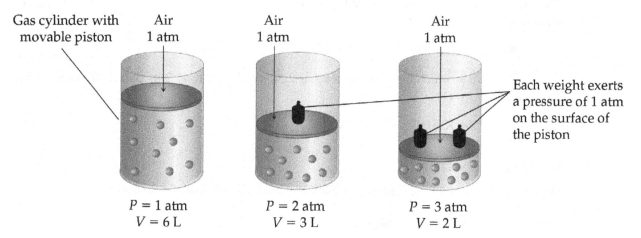

Gas cylinder with movable piston

Air 1 atm Air 1 atm Air 1 atm

Each weight exerts a pressure of 1 atm on the surface of the piston

$P = 1$ atm
$V = 6$ L

$P = 2$ atm
$V = 3$ L

$P = 3$ atm
$V = 2$ L

Example 11.1

A gas sample occupies 0.80 L at a pressure of 2.1 atm. What pressure must be exerted on the gas to decrease the volume to 0.20 L?

Solution

The volume has been decreased by a factor of 4. Therefore, the pressure must be increased by a factor of 4. The new pressure would be 2.1 atm $\times$ 4 = 8.4 atm.

Practice Exercise 11.1

A gas sample occupies 3.0 L at a pressure of 1.5 atm. What volume will the gas occupy if the pressure exerted on the gas is decreased to 0.75 atm?

Pressure and Temperature

The pressure of a gas *increases* when its temperature *increases*. In other words, the pressure of a gas is directly proportional to its temperature ($P \propto T$). This means that when the temperature of a gas is increased, the gas molecules are moving faster and hit the walls of the container harder, resulting in an increase in pressure.

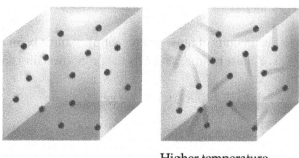

Higher temperature— molecules move faster

Example 11.2

A gas sample at 25 °C exerts a pressure of 1.25 atm. At what temperature (in °C) will the pressure be equal to 2.50 atm?

Solution

The temperature in K must be doubled to increase the pressure from 1.25 to 2.50 atm. We must therefore convert 25 °C to K.

$$K = 25 \,°C + 273.15 = 298 \,K$$

The temperature at which the pressure reaches 2.50 atm is 298 K × 2 = 596 K. We then convert this result back to °C.

$$°C = 596 \,K - 273.15 = 323 \,°C$$

Practice Exercise 11.2

What is the pressure of a gas sample at 50 K if it exerts a pressure of 1.20 atm at 200 K?

Pressure and Number of Moles

The pressure of a gas *increases* when the number of moles of gas *increases*. In other words, the pressure of a gas is directly proportional to the number of moles present ($P \propto n$). This means that when the number of moles of a gas is increased, there are more gas molecules hitting the walls of the container, resulting in an increase in pressure.

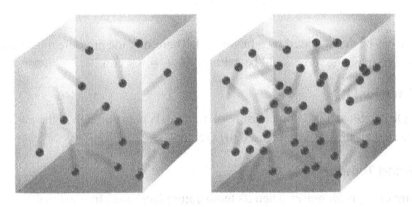

The three relationships presented above can be combined into one summary equation, the **ideal gas equation (or ideal gas law)**:

$$PV = nRT$$

where P must be in units of atmospheres (atm), V must be in units of liters (L), n must be in units of moles, T must be in K, and R is the proportionality constant, called the **ideal gas constant**.

If we solve $PV = nRT$ for R, we get

$$R = \frac{PV}{nT}$$

When we do this, we see that the units on R are

$$\frac{atm \cdot L}{mol \cdot K}, \text{ which is more often written } \frac{L \cdot atm}{K \cdot mol}$$

You should notice that the ideal gas law contains five terms: P, V, n, R, and T. Four of the terms are variables (P, V, n, and T), and one is a constant, R. The value of R is

$$0.0821 \frac{L \cdot atm}{K \cdot mol}.$$

Problems involving the ideal gas law will generally specify three of the four variables, and you will be asked to solve for the fourth one. When using the ideal gas law, you should always make sure that the variables are expressed in the units that are compatible with the units of R, the ideal gas constant. If they are not in the correct units, you must convert them to the correct units before substituting them into the equation.

Example 11.3

How many moles of CH_4 are there in 44.5 L of gas at 22 °C and a pressure of 73.0 mm Hg? How many grams of CH_4 is this?

Solution

We will use the ideal gas law, solving for the number of moles (n). Because the pressure must be in atmospheres and the temperature in K, we first convert the given units to these units.

$$atm = 73.0 \text{ mm Hg} \times \frac{1 \text{ atm}}{760 \text{ mm Hg}}$$

We then solve $PV = nRT$ for n by substituting in the known values.

$$n = \frac{PV}{RT} = \frac{(0.0961 \text{ atm})(44.5 \text{ L})}{(0.0821 \text{ L} \times \text{atm}/\text{K} \times \text{mol})(295 \text{ K})} = 0.177 \text{ mol CH}_4$$

Finally, we convert moles to grams to answer the second part of the problem.

$$0.177 \text{ mol CH}_4 \times \frac{16 \text{ g CH}_4}{1 \text{ mol CH}_4} = 2.83 \text{ g CH}_4$$

There are 0.177 moles of CH_4, which is 2.83 g of CH_4.

Practice Exercise 11.3

What volume will be occupied by 1.00 mole of CO_2 gas at 0 °C and 1.20 atm of pressure?

When using the ideal gas law to solve for the value of variables, we must keep in mind that our answers will not be exactly correct. This is because we have used the ideal gas law to solve problems involving real gases. The difference between an ideal gas and a real gas is that ideal gases experience no intermolecular attractive forces, whereas real gases do. However, real gases do not deviate too much from ideal-gas behavior as long as the temperature is not extremely low and the pressure is not extremely high. Because you will not encounter problems involving extreme temperatures and pressures either in your textbook or in this *Study Guide*, the answers you get using the ideal gas law will be valid.

11.3 Getting the Most from the Ideal Gas Law

The ideal gas law can be used to determine the molar mass, density, and molar volume of a gas when the conditions of the gas are specified. In addition, stoichiometry problems dealing with reactions involving gases can be solved using the ideal gas law. Examples and practice problems involving these types of calculations are presented in Chapter 10. The versatility of the ideal gas law makes it ideal for solving a wide variety of problems. Consult the textbook and this guide for a complete presentation of all of the types of gas problems you will be expected to know how to solve.

11.4 Deviations from Ideality

When conditions under which a gas is contained become extreme, meaning a very low temperature and a very high pressure, the assumption of the lack of intermolecular forces among gas particles cannot be made anymore. This is when we no longer can assume that gases behave ideally. The mathematical description of gas behavior also changes to account for the nonzero volumes of gas particles (assumed by the ideal gas law). The van der Waals equation is the ideal gas equation modified to account for the nonideal behavior.

$$\left[P_{real} + a\left(\frac{n}{V_{real}}\right)^2 \right]\left(\frac{V_{real}}{n} - b\right) = RT$$

The two new constants, **a** and **b** account for the intermolecular forces and nonzero volumes of gas particles, respectively. Since London forces are particularly strong for large molecules, both **a** and **b** are expected to be more substantial for larger molecules. This can be clearly seen in the table in Section 11.4 of your textbook. When the conditions become more "normal," the van der Waals equation reverts to the ideal gas equation.

Example 11.3

Using the van der Waals equation and the table in Section 11.4 of your textbook, estimate the pressure exerted by 1.000 mole of $O_2(g)$ in a 1.000 L container at 273 K. How does this pressure compare with the pressure calculated using the ideal gas equation?

Solution

We will use the van der Waals equation, solving for the pressure in atmospheres (P). The temperature is already in K. The value of **a** is 1.36 atm L^2/mol^2, and the value of **b** is 0.0318 L/mol. While normally it would be best to manipulate an equation to solve for the unknown, the van der Waals equation is complicated enough to first plug in the given information (V, n, T, R, **a**, and **b**, including units!), and then evaluate P.

$$\left[P_{real} + \underbrace{1.36\frac{atm\ L^2}{mol^2}\left(\frac{1.000\ mol}{1.000\ L}\right)^2}_{1.36\ atm} \right]\left(\underbrace{\frac{1.000\ L}{1.000\ mol} - 0.0318\frac{L}{mol}}_{9.68\frac{L}{mol}} \right) = \underbrace{\left(0.0821\frac{L\ atm}{mol\ K}\right)(273\ K)}_{22.4\frac{L\ atm}{mol}}$$

$$[P_{real} + 1.36\text{ atm}]0.968\frac{\text{L}}{\text{mol}} = 22.4\frac{\text{L atm}}{\text{mol}}$$

$$P_{real} + 1.36\text{ atm} = \frac{22.4\dfrac{\text{L atm}}{\text{mol}}}{0.968\dfrac{\text{L}}{\text{mol}}} = 23.1\text{ atm}$$

$$P_{real} = 23.1\text{ atm} - 1.36\text{ atm} = 21.7\text{ atm}$$

We now need to solve for pressure from the ideal gas equation, $PV = nRT$:

$$P = \frac{nRT}{V} = \frac{(1.000\text{ mol})(0.0821\text{ L}\cdot\text{atm/K}\cdot\text{mol})(273\text{ K})}{1.000\text{ L}} = 22.4\text{ atm O}_2$$

The pressure of the "real" O_2 is (22.4 atm − 21.7 atm) = 0.7 atm lower than that of the "ideal" O_2.

Practice Exercise 11.4

Using the van der Waals equation and the table in Section 11.4 of your textbook, estimate the pressure exerted by 1.00 mole of C_2H_4 gas in a 4.00 L container at 45 °C.

Strategies for Working Problems in Chapter 11

Overview of Problem Types: What You Should Be Able to Do

Chapter 11 provides a description of the gaseous state and presents the mathematical description of a gas in the form of the ideal gas equation. The types of problems that you should be able to solve after mastering Chapter 11 are

1. Use the ideal gas equation to find the value of P, V, T, or n when the values of the other three are given.
2. Solve for the molar mass of a gas using the molar mass form of the ideal gas equation.
3. Solve initial-condition, final-condition problems.
4. Solve gas stoichiometry problems using the ideal gas law and a balanced chemical equation.

Flowcharts

Flowchart 11.1 **Using the ideal gas equation to find the value of P, V, T, or n when the values of the other three are given**

Flowchart 11.1

Summary

Method for using the ideal gas equation to find the value of P, V, T, or n when the values of the other three are given
Step 1
• **Algebraically solve the ideal gas equation for the value to be determined.**

$$\downarrow$$

Step 2
• **Make sure that the given values are in the correct units for use in the ideal gas equation (P in atmospheres, n in moles, V in liters, T in K). Convert any values that are not in the correct units.**

Step 3
• **Substitute the given values into the equation obtained in step 1 and perform the arithmetic operation to arrive at the solution.**

Example 1

A gas sample contains 3.20 moles of gas at a temperature of 30.0 °C and a pressure of 832 mm Hg. What volume does the gas occupy?

Solution

Perform Step 1

$$V = \frac{nRT}{P}$$

Perform Step 2

P must be converted to atm and T must be converted to K.

$$atm = 832 \text{ mm Hg} \times \frac{1 \text{ atm}}{760 \text{ mm Hg}} = 1.09 \text{ atm}$$

$$K = 30.0\,°C + 273.15 = 303.2 \text{ K}$$

Perform Step 3

$$n = \frac{(53.7 \text{ atm})(12.0 \text{ L})}{(0.0821 \text{ L} \cdot \text{atm/mol} \cdot \text{K})(301.6 \text{ K})} = 26.0 \text{ mol of gas}$$

Example 2

A container of compressed oxygen gas has a volume of 12.0 L. The pressure of the compressed gas is 53.7 atm, and the temperature of the gas is 28.4 °C. How many moles of gas are present?

Solution

Perform Step 1

$$n = \frac{PV}{RT}$$

Perform Step 2

T must be converted to K.

$$K = 28.4\,°C + 273.15 = 301.6 \text{ K}$$

Perform Step 3

$$n = \frac{(53.7 \text{ atm})(12.0 \text{ L})}{(0.0821 \text{ L} \cdot \text{atm/mol} \cdot \text{K})(301.6 \text{ K})} = 26.0 \text{ moles of gas}$$

Practice Problems

11.1 What is the pressure of 0.823 moles of N_2 gas in a 5.00 L container at 37.0 °C?

11.2 What is the temperature in K and °C of 4.0 moles of a gas in a 3.50 L steel container at a pressure of 98.2 atm?

11.3. Calculate the moles of gas in a 12.00 g sample of gas which occupies 7.4 L at 27 °C and 980 mm Hg.

Flowchart 11.2 Solving for the molar mass of a gas using the molar mass form of the ideal gas equation

Flowchart 11.2

Summary

Method for solving for the molar mass of a gas using the molar mass form of the ideal gas equation

Step 1

- **Write the ideal gas equation in the molar mass form, $MM = \dfrac{m_{sample}RT}{PV}$**

$$\downarrow$$

Step 2
- Make sure that the given values are in the correct units for use in the ideal gas equation (*P* in atmospheres, *n* in moles, *V* in liters, *T* in K). Convert any values that are not in the correct units.

Step 3
- Substitute the given values into the equation and perform the arithmetic operation to arrive at the molar mass.

Example 1

A 24.5 g sample of an unknown gas occupies a volume of 44.5 L at 273.0 °C and 0.100 atm. What is the molar mass of the sample?

Solution

Perform Step 1

$$MM = \frac{m_{sample}RT}{PV}$$

Perform Step 2

T must be converted to K.

$$K = 273.0\,°C + 273.15 = 546.2\ K$$

Perform Step 3

$$MM = \frac{(24.5\ g)(0.0821\ L \cdot atm/mol \cdot K)}{(0.100\ atm)(44.5\ L)} = 247\ g/mol$$

Example 2

What is the molar mass of an unknown gas if 12.04 g of the gas occupies 7.40 L at 27.0 °C and 980 mm Hg?

Solution

Perform Step 1

$$MM = \frac{m_{sample}RT}{PV}$$

Perform Step 2

T must be converted from °C to K.

$$K = 27.0\,°C + 273.15 = 300.2\ K$$

P must be converted from mm Hg to atm.

$$P = 980\ mm\ Hg \times \frac{1\ atm}{760\ mm\ Hg} = 1.29\ atm$$

Perform Step 3

$$MM = \frac{(12.04 \text{ g})(0.0821 \text{ L} \cdot \text{atm/mol} \cdot \text{K})(300.2 \text{ K})}{(1.29 \text{ atm})(7.40 \text{ L})} = 31.1 \text{ g/mol}$$

Practice Problems

11.4 Calculate the molar mass of a gas if 1.00 L of the gas has a mass of 5.38 g at 15 °C and 736 mm Hg.

11.5 What is the molar mass of a gas if 0.985 g of the gas occupies 3.00 L at a pressure of 178 mm Hg and a temperature of 22.5 °C?

Flowchart 11.3 Solving initial-condition, final-condition gas problems

Flowchart 11.3

Summary

Method for solving initial-condition final-condition gas problems
Step 1
- **List *P*, *V*, *n*, and *T* for the initial and final conditions.**
- **Write in numeric values for any known variables.**
- **Where possible, express the final variable in terms of the initial one. For example, $V_f = V_i$**

$$\downarrow$$

Step 2

- **Write the expression $\dfrac{P_i V_i}{n_i T_i} = \dfrac{P_f V_f}{n_f T_f}$**

- **Whenever possible, rewrite this expression to show final variables in terms of initial ones (from Step 1).**
- **Cancel factors that are identical on the two sides of the expression.**

$$\downarrow$$

Step 3
- **Solve the equation algebraically for the desired variable and substitute the numerical values for the known quantities.**
- **Perform the calculation to get the answer.**

Example 1

7.62 moles of gas in a container with a moveable piston are heated from 100.0 °C to 175.0 °C. If the pressure remains constant, what volume will the gas occupy after the heating if it occupied 2.50 L initially?

Solution

Perform Step 1

Initial Conditions

$P_i = P_f$

$V_i = 2.50 \, \text{L}$

$n_i = n_f = 7.62 \, \text{mol}$

$T_i = 373.2 \, \text{K}$

Final Conditions

$P_f = P_i$

$V_f = ?$

$n_f = n_i = 7.62 \, \text{mol}$

$T_f = 448.2 \, \text{K}$

Perform Step 2

$$\frac{P_i V_i}{n_i T_i} = \frac{P_f V_f}{n_f T_f} \Rightarrow \frac{V_i}{T_i} = \frac{V_f}{T_f}$$

Perform Step 3

$$\frac{V_i}{T_i} = \frac{V_f}{T_f} \Rightarrow V_f = \frac{T_f V_i}{T_i}$$

$$V_f = \frac{(448.2 \, \text{K})(2.50 \, \text{L})}{373.2 \, \text{K}} = 3.00 \, \text{L}$$

This answer agrees with the Ideal Gas Rules. When the temperature increases, the volume will increase if the pressure and the number of moles are held constant.

Example 2

What happens to the pressure inside a steel container if the temperature of the container is decreased from 25.3 °C to 5.2 °C, and the number of moles of gas inside the container is increased from 2.32 moles to 8.63 moles? The volume of the container is 40.0 L and the initial pressure within the container is 170.0 atm.

Solution

Perform Step 1

Initial Conditions

$P_i = 170.0 \, \text{atm}$

$V_i = V_f = 40.0 \, \text{L}$

$n_i = 2.32 \, \text{moles}$

$T_i = 298.5 \, \text{K}$

Final Conditions

$P_f = ?$

$V_f = V_i = 40.0 \, \text{L}$

$n_f = 8.63 \, \text{moles}$

$T_f = 278.4 \, \text{K}$

Perform Step 2

$$\frac{P_i V_i}{n_i T_i} = \frac{P_f V_f}{n_f T_f} \Rightarrow \frac{P_i}{n_i T_i} = \frac{P_f}{n_f T_f}$$

Perform Step 3

$$\frac{P_i}{n_i T_i} = \frac{P_f}{n_f T_f} \Rightarrow P_f = \frac{P_i n_f T_f}{n_i T_i}$$

$$P_f = \frac{(170.0 \text{ atm})(8.63 \text{ moles})(278.4 \text{ K})}{(2.32 \text{ moles})(298.5 \text{ K})} = 590 \text{ atm}$$

Practice Problems

11.6 A quantity of gas occupying 2.10 L at a pressure of 750 mm Hg is compressed to a volume of 1.20 L. Calculate the resulting pressure (atm) of the gas, assuming that the temperature and the number of moles are held constant.

11.7 A sample of gas occupying a volume of 8.20 L at a temperature of 20.0 °C and a pressure of 0.832 atm is heated to 50.0 °C and its pressure is increased to 9.47 atm. Calculate the final volume of the gas.

11.8 If 3.00 moles of a gas occupy 65.3 L at a 25 °C and 1.05 atm pressure, what volume will the gas occupy at 10 °C and a pressure of 0.83 atm pressure?

Flowchart 11.4 Solving gas stoichiometry problems using the ideal gas law and a balanced chemical equation

Flowchart 11.4

Summary

Method for solving gas stoichiometry problems using the ideal gas law and a balanced chemical equation

Step 1
- **Use the ideal gas equation to determine the number of moles of gas.**

↓

Step 2
- **Use the coefficients in the balanced equation from the reaction to determine the number of moles of any other substance in the equation.**

↓

Step 3
- **Use the molar mass of the substance if asked to determine the number of grams of the other substance.**

Example 1

Pure oxygen can be produced by heating mercury(II) oxide, HgO. The balanced equation is given below:

$$2HgO(s) \rightarrow 2Hg(l) + O_2(g)$$

How many moles of HgO will be needed to produce 45.0 L of oxygen gas at temperature of 22.4 °C and a pressure of 1.23 atm?

Solution

Perform Step 1

Use the ideal gas equation to find the moles of oxygen. (Remember to convert °C to K.)

$$n = \frac{PV}{RT}$$

$$n = \frac{(1.23 \text{ atm})(45.0 \text{ L})}{\left(0.0821\dfrac{\text{L atm}}{\text{mol K}}\right)(295.6 \text{ K})} = 2.28 \text{ mol O}_2$$

Perform Step 2

$$\text{Moles HgO} = 2.28 \text{ moles O}_2 \times \frac{2 \text{ moles HgO}}{1 \text{ mole O}_2} = 4.56 \text{ moles HgO}$$

Example 2

How many moles of HCl will be needed to produce 300.0 L of H_2 gas at a temperature of 25.0 °C and a pressure of 1.00 atm? The balanced equation for the reaction is shown below.

$$Zn(s) + 2HCl(aq) \rightarrow H_2(g) + ZnCl_2(aq)$$

Solution

Perform Step 1

Use the ideal gas equation to find the moles of hydrogen. (Remember to convert °C to K.)

$$n = \frac{PV}{RT}$$

$$n = \frac{(1.00 \text{ atm})(300.0 \text{ L})}{\left(0.0821\dfrac{\text{L atm}}{\text{mol K}}\right)(298.2 \text{ K})} = 12.3 \text{ mol H}_2$$

Perform Step 2

$$\text{Moles HCl} = 12.3 \text{ moles H}_2 \times \frac{2 \text{ moles HCl}}{1 \text{ mole H}_2} = 24.6 \text{ moles HCl}$$

Example 3

How many grams of C_8H_{18} will be needed to produce 250.0 L of CO_2 gas at a temperature of 0.00 °C and a pressure of 752 mm Hg? The balanced equation for the reaction is shown below.

$$2C_8H_{18}(l) + 25O_2(g) \rightarrow 16CO_2(g) + 25H_2O(g)$$

Solution

Perform Step 1

Use the ideal gas equation to find the moles of CO_2. (Remember to convert °C to K and mm Hg to atm.)

$$n = \frac{PV}{RT}$$

$$n = \frac{(1.00 \text{ atm})(300.0 \text{ L})}{\left(0.0821\dfrac{\text{L atm}}{\text{mol K}}\right)(298.2 \text{ K})} = 12.3 \text{ mol H}_2$$

$$n = \frac{PV}{RT}$$

$$n = \frac{(0.989 \text{ atm})(250.0 \text{ L})}{\left(0.0821\dfrac{\text{L atm}}{\text{mol K}}\right)(273.15 \text{ K})} = 11.0 \text{ mol CO}_2$$

Perform Step 2

$$\text{Moles C}_8\text{H}_{18} = 11.0 \text{ moles CO}_2 \times \frac{2 \text{ moles C}_2\text{H}_{18}}{16 \text{ mole CO}_2} = 1.38 \text{ moles C}_8\text{H}_{18}$$

Perform Step 3

$$\text{grams of C}_8\text{H}_{18} = 1.38 \text{ moles C}_8\text{H}_{18} \times \frac{114.2 \text{ g C}_8\text{H}_{18}}{\text{mol C}_8\text{H}_{18}} = 158.0 \text{ g of C}_8\text{H}_{18}$$

Practice Problems

11.9 How many moles of P_4 are needed to produce 50.0 L of PH_3 gas at 28.0 °C and 2.0 atm pressure? The balanced equation for the process is shown below.

$$P_4(s) + 6H_2(g) \rightarrow 4PH_3(g)$$

11.10 How many grams of CO_2 will be produced from the reaction of 5175.0 L of C_3H_8 with oxygen gas at 35.0 °C and 1.73 atm pressure? The balanced equation for the process is shown below.

$$C_3H_8(g) + 5O_2(g) \rightarrow 3CO_2(g) + 4H_2O(g)$$

11.11 How many moles of O_2 will be needed to produce 179.0 L of CO_2 gas at 25.0 °C and 1.03 atm pressure? The balanced equation for the process is shown below.

$$C_3H_8(g) + 5O_2(g) \rightarrow 3CO_2(g) + 4H_2O(g)$$

Quiz for Chapter 11 Problems

1. What volume will 3.64 moles of N_2 gas occupy at 25.0 °C and 4.32 atm?

2. How many moles of gas are present in a 5.0 L container that is being held at a temperature of 10.0 °C and a pressure of 109.0 atm pressure?

3. Calcium carbonate can be heated to produce carbon dioxide gas according to the following reaction. How many grams of $CaCO_3$ are required for the production of 1000.0 L of CO_2 at 37 °C and 1.06 atm pressure?

$$CaCO_3(s) \rightarrow CaO(s) + CO_2(g)$$

4. Fill in the blanks below.
   (a) When the pressure is doubled on a sample of gas occupying 6.80 L, the new volume (holding the temperature constant) will be _____.
   (b) When the temperature is doubled on a sample of gas occupying 6.80 L, the new volume (holding the pressure constant) will be _____.

5. What volume will 8.92 g of Ne gas occupy at 22.0 °C and 782 mm Hg?

6. At what temperature (in °C) will a 4.0-mole sample of gas exert a pressure of 39.0 atm in a 2.5-L container?

7. If a gas occupies a volume of 4.21 L at 50 °C, what volume will the gas occupy at 100 °C? Assume the number of moles and the pressure are held constant.

8. What pressure will be exerted by 5.00 moles of a gas at 32.0 °C, at an initial pressure of 8.2 atm, if the temperature of this quantity of gas is decreased to 20.0 °C?

9. Oxygen gas can be produced by the decomposition of hydrogen peroxide according to the following reaction: $H_2O_2(l) \rightarrow H_2O(l) + O_2(g)$
   (a) Balance the equation for the reaction.
   (b) How many moles of H_2O_2 are required to produce 15.0 L of O_2 gas at a temperature of 295 K and a pressure of 740.0 mm Hg?

10. Which sample contains more moles of gas, (a) 5.00 L of He at 22.0 °C and 2.0 atm pressure, or (b) 2.50 g of He?

11. How many grams of $KClO_3$ are needed to produce 780.0 L of O_2 at 24.30 °C and 1.200 atm pressure according to the following reaction? (Hint: Make sure the equation is balanced.)

    $KClO_3 \rightarrow KCl + O_2$

12. Calculate the molar mass of a gas if 2.68 g of the gas occupies 2.00 L at 0 °C and 764 mm Hg.

Exercises for Self-Testing

A. Completion. *Write the correct word(s) to complete each statement below.*

1. A(n) _____ _____ can be thought of as consisting of pointlike particles that do not attract or repel one another at all.
2. _____ are compressible, but _____ and _____ are not.
3. When the volume of a gas is halved, holding the temperature constant, the pressure of the gas will _____.
4. In the ideal gas law, $PV = nRT$, n represents the number of _____.
5. If a sample of gas occupies 3.50 L at 200 K, it will occupy a volume of 7.00 L at _____ K.
6. Real gases deviate most from ideal-gas behavior at extreme _____ and _____.
7. One atmosphere equals _____ mm Hg.
8. Atmospheric pressure is measured by using a _____.
9. The pressure of a gas _____ when the volume it occupies is decreased and the temperature is held constant.
10. The pressure of a gas _____ when the number of moles is increased and the volume is held constant.

B. True-False. *Indicate whether each statement below is true or false, and explain why the false statements are false.*

_____ 1.   The speed of gas molecules will increase whenever the temperature is increased.

_____ 2.   When the temperature of a gas is increased, the pressure will increase.

_____ 3.   If a sample of gas in a 2.0-L container is transferred to a 4.0-L container at the same temperature, the pressure will increase.

_____ 4.   The molecules of an ideal gas experience no intermolecular attractive forces.

_____ 5. If the temperature of a gas in a container is increased, the number of moles of gas will increase also.

_____ 6. Gases are more compressible than solids.

_____ 7. If the volume of a container filled with gas is doubled at a constant temperature, the pressure of the gas also doubles.

_____ 8. Gases composed of large molecules exhibit greater deviations from the ideal behavior.

C. Questions and Problems

1. A radial tire has an air pressure of 2.20 atm at 298 K. After hours of high-speed traveling, the air in the tire reaches a temperature of 325 K. Will the driver need to add air to the tire or let some of the air out of the tire to restore it to its original pressure? Explain.

2. Use the molecular description of solids, liquids, and gases to explain the following phenomena:
   (a) Liquids take the shape of their container, but solids are rigid.
   (b) Gases are compressible, but liquids and solids are not.
   (c) Gases fill the volume of their container, but liquids and solids do not.

3. A gas sample occupies a volume of 3.2 L at a pressure of 1.3 atm and a temperature of 25 °C. How many molecules of gas are present in the sample?

4. Using the molecular model of gases, explain the following phenomena:
   (a) Increasing the temperature causes the pressure of a gas to increase.
   (b) Increasing the volume causes the pressure of a gas to decrease.
   (c) Increasing the number of moles causes the pressure of a gas to increase.

5. We have said that real gases deviate most from ideal-gas behavior at extreme temperatures and pressures. Explain why this would be the case at very low temperature and very high pressure.

6. Assuming the same T and P conditions, which one of the following gases is expected to deviate the most from the ideal gas behavior and which one is expected to deviate the least?
   (a) H_2 (b) F_2 (c) Cl_2 (d) O_2 (e) CH_4

12

Solutions

Learning Outcomes

1. Describe a solution in terms of solvent, solute, and homogeneity.

2. Describe the energetics and intermolecular attractions associated with creating a solution.

3. Define entropy and its natural tendency.

4. Describe the role of entropy in forming solutions.

5. Describe how temperature and pressure affect solubility of gaseous and solid solutes.

6. Calculate the solubility of a solute or the grams of solute required for saturation.

7. Calculate and then describe how to prepare a specific volume of a solution that has a specified molar concentration.

8. Perform calculations on solutions that make use of molarity.

9. Perform calculations on solutions that involve their concentrations in terms of percent composition.

10. Perform calculations regarding reactions that occur in solution that involve reaction stoichiometry and reactant concentrations.

11. Describe the vapor pressure of a liquid and how it depends on temperature.

12. Calculate vapor-pressure lowering, freezing-point depression, boiling-point elevation, and solute molar mass for solutions via colligative property laws.

13. Describe the two parts of a soap molecule and how it functions to make unlikes dissolve.

Chapter Outline

D. Solutions
   1. Boiling-point elevation
   2. Freezing-point depression
   3. Vapor-pressure lowering
   4. Colligative properties—properties that depend on the number of dissolved solute particles and not on their identity
   5. Calculation of freezing-point depression, ΔT_f, and boiling-point elevation, ΔT_b

12.9 One More Thing: Getting Unlikes to Dissolve—Soaps and Detergents
   A. Hydrophilic and hydrophobic regions in soap molecules
   B. Micelles

Review of Key Concepts from the Text

12.1 What Is a Solution?

A solution is a homogeneous mixture of two or more substances. Homogeneous means that the composition of the mixture is the same throughout the solution, even down to the molecular level. Homogeneous solutions appear to contain only one component when viewed with the naked eye. For example, a solution of sugar and water appears to contain only water. A solution of carbon in iron (steel) appears to contain only one metal. Heterogeneous mixtures, on the other hand, obviously contain more than one substance. It is obvious to the naked eye that a mixture of sand and water contains a solid and a liquid, just as it is obvious that a mixture of sand and sugar contains two solids. Although solutions *appear* to contain only one component, physical methods (such as evaporation) can be used to separate the components of a solution. Recall that whereas the physical combination of two substances results in a mixture, the chemical combination of two substances results in a compound.

The components of a solution that are present in the smaller amounts are the *solutes*, and the component present in the largest amount is the *solvent*. The most common liquid solvent for solutions is water, often called the universal solvent. Most of the solutions we are familiar with contain a solid solute dissolved in a liquid solvent, such as salt in water or sugar in coffee. However, the solute can be a solid, a liquid, or a gas, and the solvent can be a solid, a liquid, or a gas. (If the solvent is a solid, the solution was made by first melting the solvent, then dissolving the solute and allowing the solution to solidify.) Solutions in which water is the solvent are called *aqueous solutions*. Examples of each of the possible types of solutions are presented below.

Type of Solution	Example
Solid solute in solid solvent	Metal alloy such as brass (zinc in copper)
Liquid solute in solid solvent	Dental amalgam (mercury in silver)
Gaseous solute in solid solvent	Hydrogen in palladium
Solid solute in liquid solvent	Salt water

Type of Solution	Example
Liquid solute in liquid solvent	Rubbing alcohol (isopropyl alcohol in water)
Liquid solute in gaseous solvent	
Gaseous solute in liquid solvent	Seltzer water (carbon dioxide in water)
Gaseous solute in gaseous solvent	Air (oxygen in nitrogen)

Example 12.1

Indicate which of the following mixtures are solutions:

(a) gold   (b) chunky chicken soup   (c) clear chicken broth   (d) liquid soap

(e) oil-and-vinegar salad dressing   (f) steel   (g) blood

Solutions

The solutions (homogeneous mixtures) are the mixtures that appear to contain only one component.

(c), (d), and (f) are solutions.

(a) is not a mixture (gold is an element).

(b), (e), and (g) are heterogeneous mixtures.

Practice Exercise 12.1

Indicate which of the following mixtures are solutions:

(a) gasoline   (b) rocky soil   (c) brass   (d) apple juice

(e) 18-karat gold   (f) clam chowder

12.2   Energy and the Formation of Solutions

Solute and solvent molecules must intermix at the molecular level in order for a solution to form. This can happen only if the solute molecules are separated and the solvent molecules are separated so that they can intermingle in the solution. Therefore, when we go from separate solute and solvent to solution, all of the intermolecular forces holding the solute molecules together must be broken, and all of the intermolecular forces holding the solvent molecules together must be broken. Both of these processes require energy (ΔE is positive), because breaking any attractive forces always requires an input of energy. We might ask why a solution would form if energy has to be put into the system in order for that to happen. (Remember that any time we must add energy, we get a less stable, higher-energy product that generally will not form spontaneously.) However, there is one other process that occurs. After the solute molecules are separated (Step 1) and the solvent molecules are separated (Step 2), the intermingled solute and solvent molecules feel an attraction for each other through intermolecular attractive forces (Step 3). This third process gives off energy (ΔE is negative) because the intermolecular attractions are a stabilizing influence. (The intermolecular attractive forces are ion–dipole forces when the solvent is an ionic compound and the solvent is water.) The three steps are shown diagramatically below.

Step 1: Freeing ions from the crystal lattice of the solute   **Step 2:** Making room in the solvent

Energy must be supplied to break the ionic bonds holding the ions in the lattice

Energy must be supplied to overcome the attractive interactions between solvent molecules

Step 3: Interaction between solvent and solute

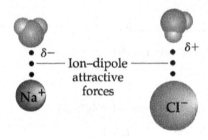

The total energy change for the dissolution process is the sum of the energy changes for the individual steps.

$$\Delta E_{total} = \Delta E_{step\ 1} + \Delta E_{step\ 2} + \Delta E_{step\ 3}$$

The amount of energy released in Step 3, called the *hydration energy* for aqueous solutions, is the determining factor in whether a substance will dissolve in a solvent because it is the only step that releases energy. If more energy is released in Step 3 than is required by Steps 1 and 2 combined, the overall energy change is negative, and the substance will dissolve.

If $\Delta E_{total} < 0$, the substance will dissolve.

You might think that if $\Delta E_{total} > 0$, the substance will not dissolve, but this is not necessarily the case. If ΔE_{total} is only slightly positive (<150 kJ), the solute might still be soluble. There is yet another factor we have to consider, *entropy*, which we will discuss in the next section. So we cannot say with certainty that if the overall energy change is positive, the substance will not dissolve. We can only say that it probably will not dissolve.

If $\Delta E_{total} > 0$, the substance probably will not dissolve (is probably insoluble).

Example 12.2

For a given solute in water, the energy change for each step is presented below. Is this solute soluble or insoluble in water? Explain.

$$\Delta E_{step\ 1} = 786\ \text{kJ}, \Delta E_{step\ 2} = 95\ \text{kJ, and } \Delta E_{step\ 3} = -783\ \text{kJ}$$

Solution

To determine whether the solute is soluble, we must calculate the total energy change.

$$\Delta E_{total} = 786\,kJ + 95\,kJ + (-783\,kJ) = 98\,kJ$$

Because ΔE_{total} is not negative, we cannot say that the solute is soluble in water. However, since ΔE_{total} is positive but less than 150 kJ, we cannot say with certainty that the solute is not soluble in water. Hence, we can only say that the solute is probably soluble in water.

Practice Exercise 12.2

For a given solute in water, the energy change for each step is presented below. Is this solute soluble or insoluble in water? Explain.

$$\Delta E_{step\ 1} = 583\,kJ, \Delta E_{step\ 2} = 98\,kJ, \text{ and } \Delta E_{step\ 3} = -823\,kJ$$

A general rule for the solubility of substances is "like dissolves like." This means that polar substances will most likely dissolve in polar solvents, and nonpolar solutes will most likely dissolve in nonpolar solvents. This rule makes sense to us if we think about it in terms of the three steps in the solution process—breaking intermolecular attractive forces between solute molecules, breaking intermolecular attractive forces between solvent molecules, and forming intermolecular attractive forces between solute and solvent molecules. If the solute and solvent have similar intermolecular attractive forces (which is the case if they are "like" substances), the strength of the intermolecular attractive forces in Step 3 will be similar to the strength of the forces in Steps 1 and 2, and the solution should form. The solute molecules will have no problem intermingling with the solvent molecules, because the types of intermolecular forces are the same in both compounds.

Example 12.3

Predict which of the following compounds will be soluble in hexane, C_6H_{14}. Explain your answers.

(a) CH_3OH (b) CCl_4 (c) CH_3CH_3 (d) NaI   (e) HCl

Solution

C_6H_{14} is a nonpolar solvent and therefore will dissolve nonpolar solutes. Only (b) and (c) are nonpolar solutes, so only (b) and (c) are soluble in hexane.

Practice Exercise 12.3

Predict which of the following compounds will be soluble in H_2O. Explain your answer.

(a) CH_3OCH_3 (b) CBr_4 (c) $CH_3CH_2CH_3$ (d) NaI   (e) HBr

We have said that a solution may form even if $\Delta E_{total} > 0$. This is true because of another factor that determines the solubility of a substance—entropy. If ΔE_{total} is only slightly positive, the entropy can favor the dissolving process, as we shall see in the next section.

12.3   Entropy and the Formation of Solutions

Entropy is a measure of the amount of disorder (or randomness) in a system. Systems tend toward a state of high entropy, and changes that result in an increase in entropy tend to be spontaneous. This is evident when you consider the state of your room. You must put energy in to decrease the amount of disorder in the room, (i.e., put everything in its place), but an increasing amount of disorder seems to happen with no input of energy on your part at all. The second law of thermodynamics addresses the concept of increasing entropy by stating that for anything to happen spontaneously anywhere in the universe, the entropy of the universe must increase. Let's see how this relates to solutions.

There is less entropy (more order) when the solute and solvent are separate than there is when they have formed a solution. This is because the atoms are arranged less randomly in the pure solute and solvent than they are in the solution. (See the figure below.)

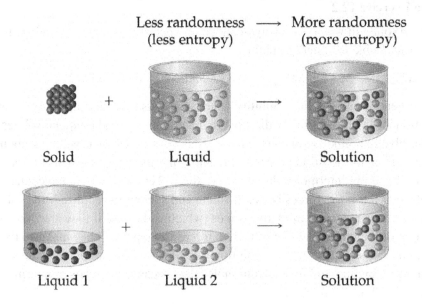

Less randomness $\longrightarrow$ More randomness
(less entropy)   (more entropy)

Solid   Liquid   Solution

Liquid 1   Liquid 2   Solution

The increase in entropy favors the formation of solutions and can overcome a positive ΔE_{total} if ΔE_{total} is small enough. This is usually the case when a nonpolar substance is being dissolved in a nonpolar solvent because the intermolecular attractive forces are weak London forces.

12.4   Solubility, Temperature, and Pressure

Solubility is the term used to describe how much solute can be dissolved in a given amount of solvent. When the maximum amount of solute is dissolved in solution, the solution is *saturated*. We typically speak of solubility in terms of the grams of solute that can be dissolved in 100 g of solvent. For example, the solubility of NaCl in water is approximately 37 g per 100 g of water at 25 °C. We must specify the temperature when giving solubility values because the solubility of substances is usually different at different temperatures. We are familiar with this phenomenon because we have experienced that it is easy to dissolve sugar in a cup of hot coffee, but it is much more difficult to get sugar to dissolve in a glass of iced tea. In fact, we would find that we can get more sugar to dissolve in the hot coffee than we can in the iced tea. Thus, we have firsthand experience with the effect of temperature on solubility.

The effect of temperature on solubility is different for solids dissolved in liquids than for gases dissolved in liquids. In the case of solids, it is generally true that the solubility *increases* with an increase in temperature (although there are a few exceptions to this). In the case of gases, the solubility generally *decreases* when the temperature is increased. We have seen this phenomenon when we have observed that a warm soft drink will fizz a lot more when opened than a cold one will. The fizzing is simply the dissolved carbon dioxide escaping from the liquid.

Example 12.4

Indicate which of the following solutions will probably allow more solute to be dissolved at higher temperature than at lower temperature. Explain your answers.

(a) N_2 in the bloodstream   (b) Salt water   (c) Sugar in iced tea

Solution

Solid solutes will likely be more soluble at higher temperatures, whereas gaseous solutes will be less soluble at higher temperatures. Therefore, (b) and (c) will allow more solute to be dissolved at higher temperature, and (a) will allow less solute to be dissolved at higher temperature.

Practice Exercise 12.4

Indicate which of the following solutions will probably allow more solute to be dissolved at higher temperature than at lower temperature. Explain your answers.

(a) Iodine crystals in alcohol (tincture of iodine)   (b) Carbonated water

(c) Sugar in lemonade

Pressure affects the solubility of a gaseous solute in a liquid solvent. Gases are more soluble in liquids when the pressure above the liquid is greater. We have also witnessed this phenomenon when we have opened a can or bottle of a soft drink. Generally, the soft drink will fizz when we take the top off. Removing the cap reduces the pressure above the liquid, allowing some of the carbon dioxide to escape.

Henry's law is a statement of the effect of pressure on the solubility of gases. It simply states that the solubility of a gas in a liquid increases as the pressure of the gas over the solution increases.

12.5 Molarity

When we are speaking about solutions, it is often helpful to know how much solute is present in the solution. There are many ways of indicating this, some more quantitative than others. For example, two terms that provide a rough idea of the relative amounts of solute and solvent in solution are *dilute* and *concentrated*. A dilute solution contains a relatively small amount of solute, and a concentrated solution contains a large amount of solute. We might also speak of a solution as being *saturated*. We know from our previous discussion that a saturated solution contains the maximum amount of solute that the solvent can dissolve at a particular temperature.

We would not know the exact amount (moles or grams) of solute present in a saturated solution, however, unless we knew the solubility of the particular solute in the given solvent. Usually a more exact measure of the relative amounts of solute and solvent is needed, and we express the concentration of a solution in terms that tell us exactly how much solute there is in a certain amount of solution. When describing the amount of solute dissolved, chemists most often think in terms of moles and express the concentration in terms of *molarity*. The molarity of a solution is the number of moles of solute that are present in one liter of solution.

$$\text{Molarity} = \frac{\text{Moles of solute}}{\text{Liters of solution}}$$

To calculate the molarity of any solution, we simply divide the number of moles of the solute by the number of liters of solution. We may have to change grams to moles, or milliliters to liters before performing the calculations, but we know how to do both of these conversions.

We can rearrange the molarity equation to obtain two alternate forms.

(1) Moles of solute = Molarity × Liters of solution

$$(2) \text{ Liters of solution} = \frac{\text{Moles of solute}}{\text{Molarity}}$$

We can use these relationships to solve for variables in the molarity equation or we can use the conversion-factor method described in Section 12.5 of your textbook. Either method, if done correctly, produces the same result.

Example 12.5

A solution is prepared by mixing 100.0 g of glucose $(C_6H_{12}O_6)$ with enough water to make 500.0 mL of solution. Calculate the molarity of the solution.

Solution

$$\text{Molarity} = \frac{\text{Moles of solute}}{\text{Liters of solution}}$$

$$\text{Moles of solute} = 100.0 \text{ g } C_6H_{12}O_6 \times \frac{1 \text{ mol } C_6H_{12}O_6}{180.2 \text{ g } C_6H_{12}O_6} = 0.5549 \text{ mol } C_6H_{12}O_6$$

Liters of solution = 0.5000 L

$$\text{Therefore, molarity} = \frac{0.5549 \text{ mol } C_6H_{12}O_6}{0.5000 \text{ L solution}} = 1.110 \text{ M.}$$

Practice Exercise 12.5

A solution is prepared by mixing 6.8 g of sodium chloride (NaCl) with enough water to make 250.0 mL of solution. Calculate the molarity of the solution.

Example 12.6

How many milliliters of the solution in Example 12.5 are needed to provide 35.0 g of glucose?

Solution

We must first convert the 35.0 g to moles because our molarity equation involves moles and liters of solution.

$$\text{Moles of glucose} = 35.0 \text{ g C}_6\text{H}_{12}\text{O}_6 \times \frac{1 \text{ mol C}_6\text{H}_{12}\text{O}_6}{180.2 \text{ g C}_6\text{H}_{12}\text{O}_6} = 0.194 \text{ mol C}_6\text{H}_{12}\text{O}_6$$

We then calculate the liters of solution containing 0.194 mole of $C_6H_{12}O_6$.

$$\text{Liters of solution} = \frac{\text{Moles of solute}}{\text{Molarity}} = \frac{0.194 \text{ mol C}_6\text{H}_{12}\text{O}_6}{1.110 \text{ mol/L}} = 0.175 \text{ L solution}$$

Finally, we convert the liters of solution to milliliters.

$$0.175 \text{ L} \times \frac{1000 \text{ mL}}{1 \text{ L}} = 175 \text{ mL}$$

Practice Exercise 12.6

How many milliliters of the solution in Practice Exercise 12.5 are needed to provide 100.0 g of NaCl?

It is important to note that the molarity is the number of moles of solute per liter of *solution*, not per liter of *solvent*. This is an important distinction that has particular relevance when you are preparing solutions, as we shall see shortly. There are two methods commonly used to prepare solutions. One method involves combining the solute and solvent (from scratch), and the other method involves diluting a more concentrated stock solution.

Preparing a Solution from Solute and Solvent (from Scratch)

To prepare a solution from solute and solvent, we must do three things:

1. Determine the mass of solute needed. We calculate the mass from the number of moles of solute needed.
2. Completely dissolve that amount of solute in a small volume of the solvent.
3. Add enough solvent to bring the total volume of the solution up to the desired volume.

Notice that we did not just take the grams of solute and add an amount of solvent equal to the volume of solution we need to prepare. Molarity is moles of solute per liter of *solution*, not per liter of *solvent*. If we added an amount of solvent equal to the desired volume of the solution, the volume we obtained would be greater than the desired volume of solution because the solute would occupy some volume.

Example 12.7

Describe how you would prepare 250.0 mL of a 0.300 M solution of KCl from solid KCl and water.

Solution

First we calculate the moles of KCl needed.

Moles of solute = Molarity × Volume (in liters)
Moles of KCl = 0.300 mol/L × 0.250 L = 0.0750 mol KCl

Next we calculate the grams of KCl needed.

$$\text{Grams of KCl} = 0.0750 \text{ mol KCl} \times \frac{74.55 \text{ g KCl}}{1 \text{ mol KCl}} = 5.59 \text{ g KCl}$$

To prepare the solution, mix 5.59 g of KCl with enough water to completely dissolve the KCl. After all of the KCl is dissolved, add enough water to bring the total volume of solution to 250.0 mL.

Practice Exercise 12.7

Describe how you would prepare 500.0 mL of a 0.100 M solution of $NaNO_3$ from solid $NaNO_3$ and water.

Preparing a Solution from a More Concentrated Stock Solution

Sometimes we are required to make a solution by diluting a currently existing solution. To prepare a solution from a concentrated stock solution, we need to do three things:

1. Calculate the number of moles of solute in the volume of the solution we are preparing.
2. Calculate the volume of the concentrated solution that contains this number of moles of solute, and place this volume in a container.
3. Add enough solvent to bring the total volume of the solution up to the desired volume.

An alternate method is to use the relationship $M_1V_1 = M_2V_2$, where one of the solutions is the more concentrated solution and the other is the more dilute solution. The volume of the concentrated solution needed can be found by solving for V_1.

$$V_1 = \frac{M_2V_2}{M_1}$$

Both methods are illustrated in the example below.

Example 12.8

How many milliliters of 18 M H_2SO_4 are needed to prepare 500.0 mL of 2.00 M H_2SO_4? How would you prepare the solution?

Solution

First we calculate the number of moles of H_2SO_4 in 500.0 mL of 2.00 M H_2SO_4.

$$\text{Moles of } H_2SO_4 = 2.00 \text{ mol/L} \times 0.500 \text{ L} = 1.00 \text{ mol } H_2SO_4$$

The volume of the concentrated solution that contains 1.00 mole of H_2SO_4 is:

$$\text{Liters of solution} = \frac{\text{Moles of solute}}{\text{Molarity}}$$

$$\text{Liters of solution} = \frac{1.00 \text{ mol } H_2SO_4}{18.0 \text{ mol/L}} = 0.0556 \text{ L of } 18.0 \text{ M } H_2SO_4$$

Finally, we convert the liters of solution to milliliters.

$$0.0556 \text{ L of solution} = \frac{1000 \text{ mL}}{1 \text{ L}} = 55.6 \text{ mL of } 18.0 \text{ M } H_2SO_4$$

To prepare the solution, we would *dilute* 55.6 mL of 18.0 M H_2SO_4 *to* 500.0 mL.

Notice that we can confirm that our result is correct by calculating the molarity of the solution we have just made. Do this now, and you will find that the molarity of a solution containing 55.6 mL of 18.0 M H_2SO_4 (calculate the moles of H_2SO_4) in 500.0 mL of solution is 2.00 M.

Alternatively, we could have used the relationship $M_1V_1 = M_2V_2$.

$$18.0 \text{ M} \times V_1 = 2.00 \text{ M} \times 500.0 \text{ mL}$$

Solving for V_1, we get

$$V_1 = \frac{2.00\ \text{M} \times 500.0\ \text{mL}}{18.0\ \text{M}} = 55.6\ \text{mL of }18.0\ \text{M H}_2\text{SO}_4$$

Both methods provide the correct answer, so use whichever method you prefer.

Practice Exercise 12.8

How many milliliters of 12.0 M HCl are needed to prepare 250.0 mL of 1.00 M HCl?

12.6   Percent Composition for Solutions

In addition to molarity, we can specify the exact amount of solute in a given amount of solution by three other units, all involving percents. Each is presented below.

$$1.\ \text{Percent by mass (mass \%)} = \frac{\text{Mass of solute}}{\text{Mass of solution}} \times 100\%$$

which can also be written as

$$\text{Percent by mass (mass \%)} = \frac{\text{Mass of solute}}{\text{Mass of solute + Mass of solvent}} \times 100\%$$

Example 12.9

Calculate the percent by mass of a solution that is prepared by mixing 35.0 g of ethanol (C_2H_5OH) and 150.0 g of water.

Solution

$$\text{Percent by mass (mass \%)} = \frac{\text{Mass of solute}}{\text{Mass of solute + Mass of solvent}} \times 100\%$$

$$= \frac{35.0\ \text{g C}_2\text{H}_5\text{OH}}{35.0\ \text{g C}_2\text{H}_5\text{OH} + 150.0\ \text{g H}_2\text{O}} \times 100\% = 18.9\%$$

Practice Exercise 12.9

Calculate the percent by mass of a solution that is prepared by mixing 40.0 g of sucrose $(C_{12}H_{22}O_{11})$ and 250.0 g of water.

$$2.\ \text{Percent by volume (vol \%)} = \frac{\text{Volume of solute}}{\text{Volume of solution}} \times 100\%$$

Notice that in percent by volume problems you must be given the volume of solution because the volume of solution is generally not equal to the sum of the volumes of the solute and the solvent, especially when the solute is a solid. If you need some convincing, combine a cup of sugar and a cup of water. You will surely not obtain two cups of sugar water!

Example 12.10

Calculate the percent by volume of a solution that contains 20.0 mL of ethanol (C_2H_5OH) in 100.0 mL of solution.

Solution

$$\text{Percent by volume (vol \%)} = \frac{\text{Volume of solute}}{\text{Volume of solution}} \times 100\%$$

$$= \frac{20.0 \text{ mL}}{100.0 \text{ mL}} \times 100\% = 20.0\%$$

Practice Exercise 12.10

Calculate the percent by volume of a solution that contains 125.0 mL of methanol (CH_3OH) in 500.0 mL of solution.

$$3. \text{ Percent by mass/volume (mass/vol \%)} = \frac{\text{Mass of solute}}{\text{Volume of solution}} \times 100\%$$

Example 12.11

Calculate the percent by mass/volume of a solution prepared by dissolving 2.50 g of NaCl in enough water to make 50.0 mL of solution.

Solution

$$\text{Percent by mass/volume (mass/vol \%)} = \frac{\text{Mass of solute}}{\text{Volume of solution}} \times 100\%$$

$$= \frac{2.50 \text{ g}}{50.0 \text{ mL}} \times 100\% = 5.00\%$$

Practice Exercise 12.11

Calculate the percent by mass/volume of a solution prepared by dissolving 45.0 g of KBr in enough water to make 350.0 mL of solution.

12.7 Reactions in Solution

Most chemical reactions occur in solution. Therefore, the stoichiometric calculations involving most reactions must be performed based on volumes of solutions rather than on grams of reactants. This change, however, requires us to make a few minor adjustments to the procedures we have used in our previous stoichiometry calculations. When dealing with solutions, we will generally need to calculate moles from the molarity and volume of solution:

$$\text{Moles} = \text{Molarity (mol/L)} \times \text{Volume (L)}$$

After we have calculated the moles of reacting substance, the procedure is the same as before to determine the moles of product.

If we want to determine the volume of a solution needed to provide a given number of moles of reactant after we have used stoichiometry to obtain the moles of substance, we use the molarity and the moles to calculate the volume:

$$\text{Volume} = \text{Moles/Molarity (mol/L)}$$

Example 12.12

How many grams of $Al_2(SO_4)_3$ will be produced from the reaction of 350.0 mL of 0.115 M H_2SO_4 and an excess of Al? The other product of the reaction is H_2 gas.

Solution

The balanced equation for the reaction is:

$$2Al + 3H_2SO_2 \rightarrow Al_2(SO_4)_3 + 6H_2$$

Moles of H_2SO_2 = 0.115 mol/L $\times$ 0.350 L = 0.0403 mol H_2SO_4

Moles of $Al_2(SO_4)_3$ = 0.0403 mol H_2SO_4 $\times$ $\dfrac{1 \text{ mol } Al_2(SO_4)_3}{3 \text{ mol } H_2SO_4}$ = 0.0134 mol $Al_2(SO_4)_3$

Grams of $Al_2(SO_4)_3$ = 0.0134 mol $Al_2(SO_4)_3$ $\times$ $\dfrac{342.15 \text{ g } Al_2(SO_4)_3}{1 \text{ mol } Al_2(SO_4)_3}$ = 4.58 g $Al_2(SO_4)_3$

Example 12.13

How many milliliters of 0.237 M $MgCl_2$ would be needed to react with an excess of $AgNO_3$ to produce 10.0 g of AgCl?

Solution

The balanced equation for the reaction is

$$MgCl_2 + 2AgNO_3 \rightarrow 2AgCl + Mg(NO_3)_2$$

Moles of AgCl = 10.0 g AgCl $\times$ $\dfrac{1 \text{ mol AgCl}}{143.32 \text{ g AgCl}}$ = 0.0698 mol AgCl

Moles of $MgCl_2$ = 0.0698 mol AgCl $\times$ $\dfrac{1 \text{ mol } MgCl_2}{2 \text{ mol AgCl}}$ = 0.0349 mol $MgCl_2$

Volume of $MgCl_2$ = $\dfrac{0.0349 \text{ mol } MgCl_2}{0.237 \text{ mol/L}}$ = 0.147 L $MgCl_2$ = 147 mL $MgCl_2$

Therefore, 147 mL of 0.237 M $MgCl_2$ would produce 10.0 g of AgCl.

Practice Exercise 12.12

Consider the *unbalanced* equation below. How many grams of SiH_4 can be prepared by reacting 50.0 mL of 1.25 M HCl with an excess of Mg_2Si?

$$Mg_2Si + HCl \rightarrow MgCl_2 + SiH_4$$

Practice Exercise 12.13

How many milliliters of 0.125 M Na_2CO_3 would be needed to react with an excess of CaI_2 to produce 75.0 g of $CaCO_3$?

12.8 Colligative Properties of Solutions

Colligative properties are properties of a solution that depend on the number of solute particles present in solution, not on the identity of the particles. The three colligative properties presented in the textbook are freezing-point depression, boiling-point elevation, and vapor-pressure lowering. (The solute in each case is a nonvolatile solute—one that does not readily evaporate.)

Freezing-Point Depression

The freezing point of a solution will be lower than the freezing point of the pure solvent. This change in freezing point is called the freezing-point depression and is denoted by ΔT_f.

The change in the freezing point is calculated by multiplying the freezing-point constant K_f by the moles of solute particles and dividing by the kilograms of solvent, as shown in the following equation:

$$\Delta T_f = \frac{K_f \times \text{Moles of solute particles}}{\text{Kilograms of solvent}}$$

Boiling-Point Elevation

The boiling point of a solution will be higher than the boiling point of the pure solvent. The amount by which the boiling point increases is called the boiling-point elevation, ΔT_b. ΔT_b is calculated from the boiling-point constant, the moles of solute particles, and the kilograms of solvent in the same way as the freezing-point depression was calculated from the freezing-point constant. The equation is shown below.

$$\Delta T_b = \frac{K_b \times \text{Moles of solute particles}}{\text{Kilograms of solvent}}$$

Values of K_f and K_b for many substances are provided in Table 12.2 in the textbook.

Vapor Pressure

The vapor pressure of a solution containing a nonvolatile solute is always less than the vapor pressure of the pure solvent. Section 12.8 in the textbook provides a more thorough discussion of colligative properties.

Example 12.14

What is the boiling point of a solution consisting of 450 g of KCl and 750 g of water?

Solution

KCl dissociates in water to produce K^+ and Cl^- ions. Therefore, a solution that contains KCl will have twice that number of ions in 1 L of solution.

$$\text{Moles of KCl} = 450.0 \text{ g KCl} \times \frac{1 \text{ mol KCl}}{74.551 \text{ g KCl}} = 6.04 \text{ mol KCl}$$

$$\text{Moles of ions} = 6.04 \text{ mol KCl} \times \frac{2 \text{ mol ions}}{1 \text{ mol KCl}} = 12.08 \text{ mol ions}$$

$$\Delta T_b = \frac{K_b \times \text{Moles of solute particles}}{\text{Kilograms of solvent}}$$

$$\Delta T_b = \frac{0.52 \,^\circ\text{C} \cdot \text{kg/mol} \times 12.08 \text{ mol solute particles}}{0.750 \text{ kg water}} = 8.4 \,^\circ\text{C}$$

The boiling point of the solution will be $100.0\,^\circ\text{C} + 8.38\,^\circ\text{C} = 108.4\,^\circ\text{C}$.

Example 12.15

Calculate the freezing-point depression for a solution that consists of 275.0 g of NaF in 5250 g of water.

$$\Delta T_f = \frac{K_f \times \text{Moles of solute particles}}{\text{Kilograms of solvent}}$$

$$\text{Moles of NaF} = 275.0 \text{ g NaF} \times \frac{1 \text{ mol NaF}}{41.99 \text{ g NaF}} = 6.549 \text{ mol NaF}$$

One mole of NaF produces 2 moles of particles in solution: $NaF \rightarrow Na^+ + F^-$

Moles of particles $= 2 \times 6.549 = 13.10$ mol particles

$$\Delta T_f = \frac{1.86 \text{ °C} \cdot \text{kg/mol} \times 13.10 \text{ mol particles}}{5.250 \text{ kg solvent}} = 4.64 \text{ °C}$$

The freezing point of the solution will be $0.00 \text{ °C} - 4.64 \text{ °C} = -4.64 \text{ °C}$.

Practice Exercise 12.14

Calculate the boiling point of a solution composed of 25.0 g of ethylene glycol $(C_2H_6O_2)$ dissolved in 75.0 g of water.

Practice Exercise 12.15

Calculate the freezing point of a solution composed of 50.0 g of sucrose $(C_{12}H_{22}O_{11})$ dissolved in 650.0 g of water.

A common use of freezing-point depression and boiling-point elevation data is the determination of the molar masses of an unknown compound. If we take a specific mass of the unknown, dissolve it in a specific number of grams of a solvent, and determine the new freezing point (or boiling point) of the solution, we can calculate the molar mass of the unknown. First, we use the freezing-point depression equation to find the moles of solute. Because molar mass = grams/moles, we simply divide the number of moles of unknown into the number of grams that we dissolved in the solvent. This will provide us with the molar mass of the compound.

Example 12.16

When 60.0 g of an unknown substance is dissolved in 200.0 g of water, the freezing point of the solution is -3.10 °C. What is the molar mass of the unknown?

Solution

$$\Delta T_f = \frac{K_f \times \text{Moles of solute particles}}{\text{Kilograms of solvent}}$$

$$3.10 \text{ °C} = \frac{1.86 \text{ °C} \cdot \text{kg/mol solute} \times \text{Moles of solute particles}}{0.2000 \text{ kg of solvent}}$$

Solving the above equation algebraically for moles of solute particles, we obtain:

$$\text{Moles of solute particles} = \frac{3.10 \text{ °C} \times 0.2000 \text{ kg solvent}}{1.86 \text{ °C} \cdot \text{kg/mol solute}} = 0.333 \text{ mol solute particles}$$

$$\text{Molar mass of solute} = \frac{\text{Grams of solute}}{\text{Moles of solute}} = \frac{60.0 \text{ g}}{0.333 \text{ mol}} = 180 \text{ g/mol}$$

Practice Exercise 12.16

When 46.0 g of an unknown substance is dissolved in 250.0 g of water, the freezing point of the solution is -3.72 °C. What is the molar mass of the unknown?

12.9   Getting Unlikes to Dissolve—Soaps and Detergents

We established the general rule that "like dissolves like," but it is sometimes necessary to create conditions in which "unlikes" can form solutions. One example is the dissolving of grease and oils in water when we wash our clothing or hands. We need a substance that has a nonpolar section, which will dissolve the nonpolar oils and grease, as well as a polar section, which will allow it to be soluble in water. Soaps and detergents are such substances.

Soaps and detergents consist of molecules that contain a long nonpolar "tail" of carbon and hydrogen and a highly polar "head" group. The nonpolar tail is said to be *hydrophobic* (water-fearing) because it will not dissolve in water. The polar head is said to be *hydrophilic* (water-loving) because it is highly water soluble. The soap molecules arrange themselves in spheres, called *micelles*, in which the nonpolar tails are in the interior of the sphere and the polar heads are on the surface, as shown in the figure below.

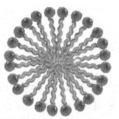

Cross section of a soap micelle

When greasy dirt comes into contact with the soap micelles, the hydrocarbon tails penetrate the grease, and the ionic heads dissolve in water. The greasy dirt is therefore dispersed throughout the water inside the micelles and can be rinsed away in the wash. Hence, we have succeeded in devising a way for "unlikes" to dissolve.

Water   Water

Water   Water

Strategies for Working Problems in Chapter 12

Overview of Problem Types: What You Should Be Able to Do

Chapter 12 provides a discussion of solutions, including solution formation and concentration units. The types of problems that you should be able to solve after mastering Chapter 12 are:

1. Prepare a solution from solute and solvent (from scratch).
2. Prepare a solution from a more concentrated stock solution.
3. Calculate the percent composition of a solution.
4. Solve solution stoichiometry problems.
5. Solve acid–base titration problems.
6. Determine molar mass using boiling-point elevation (ΔT_b) or freezing-point depression (ΔT_f) data.

Flowcharts

Flowchart 12.1 Preparing a solution from solute and solvent (from scratch)

Flowchart 12.1

Summary:

Method for preparing a solution from solute and solvent (from scratch)

Step 1
- **Determine the moles of solute needed to prepare the solution.**
- **Moles of solute = volume of solution (in liters) × molarity of solution**

$\downarrow$

Step 2
- **Convert the moles of solute to grams of solute.**
- **Use the molar mass as a conversion factor.**

$\downarrow$

Step 3
- **Place the number of grams from Step 2 in a volumetric flask and add enough water to bring the total volume of the solution to the desired volume.**
- **Do not simply add an amount of water equal to the desired volume of solution.**

Example 1:

Explain how you would prepare 2.50 L of a 0.350 M solution of glucose, $C_6H_{12}O_6$.

Solution:

Perform Step 1:

$$\text{Moles of glucose needed} = \text{Liters of solution} \times \text{molarity of solution}$$
$$= 2.50 \, \text{L} \times 0.350 \, \text{mol/L}$$
$$= 0.875 \, \text{mol of glucose}$$

Perform Step 2:

$$\text{grams of glucose} = 0.875 \, \text{mol glucose} \times \frac{180.15 \, \text{g}}{1 \, \text{mol glucose}} = 158 \, \text{g glucose}$$

Perform Step 3:

Place 158 g of glucose in a 2.50 L volumetric flask, and add enough water to bring the level of the solution to 2.50 L.

Example 2:

Explain how you would prepare 500.0 mL of a 1.20 M solution of potassium nitrate, KNO_3.

Solution:

Perform Step 1:

$$\text{Moles of } KNO_3 \text{ needed} = \text{Liters of solution} \times \text{molarity of solution}$$
$$= 0.500\,\text{L} \times 1.20\,\text{mol/L}$$
$$= 0.600 \text{ moles of } KNO_3$$

Perform Step 2:

$$\text{grams of glucose} = 0.600 \text{ mol } KNO_3 \times \frac{101.1\,\text{g}}{1 \text{ mol } KNO_3} = 60.6 \text{ g } KNO_3$$

Perform Step 3:

Place 60.6 g of KNO_3 in a 500.0 mL volumetric flask, and add enough water to bring the level of the solution to 500.0 mL.

Practice Problems:

12.1 Explain how you would prepare 1.00 L of a 4.50 M solution of NaCl.

12.2 Explain how you would prepare 250.0 mL of a 0.250 M solution of sucrose, $C_{12}H_{22}O_{11}$.

Flowchart 12.2 Preparing a solution from a more concentrated stock solution. (Two methods are presented.)

Flowchart 12.2—Method A

Step 1
- Determine the moles of solute needed to prepare the solution.
- Moles of solute = volume of solution (in liters) × molarity of solution

$\downarrow$

Step 2
- Convert the moles of solute to volume of stock solution by using the molarity of the stock solution as a conversion factor. (Multiply the moles of solute by the inverse of the stock solution's molarity.)

$\downarrow$

Step 3
- Place the volume of stock solution from Step 2 in a volumetric flask and add enough water to bring the total volume of the solution to the desired volume.

Flowchart 12.2—Method B

Step 1

- **Use the dilution equation shown below to determine the volume of the stock solution needed:**

$$\text{Molarity}_{\substack{\text{stock} \\ \text{solution}}} \times \text{Volume}_{\substack{\text{stock} \\ \text{solution}}} = \text{Molarity}_{\substack{\text{diluted} \\ \text{solution}}} \times \text{Volume}_{\substack{\text{diluted} \\ \text{solution}}}$$

- **Algebraically solve the equation for the volume of the stock solution.**

$$\downarrow$$

Step 2

- **Place the volume of stock solution from Step 2 in a volumetric flask and add enough water to bring the total volume of the solution to the desired volume.**

Example 1:

Explain how you would prepare 500.0 mL of a 0.100 M solution of glucose from a stock solution that is 0.250 M.

Solution (By Method A)

Perform Step 1:

$$
\begin{aligned}
\text{Moles of glucose needed} &= \text{volume of solution (L)} \times \text{molarity of solution} \\
&= 0.500 \, \text{L} \times 0.100 \, \text{mol/L} \\
&= 0.0500 \text{ moles of glucose}
\end{aligned}
$$

Perform Step 2:

$$
\begin{aligned}
\text{Volume of stock solution} &= \frac{\text{moles of solute needed}}{\text{molarity of stock solution}} \\
&= 0.0500 \, \text{mol} \times 1/0.250 \, \text{mol/L} \\
&= 0.200 \, \text{L} = 2.00 \times 10^2 \, \text{mL of stock solution}
\end{aligned}
$$

Perform Step 3:

Place 2.00×10^2 (200.) mL of stock solution in a 500.0 mL volumetric flask and bring the solution level up to the desired volume.

Solution (By Method B)

Perform Step 1:

$$\text{Molarity}_{\text{stock solution}} \, \text{Volume}_{\text{stock solution}} = \text{Molarity}_{\text{diluted solution}} \, \text{Volume}_{\text{diluted solution}}$$

$$(0.250 \, \text{mol/L})(V_{\text{stock solution}}) = (0.500 \, \text{mol/L})(0.100 \, \text{L})$$

$$\text{Volume}_{\text{stock solution}} = \frac{(0.500 \, \text{mol/L})(0.100 \, \text{L})}{(0.250 \, \text{mol/L})} = 0.200 \, \text{L} = 2.00 \times 10^2 \, \text{mL}$$

Perform Step 2:

Place 2.00×10^2 (200.) mL of stock solution in a 500.0 mL volumetric flask and bring the solution level up to the desired volume.

Example 2:

Explain how you would prepare 2.00 L of a 0.750 M HCl solution from a stock solution that is 6.0 M.

Solution (By Method A)

Perform Step 1:

Moles of HCl needed = volume of solution (L) × molarity of solution
= 2.00 L × 0.750 mol/L = 1.50 mol of HCl

Perform Step 2:

$$\text{Volume of stock solution} = \frac{\text{moles of solute needed}}{\text{molarity of stock solution}}$$
$$= 1.50 \text{ mol}/6.00 \text{ mol}/L$$
$$= 0.250 \text{ L} = 2.50 \times 10^2 \text{ mL of stock solution}$$

Perform Step 3:

Place 2.50×10^2 (250.) mL of stock solution in a 2.00 L volumetric flask and bring the solution level up to the desired volume.

Solution (By Method B)

Perform Step 1:

$$\text{Molarity}_{\text{stock solution}} \text{ Volume}_{\text{stock solution}} = \text{Molarity}_{\text{diluted solution}} \text{ Volume}_{\text{diluted solution}}$$
$$(6.00 \text{ mol}/L)(V_{\text{stock solution}}) = (0.750 \text{ mol}/L)(2.00 \text{ L})$$

$$\text{Volume}_{\text{stock solution}} = \frac{(0.750 \text{ mol}/L)(2.00 \text{ L})}{6.00 \text{ mol}/L} = 0.250 \text{ L} = 2.50 \times 10^2 \text{ mL}$$

Perform Step 2:

Place 2.50×10^2 (250.) mL of stock solution in a 2.00 L volumetric flask and bring the solution level up to the desired volume.

Practice Problems:

12.3 Explain how you would prepare 1.00 L of a 2.50 M NaCl solution from a stock solution that is 4.50 M.

12.4 Explain how you would prepare 250.0 mL of a 12.00 M H_2SO_4 solution from a stock solution that is 18.00 M.

Flowchart 12.3 Calculating the percent composition of a solution

Flowchart 12.3

Summary:

Method for calculating the percent composition of a solution

Step 1
- **Determine the type of percent composition calculation the problem involves (i.e., percent by mass, volume percent by volume, or percent by mass/volume).**

↓

Step 2

• **Calculate the percent composition using the appropriate formula.**

$$\text{Percent by mass (wt\%)} = \frac{\text{grams of solute}}{\text{grams of solution}} \times 100\%$$

$$\text{Percent by volume (vol \%)} = \frac{\text{volume of solute}}{\text{grams of solution}} \times 100\%$$

$$\text{Percent by mass/volume (vol \%)} = \frac{\text{grams of solute}}{\text{volume of solution}} \times 100\%$$

Example 1:

Calculate the percent by mass of a solution that contains 125.0 g of $KClO_3$ in 450.0 g of solution.

Solution:

Perform Step 1:

The problem asks for percent by mass, so we will use:

$$\text{Percent by mass (wt\%)} = \frac{\text{grams of solute}}{\text{grams of solution}} \times 100\%$$

Perform Step 2:

$$\%KClO_3 \text{ by mass} = \frac{125.0 \text{ g of } KClO_3 \times 100\%}{450.0 \text{ g of solution}} = 27.78\% \ KClO_3$$

Example 2:

Calculate the percent by mass of a solution that contains 55.0 g of NaBr in 150.0 g of water.

Solution

Perform Step 1:

The problem asks for percent by mass, so we will use

$$\text{Percent by mass (wt\%)} = \frac{\text{grams of solute}}{\text{grams of solution}} \times 100\%$$

we must add the mass of the solute, NaBr, and the mass of the solvent, water, to get the total mass of the solution. Therefore, the mass of the solution equals 55.0 g (NaBr) + 150.0 g (water), or 205.0 g of solution.

Perform Step 2:

$$\% \text{ NaBr by mass} = \frac{55.0 \text{ g of } KClO_3}{205.0 \text{ g of solution}} \times 100\% = 26.8\% \ NaBr$$

Example 3:

Calculate the percent by mass/volume of a solution that contains 80.0 g of $LiNO_3$ in 550.0 mL of solution.

Solution:

Perform Step 1:

The problem asks for percent by mass/volume, so we will use

$$\text{Percent by mass/volume (wt/vol \%)} = \frac{\text{grams of solute}}{\text{mL of solution}} \times 100\%$$

Perform Step 2:

$$\% \ LiNO_3 \ \text{by mass/vol} = \frac{80.0 \ \text{g of } LiNO_3}{550.0 \ \text{mL of solution}} \times 100\% = 14.5\% \ LiNO_3$$

Practice Problems

12.5 Calculate the percent by volume of a solution that contains 15.0 mL of methanol, CH_3OH, in 120.0 mL of solution.

12.6 Calculate the percent by mass/volume of a solution that contains 65.0 g of sucrose, $C_{12}H_{22}O_{11}$, in 550.0 mL of solution.

Flowchart 12.4 Solving solution stoichiometry problems

Flowchart 12.4

Summary:

Procedure for solving solution stoichiometry problems

Step 1
- **Write the balanced equation for the reaction that is occurring in the solution.**

↓

Step 2
- **Convert all given amounts to moles using molarities (if volumes are given) and/or molar masses (if grams are given) as conversion factors.**

↓

Step 3
- **Use the coefficients from the balanced equation to find the relationship between the number of moles of the given substance and the number of moles of the unknown substance.**
- **The ratio of the coefficients is used as a conversion factor, with the coefficient in front of the unknown as the numerator and the coefficient in front of the given as the denominator.**
- **The conversion factor is multiplied by the number of moles of the given substance.**

↓

Step 4
- **Convert the moles of the unknown, calculated in Step 3, to appropriate units using molarities (if volumes are asked for) and/or molar masses (if grams are asked for) as conversion factors.**

Example 1:

How many milliliters of 0.300 M NaOH are required to react exactly with 50.0 mL of 0.200 M H_2SO_4 according to the reaction represented by the unbalanced equation below?

$$H_2SO_4 + NaOH \rightarrow Na_2SO_4 + H_2O$$

Solution:

Perform Step 1:

$$H_2SO_4 + 2NaOH \rightarrow Na_2SO_4 + 2H_2O$$

Perform Step 2:

Moles of H_2SO_4 = Molarity $\times$ Volume (L) = 0.200 mol/L H_2SO_4 $\times$ 0.0500 L H_2SO_4
= 0.0100 mol H_2SO_4

Perform Step 3:

Moles NaOH = 0.0100 mol H_2SO_4 $\times$ $\dfrac{2 \text{ mol NaOH}}{1 \text{ mol } H_2SO_4}$ = 0.0200 mol NaOH

Perform Step 4:

Liters of NaOH = $\dfrac{0.0200 \text{ mol NaOH}}{0.300 \text{ mol/L NaOH}}$ = 0.0667 L NaOH

Volume of NaOH in mL = 66.7 mL NaOH

Example 2:

What is the maximum number of grams of $MgCl_2$ that can be prepared from reacting 350.0 mL of 0.125 M NaCl with an excess amount of $Mg(NO_3)_2$ according to the unbalanced equation below for the reaction?

$$NaCl + Mg(NO_3)_2 \rightarrow MgCl_2 + NaNO_3$$

Solution:

Perform Step 1:

$$2NaCl + Mg(NO_3)_2 \rightarrow MgCl_2 + 2NaNO_3$$

Perform Step 2:

Moles of NaCl = Molarity $\times$ Volume = 0.125 mol/L NaCl $\times$ 0.350 L NaCl = 0.0438 mol NaCl

Perform Step 3:

Moles $MgCl_2$ = 0.0438 mol NaCl $\times$ $\dfrac{1 \text{ mol } MgCl_2}{2 \text{ mol NaCl}}$ = 0.0219 mol $MgCl_2$

Perform Step 4:

grams $MgCl_2$ = 0.0219 mol $MgCl_2$ $\times$ $\dfrac{95.211 \text{ g } MgCl_2}{1 \text{ mol } MgCl_2}$ = 2.09 g $MgCl_2$

Example 3:

How would you prepare 25.00 g of $CaCO_3$ from a 0.400 M solution of Ca_3N_2 and a 0.100 M solution of Na_2CO_3? The unbalanced reaction is given below:

$$Ca_3N_2 + Na_2CO_3 \rightarrow CaCO_3 + NaNO_3$$

Solution:

Perform Step 1:

$$Ca_3N_2 + 3Na_2CO_3 \rightarrow 3CaCO_3 + 2Na_3N$$

Perform Step 2:

$$\text{Moles of } CaCO_3 = 25.00 \text{ g } CaCO_3 \times \frac{1 \text{ mol}}{100.09 \text{ g}} = 0.2498 \text{ mol } CaCO_3$$

Perform Step 3:

$$\text{Moles of } Ca_3N_2 = 0.2498 \text{ mol } CaCO_3 \times \frac{1 \text{ mol } Ca_3N_2}{3 \text{ mol } CaCO_3} = 0.08327 \text{ mol } Ca_3N_2$$

$$\text{Moles of } Na_2CO_3 = 0.2498 \text{ mol } CaCO_3 \times \frac{3 \text{ mol } Na_2CO_3}{3 \text{ mol } CaCO_3} = 0.2498 \text{ mol } Na_2CO_3$$

Perform Step 4:

$$\text{Volume of } Ca_3N_2 = \frac{\text{Moles of } Ca_3N_2}{\text{Molarity}} = \frac{0.08327 \text{ mol } Ca_3N_2}{0.400 \text{ mol/L}} = 0.208 \text{ L} = 208 \text{ mL}$$

$$\text{Volume of } Na_2CO_3 = \frac{\text{Moles of } Na_2CO_3}{\text{Molarity}} = \frac{0.2498 \text{ mol } Na_2CO_3}{0.100 \text{ mol/L}} = 2.50 \text{ L} = 2500 \text{ mL}$$

To prepare 25.00 g of $CaCO_3$:

Combine 208 mL of 0.400 M Ca_3N_2 and 2500 mL of 0.100 M Na_2CO_3.

Practice Problems:

12.7 How many grams of SiH_4 can be prepared by reacting 76.0 mL of 0.750 M HCl with an excess of Mg_2Si, according to the reaction represented by the unbalanced equation below.

$$Mg_2Si + HCl \rightarrow MgCl_2 + SiH_4$$

12.8 How would you prepare 45.0 g of PbC_2O_4 from a 0.250 M solution of $Pb(NO_3)_2$ and a 0.125 M solution of $K_2C_2O_4$? The unbalanced reaction is given below.

$$Pb(NO_3)_2 + K_2C_2O_4 \rightarrow KNO_3 + PbC_2O_4$$

Flowchart 12.5 Solving acid–base titration problems

Flowchart 12.5

Summary:

Method for solving acid–base titration problems

Step 1
- Write the balanced equation for the reaction.

$$\downarrow$$

Step 2
- Use the molarity to convert the volume of the base used to reach the equivalence point to moles.
- Volume (in liters) × Molarity = moles

$$\downarrow$$

Step 3
- Use the coefficients in the balanced equation to convert moles of based added to moles of acid that reacted.
- The number of moles of acid that reacted is the number of moles of acid originally in the flask.

$$\downarrow$$

Step 4
- Convert the moles of the acid, calculated in Step 3, to the appropriate units.
- If the molarity of the acid is asked for, divide the moles of acid by the volume of the original acid solution (before the titration began) because molarity = moles/liters

Example 1:

50.0 mL of an H_2SO_4 solution of unknown concentration is titrated with 0.100 M NaOH. 32.30 mL of base are required to neutralize the acid. What is the concentration of the H_2SO_4?

Solution:

Perform Step 1:

$$H_2SO_4 + 2NaOH \rightarrow Na_2SO_4 + 2H_2O$$

Perform Step 2:

$$
\begin{aligned}
\text{Moles of NaOH} &= \text{Volume of NaOH} \times \text{Molarity of NaOH} \\
&= 0.03230 \, \text{L} \times 0.100 \, \text{mol/L} \\
&= 0.003230 \, \text{mol NaOH}
\end{aligned}
$$

Perform Step 3:

$$\text{Moles of } H_2SO_4 = 0.003230 \, \text{mol NaOH} \times \frac{1 \, \text{mol } H_2SO_4}{2 \, \text{mol NaOH}} = 0.001\,615 \, \text{mol } H_2SO_4$$

Perform Step 4:

$$\text{Molarity of } H_2SO_4 = \frac{0.001615 \, \text{mol } H_2SO_4}{0.0500 \, \text{L } H_2SO_4} = 0.0323 \, \text{M } H_2SO_4$$

Example 2:

50.0 mL of an HCl solution of unknown concentration is titrated with 0.200 M NaOH. 28.92 mL of base are required to neutralize the acid. What is the concentration of the HCl?

Solution:

Perform Step 1:

$$HCl + NaOH \rightarrow NaCl + H_2O$$

Perform Step 2:

$$
\begin{aligned}
\text{Moles of NaOH} &= \text{Volume of NaOH} \times \text{Molarity of NaOH} \\
&= 0.028\,92 \text{ L} \times 0.200 \text{ mol/L} \\
&= 0.005\,784 \text{ mol NaOH}
\end{aligned}
$$

Perform Step 3:

$$\text{Moles of HCl} = 0.005\,784 \text{ mol NaOH} \times \frac{1 \text{ mol HCl}}{1 \text{ mol NaOH}} = 0.005\,78 \text{ mol HCl}$$

Perform Step 4:

$$\text{Molarity of HCl} = \frac{0.005\,784 \text{ mol HCl}}{0.0500 \text{ L HCl}} = 0.116 \text{ M HCl}$$

Practice Problems:

12.9 50.0 mL of an HCl solution of unknown concentration is titrated with 0.100 M NaOH. 32.33 mL of base are required to neutralize the acid. What is the concentration of the HCl?

12.10 50.0 mL of an H_2SO_4 solution of unknown concentration is titrated with 0.250 M NaOH. 39.36 mL of base are required to neutralize the acid. What is the concentration of the H_2SO_4?

Flowchart 12.6 Determining molar mass using boiling point elevation (ΔT_b) or freezing point depression (ΔT_f) data

Flowchart 12.6

Summary:

Procedure for determining molar mass using boiling point elevation (ΔT_b) or freezing point depression (ΔT_f) data

Step 1

- Use the boiling point elevation equation (or the freezing point depression equation), shown below in terms of the boiling point elevation, to determine the moles of solute

$$\Delta T_b = K_b \times \frac{\text{moles of solute}}{\text{kilograms of solvent}}$$

$$\text{moles of solute} = \frac{\Delta T_b \times \text{kilograms of solvent}}{K_b}$$

- The value of K_b or K_f will be given in the problem if the solvent is not water.
- Remember to convert the mass of solvent to kilograms if it is given in grams.

$\downarrow$

Step 2
  • **Divide the grams of solute (given in the problem) by the moles of solute (calculated in Step 1) to determine the molar mass.**

Example 1:

A solution of an unknown compound was prepared by dissolving 1.50 g of the compound in 30.0 g of water. The freezing point of the solution was determined to be $-1.02\,°C$. What is the molar mass of the unknown compound?

Solution:

Perform Step 1:

$$\text{moles of solute} = \frac{\Delta T_f \times \text{kilograms of solvent}}{K_f}$$

$$\text{moles of solute} = \frac{1.02\,°C \times 0.030\,\text{kg of water}}{1.86\,°C \cdot \text{kg solvent/moles solute}}$$

$$= 0.0165\,\text{moles of solute}$$

Perform Step 2:

$$\text{Molar mass of solute} = \frac{1.50\,\text{g of compound}}{0.0165\,\text{moles of compound}} = 90.1\,\text{g/mol}$$

Example 2:

44.00 g of an unknown compound were dissolved in 100.0 g of water. The boiling point of the solution was determined to be $101.27\,°C$. What is the molar mass of the unknown compound?

Solution:

Perform Step 1:

$$\text{moles of solute} = \frac{\Delta T_b \times \text{kilograms of solvent}}{K_b}$$

$$\text{moles of solute} = \frac{1.27\,°C \times 0.100\,\text{kg of water}}{0.52\,°C \cdot \text{kg solvent/moles solute}}$$

$$= 0.244\,\text{moles of solute}$$

Perform Step 2:

$$\text{Molar mass of solute} = \frac{44.00\,\text{g of compound}}{0.244\,\text{moles of compound}} = 180.\,\text{g/mol}$$

Practice Problem:

12.11 2.20 g of an unknown compound were dissolved in 10.0 g of water. The freezing point of the solution was determined to be $-2.33\,°C$. What is the molar mass of the unknown compound?

Quiz for Chapter 12 Problems

1. How many grams of $BaCl_2$ can be produced from the reaction of 82.5 mL of a 0.175 M NaCl solution with an excess of $Ba(OH)_2$ according to the reaction represented by the unbalanced reaction below:

   $$NaCl + Ba(OH)_2 \rightarrow BaCl_2 + NaOH$$

2. Describe how you would prepare 100.0 mL of a 0.420 M $C_2H_2O_6$ solution from a stock solution whose concentration is 5.00 M.

3. Calculate the volume percent of a solution that contains 35.0 mL of ethanol in 750.0 mL of solution.

4. How many milliliters of a 0.123 M HCl solution are required to react exactly with 35.32 mL of a 0.282 M NaOH solution?

5. 50.00 mL of H_2SO_4 solution is titrated to neutralization with 22.34 mL of 0.120 M NaOH. Calculate the concentration of the H_2SO_4 solution.

6. Calculate the mass percent of a solution that contains 80.0 g of KNO_3 and 250.0 g of water.

7. When 10.00 g of an unknown compound are dissolved in 250.0 g of water, the freezing point of the solution is $-3.70\,°C$. What is the molar mass of the compound?

8. Calculate the mass percent of a solution that contains 400.0 g of KBr in 1.000 kg of solution.

9. Calculate the mass/volume percent in a solution that contains 8.50 g of C_3H_6O in 125.0 mL of solution.

10. Describe how you would prepare 500.0 mL of a 0.631 M Na_2SO_4 solution from solid Na_2SO_4 and water.

Exercises for Self-Testing

A. Completion. *Write the correct word(s) to complete each statement below.*

1. All solutions are _____ mixtures.
2. In a solution prepared by dissolving 2.0 g of naphthalene in 25.0 g of benzene, naphthalene is called the _____, and benzene is called the _____.
3. _____ is the term used to describe the maximum amount of solute that can be dissolved in a given amount of solvent.
4. The general rule for solubility is that _____ dissolves _____.
5. The solubility of a _____ solute in a liquid solvent will likely increase with temperature, whereas the solubility of a _____ solute in a liquid solvent will decrease with temperature.
6. The solubility of a _____ solute in a liquid solute will increase if the pressure above the solution is increased.
7. _____ is a measure of the disorder in a system.
8. A _____ solution is one in which the maximum amount of solute is dissolved in solution.
9. The ΔE_{total} for the solution process involves adding the ΔE values for _____ steps.
10. The ΔE value for step _____ in the solution process is the only negative value.
11. The nonpolar tail of a soap molecule will not dissolve in water and is said to be _____, whereas the polar head is said to be _____ because it has an affinity for water.
12. Soap molecules form spherical _____, which allow greasy dirt to be dissolved in water.

13. The molarity of a solution is defined as the number of _____ of solute dissolved in one _____ of solution.

14. Solutions can either be made from scratch (combining solute plus solvent) or from a more _____ stock solution.

15. The percent by mass of a solution containing 25.0 g of solute and 100.0 g of solvent is _____.

16. The amount by which the boiling point of a solution is higher than the boiling point of the pure solvent is the _____ _____ _____ .

17. The vapor pressure above a solution is _____ than the vapor pressure above the pure solvent.

18. When performing solution stoichiometry problems, we can convert the volume of a solution to moles by using the _____.

19. A 1.0 M KCl solution will have a _____ (higher or lower) boiling point than a 1.0 M solution of $CaCl_2$.

20. Three colligative properties are boiling-point elevation, freezing-point depression, and _____ _____ .

B. True-False. *Indicate whether each statement below is true or false, and explain why the false statements are false.*

_____ 1.   Water is the solvent in a solution containing 50.0 mL of ethanol and 25.0 mL of water.

_____ 2.   The ΔE value for Step 1 of the solution process may be positive or negative.

_____ 3.   A solute will not be soluble in a solvent if the ΔE_{total} is a positive number.

_____ 4. BF_3 is insoluble in nonpolar solvents because of the polar BF bonds.

_____ 5.   The outer portion of a soap micelle is hydrophilic, whereas the inner portion is hydrophobic.

_____ 6.   The solvent and solute must be melted in order for two solids to form a solution.

_____ 7.   The percent by volume of a solution made by mixing 15.0 mL of solute and 100.0 mL of solvent is 15.0%.

_____ 8. CH_3OH should dissolve in H_2O.

_____ 9. CO_2 dissolves better in water at higher temperatures.

_____ 10. Once a liquid solution is saturated with a solid solute, it will be impossible to get any more solute to dissolve in the solvent.

_____ 11. The volume of a solution cannot be determined by adding the volumes of the solute and solvent.

_____ 12. The boiling-point elevation for a 1.50 M NaCl solution will be less than that of a 3.00 M solution of glucose, $C_6H_{12}O_6$.

_____ 13. The vapor pressure of a solution is higher than the vapor pressure of the pure solvent.

_____ 14. An addition of 1 g of NaCl will lower the freezing point of 150 mL of water more than an addition of 5 g of sucrose, $C_{12}H_{22}O_{11}$.

C. Questions and Problems

1. Explain why a pot of water will form bubbles as it is being heated.

   Use the following information for Problems 2–6: A solution is prepared by mixing 45.0 g of glucose $(C_6H_{12}O_6)$ with enough water to make 300.0 mL of solution.

2. Calculate the percent by mass/volume of the solution.

3. Calculate the molarity of the solution.
4. Could this solution be used as a stock solution to make a glucose solution with a molarity of 1.5 M? Why or why not?
5. How would you prepare 500.0 mL of a 0.100 M glucose solution from this solution?
6. How many milliliters of this solution would be needed to obtain 100.0 g of glucose?
7. Explain why the signs of ΔE for Steps 1 and 2 of the solution process are positive but the sign of ΔE for Step 3 is negative.
8. Explain the molecular basis for the general rule that "like dissolves like."
9. A solution is made by combining 5.0 g of a solute whose density is 1.0 g/mL with 95.0 g of a solvent whose density is 1.0 g/mL. Will the percent by mass of the solution be equal to, less than, or greater than the percent by volume of the solution? Explain fully. (Assume in this case that the total volume is equal to the volume of solute plus the volume of solvent.)
10. Explain with the use of a diagram how soapy water dissolves greasy dirt.
11. Calculate the number of milliliters of 0.0160 M H_2SO_4 needed to react with 250.0 mL of 0.0210 M NaOH to produce Na_2SO_4. (*Hint:* Write the balanced equation for the reaction.)
12. Determine the freezing point of a solution that contains 1 mole of each of the following in 1 kg of solvent:

    (a) LiCl   (b) K_2SO_4 (c) $C_6H_{12}O_6$
13. When 250.0 g of an unknown substance is dissolved in 750.0 g of water, the freezing point of the solution is $-10.00\,°C$. What is the molar mass of the unknown?
14. Arrange the following solutions in order of increasing vapor pressure. (Assume each solution contains 1.00 mole of the compound given in 1 kg of solvent.) Explain your answer.

    (a) $C_6H_{12}O_6$ (b) K_2CO_3 (c) $Al_2(SO_4)_3$ (d) LiBr
15. How many moles of H_3PO_4 are needed to neutralize 250.0 mL of 0.100 M NaOH? How many milliliters of 0.0500 M H_3PO_4 are required?

13

When Reactants Turn Into Products

Learning Outcomes

1. Describe what a chemical reaction is in terms of mechanism, bonds broken, and bonds formed.

2. Calculate ΔE_{rxn} and identify a reaction as exothermic (downhill in energy) or endothermic (uphill in energy).

3. Describe the relationship between activation energy, E_a, and reaction rate.

4. Sketch a reaction energy profile.

5. Describe what a catalyst is and how it affects a reaction's rate.

6. Describe in words and mathematically how concentration and inherent factors affect reaction rates.

7. Understand the meaning of reaction orders in a rate law.

8. Derive a reaction's rate law from kinetic data.

9. Understand what a reaction mechanism is and its relationship to reaction rate.

10. Postulate a reaction mechanism from rate data or the rate law.

Chapter Outline

13.1 Chemical Kinetics

    A. Collisions necessary for reactions to occur

    B. Importance of understanding mechanisms

13.2 Energy Changes and Chemical Reactions

    A. Net energy changes between reactants and products

        1. Exothermic and endothermic reactions

        2. Reaction-energy profiles

    B. Energy changes for reverse reactions

    C. The energetics of bond breaking and bond making in determining net energy change

Review of Key Concepts from the Text

13.1 Chemical Kinetics

Chemical kinetics is the branch of chemistry that deals with the rates at which chemical reactions occur and the factors that affect those rates. Much of what determines the rate of a reaction is what happens to the reactants during the time they are being converted to products. The exact sequence of steps that molecules go through on changing from reactants to products is called the *reaction mechanism*. It is impossible to "see" a reaction as it is occurring because the process is too fast. However, we can determine the mechanism of a reaction by studying the rate of the reaction and by observing how the rate changes when we change the concentration of the reactants or the temperature. We will see how this is done later in the chapter, but first we must look more closely at the energy changes that occur during chemical reactions.

13.2 Energy Changes and Chemical Reactions

Most chemical reactions are accompanied by a change in energy. Energy might be given off during the reaction (as in combustion reactions), or energy might be consumed (as in the decomposition of NH_3 into nitrogen and hydrogen). Reactions that give off energy (more specifically, *heat* energy) are called *exothermic* reactions, and reactions that consume energy are called *endothermic* reactions. The container in which an exothermic reaction is occurring will feel hot to the touch (from the energy being released), and the container in which an endothermic reaction is occurring will feel cold to the touch (because it is taking energy from the surroundings to be used in the reaction).

We can also determine the relative energies of the reactants and products if we know whether a reaction is endothermic or exothermic. In an endothermic reaction, the products will have more energy than the reactants (because energy has been absorbed during the reaction), and in an exothermic reaction, the reactants will have more energy than the products (because energy has been released during the reaction).

We can graphically represent the energy changes that occur during a chemical reaction with a *reaction-energy profile* (sometimes called a reaction-energy diagram), which plots the relative energies of the reactants and products. The x-axis is the reaction coordinate, sometimes called reaction progress; it traces the state of the reaction from start to finish. The energy is graphed on the y-axis. The reaction-energy profiles for an endothermic reaction and an exothermic reaction are shown below.

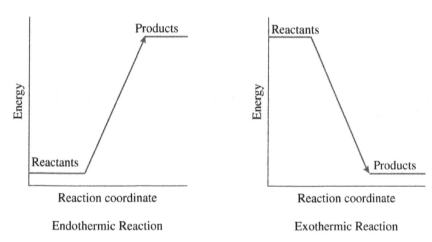

The difference in energy between the reactants and the products is called the change in energy of the reaction (ΔE_{rxn}). ΔE_{rxn} is equal to the energy of the products minus the energy of the reactants. ΔE_{rxn} is therefore positive for an endothermic reaction (in which the products have more energy) and negative for an exothermic reaction (in which the reactants have more energy).

$$\Delta E_{rxn} = E_{products} - E_{reactants}$$

Example 13.1

Indicate which of the following reactions are endothermic and which are exothermic. Draw a reaction-energy profile for each reaction.

(a) $2Fe + Al_2O_3 \rightarrow 2Al + Fe_2O_3$ $\Delta E_{rxn} = +852 \, kJ$

(b) A reaction in which the products contain 479 kJ more energy than the reactants.

(c) A reaction whose container feels hot to the touch while the reaction is occurring.

Solution

(a) This reaction has a positive ΔE_{rxn} and is therefore an endothermic reaction. Its reaction-energy profile is shown at the top of the next page.

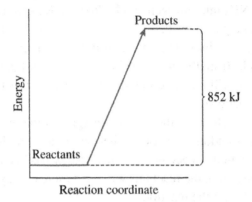

(b) This reaction is also endothermic since the products contain more energy than the reactants. Its reaction-energy profile will be similar to that in (a).

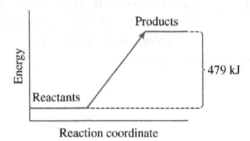

(c) This reaction is an exothermic reaction, giving off heat as it occurs. Its reaction-energy profile is shown below.

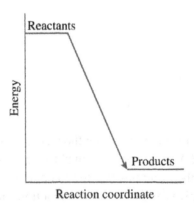

Practice Exercise 13.1

Indicate which of the following reactions are endothermic and which are exothermic. Draw a reaction-energy profile for each reaction.

(a) $C + O_2 \rightarrow CO_2$ $\Delta E_{rxn} = +394$ kJ

(b) A reaction whose container feels cool to the touch.

(c) A reaction in which the reactants contain 238 kJ more energy than the products.

STUDENT WORKBOOK AND SELECTED SOLUTIONS

It should be evident from the reaction-energy profile that the size and sign of the ΔE_{rxn} for the reverse reaction are related to those of the forward reaction. Because the energy difference between the reactants and products (or products and reactants in the reverse reaction) does not change, the size of the ΔE_{rxn} remains the same. However, if the forward reaction is exothermic, the reverse reaction will be endothermic, and vice versa. Hence the signs of the ΔE_{rxn} will be opposite for forward and reverse reactions.

Example 13.2

A reaction consumes 380 kJ of energy.

(a) Is the reaction endothermic or exothermic?

(b) Do the reactants or products have more energy?

(c) What is the ΔE_{rxn} for the reverse reaction?

(d) Does the reaction container feel hot or cold to the touch?

Solutions

(a) The reaction is endothermic because it consumes energy.

(b) The products have more energy than the reactants.

(c) The ΔE_{rxn} for the reverse reaction is -380 kJ.

(d) The reaction container feels cold to the touch.

Practice Exercise 13.2

A reaction gives off 560 kJ of energy.

(a) Is ΔE_{rxn} positive or negative for the reaction?

(b) Do the reactants or products have more energy?

(c) What is the ΔE_{rxn} for the reverse reaction?

(d) Does the reaction container feel hot or cold to the touch?

Another way to determine the value of ΔE_{rxn} is by comparing the amount of energy required to break reactant bonds with the amount of energy released when product bonds are formed.

$$\Delta E_{rxn} = \Delta E_{\text{required by bond breaking}} - \Delta E_{\text{released by bond forming}}$$

If more energy is used up in bond breaking than is released in bond making, the reaction will be endothermic. On the other hand, if the reaction releases more energy on bond making than it requires for bond breaking, the reaction will be exothermic.

Example 13.3

In the reaction of CH_4 with Cl_2 to produce $CHCl_3$ and HCl, the total energy required to break the bonds in the reactants is 1950 kJ. The total energy released in forming the bonds in the products is 2290 kJ.

(a) Is this reaction endothermic or exothermic?

(b) What is the ΔE_{rxn} for the reaction?

(c) Will the reaction container feel hot or cold to the touch? Explain your answer.

Solutions

(a) The energy released upon bond formation (2290 kJ) is greater than the energy consumed in bond breaking (1950 kJ). Therefore, the reaction is exothermic.

(b) The ΔE_{rxn} will equal the difference between the energy consumed and the energy released: $\Delta E_{rxn} = 1950\,\text{kJ} - 2290\,\text{kJ} = -340\,\text{kJ}$.

(c) The reaction container will feel hot to the touch because energy is being released during the reaction.

Practice Exercise 13.3

In the reaction of H_2 with Cl_2 to produce HCl, the total energy required to break the bonds in the reactants is 680 kJ. The total energy released in forming the bonds in the products is 860 kJ.

(a) Is this reaction endothermic or exothermic?

(b) What is the ΔE_{rxn} for the reaction?

(c) What is the ΔE_{rxn} for the reverse reaction?

(d) Will the reaction container feel hot or cold to the touch? Explain your answer.

13.3 Reaction Rates and Activation Energy— Getting Over the Hill

When we look at reaction-energy profiles of endothermic and exothermic reactions, it seems that endothermic reactions would be slow reactions (trudging up an energy barrier) whereas exothermic reactions would be fast reactions (sliding down an energy hill). However, it happens that some endothermic reactions are faster than some exothermic reactions. Evidently, the energy change for the reaction is not the relevant consideration in determining reaction rates. There is another entity, the *activation energy* of the reaction (E_a), that determines the rate at which a reaction will occur. The activation energy is the energy hill that exists between the reactants and the products, which every reaction must climb—endothermic and exothermic reactions alike.

Whenever a reaction occurs, reactant molecules must be converted to product molecules, and bonds are broken and formed in the process. The activation energy represents the energy that must be added to the reactant molecules in order for them to collide hard enough to begin the bond-breaking process. If the reactant molecules have less energy than this, they will only bounce off one another, and no product will be formed. However, when the reactant molecules collide with the necessary amount of energy (the activation energy), they can proceed to products. This phenomenon is illustrated graphically in the reaction-energy profile shown at the top of the next page.

All reactions must begin by absorbing enough energy to get over the activation energy barrier. It is the height of this barrier that determines how fast a reaction will occur. The higher the activation energy barrier, the slower the reaction rate will be.

At the top of the activation energy barrier, the reactant molecules have been transformed into a special, high-energy species, called the *transition state*. In the transition state, the reactant molecule bonds that will be broken are partially broken, and the product molecule bonds that will be formed are partially formed. Because the transition state contains partially broken and partially formed bonds, it is extremely unstable. It is called the transition state because it represents molecules in transition on the way from reactants to products.

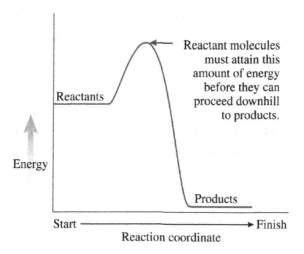

Reactant molecules
must attain this
amount of energy
before they can
proceed downhill
to products.

Reactants

Energy

Products

Start ⟶ Finish

Reaction coordinate

Example 13.4

In the reaction of A + B ⟶ C, the energy of the reactants is 90 kJ/mol, and the energy of the products is 50 kJ/mol. The energy of the transition state is 120 kJ/mol.

(a) Is the reaction endothermic or exothermic?

(b) What is the ΔE_{rxn}?

(c) What is the E_a?

Solutions

(a) The reaction is exothermic since the products have less energy than the reactants.

(b) $\Delta E_{rxn} = E_{products} - E_{reactants} = 50\,kJ/mol - 90\,kJ/mol = -40\,kJ/mol$

(c) The E_a is the difference in energy between the reactants and the transition state.
$E_a = E_{transition\ state} - E_{reactants} = 120\,kJ/mol - 90\,kJ/mol = 30\,kJ/mol$

Practice Exercise 13.4

In the reaction of X + Y ⟶ Z, the energy of the reactants is 60 kJ/mol, and the energy of the products is 90 kJ/mol. The energy of the transition state is 130 kJ/mol.

(a) Is the reaction endothermic or exothermic?

(b) What is the ΔE_{rxn}?

(c) What is the E_a?

Temperature and the Collision Theory of Reaction Rates

We mentioned in the previous section that reactant molecules have to collide in order for a reaction to occur. In addition, the collisions have to be energetic enough to get over the activation energy barrier (E_a) in order for the reactants to be converted to products. The more collisions that occur with sufficient energy, the faster a reaction will occur. The speed of the reaction depends on the height of the activation energy barrier, because the higher the barrier, the fewer the number of molecules with energy equal to or greater than the activation energy.

It turns out that increasing the temperature always increases the rate of a chemical reaction. We can explain this phenomenon when we consider the effect of temperature on the energy of the molecules. Increasing the temperature of the reaction causes the reactant molecules to move faster and consequently have more kinetic energy. As a result, a greater number of collisions will occur with energy equal to or greater than the activation energy, thereby allowing the reaction to proceed at a faster rate. In fact, the reaction rate is approximately doubled for every 10 °C rise in temperature.

Another factor to consider in determining whether a reaction will occur is the relative orientation of the colliding molecules. The reactant molecules must be oriented properly, or no reaction can occur. We will consider the reaction of H_2 with I_2 to form HI as an example. The reaction that occurs is $H_2 + I_2 \rightarrow 2HI$. In order for the reaction to occur, the H_2 and I_2 molecules have to be perfectly aligned for the H—I bonds to form, as shown below.

H—H	H·····H	H H
+	⋮ ⋮	| + |
I—I	I·····I	I I
Reactants	Transition state	Products

If the H_2 and I_2 molecules collide end to end, the H—I bonds cannot form, and no reaction will take place.

We can express the dependence of reaction rate on both the number of collisions with sufficient energy and the proper orientation in the following equation:

$$\text{Reaction rate} = \underbrace{\begin{array}{c}\text{Number of collisions}\\\text{with energy greater than } E_a\\\text{that occur per unit time}\end{array}}_{\text{Energy factor}} \times \underbrace{\begin{array}{c}\text{Fraction of collisions with}\\\text{proper orientation}\end{array}}_{\text{Orientation factor}}$$

The orientation factor is always a number between 0 and 1, with 1 meaning that any orientation will work (100% of the sufficiently energetic collisions are properly oriented), and 0 meaning that no orientation will work (0% of the sufficiently energetic collisions are properly oriented). Orientation factors are typically around 0.1 (meaning around 10% of the sufficiently energetic collisions are properly oriented).

Example 13.5

A chemical reaction has an orientation factor of 0.40. Will this reaction occur faster or slower than a reaction with an orientation factor of 0.50, if the numbers of collisions per second are equal in the two reactions?

Solution

The reaction with an orientation factor of 0.40 will occur more slowly than the one with a collision factor of 0.50. Only 40% of the sufficiently energetic reactants have proper orientation in the first reaction, whereas 50% have the proper orientation in the second reaction.

Practice Exercise 13.5

Will a reaction that has 500 collisions per second and an orientation factor of 0.3 occur faster, slower, or at the same rate as a reaction that has 400 collisions per second and an orientation factor of 0.4? Explain.

Catalysts

Catalysts are substances that speed up the rate of a chemical reaction without being consumed in the process. A catalyst speeds up the rate of reaction by providing a new pathway for the reaction to occur. The new pathway has a lower activation energy than the uncatalyzed pathway. This is shown in the adjacent figure. Because the new pathway has a lower activation energy, more reactant molecules will have the energy necessary to climb the activation energy barrier than would have been the case without the catalyst. Catalysts often do this by somehow weakening a reactant bond that must be broken, thereby decreasing the activation energy.

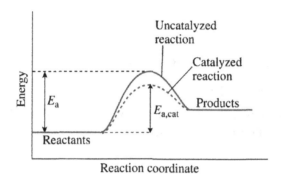

Biological reactions utilize catalysts routinely. These catalysts are known as *enzymes*, and the reactants upon which they work are called *substrates*. The enzyme acts on the substrate in such a way that the activation energy of the reaction is lowered. Enzymes exhibit a high degree of specificity, meaning that an enzyme will generally act on only one substrate (or a small number of substrates) and will catalyze only one reaction (or a very small number of reactions). Their specificity is due to the fact that enzymes contain specially shaped pockets or depressions that fit only specific reactant molecules. This pocketed region, called the active site, fits the shape of the reactant molecule similarly to the way a lock fits a key. This is why the mechanism of enzyme action is often called the "lock and key" mechanism. Once the lock and key are in place, the enzyme weakens some of the substrate bonds and brings the substrate into the proper orientation to react, with a lower E_a. Without enzymes, most biological reactions would occur so slowly that the organism would die due to insufficient amounts of the products of the reaction.

13.4   How Concentration Affects Reaction Rate

Changing the concentration of the reactants has an effect on the rate of the reaction because the concentration affects the number of collisions between reactant molecules. The effect of concentration on rate is used to determine the mechanism of a chemical reaction. We can rewrite the reaction rate equation in the following manner:

$$\text{Reaction rate} = \begin{array}{c}\text{Total number of}\\ \text{collisions that occur}\\ \text{per unit time}\end{array} \times \begin{array}{c}\text{Fraction of collisions}\\ \text{with energy greater}\\ \text{than } E_a\end{array} \times \begin{array}{c}\text{Fraction of collisions}\\ \text{with proper orientation}\end{array}$$

Notice that the first two terms are what we previously called the energy factor and the third term is what we called the orientation factor.

In the reaction rate equation above, the last two factors are characteristics of the reacting molecules and will not change with changing concentrations of reactants. Chemists combine these two factors and represent their combination with the letter k, called the rate constant for the reaction. The equation, called the *rate law*, can therefore be rewritten as

$$\text{Reaction rate} = k \times \begin{array}{c}\text{Total number of collisions}\\ \text{that occur per unit time}\end{array}$$

We can now calculate the rate of a chemical reaction if we know how many collisions are occurring per unit time. The number of collisions occurring per unit time is proportional to the product of the concentrations of the reactants, each raised to some power or exponent. The power is called the *order* of the reaction with respect to that reactant. An exponent of 1 is called first-order, an exponent of 2 is called second-order, and so on. The sum of all of the exponents in the rate law is the *overall order of the reaction.*

Example 13.6

The reaction rate for a reaction $A + B \rightarrow C$ is expressed as:

$$\text{Reaction rate} = k[A][B]^3$$

What is the order of the reaction with respect to A? What is the order of the reaction with respect to B? What is the overall order of the reaction?

Solution

The order of the reaction with respect to A is 1 (first-order). The order of the reaction with respect to B is 3 (third-order). The overall order of the reaction is 4 (fourth-order).

Practice Exercise 13.6

The reaction rate for the reaction $2NO + O_2 \rightarrow 2NO_2$ is expressed as:

$$\text{Reaction rate} = k[NO]^2[O_2]$$

What is the order of the reaction with respect to NO? What is the order of the reaction with respect to O_2? What is the overall order of the reaction?

13.5 Reaction Order

In the reaction rate equation for a general reaction, $aA + bB \rightarrow P$, we can substitute the term involving the total number of collisions per unit time with the concentrations raised to certain powers. Each power is the order of the reaction with respect to that particular reactant. The general rate law for the reaction $aA + bB \rightarrow P$ is:

$$\text{Reaction rate} = k[A]^x[B]^y$$

Notice that in the general rate law the exponents on the concentration values are x and y, and are *not* the coefficients in the balanced equation, a and b. The values of x and y are the orders of the reaction with respect to A and B, respectively, and must be determined experimentally by a procedure we will describe shortly. When the values of x and y are included in the rate law, the rate law is called the experimentally determined rate law, or the experimental rate law.

To determine the values of x and y, we must determine how changing the concentrations of A and B, respectively, changes the rate of the reaction. Kinetics experiments are experiments in which the concentration of one reactant is changed by a certain factor (such as doubled or tripled) while keeping the concentrations of the other reactants constant. The factor by which the reaction rate changes as a result of the concentration change is noted and used to determine the values of the exponents. The exponent on a particular reactant's concentration will be equal to the power to which the change in concentration must be raised in order to get the observed change in the rate. For example, if [A] is doubled, and the rate doubles as a result, the value of x will be 1. This is because 2 (the factor by which the concentration is changed) must be raised to the first power to get 2 (the resulting change in the rate). If there is no change in rate when there is a concentration change, the factor by which the rate changes is 1, and the value of the exponent is 0, because any number raised to the 0 power equals 1 (e.g., $2^0 = 1$). The following table indicates the appropriate exponent for particular changes in concentration and resulting rate changes.

Factor by Which the Concentration Changes	Factor by Which the Reaction Rate Changes	Value of the Exponent in the Experimental Rate Law
2	1 (no change)	$0(2^0 = 1)$
2	2	$1(2^1 = 2)$
2	4	$2(2^2 = 4)$
2	8	$3(2^3 = 8)$
2	16	$4(2^4 = 16)$
3	1 (no change)	$0(3^0 = 1)$
3	3	$1(3^1 = 3)$
3	9	$2(3^2 = 9)$
3	27	$3(3^3 = 27)$
3	81	$4(3^4 = 81)$

In order to find the exponent on the concentration term for a specific reactant, we must locate two experiments in which the concentration of the reactant we are interested in changes but all other concentrations remain the same. This is important because we want to know how the rate changes when we change the reactant in question. Therefore, it must be the only reactant whose concentration is changing.

Once we know which experiments to consider in determining the exponent for a particular reactant, we use the change in its concentration and the resulting change in the rate to determine the exponent. The following example demonstrates how this procedure is used to determine the exponents in an experimentally determined rate law (experimental rate law) for a chemical reaction.

Example 13.7

A kinetics experiment on the reaction of $bB + cC \rightarrow dD$ yielded the data shown in the table below.

Experiment	[B]	[C]	Rate (M/s) of D Formation
1	0.100	0.100	0.032
2	0.100	0.200	0.128
3	0.200	0.100	0.064

(a) Write the general rate law for the reaction.

(b) Which two experiments would you compare to determine the exponent on B? On C?

(c) What are the exponents on B and C in the experimentally determined rate law?

(d) Write the experimentally determined rate law for the reaction.

(e) What is the order of the reaction with respect to B? With respect to C? What is the overall order of the reaction?

Solutions

(a) The general rate law for the reaction is reaction rate $= k[B]^x[C]^y$.

(b) To get the exponent on B, we consider two experiments where [B] is changing but [C] remains the same. This is true of experiments 1 and 3. To get the exponent on C, we consider two experiments where [C] is changing but [B] remains the same. This is true of experiments 1 and 2.

(c) To get the values of x and y, we use the two sets of experiments we have identified in part (b) above. *Exponent on B:* In experiments 1 and 3, we find that doubling the concentration of B doubles the reaction rate. Thus, the exponent on B is 1. *Exponent on C:* In experiments 1 and 2, we find that doubling the concentration of C quadruples the reaction rate. Thus, the exponent on C is 2.

(d) Reaction rate $= k[B][C]^2$

(e) The order of the reaction with respect to B is 1. The order of the reaction with respect to C is 2. The overall order of the reaction is 3.

Practice Exercise 13.7

A kinetics experiment on the reaction of $rR + sS \rightarrow tT$ yielded the data shown in the table below.

Experiment	[R]	[S]	Rate (M/s) of T Formation
1	0.120	0.120	0.030
2	0.360	0.120	0.090
3	0.120	0.240	0.060

(a) Write the general rate law for the reaction.

(b) Which two experiments would you compare to determine the exponent on R? On S?

(c) What are the exponents on R and S in the experimentally determined rate law?

(d) Write the experimentally determined rate law for the reaction.

(e) What is the order of the reaction with respect to R? With respect to S? What is the overall order of the reaction?

13.6 Why Reaction Orders Have the Values They Do—Mechanisms

Although some chemical reactions occur in one step, most reactions occur in a series of steps. The elementary step(s) that indicate the pathway from reactants to products are called the *reaction mechanism*. The individual steps generally involve one, two, or three reactant molecules. The steps are given specific designations, depending on the number of reactant molecules that are involved in the step. The designations are shown at the top of the next page.

Number of Reactant Molecules in Elementary Steps	Designation of Step
1	Unimolecular
2	Bimolecular
3	Termolecular

Because the elementary steps in a mechanism represent the actual collisions, the coefficients in the balanced equations are the exponents in the rate law for the individual steps in the mechanism.

A reaction cannot proceed any faster than the slowest step in its mechanism, so the slowest step is called the *rate-determining step* of the reaction, or simply the rds. We can write the rate law for a chemical reaction if we know the stoichiometry of the rate-determining step. This rate law, predicted from the slow step in the mechanism, is called the *predicted rate law* because it was not determined by experiment.

Example 13.8

The slow step in the mechanism for the production of NO_2F from NO_2 and F_2 is $2NO_2 + F_2 \rightarrow NO_2F + F$. Write the predicted rate law for the reaction.

Solution

The predicted rate law for the reaction contains the reactants with their respective coefficients as exponents. The predicted rate law is therefore:

Reaction rate $= k[NO_2]^2[F_2]$.

Practice Exercise 13.8

The slow step in the mechanism for the production of C from A and B is $A + 2B \rightarrow AB_2$. Write the predicted rate law for the reaction.

We can also proceed in the other direction. If we know the experimentally determined rate law and the overall reaction, we can postulate a mechanism that is consistent with the information we have. We use the rate law to determine the reactants in the slowest step of the mechanism and devise additional steps whose overall sum equals the overall chemical reaction. This process is shown in the example below. (See Section 13.6 of the textbook for a complete discussion of this procedure, with examples.)

As you study reaction mechanisms, you will find that there will be species that appear in the mechanism but not in the overall reaction. Consider the following mechanism, for the reaction of A_2 with B to produce A_2B:

Step 1: $A_2 + Z \rightarrow AZ + A$ (slow)

Step 2: $A + B \rightarrow AB$ (fast)

Step 3: $AZ + AB \rightarrow Z + A_2B$ (fast)

When we add the three steps in the mechanism, we get the overall reaction. Species that appear on both sides of the equations cancel each other out, as shown at the top of the next page.

Step 1: $A_2 + Z \rightarrow AZ + A$ (slow)

Step 2: $A + B \rightarrow AB$ (fast)

Step 3: $AZ + AB \rightarrow Z + A_2B$ (fast)

Overall reaction: $A_2 + B \rightarrow A_2B$

There are some species that appear in the mechanism but not in the overall reaction. They are canceled out when the individual steps are added together. The first is Z, which appeared first as a reactant in step 1, the rate-determining step, and then as a product in step 3. Species that appear first as a reactant in the rate-determining step and then as a product in a later step are catalysts. They speed up the rate-determining step but are regenerated later as a product. This is in keeping with the requirement that a catalyst not be consumed during a reaction.

The other species that appear in the mechanism but not in the overall reaction are AZ and AB. Species that are produced during an earlier step of the reaction and then consumed in a later step are called *reaction intermediates*. While relatively unstable (they disappear almost immediately after they are produced) and of high-energy, intermediates can be detected during the course of a reaction and subsequently characterized. Their appearance and disappearance provide valuable clues in regards to plausible reaction mechanisms.

Example 13.9

The mechanism for the reaction of A with A to produce A_2 is shown below.

Step 1: $A + BC \rightarrow AB + C$ (slow)

Step 2: $A + AB \rightarrow B + A_2$ (fast)

Step 3: $B + C \rightarrow BC$ (fast)

(a) Write the equation for the overall reaction that is occurring.

(b) What is the predicted rate law for the reaction?

(c) Are there any catalysts present? If so, identify them.

(d) Are there any reaction intermediates present? If so, identify them.

Solutions

(a) We add the steps to get the overall reaction that is occurring.

Step 1: $A + BC \rightarrow AB + C$ (slow)

Step 2: $A + AB \rightarrow B + A_2$ (fast)

Step 3: B + C → BC (fast)

Overall reaction: A + A → A$_2$

(b) The predicted rate law can be written from the slow step in the mechanism:
Reaction rate = k[A][BC].

(c) Any catalyst(s) present would appear in the mechanism first as a reactant in the slow step and then as a product in a subsequent step. When we examine the mechanism, we see that BC appears first as a reactant in step 1 and then as a product in step 3. Therefore, BC is a catalyst in the reaction.

(d) Any reaction intermediate(s) present would appear in the mechanism first as a product and then as a reactant in a subsequent step. When we examine the mechanism, we see that AB appears first as a product in step 1 and then as a reactant in step 2. We also see that B appears first as a product in step 2 and then as a reactant in step 3. Therefore, AB and B are both intermediates in the reaction.

Practice Exercise 13.9

The two-step mechanism for the decomposition of H_2O_2 to produce H_2O and O_2 is shown below.

Step 1: H_2O_2 + I$^-$ → H_2O + IO$^-$ (slow)

Step 2: H_2O_2 + IO$^-$ → H_2O + O_2 + I$^-$ (fast)

(a) Write the equation for the overall reaction that is occurring.

(b) What is the predicted rate law for the reaction?

(c) Are there any catalysts present? If so, identify them.

(d) Are there any reaction intermediates present? If so, identify them.

When you are asked to postulate a mechanism from an overall reaction and the slow step in a mechanism, there are several generalizations you should bear in mind.

1. Elementary steps usually have no more than two reactants, but there are exceptions.
2. The elementary steps in the mechanism must add up to give the overall reaction.
3. The slow step in the mechanism must yield a rate law that is consistent with the experimentally determined rate law.
4. When your postulated mechanism agrees with the experimentally determined rate law, you can say only that the mechanism is plausible. There may be other mechanisms that are also consistent with the data and cannot be ruled out. It *is* possible, however, to rule out a postulated mechanism if it does not agree with the experimentally determined rate law.

The above generalizations will be invaluable when you are asked to postulate a reaction mechanism. (See Section 13.6 of the textbook for a complete discussion of these concepts.)

Strategies for Working Problems in Chapter 13

Overview of Problem Types: What You Should Be Able to Do

Chapter 13 provides a discussion of the factors that influence the rate at which chemical reactions will occur. The types of problems that you should be able to solve after mastering Chapter 13 are:

1. Draw a reaction energy profile and calculating ΔE_{rxn} when given the energy of the reactants and products, and the activation energy, E_a, of a reaction.

2. Write the rate law for a chemical reaction and determine the order of the reaction when given experimental data.

3. Determine a possible mechanism of a reaction when provided with the overall reaction and the rate law.

Flowcharts

Flowchart 13.1 Drawing a reaction-energy profile and calculating ΔE_r when given the energy of the reactants and products, and the activation energy, E_a of a reaction

Flowchart 13.1

Summary:

Method for drawing a reaction-energy profile when given the energy of the reactants and products (or the ΔE_{rxn} for the reaction) and the activation energy, E_a of the reaction

Step 1
- **Draw the x- and y-axis of a graph and label the x-axis "Reaction Coordinate" and the y-axis "Energy".**

$\downarrow$

Step 2
- **Place the energy of the reactants on the y-axis (on the left portion of the x-axis), and place the energy of the products on the right portion of the x-axis.**

$\downarrow$

Step 3
- **Place the energy of the transition state between the energy of the reactants and the products.**

$\downarrow$

Step 4
- **Draw a curve connecting the energy of the reactants to that of the transition state and the energy of the transition state to that of the products.**
- **Determine the ΔE_{rxn} (if it is not provided in the problems) by subtracting the energy of the reactants from the energy of the products of the reaction. $(E_{products} - E_{reactants})$.**
- **The reaction is endothermic if the energy of the products is greater than the energy of the reactants, and the reaction is exothermic if the energy of the products is less that the energy of the reactants.**

Example 1

In a hypothetical chemical reaction, the energy of the reactants is 200 kJ/mol, the energy of the products is 250 kJ/mol, and the energy of the transition state is 400 kJ/mol. Draw the reaction-energy profile for the reaction. Is the reaction endothermic or exothermic?

Solution

Perform Step 1

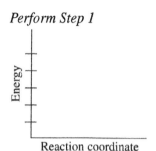

Perform Step 2

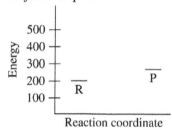

Perform Step 3

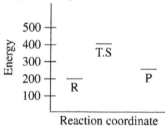

Perform Step 4

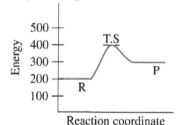

The reaction is endothermic.

Example 2

In a hypothetical chemical reaction the energy of the reactants is 200 kJ/mol, the reactants of the products is 150 kJ/mol, and the energy of the transition state is 250 kJ/mol. Draw the reaction energy profile for the reaction. Is the reaction endothermic or exothermic?

Solution

Perform Step 1

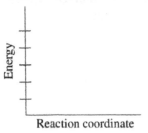

Perform Step 2

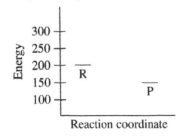

Perform Step 3

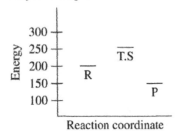

Perform Step 4

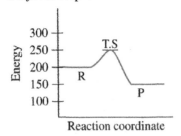

The reaction is exothermic.

Practice Problems

13.1 In the reaction X → Y, the energy of X is 50 kJ/mol, the energy of Y is 30 kJ/mol, and the energy of the transition state is 80 kJ/mol. Draw the reaction energy profile for the reaction. Is the reaction endothermic or exothermic?

13.2 In a certain chemical reaction, the energy of the reactants is 35 kJ/mol, the energy of the products is 45 kJ/mol, and the energy of the transition state is 50 kJ/mol. Draw the reaction energy profile for the reaction. Is the reaction endothermic or exothermic?

Flowchart 13.2 Writing the rate law for a chemical reaction and determining the order of the reaction when given experimental data

Flowchart 13.2

Summary:

Method for writing the rate law for a chemical reaction and determining the order of the reaction when given experimental data

Step 1
- Write the general rate law for the reaction.
- The general rate law for a reaction $A + B + C \rightarrow D$ is
  Rate $= k[A]^x[B]^y[C]^z$, where x, y, and z and are determined from the experimental data.

$\downarrow$

Step 2
- Determine the values of x, y, and z as indicated below.
- Determine the value of x by identifying two experiments for which [A] changes but the [B] and [C] remain the same.
- Determine how the rate changes with the change in [A] to get the value of x.
- Determine the value of y by identifying two experiments for which [B] changes but the [A] and [C] remain the same. Determine how the rate changes with the change in [B] to get the value of y.
- Determine the value of z by identifying two experiments for which [C] changes but the [A] and [B] remain the same. Determine how the rate changes with the change in [C] to get the value of z.

$\downarrow$

Step 3
- Determine the order of the reaction by adding the values of x, y and z.

Example 1

In a kinetics study of the reaction $A + B + C \rightarrow D$, the following data were obtained. Write the rate law for the reaction, and determine the overall order of the reaction.

Experiment	[A]	[B]	[C]	Rate (M/s) of D formation
1	0.100	0.100	0.100	0.032
2	0.100	0.200	0.100	0.128
3	0.200	0.100	0.100	0.064
4	0.100	0.100	0.200	0.064

Solution

Perform Step 1

General rate law: Rate $= k[A]^x[B]^y[C]^z$

Perform Step 2

(a) Use experiments 1 and 3 to find the value of x. When [A] doubles, holding [B] and [C] constant, the rate also doubles. Because $2^1 = 2$, the value of $x = 1$.

(b) Use experiments 1 and 2 to find the value of y. When [B] doubles, holding [A] and [C] constant, the rate increases by a factor of 4. Because 2^2 is 4, the value of $y = 2$.

(c) Use experiments 1 and 4 to find the value of z. When [C] doubles, holding [A] and [B] constant, the rate increases by a factor of 2. Because 2^1 is 2, the value of $z = 1$.

The rate law is therefore:

$$\text{Reaction rate} = k[A]^1[B]^2[C]^1 = k[A]^1[B]^2[C]$$

Perform Step 3

The sum of x, y, and $z = 1 + 2 + 1 = 4$. Therefore, the overall order of the reaction is 4.

Example 2

The rate data for the reaction:

$$E + F \rightarrow G$$

are shown below. Write the rate law for the reaction and determine the overall order of the reaction.

Experiment	[E]	[F]	Rate (M/s) of G formation
1	0.010	0.015	0.001 38
2	0.010	0.045	0.004 14
3	0.030	0.015	0.012 42

Solution

Perform Step 1

General rate law: Reaction rate $= k[E]^x[F]^y$

Perform Step 2

(a) Use experiments 1 and 3 to find the value of x. When [E] is tripled $(0.030/0.010 = 3)$, holding [F] constant, the rate increases by a factor of 9 $(0.012\ 42/0.001\ 38 = 9)$. Because $3^2 = 9$, the value of $x = 2$.

(b) Use experiments 1 and 2 to find the value of y. When [F] triples, holding [E] constant, the rate triples $(0.004\ 14/0.001\ 38 = 3)$. Because $3^1 = 3$, the value of $y = 1$.

Therefore, Reaction rate $= k[E]^2[F]^1 = k[E]^2[F]$

Perform Step 3

The sum of x and $y = 1 + 2 = 3$. Therefore, the overall order of the reaction is 3.

Practice Problem

13.3 The results of kinetics experiments for the reaction:

$$R + S \rightarrow Z$$

are shown below. Write the rate law for the reaction and indicate the overall order of the reaction.

Experiment	[R]	[S]	Rate (M/s) of Z formation
1	0.240	0.240	0.060
2	0.720	0.240	0.180
3	0.240	0.480	0.120

Flowchart 13.3 **Determining a possible mechanism of a reaction when provided with the overall reaction and the rate law**

Flowchart 13.3

Summary:

Method for determining a possible mechanism of a reaction when provided with the overall reaction and the rate law

Step 1
- **Write the slow step in the mechanism when provided with the rate law and the overall reaction.**
- **The species whose concentrations are represented in the rate law are the reactants in the slow step of the mechanism.**
- **The exponents on the concentrations are the coefficients in the slow step.**
- **The products of the slow step should reflect the same number and kind of atoms as the reactants in the slow step, but arranged to make different substances.**

Step 2
- **Add additional steps to the mechanism such that the sum of all steps equals the overall reaction.**
- **No elementary steps should have more than two reactants.**

Example 1

Determine a possible mechanism for the reaction of $A_2 + B \rightarrow A_2B$. The rate law for the reaction is:

Reaction rate $= k[A_2][Z]$.

Solution

Perform Step 1

The reactants in the slow step of the reaction are A_2 and Z because they appear in the rate law. The coefficients for both species are 1 because the exponent of each is 1 in the rate law.

A possible first step in the mechanism is therefore

$A_2 + Z \rightarrow AZ + A$

Perform Step 2

Adding additional steps such that the sum of all steps equals the overall reaction, a possible mechanism is:

Step 1)	$A_2 + Z \rightarrow AZ + A$	(slow)
Step 2)	$A + B \rightarrow AB$	(fast)
Step 3)	$AZ + AB \rightarrow Z + A_2B$	(fast)
Overall reaction	$A_2 + B \rightarrow A_2B$	

Practice Problem

13.4 Determine a possible mechanism for the reaction of

$$Br_2 + CHBr_3 \rightarrow HBr + CBr_4$$

The rate law for the reaction is:

$$\text{Reaction rate} = k[Br_2].$$

Quiz for Chapter 13 Problems

1. Fill in the blanks in the following statement: When the energy of the _____ is greater than the energy of the _____ the reaction is exothermic, and when the energy of the _____ is greater than the energy of the _____ the reaction is endothermic.

2. The reaction producing CO_2 and NO from CO and NO_2 is shown below:

   $$CO + NO_2 \rightarrow CO_2 + NO$$

   The reaction is first order with respect to both CO and NO_2.
   (a) Write the rate law for the reaction.
   (b) What is the overall order of the reaction?
   (c) What happens to the reaction rate if the concentration of CO is doubled?

3. Draw a reaction energy profile diagram for hypothetical chemical reaction with the following characteristics. The energy of the reactants is 20 kJ/mol, the energy of the products is 30 kJ/mol, and the energy of the transition state is 40 kJ/mol. Is the reaction endothermic or exothermic? Explain.

4. Write the general rate law for the following reaction, using x and y as orders.

   $$2NO + O_2 \rightarrow 2NO_2$$

5. The rate law for the reaction: $2NO + Br_2 \rightarrow 2NOBr$ is Reaction rate $= k[NO]^2[Br_2]$ How will the rate of the reaction change when the following changes occur?
   (a) [NO] is doubled   (b) $[Br_2]$ is tripled   (c) [NO] is tripled
   (d) [NO] doubles and $[Br_2]$ triples

6. A reaction releases 700 kJ of energy.
   (a) Is the reaction endothermic or exothermic?
   (b) In the reaction energy profile are the reactants or products higher? Explain.

7. Determine a possible mechanism for the reaction: $2NO + Cl_2 \rightarrow 2NOCl$
   The rate law for the reaction is: Reaction rate $= k[NO][Cl_2]$.

8. The rate law for the reaction: $3A + B \rightarrow D$ is: Reaction rate $= k[A][B]^2$. What is the overall order of the reaction?

STUDENT WORKBOOK AND SELECTED SOLUTIONS

9. The slow step in the mechanism of the reaction of H_2 with ICl to produce HCl and I_2 is shown below. Write the rate law for the reaction:

$$H_2 + ICl \rightarrow HI + HCl$$

10. Write the rate law for the reaction: $A + 2B \rightarrow D$ if the following is true. The rate of the reaction quadruples when the concentration of A is doubled, and the rate of the reaction triples when the concentration of B is tripled.

Exercises for Self-Testing

A. Completion. *Write the correct word(s) to complete each statement below.*

1. The series of elementary steps that represent the pathway of a reaction as reactants are converted to products is the _____ _____.
2. In the reaction-energy profile for a(n) _____ reaction, the products are higher than the reactants.
3. A(n) _____ participates in a chemical reaction by speeding it up but is not consumed during the reaction.
4. The species that exists at the top of a reaction-energy profile is the _____ _____.
5. The rate of a chemical reaction will approximately double for every _____ rise in temperature.
6. A catalyst that operates in biological systems is called a(n) _____.
7. The slow step in a reaction mechanism is always the _____ _____ _____.
8. A catalyst speeds up a chemical reaction by providing an alternate pathway that has a lower _____ _____ than the uncatalyzed reaction.
9. If a reaction is second-order with respect to a reactant A, quadrupling the concentration of A will cause the rate of the reaction to increase by a factor of _____.
10. In the _____ _____, the reactant bonds are partially broken and the product bonds are partially formed.
11. A high-energy species that is produced in one step of a reaction mechanism but quickly used up in a subsequent step is called a(n) _____ _____.
12. If the energy required to break the reactant bonds is greater than the energy released in forming the product bonds, the sign of the ΔE_{rxn} for the chemical reaction will be _____.
13. The container in which an exothermic reaction is occurring will feel _____ to the touch, whereas the container in which an endothermic reaction is occurring will feel _____ to the touch.
14. Increasing the concentration of reactants will generally _____ the rate of a chemical reaction.
15. The minimum energy required for the bonds in the reactant molecules to begin breaking is known as the _____ _____.

B. True-False. *Indicate whether each statement below is true or false, and explain why the false statements are false.*

_____ 1.   Exothermic reactions will be faster than endothermic reactions because the reaction goes downhill in energy.

_____ 2.   Although ΔE_{rxn} values can be either positive or negative, all activation energy values are positive.

_____ 3. The slow step in a reaction mechanism is the rate-determining step.
_____ 4. The reaction that has the rate law Reaction rate $= k[A]^{1/2}[B]^2$ has an overall order of $2\frac{1}{2}$.
_____ 5. The orientation factor for a chemical reaction gives the proportion of the collisions that will be of sufficient energy to cause a reaction to occur.
_____ 6. Reaction intermediate appears in the proposed reaction mechanism.
_____ 7. The rate of a chemical reaction involving gaseous reactants is dependent on the size of the container in which the reaction is taking place.
_____ 8. The rate law for a chemical reaction can be used to prove that a mechanism is correct.
_____ 9. If a reaction is second-order in a reactant A, tripling the concentration of A will double the rate.
_____ 10. The exponents in the rate law for a chemical reaction are the same as the coefficients of the reactants in the balanced equation.

C. Questions and Problems

1. Explain why the slow step in a mechanism is the rate-determining step.
2. A kinetics experiment on the reaction of $A + B \rightarrow E$ yielded the data shown in the table below.

Experiment	[A]	[B]	Rate (M/s) of E Formation
1	0.010	0.020	0.150
2	0.010	0.060	0.450
3	0.020	0.020	0.600

   (a) Write the general rate law for the reaction.
   (b) Write the experimentally determined rate law for the reaction.
   (c) What is the order of the reaction with respect to A? With respect to B? What is the overall order of the reaction?

3. Explain the difference between a transition state and a reaction intermediate.

4. The ΔE_{rxn} for a certain reaction is -983.5 kJ.
   (a) Is the reaction endothermic or exothermic?
   (b) Draw a reaction-energy profile for the reaction, labeling the axes, the reactants and products, the activation energy, and the transition state.
   (c) What is the ΔE_{rxn} for the reverse reaction?
   (d) Will the container in which the reaction is occurring feel hot or cold to the touch?

5. The reaction producing CO_2 and NO from CO and NO_2 is shown below.

   $$CO + NO_2 \rightarrow CO_2 + NO$$

   The reaction is first-order with respect to both CO and NO_2.
   (a) Write the rate law for the reaction.
   (b) What is the overall order of the reaction?
   (c) What will happen to the reaction rate if the concentration of CO is tripled?

6. Explain why enzymes are so valuable in speeding up processes in biological systems.

7. Does a catalyst affect the rate of the reverse reaction? Explain your answer.

14

Chemical Equilibrium

Learning Outcomes

1. Describe equilibrium for a chemical reaction in terms of forward and reverse reaction rates.
2. Describe equilibrium for a chemical reaction in terms of its position.
3. Explain why forward and reverse reaction rates change so as to obtain equilibrium.
4. Write the equilibrium constant expression for a chemical reaction.
5. Calculate the value of the equilibrium constant for a reaction.
6. Interpret the meaning of an equilibrium constant based on its magnitude.
7. Describe an equilibrium constant in terms of the rate constants for the forward and reverse reactions.
8. Predict which way an equilibrium will shift upon a disturbance of adding or removing one or more reactants and/or products.
9. Predict which way an equilibrium will shift upon heating or cooling a reaction and if the reaction is exothermic or endothermic.
10. Predict the effect of K_{eq} when a reaction is heated or cooled.
11. Write an equilibrium constant expression for a reaction that contains pure liquids and solids.
12. Calculate solubility given a K_{sp} value.
13. Calculate equilibrium concentrations given a K_{eq} value.

Chapter Outline

14.1 Dynamic Equilibrium—My Reaction Seems to Have Stopped!
   A. Three steps of the contact process for sulfuric acid production
   B. State of dynamic equilibrium
   C. Possible equilibrium positions
   D. "One-way" reactions

Review of Key Concepts from the Text

14.1 Dynamic Equilibrium—My Reaction Seems to Have Stopped!

All chemical reactions reach what chemists call dynamic equilibrium (or equilibrium), at which point the reaction seems to have stopped completely. That is, the reaction seems to be forming no more products. However, this is not the case. The reaction is still producing product, but at the same time an equal amount of product is being converted back to reactants, by the reverse reaction.

Many reactions are reversible, and a double arrow is used to indicate those reactions in which the reverse reaction occurs to an appreciable extent.

$$A + B \rightleftharpoons C$$

When the rate of the forward reaction, $A + B \rightarrow C$, is equal to the rate of the reverse reaction, $C \rightarrow A + B$, the reaction is at dynamic equilibrium. It will appear to have stopped because there will be no more change in the amounts of the reactants or products. (A common misconception is that the *amounts* of reactants and products are equal at equilibrium. This is not the case. It is the *rates* of the forward and reverse reaction that are equal, not the amounts of reactants and products.)

Some reactions reach dynamic equilibrium soon after the reaction has begun, some reach equilibrium when about half of the reactants have been converted to products, and still others do not reach equilibrium until most of the reactants have been converted to products. In the first case (equilibrium reached soon after the reaction has begun), we say that the equilibrium lies to the left. In the second case (equilibrium reached when about half of the reactants have been converted to products), we say that the equilibrium lies near the middle. In the last case (equilibrium reached when most of the reactants have been converted to products), we say that the equilibrium lies to the right.

Although all chemical reactions reach an equilibrium state, there are two special cases in which we don't treat reactions as if they are equilibrium reactions. Those two cases are described below.

1. Reactions that reach equilibrium almost immediately after the reaction has begun are designated "no reaction" (NR). An "x" is drawn through the arrow that separates the reactants from the products, as shown below.

    $$A + B \xrightarrow{\times} C \text{ no reaction (NR)}$$

2. Reactions that do not reach equilibrium until essentially all of the reactants have been converted to products are said to have "gone to completion." Because there is almost no reverse reaction occurring, we use only a single forward arrow in the equation, as shown below.

    $$A + B \rightarrow C$$

Example 14.1

A chemical reaction, $X + Y \rightarrow Z$, reaches equilibrium after essentially all of the reactants have been converted to products. Is it appropriate to write the reaction with a double arrow? Explain.

Solution

No. If the reaction reaches equilibrium after essentially all reactants have been converted to products, the reaction is one that "goes to completion" and should be written with a single forward arrow.

Practice Exercise 14.1

A reaction, $A + B \rightarrow C + D$, reaches equilibrium almost immediately after the reaction begins. What type of arrow (single arrow to the right, double arrow, or single arrow with an "x" through it) should separate the reactants from the products in the reaction equation?

14.2 Why Do Chemical Reactions Reach Equilibrium?

We have seen that when a system is at chemical equilibrium, the rates at which the forward and reverse reactions occur are equal. Why does the equalization of rates occur? We will consider a reaction, A + B → C, which begins with only A and B present. Initially, only the forward reaction, A + B → C, occurs because there is no C present. However, after some C has been formed, the reverse reaction, C → A + B, begins to occur. As the amounts of A and B decrease, the forward reaction slows down. As the amount of C increases, the reverse reaction speeds up. At some point the two rates will become equal, and the reaction will be at equilibrium. (See Section 14.2 of the textbook for a detailed discussion of this concept.)

Example 14.2

If the reaction X + Y ⇌ Z begins with only Y and Z in the reaction mixture, will the forward reaction or the reverse reaction take place initially?

Solution

The reverse reaction will occur initially. Because there is no X present, the forward reaction cannot occur.

Practice Exercise 14.2

The reaction C + D ⇌ E has an equilibrium that lies *very* far to the left. If the reaction is begun with only E present, will the reverse reaction proceed to a small extent or to a large extent? Explain.

14.3 The Position of Equilibrium — The Equilibrium Constant, K_{eq}

We have seen that the equilibrium position of a reaction can lie to the left, to the right, or toward the middle. We quantify (indicate by assigning a numerical value) the position of equilibrium by the equilibrium constant, K_{eq}. The K_{eq} shows the relationship between the equilibrium concentrations of reactants and products. Because the relative amounts (or concentrations) of reactants and products remain unchanged once equilibrium is reached (assuming that the reaction is not subjected to a stress such as a concentration change or a temperature change), we can write an expression showing the relationship between the equilibrium concentrations of reactants and products. This expression is called the equilibrium constant expression and is discussed in detail in Section 14.3 of the textbook.

For the general reaction aA + bB ⇌ cC + dD,

$$K_{eq} = \frac{[C]^c[D]^d}{[A]^a[B]^b}$$

Notice that plus signs are never included in the equilibrium constant expression and that the coefficients from the balanced equation are included only as exponents in this expression (they never appear inside the brackets).

The value of K_{eq} is related to the values of the forward rate constant k_f and the reverse rate constant k_r in the following manner:

$$K_{eq} = \frac{k_f}{k_r}$$

If the forward rate constant is much larger than the reverse rate constant, K_{eq} will be a very large number. If the reverse rate constant is much larger, K_{eq} will be a very small fraction.

We can determine the value of the K_{eq} if we know the equilibrium concentrations of all species present, as demonstrated in the following example.

Example 14.3

For the reaction $N_2 + 3H_2 \rightleftharpoons 2NH_3$, calculate the value of K_{eq} if the following concentrations are present at equilibrium: $[N_2] = 0.22$ M; $[H_2] = 0.14$ M; and $[NH_3] = 0.055$ M.

Solution

$$K_{eq} = \frac{[NH_3]^2}{[N_2][H_2]^3} = \frac{(0.055)^2}{(0.22)(014)^3} = 5.0$$

Practice Exercise 14.3

For the reaction $2A + B \rightleftharpoons 2C$, calculate the value of K_{eq} if the following concentrations are present at equilibrium: $[A] = 0.45$ M; $[B] = 0.31$ M; and $[C] = 0.055$ M.

Because the K_{eq} for a reaction at equilibrium represents $\dfrac{[\text{products}]}{[\text{reactants}]}$, its value can tell us roughly where the equilibrium lies—to the left, to the right, or near the middle.

- If K_{eq} is large ($>10^3$), the equilibrium lies to the right, and there are mostly products in the equilibrium mixture.
- If K_{eq} is small ($<10^{-3}$), the equilibrium lies to the left, and there are mostly reactants in the equilibrium mixture.
- If K_{eq} is close to 1 (between 10^{-3} and 10^3), the equilibrium lies near the middle.

Two special cases have K_{eq} values that are reflective of their characteristics.

- If K_{eq} is very small ($\leq 10^{-5}$), there is essentially no reaction (NR).
- If K_{eq} is very large ($\geq 10^5$), the reaction essentially goes to completion.

Example 14.4

Given the following K_{eq} values, classify the direction of equilibrium for each reaction as no reaction (NR), to the left, near the middle, to the right, or goes to completion.

(a) $K_{eq} = 1.0 \times 10^5$ (b) $K_{eq} = 0.0003$ (c) $K_{eq} = 8.3$
(d) $K_{eq} = 2 \times 10^3$ (e) $K_{eq} = 360$

Solutions

(a) goes to completion (b) to the left (c) near the middle
(d) to the right (e) near the middle

Practice Exercise 14.4

Given the following K_{eq} values, classify the direction of equilibrium for each reaction as no reaction (NR), to the left, near the middle, to the right, or goes to completion.

(a) $K_{eq} = 1.0 \times 10^{-9}$ (b) $K_{eq} = 3500$ (c) $K_{eq} = 0.67$
(d) $K_{eq} = 8.2 \times 10^3$ (e) $K_{eq} = 4 \times 10^{-3}$

14.4 Disturbing a Reaction Already at Equilibrium — Le Châtelier's Principle

When a system at equilibrium is subjected to a stress, such as a change in the concentration of one of the components of the equilibrium mixture or a change in the temperature, the system will react in such a way as to relieve the stress. This principle, known as *Le Châtelier's principle*, can be used to predict shifts in the position of equilibrium. First we will consider what happens to the equilibrium when a change is made to the concentration of any of the components of the equilibrium system.

When there is an increase in the concentration of one of the components of an equilibrium mixture, an equilibrium system will always shift *away from* the increase in concentration (to remove the added substance). When there is a decrease in the concentration of one of the components, an equilibrium system will always shift *toward* the decrease in concentration (to produce more of the substance that was removed).

Example 14.5

Consider the equilibrium reaction $H_2 + Cl_2 \rightleftharpoons 2HCl$. Which way will the equilibrium shift if the following stresses are placed on the system:

(a) $[H_2]$ is decreased   (b) [HCl] is increased

(c) $[Cl_2]$ is increased   (d) [HCl] is decreased

Solutions

(a) to the left   (b) to the left

(c) to the right   (d) to the right

Practice Exercise 14.5

Consider the equilibrium reaction $N_2 + O_2 \rightleftharpoons 2NO$. Which way will the equilibrium shift if the following stresses are placed on the system:

(a) [NO] is decreased   (b) $[O_2]$ is increased

(c) $[O_2]$ is decreased   (d) $[N_2]$ is increased

14.5 How Equilibrium Responds to Temperature Change

We learned earlier that chemical reactions either absorb (consume) heat or give off (release) heat when they occur. Reactions that absorb heat are *endothermic*, and their energy change (ΔE) is positive. Reactions that give off heat are *exothermic*, and their energy change (ΔE) is negative.

A temperature change affects endothermic and exothermic reactions in opposite ways. It is helpful to assign heat as either a reactant or a product when we need to determine the equilibrium shift caused by a temperature change.

- Endothermic reactions absorb heat, so heat is placed on the reactant side.
- Exothermic reactions give off heat, so heat is placed on the product side.

Example 14.6

Given the following reaction and value of ΔE, assign heat as either a reactant or a product:

$$2HBr \rightleftharpoons H_2 + Br_2 \qquad \Delta E = +72.8 \text{ kJ}$$

Solution

Because ΔE is positive, this is an endothermic reaction, and heat is a reactant:

$$\text{Heat} + 2\text{HBr} \rightleftharpoons \text{H}_2 + \text{Br}_2.$$

Practice Exercise 14.6

Given the following reaction and value of ΔE, assign heat as either a reactant or a product:

$$\text{N}_2 + 3\text{H}_2 \rightleftharpoons 2\text{NH}_3 \quad \Delta E = -92.0 \text{ kJ}$$

We can now treat a temperature change just like concentration changes. When we do this, the following generalizations are apparent:

1. For an endothermic reaction, an increase in temperature will cause the equilibrium to shift to the right; a decrease in temperature will cause the equilibrium to shift to the left.
2. For an exothermic reaction, an increase in temperature will cause the equilibrium to shift to the left; a decrease in temperature will cause the equilibrium to shift to the right.

Note that it is *never* correct to say that heat has been added to the right or left side of the equation. The heat is added to the *entire* reaction mixture.

Example 14.7

Given the reaction $2\text{HBr} \rightleftharpoons \text{H}_2 + \text{Br}_2$, for which $\Delta E = +72.8 \text{ kJ}$, in which direction will the equilibrium shift if the temperature is decreased?

Solution

The reaction is an endothermic reaction, with heat on the left:

$$\text{Heat} + 2\text{HBr} \rightleftharpoons \text{H}_2 + \text{Br}_2.$$

Therefore, decreasing the temperature will shift the equilibrium to the left.

Practice Exercise 14.7

Given the reaction $\text{N}_2 + 3\text{H}_2 \rightleftharpoons 2\text{NH}_3$, for which $\Delta E = -92.0 \text{ kJ}$, in which direction will the equilibrium shift if the temperature is increased?

One final stress that might be imposed on a system at equilibrium is the addition of a catalyst. We saw in Chapter 13 that a catalyst speeds up the rate of reaction by providing an alternate path that will lower the activation energy of the reaction, as shown in the energy diagram below.

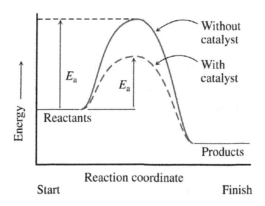

Copyright © 2015 Pearson Education, Inc.

We can see from the figure that the activation energy of the reverse reaction is lowered by exactly the same amount as the activation energy of the forward reaction. This means that the speed of both the forward and reverse reactions will be increased by exactly the same amount when a catalyst is added, and there will be no shift in the position of equilibrium.

So why add a catalyst to an equilibrium system? Because the rates of the reactions will be increased, the system will reach the equilibrium state sooner with a catalyst than without one. The time savings will usually translate into financial savings.

14.6 Equilibria for Heterogeneous Reactions, Solubility, and Equilibrium Calculations

Many chemical reactions involve more than one phase, and are referred to as heterogeneous reactions. These reactions are easily identified because they involve reactants or products that are in the solid or liquid phase, as indicated by the phase designations (s) for solid or (l) for liquid. An example of an equation for a heterogeneous equilibrium reaction is shown below.

$$AgCl(s) \rightleftharpoons Ag^+(aq) + Cl^-(aq)$$

When the K_{eq} expression for a heterogeneous equilibrium reaction is written the pure liquids and pure solids are omitted from the expression. Thus, the K_{eq} expression for the reaction above would be:

$$K_{eq} = [Ag^+][Cl^-]$$

Another example of a heterogeneous equilibrium reaction and its K_{eq} expression are shown below.

$$2FeS(s) + 3H_2O(g) \rightleftharpoons Fe_2O_3(s) + 3H_2(g)$$

$$K_{eq} = \frac{[H_2]^3}{[H_2O]^3}$$

The concentrations of pure solids and pure liquids are left out of the equilibrium constant expressions because the concentrations of liquids and solids remain constant during the course of the reaction. Their molar concentrations remain constant (and are related to the density of the substance) because these substances do not expand or contract as do gases. The constants that represent the concentrations of solids and liquids are included in the value of the equilibrium constant K_{eq}, so their concentrations do not appear on the right side of the K_{eq} expression. Section 14.6 of the textbook provides a complete discussion of heterogeneous equilibrium reactions.

Example 14.8

Write the K_{eq} expression for the following reactions:

(a) $SnO_2(s) + 2CO(g) \rightleftharpoons Sn(s) + 2CO_2(g)$

(b) $Ti(s) + 2Cl_2(g) \rightleftharpoons TiCl_4(l)$

Solution

(a) $K_{eq} = \dfrac{[CO_2]^2}{[CO]^2}$ (b) $K_{eq} = \dfrac{1}{[Cl_2]^2}$

Practice Exercise 14.8

Write the K_{eq} expression for the following reactions:

(a) $Fe(s) + H_2O(g) \rightleftharpoons FeO(s) + H_2(g)$

(b) $NH_4HS(s) \rightleftharpoons NH_3(g) + H_2S(g)$

A special kind of heterogeneous equilibrium reaction involves the equilibrium between insoluble ionic solids and the ions of which they are composed. Actually, even though we say that these solids are "insoluble," a very tiny amount does dissolve, breaking into ions in the solution. This process is illustrated below for $AgCl(s)$.

$$AgCl(s) \rightleftharpoons Ag^+(aq) + Cl^-(aq)$$

The K_{eq} for this reaction is called a solubility product constant and designated K_{sp}. Because very little of the soluble ionic salt dissolves, the concentration of ions in solution is quite small and the K_{sp} values are extremely small, sometimes as small as 10^{-54} or smaller. Table 14.1 in the textbook provides K_{sp} values for some slightly soluble salts.

If we know the equilibrium concentrations of the ions in this type of reaction, we can calculate the value of K_{sp}. We can use the solubility of the salt to determine the equilibrium concentration of the ions, as shown in Example 14.9.

Example 14.9

Sparingly soluble $Fe(OH)_2$ dissolves in water to produce $[Fe^{2+}] = 1.6 \times 10^{-5}$ M. What is the value of K_{sp} for the reaction?

Solution

The equation and K_{sp} expression for the solubility reaction are shown below.

$$Fe(OH)_2(s) \rightleftharpoons Fe^{2+}(aq) + 2OH^-(aq); K_{sp} = [Fe^{2+}][OH^-]^2$$
$$[Fe^{2+}] = 1.6 \times 10^{-5}$$

Use the coefficients in the balanced equation to determine the $[OH^-]$.

$$[OH^-] = \frac{1.6 \times 10^{-5}\,\text{mol Fe}^{2+}}{1\,\text{L}} \times \frac{2\,\text{mol OH}^-}{1\,\text{mol Fe}^{2+}} = 3.2 \times 10^{-5}\,\text{M}$$

Substituting the ion values into the K_{sp} expression, we obtain

$$K_{sp} = [Fe^{2+}][OH^-]^2 = (1.6 \times 10^{-5})(3.2 \times 10^{-5})^2 = 1.6 \times 10^{-14}$$

Practice Exercise 14.9

Sparingly soluble CuI dissolves in water to produce $[Cu^+] = 2.26 \times 10^{-6}$ M. What is the value of K_{sp} for the reaction?

If the K_{sp} value is known, we can use this value to calculate the solubility of the slightly soluble salt. To do this, we first write the equation for the equilibrium reaction. We express the solubility of the slightly soluble salt in terms of s, using the coefficients in the balanced equation. We then substitute the equilibrium concentrations, expressed in terms of s, into the K_{sp} expression and solve for the value of s. For example:

Example 14.10

Calculate the solubility of lead(II) sulfide, PbS, in water at 25 °C. K_{sp} is 3.4×10^{-28}.

Solution

The balanced equation and the K_{sp} expression for the reaction are:

$$PbS(s) \rightleftharpoons Pb^{2+}(aq) + S^{2-}(aq); K_{sp} = [Pb^{2+}][S^{2-}] = 3.4 \times 10^{-28}$$

If s = solubility of PbS, s moles/liter of PbS dissolve in solution. Therefore, s moles/liter of Pb^{2+} and s moles/liter of S are produced, and the K_{sp} expression can be written in terms of s.

$$K_{sp} = [Pb^{2+}][S^{2-}]$$
$$3.4 \times 10^{-28} = (s)(s) = s^2$$

Taking the square root of both sides we obtain $s = 1.84 \times 10^{-14}$ M. The solubility of PbS is 1.84×10^{-14} M.

Practice Exercise 14.10

Calculate the solubility of cadmium sulfide, CdS, in water at 25 °C. K_{sp} is 8.0×10^{-28}.

We have just seen that knowing the value of the K_{sp} allows us to calculate the equilibrium in concentrations. This is true not just for equilibria of sparingly soluble salts but for all equilibrium reactions. If we know the value of K_{eq} and all equilibrium concentrations except one, we can write the K_{eq} expression, substitute all known values into the expression, and solve for the value of the missing concentration. For example:

Example 14.11

For the reaction $2SO_2(g) + O_2(g) \rightleftharpoons 2SO_3(g)$, $K_{eq} = 8.10 \times 10^2$. If the equilibrium concentration of $SO_3 = 4.78$ M and the equilibrium concentration of $SO_2 = 0.216$ M, calculate the equilibrium concentration of O_2.

Solution

The equilibrium constant expression for the reaction is:

$$K_{eq} = \frac{[SO_2]^2}{[SO_2]^2[O_2]}$$

Algebraically solving the expression for $[O_2]$ we obtain:

$$[O_2] = \frac{[SO_2]^2}{K_{eq}[SO_2]^2} = \frac{(4.78)^2}{8.10 \times 10^2 \times (0.216)^2} = 6.05 \times 10^{-1} \text{ M}$$

Practice Exercise 14.11

For the reaction $N_2O_4(g) \rightleftharpoons 2NO(g)$, $K_{eq} = 4.64 \times 10^{-3}$. If the equilibrium concentration of NO = 0.156 M, calculate the equilibrium concentration of N_2O_4.

Strategies for Working Problems in Chapter 14

Overview of Problem Types: What You Should Be Able to Do

Chapter 14 presents an introduction to the state of chemical equilibrium, a state in which the rate of the forward reaction equals the rate of the reverse reaction. The types of problems that you should be able to solve after mastering Chapter 14 are:

1. Determine the value of the equilibrium constant, K_{eq} by using the equilibrium concentrations.
2. Calculate the K_{sp} from the solubility.
3. Calculate the solubility from the K_{sp} value.

   (Note that the solubility is also sometimes referred to as the saturation solubility or the maximum solubility.)
4. Determine the missing equilibrium concentration from the K_{eq} value and the values of all other equilibrium concentrations.

Flowcharts

Flowchart 14.1 Determining the value of the equilibrium constant, K_{eq} by using the equilibrium concentrations

Flowchart 14.1

Summary:

Method for determining the value of the equilibrium constant, K_{eq} by using the equilibrium concentrations

Step 1
- **Write the equilibrium constant expression for the reaction.**

$\downarrow$

Step 2
- **Substitute the molar concentrations for all species in the reaction.**

$\downarrow$

Step 3
- **Perform the calculation to determine the value of K_{eq}.**

Example 1

The reaction involving the production of ammonia gas is

$$N_2 + 3H_2 \rightleftharpoons 2NH_3$$

At a certain temperature, when the reaction is at equilibrium the concentrations of the species are the following.

$$[N_2] = 0.005\,61\,M;\, [H_2] = 0.813\,M;\, [NH_3] = 0.241\,M$$

Calculate the value of K_{eq}.

Solution

Perform Step 1

$$K_{eq} = \frac{[NH_3]^2}{[N_2][H_2]^3}$$

Perform Step 2

$$K_{eq} = \frac{[NH_3]^2}{[N_2][H_2]^3} = \frac{(0.241)^2}{(0.00561)(0.813)^3}$$

Perform Step 3

$$K_{eq} = \frac{(0.241)^2}{(0.00561)(0.813)^3} = 19.3$$

Example 2

For the reaction $H_2 + I_2 \rightleftharpoons 2HI$, the equilibrium concentrations are as follows:
$[H_2] = 0.022\ M; [I_2] = 0.36\ M; [HI] = 0.62\ M$
Calculate the value of K_{eq} for the reaction.

Solution

Perform Step 1

$$K_{eq} = \frac{[HI]^2}{[H_2][I_2]}$$

Perform Step 2

$$K_{eq} = \frac{[HI]^2}{[H_2][I_2]} = \frac{(0.62)^2}{(0.022)(0.36)}$$

Perform Step 3

$$K_{eq} = \frac{(0.62)^2}{(0.022)(0.36)} = 49$$

Practice Problems

14.1 For the hypothetical reaction $2A + B \rightleftharpoons 2C$, calculate the value of K_{eq} if the following concentrations are present at equilibrium:

$[A] = 0.090\ M; [B] = 0.072\ M; [C] = 0.011\ M$

14.2 For the reaction of $2SO_2 + O_2 \rightleftharpoons 2SO_3$, the following concentrations are present at equilibrium:

$[SO_2] = 0.014\ M; [O_2] = 0.031; [SO_3] = 0.24\ M$

Calculate the value of K_{eq} for the reaction.

Flowchart 14.2 Calculating the K_{sp} from the solubility

Flowchart 14.2

Summary:

Method for calculating the K_{sp} from the solubility

Step 1
- Write the equation for the equilibrium reaction of the salt dissolving in water.
- Write the equilibrium constant expression for the reaction from the equation.

$\downarrow$

Step 2
- Substitute the molar concentrations of the ions in solution into the K_{sp} expression.
- Use the coefficients in the solubility reaction to determine the concentration of species whose concentrations are not given in the problem.

$\downarrow$

Step 3
- Perform the calculation to determine the value of K_{sp}.

Example 1

Sparingly soluble $PbCl_2$ dissolves in water to yield a $[Pb^{2+}]$ of 0.039 M. Calculate the value of K_{sp} for the reaction.

Solution

Perform Step 1

$PbCl_2(s) \rightleftharpoons Pb^{2+}(aq) + 2\,Cl^-(aq)$

$K_{sp} = [Pb^{2+}][Cl^-]^2$

Perform Step 2

$[Pb^{2+}] = 0.039$ M.

$[Cl^-] = 2\,[Pb^{2+}] = 0.078$ M

$K_{sp} = [Pb^{2+}][Cl^-]^2 = (0.039)(0.078)^2$

Perform Step 3

$K_{sp} = (0.039)(0.078)^2 = 2.4 \times 10^{-4}$

Example 2

Sparingly soluble CuI dissolves in water to yield a $[Cu^+]$ of 2.26×10^{-6} M. Calculate the value of K_{sp} for the reaction.

Solution

Perform Step 1

$CuI\,(s) \rightleftharpoons Cu^+(aq) + I^-(aq)$

$K_{sp} = [Cu^+][I^-]$

Perform Step 2

$[Cu^+] = 2.26 \times 10^{-6}\,M.$

$[I^-] = [Cu^+] = 2.26 \times 10^{-6}\,M.$

$K_{sp} = [Cu^+][I^-] = (2.26 \times 10^{-6})(2.26 \times 10^{-6}) = 5.1 \times 10^{-12}$

Perform Step 3

$K_{sp} = (2.26 \times 10^{-6})(2.26 \times 10^{-6}) = 5.1 \times 10^{-12}$

Practice Problems

14.3 When the sparingly soluble salt BaF_2 dissolves in water, the $[Ba^{2+}] = 7.5 \times 10^{-3}\,M.$ Calculate the K_{sp} for BaF_2.

14.4 When sparingly soluble $BaCO_3$ dissolves in water (to produce Ba^{2+} and CO_3^{2-} ions) the $[Ba^{2+}] = 8.9 \times 10^{-5}\,M.$ Calculate the K_{sp} for $BaCO_3$.

Flowchart 14.3 Calculating the solubility from the K_{sp} value

Flowchart 14.3

Summary:

Method for calculating the solubility from the K_{sp} value

Step 1
- **Write the equation for the equilibrium reaction of the salt dissolving in water.**
- **Write the equilibrium constant expression for the reaction from the equation.**

$\downarrow$

Step 2
- **Set the solubility of the compound equal to the variable s.**
- **Express the equilibrium concentrations of the ions in terms of s using the coefficients in the balanced equation.**
- **Table 14.2 in the textbook provides the ion solubilities in terms of s for various salts.**

$\downarrow$

Step 3
- **Substitute the equilibrium ion concentrations, expressed in terms of s, into the K_{sp} expression.**
- **Solve for the value of s.**

Example 1

What is the maximum solubility of $CdCO_3$ in water at 25 °C? $K_{sp} = 1.8 \times 10^{-14}$.

Solution

Perform Step 1

$CdCO_3(s) \rightleftharpoons Cd^{2+}(aq) + CO_3^{2-}(aq)$

$K_{sp} = [Cd^{2+}][CO_3^{2-}]$

Perform Step 2

s = solubility of $CdCO_3$

Because 1 mole of $CdCO_3$ produces 1 mole each of Cd^{2+} and CO_3^{2-},
$[Cd^{2+}] = [CO_3^{2-}] = s$.

Perform Step 3

$K_{sp} = [Cd^{2+}][CO_3^{2-}]$

$1.8 \times 10^{-14} = s \times s = s^2$

Taking the square root of both sides to solve for s, we obtain $s = 1.3 \times 10^{-7}$.

Therefore, the maximum solubility of $CdCO_3$ in water at 25 °C is 1.3×10^{-7} M.

Example 2

What is the maximum solubility of CuBr in water at 25 °C? The K_{sp} for CuBr at 25 °C is 4.2×10^{-8}.

Solution

Perform Step 1

$CuBr(s) \rightleftharpoons Cu^+(aq) + Br^-(aq)$

$K_{sp} = [Cu^+][Br^-]$

Perform Step 2

s = solubility of CuBr

Because 1 mole of CuBr produces 1 mole each of Cu^+ and Br^-,
$[Cu^+] = [Br^-] = s$.

Perform Step 3

$K_{sp} = [Cu^+][Br^-]$

$4.2 \times 10^{-8} = s \times s = s^2$

Taking the square root of both sides to solve for s, we obtain $s = 2.0 \times 10^{-4}$.

Therefore, the maximum solubility of CuBr in water at 25 °C is 2.0×10^{-4} M.

Practice Problems

14.5 Calculate the maximum solubility for AgBr at 25 °C. The K_{sp} for AgBr at 25 °C is 7.7×10^{-13}.

Flowchart 14.4 Determining the missing equilibrium concentration from the K_{eq} value and the values of all other equilibrium concentrations

Flowchart 14.4

Summary:

Method for determining the missing equilibrium concentration from the K_{eq} value and the values of all other equilibrium concentrations

Step 1
- **Write the equilibrium constant expression for the reaction.**

Step 2
• Substitute the value of K_{eq} and the molar concentrations for all known species into the K_{sp} expression.

$\downarrow$

Step 3
• Algebraically solve the equation to determine the value of the missing equilibrium concentration.

Example 1

For the reaction $2A_2 + B_2 \rightleftharpoons 2A_2B$, the $K_{eq} = 3.36 \times 10^{18}$ and the equilibrium concentrations of A_2 and A_2B are as follows: $[A_2] = 0.0224$ M; $[A_2B] = 0.596$ M. Calculate the $[B_2]$.

Solution

Perform Step 1

$$K_{eq} = \frac{[A_2B]^2}{[A_2]^2[B_2]}$$

Perform Step 2

$$K_{eq} = \frac{[A_2B]^2}{[A_2]^2[B_2]}$$

$$3.36 \times 10^{18} = \frac{(0.596)^2}{(0.0224)^2[B_2]}$$

Perform Step 3

$$3.36 \times 10^{18} = \frac{(0.596)^2}{(0.0224)^2[B_2]}$$

Multiplying both sides of the equation by $[B_2]$, and dividing both sides of the equation by 3.36×10^{18}, we obtain $[B_2] = 2.11 \times 10^{-16}$ M.

Practice Problems

14.6 For the reaction $N_2 + O_2 \rightleftharpoons 2NO$, the $K_{eq} = 4.03 \times 10^{-2}$ and the equilibrium concentrations of N_2 and NO are as follows: $[N_2] = 0.036$ M, $[NO] = 0.017$ M. Calculate the equilibrium concentration of O_2.

Quiz for Chapter 14 Problems

1. When the reaction of $CoCl_2(aq) \rightleftharpoons Co^{2+}(aq) + 2Cl^-(aq)$ occurs at 25 °C, the following equilibrium concentrations are observed: $[CoCl_2] = 3.2 \times 10^{-2}$ M; $[Co^{2+}] = 1.0 \times 10^{-5}$ M; $[Cl^-] = 2.38 \times 10^{-4}$ M. Calculate K_{eq} for the reaction.

2. In the equilibrium constant, K_{eq}, the concentration(s) of the _____ of the reaction appear in the numerator and the concentration(s) of the _____ appear in the denominator.

3. The slightly soluble salt, AB, has a solubility of 5.32×10^{-6} M. Calculate the K_{sp} for AB.

4. Write the equilibrium constant expression, K_{eq}, for the following gaseous reaction:

$2SO_2 + O_2 \rightleftharpoons 2SO_3$

5. For the gaseous reaction $X \rightleftharpoons 2$ Y, $K_{eq} = 2.3 \times 10^{-4}$. If the concentration of Y at equilibrium $= 8.5 \times 10^{-3}$ M, calculate the equilibrium concentration of X.

6. In a saturated solution of the slightly soluble salt CuBr, the $[Cu^+] = [Br^-] = 2.0 \times 10^{-4}$ M. Calculate the K_{sp} for CuBr.

7. For the hypothetical gaseous reaction $3A \rightleftharpoons 2B$, the equilibrium concentrations are:

$[A] = 1 \times 10^{-3}$ M and $[B] = 2 \times 10^{-1}$ M. Calculate K_{eq} for the reaction.

8. For the gaseous reaction $2SO_2 + O_2 \rightleftharpoons 2SO_3$, $K_{eq} = 810$. The equilibrium concentration of SO_2 is 0.2 M, and the equilibrium concentration of $SO_3 = 1.61$ M. What is the equilibrium concentration of O_2?

9. The K_{sp} value for the slightly soluble salt AgI(s) is 1.5×10^{-16} at 25 °C. What is the solubility of AgI at 25 °C?

10. If $[Ag^+]$ in a saturated solution of AgBr is 9×10^{-7} M and the $[Ag^+]$ in a saturated solution of AgCl is 1×10^{-5} M, which salt is more soluble in water? Explain.

11. The K_{eq} for three different reactions are shown below:

$A \rightleftharpoons B: K_{eq} = 1 \times 10^{-2}$

$E \rightleftharpoons F: K_{eq} = 1 \times 10^{-18}$

$X \rightleftharpoons Y: K_{eq} = 1 \times 10^5$

Which reaction will contain more reactants in its equilibrium state? Explain.

12. Write the K_{eq} expression for the following reaction: $NH_4HS(s) \rightleftharpoons NH_3(g) + H_2S(g)$.

Exercises for Self-Testing

A. **Completion.** *Write the correct word(s) to complete each statement below.*

1. At the point of _____ _____, chemical reactions appear to have stopped.
2. Reactions that reach equilibrium almost immediately after they have started can be designated NR, which means _____ _____.
3. Reactions that do not reach equilibrium until essentially all reactants have been converted to products are said to have _____ _____ _____.
4. At the point of dynamic equilibrium, the _____ of the forward reaction equals the _____ of the reverse reaction.
5. For a reaction with a K_{eq} of 1×10^{-14}, the equilibrium lies far to the _____.
6. When a change in concentration is made on an equilibrium system, the equilibrium will always shift _____ _____ an increase and _____ a decrease in concentration.
7. A temperature increase will cause a shift to the left for an _____ reaction.
8. Addition of a _____ causes no shift in the position of equilibrium for a reaction.
9. When a system with a very large K_{eq} value reaches equilibrium, there will be many more _____ than _____ in the equilibrium mixture.

10. A catalyst speeds up the _____ _____ by the same amount it speeds up the _____ _____

11. When the solubility of a slightly soluble salt is calculated from the value of its K_{sp}, the solubility of the salt is represented by the letter _____.

B. True/False. *Indicate whether each statement below is true or false, and explain why the false statements are false.*

_____ 1. Changing the initial concentrations of the reactants will change the value of K_{eq}.

_____ 2. A very large value of K_{eq} means that there are many more reactants present at equilibrium than there are products.

_____ 3. A reaction that has gone to completion has some reactants present in the mixture.

_____ 4. A reaction that has a very small K_{eq} will be mostly reactants at equilibrium.

_____ 5. When heat is added to an equilibrium system, it will always produce more products.

_____ 6. If k_r is much larger than k_f, the value of K_{eq} will be much smaller than 1.

_____ 7. Adding a catalyst to an equilibrium system does no good because it does not shift the position of equilibrium.

_____ 8. The amount of product produced in an exothermic reaction can be increased if the temperature of the reaction mixture is increased.

_____ 9. The value of K_{eq} is a constant and will not change if the temperature at which the reaction occurs is changed.

_____ 10. Increasing the temperature of an endothermic reaction will shift the equilibrium to the right.

_____ 11. If AgCl has a solubility of 1.26×10^{-5} M, the K_{sp} for AgCl is 2.56×10^{-10}.

_____ 12. If $PbI_2(s)$ has the K_{sp} of 1.4×10^{-8}, $[I^-] = 1.5 \times 10^{-3}$ M is in the saturated solution?

_____ 13. If a poorly soluble ionic compound has a generic formula AB(s), then $K_{sp} = s^2$.

_____ 14. If a poorly soluble ionic compound has a generic formula $A_2B(s)$, then $K_{sp} = 2s^2$.

C. Questions and Problems

1. Write equilibrium constant expressions for the following reactions:
   (a) $2SO_2(g) + O_2(g) \rightleftharpoons 2SO_3(g)$
   (b) $N_2O_4(g) \rightleftharpoons 2NO_2(g)$
   (c) $2NO(g) \rightleftharpoons N_2(g) + O_2(g)$
   (d) $CH_4(g) + H_2O(g) \rightleftharpoons CO(g) + 3H_2(g)$

2. Calculate the value of K_{eq} for the reaction $PCl_5(g) \rightleftharpoons PCl_3(g) + Cl_2(g)$ if the equilibrium concentrations are as follows: $[PCl_5] = 0.083$ M; $[PCl_3] = 0.15$ M; and $[Cl_2] = 0.32$ M.

3. For the reaction $N_2O_4(g) \rightleftharpoons N_2(g) + O_2(g)$, for which $\Delta E = -9.2$ kJ, answer the following questions:
   (a) If oxygen is added to the system, which way will the equilibrium shift?
   (b) If nitrogen is added to the system, which way will the equilibrium shift?
   (c) If the temperature is increased, which way will the equilibrium shift?
   (d) If some nitrogen is removed from the system, which way will the equilibrium shift?
   (e) If a catalyst is added to the system, which way will the equilibrium shift?

4. Draw an appropriate diagram to explain the effect of a catalyst on an equilibrium system.

5. The reaction that produces ammonia from hydrogen and nitrogen gases is exothermic. Should the temperature of the reaction mixture be increased or decreased in order to make more product? Explain your answer.

6. Write the K_{eq} expression for the reaction:

   $Fe_2O_3(s) + 6HCl(g) \rightleftharpoons 2FeCl_3(s) + 3H_2O(g)$.

7. For the gaseous reaction $A_2 + B_2 \rightleftharpoons 2AB$, the $K_{eq} = 9.37 \times 10^{-3}$ and the equilibrium concentrations of A_2 and AB are as follows: $[A_2] = 0.981$ M; $[AB] = 0.17$ M. Calculate the $[B_2]$.

8. Write the K_{sp} expression for each of the following ionic compounds:
   (a) $Ca(OH)_2$ (b) $MgCO_3$ (c) PbF_2

9. Can the theoretical yield of a reaction be calculated in the same way for reactions that run to completion (no reactants left) and for reactions that reach equilibrium when there is still a non-negligible amount of reactants present? Explain your answer.

10. Consider the following two reactions. Write the corresponding equilibrium constant expressions and comment on the presence/absence of the water concentration term in them.
    (i) $2H_2(g) + O_2(g) \rightleftharpoons 2H_2O(g)$
    (ii) $HF(aq) + H_2O(l) \rightleftharpoons F^-(aq) + H_3O^+(aq)$

15

Electrolytes, Acids, and Bases

Learning Outcomes

1. Describe the relationship between being an electrolyte and dissociation.

2. Recognize whether a compound is an electrolyte or not from its formula.

3. Describe the difference between a weak and strong electrolyte in terms of equilibrium.

4. Recognize acids and their properties from their formula.

5. Describe the species present in a solution of a strong acid versus a weak acid.

6. Recognize bases and their properties from their formula.

7. Understand the chemistry and calculate an amount of base required for a base to neutralize a given amount of acid.

8. Describe ammonia (NH_3) as a base in terms of Brønsted–Lowry definition.

9. Describe carbonate and hydride as bases in terms of Brønsted–Lowry definition.

10. Describe a solution of a weak base in terms of Brønsted–Lowry definition.

11. Describe what is meant by the autoionization of water in terms of equilibrium.

12. Describe both qualitatively and quantitatively whether an aqueous solution is acidic, neutral, or basic.

13. Understand base-10 logarithms.

14. Understand and calculate the pH of a solution.

15. Calculate hydronium and hydroxide concentrations given the pH.

16. Understand what a conjugate acid and a conjugate base is.

17. Describe how to make a buffer and how it works.

Chapter Outline

B. Definition of a buffered solution

C. Conjugate acid–base pairs

D. Components of a buffered solution

E. Reactions that occur when acid or base is added to a buffered solution

Review of Key Concepts from the Text

15.1 Electrolytes and Nonelectrolytes

Electrolytes are substances whose aqueous solutions conduct electricity. The aqueous solutions of nonelectrolytes do not conduct electricity. In order for a substance to be classified as an electrolyte, two things must be true.

1. The substance must be soluble in water (so that it can form an aqueous solution).
2. An aqueous solution of the substance must conduct electricity.

The substances that fall into this category are acids, bases, and salts. Aqueous solutions of acids, bases, and salts are therefore electrolytes. It is extremely important for you to be able to recognize acids, bases, and salts from their formulas. We will take a closer look at each type of substance.

Before the chemistry of acids and bases was known, these substances were described by three characteristics, as follows:

- *Acids* taste sour, turn litmus red, and dissolve active metals, producing hydrogen gas.
- *Bases* taste bitter, turn litmus blue, and feel slippery to the touch.

Salts, on the other hand, were described as follows:

- *Salts* are compounds that are produced from the neutralization reaction of acids and bases.

We'll learn more about salts later.

As previously stated, electrolytes conduct electricity when dissolved in water, but nonelectrolytes do not.

Electrolytes can be divided into two categories—strong electrolytes and weak electrolytes. In the light bulb test, strong electrolytes light the bulb brightly, and weak electrolytes light the bulb only weakly.

It is easy for us to do the light bulb test to see if a substance is an electrolyte, but how do we recognize electrolytes from the formula of the compound?

Arrhenius was the first to recognize that electrolytes are substances that dissociate into ions in solution. It is the ions that serve as the mobile charge carriers that conduct electricity. We learned in Chapter 12 that salts are ionic compounds that dissociate into ions when dissolved in water. We can easily recognize salts because they are composed of a metal and a nonmetal, or they contain a polyatomic ion.

Molecular compounds, on the other hand, do not usually produce ions when dissolved in water, and are therefore nonelectrolytes.

Example 15.1

Classify the following compounds as electrolytes or nonelectrolytes:

(a) KCl   (b) $C_6H_{12}O_6$ (c) NH_4NO_3 (d) NaOH   (e) CO_2

Solutions

(a), (c), and (d) are electrolytes; (b) and (e) are nonelectrolytes.

Practice Exercise 15.1

Classify the following compounds as electrolytes or nonelectrolytes:

(a) SO_3 (b) Ne   (c) Na_3PO_4 (d) LiBr   (e) $MgSO_4$

We said earlier that molecular compounds do not usually produce ions. The class of compounds that are the exception to this rule are the covalent compounds whose formula is HX, when X is a halogen (F, Cl, Br, or I). These compounds dissociate into H^+ and X^- ions when they are dissolved in water. Thus, they must be added to our list of electrolytes.

The HX compounds are actually a subgroup of a larger class of compounds, HA, where A is a halogen or a polyatomic ion (NO_3^-, CN^-, HSO_4^-, etc.). The HA compounds are acids, and all acids are electrolytes.

15.2 Electrolytes Weak and Strong

Electrolytes are substances that break into ions and conduct electricity when dissolved in water. However, different compounds break up to different extents. Some break completely into ions. (We say that they are 100% dissociated in solution.) Others break up to an extent less than 100% (some *much* less than 100%).

The electrolytes that dissociate 100% are called *strong electrolytes*, and those that dissociate less than 100% are called *weak electrolytes*. As you might expect, weak electrolytes will light a light bulb to a lesser extent than strong electrolytes, so a simple light bulb test will allow us to classify electrolytes as weak or strong.

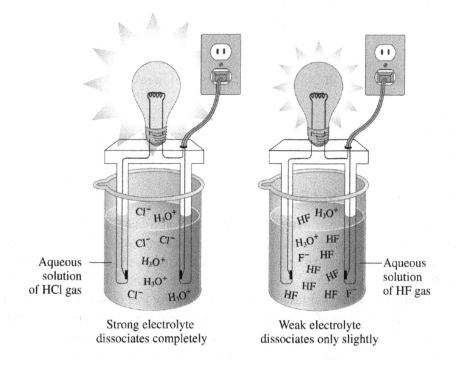

Aqueous solution of HCl gas

Strong electrolyte dissociates completely

Aqueous solution of HF gas

Weak electrolyte dissociates only slightly

We can also tell whether a substance is a weak or a strong electrolyte by knowing what type of compound it is and by remembering the following statements:

1. All soluble salts are strong electrolytes.
2. All strong acids are strong electrolytes.
3. All weak acids are weak electrolytes.

(We will add to this list later in the chapter.)

If we know whether a compound is a salt, a strong acid, or a weak acid, we can classify the substance as either a strong electrolyte or a weak electrolyte. How can you tell if a substance is a salt, a strong acid, or a weak acid? We'll use the following definitions and/or rules:

1. A *salt* is any ionic compound that contains a metal ion with a nonmetal ion, a polyatomic ion with a nonmetal ion, a metal ion with a polyatomic ion, or two polyatomic ions (i.e., any compound containing positive and negative ions).
2. In general, an *acid* is a compound with the general formula HA.

   (a) A *strong acid* is an acid that completely dissociates in water. There are relatively few strong acids, and you should commit the ones listed here to memory: HCl, HBr, HI, HNO_3, H_2SO_4, and $HClO_4$.

   (b) A *weak acid* is an acid that partially dissociates in water. The weak acids are all acids that are not strong! Because we have memorized the list of strong acids, we can assume that any acid that is not on this list is weak. If you identify a substance as an acid (HA) and it is not one of the strong acids you have memorized, it is most probably a weak acid.

Example 15.2

Identify the following compounds as either a salt, a strong acid, or a weak acid:

(a) HNO_3 (b) $CaCl_2$ (c) HF   (d) HCN   (e) H_2SO_4

Solutions

(a) strong acid   (b) salt   (c) weak acid   (d) weak acid   (e) strong acid

Practice Exercise 15.2

Identify the following compounds as either a salt, a strong acid, or a weak acid:

(a) $HClO_2$ (b) KF   (c) HCl   (d) NH_4Cl (e) $HClO_4$

15.3   Acids, Weak and Strong

We know that acids are electrolytes that produce H^+ ions (which really exist as H_3O^+ ions) in solution. Acids can be strong or weak. The strong acids are strong electrolytes, and the weak acids are weak electrolytes. Arrhenius defined an acid as any electrolyte that, when added to water, dissociates to produce H^+ ions in solution. We now know that the H^+ (H_3O^+) is the ion that is responsible for the sour taste of acids, the red color that is produced when acids react with litmus, and the production of hydrogen gas when an acid reacts with an active metal such as zinc.

Although we wrote the general acid formula as HA, some acids can have more than one "acidic" hydrogen. (An acidic hydrogen is one that can dissociate when the acid is placed in water.) There are certain cases in which not all H atoms in a molecule are dissociable. For example, acetic acid, $HC_2H_3O_2$, has only one dissociable hydrogen—the one that is written first in the formula. Acids that have only one dissociable hydrogen are said to be *monoprotic*.

Other acids with the general formula H_2A, such as H_2SO_4, and the general formula H_3A, such as H_3PO_4, can dissociate more than one H. These acids are called *polyprotic*, and a prefix is used to specify the number of dissociable hydrogens present.

1. An acid with one dissociable hydrogen is called monoprotic.
2. An acid with two dissociable hydrogens is called diprotic.
3. An acid with three dissociable hydrogens is called triprotic.

The general chemical reactions responsible for the production of H_3O^+ in an acid solution can be written as follows for the three types of acids:

Monoprotic: $HA + H_2O \rightarrow H_3O^+ + A^-$

Diprotic: $H_2A + 2H_2O \rightarrow 2H_3O^+ + A^{2-}$

Triprotic: $H_3A + 3H_2O \rightarrow 3H_3O^+ + A^{3-}$

Although we wrote one reaction for each type of acid, diprotic and triprotic acids dissociate in more than one step. A separate reaction can be written for each step in the dissociation process, as in the three-step dissociation of H_3PO_4, shown below.

First dissociation: $H_3PO_4 + H_2O \rightarrow H_3O^+ + H_2PO_4^-$

Second dissociation: $H_2PO_4^- + H_2O \rightarrow H_3O^+ + HPO_4^{2-}$

Third dissociation: $HPO_4^{2-} + H_2O \rightarrow H_3O^+ + PO_4^{3-}$

Notice that the magnitude of the negative charge on the anion increases by one with each successive dissociation (i.e., $1-, 2-, 3-$).

For a weak acid, the dissociation reaction is an equilibrium reaction, and the value of K_{eq} is a measure of the acid strength. The weaker the acid, the smaller the K_{eq}. A very large K_{eq} means that the H^+ is easily removed from the acid. A small K_{eq} indicates that the H^+ is not easily removed from the acid.

The K_{eq} for an acid dissociation is given the special designation K_a. Notice that in the case of polyprotic acids, the K_a for the loss of the first proton is greater than the K_a for the loss of the second proton, which is greater than the K_a for the loss of the third proton. (See Section 15.3 in the textbook.) The reason for the decrease in the size of the K_a values (or the decrease in the dissociability of the second and third hydrogens) is based on the fact that unlike charges attract. After the first proton is removed, the subsequent hydrogen(s) must be removed from a negatively charged species. Because H^+ ions are positive, it will be more difficult to remove an H^+ from an already negatively charged species than from a neutral one.

Example 15.3

Would you expect H_3PO_4 or HPO_4^{2-} to have a larger K_a?

Solutions

We would expect H_3PO_4 to have a larger K_a because it is neutral. It will be harder to remove the H^+ from the negatively charged HPO_4^{2-}, so it will have the smaller of the two K_a values.

Practice Exercise 15.3

Would you expect HCO_3^- to have a larger or smaller K_a than H_2CO_3?

15.4 Bases—The Anti-Acids

Bases are substances that were found to have the power to neutralize the hydronium ions (H_3O^+) produced in acidic solutions. The hydroxide ions (OH^-) from the base react with the hydronium ions (H_3O^+) to turn H_3O^+ into water, according to the following reaction:

$$H_3O^+ + OH^- \rightarrow 2H_2O$$

Based on the role of the OH^- ions in the neutralization process, Arrhenius defined a base as any electrolyte that, when added to water, dissociates to produce OH^-.

Because bases produce ions when dissolved in water, they are electrolytes. As was the case with acids, bases can be either strong electrolytes or weak electrolytes. To recognize a strong base from its formula, we must know that the strong bases are the hydroxides of the group IA and IIA metals (except for Be). The neutralization of a strong acid with a strong base can be written as:

$$HA + MOH \rightarrow MA + H_2O$$

You should recognize MA as a salt, because it is made of a metal ion and a nonmetal ion.

Example 15.4

Write the neutralization equation for the reaction of HI and KOH.

Solution

$$HI + KOH \rightarrow KI + H_2O$$

Practice Exercise 15.4

Write the neutralization equation for the reaction of HF and LiOH.

15.5 Help! I Need Another Definition of Acid and Base

The Arrhenius definitions of acids and bases worked well to describe most acids and bases known at the time. However, some substances, such as NH_3, that did not have an OH^- in their formula exhibited all of the properties of bases. Because these substances did not even *have* an OH^- in their formula, they could not have dissociated to produce OH^- in water. Furthermore, the Arrhenius definitions were applicable to aqueous solutions only. It was known that acid–base reactions could occur in a nonaqueous environment as well. Hence, it was clear that broader definitions of acids and bases were needed—ones that would apply to all substances exhibiting the properties of acids and bases, in any type of environment. The Brønsted–Lowry definitions fit the requirement:

- An acid is any substance that can serve as a proton donor.
- A base is any substance that can serve as a proton acceptor.

The new definition of bases meant that most substances that have a lone pair of electrons could accept a proton and therefore be classified as a base. This definition of bases explains why NH_3, with its lone pair of electrons on nitrogen, acts as a base. (See Section 15.5 of the textbook for a complete discussion of this concept.)

Because the Brønsted–Lowry definitions speak of acids and bases in terms of donating and accepting protons, acids and bases can be identified based on the specific reactions in which they are participating. The same substance can act as an acid in one reaction and a base in another. One such substance is water, as shown in the reactions below.

Water as an acid: $H_2O + NH_3 \rightarrow OH^- + NH_4^+$

Water as a base: $H_2O + HCl \rightarrow H_3O^+ + Cl^-$

In the first reaction, H_2O *donates* a proton to NH_3, and in the second reaction, H_2O *accepts* a proton from HCl.

Example 15.5

Write a reaction showing ClO^- acting as a base in a reaction with HNO_3.

Solution

$ClO^- + HNO_3 \rightarrow HClO + NO_3^-$

Practice Exercise 15.5

Write a reaction showing $HClO$ acting as an acid in a reaction with NH_3.

15.6   Weak Bases

As stated earlier, bases can be either strong electrolytes or weak electrolytes, as was the case with acids. The strong bases (hydroxides of group IA and IIA metals, except Be) are strong electrolytes, and weak bases are weak electrolytes. Although it is a little more difficult to recognize weak bases than it is to recognize weak acids, you should be aware that one of the most commonly used weak bases is ammonia, NH_3. Some other common weak bases contain nitrogen bonded to three groups, such as CH_3NH_2. We can now rewrite our list of substances that act as electrolytes in a slightly different way.

1. All soluble salts are strong electrolytes.
2. All strong acids are strong electrolytes.
3. All weak acids are weak electrolytes.
4. All strong bases are strong electrolytes.
5. All weak bases are weak electrolytes.

Weak bases produce small amounts of OH^- in water solution (by accepting a proton from water). An equilibrium reaction that appropriately describes the reaction of weak bases with water is that of ammonia.

$NH_3 + H_2O \rightleftharpoons NH_4 + OH^-$

The size of the K_{eq} determines how weak the base is. The smaller the K_{eq}, the weaker the base. (The K_{eq} for a base reaction is given the special name K_b.)

Note that in speaking of acids and bases, we can say that an acid or base is either strong or weak. There are different degrees to which they can be weak, depending on the size of the K_{eq}, but there is only one level of strong (complete dissociation) in aqueous solutions. Therefore, we can say that an acid or base is very weak, but we do not need to say that it is very strong.

15.7 Is This Solution Acidic or Basic? Understanding Water, Autodissociation, and K_w

The distinction between whether a substance is acidic or basic is made on the basis of the relative amounts of H_3O^+ and OH^- in a solution of the substance. Acidic solutions have more H_3O^+ than OH^-; basic solutions have more OH^- than H_3O^+. Neutral solutions have an equal amount of H_3O^+ and OH^-. The amounts of H_3O^+ and OH^- are considered in terms of molarity, or molar concentrations.

For a neutral solution, $[H_3O^+]$ and $[OH^-]$ are both equal to 1×10^{-7} M. These small amounts of H_3O^+ and OH^- in pure water come from the reaction of water with itself, called autodissociation. The autodissociation reaction is given below.

$$H_2O + H_2O \rightleftharpoons H_3O^+ + OH^-$$

A special K_{eq}, called K_w is defined as follows:

$$K_w = [H_3O^+][OH^-]$$

In all aqueous solutions, $K_w = 1 \times 10^{-14}$ (at 25 °C). In neutral solutions, $[H_3O^+] = [OH^-] = 1 \times 10^{-7}$ M.

Because the value of K_w is *always* equal to 1×10^{-14}, if we know the $[H_3O^+]$, we can calculate the $[OH^-]$, and vice versa.

Example 15.6

Calculate the $[OH^-]$ in solution if the $[H_3O^+] = 1 \times 10^{-6}$ M.

Solution

$$[H_3O^+][OH^-] = 1 \times 10^{-14}$$

$$1 \times 10^{-6}[OH^-] = 1 \times 10^{-14}$$

$$[OH^-] = \frac{1 \times 10^{-14}}{1 \times 10^{-6}} = 1 \times 10^{-8} \text{ M}$$

Practice Exercise 15.6

Calculate the $[H_3O^+]$ in solution if the $[OH^-] = 1 \times 10^{-2}$ M.

Notice that in acidic solutions $[H_3O^+] > [OH^-]$, and in basic solutions $[OH^-] > [H_3O^+]$. This means that in acidic solutions the $[H_3O^+]$ will be greater than 1×10^{-7} M, and in basic solutions the $[OH^-]$ will be greater than 1×10^{-7} M.

Example 15.7

Is the solution in Example 15.6 acidic or basic?

Solution

The $[H_3O^+] = 1 \times 10^{-6}$ M (which is $> 1 \times 10^{-7}$ M). Therefore, the solution is acidic.

Practice Exercise 15.7

Is the solution in Practice Exercise 15.6 acidic or basic?

15.8 The pH Scale

We saw in the previous section that we can tell whether a solution is acidic or basic if we know its $[H_3O^+]$ or its $[OH^-]$. Chemists have developed a way of expressing the concentration of H_3O^+ in a solution using numbers that are not written as a negative power of 10, but rather as numbers that are generally between 1 and 14. This method is based on taking the negative logarithm of the $[H_3O^+]$, or $-\log[H_3O^+]$. This value is defined as the **pH**.

$$pH = -\log[H_3O^+]$$

The small letter p in front of the H has the meaning "minus the logarithm of." Whenever you encounter "p" in your study of chemistry, it will always mean "−log." (See Section 15.8 of the textbook for a detailed discussion of logarithms and calculating logarithms.)

We can calculate the pH of an aqueous solution of any substance if we know the $[H_3O^+]$ in the solution.

Example 15.8

Calculate the pH of a solution for which the $[H_3O^+] = 1 \times 10^{-4}$ M. Is the solution acidic or basic?

Solution

$$pH = -\log[H_3O^+] = -\log(1 \times 10^{-4}) = 4.$$

The solution is acidic because the $[H_3O^+] > 1 \times 10^{-7}$M.

Practice Exercise 15.8

Calculate the pH of a solution for which the $[H_3O^+] = 1 \times 10^{-9}$ M. Is the solution acidic or basic?

We see now that we can tell if a solution is acidic, basic, or neutral by the value of its pH.

- In acidic solutions, pH < 7.
- In basic solutions, pH > 7.
- In neutral solutions, pH = 7.

We have seen how to calculate the pH from the $[H_3O^+]$. We can also calculate the $[H_3O^+]$ if we know the pH.

$$[H_3O^+] = 10^{-pH}$$

Example 15.9

Calculate the $[H_3O^+]$ and the $[OH^-]$ of a solution in which pH = 9. Is the solution acidic, basic, or neutral?

Solution

$$[H_3O^+] = 10^{-pH}$$
$$[H_3O^+] = 10^{-9} \text{ M (or } 1 \times 10^{-9} \text{ M)}$$
$$[OH^-] = 10^{-5} \text{ M (or } 1 \times 10^{-5} \text{ M)}$$

The solution is basic because the pH > 7.

Practice Exercise 15.9

Calculate the $[H_3O^+]$ and the $[OH^-]$ of a solution in which pH = 4. Is the solution acidic, basic, or neutral?

When doing pH problems, you should always check to see that your answers are consistent. If the solution is acidic, its pH < 7 and the $[H_3O^+]$ must be greater than 1×10^{-7} M. You will find inadvertent mistakes if you always check to see that your answers are consistent.

15.9 Resisting pH Changes—Buffers

Many biological systems, such as blood, contain natural buffers. These solutions have the ability to resist large changes in pH when strong acids or strong bases are added to the solution. These solutions are called *buffered solutions* (or buffers). Note that the pH of a buffered solution does not remain unchanged when an acid or base is added, but the solutions do not experience a *large* change in pH. The pH does change slightly, but not as much as when strong acids or bases are added to unbuffered solutions.

What makes some solutions buffers, while others are not? It is the presence of a weak acid and its conjugate base, or a weak base and its conjugate acid. Let's now discuss the meaning of the terms conjugate acid and conjugate base.

When we discussed the Brønsted–Lowry definitions of acids and bases, we defined acids as proton donors and bases as proton acceptors. The conjugate base of an acid is the species left after the acid donates a proton. The conjugate acid of a base is the species produced after the base accepts a proton. These relationships are shown for HF and NH_3 below.

For the acid HF, the conjugate base is F^-.

For the base F^-, the conjugate acid is HF.

For the base NH_3, the conjugate acid is NH_4^+.

For the acid NH_4^+, the conjugate base is NH_3.

HF and F^- are a weak acid and its conjugate base (called a conjugate pair).

NH_3 and NH_4^+ are a weak base and its conjugate acid (also called a conjugate pair).

Notice that the relationship between the members of a conjugate pair is that *the acid has one more proton than the base.*

Example 15.10

Indicate the conjugate base for each acid below. Which of the acids could not be used in a buffered solution?

(a) HCN   (b) H_2CO_3 (c) $H_2PO_4^-$ (d) HCl

Solutions

(a) CN^- (b) HCO_3^- (c) HPO_4^{2-} (d) Cl^-

HCl could not be used in a buffered solution. The acid in a buffered solution must be a weak acid, and HCl is a strong acid.

Practice Exercise 15.10

Indicate the conjugate acid for each base below. Which of the bases could not be used in a buffered solution?

(a) NH_2^- (b) Br^- (c) $H_2PO_4^-$ (d) SO_4^{2-}

Now we are ready to return to our discussion on buffered solutions. A buffered solution consists of a conjugate weak acid–base pair in water. An example would be a solution of HF and F^- or a solution of NH_3 and NH_4^+. The F^- and NH_4^+ in these solutions would have to come from a salt containing these ions, such as NaF and NH_4Cl. Because the conjugate species often come from a salt, buffered solutions are sometimes defined as solutions that contain a weak acid and the salt of a weak acid, or a weak base and the salt of a weak base. We'll now consider how the solution of the weak acid HF and the conjugate base F^- acts as a buffer.

If we add some strong acid, it will react with the F^- to produce HF, according to the following reaction:

$$F^- + H_3O^+ \rightleftharpoons HF + H_2O$$

If we add some strong base, it will react with the HF to produce F^-, according to the following reaction:

$$HF + OH^- \rightleftharpoons F^- + H_2O$$

We see that the F^- changes the added strong acid, H_3O^+, into the weak acid HF. The HF changes the added strong base, OH^-, into the weak base F^-. The reason that the pH changes slightly is that we are still left with a weak acid, HF, or a weak base, F^-, after the addition of strong acid or strong base. (See Section 15.9 of the textbook for a more detailed discussion of this concept.)

Example 15.11

Which of the following conjugate pairs could *not* be used to make a buffered solution?

(a) HCN/CN^- (b) HI/I^- (c) $CH_3NH_2/CH_3NH_3^+$

Solution

(b) could not be used as a buffer because HI is a strong acid.

Practice Exercise 15.11

Which component of the buffered solution containing NH_3 and NH_4^+ will react with added base? Which component will react with added acid?

Strategies for Working Problems in Chapter 15

Overview of Problem Types: What You Should Be Able to Do

Chapter 15 provides a discussion of the properties and reactions of acids, bases, and salts. These compounds are known as electrolytes because solutions made with them conduct electricity. The types of problems that you should be able to solve after mastering Chapter 15 are:

1. Classify a given compound as a strong acid, weak acid, strong base, weak base, or salt, and determine whether the compound is a strong electrolyte, a weak electrolyte, or a nonelectrolyte.

2. Write the formula of the salt that is produced from the reaction of any given acid and base.

3. Identify Brønsted–Lowry acids and bases, and conjugate acid–base pairs.

4. Use the K_w to solve for the $[OH^-]$ when given the $[H_3O^+]$, or the $[H_3O^+]$ when given the $[OH^-]$.

5. Write the reaction that occurs when acid or base is added to a buffer solution.

Flowcharts

Flowchart 15.1 **Classifying a given compound as a strong acid, weak acid, strong base, weak base, or salt, and determining whether the compound is a strong electrolyte, a weak electrolyte, or a nonelectrolyte**

Flowchart 15.1

Summary

Method for classifying a given compound as a strong acid, weak acid, strong base, weak base, or salt, and determining whether the compound is a strong electrolyte, a weak electrolyte, or a nonelectrolyte

Step 1
- **Determine whether the compound is an acid, base, salt, or neither.**
- **An acid has the general formula HX (subscripts not included), where X is a monatomic or polyatomic ion.**
- **Many bases have the general formula $M(OH)_n$ where M is a metal ion.**
- **NH_3 is a common weak base. Other weak bases are derivatives of NH_3 in which one or more of the H atoms on NH_3 have been replaced by other atoms or groups of atoms (e.g., CH_3NH_2).**
- **If the compound is an acid (HX) proceed to step 2.**
- **If the compound is a base $(M(OH)_n)$ or an NH_3 type compound, proceed to step 3.**
- **If the compound is not an acid or base, and contains positive and negative ions, it is a salt. Salts are strong electrolytes.**
- **If the compound is not an acid, base, or salt, it is most probably a nonelectrolyte.**

$\downarrow$

Step 2
- **Determine if the acid is one of the strong acids—HCl, HBr, HI, HNO_3, H_2SO_4, $HClO_4$. Strong acids are strong electrolytes.**
- **If the acid is not one of the strong acids, it is a weak acid. Weak acids are weak electrolytes.**

$\downarrow$

Step 3
- **Determine if the base is one of the strong bases—hydroxides of the Group IA and IIA metals. Strong bases are strong electrolytes.**
- **If the base is not one of the strong bases, it is a weak base. Weak bases are weak electrolytes.**
- **Remember that NH_3 and NH_3 derivatives are weak bases.**

Example 1

Classify $Mg(OH)_2$ as a strong acid, weak acid, strong base, weak base, or salt, and determine whether the compound is a strong electrolyte, a weak electrolyte, or a nonelectrolyte.

Solution

Perform Step 1

$Mg(OH)_2$ is a base because it has the general formula of a base, $M(OH)_2$.

Perform Step 2

$Mg(OH)_2$ is a Group IIA hydroxide, therefore $Mg(OH)_2$ is a strong base.

Therefore, $Mg(OH)_2$ is a strong electrolyte because strong bases are strong electrolytes.

Example 2

Classify HF as a strong acid, weak acid, strong base, weak base, or salt, and determine whether the compound is a strong electrolyte, a weak electrolyte, or a nonelectrolyte.

Solution

Perform Step 1

HF is an acid because it has the general formula of an acid, HX.

Perform Step 2

HF is not one of the strong acids, therefore it is a weak acid.

Therefore, HF is a weak electrolyte because weak acids are weak electrolytes.

Example 3

Classify LiBr as a strong acid, weak acid, strong base, weak base, or salt, and determine whether the compound is a strong electrolyte, a weak electrolyte, or a nonelectrolyte.

Solution

Perform Step 1

LiBr is a salt because it has neither the general formula of an acid, HX, nor the general formula for a base, $M(OH)_n$.

Perform Step 2

LiBr is a salt, and therefore, it is a strong electrolyte.

Practice Problems

15.1 Classify each of the following as a strong acid, weak acid, strong base, weak base, or salt, and determine whether the compound is a strong electrolyte, a weak electrolyte, or a nonelectrolyte.

   (a) HBr   (b) H_2CO_3 (c) $HC_2O_3O_2$

15.2 Classify each of the following as a strong acid, weak acid, strong base, weak base, or salt, and determine whether the compound is a strong electrolyte, a weak electrolyte, or a nonelectrolyte.

   (a) $Ca(OH)_2$ (b) $NH_2CH_2CH_3$ (c) H_2SO_4

Flowchart 15.2 **Writing the formula of the salt that is produced from the reaction of any given acid and base**

Flowchart 15.2

Summary

Method for writing the formula of the salt that is produced from the reaction of any given acid and base

Step 1
• **Determine the anion that appears in the acid.**

↓

Step 2
• **Determine the cation that appears in the base.**

↓

Step 3
• **Combine the cation from Step 2 with the anion from Step 1 to determine the formula of the salt.**

Example 1

Write the formula of the salt formed from the combination of HI and $Mg(OH)_2$.

Solution

Perform Step 1
The anion that appears in the acid is I^-.

Perform Step 2
The cation that appears in the base is Mg^{2+}.

Perform Step 3
Mg^{2+} and I^- combine to form the compound MgI_2.

Example 2

Write the formula of the salt formed from the combination of H_3PO_4 and LiOH.

Perform Step 1
The anion that appears in the acid is PO_4^{3-}.

Perform Step 2

The cation that appears in the base is Li^+.

Perform Step 3

Li^+ and PO_4^{3-} combine to form the compound Li_3PO_4.

Practice Problems

15.3 Write the formula of the salt formed from the combination of H_2SO_4 and KOH.

Flowchart 15.3 Identifying conjugate acid–base pairs

Flowchart 15.3

Summary

Method for identifying conjugate acid–base pairs

Step 1
- **Identify two species whose formulas differ by only the presence of one H^+. These are the conjugate acid–base pairs.**

$$\downarrow$$

Step 2
- **Label the species with one more H^+ as the acid.**
- **Label the species with one less H^+ as the base.**

Example 1

Identify the conjugate acid–base pairs in the reaction below. For each pair, indicate which species is the acid and which species is the base.

$HNO_2 + H_2O \rightarrow NO_2^- + H_3O^+$.

Solution

Perform Step 1

    The HNO_2 and NO_2^- differ by only one H^+.

    The H_2O and H_3O^+ differ by only one H^+.

    These are the two sets of acid–base pairs.

Perform Step 2

In the $HNO_2-NO_2^-$ pair:

    The HNO_2 has one more H^+ and is therefore, the acid of the pair.

    The NO_2^- has one less H^+ and is therefore, the base of the pair.

In the H_2O and H_3O^+ pair:

    The H_3O^+ has one more H^+ and is therefore, the acid of the pair.

    The H_2O has one less H^+ and is therefore, the base of the pair.

Example 2

Identify the conjugate acid–base pairs in the reaction below. For each pair, indicate which species is the acid and which species is the base.

$HCl + S^{2-} \rightarrow HS^- + Cl^-$

Solution

Perform Step 1

The HCl and Cl⁻ differ by only one H^+.

The S^{2-} and HS^- differ by only one H^+.

These are the two sets of acid–base pairs.

Perform Step 2

In the HCl and Cl⁻ pair:

> The HCl has one more H^+ and is therefore, the acid of the pair.

> The Cl⁻ has one less H^+ and is therefore, the base of the pair.

In the S^{2-} and HS^- pair:

> The HS^- has one more H^+ and is therefore, the acid of the pair.

> The S^{2-} has one less H^+ and is therefore, the base of the pair.

Practice Problems

15.4 Identify the conjugate acid–base pairs in the reaction below. For each pair, indicate which species is the acid and which species is the base.

$$HBr + NH_3 \rightarrow NH_4^+ + Br^-$$

15.5 Identify the conjugate acid–base pairs in the reaction below. For each pair, indicate which species is the acid and which species is the base.

$$CH_3NH_2 + H_2O \rightarrow CH_3NH_3^+ + OH^-$$

Flowchart 15.4 **Using the K_w to solve for the $[OH^-]$ when given the $[H_3O^+]$, or the $[H_3O^+]$ when given the $[OH^-]$**

Flowchart 15.4

Summary

Method for using the K_w to solve for the $[OH^-]$ when given the $[H_3O^+]$, or the $[H_3O^+]$ when given the $[OH^-]$

Step 1
• **If given $[H_3O^+]$, plug the $[H_3O^+]$ into the K_w expression,**
$K_w = 10^{-14} = [H_3O^+] \times [OH^-]$

or

• **If given $[OH^-]$, plug the $[OH^-]$ into the K_w expression,**
$K_w = 10^{-14} = [H_3O^+] \times [OH^-]$

↓

Step 2
• **Divide both sides of the expression by $[H_3O^+]$ to isolate $[OH^-]$.**

or

• **Divide both sides of the expression by $[OH^-]$ to isolate $[H_3O^+]$.**

↓

Step 3
- Solve for $[OH^-]$.

or

- Solve for $[H_3O^+]$.

Example 1

Calculate the $[OH^-]$ in a solution in which $[H_3O^+] = 2.3 \times 10^{-4}$ M.

Solution

Perform Step 1

$K_w = 10^{-14} = [H_3O^+] \times [OH^-]$

$10^{-14} = 2.3 \times 10^{-4} \times [OH^-]$

Perform Step 2

$$\frac{10^{-14}}{2.3 \times 10^{-4}} = \frac{2.3 \times 10^{-4} \times [OH^-]}{2.3 \times 10^{-4}}$$

Perform Step 3

$4.3 \times 10^{-11} M = [OH^-]$

Example 2

Calculate the $[H_3O^+]$ in a solution in which $[OH^-] = 2.7 \times 10^{-9}$ M.

Solution:

Perform Step 1

$K_w = 10^{-14} = [H_3O^+][OH^-]$

$10^{-14} = [H_3O^+] \times 2.7 \times 10^{-9}$

Perform Step 2

$$\frac{10^{-14}}{2.7 \times 10^{-9}} = \frac{2.7 \times 10^{-4} \times [OH^-]}{2.7 \times 10^{-9}}$$

Perform Step 3

$3.7 \times 10^{-6} M = [H_3O^+]$

Practice Problems

15.6 Calculate the $[H_3O^+]$ in a solution in which $[OH^-] = 1.3 \times 10^{-2}$ M.

15.7 Calculate the $[OH^-]$ in a solution in which $[H_3O^+] = 5.8 \times 10^{-8}$ M.

Flowchart 15.5 Writing the reaction that occurs when strong acid (H_3O^+) or strong base (OH^-) is added to a buffer solution

Flowchart 15.5

Summary

Method for writing the reaction that occurs when acid or base is added to a buffer solution

Step 1
- Identify the acid and base components of the buffer solution. (The acid component is the species with one more H^+.)
- If acid (H_3O^+) is added to the buffer solution, go to step **2**.
- If base (OH^-) is added to the buffer solution, go to step **3**.

$\downarrow$

Step 2
- React the added acid (H_3O^+) with the base component of the buffer solution. (The H_3O^+ will add an H^+ to the base component of the buffer, producing water and the acid component of the buffer.)

$\downarrow$

Step 3
- React the added base (OH^-) with the acid component of the buffer solution. (The OH^- will take an H^+ from the acid component of the buffer, producing water and the base component of the buffer.)

Example 1

A buffer solution is prepared using HF and F^-. Write the reaction that occurs when strong acid is added to this buffer solution.

Solution

Perform Step 1

The acid component of the buffer solution is HF; the base component is F^-. Because acid is being added to the solution, we proceed to step 2.

Perform Step 2

The added H_3O^+ will react with the F^- according to the following equation.

$$F^- + H_3O^+ \rightarrow HF + H_2O$$

Example 2

A buffer solution is prepared using H_2CO_3 and HCO_3^-. Write the reaction that occurs when strong base is added to this buffer solution.

Solution

Perform Step 1

The acid component of the buffer solution is H_2CO_3; the base component is HCO_3^-. Because base is being added to the solution, we proceed to step 3.

Perform Step 3

The added OH^- will react with the H_2CO_3 according to the following equation:

$$H_2CO_3 + OH^- \rightarrow HCO_3^- + H_2O$$

Practice Problems

15.8 A buffer solution is prepared using H_3BO_3 and $H_2BO_3^-$. Write the reaction that occurs when strong acid is added to this buffer solution.

15.9 A buffer solution is prepared using HCN and CN^-. Write the reaction that occurs when strong base is added to this buffer solution.

Quiz for Chapter 15 Problems

1. Classify the following compounds as electrolytes or nonelectrolytes:
   (a) HBr   (b) CH_3CH_3 (c) NH_3

2. Write the formula of the salt produced in the reaction of the following acids and bases:
   (a) $H_3PO_4 + NaOH$ (b) $HNO_3 + Mg(OH)_2$

3. Write the formula for the conjugate base for the following acids:
   (a) H_3BO_3 (b) $HC_2H_3O_2$ (c) NH_4^+

4. Classify the following as strong acids or weak acids:
   (a) HCl   (b) HF   (c) HI   (d) HBr

5. Complete the sentence. Of the species NaF, HF, NH_3, and $HC_2H_3O_2$, the strong electrolyte is
   _____.

6. Calculate the $[OH^-]$ and $[H_3O^+]$ in a solution in which the ions have equal concentration.

7. Classify the following as strong bases or weak bases:
   (a) NH_3 (b) CH_3NH_2 (c) LiOH   (d) $Ca(OH)_2$

8. Write the formula of the conjugate acid for the following bases:
   (a) NH_3 (b) NO_3^- (c) HPO_4^{2-}

9. Calculate the $[OH^-]$ in a solution in which the $[H_3O^+] = 8.9 \times 10^{-5}\, M$.

10. Indicate which of the following species represent acid–base pairs.
    (a) HNO_3 (b) $H_2PO_4^-$ (c) H_2SO_4 (d) NO_3^-
    (e) NH_4^+ (f) H_3PO_4 (g) NH_3 (h) HSO_4^-

11. A buffer solution is made up of H_2S and HS^-.
    (a) Write the reaction that occurs when strong acid is added to the solution.
    (b) Write the reaction that occurs when strong base is added to the reaction.

12. Calculate the $[H_3O^+]$ in a solution in which the $[OH^-] = 1 \times 10^{-5}\, M$.

Exercises for Self-Testing

A. Completion. *Write the correct word(s) to complete each statement below.*

1. Acids react with active metals, such as zinc, to produce _____ gas.

2. _____ electrolytes are substances that dissociate completely in aqueous solution, whereas _____ electrolytes dissociate partially.

3. Acids turn litmus dye _____, and bases turn litmus dye _____.

4. Most molecular compounds do not dissociate when placed in water and are therefore _____.

5. According to the Brønsted–Lowry theory of acids and bases, acids are _____ _____ and bases are _____ _____.

6. An acid that has two dissociable hydrogen atoms is a _____ acid.

7. The neutralization reaction between an acid and a base produces a _____ plus _____.

8. A solution in which $[H_3O^+] = 1 \times 10^{-5}$ M has a pH of _____.

9. An acid with a K_a of 1×10^{-8} is _____ than an acid with a K_a of 1×10^{-2}.

10. A solution that resists a change in pH is called a _____ _____.

11. An acidic solution has a pH less than _____.

12. HCl, HI, and HBr are all examples of _____ acids.

13. The conjugate acid of HS^- is _____, and the conjugate base of HS^- is _____.

14. The pH of a solution in which $[OH^-] = 1 \times 10^{-4}$ M is _____.

15. The strong bases are the hydroxides of group _____ and _____ metals.

B. True/False. *Indicate whether each statement below is true or false, and explain why the false statements are false.*

_____ 1.   Bases turn litmus dye blue.

_____ 2.   HCl is an example of a nonelectrolyte because it is a molecular compound.

_____ 3.   All weak acids and weak bases are weak electrolytes.

_____ 4. NH_3 is an example of an Arrhenius acid because it has three hydrogens.

_____ 5.   The pH of a solution in which $[H_3O^+] = 10^{-2}$ is -2.

_____ 6.   There is some H_3O^+ present in distilled water even though it is neutral.

_____ 7.   When a strong acid or base is added to a buffered solution, the pH does not change at all.

_____ 8. H_2SO_4 is the conjugate acid of SO_4^{2-}.

_____ 9.   HCN and its conjugate base, CN^-, can be used to make a buffered solution.

_____ 10.   When an acid reacts with water, the water acts as an Arrhenius base.

C. Questions and Problems

1. Explain why the Arrhenius definition of acids and bases needed to be broadened.

2. Predict the salt that will be produced from the following acid–base neutralization reaction:

   $HNO_3 + KOH \rightarrow$ ____ $+ H_2O$

3. Calculate the pH of a solution in which $[OH^-] = 1 \times 10^{-10}$ M.

4. Explain what the value of K_a tells us about the strength of an acid.

5. A student is asked to prepare a buffered solution in which the acid will be HCN. She is told that the other ingredient will be either CH_3CN or KCN, and that she must choose. Which one should she choose? Why?

6. NH_3 is a Brønsted–Lowry base but not an Arrhenius base. Explain why this is so.

7. Autodissociation of water in an endothermic process. How do you expect K_w to change as the temperature increases? Explain your answer.

8. Would you expect pure water to have pH = 7 at temperatures above 25 °C?

16

Nuclear Chemistry

Learning Outcomes

1. Understand what an atom's mass defect is and its relationship to binding energy.

2. Understand what the band of stability is and its relationship to the neutron-to-proton ratio.

3. Understand what half-life is.

4. Understand what radioactivity is and the symbols used to represent it.

5. Determine the products of the various forms of radioactive decay.

6. Write and balance nuclear reactions.

7. Understand half-life and use it to calculate the age of an object.

8. Understand nuclear fission, nuclear fusion, and their impact on society.

9. Understand how radioactivity and radiation can both hurt and be helpful to living organisms.

Chapter Outline

16.1 The Case of the Missing Mass: Mass Defect and the Stability of the Nucleus

    A. Calculation of mass defect

    B. Conversion of mass defect to energy using $E = mc^2$

    C. Mass defect and binding energy per nucleon

    D. Unstable nuclei—Radioactivity

16.2 Half-Life and the Band of Stability

    A. Neutron-to-proton (n/p) ratio

    B. The band of stability

16.3 Spontaneous Nuclear Changes—Radioactivity

    A. Unstable nuclei and the "ocean of instability"

    B. Relationship between n/p and radioactive decay mode

        1. β^- emission

        2. Positron emission or electron capture

3. α particle emission

C. Gamma (γ) Radiation

16.4 Using Radioactive Isotopes to Date Objects

A. Radiocarbon dating

B. Age calculations from radioactive isotope content

16.5 Nuclear Energy—Fission and Fusion

A. Nuclear fission

1. Fission reaction

2. Nuclear explosion

3. Nuclear reactors

4. Problems with nuclear fission

B. Nuclear fusion

1. Fusion reaction

2. Problem with reaching break-even point with fusion

3. The hydrogen bomb

16.6 One More Thing: Biological Effects and Medical Applications of Radioactivity

A. Magnetic resonance imaging (MRI)

B. Biological effects of ionizing radiation

C. Use of radiation imaging in medical diagnosis

D. Radioimmunoassays

E. Irradiation of food products to kill bacteria

Review of Key Concepts from the Text

16.1 The Case of the Missing Mass: Mass Defect and the Stability of the Nucleus

Because we will be using the symbols for isotopes throughout this chapter, we need to briefly review the symbols we use to describe isotopes, which we first encountered in Chapter 3. The general symbol for an isotope can be written as:

$$^A_Z X$$

where X is the elemental symbol, $Z =$ atomic number (# of protons), and $A =$ mass number (protons + neutrons). We can therefore determine the number of nucleons (protons and neutrons) in a nucleus from the isotopic symbol.

Number of protons $= Z$

Number of nucleons (protons + neutrons) $= A$

Number of neutrons $= A - Z$

All atoms are composed of a nucleus in the center and electrons surrounding the nucleus. Nuclei (plural for nucleus) are made up of protons and neutrons. (The only exception to this is the $^1_1$H nucleus, which contains one proton and no neutrons.) It is logical to assume that the sum of the masses of the components of an atom would be equal to the actual mass of the atom. This, however, is not the case. It happens that the mass of an atom is always slightly *lower than* the sum of the masses of the components. The missing mass is called the mass defect. The mass defect of an isotope can be calculated by adding the masses of the protons, neutrons, and electrons that make up a mole of the isotope and subtracting the molar mass of the isotope.

Mass defect = Mass of subatomic particles in one mole of atoms − Mass of one mole of atoms

This procedure is illustrated in the example below.

Example 16.1

Calculate the mass defect of one mole of $^{32}_{16}S$ atoms. The mass of one mole of $^{32}_{16}S$ is 31.972 07 g.

Solution

Mass defect = Mass of subatomic particles in one mole of atoms − Mass of one mole of atoms
First, we calculate the mass of one mole of each subatomic particle. To do so, we need to look up the mass of a proton ($1.672\,623 \times 10^{-27}$ kg), neutron ($1.674\,6929 \times 10^{-27}$ kg), and an electron ($9.109\,390 \times 10^{-31}$ kg), on the inside back cover of your textbook under *Fundamental Constants*.

$$\frac{1.672\,623 \times 10^{-27}\,\text{kg}}{1\,\text{proton}} \times \frac{1000\,\text{g}}{1\,\text{kg}} \times \frac{6.022 \times 10^{23}\,\text{protons}}{1\,\text{mol protons}} = 1.0073\frac{\text{g}}{\text{mol}}$$

$$\frac{1.674\,929 \times 10^{-27}\,\text{kg}}{1\,\text{neutron}} \times \frac{1000\,\text{g}}{1\,\text{kg}} \times \frac{6.022 \times 10^{23}\,\text{neutrons}}{1\,\text{mol neutrons}} = 1.0087\frac{\text{g}}{\text{mol}}$$

$$\frac{9.109\,390 \times 10^{-31}\,\text{kg}}{1\,\text{electron}} \times \frac{1000\,\text{g}}{1\,\text{kg}} \times \frac{6.022 \times 10^{23}\,\text{electrons}}{1\,\text{mol electrons}} = 0.000\,55\frac{\text{g}}{\text{mol}}$$

Now, we calculate the mass of subatomic particles in one mole of atoms:

$$16\,\text{mol protons} \times \frac{1.0073\,\text{g}}{1\,\text{mol protons}} = 16.1168\,\text{g}$$

$$16\,\text{mol neutrons} \times \frac{1.0087\,\text{g}}{1\,\text{mol neutrons}} = 16.1392\,\text{g}$$

$$16\,\text{mol electrons} \times \frac{0.000\,55\,\text{g}}{1\,\text{mol electrons}} = 0.0088\,\text{g}$$

Then we add up the results to get the sum of the masses of the components.

$$16.1168\,\text{g} + 16.1392\,\text{g} + 0.0088\,\text{g} = 32.2648\,\text{g}$$

Finally, we subtract the mass of one mole of $^{32}_{16}S$ atoms from this total to get the mass defect.

$$32.2648\,\text{g} - 31.972\,07\,\text{g} = 0.2927\,\text{g/mol}$$

Practice Exercise 16.1

Calculate the mass defect of one mole of $^{63}_{29}Cu$. The mass of one mole of $^{63}_{29}Cu$ is 62.939 60 g.

The missing mass is converted to energy upon formation of the atoms from the component parts. This energy is called the binding energy and can be thought of as the energy that holds the nucleus together. The binding energy can be calculated using Einstein's equation $E = mc^2$, where *m* is the mass in kilograms and *c* is the speed of light, which is equal to 3.00×10^8 m/s. The following example illustrates the calculation of the binding energy for a mole of $^{32}_{16}S$ atoms.

Example 16.2

Calculate the binding energy for one mole of $^{32}_{16}S$ atoms.

Solution

The mass defect for $^{32}_{16}S$ is 0.2927 g/mol (from Example 16.1). First we convert this to kilograms/mole.

$$\text{mass defect in kg/mol} = 0.2927 \text{ g} \times \frac{1 \text{ kg}}{1000 \text{ g}} = 0.000\,2927 \text{ kg/mol}$$

Then we use Einstein's equation to calculate the binding energy.

$$\begin{aligned} E &= mc^2 \\ &= (0.000\,2927 \text{ kg/mol}) \times (3.00 \times 10^8 \text{ m/s})^2 \\ &= 2.63 \times 10^{13} \text{ kg} \cdot \text{m}^2/\text{s}^2 \cdot \text{mol} \\ &= 2.63 \times 10^{13} \text{ J/mol} = 2.63 \times 10^{10} \text{ kJ/mol} \end{aligned}$$

Practice Exercise 16.2

Calculate the binding energy for one mole of $^{63}_{29}Cu$ atoms.

The binding energy is very important in determining the stability of a nucleus. As you might expect, the greater the binding energy, the more stable the nucleus because the nucleus is being held together more tightly. However, to directly compare the stability of different atoms, we must take into consideration not only the binding energy but also the number of nucleons that are being held together. Thus, we must calculate the binding energy per mole of nucleons, a quantity called the binding energy per nucleon. The binding energy per nucleon is calculated by dividing the binding energy by the number of nucleons (protons and neutrons) in the nucleus.

$$\text{Binding energy per mole of nucleons} = \frac{\text{Binding energy per mol}}{\text{Number of nucleons}}$$

The binding energy per nucleon is often expressed in millions of electron volts, or megaelectron volt (MeV) instead of in kilojoules (kJ). We can convert kJ to MeV by using the following conversion factor:

$$1 \text{ MeV} = 96.5 \times 10^6 \text{ kJ (per mole of nucleons)}$$

A calculation of the binding energy per mole of nucleons is shown in the example below.

Example 16.3

Calculate the binding energy per nucleon of $^{32}_{16}S$ and $^{63}_{29}Cu$. Which nucleus is more stable?

Solution

To determine which nucleus is more stable, we need to calculate the binding energy per nucleon for each one. We will use the binding energies calculated in Example 16.2 and Practice Exercise 16.2.

$$\begin{aligned} \text{Binding energy per mole of nucleons for } ^{32}_{16}S &= \frac{2.63 \times 10^{10} \text{ kJ}}{1 \text{ mol}} \times \frac{1 \text{ mol}}{32 \text{ mol nucleons}} \\ &= 8.22 \times 10^8 \text{ kJ/mol of nucleons} \end{aligned}$$

Converting to MeV:

$$8.22 \times 10^8 \text{ kJ/mol of nucleons} \times \frac{1 \text{ MeV}}{96.5 \times 10^6 \text{ kJ}} = 8.52 \text{ MeV/mol of nucleons}$$

$$\text{Binding energy per mol of nucleons for } {}^{63}_{29}\text{Cu} = \frac{5.254 \times 10^{10} \text{ kJ}}{1 \text{ mol}} \times \frac{1 \text{ mol}}{63 \text{ mol nucleons}}$$

$$= 8.34 \times 10^8 \text{ kJ/mol of nucleons}$$

Converting to MeV:

$$8.34 \times 10^8 \text{ kJ/mol of nucleons} \times \frac{1 \text{ MeV}}{96.5 \times 10^6 \text{ kJ}} = 8.64 \text{ MeV/mol of nucleons}$$

The ${}^{63}_{29}\text{Cu}$ is the more stable nucleus because it has the greater binding energy per mole of nucleons.

Practice Exercise 16.3

Calculate the binding energy per nucleon of ${}^{40}_{20}\text{Ca}$ and ${}^{24}_{12}\text{Mg}$. Which nucleus is more stable? The mass of one mole of ${}^{40}_{20}\text{Ca}$ is 39.962 59 g, and the mass of one mole of ${}^{24}_{12}\text{Mg}$ is 23.985 04 g.

    A plot of the binding energy per nucleon for the most stable isotope of each element is shown in Section 16.1 of the textbook and reprinted here. (You can verify your answers to Practice Exercise 16.3 from the figure.) The highest point on the plot is occupied by the ${}^{56}_{26}\text{Fe}$ nucleus, the most stable nucleus that exists. You can always compare the relative stabilities of isotopes by their positions on this plot.

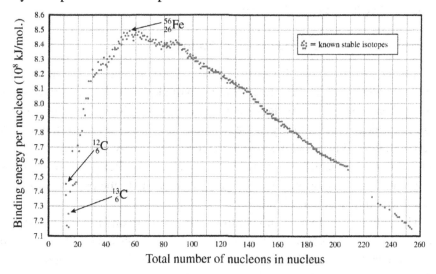

Example 16.4

Use the figure above to arrange the following isotopes in the order of decreasing stability.

$${}^{40}_{20}\text{Ca} \quad {}^{60}_{27}\text{Co} \quad {}^{235}_{92}\text{U} \quad {}^{80}_{35}\text{Br}$$

Solution

The isotopes with the greater binding energy per nucleon values have greater stability than those with smaller values. So the order of decreasing stability is as follows:

$${}^{60}_{27}\text{Co} > {}^{80}_{35}\text{Br} > {}^{40}_{20}\text{Ca} > {}^{235}_{92}\text{U}$$

Practice Exercise 16.4

Use the plot above to arrange the following isotopes in the order of decreasing stability:

$$^{200}_{79}\text{Au} \quad ^{108}_{47}\text{Ag} \quad ^{195}_{78}\text{Pt} \quad ^{52}_{24}\text{Cr}$$

The enormous amount of energy that binds nucleons together makes it very difficult to change one nucleus into another. In fact, the types of reactions that involve changing nuclei (called nuclear reactions) involve much more energy than the energy associated with the chemical reactions we have been studying so far.

Chemical reactions involve only the redistribution of electrons, whereas nuclear reactions involve either breaking apart or combining nuclei. Most chemical reactions involve energies ranging from less than 100 kJ/mol to a few thousand kJ/mol. Nuclear reactions, however, involve energies of billions of kJ/mol. When we understand the large differences in energy requirements for chemical reactions and nuclear reactions, we are not surprised that the ancient alchemists failed in their attempts to convert lead into gold. The chemical reactions they were attempting to use did not produce anywhere near enough energy to cause the nuclear reaction that would have been necessary to change the identity of the elements.

Although the alchemists could not convert one nucleus into another, nature does this spontaneously in the case of some nuclei, called *radioactive nuclei*. Radioactive nuclei are unstable and spontaneously eject pieces of themselves, transforming themselves into a new element. A *radioactive atom* is one that contains an unstable nucleus that undergoes spontaneous change. Although a complete study of the reasons for nuclear instability will not be undertaken here, it is possible for us to determine what types of nuclei will be unstable. We can even predict how they will change in order to become more stable.

16.2 Half-Life and the Band of Stability

Because nuclei are composed of positively charged protons and neutral neutrons, one might wonder what holds all of the positively charged particles in such close proximity given that like charges repel each other. The nucleons in the nucleus are held together by a force called the *strong force*, which is present only in the nuclei of atoms. However, even the strong force is not strong enough to hold two protons together if they are directly touching each other. This is where the neutrons come in. Neutrons bind the whole nucleus together by means of the strong force. They can almost be thought of as neutral particles that shield the protons from the repulsive forces they would feel from each other.

The only isotope that has no neutrons, ^1_1H, has only one proton. It doesn't need any neutrons to shield protons from each other. The more protons a nucleus has, the more neutrons it needs to insulate the protons from each other. This is why the smaller elements, like $^{14}_7\text{N}$, usually have about one neutron for every proton, but the larger elements, like $^{238}_{92}\text{U}$, have many more neutrons than protons.

The number of neutrons compared to the number of protons is expressed as a ratio, the neutron-to-proton (n/p) ratio. In general, the elements with an atomic number between 1 and 20 have an n/p ratio of 1. For the larger nuclei, the n/p ratios are ~1.5.

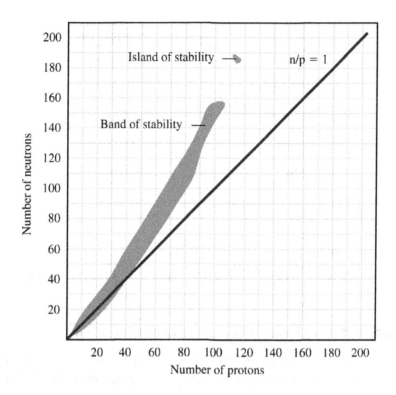

It is possible for a nucleus to have too many or too few neutrons. A nucleus with fewer than 20 protons needs an n/p ratio about equal to 1 and would have too many neutrons if the ratio were much bigger than that. On the other hand, a nucleus with more than 80 protons needs an n/p ratio of about 1.5 and would have too few neutrons if the ratio were much smaller than that. A plot of the number of protons (*x*-axis) versus the number of neutrons (*y*-axis) for all known stable isotopes yields an area known as the *band of stability*, shown in the figure above. All of the isotopes within the band are considered stable. Stable isotopes are those that are either nonradioactive, or radioactive with a measurable *half-life*.

The half-life of an isotope is defined as the amount of time it takes for the number of radioactive nuclei to drop to exactly half of the initial number due to spontaneous nuclear change. So, 25.0×10^{23} atoms of a radioactive isotope with a half-life of 4 minutes would decrease to 12.5×10^{23} atoms after 4 minutes, 6.25×10^{23} atoms after another 4 minutes (8 minutes total), 3.125×10^{23} atoms after another 4 minutes (12 minutes total), and so on.

Example 16.5

The half-life of tritium, ^3_1H, is 12.3 years. How old is a sample of tritium that originally contained 0.0400 g of tritium and now contains 0.001 25 g?

Solution

If the original sample was 0.0400 g, we can determine the mass of tritium remaining after each half-life.

Half-life	Amount remaining (in grams)
First	0.0200
Second	0.0100
Third	0.005 00
Fourth	0.002 50
Fifth	0.001 25

A mass of 0.001 25 g would be remaining after five half-lives. The age of the tritium would therefore be $5 \times 12.3 = 61.5$ years.

Practice Exercise 16.5

The half-life of $^{40}_{15}P$ is 14.28 days. How many atoms of a 1.00-mole sample of $^{40}_{15}P$ will be present after 100 days?

16.3 Spontaneous Nuclear Changes—Radioactivity

All of the isotopes located inside the band of stability are stable enough to exist indefinitely or for a measurable amount of time. If an isotope lies outside of the band of stability, it will spontaneously decay in order to convert to an isotope within the band of stability. This spontaneous decay is called *radioactivity*, or radioactive decay. Isotopes that are located above the band of stability would be unstable because they would have too many neutrons. For example, $^{18}_{6}C$ would have 6 protons and 12 neutrons. If this isotope were plotted on the graph on page 329, its point would be well above the band of stability. On the other hand, isotopes that are located below the band of stability would be unstable because they would have too few neutrons. For example, $^{25}_{20}Ca$ would have 20 protons and 5 neutrons. If this isotope were plotted on the graph on page 329, its point would be well below the band of stability.

You should notice that the band of stability ends at approximately 108 protons. Then there is a small circular region (an "island" of stability) in the region around 112–118 protons. These isotopes are predicted to be stable but have not yet been observed.

To briefly summarize, the regions above, below, and beyond the band of stability can be thought of as regions of instability. Any isotopes that would exist in these regions will decay spontaneously to be converted to an isotope that lies within the band of stability. Isotopes that are above the band of stability have too many neutrons for the number of protons; elements that are below the band of stability have too few neutrons for the number of protons. Elements whose atomic numbers are beyond the island of stability (atomic numbers greater than 118) contain too many protons to exist. The unstable isotopes will decay in a manner that will allow them to be converted to a stable isotope (within the band of stability). The different mechanisms by which radioactive decay can occur will be discussed shortly.

Example 16.6

Predict whether each of the following isotopes will lie within, below, above, or beyond the band of stability. Explain your reasoning.

(a) $^{30}_{20}Ca$ (b) $^{80}_{27}Co$ (c) $^{9}_{3}Li$ (d) $^{206}_{109}Mt$

Solution

We use the generalization that stable isotopes of elements with atomic numbers between 1 and 20 have an n/p ratio of about 1, while larger isotopes have greater n/p ratios, up to about 1.5.

(a) $^{30}_{20}$Ca has too few neutrons and will lie below the band of stability.

(b) $^{80}_{27}$Co has too many neutrons and will lie above the band of stability.

(c) $^{9}_{3}$Li has too many neutrons and will lie above the band of stability.

(d) $^{206}_{109}$Mt has too many protons and will lie beyond the band of stability.

Practice Exercise 16.6

Predict whether each of the following isotopes will lie within, below, above, or beyond the band of stability. Explain your reasoning.

(a) $^{12}_{6}$C   (b) $^{12}_{7}$N   (c) $^{30}_{12}$Mg   (d) $^{235}_{92}$U

Knowing Your Subatomic Particles

There are several subatomic particles that are involved in radioactive decay. Their names and symbols are presented in the table below.

Particle name	Symbol
Neutron	$^{1}_{0}$n
Proton	$^{1}_{1}$p
Electron (also called a beta particle)	$^{0}_{-1}$e or $^{0}_{-1}\beta$
Positron (also called antimatter)	$^{0}_{+1}$e or $^{0}_{+1}\beta$

Conversion of a Neutron to a Proton: Radioactive Decay via Beta Emission

Isotopes that lie above the band of stability have too many neutrons for the number of protons and need to convert a neutron into a proton to be brought into (or closer to) the band of stability. This conversion can take place if the nucleus ejects an electron (or beta particle). This is a little easier to understand if we think of a neutron as the combination of a proton and an electron. When the neutron (proton plus electron) ejects an electron, a proton will be left. The conversion of a neutron to a proton decreases the n/p ratio and brings the unstable nucleus into (or closer to) the band of stability.

Example 16.7

The $^{16}_{7}$N nucleus is unstable because it has too many neutrons. What isotope will be produced if the $^{16}_{7}$N nucleus ejects an electron (beta particle)? Write a reaction illustrating this process.

Solution

When the $^{16}_{7}$N nucleus ejects an electron, it will contain one more proton and one less neutron. The new isotope will therefore have 8 protons and 8 neutrons, making it $^{16}_{8}$O, a stable isotope. The reaction is written as follows:

$$^{16}_{7}\text{N} \rightarrow {}^{16}_{8}\text{O} + {}^{0}_{-1}\text{e}$$

Practice Exercise 16.7

The $^{198}_{79}$Au nucleus is unstable because it has too many neutrons. What isotope will be produced if the $^{198}_{79}$Au nucleus ejects an electron (beta particle)?

The reaction in Example 16.7 is called a *nuclear reaction* because it represents a change to the atom's nucleus. In nuclear reactions that represent decay processes, the reactant isotope is called the *parent*, and the resulting isotope is called a *daughter*.

Nuclear reactions are balanced by making sure that the sum of the mass numbers on the reactants' side equals the sum of the mass numbers on the products side, and that the sum of the atomic numbers is also the same on both sides of the equation. We can use this principle to complete nuclear reactions that are missing either a reactant or a product.

Example 16.8

Fill in the blank in the nuclear reaction below.

$$^{277}_{89}\text{Ac} \rightarrow \underline{} + {}^{0}_{-1}\text{e}$$

Solution

To fill in the blank, we determine the mass number and charge of the missing piece. The mass number must be 227, and the charge must be 90 for the equation to balance. Therefore, the missing piece is $^{227}_{90}\text{Th}$.

Practice Exercise 16.8

Fill in the blank in the nuclear reaction below.

$$^{131}_{53}\text{I} \rightarrow \underline{} + {}^{0}_{-1}\text{e}$$

Radioactive Decay via Positron Emission and Electron Capture

Isotopes that lie below the band of stability have too few neutrons for the number of protons and need to convert a proton into a neutron to be brought into (or closer to) the band of stability. This conversion can take place if the nucleus ejects a positron, as shown in the reaction below.

$$^{12}_{7}\text{N} \rightarrow {}^{12}_{6}\text{C} + {}^{0}_{1}\text{e} \text{ (positron emission)}$$

An alternative mechanism for converting a proton into a neutron is for the nucleus to capture an electron. (Recall our thinking of a neutron as a proton plus an electron.) If an electron joins a proton in the nucleus, the proton will be converted to a neutron, thereby increasing the n/p ratio of the isotope. This is shown in the reaction below.

$$^{12}_{7}\text{N} + {}^{0}_{-1}\text{e} \rightarrow {}^{12}_{6}\text{C} \text{ (electron capture)}$$

Both positron emission and electron capture occur for radioactive isotopes that are below the band of stability (where the n/p ratio is too low). The specific mode that actually occurs depends on the particular radioactive isotope.

Radioactive Decay via Alpha Particle Emission

A radioactive isotope that is beyond the band of stability is too large to be nonradioactive and must emit a chunk of its nucleus to enter the band of stability. These isotopes emit an alpha (α) particle to achieve this. You will recall (from our discussion of the Rutherford experiment in

Chapter 3) that an alpha particle is equivalent to a helium nucleus. It consists of two protons and two neutrons and has the symbol $_2^4$He. (Although an alpha particle has a charge of 12, this is not indicated in the symbol.) When a nucleus emits an alpha particle, it gets rid of two protons and two neutrons all at once, decreasing its atomic number by 2 and its mass number by 4. This is shown in the reaction below.

$$_{84}^{210}\text{Po} \rightarrow _{82}^{206}\text{Pb} + _2^4\text{He}$$

Up to this point, we have discussed the radioactive decay processes as if only one process happens to an isotope, with no further reaction. In reality, however, some radioactive isotopes undergo several decays before arriving at a stable isotope. The sum of steps in a multistep decay is called a *radioactive decay pathway*. The radioactive decay pathway for $_{92}^{238}$U is shown in the figure in Section 16.3 of the textbook.

Gamma Radiation

When radioactive nuclei undergo decay, they emit energy in the process. Some of the energy is carried away by the emitted particles in the form of kinetic energy. Many radioactive nuclei, however, release energy in the form of electromagnetic radiation, usually gamma rays (γ rays). Gamma rays are very high energy radiation (more energetic than X-rays) and can be quite harmful to living organisms. Because gamma rays are electromagnetic radiation and not particles, their emission does not change the atomic number or the mass number of the isotope emitting the gamma radiation. Gamma radiation emission often accompanies the decay of particles that we discussed earlier.

16.4 Using Radioactive Isotopes to Date Objects

Naturally occurring radioactive isotopes have a characteristic half-life and a characteristic decay product that can be used to determine the age of an object. (Carbon-14 is the isotope that is used to date objects that were once alive.) We use the half-life of the isotope and the fraction of original isotope remaining at the time of dating to determine the age of the object. The following example demonstrates how this can be done.

Example 16.9

$_{92}^{238}$U has a half-life of 4.46×10^9 years as it decays to $_{82}^{206}$Pb. If a rock contains 25 atoms of $_{92}^{238}$U for every 75 atoms of $_{82}^{206}$Pb, how old is the rock?

Solution

We first assume that all of the $_{82}^{206}$Pb comes from the decay of the $_{92}^{238}$U. (For these types of problems, we always assume that all of the decay product has come from the radioactive isotope we are using to date the object.) The original amount of $_{92}^{238}$U would have thus been 100 out of every 100 atoms. Twenty-five percent of the $_{92}^{238}$U is remaining at the time of dating. This means that two half-lives have passed ($100 \rightarrow 50 \rightarrow 25$). The rock is therefore $2 \times 4.46 \times 10^9 = 8.92 \times 10^9$ years old.

Practice Exercise 16.9

$_6^{14}$C has a half-life of 5715 years. If a piece of wood contains 1/16 the amount of carbon-14 that it contained while the tree from which it came was alive, what is the age of the wood?

If we are trying to date an object for which the amount of radioactive isotope remaining is not equivalent to an exact number of half-lives, we use another method. We will still use the fraction of radioactive isotope remaining at the time of dating, and we use it in the following equation:

$$\text{Age} = \frac{-2.303 \times \log(\% \text{ Radioactive isotope remaining}/100\%)}{0.693} \times \text{Half-life}$$

Example 16.10

The half-life of $^{40}_{19}K$ is 1.2×10^9 years, as it decays to $^{40}_{18}Ar$. How old is a sample of moon rock that contains 18% $^{40}_{19}K$ and 82% $^{40}_{18}Ar$ by mass?

Solution

We substitute the values into the equation for determining age.

$$\text{Age} = \frac{-2.303 \times \log(\% \text{ Radioactive isotope remaining}/100\%)}{0.693} \times \text{Half-life}$$

$$= \frac{-2.303 \times \log(18/100)}{0.693} \times 1.2 \times 10^9 \text{ years} = 3.0 \times 10^9 \text{ years}$$

Practice Exercise 16.10

What is the age of a piece of cloth that contains 5.2% of the amount of $^{14}_{6}C$ present in living organisms?

16.5   Nuclear Energy—Fission and Fusion

We saw earlier that the $^{56}_{26}Fe$ isotope is the most stable isotope that exists. It is at the top of the plot of binding energy per nucleon. (See the figure on page 327) It is clear from the diagram that isotopes smaller or larger than the $^{56}_{26}Fe$ isotope are less stable. Very light isotopes, such as $^{2}_{1}H$, and very heavy isotopes, such as $^{235}_{92}U$, can gain stability and release energy if they can somehow be converted to isotopes that are closer to $^{56}_{26}Fe$ on the curve.

The heavier isotopes need to be converted to lighter isotopes, and the lighter isotopes need to be converted to heavier isotopes. There are mechanisms for performing both of these conversions—fission and fusion.

- Nuclear fission is the breaking up of a large nucleus into smaller ones.
- Nuclear fusion is the combining of small nuclei into larger ones.

Both processes release energy because the combined mass defects of the products is greater than the combined mass defects of the reactants. In other words, the binding energy per nucleon has increased, so mass has been converted to energy.

A small amount of mass converts to a large amount of energy, because of the $E = mc^2$ relationship. Both nuclear fission and nuclear fusion are capable of producing large amounts of energy. To date, producing large amounts of energy for mass consumption has been achieved only through nuclear fission. This is the process that is occurring in the nuclear reactors that provide some of the world's energy.

In one fission reaction commonly used, $^{235}_{92}U$ is the fuel. The reaction that occurs is shown in the figure below.

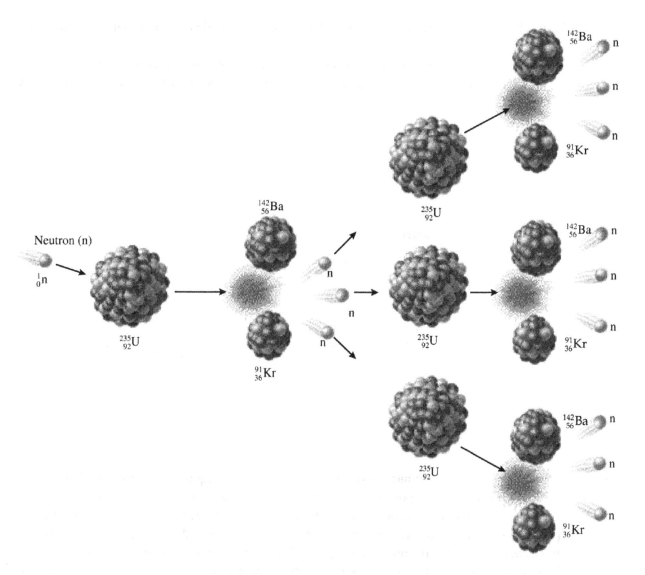

It is important to notice that three neutrons are produced each time the reaction occurs. These three neutrons cause fission of three more atoms, which in turn produce nine more neutrons. The process continues, producing more and more neutrons to cause the fission of more and more atoms. The reaction itself produces the neutrons it needs to continue the reaction. If the mass of uranium is small, many of the neutrons will escape before encountering additional uranium atoms. However, if the mass of uranium is large enough, the neutrons produced will encounter other uranium atoms and keep the reaction going. This type of reaction (which keeps going even if no additional neutrons are supplied) is called a *chain reaction* because it is essentially self-sustaining. The amount of uranium needed to produce a self-sustaining reaction is called the *critical mass*. In a nuclear reactor, the chain reaction is moderated with control rods containing material that absorbs some of the neutrons. When the fission reaction is not moderated, a nuclear explosion could result.

Although nuclear fission reactions can produce enormous amounts of energy, nuclear reactors currently supply less than 20% of the world's energy needs. There is a real concern about contaminating the environment with radioactive material in the case of an accident. In addition, the problem of how to store the nuclear wastes generated by nuclear reactors has not been solved. Until these issues are resolved, nuclear fission will not become a major source of the world's energy.

Unlike nuclear fission, which uses highly radioactive material and generates radioactive waste, nuclear fusion is a very safe process. It converts hydrogen to helium, by the process shown in the figure below.

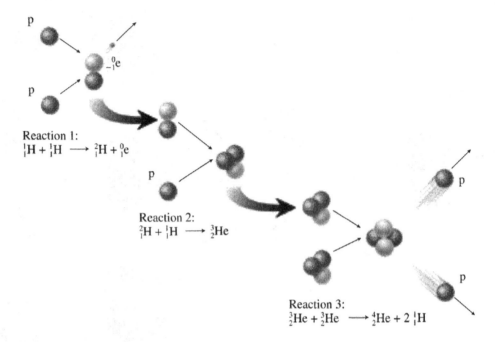

Reaction 1:
$${}_{1}^{1}H + {}_{1}^{1}H \longrightarrow {}_{1}^{2}H + {}_{1}^{0}e$$

Reaction 2:
$${}_{1}^{2}H + {}_{1}^{1}H \longrightarrow {}_{2}^{3}He$$

Reaction 3:
$${}_{2}^{3}He + {}_{2}^{3}He \longrightarrow {}_{2}^{4}He + 2\,{}_{1}^{1}H$$

The problem with nuclear fusion, however, is that it requires temperatures of up to 15,000,000 K to force the positively charged hydrogen nuclei to overcome their repulsion and combine. Scientists are working on new ways to heat the hydrogen and have made considerable progress over the past decade. However, we have not yet reached the point where we can get more energy out of the fusion reaction than we have to put into it to make it occur. In fact, we have not even reached the break-even point—the point at which we get as much energy out as we put in. The research will continue, however, and there is optimism that nuclear fusion will be a reliable source of clean energy in the not-too-distant future.

16.6 Biological Effects and Medical Applications of Radioactivity

The radiation emitted by radioactive isotopes is used medically primarily in two ways: as a diagnostic tool for identifying disorders, and as a therapeutic treatment for cancer and other disorders.

Magnetic resonance imaging (MRI) is used routinely to produce three-dimensional images of the internal organs of the body, and allows physicians to locate abnormalities due to tumors or other causes. Originally, the process of MRI, which uses radio waves to stimulate certain nuclei in the presence of a very strong magnetic field, was called nuclear magnetic resonance. Although the process does not actually involve radioactivity (the nuclei stimulated by the radio waves are not radioactive), the name nuclear magnetic resonance (NMR) had to be changed to magnetic resonance imaging because of the public fear of anything with the word *nuclear* in it.

The public fear of anything related to nuclear processes arose because of the damage that nuclear radiation exposure causes to living tissue. Nuclear emissions (α particles, β particles, and γ rays) are very high energy and can cause harm ranging from simple burns to death. In addition to the burns caused by the high energy of nuclear radiation, the radiation can cause biological molecules to ionize. When biological molecules ionize, an electron is knocked out of a bond, causing the bond to weaken or break.

Not all radiation has the same effect, however. In general, α particles are less dangerous than β particles when the source of radiation is outside of the body. Alpha particles are more massive than β particles and therefore move more slowly and have less ability to penetrate. In fact, α particles are stopped by a few sheets of paper, by clothing, or even by the top layer of skin. Beta particles are smaller and move more rapidly than α particles, but they can still be stopped by a block of wood or heavy clothing. When the source of radiation is ingested, both α and β particles are extremely damaging. The most dangerous form of external radiation is γ radiation, which has great penetrating power. A lead block several inches thick is required to stop it.

There are several therapeutic uses of radiation. These include external radiation therapy (in which radiation is used to destroy cancerous tumors), radiation imaging (in which a radiation emitter is injected or fed to a patient to analyze organ function), and radioimmunoassays (in which radioactive substances are bound to substances in the blood to determine the concentration of the substance).

Radiation is also used advantageously in the preservation of fruits, vegetables, meats, and poultry. Although there is still widespread debate about the use of radiation in this manner, many persons have been saved from fatal food poisoning through this use of nuclear radiation.

Strategies for Working Problems in Chapter 16

Overview of Problem Types: What You Should Be Able to Do

Chapter 16 provides a discussion of the reactions that involve the nuclei of atoms and the energy that these reactions produce. The types of problems that you should be able to solve after mastering Chapter 16 are:

1. Determine the mass defect associated with an atom.

2. Determine the binding energy per mole of nucleons and the binding energy per nucleon for an atom.

3. Use the half-life of an isotope to determine how much of the isotope will be left after a given amount of time.

4. Determine the age of objects (date objects) using radioactive objects.

Flowcharts

Flowchart 16.1 Determining the mass defect associated with an atom

Flowchart 16.1

Summary

Method for determining the mass defect associated with an atom

Step 1
- **Determine the sum of the masses of the moles of protons, neutrons, and electrons in the atom.**

$\downarrow$

Step 2
- **Identify the mass of one mole of atoms of the element (given in the problem).**

$\downarrow$

Step 3
- **Subtract the mass of one mole of atoms of the element (step 2) from the sum of the masses of the moles of protons, neutrons, and electrons (step 1) in the atom.**

Example 1

Calculate the mass defect of $^{14}_{6}C$. The mass of one mole is 14.003 24 g.

Solution

Perform Step 1

$$6 \text{ mol protons} \times \frac{1.0073 \text{ g}}{1 \text{ mol protons}} = 6.0438 \text{ g}$$

$$8 \text{ mol neutrons} \times \frac{1.0087 \text{ g}}{1 \text{ mol neutrons}} = 8.0696 \text{ g}$$

$$6 \text{ mol electrons} \times \frac{0.000\,55\text{ g}}{1\text{ mol electrons}} = 0.0033\text{ g}$$
$$\text{Total Mass} = 14.1167\text{ g}$$

Perform Step 2

The mass of $^{14}_{6}\text{C}$ is 14.003 24 g/mol.

Perform Step 3

The mass defect for $^{14}_{6}\text{C}$ is

14.1167 g $-$ 14.003 24 g $= 0.1135$ g/mol

Example 2

Calculate the mass defect of one mole of $^{60}_{27}\text{Co}$ atoms.
The mass of $^{60}_{27}\text{Co}$ is 59.933 82 g/mol.

Solution

Perform Step 1

$$27 \text{ mol protons} \times \frac{1.0073\text{ g}}{1\text{ mol protons}} = 27.1971\text{ g}$$

$$33 \text{ mol neutrons} \times \frac{1.0087\text{ g}}{1\text{ mol neutrons}} = 33.2871\text{ g}$$

$$27 \text{ mol electrons} \times \frac{0.000\,55\text{ g}}{1\text{ mol protons}} = 0.0148\text{ g}$$
$$\text{Total Mass} = 60.4990\text{ g}$$

Perform Step 2

The mass of $^{60}_{27}\text{Co}$ is 59.933 82 g/mol.

Perform Step 3

The mass defect for $^{60}_{27}\text{Co}$ is 60.4990 g $-$ 59.933 82 g $= 0.5652$ g/mol

Practice Problems

16.1 Calculate the mass defect of a mole of $^{56}_{26}\text{Fe}$ if the mass of one mole of $^{56}_{26}\text{Fe}$ is 55.920 68 g.

16.2 Which has a greater mass defect, $^{12}_{6}\text{C}$ with a mass of 12.000 g/mol, or $^{10}_{5}\text{B}$, with a mass of 10.0129 g/mol?

Flowchart 16.2 Determining the binding energy per mole of nucleons and the binding energy per mole per nucleon for an atom

Flowchart 16.2

Summary

Method for determining the binding energy per mole of nucleons for an atom

Step 1

• **Determine the mass defect using the procedure in Chart 16.1.**

Step 2
- Use $E = mc^2$ to calculate the binding energy per mole of nucleons.
- The mass must be in units of kg and c must be in units of m/s
- The units on the answer will be joules, J, because
  $1 \text{ kg m}^2/\text{s}^2 \text{ mol} = 1 \text{ joule/mol}$.

Step 3
- Calculate the binding energy per mole of nucleons per nucleon by dividing the binding energy per mole of nucleons by the total number of nucleons in the atom.

Example 1

Calculate the binding energy of $^{14}_{6}\text{C}$. The mass of one mole is 14.003 24 g. What is the binding energy per nucleon for this atom?

Solution

Perform Step 1

The mass defect of $^{14}_{6}\text{C}$ is 0.1135 g/mol. (See Flowchart 16.1.)

Perform Step 2

Binding Energy: $E = mc^2$

$\qquad = (0.000\ 1135 \text{ kg/mol}) \times (3.00 \times 10^8 \text{ m/s})^2$

$\qquad = 1.02 \times 10^{13} \text{ kg m}^2/\text{s}^2 \text{ mol}$

$\qquad = 1.02 \times 10^{13} \text{ J/mol}$

$\qquad = 1.02 \times 10^{10} \text{ kJ/mol}$

Perform Step 3

$$\text{Binding Energy per nucleon} = \frac{1.02 \times 10^{10} \text{ kJ/mol}}{14 \text{ nucleons}} = 7.29 \times 10^8 \text{ kJ/mol/nucleon}$$

Example 2

Calculate the binding energy per nucleon of $^{63}_{29}\text{Cu}$. The mass of one mole is 62.939 60 g.

Solution

Perform Step 1

The mass defect of $^{63}_{29}\text{Cu}$ is calculated to be 0.5838 g/mol.

(The procedures outlined in Flowchart 16.1 were used to get this number. You should go through the steps yourself to calculate this result.)

Perform Step 2

Binding Energy: $E = mc^2$

$\qquad = (0.000\ 5838 \text{ kg/mol}) \times (3.00 \times 10^8 \text{ m/s})^2$

$\qquad = 5.25 \times 10^{13} \text{ kg m}^2/\text{s}^2 \text{ mol}$

$\qquad = 5.25 \times 10^{13} \text{ J/mol}$

$\qquad = 5.25 \times 10^{10} \text{ kJ/mol}$

Perform Step 3

$$\text{Binding Energy per mole of nucleons} = \frac{5.25 \times 10^{10}\,\text{kJ}}{1\,\text{mol}} \times \frac{1\,\text{mol}}{63\,\text{mol nucleons}}$$
$$= 8.33 \times 10^8\,\text{kJ/mol of nucleons}$$

Practice Problems

16.3 Calculate the binding energy per nucleon of $^{19}_{9}\text{F}$. The mass of one mole is 18.9984 g.

16.4 Calculate the binding energy per nucleon of $^{16}_{8}\text{O}$. The mass of one mole is 15.9949 g.

Flowchart 16.3 Using the half-life of an isotope to determine how much of the isotope will be left after a given amount of time

Flowchart 16.3

Summary

Method for using the half-life of an isotope to determine how much of the isotope will be left after a given amount of time, or how long it will take for a given amount of the isotope to decay

Step 1

• Use the half-life to construct a table of the amount remaining after successive half-lives.

↓

Step 2

• Use the table constructed in step 1 to determine how much of the substance will be left after a given amount of time.

• Values in the table will also allow you to determine how long it will take for a given amount of the isotope to decay to a specified amount.

Example 1

The half-life of iodine-123 is 13.1 hours. How much of a 25.00 g sample will remain after 65.5 hours?

Solution

Perform Step 1

Construct the table of amount remaining after successive half-lives.

Half-Lives Elapsed	Time Elapsed	Amount Remaining
0	0.0 hours	25.00 g
1	13.1 hours	12.50 g
2	26.2 hours	6.25 g
3	39.3 hours	3.13 g
4	52.4 hours	1.57 g
5	65.5 hours	0.781 g

Perform Step 2

Identify the amount remaining after the specified time has elapsed. After 65.5 hours, 0.781 g will remain.

Example 2

The half-life of phosphorus-32 is approximately 14 days. How much of a 100.00 g sample will remain after 42 days?

Solution

Perform Step 1

Construct the table of amount remaining after successive half-lives.

Half-Lives Elapsed	Time Elapsed	Amount Remaining
0	0 days	100.00 g
1	14 days	50.00 g
2	28 days	25.00 g
3	42 days	12.5 g

Perform Step 2

Identify the amount remaining after the specified time has elapsed. After 42 days, 12.5 g will remain.

Example 3

The half-life of $^{55}_{27}Co$ is approximately 17.5 hours. How long will it take for a 1500.0 g sample of the isotope to decay to approximately 50 g?

Solution

Perform Step 1

Construct the table to indicate the amount remaining after successive half-lives.

Half-Lives Elapsed	Time Elapsed	Amount Remaining
0	0 hours	1500.0 g
1	17.5 hours	750.0 g
2	35 hours	375.0 g
3	52.5 hours	187.5 g
4	70 hours	93.8 g
5	87.5 hours	46.9 g

Perform Step 2

It will take about 87.5 hours (five half-lives) for the isotope to decay to approximately 50 g.

Practice Problems

16.5 The half-life of $^{60}_{27}\text{Co}$ is approximately 15.27 years. How much of a 100.00 g sample will remain after 45.8 years?

16.6 $^{19}_{8}\text{O}$ has a half-life of approximately 27 seconds. Approximately how long will it take for 250 g of the isotope to decay to 4 g?

Flowchart 16.4 Determining the age of objects (date objects) using radioactive objects

Flowchart 16.4

Summary

Method for determining the age of objects (date objects) using radioactive objects

Step 1
- Use the formula shown below to calculate the age of the object.

$$\text{Age} = -\frac{-2.303 \times \log\left(\%\text{ radioactive isotope remaining}/100\%\right)}{0.693} \times \text{half-life}$$

$\downarrow$

Step 2
- Substitute the given values into the equation and solve for the unknown variable.

Example 1

$^{238}_{92}\text{U}$ has a half-life of 4.46×10^9 years as it decays to $^{206}_{82}\text{Pb}$. If a rock contains 45 atoms of $^{238}_{92}\text{U}$ for every 55 atoms of $^{206}_{82}\text{Pb}$, how old is the rock? (Note that the total number of atoms is 100. This means that the % radioactive isotope remaining is 45%.)

Solution

Perform Step 1

$$\text{Age} = \frac{-2.303 \times \log\left[\%\text{ radioactive isotope remaining}/100\%\right]}{0.693} \times \text{half-life}$$

Perform Step 2

The equation becomes

$$\text{Age} = \frac{-2.303 \times \log\left(45/100\right)}{0.693} \times 4.46 \times 10^9$$
$$= 5.14 \times 10^9 \text{ years}$$

Example 2

$^{14}_{6}\text{C}$ has a half-life of 5.715×10^3 years as it decays to $^{14}_{7}\text{N}$. If a cloth contains 20 atoms of $^{14}_{6}\text{C}$ for every 80 atoms of $^{14}_{7}\text{N}$, how old is the cloth?

Solution

Perform Step 1

$$\text{Age} = \frac{-2.303 \times \log\left[\,\%\text{ radioactive isotope remaining}/100\%\,\right]}{0.693} \times \text{half-life}$$

Perform Step 2

The equation becomes

$$\text{Age} = \frac{-2.303 \times \log\left[\,20/100\,\right]}{0.693} \times 5.715 \times 10^3$$

$$= 1.33 \times 10^4 \text{ years}$$

Practice Problems

16.7 $^{238}_{92}$U has a half-life of 4.46×10^9 years as it decays to $^{206}_{82}$Pb. If a rock contains 85 atoms of $^{238}_{92}$U for every 15 atoms of $^{206}_{82}$Pb, how old is the rock?

16.8 $^{14}_6$C has a half-life of 5.715×10^3 years as it decays to $^{14}_7$N. If a fossil contains 5 atoms of $^{14}_6$C for every 95 atoms of $^{14}_7$N, how old is the fossil?

Quiz for Chapter 16 Problems

1. $^{14}_6$C has a half-life of 5.715×10^3 years as it decays to $^{14}_7$N. If a scroll contains 10 atoms of $^{14}_6$C for every 90 atoms of $^{14}_7$N, how old is the scroll?

2. $^{238}_{92}$U has a half-life of 4.46×10^9 years as it decays to $^{206}_{82}$Pb. If a rock contains 35 atoms of $^{238}_{92}$U for every 65 atoms of $^{206}_{82}$Pb, how old is the rock?

3. Calculate the mass defect of $^9_4$Be. The mass of one mole is 9.0122 g.

4. Calculate the binding energy per nucleon for $^9_4$Be. Use the result from Problem 3 to perform the calculation.)

5. Calculate the binding energy per nucleon of $^{40}_{20}$Ca. The mass of one mole is 39.9626 g.

6. The half-life of $^{207}_{83}$Bi is 35 years. How much of a 500.0-g sample of this isotope of bismuth will remain after 175 years?

7. The half-life of $^{93}_{39}$Y is approximately 10 hours. How long will it take for a 25 g sample of the isotope to decay to approximately 3 g?

8. Determine which has a greater mass defect per nucleon: $^{40}_{18}$Ar, with a mass of 39.9624 g/mol, or $^{27}_{13}$Al, with a mass of 26.9815 g/mol.

9. Complete the following sentence. The difference between the sum of the masses of the atomic particles in an atom and the mass of the atom is called the _____ _____.

10. What fraction of a radioactive isotope will remain after 200 years if the isotope has a half-life of 25 years?

Exercises for Self-Testing

A. Completion. *Write the correct word(s) to complete each statement below.*

1. The difference between the sum of the masses of the atomic particles and the mass of the atom is called the _____ _____.

2. The most stable of all isotopes is _____ because it has the highest binding energy per nucleon.
3. The nuclear particle that acts as the "glue" to keep the nucleus from flying apart is the _____.
4. A particle that is considered to be antimatter is the _____.
5. An isotope whose nuclei contain too many neutrons lies _____ the band of stability.
6. When an alpha particle is emitted from a nucleus, the atomic number decreases by _____, and the mass number decreases by _____.
7. An isotope that is _____ the band of stability will emit an electron to become stable.
8. The type of nuclear radiation that is a high-energy form of electromagnetic radiation is _____ _____.
9. In nuclear _____ a large nucleus is split into smaller ones, and in nuclear _____ two small nuclei are combined.
10. The minimum amount of fuel necessary for a nuclear reaction to become self-sustaining is called the _____ _____.

B. True/False. *Indicate whether each statement below is true or false, and explain why the false statements are false.*

_____ 1. The protons and electrons in an isotope are the nucleons.
_____ 2. Different isotopes of the same element must have different mass numbers.
_____ 3. Much more energy is involved in a nuclear reaction than in a chemical reaction.
_____ 4. A comparison of the binding energies of two isotopes will allow you to determine which one is more stable.
_____ 5. An isotope that lies below the band of stability can capture a positron to become stable.
_____ 6. $^{28}_{8}O$ lies above the band of stability.
_____ 7. Alpha particles move faster than beta particles and require more shielding to stop them.
_____ 8. Magnetic resonance imaging provides a physician with more information than nuclear magnetic resonance.
_____ 9. All isotopes of elements with atomic numbers greater than 84 are radioactive.
_____ 10. The cleaner source of nuclear energy is nuclear fusion.

C. Questions and Problems

1. Why is the mass of an atom less than the sum of the masses of the particles that make up the atom?
2. One isotope of nickel has a mass number of 57; another isotope has a mass number of 66. Both of the isotopes are unstable. One isotope decays by beta emission and one decays by positron emission. Which is which? Explain your answer.
3. A 10.36-kg sample of a radioactive substance was found to have decayed to 1.295 kg in a period of 18 years. Calculate the half-life of the substance.
4. In the production of element 118, a nuclear fusion reaction involving an isotope of krypton and one other element might be carried out. The two nuclei would fuse to produce element 118. What is the other element that would be used in the reaction?
5. Describe two medical uses of radioactivity.

The Chemistry of Carbon

Learning Outcomes

1. Understand the allotropes of carbon and the ability of the carbon atom to catenate to form hydrocarbons.

2. Draw structural formulas of hydrocarbon molecules.

3. Draw and interpret line drawings of hydrocarbon molecules.

4. Understand and recognize isomers.

5. Assign hydrocarbon molecules an IUPAC name.

6. Understand the basic properties of hydrocarbon molecules.

7. Recognize the various functional groups often present in functionalized hydrocarbons and how they are named.

8. Assign functionalized hydrocarbons a name.

9. Understand some of the properties bestowed on functionalized hydrocarbons by the functional group.

10. Understand why one functional group might react with another.

Chapter Outline

17.1 Carbon—A Unique Element

    A. Long-chain and ring formation by carbon atoms

    B. Three allotropes of carbon

    C. Different ways in which carbon forms four covalent bonds

17.2 Naturally Occurring Compounds of Carbon with Hydrogen—Hydrocarbons

    A. Petroleum—The source of hydrocarbons

    B. Uses of petroleum

    C. Chains of carbon

1. Chain length
    a. Linear hydrocarbons
    b. Branched hydrocarbons
   2. Drawing structures of hydrocarbons
    a. Full structures
    b. Line drawings
  D. Saturated versus unsaturated hydrocarbons
   1. Definition of saturated and unsaturated
   2. Saturation in fats and unsaturation in oils
17.3 Naming Hydrocarbons
  A. Common names
  B. IUPAC nomenclature system
   1. Suffixes *-ane, -ene*, and *-yne*
   2. Root names
  C. Homologous series
  D. General formula for alkanes, alkenes, and alkynes
  E. IUPAC rules for naming alkanes, alkenes, and alkynes
   1. Linear chains
   2. Branched chains
17.4 Properties of Hydrocarbons
  A. Solubility
  B. Boiling points
17.5 Functionalized Hydrocarbons—Bring on the Heteroatoms
  A. Halogen functionalized hydrocarbons
   1. Preparation
   2. Reactions
  B. Alcohols
   1. Ethanol
   2. Methanol
   3. Solubility of alcohols
  C. Ethers
   1. General structure
   2. Naming ethers
  D. Organic acids—Carboxylic acids
   1. General formula and structure
   2. Reaction of carboxylic acids with water and K_a values
   3. Acetic acid and formic acid
  E. Organic bases—Amines
   1. General structure of primary, secondary, and tertiary amines
   2. Reaction of amines with water and K_b values

F. Ketones and aldehydes
  1. General structure
  2. Naming ketones and aldehydes
17.6 One More Thing: Functional Groups and Organic Synthesis
  A. Active pharmaceutical ingredients
  B. Multi-step organic synthesis

Review of Key Concepts from the Text

17.1 Carbon—A Unique Element

The chemistry of carbon compounds is the subject of the branch of chemistry known as *organic chemistry*. Carbon is the only element that is the subject of an entire branch of chemistry, because of its unique characteristics. Carbon is the only atom that has the ability to form very long chains, rings, and other frameworks. The ability of carbon to connect and link together is called *catenation*. Several basic structures of the carbon frameworks, such as various chains and rings, are shown in Section 17.1 of your textbook.

Elemental carbon comes in three different structural forms, called *allotropes*. The three carbon allotropes are diamond, graphite, and fullerene, shown in the figure below.

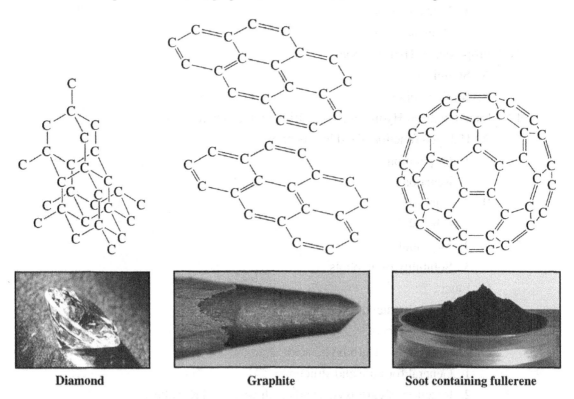

Diamond   **Graphite**   **Soot containing fullerene**

The allotropes have very different physical properties due to different bonding in the three forms. Diamond contains carbons that are all bonded to four other carbons by single bonds in a tetrahedral arrangement. It has a very rigid structure and is the hardest natural substance known. It is used to cut through other hard materials such as steel.

Graphite, on the other hand, consists of sheets of carbon atoms. The atoms in each sheet are bonded to three other carbons in a trigonal planar arrangement. The sheets are held together by weak London forces, allowing the layers to slide easily over one another. This characteristic makes graphite particularly well suited for use in pencils and as a lubricant.

Fullerene, the third allotrope of carbon, was discovered in 1990. Like graphite, fullerene contains carbons that are trigonal and bonded to three other carbon atoms. The difference between graphite and fullerene, however, is that fullerene exists as spheres (as opposed to the sheets in graphite) that contain five- and six-membered rings joined together. Fullerene molecules have the shape of a soccer ball, with each sphere containing 60 carbon atoms.

Diamond and graphite are both network covalent solids, whereas fullerene is a molecular solid. The common feature of all three allotropes is that every carbon atom forms four covalent bonds to other carbons. The carbon atoms in diamond form four single bonds to other carbons, whereas the carbon atoms in graphite and fullerene form two single bonds and one double bond to other carbon atoms.

17.2   Naturally Occurring Compounds of Carbon with Hydrogen—Hydrocarbons

Hydrocarbons, found naturally in petroleum (crude oil), are compounds that contain only carbon and hydrogen. Because carbon atoms have the special ability to link together in long chains (catenation), carbon and hydrogen form countless numbers of different compounds. The carbon atoms are bonded together by single, double, or triple bonds, and the hydrogen atoms are always bonded to the carbon atoms by single bonds (because hydrogen can only form one bond).

Hydrocarbons are nonpolar molecules that burn well and are used as a source of fuel. Petroleum is also used as the source from which fertilizers, detergents, medicines, and numerous other chemicals are produced.

Chains of Carbon

Hydrocarbon molecules consist of chains of carbon atoms and are often described in terms of their chain length. The chain length of a hydrocarbon is the number of carbon atoms that are bound together to form the longest continuous chain in the molecule. Hydrocarbons can be *linear* or *branched*. In linear hydrocarbons, all of the carbons are attached in a straight line with no branches. All carbons are attached to two other carbons (except the two on the end, which are only attached to one other carbon). Branched hydrocarbons, on the other hand, contain some carbons in a linear chain and some that are branched off of the linear chain.

For example, a molecule with six carbons and 14 hydrogens, C_6H_{14}, can be arranged in a number of different ways. The molecule $CH_3CH_2CH_2CH_2CH_2CH_3$ has a linear chain of six carbons. We can, however, arrange six carbons in other ways. One way involves having a linear chain of five carbons, with a one-carbon branch off of the straight chain, as shown below.

$$CH_3 - CH - CH_2 - CH_2 - CH_3$$
$$| $$
$$CH_3$$

Yet another way involves having a linear chain of only four carbons, with a one-carbon branch off of two of the carbons on the straight chain, as shown below.

$$CH_3 - CH - CH - CH_3$$
$$|\qquad|$$
$$CH_3\quad CH_3$$

The linear and branched hydrocarbons discussed above are *isomers* of each other. Isomers are compounds that have the same molecular formula (same number and kind of atoms) but whose molecules have a different arrangement of atoms. To determine whether two compounds are isomers of each other, we have to determine whether the molecular formulas are the same and whether the arrangement of atoms is indeed different.

Example 17.1

Identify the following pairs of molecules as the same, isomers, or coming from different compounds:

(a) $CH_3-CH-CH_2-CH_3$ and $CH_3-CH_2-CH_2-CH_3$
              |
            CH_3

(b) $CH_3-CH_2-CH_2-CH_2$ and $CH_3-CH_2-CH_2-CH_2-CH_3$
                     |
                  CH_3

(c) $CH_3-CH_2-CH-CH_3$ and $CH_3-CH_2-CH_2-CH_2-CH_3$
                 |
              CH_3

Solution

To determine whether the compounds are different, we must determine the molecular formula for each compound. If the molecular formulas are different, the molecules come from different compounds. If the molecular formulas are the same, the molecules are either identical or isomers of each other. To determine whether the molecules are identical, it is necessary to determine if the arrangement of atoms is really different or only appears to be different based on the way it is drawn. Using these general principles, we arrive at the following answers for Example 17.1:

(a) The two molecules are from different compounds. The first compound has the molecular formula C_5H_{12}, and the second compound has the molecular formula C_4H_{10}.

(b) The two molecules have the same molecular formula, C_5H_{12}, so they are either identical or isomers. It may appear that the first compound is a branched hydrocarbon and the second one a linear compound. But don't be fooled. An end carbon drawn up or down does not constitute a branch, because all four positions on a tetrahedral carbon atom are equivalent. Therefore, both of the compounds are really linear hydrocarbons containing five carbons. The two molecules are identical.

(c) These two compounds are isomers. The molecular formulas are the same, C_5H_{12}, and the arrangement of atoms is really different. The first compound is a branched hydrocarbon, and the second one is linear.

Practice Exercise 17.1

Identify the following pairs of molecules as the same, isomers, or coming from different compounds.

(a) 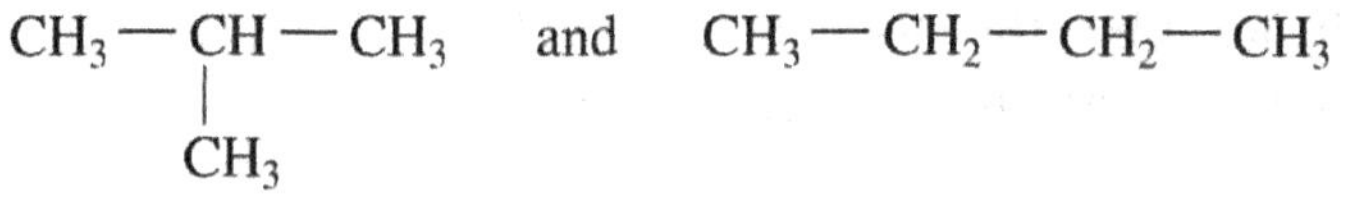 CH_3—$\underset{\underset{\displaystyle CH_3}{|}}{CH}$—$CH_3$ and CH_3—CH_2—CH_2—CH_3

(b) 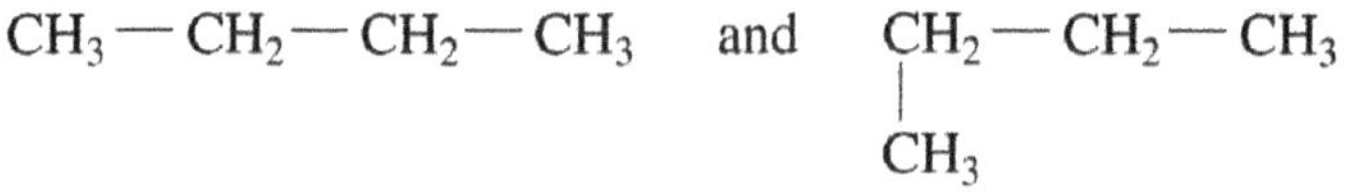 CH_3—CH_2—CH_2—CH_3 and $\underset{\underset{\displaystyle CH_3}{|}}{CH_2}$—$CH_2$—$CH_3$

(c) 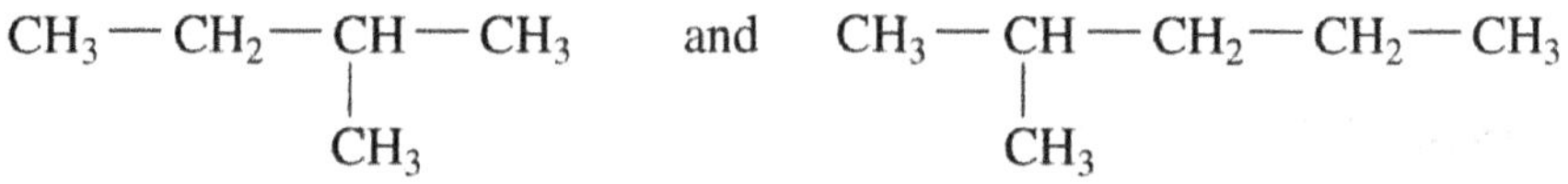 CH_3—CH_2—$\underset{\underset{\displaystyle CH_3}{|}}{CH}$—$CH_3$ and CH_3—$\underset{\underset{\displaystyle CH_3}{|}}{CH}$—$CH_2$—$CH_2$—$CH_3$

Although it is easy to determine the chain length of a linear hydrocarbon, the chain length of a branched hydrocarbon is not always obvious because the longest continuous chain of carbons may not be in a straight line, as shown in the example below.

Example 17.2

Indicate the chain length (number of carbon atoms that are bound together to form the longest continuous chain) in each of the following molecules:

(a) CH_3—CH_2—CH_2—CH_2—CH_2—CH_3

(b) 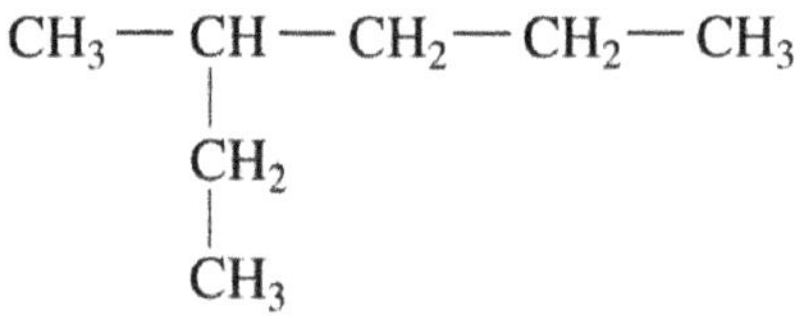 CH_3—$\underset{\underset{\displaystyle CH_2}{|}}{CH}$—$CH_2$—$CH_2$—$CH_3$
$\underset{\underset{\displaystyle CH_3}{|}}{}$

(c) 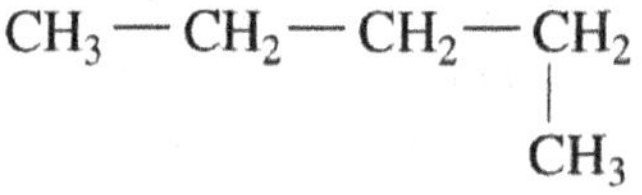 CH_3—CH_2—CH_2—$\underset{\underset{\displaystyle CH_3}{|}}{CH_2}$

Solution

To find the longest chain, we will start at an end carbon and count all of the carbons that are connected in the chain. You may find it useful to circle the carbons in the longest chain, because some groups may appear to be branches but are actually members of the longest chain. We'll use this method for the compounds above.

(a) This is a linear hydrocarbon, so the longest chain is obvious. It contains six carbons.

(b) This compound is a little trickier. If we count the atoms in the chain going left to right, there are five carbons. However, when we look more carefully, we see that the longest continuous chain contains the two carbons attached to the second carbon from the left. This chain contains six carbons and is the longest continuous chain in the molecule. Therefore, the molecule has a chain length of six, which we have circled in the figure below.

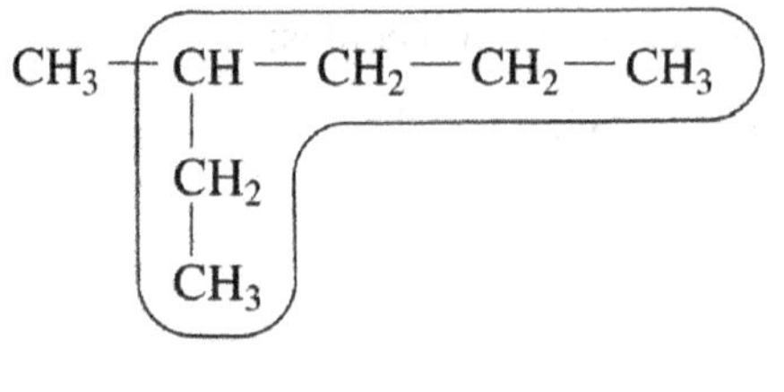

CH_3—$\begin{array}{l} CH—CH_2—CH_2—CH_3 \\ | \\ CH_2 \\ | \\ CH_3 \end{array}$

(c) This is really a linear hydrocarbon (like the compound in Example 17.1b). The longest continuous chain contains five carbons has been circled:

$$CH_3 — CH_2 — CH_2 — CH_2$$
$$| $$
$$CH_3$$

Practice Exercise 17.2

Indicate the chain length (number of carbon atoms that are bound together to form the longest continuous chain) in each of the following molecules:

(a) $CH_3 — CH_2 — CH_2 — CH — CH_2 — CH_3$
$\quad\quad\quad\quad\quad\quad\quad\quad | $
$\quad\quad\quad\quad\quad\quad CH_2 — CH_2 — CH_3$

(b) $CH_3 — CH — CH_2 — CH_3$
$\quad\quad\quad\quad | $
$\quad\quad\quad\quad CH_2$
$\quad\quad\quad\quad | $
$\quad\quad\quad\quad CH_3$

(c) $CH_3 — CH_2 — CH_2 — CH — CH_2 — CH_3$
$\quad\quad\quad\quad\quad\quad\quad\quad\quad | $
$\quad\quad\quad\quad\quad\quad\quad\quad\quad CH_3$

Up to this point, we have written the structure of hydrocarbon molecules by writing in each carbon and each hydrogen in the molecule. However, there is a shorthand method for writing the structure of hydrocarbons. This shorthand notation, called a *line drawing*, involves showing only the bonds between carbon atoms. The symbols C and H and all of the bonds between the C and H atoms are excluded. The lines are drawn in a zigzag pattern to represent the tetrahedral shape of the carbon atoms in the molecule. For example, the six-carbon linear hydrocarbon discussed earlier, $CH_3CH_2CH_2CH_2CH_2CH_3$, would be shown in its line drawing as

The isomer that has four carbons in a straight chain and a two-carbon branch would be drawn as

Section 17.2 of the textbook provides additional examples.

Example 17.3

Draw the line drawing notation for each of the following molecules:

(a) $CH_3 — CH_2 — CH_2 — CH_2 — CH_2 — CH_2 — CH_2 — CH_3$

(b) $CH_3 — CH — CH_2 — CH_2 — CH_3$
$\quad\quad\quad\quad | $
$\quad\quad\quad\quad CH_3$

(c) $CH_3 — CH — CH — CH_3$
$\quad\quad\quad\quad | \quad\quad | $
$\quad\quad\quad\quad CH_3 \quad CH_3$

Solutions

(a)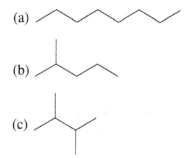

(b)

(c)

Practice Exercise 17.3

Draw the line drawing notation for each of the following molecules:

(a) $CH_3 - CH_2 - CH_2 - CH_2 - CH_2 - CH_2 - \overset{\overset{\displaystyle CH_3}{|}}{\underset{\underset{\displaystyle CH_3}{|}}{C}} - CH_3$

(b) $CH_3 - \overset{|}{\underset{\underset{\displaystyle CH_3}{|}}{CH}} - \overset{|}{\underset{\underset{\displaystyle CH_3}{|}}{CH}} - \overset{|}{\underset{\underset{\displaystyle CH_3}{|}}{CH}} - CH_3$

(c) $CH_3 - \overset{|}{\underset{\underset{\displaystyle CH_3}{|}}{CH}} - \overset{|}{\underset{\underset{\displaystyle CH_3}{|}}{CH}} - CH_2 - CH_3$

Saturated Versus Unsaturated Hydrocarbons

All of the hydrocarbons we have considered so far have had only single bonds. However, hydrocarbons can have double or triple bonds. An example of a hydrocarbon containing a double bond is propene, $H_2C=CH-CH_3$. Notice that propene has two fewer hydrogen atoms than its equivalent three-carbon hydrocarbon containing only single bonds, propane, $H_3C-CH_2-CH_3$. The three-carbon hydrocarbon that contains a triple bond is propyne, $HC\equiv CCH_3$. Ethyne, C_2H_2, contains two fewer hydrogen atoms than ethene.

Hydrocarbons containing fewer hydrogen atoms than the maximum number possible (due to the presence of double or triple bonds) are called *unsaturated* hydrocarbons. Hydrocarbons that contain the maximum number of hydrogens possible for the number of carbons contain only single bonds and are said to be *saturated*. A saturated hydrocarbon will have the general formula C_nH_{2n+2}, a hydrocarbon with one double bond will have the general formula C_nH_{2n}, and a hydrocarbon with one triple bond will have the general formula C_nH_{2n-2}.

Example 17.4

Indicate which of the following compounds are saturated and which are unsaturated. If the hydrocarbon is unsaturated, indicate the formula of the saturated hydrocarbon with the same number of carbons.

(a) C_3H_6 (b) C_5H_{12} (c) $C_{10}H_{18}$ (d) $C_{40}H_{82}$

Solutions

(a) C_3H_6 is unsaturated because its formula fits the pattern C_nH_{2n}. The formula for a saturated hydrocarbon with three carbons would be C_3H_8.

(b) C_5H_{12} is saturated because its formula fits the pattern C_nH_{2n+2}.

(c) $C_{10}H_{18}$ is unsaturated because its formula fits the pattern C_nH_{2n-2}. The formula for a saturated hydrocarbon with 10 carbons would be $C_{10}H_{22}$.

(d) $C_{40}H_{82}$ is saturated because its formula fits the pattern C_nH_{2n+2}.

Practice Exercise 17.4

Indicate which of the following compounds are saturated and which are unsaturated. If the hydrocarbon is unsaturated, indicate the formula of the saturated hydrocarbon with the same number of carbons.

(a) C_8H_{18} (b) C_3H_4 (c) C_9H_{18} (d) $C_{16}H_{34}$

17.3 Naming Hydrocarbons

The International Union of Pure and Applied Chemistry (IUPAC) developed a system for naming compounds, including hydrocarbons. The hydrocarbons are classified as alkanes if they contain only single bonds, as alkenes if they contain at least one double bond, and as alkynes if they contain at least one triple bond. The name of the hydrocarbon includes a prefix to indicate the number of carbon atoms in the longest continuous chain of carbons in the molecule, and the suffix *-ane*, *-ene*, or *-yne* to indicate whether the compound is an alkane, alkene, or alkyne, respectively.

The prefixes used to indicate the chain lengths of from 1 to 10 carbon atoms are shown below.

Chain Length	Prefix	Chain Length	Prefix
1	meth-	6	hex-
2	eth-	7	hept-
3	prop-	8	oct-
4	but-	9	non-
5	pent-	10	dec-

The table on page 356 provides the names of the linear alkanes, alkenes, and alkynes containing from 1 to 10 carbons. Each column in the table represents a *homologous* series—that is, a series that consists of compounds that are similar to one another.

The names in the table are for the linear hydrocarbons. Naming branched hydro carbons is a little more challenging but can be accomplished by applying the appropriate rules. The branches are called alkyl groups and are named according to the number of carbons present. The prefixes listed above are used with the ending *-yl*. For example, a branch containing three carbon atoms would be named propyl, one containing six carbons would be named hexyl, and so on. The rules for naming hydrocarbons are listed below.

1. Identify the longest continuous chain in the molecule. The number of carbons in this chain is the basis for the name of the hydrocarbon.
2. Identify the branches according to the number of carbons in each branch and the position of the branch(es) on the chain. The carbons on the chain should be numbered in such a way as to give the lowest combination of numbers for the branches.

3. If more than one branch with the same structure is present, use the prefixes *di-*, *tri-*, *tetra-*, *penta-*, and so on, and give the number of the position of each of the branches on the chain.
4. Give the names of the branches in alphabetical order.
5. If the compound is an alkene or an alkyne, the longest continuous chain must contain the double or triple bond.
6. In an alkene or alkyne, the longest continuous chain is numbered from the end that gives the lowest number(s) for the position(s) of the multiple bond(s).

Example 17.5

Name the following compounds:

(a) $CH_3-CH_2-CH_2-CH_2-CH_2-CH_2-CH_3$

(b) 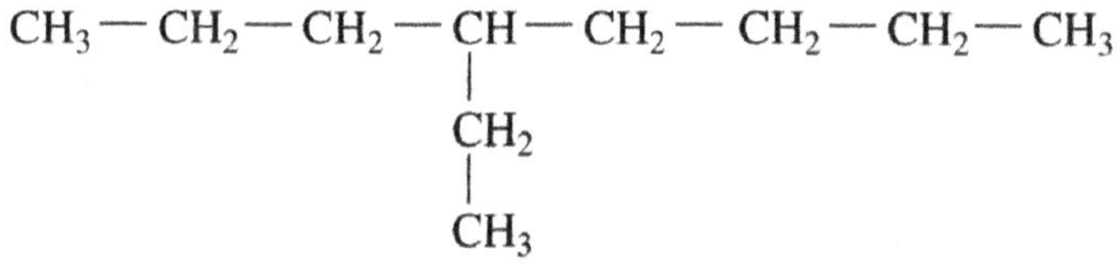 $CH_3-CH_2-CH_2-\underset{\underset{\displaystyle CH_3}{\overset{\displaystyle |}{\underset{\displaystyle |}{CH_2}}}}{CH}-CH_2-CH_2-CH_2-CH_3$

(c) 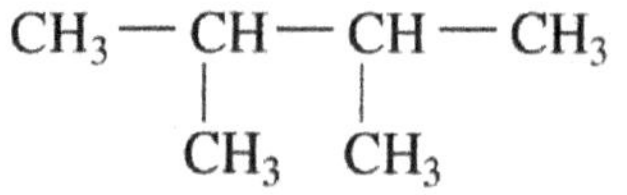 $CH_3-\underset{\underset{\displaystyle CH_3}{|}}{CH}-\underset{\underset{\displaystyle CH_3}{|}}{CH}-CH_3$

(d) $CH_2\!=\!CH-CH_2-CH_3$

(e) 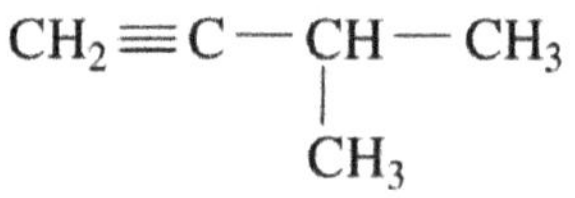 $CH_2\!\equiv\!C-\underset{\underset{\displaystyle CH_3}{|}}{CH}-CH_3$

Solution

(a) heptane   (b) 4-ethyloctane   (c) 2,3-dimethylbutane   (d) 1-butene

(e) 3-methyl-1-butyne

Practice Exercise 17.5

Name the following compounds:

(a) 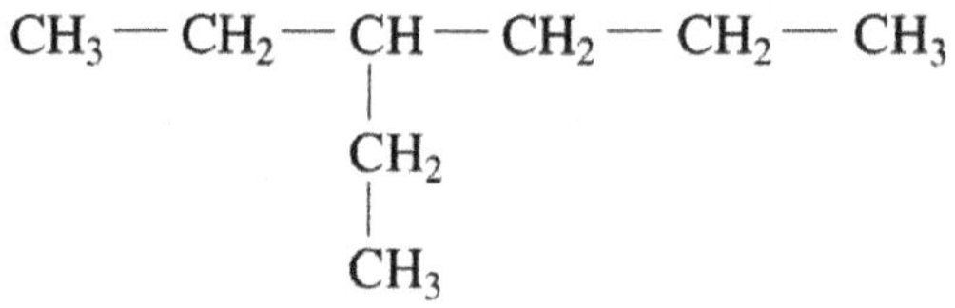 $CH_3-CH_2-\underset{\underset{\displaystyle CH_3}{\overset{\displaystyle |}{\underset{\displaystyle |}{CH_2}}}}{CH}-CH_2-CH_2-CH_3$

(b) 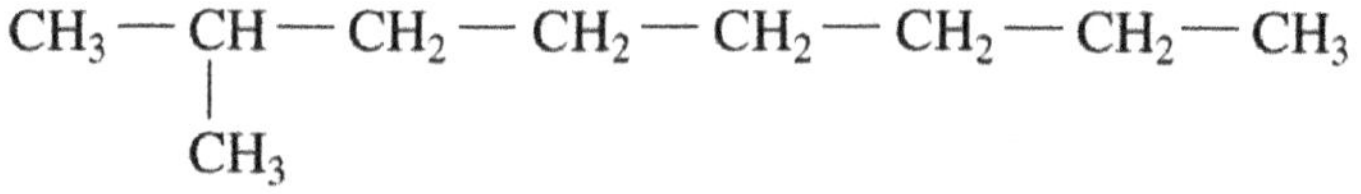 $CH_3-\underset{\underset{\displaystyle CH_3}{|}}{CH}-CH_2-CH_2-CH_2-CH_2-CH_2-CH_3$

(c) 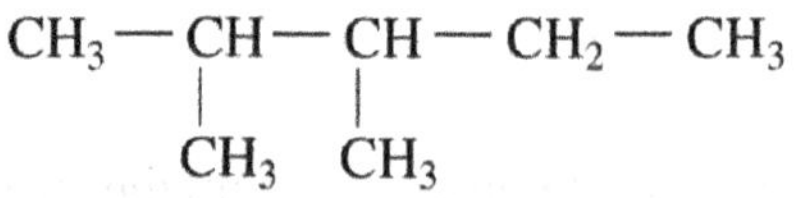 $CH_3-\underset{\underset{\displaystyle CH_3}{|}}{CH}-\underset{\underset{\displaystyle CH_3}{|}}{CH}-CH_2-CH_3$

(d) $HC\!\equiv\!C-CH_2-CH_2-CH_2-CH_2-CH_3$

(e) 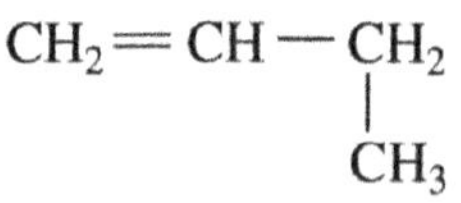 $CH_2\!=\!CH-\underset{\underset{\displaystyle CH_3}{|}}{CH_2}$

Chain length (n)	Alkanes	Alkenes	Alkynes
1	Methane	—	—
2	Ethane	Ethene	Ethyne
3	Propane	Propene	Propyne
4	Butane	Butene	Butyne
5	Pentane	Pentene	Pentyne
6	Hexane	Hexene	Hexyne
7	Heptane	Heptene	Heptyne
8	Octane	Octene	Octyne
9	Nonane	Nonene	Nonyne
10	Decane	Decene	Decyne
n	C_nH_{2n+2}	C_nH_{2n}	C_nH_{2n-2}

Just as the IUPAC system rules allow you to name a compound from the structure of its molecules, knowing the rules also allows you to draw the molecule from the name of the compound.

Example 17.6

Draw the molecular structure of the following compounds:

(a) pentane   (b) 2,2,4-trimethyldecane   (c) 2-heptyne   (d) 2-methyl-1-butene

Solutions

(a) $CH_3 - CH_2 - CH_2 - CH_2 - CH_3$

(b) 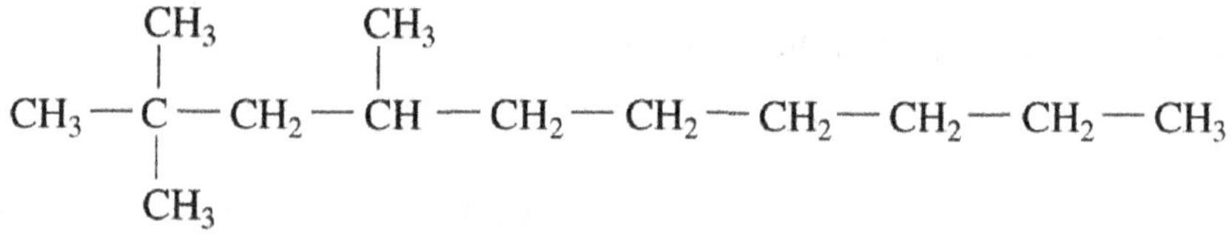
$$
\begin{array}{c}
\quad\quad CH_3 \quad\quad\quad CH_3 \\
\quad\quad | \quad\quad\quad\quad | \\
CH_3 - C - CH_2 - CH - CH_2 - CH_2 - CH_2 - CH_2 - CH_2 - CH_3 \\
\quad\quad | \\
\quad\quad CH_3
\end{array}
$$

(c) $CH_3 - C \equiv C - CH_2 - CH_2 - CH_2 - CH_3$

(d) 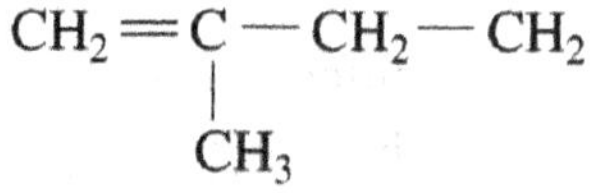
$$
\begin{array}{c}
CH_2 = C - CH_2 - CH_2 \\
\quad\quad | \\
\quad\quad CH_3
\end{array}
$$

Practice Exercise 17.6

Draw the molecular structure of the following compounds:

(a) 3-ethylnonane   (b) 2,5-dimethyloctane   (c) 2-butyne   (d) 2-methyl-2-pentene

17.4   Properties of Hydrocarbons

The hydrocarbons are nonpolar compounds and have the properties we would predict for nonpolar molecules. They are insoluble in water and have relatively low boiling points. The first four alkanes are all gases at room temperature, the next 15 (C_5 through C_{19}) are liquids, and those containing at least 20 carbon atoms are solids. As we would expect, the boiling points steadily increase as the chain length increases (due to increasing London forces).

17.5   Functionalized Hydrocarbons—Bring On the Heteroatoms

The alkanes are generally unreactive compounds because there are no regions in the molecule that contain atoms with partial charges (as would be the case if the molecule contained atoms of different electronegativities). Having an uneven distribution of charge in a molecule is generally a prerequisite for most chemical reactions to occur. Hydrocarbons can be converted to molecules with regions of partial charges by replacing a hydrogen with an atom (or group of atoms) that is more electronegative than carbon. The presence of atoms such as the halogens (F, Cl, Br, I), oxygen, and nitrogen will make the molecule more reactive. By introducing one of these atoms into the hydrocarbon, we will have transformed the hydrocarbon into what is called a *functionalized hydrocarbon*. The functional group is the section of the molecule that is responsible for its reactivity and gives the molecule unique properties.

Replacing a hydrogen with another atom or group of atoms produces functionalized hydrocarbons. The functional groups possible, along with the class of functionalized hydrocarbon produced, are shown below.

Functional Group	Functionalized Hydrocarbon
Halogen (F, Cl, Br, I)	Haloalkane (e.g., chloropentane)
	(*or* alkyl halide; e.g., pentyl chloride)
—OH	Alcohol (e.g., ethanol)
—NH$_2$, —NHR, —NR$_2$	Amine (e.g., ethylamine)
(R stands for a hydrocarbon part of the molecule)	

$$-C\overset{\displaystyle O}{\underset{\displaystyle OH}{\big\|}}$$

Carboxylic acid (e.g., butanoic acid)

The properties of the functionalized hydrocarbons are quite different from those of the hydrocarbon from which they were produced. The functionalized hydrocarbons are polar, and some (alcohols, carboxylic acids, and amines) can participate in hydrogen bonding. The boiling points of the functionalized hydrocarbons are higher, and they are typically more soluble in water than the corresponding hydrocarbon. We will consider some of the functionalized hydrocarbons in more detail in the following sections. When we discuss the functionalized hydrocarbons, the hydrocarbon portion of the molecule is abbreviated by an R. The R simply represents the C and H atoms that make up the hydrocarbon section of the molecule. Halogens are abbreviated by letter X.

Example 17.7

Indicate the type of functionalized hydrocarbon group to which each of the following compounds belongs.

(a) $CH_3-C\overset{\displaystyle O}{\underset{\displaystyle OH}{\big\|}}$ (b) CH_3CH_2Cl (c) $CH_3CH_2NH_2$ (d) $CH_3CH_2CH_2OH$

Solutions

(a) carboxylic acid   (b) haloalkane   (c) amine   (d) alcohol

Practice Exercise 17.7

Indicate the type of functionalized hydrocarbon group to which each of the following compounds belongs.

(a) CH_3NHCH_2 (b) CH_3CH_2OH (c) $CH_3CH_2CH_2Br$

(d) $CH_3-CH_2-CH_2-C\overset{\displaystyle O}{\underset{\displaystyle OH}{\big\|}}$

Halogen Functionalized Hydrocarbons (R–X)

Halogen functionalized hydrocarbons contain one or more halogen atoms in the molecule. They do not occur naturally but are prepared by reacting hydrocarbons with halogens in the presence of ultraviolet light. (See the reaction below.) They are named as haloalkanes (or haloalkenes or haloalkynes), with the name of the halogen (with an *-o* replacing the *-ine*) written before the name of the hydrocarbon. The position of the halogen is specified by a number, just as in alkyl branches.

$$H-\underset{\underset{H}{|}}{\overset{\overset{H}{|}}{C}}-H + Cl_2 \xrightarrow{UV} H-\underset{\underset{H}{|}}{\overset{\overset{H}{|}}{C}}-Cl + H-\underset{\underset{H}{|}}{\overset{\overset{Cl}{|}}{C}}-Cl + Cl-\underset{\underset{H}{|}}{\overset{\overset{Cl}{|}}{C}}-Cl + Cl-\underset{\underset{Cl}{|}}{\overset{\overset{Cl}{|}}{C}}-Cl$$

Example 17.8

Name the following haloalkanes:

(a) $CH_3CH_2CH_2Br$ (b) $CH_3CH_2CH_2CH_2CH_2CH_2I$ (c) $ClCHCH_2CH_2CH_3$
$$\qquad\qquad\qquad\qquad\qquad\qquad\qquad\qquad\qquad\qquad\qquad\qquad\qquad\quad |$$
$$\qquad\qquad\qquad\qquad\qquad\qquad\qquad\qquad\qquad\qquad\qquad\qquad\quad CH_3$$

Solutions

(a) 1-bromopropane   (b) 1-iodohexane   (c) 2-chloropentane

Practice Exercise 17.8

Name the following haloalkanes:

(a) CH_3CH_2F (b) $CH_3CH_2CHCH_2CH_2CH_3$
$$\qquad\qquad\qquad\qquad\qquad\qquad\quad |$$
$$\qquad\qquad\qquad\qquad\qquad\qquad\quad Br$$

(c) $CH_3CH_2CHCH_2CH_2$
$$\qquad\quad |$$
$$\qquad\quad I$$

The halogen functionalized hydrocarbons are relatively unreactive, and some of them (the chlorofluoro hydrocarbons) can cause many environmental problems. (See Section 17.2 of the textbook for a more complete discussion of these concepts.)

Although generally unreactive, the halogen functionalized hydrocarbons can be made to undergo a variety of reactions to produce other functionalized hydrocarbons, such as the alcohols, which we discuss in the next section.

OH Functionalized Hydrocarbons—Alcohols (R–OH)

Alcohols contain an —OH group in place of one of the hydrogen atoms in the hydrocarbon. Alcohols are named according to the number of carbons in the hydrocarbon part of the molecule (often designated as the R group). The prefix for the number of carbons is followed by the suffix *-ol.* For example, CH_3CH_2OH is called ethanol. Ethanol is the alcohol found in beer, wine, and other alcoholic beverages. The concentration of ethanol in beverages is given by the concentration unit *proof,* which is twice the percent by volume of alcohol. The sources of ethanol and the production of ethanol via fermentation are discussed in detail in the textbook.

Alcohol molecules can form hydrogen bonds to each other and also to water molecules. For this reason, the alcohols have relatively high boiling points and are soluble in water (up to five carbons).

Alcohols other than ethanol are toxic to humans. This is due to the differences in the products of the biological reactions that metabolize alcohols.

Although alcohols can be quite toxic, they are very useful compounds that serve as solvents for a number of other substances. They are also quite useful in making other functionalized hydrocarbons, such as ethers, which we discuss in the next section.

Example 17.9

Arrange the following compounds in order of increasing boiling point:

(a) CH_3CH_2OH (b) CH_3CH_2Cl (c) $CH_3CH_2CH_3$ (d) $CH_3CH_2CH_2OH$

Solution

The intermolecular attractive forces operating in each compound must be considered. Only London forces are operating in the alkane, and therefore it will boil at the lowest temperature. Dipole–dipole interactions and London forces are operating in the haloalkane, and it will boil next. The alcohols contain an —OH group, which gives rise to hydrogen bonding. They will boil highest. The larger alcohol will boil at a higher temperature than the smaller alcohol due to more London forces in the larger molecule. The order of increasing boiling points is therefore as follows:

$$CH_3CH_2CH_3 < CH_3CH_2Cl < CH_3CH_2OH < CH_3CH_2CH_2OH$$

Practice Exercise 17.9

Arrange the following compounds in order of increasing boiling point:

(a) CH_3CH_2Br (b) CH_3Br (c) CH_3CH_2OH (d) CH_3CH_3

Ethers (R–O–R)

Ethers have the general formula $R-O-R$ and can be thought of as an alcohol $(R-OH)$ in which the H was replaced by another R group. Ethers are named by naming the two alkyl groups attached to the oxygen and ending with the word "ether." The names of the alkyl groups are always given in alphabetical order, not simply reading from left to right, which is a common mistake. For example, $CH_3-O-CH_2CH_3$ would be named ethyl methyl ether, not methyl ethyl ether. If the alkyl (R) groups are the same, the prefix *di-* is used to name the compound.

Ethers are commonly used as solvents but are no longer used as anesthetics.

Example 17.10

Name the following ethers:

(a) $CH_3-O-CH_2CH_2CH_3$ (b) CH_3-O-CH_3

Solutions

(a) methyl propyl ether   (b) dimethyl ether

Practice Exercise 17.10

Name the following ethers:

(a) $CH_3CH_2CH_2CH_2-O-CH_2CH_2CH_2CH_3$

(b) $CH_3CH_2-O-CH_2CH_2CH_3$

Organic Acids (Carboxylic Acids: R—COH)

Carboxylic acids contain the —COOH functional group, which has both a C—O single bond and a C=O double bond. They act as weak acids in the same manner as the weak acids we studied earlier; they donate a proton (H^+). As with the weak acids studied previously, the dissociation occurs to a small extent and is written as an equilibrium reaction where the equilibrium lies to the left. The reaction of carboxylic acids with water can be represented as follows:

$$RCOOH + H_2O \rightleftharpoons RCOO^- + H_3O^+$$

Carboxylic acids are named according to the number of carbons in the longest continuous chain that includes the —COOH group. The final -e in the alkane name is replaced by the suffix -oic acid. For example, the carboxylic acid shown below is named butanoic acid, because the longest chain contains four carbons. (Always include the C in the —COOH group when determining the prefix for carboxylic acids.)

$$CH_3CH_2CH_2 \overset{\overset{\displaystyle O}{\|}}{-C} -OH$$

One of the most widely known carboxylic acids is acetic acid (ethanoic acid by IUPAC naming rules), the acid in vinegar. Carboxylic acids are very important as starting materials for other organic compounds.

Example 17.11

Arrange the following compounds in order of increasing solubility in water:

(a) CH_3CH_2COOH (b) CH_3CH_2Cl (c) $CH_3CH_2CH_3$ (d) $CH_3CH_2CH_2COOH$

Solution

The carboxylic acids will be the most soluble because they can form hydrogen bonds with water. The larger carboxylic acid will be less soluble than the smaller one because it has a larger nonpolar R group. The polar haloalkane will be next in solubility, and the nonpolar alkane will be the least polar. The order of increasing solubility in water is therefore as follows:

$$CH_3CH_2CH_3 < CH_3CH_2Cl < CH_3CH_2CH_2COOH < CH_3CH_2COOH$$

Practice Exercise 17.11

Arrange the following compounds in order of increasing solubility in water:

(a) CH_3Br (b) CH_3COOH (c) CH_3CH_3 (d) $CH_3CH_2CH_2CH_2CH_2COOH$

Organic Bases (Amines: H_2NR, HNR_2, NR_3)

Amines contain a nitrogen atom bonded to one, two, or three alkyl groups. They can be thought of as ammonia, NH_3, with one, two, or three of the hydrogens replaced by alkyl groups. Just as ammonia is a Brønsted–Lowry base due to the lone pair of electrons on the nitrogen, the amines are also Brønsted–Lowry bases because of the lone pairs of electrons on the nitrogens in their molecules. The reaction of an amine with an acid is shown below.

$$R\ddot{N}H_2 + H^+ \longrightarrow \left[R\overset{H}{\underset{}{\ddot{N}}}H_2 \right]^+$$

Like the alcohols, amine molecules can form hydrogen bonds with each other or with water. Therefore, their boiling points are relatively high, and the smaller amines are soluble in water.

Amines are named by naming the alkyl (R) groups attached to the nitrogen and ending with the word "amine." For example, $CH_3NHCH_2CH_3$ is named ethylmethyl amine. If the alkyl groups are the same, the prefix *di-* or *tri-* is used to name the compound.

Amines have a very strong odor and are often produced by rotting flesh. They are also quite useful in the production of other organic compounds.

Ketones (R_2C=O) and Aldehydes (RHC=O)

Ketones and aldehydes contain the C=O group, known as the *carbonyl group*. (Carboxylic acids are the other group of compounds we have discussed that contain a carbonyl group.) The only difference between aldehydes and ketones is that aldehydes have an H and one alkyl (R) group attached to the carbonyl, whereas ketones have two R groups attached. The aldehyde group is often indicated —CHO.

Ketones and aldehydes are named according to the number of carbons in the chain (including the carbonyl carbon). Ketones end in the suffix *-one*, and aldehydes end in the suffix *-al*. The position of the carbonyl carbon must be specified in ketones but not in aldehydes. In aldehydes the carbonyl carbon is always carbon number one. The following example illustrates this. (More examples are found in Section 17.4 of the textbook.)

Example 17.12

Name the following compounds:

$$\text{(a) } CH_3\overset{\displaystyle O}{\overset{\displaystyle \|}{\underset{\displaystyle CH_3}{\underset{|}{CH}}CH}} \qquad \text{(b) } CH_3CH_2\overset{\displaystyle O}{\overset{\displaystyle \|}{C}}CH_2CH_3 \qquad \text{(c) } CH_3\overset{\displaystyle O}{\overset{\displaystyle \|}{C}}CH_2CH_3$$

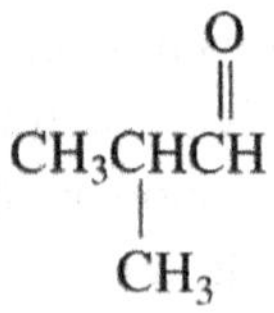

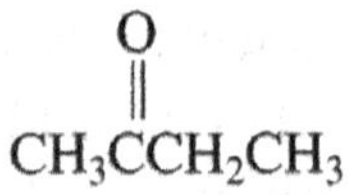

Solutions

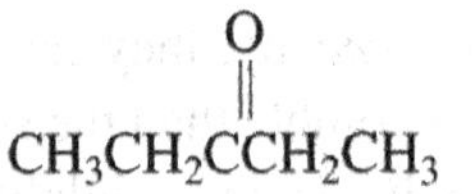

(a) 2-methylpropanal   (b) 3-pentanone   (c) 2-butanone

Practice Exercise 17.12

Name the following compounds:

$$\text{(a) } CH_3\overset{\displaystyle O}{\overset{\displaystyle \|}{C}}CH_3 \qquad \text{(b) } CH_3CH_2CH_2CH_2\overset{\displaystyle O}{\overset{\displaystyle \|}{C}}H \qquad \text{(c) } CH_3\overset{\displaystyle O}{\overset{\displaystyle \|}{\underset{\displaystyle CH_3}{\underset{|}{CH}}CH_2}}CH$$

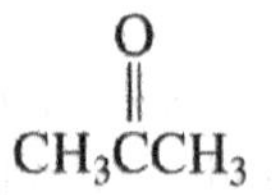

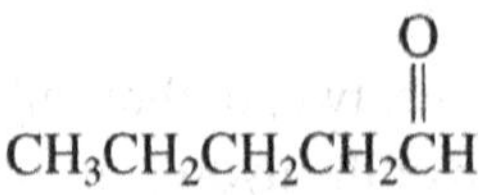

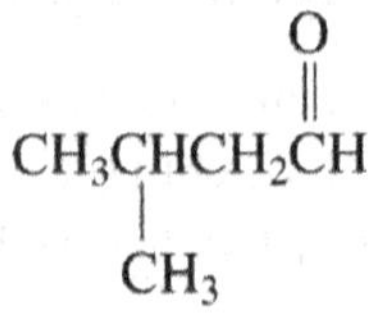

Strategies for Working Problems in Chapter 17

Overview of Problem Types: What You Should Be Able to Do

Chapter 17 presents an overview of the field of organic chemistry, the chemistry of carbon compounds. The types of problems that you should be able to solve after mastering Chapter 17 are:

1. Determine whether a hydrocarbon is saturated or unsaturated.

2. Name branched hydrocarbons.

3. Classify organic compounds according to the functional group present in the structure of the compound.

Flowcharts

Flowchart 17.1 Determining whether a hydrocarbon is saturated or unsaturated

Flowchart 17.1

Summary

Method for determining whether a hydrocarbon is saturated or unsaturated

Step 1
*   **Draw the structure of the saturated hydrocarbon with the number of carbon atoms specified.**

$\downarrow$

Step 2
*   **Determine whether the hydrocarbon in question has the same number or a fewer number of carbon atoms than the saturated hydrocarbon.**
*   **If the number of hydrogen atoms is the same as in the structure of the saturated hydrocarbon, the hydrocarbon in question is saturated.**
*   **If the number of hydrogen atoms is less that the hydrogen atoms in the saturated hydrocarbon, the hydrocarbon in question is unsaturated.**

$\downarrow$

Step 3
*   **Use the C_nH_{2n+2} general formula for saturated hydrocarbons to check the result obtained in step 2.**

Example 1

The hydrocarbon that is commonly used to make film for packaging has the formula C_3H_6. Is this hydrocarbon saturated or unsaturated?

Solution

Perform Step 1

The structure of the saturated hydrocarbon containing three carbon atoms is:

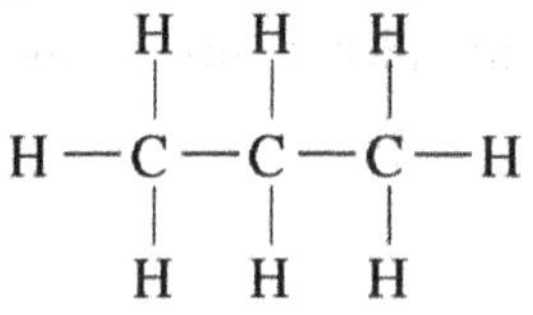

Perform Step 2

The saturated hydrocarbon has eight hydrogen atoms, and the hydrocarbon in question has only six hydrogen atoms. Therefore, C_3H_6 is unsaturated.

Perform Step 3

For a three carbon hydocarbon, $C_nH_{2n+2} = C_3H_8$

This verifies that a saturated hydrocarbon with three carbons would have eight hydrogen atoms; one with only six hydrogen atoms is unsaturated.

Example 2

The hydrocarbon that is commonly used to make synthetic rubber has the formula C_4H_6. Is this hydrocarbon saturated or unsaturated?

Solution

Perform Step 1

The structure of the saturated hydrocarbon containing four carbon atoms is

Perform Step 2

The saturated hydrocarbon has ten hydrogen atoms, and the hydrocarbon in question has only six hydrogen atoms. Therefore, C_4H_6 is unsaturated.

Perform Step 3

For a four-carbon hydrocarbon, $C_nH_{2n+2} = C_4H_{10}$.

This verifies that a saturated hydrocarbon with four carbons would have ten hydrogen atoms; one with only six hydrogen atoms is unsaturated.

Practice Problems

17.1 Is C_8H_{18} a saturated or an unsaturated hydrocarbon?

17.2 Is C_6H_{12} a saturated or an unsaturated hydrocarbon?

Flowchart 17.2 Naming branched hydrocarbons

Flowchart 17.2

Summary

Method for naming branched hydrocarbons

Step 1
- **Identify and name the main chain (longest continuous chain) in the molecule. The ending will be -ane if it is an alkane (single bonds only), -ene if an alkene (one or more double bonds), and -yne if an alkyne (one or more triple bonds).**

↓

Step 2

- Identify and name the branches off of the main chain. Name the branches according to the number of carbons in each branch.

Step 3

- Give the location of each branch on the main chain.
- Number the carbons in the longest chain in such a way as to give the lowest combination of numbers for the branches off of the main chain.
- If more than one branch with the same structure is present, use the prefixes *di*, *tri*, *tetra*, *penta*, and so on and give the number of the position of each of the branches on the chain.
- In an alkene or alkyne the longest continuous chain is numbered from the end that gives the lowest number(s) for the position of the multiple bond(s). The number of the carbon that begins the multiple bond must be specified.

Example 1

Name the following compound:

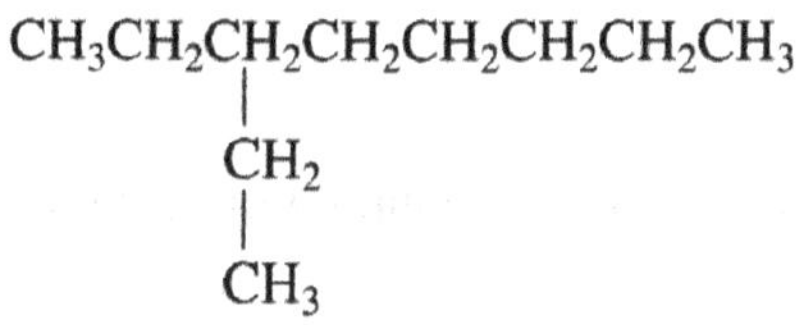

$$CH_3CH_2CH_2CH_2CH_2CH_2CH_2CH_3$$
$$|$$
$$CH_2$$
$$|$$
$$CH_3$$

Solution

Perform Step 1

The longest chain consists of eight carbon atoms, and the compound will be named as octane.

Perform Step 2

The branch is named ethyl because it has two carbon atoms in it.

Perform Step 3

The chain must be numbered from the left, in order to give the branch the position of 3 and not 6. Thus, the branch is named as 3-ethyl.

The compound is therefore named 3-ethyloctane.

Example 2

Name the following compound:

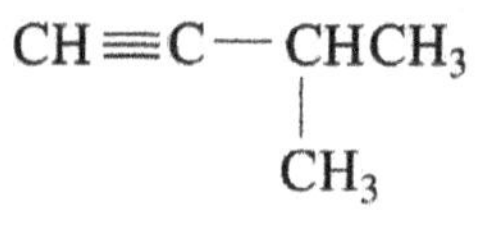

$$CH \equiv C - CHCH_3$$
$$|$$
$$CH_3$$

Solution

Perform Step 1

The longest chain consists of four carbon atoms, and the compound will be named as butyne (ending in -yne because of the triple bond).

Perform Step 2

The branch is named methyl because it has one carbon in it.

Perform Step 3

The chain must be numbered from the left in order to give the triple bond the lowest numbered position (1-butyne). Thus, the branch is named as 3-methyl.

The compound is therefore named 3-methyl-1-butyne.

Example 3

Name the following compound:

$$CH_2{=}CH-CH-CH_3$$
$$| $$
$$CH_2$$
$$| $$
$$CH_2$$
$$| $$
$$CH_3$$

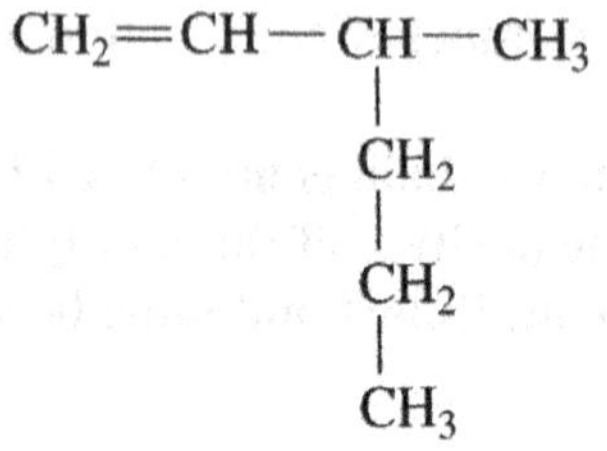

Solution

Perform Step 1

The longest chain consists of six carbon atoms, and the compound will be named as hexene (ending in -ene because of the double bond).

Perform Step 2

The branch is named methyl because it has one carbon in it.

Perform Step 3

The chain must be numbered from the left in order to give the double bond the lowest numbered position (1-hexene). Thus, the branch is named as 3-methyl.

The compound is therefore named 3-methyl-1-hexene.

Practice Problems

17.3 Name the following compound:

$$CH_3C{=}CCHCH_3$$
$$| $$
$$CH_2$$
$$| $$
$$CH_3$$

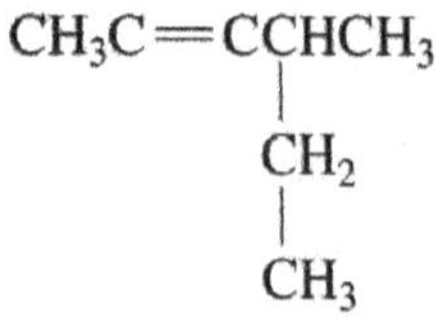

17.4 Name the following compound:

$$CH_3CH{=}CHCHCH_2CH_3$$
$$| $$
$$CH_3$$

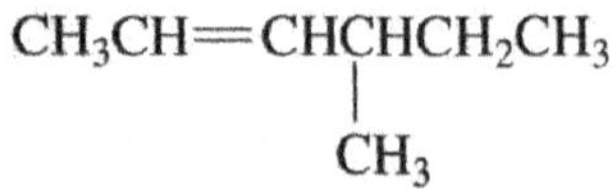

Flowchart 17.3 Classifying organic compounds according to the functional group present in the structure of the compound

Flowchart 17.3

Summary

Method for classifying organic compounds according to the functional group present in the structure of the compound

Step 1
- Use the chart of classes of organic compounds to locate the functional group found in the molecule.

$\downarrow$

Step 2
- Classify the compound based on its functional group.
- If more than one functional group is present the compound can generally be classified as both.

(e.g., $CH_2\!=\!CH\!-\!CH_2OH$ can be classified as an alkene and as an alcohol.)

Example 1

Classify the following compound according to the group in which it belongs.

$CH_3CH_2NH_2$

Solution

Perform Step 1

According to the chart, a compound with a nitrogen atom bonded to two hydrogen atoms is an amine.

Perform Step 2

The compound is an amine.

Example 2

Classify the following compound according to the group in which it belongs.

$CH_3CH_2OCH_2CH_2CH_3$

Solution

Perform Step 1

According to the chart, a compound with an oxygen atom between two alkyl groups is an ether.

Perform Step 2

The compound is an ether.

Example 3

Classify the following compound according to the group in which it belongs.

$$CH_3CH_2\overset{\overset{\displaystyle O}{\|}}{C}\!-\!H$$

Solution

Perform Step 1

According to the chart, a compound with a carbon doubly bonded to an oxygen atom and to a hydrogen atom is an aldehyde.

Perform Step 2

The compound is an aldehyde.

Practice Problems

17.5 Classify the following compound:

$$CH_3CH_2 \overset{\overset{\displaystyle O}{\|}}{-C} - CH_2CH_3$$

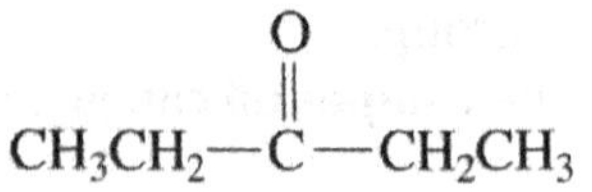

17.6 Classify the following compound:

$$CH_3 \overset{\overset{\displaystyle O}{\|}}{-C} - OH$$

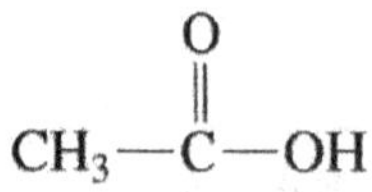

17.7 Classify the following compound:

$$CH_3F$$

Quiz for Chapter 17 Problems

1. Name the following compound.

$$\underset{\underset{\displaystyle CH_3}{|}}{CH_3CH_2CHCH_2CH_3}$$

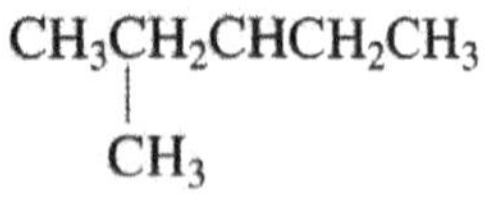

2. Fill in the blanks in the following sentence. A hydrocarbon that contains only single bonds is a(n) _____, one with a double bond is a(n) _____, and one containing a triple bond is a(n) _____.

3. Classify the following compound according to the functional group to which each belongs.
   (a) 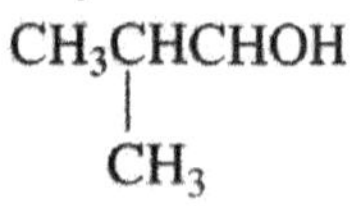 $\underset{\underset{\displaystyle CH_3}{|}}{CH_3CHCHOH}$ (b) $CH_3CH_2OCH_2CH_3$ (c) $CH_3CH_2CH_2CH_2Br$

4. Indicate whether the following compounds are saturated or unsaturated.
   (a) C_5H_{12} (b) C_7H_{14}

5. Write the functional group for each of the following classes of compounds.
   (a) alcohols   (b) ethers   (c) carboxylic acids

6. Provide the name of the alkane that contains the following number of carbons in its main chain.
   (a) three carbons   (b) seven carbons

7. Name the compound below.

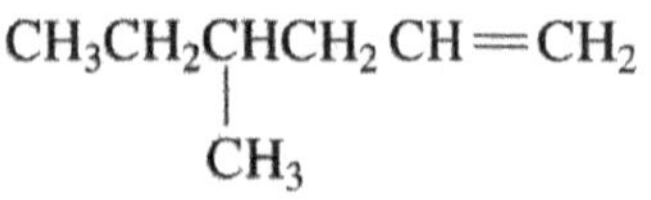

$$\underset{\underset{\displaystyle CH_3}{|}}{CH_3CH_2CHCH_2} \, CH{=}CH_2$$

8. The three classes of compounds in which there is carbon doubly bonded to an oxygen atom are _____, _____, and _____.

9. A saturated hydrocarbon that contains 12 carbons will contain _____ hydrogen atoms.

10. Draw the structure of a molecule that might be called 2-methyl-3-butene. Write the correct name of the molecule.

Exercises for Self-Testing

A. Completion. *Write the correct word(s) to complete each statement below.*

1. The three allotropes of carbon are _____, _____, and _____.
2. In its compounds the number of bonds carbon will always form to other atoms is _____.
3. _____ are hydrocarbons that contain only single bonds; _____ contain a double bond, and _____ contain a triple bond.
4. The formula of an alkane that contains nine carbon atoms will contain _____ hydrogen atoms.
5. Compounds with the same molecular formula but a different arrangement of the atoms are called _____.
6. The name of a linear hydrocarbon containing 10 carbon atoms and only single bonds is _____.
7. R — OH is the general formula for the _____.
8. The alcohol concentration unit that equals twice the percent by volume of alcohol is _____.
9. The class of functionalized hydrocarbons that contain a basic functional group are the _____.
10. The class of compounds whose structure can be abbreviated R — O — R are the _____.
11. The three classes of functionalized hydrocarbons that contain a carbonyl group are _____ _____, _____, and _____.
12. The name for a ketone always ends in the suffix _____.
13. The class of functionalized hydrocarbons that are acidic are the _____ _____.
14. Amines can form _____ _____ with water.
15. Hydrocarbons are found in the Earth's deposits of _____.

B. True/False. *Indicate whether each statement below is true or false, and explain why the false statements are false.*

_____ 1. Carbon is the only element that has the capability of forming long chains of atoms bonded to each other.
_____ 2. Alkynes contain at least one double bond.
_____ 3. Hydrocarbons are polar molecules because the electronegativity of carbon is greater than that of hydrogen.
_____ 4. 1-butene contains one double bond between carbon atoms 1 and 2.
_____ 5. 2-methylbutane and pentane are isomers of each other.
_____ 6. C_6H_{12} is the formula for an alkane.
_____ 7. The solubility of carboxylic acids in water increases as the size of the carboxylic acid increases.
_____ 8. The nitrogen atom in an amine is the source of its basic behavior.
_____ 9. Aldehydes and ketones are less soluble in water than alkanes.
_____ 10. The oxygen atom in aldehyde molecules makes it possible for aldehyde molecules to form hydrogen bonds to each other.

C. Questions and Problems

1. Write the structures of the following compounds:
   (a) 3-ethyloctane   (b) 2,4-dimethyldecane   (c) pentanal
   (d) 2-pentanone   (e) dimethyl amine   (f) propanoic acid

2. Draw a molecule of a compound from each of the following functional groups:
   (a) amine   (b) haloalkane   c) alkyne
   (d) ether   (e) carboxylic acid

3. Explain the ranking in the boiling points of the following compounds:
   $CH_3CH_2CH_3 < CH_3CH_2CH_2CH_2CH_2 < CH_3CH_2CH_2Cl < CH_3CH_2COOH < CH_3CH_2CH_2CH_2CH_2COOH$

4. Write an equation for the reaction of ethanoic acid (acetic acid) with water.

5. Explain why humans can consume ethanol in reasonable amounts with no toxicity, but cannot tolerate even small amounts of methanol without serious consequences.

6. Which of the following compounds is expected to be soluble in water? Explain your answer.
   (a) propane   (b) propanol   (c) propanone   (d) propanoic acid

7. Which organic compounds remind you the most of ammonia?

Synthetic and Biological Polymers

Learning Outcomes

1. Understand what a polymer and monomer are.
2. Understand the difference between a monomer and a monomer unit and how monomer units with $C = C$ bonds can undergo free-radical polymerization.
3. Know the monomers used to make nylon and how they polymerize via formation of an amide bond.
4. Understand what monosaccharides, disaccharides, polysaccharides, such as starch and cellulose, and carbohydrates are.
5. Understand the relationship between amino acids and proteins and the peptide bond.
6. Understand the structure and function of DNA.

Chapter Outline

C. Glucose

    1. Linear and cyclic structures of glucose

    2. Polymers of glucose

        a. Starch

        b. Cellulose

18.5 Proteins

  A. Definition of proteins

  B. Amino acids

    1. Twenty amino acids used to make proteins

    2. Ten essential amino acids

  C. Peptide linkage

  D. Distinction between polypeptides and proteins

  E. Importance of amino acid sequencing in protein function

18.6 One More Thing: DNA—The Master Biopolymer

  A. Molecular components of DNA

  B. Double-stranded DNA: The double helix

  C. DNA's role as a blueprint for protein synthesis

  D. Viruses

    1. Mechanism by which viruses kill cells

    2. Potential role of viruses in replacing defective genes in organisms and curing disease

Review of Key Concepts from the Text

18.1 Building Polymers

Polymers are very large molecules, often called macromolecules, that consist of a repeating unit. They are made from individual molecules called monomers. The monomers are connected by covalent bonds to form the polymer. The repeating structure is called the *monomer unit*.

18.2 Polyethylene and Its Relatives

Polyethylene is a polymer made from ethylene monomers, as its name suggests. Ethylene is a common name for ethene, a two-carbon hydrocarbon containing a double bond, $CH_2\!=\!CH_2$. Ethylene is the simplest alkene that exists. Its monomers join together to form the polymer. The polymer (showing three repeating units) is written as:

$$\cdots-\underset{\underset{\displaystyle H}{|}}{\overset{\overset{\displaystyle H}{|}}{C}}-\underset{\underset{\displaystyle H}{|}}{\overset{\overset{\displaystyle H}{|}}{C}}-\underset{\underset{\displaystyle H}{|}}{\overset{\overset{\displaystyle H}{|}}{C}}-\underset{\underset{\displaystyle H}{|}}{\overset{\overset{\displaystyle H}{|}}{C}}-\underset{\underset{\displaystyle H}{|}}{\overset{\overset{\displaystyle H}{|}}{C}}-\underset{\underset{\displaystyle H}{|}}{\overset{\overset{\displaystyle H}{|}}{C}}-\cdots$$

The bonds at either end of the polymer indicate that the chain continues in both directions. The repeating unit in the polymer is $-CH_2CH_2-$, even though it might appear to be $-CH_2-$. When an alkene serves as the monomer for making a polymer, the repeating unit contains both carbons of the alkene monomer. There are no double bonds in the polymer because electrons are rearranged during the formation of covalent single bonds between adjacent monomer units. The process is discussed in detail in Section 18.2 of the textbook.

Polyethylene is commonly used in everyday materials such as plastic bags, plastic bottles, and plastic pipes and tubing.

If another atom or group replaces one or more of the hydrogens in ethylene, a different polymer is produced. For example, if all four hydrogens are replaced by fluorines, the monomer is named tetrafluoroethene, and the polymer is teflon. The monomer and the polymer are shown at the top of the next page.

$$F_2C = CF_2$$

Tetrafluoroethene

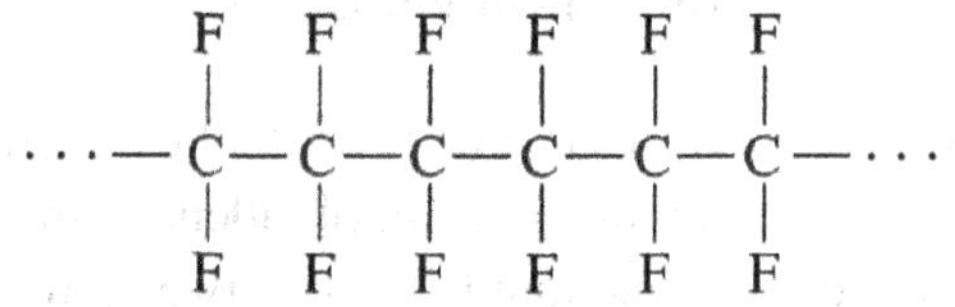

```
     F   F   F   F   F   F
     |   |   |   |   |   |
···— C — C — C — C — C — C —···
     |   |   |   |   |   |
     F   F   F   F   F   F
```

Teflon

You can always draw the polymer that is made from an alkene monomer by simply connecting the alkene monomers together and omitting the double bond between the carbons, as the following example demonstrates.

Example 18.1

Draw the monomer and the polymer (including at least three monomer units) of the following monomers:

(a) bromoethene   (b) 1,2-dichloroethene

Solution

(a) The structure of bromoethene is:

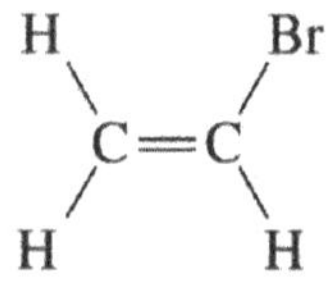

```
 H       Br
  \     /
   C = C
  /     \
 H       H
```

The structure of the polymer is, therefore:

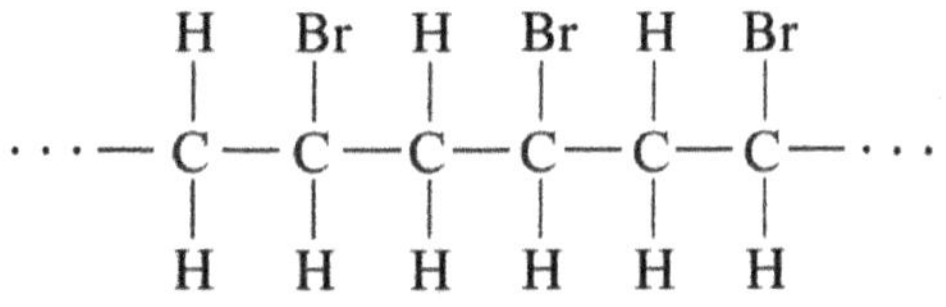

```
     H   Br  H   Br  H   Br
     |   |   |   |   |   |
···— C — C — C — C — C — C —···
     |   |   |   |   |   |
     H   H   H   H   H   H
```

(b) The structure of 1,2-dichloroethene is:

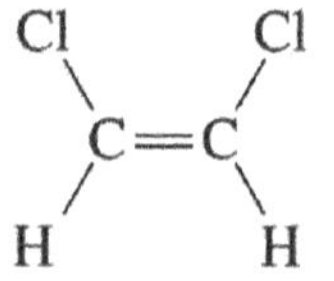

```
 Cl      Cl
  \     /
   C = C
  /     \
 H       H
```

The structure of the polymer is, therefore:

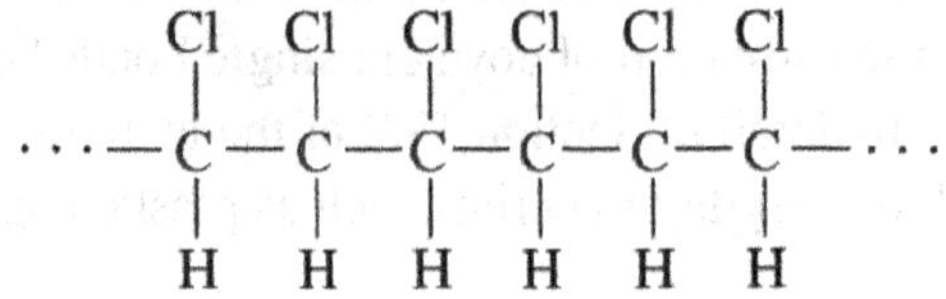

Practice Exercise 18.1

Draw the monomer and the polymer (including at least three monomer units) of the following monomers:

(a) 1-propene   (b) 1,1-difluoroethene

      You should also be able to write the structure of the monomer when given the structure of the polymer. To do this, you simply identify the repeating unit (two carbons), and write the repeating unit with a double bond between the two carbon atoms. The following example demonstrates this.

Example 18.2

Draw the monomer that was used to make the following polymers:

(a)

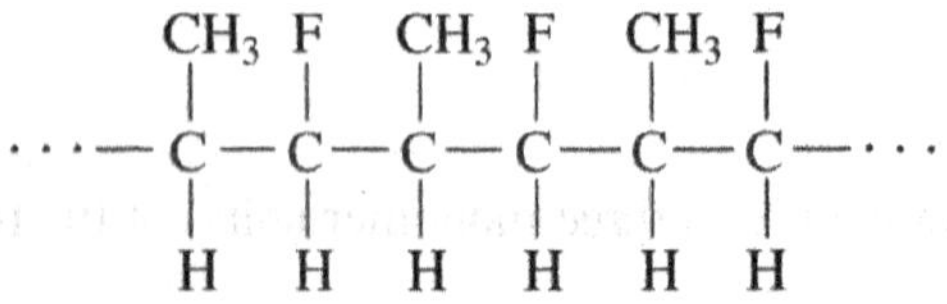

(b)

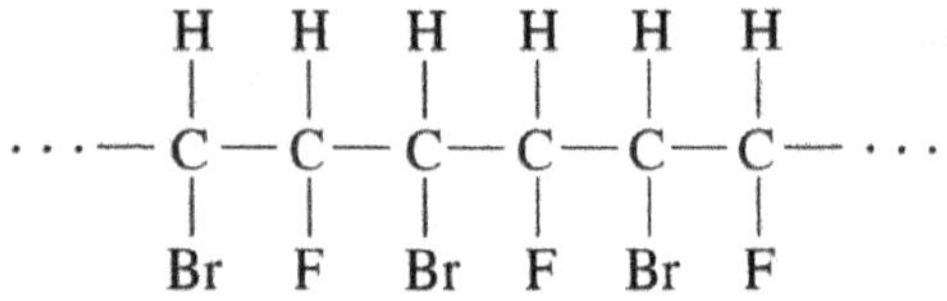

Solution

(a)

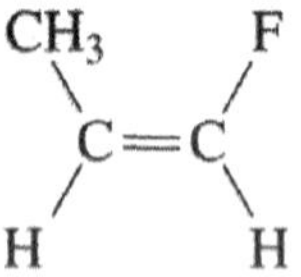

(b)

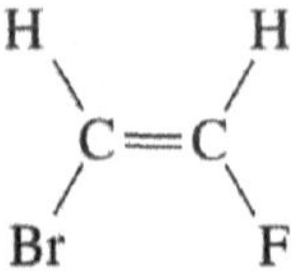

Practice Exercise 18.2

Draw the monomer that was used to make the following polymers:

(a)

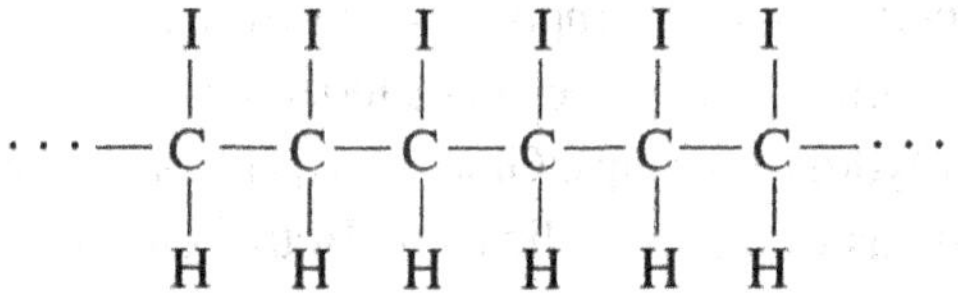

```
      I   I   I   I   I   I
      |   |   |   |   |   |
···—C —C —C —C —C —C—···
      |   |   |   |   |   |
      H   H   H   H   H   H
```

(b)

```
      H  Cl  H  Cl  H  Cl
      |   |   |   |   |   |
···—C —C —C —C —C —C— ···
      |   |   |   |   |   |
      Br  F  Br  F  Br  F
```

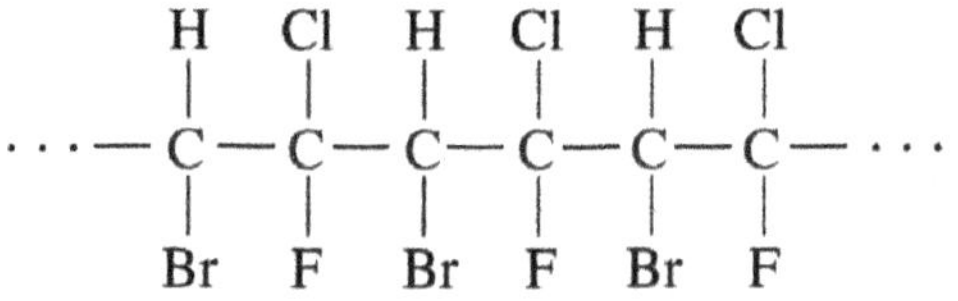

18.3 Nylon—A Polymer You Can Wear

So far we have considered only one type of polymers, addition polymers. Addition polymers are made by linking alkene monomers together. We can also form polymers by allowing two molecules with reactive ends to join together. This happens when condensation polymers are formed. Nylon is a condensation polymer, and we will look closely at the monomers used to produce it. First, however, we will look at the general mechanism of forming condensation polymers.

Condensation monomers are made by combining two monomers that each have reactive groups at both ends of the molecule. In the process of combining the monomers, a small molecule is split off from the two monomers. Sometimes this small molecule is water (hence the name *condensation polymer*), but often it is an HCl molecule, as is the case in the synthesis of nylon.

Nylon 66 is a particular type of nylon that is made by combining the acid chloride derivative of 1,6-hexanedioic acid and 1,6-diaminohexane, as shown in the figure in Section 18.3 of the textbook. In the reaction between the monomers, an HCl molecule is split off when the new bond is formed between the carbonyl of the acid chloride and the nitrogen of the amine. The new bond is called an *amide bond*, as shown below.

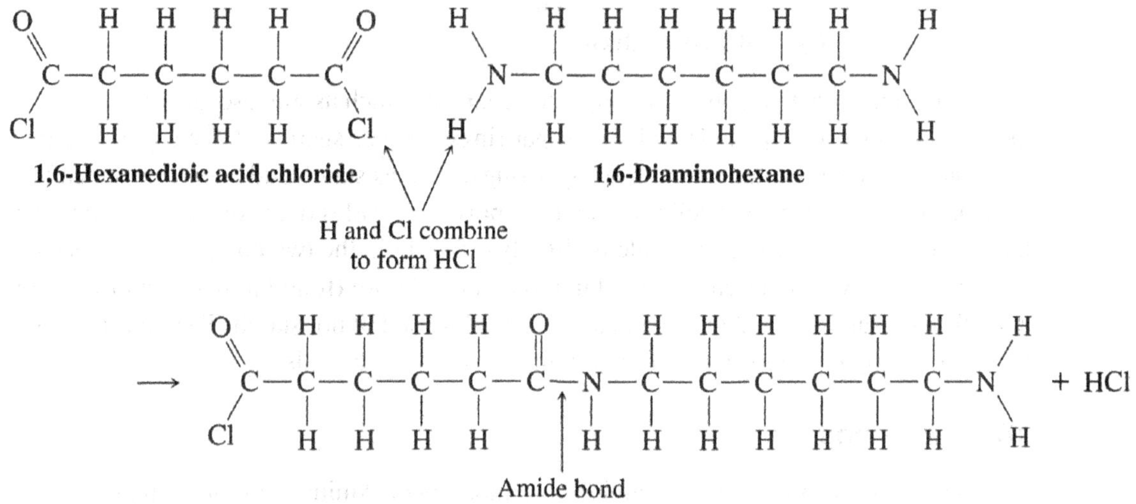

1,6-Hexanedioic acid chloride   **1,6-Diaminohexane**

H and Cl combine
to form HCl

Amide bond

When many of the monomers join together, a nylon polymer chain—held together by amide bonds—is produced. The amide linkages that occur in nylon are identical to the linkages between the amino acids in protein molecules. We will discuss this comparison in more detail in Section 18.5.

18.4 Polysaccharides and Carbohydrates

Polysaccharides are polymers made up of tens, hundreds, or even thousands of sugar molecule monomers. Sugar molecules belong to the general class of compounds known as carbohydrates, which include sugars, starches, and cellulose. Carbohydrates are characterized by having an aldehyde or ketone functional group and many —OH (hydroxy) groups. Glucose, $C_6H_{12}O_6$, is the most common sugar. It has a linear, open chain form, as well as a cyclic form. Both forms are shown below:

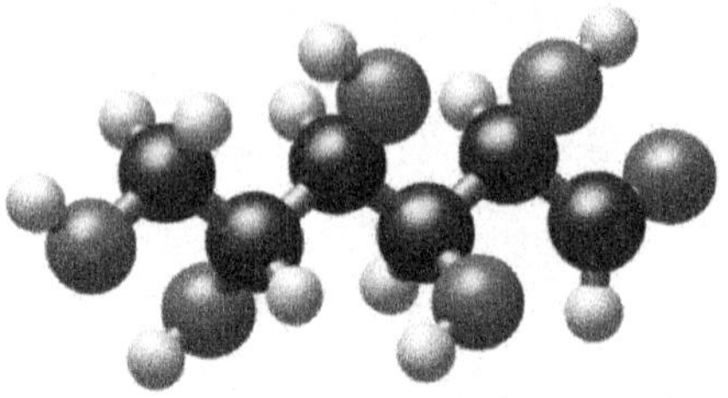

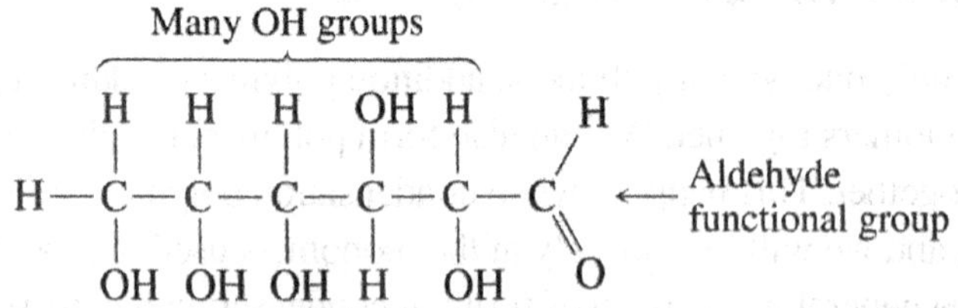

Glucose, the simplest sugar
$C_6H_{12}O_6$

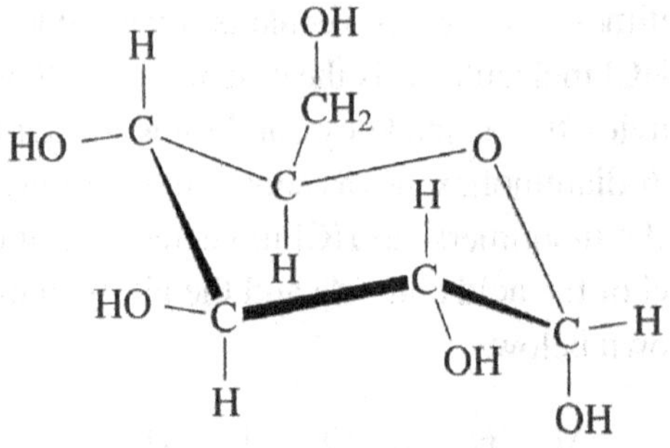

Cyclical form of glucose

Compounds that consist of only one sugar ring, such as glucose, are called monosaccharides. Compounds with two sugar rings, such as sucrose (table sugar), are called disaccharides. Compounds with many sugar rings, such as starch and cellulose, are called polysaccharides. Starch and cellulose are both polysaccharides made of glucose monomers, but the structure of the glucose molecule is slightly different in the two compounds. This structural difference is why humans can digest starch but not cellulose (found in wood) and why some animals (such as cows and termites) can digest cellulose but not starch. The digestion of cellulose is performed by microorganisms found in the gut of these animals.

18.5 Proteins

Proteins are polymers made up of amino acid monomers. Amino acids are compounds that contain both an amino group (—NH$_2$) and a carboxylic acid group (—COOH). The amino acids link together by forming amide linkages (just like the amide linkages in nylon). When protein molecules are formed, however, H_2O is split off when the amide linkage is formed, in contrast to the HCl that was split off when nylon was formed. The amide linkages in protein molecules are most often referred to as *peptide linkages*, and the individual amino acids in proteins are referred to as *peptides*.

In order to be classified as a protein, the polymer must contain at least 50 amino acids. Polymers with less than 50 amino acids are called polypeptides, and a prefix is often used to indicate the number of amino acids in polypeptides containing fewer than 20 amino acids. Section 18.5 of the textbook describes the peptide linkage and the different orders in which amino acids can link to form polypeptides.

The proteins in our bodies are made up of only specific 20 amino acids. All 20 of them are similar in that the $-COOH$ and the $-NH_2$ groups are both bonded to the same carbon atom. The carbon atom that is connected to the carboxylic acid group and the amino group is called the α carbon. The general structure of these amino acids, called α-amino acids, is shown below.

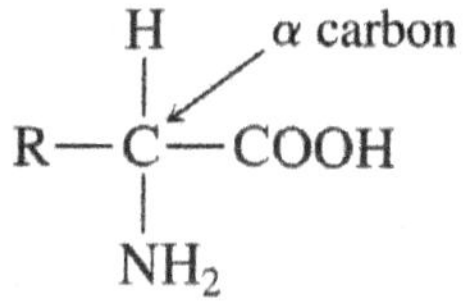

$$R-\underset{\underset{NH_2}{|}}{\overset{\overset{H}{|}}{C}}-COOH \quad \alpha \text{ carbon}$$

The 20 amino acids that make up the proteins in our bodies are shown on the next page. The body can synthesize half of the amino acids from simple molecules. The other 10 must be present in the food we eat, and are called *essential amino acids*.

The structure of the R group of the amino acid determines the properties of the acid. The simplest amino acid is glycine, which has H as its R group. When the R group is an alkyl group, the group is nonpolar and therefore hydrophobic. When the R group is acidic (as in aspartic acid), it can react with bases. When the R group is basic (as in lysine), it can react with acids. The properties of polypeptides and proteins are largely due to the structure of the R groups of the amino acid monomers. The three-dimensional shape of the protein is also due to the structure of the R groups. The protein may twist and fold to keep its hydrophobic regions inside (and out of contact with the water that is ubiquitous throughout the body), or it may arrange itself so that R groups capable of hydrogen bonding to each other are in close enough proximity to do so.

Proteins have many functions in the human body. Some proteins, such as keratin, serve a structural function in the skin, hair, and fingernails. Other proteins, such as insulin, serve as hormones that regulate the activity of different cells. Still others serve as enzymes, such as the amylase found in the digestive system to catalyze the breakdown of starch into glucose.

The function of proteins is tied directly to their three-dimensional shape. And because the shape is tied to the structure of the R groups, protein function is dependent on the protein having the correct sequence of amino acids, with each R group exactly where it is supposed to be. Any change in the position or identity of even one of the component amino acids can affect the function of the protein. This change in the identity of one amino acid is what causes the disease sickle-cell anemia. The change in the hemoglobin molecule caused by the substitution of one amino acid causes the symptoms of the disease. (A complete discussion of this mechanism appears in Section 18.5 of the textbook.)

18.6 DNA—The Master Biopolymer

The correct sequence of amino acids in proteins is vital for the proteins in the body to function as they should. The master blueprint for the sequence of amino acids in proteins is called DNA, which stands for *deoxyribonucleic acid.* DNA is a polymer consisting of a "backbone" of alternating deoxyribose sugar molecules and phosphate units. One of four amine bases is covalently bonded to each sugar molecule in the backbone of DNA, as shown below.

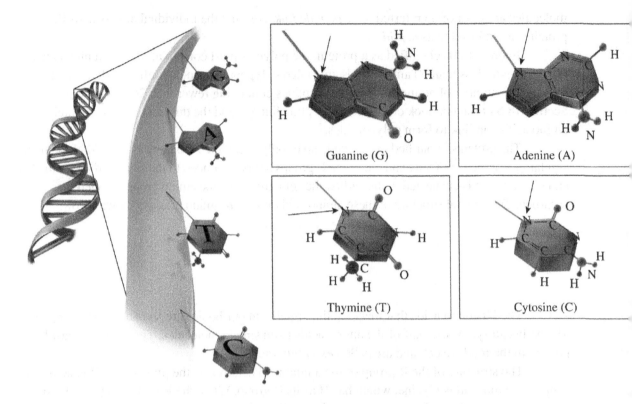

The DNA polymers form strands, two of which are associated with each other by the intermolecular attractive forces of hydrogen bonding. The two linked strands have a helical shape, and are called a *double helix*. The structure of the DNA double helix is shown to the right.

The 20 Amino Acids Used to Build Proteins
(Essential Amino Acids Are Indicated by Shading)

Neutral or polar amino acids

Alanine (Ala)

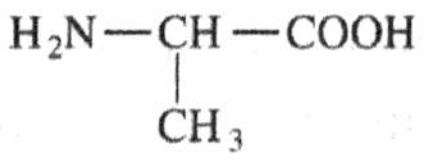

$H_2N-CH-COOH$
$|$
CH_3

Asparagine (Asn)

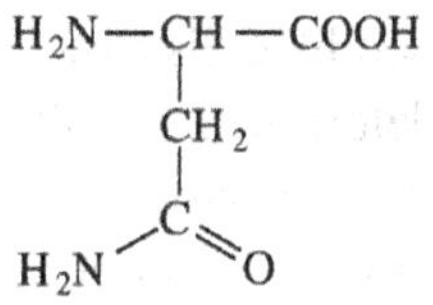

$H_2N-CH-COOH$
$|$
CH_2
$|$
$H_2N-C=O$

Cysteine (Cys)

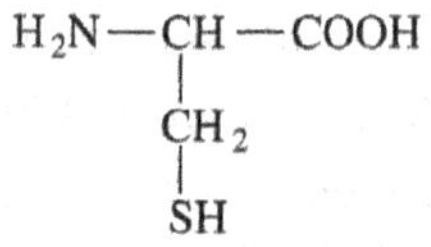

$H_2N-CH-COOH$
$|$
CH_2
$|$
SH

Glutamine (Gln)

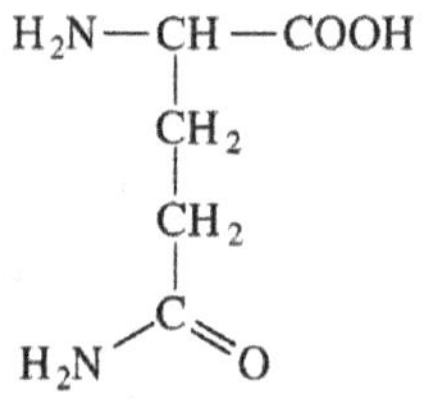

$H_2N-CH-COOH$
$|$
CH_2
$|$
CH_2
$|$
$H_2N-C=O$

Glycine (Gly)

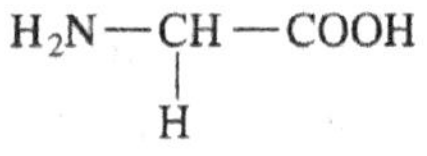

$H_2N-CH-COOH$
$|$
H

Isoleucine (Ile)

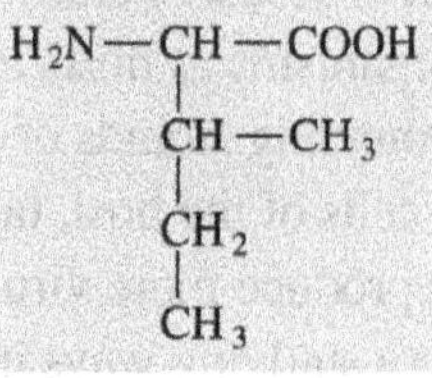

$H_2N-CH-COOH$
$|$
$CH-CH_3$
$|$
CH_2
$|$
CH_3

Leucine (Leu)

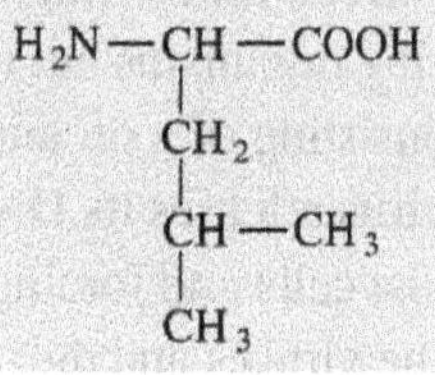

$H_2N-CH-COOH$
$|$
CH_2
$|$
$CH-CH_3$
$|$
CH_3

Methionine (Met)

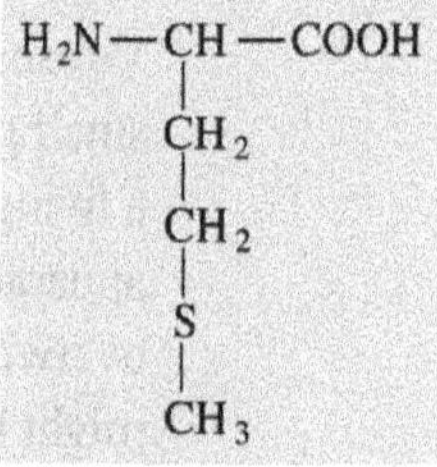

$H_2N-CH-COOH$
$|$
CH_2
$|$
CH_2
$|$
S
$|$
CH_3

Phenylalanine (Phe)

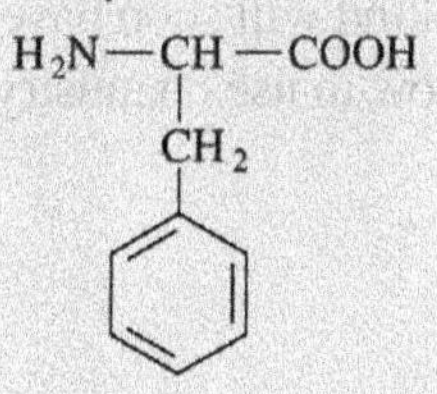

$H_2N-CH-COOH$
$|$
CH_2
$|$
(benzene ring)

Proline (Pro)

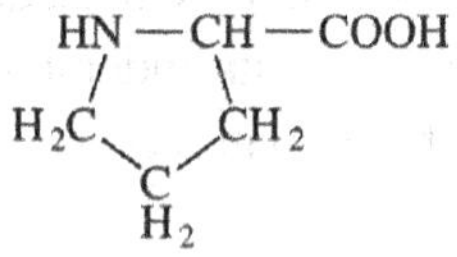

$HN-CH-COOH$
$H_2C \quad CH_2$
C
H_2

Serine (Ser)

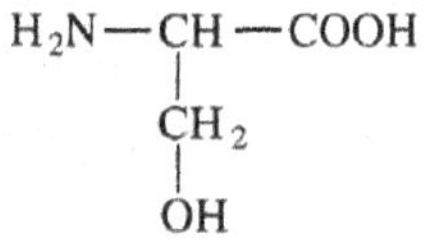

$H_2N-CH-COOH$
$|$
CH_2
$|$
OH

Threonine (Thr)

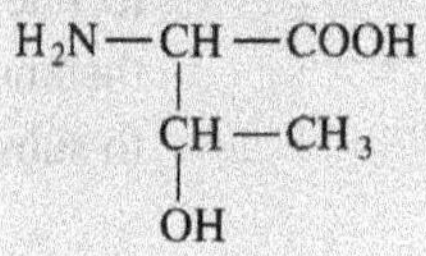

$H_2N-CH-COOH$
$|$
$CH-CH_3$
$|$
OH

Tryptophan (Trp)

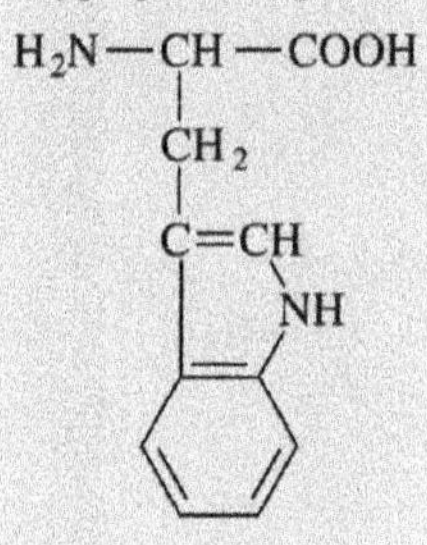

$H_2N-CH-COOH$
$|$
CH_2
$|$
$C=CH$
$\quad NH$
(indole ring)

Tyrosine (Tyr)

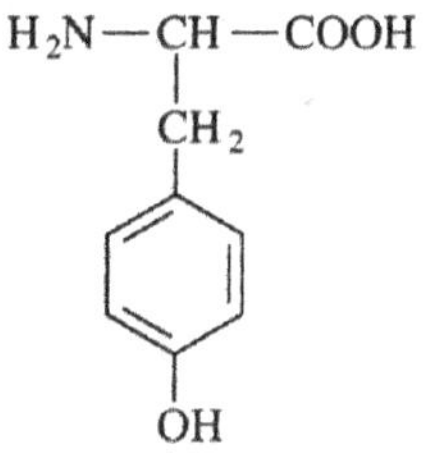

$H_2N-CH-COOH$
$|$
CH_2
$|$
(phenol ring)
OH

Valine (Val)

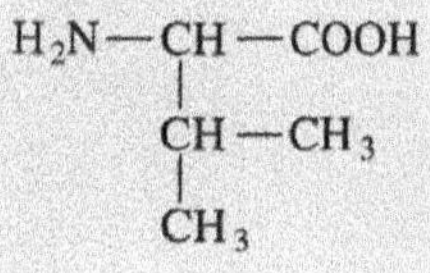

$H_2N-CH-COOH$
$|$
$CH-CH_3$
$|$
CH_3

Basic amino acids   ### Acidic amino acids

Arginine (Arg)

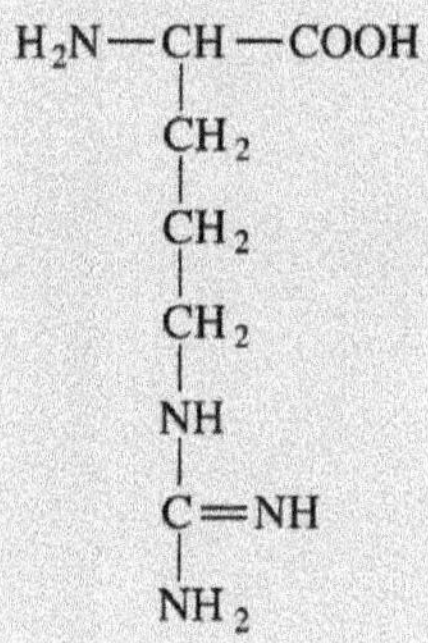

$H_2N-CH-COOH$
$|$
CH_2
$|$
CH_2
$|$
CH_2
$|$
NH
$|$
$C=NH$
$|$
NH_2

Histidine (His)

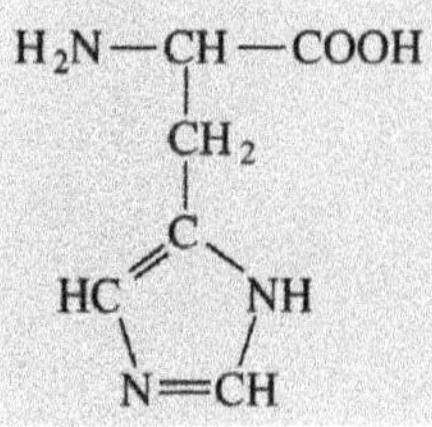

$H_2N-CH-COOH$
$|$
CH_2
$|$
C
$HC \quad NH$
$N=CH$

Lysine (Lys)

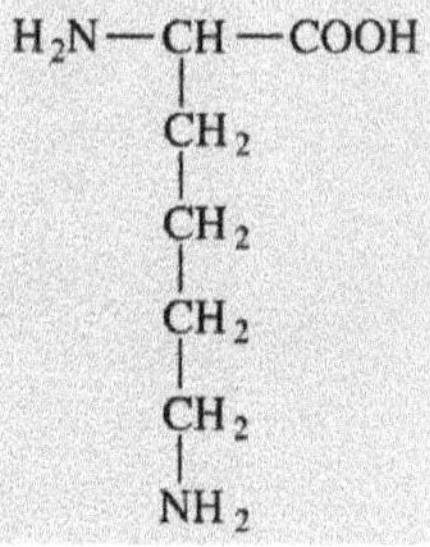

$H_2N-CH-COOH$
$|$
CH_2
$|$
CH_2
$|$
CH_2
$|$
CH_2
$|$
NH_2

Aspartic acid (Asp)

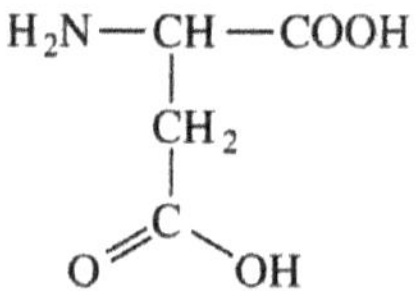

$H_2N-CH-COOH$
$|$
CH_2
$|$
C
$O \quad OH$

Glutamic acid (Glu)

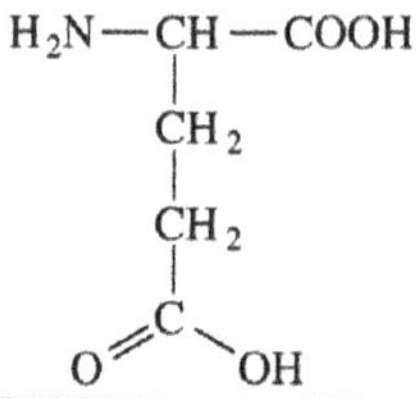

$H_2N-CH-COOH$
$|$
CH_2
$|$
CH_2
$|$
C
$O \quad OH$

The hydrogen-bonding attractions in the two strands occur between the bases, which are always paired so as to facilitate hydrogen-bond formation. Thymine is always paired with adenine, and guanine is always paired with cytosine. Therefore, in a sample of DNA, the amounts of adenine and thymine are always equal, and the amounts of guanine and cytosine are always equal.

The sequence of the bases in DNA is used by cells to specify the sequence of amino acids in proteins. The sequence of bases is the determinant of all inherited traits. Each set of three adjacent bases is called a *codon*, because the triplet of bases codes for a specific amino acid in a growing protein chain. The *genetic code* describes the compilation of all of the triplets used in the coding process and the amino acids for which they code. The DNA molecule transmits its information to RNA molecules, which serve as the template for protein synthesis.

There is currently a considerable amount of research studying the action of viruses. Viruses are not living organisms but use living organisms as hosts to reproduce. They are extremely small parasites, consisting of a protein shell surrounding strands of genetic material. Once inside a living host, the virus releases its DNA into the cells of the host, taking over the protein-making apparatus of the host cells and forcing them to reproduce more viruses. The host cells soon become overwhelmed by the viruses and die. Scientists are studying ways in which pieces of foreign DNA might be incorporated into the DNA of living cells to produce desirable results. For instance, DNA that codes for insulin production could be spliced into the DNA of persons with diabetes. The technology to produce these advances in medicine currently exists, and humankind will soon have the knowledge to produce astounding breakthroughs in our understanding of how to use chemistry to substantially improve the living conditions of the human race.

Strategies for Solving Problems in Chapter 18

Overview of Problem Types: What You Should Be Able to Do

Chapter 18 provides an introduction to the types of large organic compounds that are known as polymers. The types of problems that you should be able to solve after mastering Chapter 18 are:

1. Draw the polymer formed from a given alkene monomer.

2. Draw the polymer formed from combining a carboxylic acid or acid chloride with an amine.

3. Draw the monomer(s) used to synthesize a given alkene polymer.

Flowcharts

Flowchart 18.1 Drawing the polymer formed from a given alkene monomer

Flowchart 18.1

Summary
Method for drawing the polymer formed from a given alkene monomer.

Step 1
- **Write the alkene monomer and note the two groups connected to each carbon in the alkene.**

↓

Step 2
- **Connect pairs of alkene carbons by converting double bonds to single bonds (and with the two groups attached to each carbon intact).**

Example 1

Draw three repeating units of the polymer from the monomer shown below.

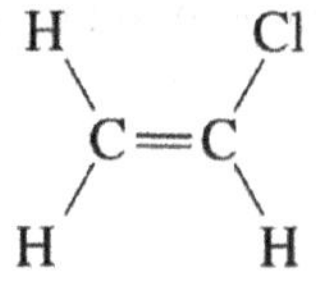

Solution

Perform Step 1
The structure of the alkene is:

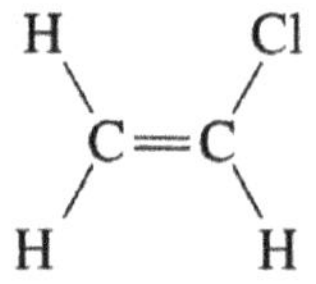

Perform Step 2

Connecting the carbons from two monomers, and converting the double bond to single bonds, we obtain the following polymer:

$$\cdots -\underset{\underset{\text{H}}{|}}{\overset{\overset{\text{H}}{|}}{\text{C}}}-\underset{\underset{\text{H}}{|}}{\overset{\overset{\text{Cl}}{|}}{\text{C}}}-\underset{\underset{\text{H}}{|}}{\overset{\overset{\text{H}}{|}}{\text{C}}}-\underset{\underset{\text{H}}{|}}{\overset{\overset{\text{Cl}}{|}}{\text{C}}}-\underset{\underset{\text{H}}{|}}{\overset{\overset{\text{H}}{|}}{\text{C}}}-\underset{\underset{\text{H}}{|}}{\overset{\overset{\text{Cl}}{|}}{\text{C}}}- \cdots$$

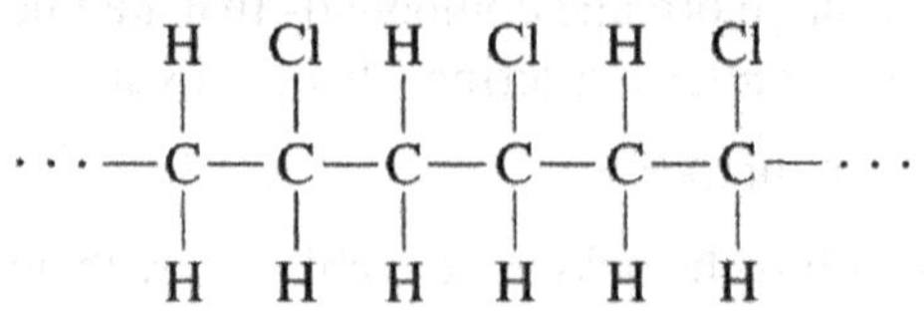

Example 2

Draw three repeating units of the polymer from the monomer shown below.

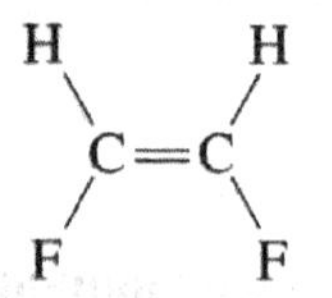

Solution

Perform Step 1

The structure of the alkene is:

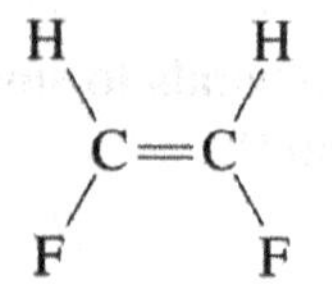

Perform Step 2

Connecting the carbons from two monomers, and converting the double bond to single bonds, we obtain the following polymer:

$$\cdots -\underset{\underset{\text{F}}{|}}{\overset{\overset{\text{H}}{|}}{\text{C}}}-\underset{\underset{\text{F}}{|}}{\overset{\overset{\text{H}}{|}}{\text{C}}}-\underset{\underset{\text{F}}{|}}{\overset{\overset{\text{H}}{|}}{\text{C}}}-\underset{\underset{\text{F}}{|}}{\overset{\overset{\text{H}}{|}}{\text{C}}}-\underset{\underset{\text{F}}{|}}{\overset{\overset{\text{H}}{|}}{\text{C}}}-\underset{\underset{\text{F}}{|}}{\overset{\overset{\text{H}}{|}}{\text{C}}}- \cdots$$

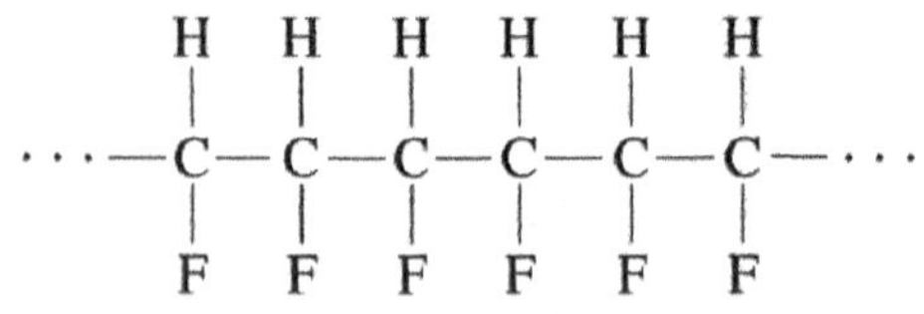

Practice Problems

18.1 Draw three repeating units of the polymer from the monomer shown below.

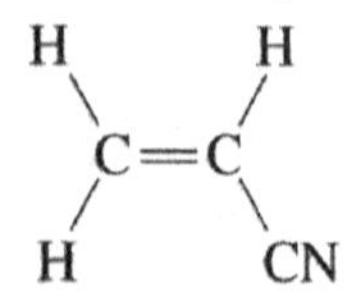

18.2 Draw three repeating units of the polymer from the monomer shown below.

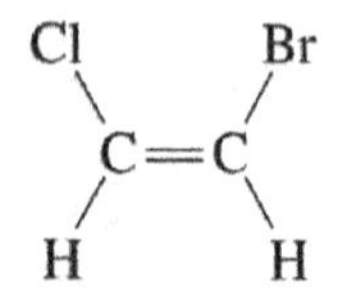

Flowchart 18.2 Drawing the polymer formed from combining a carboxylic acid or acid chloride with an amine

Flowchart 18.2

Summary

Method for drawing the polymer formed from combining a carboxylic acid or acid chloride with an amine

Step 1
- Remove the atom connected to the C=O (the OH in a carboxylic acid or the Cl in an acid chloride).
- Remove an H connected to the N in the amine.
- These groups combine to form water or HCl.

↓

Step 2
- Form the amide bond by making a bond between the C from the carboxylic acid (or acid chloride) and the N from the amine.

Example 1

Draw three repeating units of the polymer formed from the combination of the following monomers:

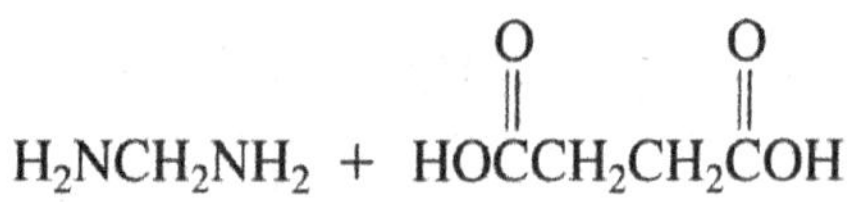

$$H_2NCH_2NH_2 \ + \ HOCCH_2CH_2COH$$
$$(\text{each C=O})$$

Solution

Perform Step 1

Removing the —OH groups from the C=O and an H from the NH_2 (to form water) we obtain

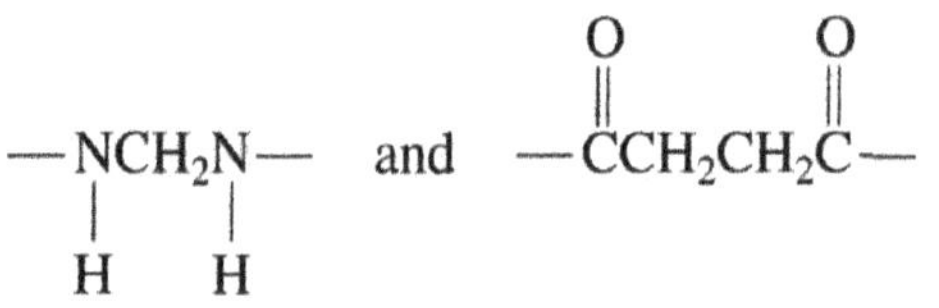

$$-NCH_2N- \quad \text{and} \quad -CCH_2CH_2C-$$

Perform Step 2

Connecting the N atoms with the C=O, we get the polymer

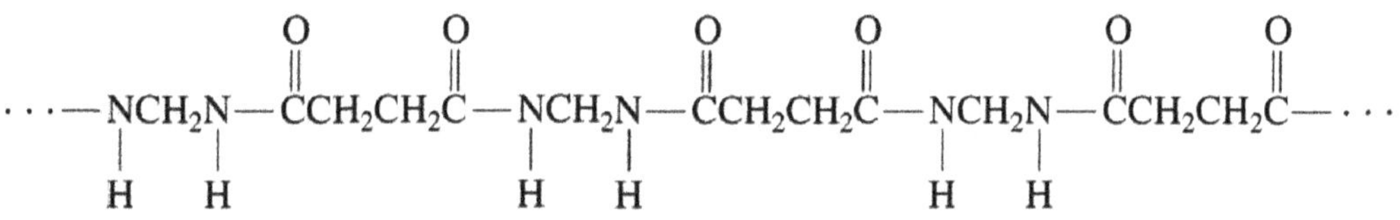

$$\cdots-NCH_2N-CCH_2CH_2C-NCH_2N-CCH_2CH_2C-NCH_2N-CCH_2CH_2C-\cdots$$

Example 2

Draw three repeating units of the polymer formed from the combination of the following monomers:

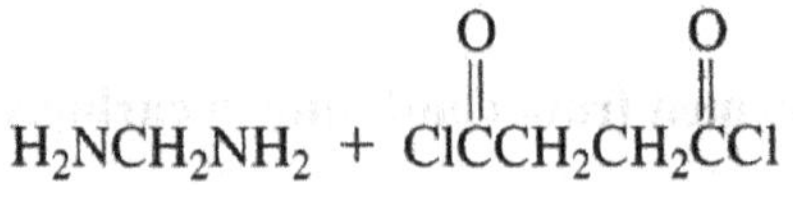

$$\text{H}_2\text{NCH}_2\text{NH}_2 \ + \ \overset{\overset{\displaystyle O}{\|}}{\text{ClCCH}_2\text{CH}_2}\overset{\overset{\displaystyle O}{\|}}{\text{CCl}}$$

Solution

Perform Step 1

Removing the Cl atoms from the $C{=}O$ and an H from the NH_2 (to form HCl) we obtain:

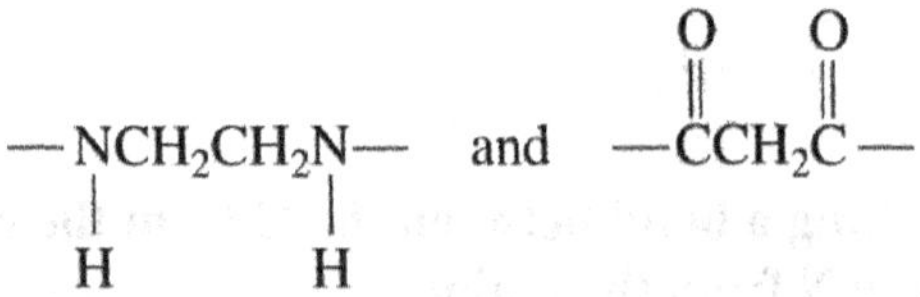

$$-\underset{\underset{\displaystyle H}{|}}{\text{NCH}_2\text{CH}_2}\underset{\underset{\displaystyle H}{|}}{\text{N}}- \quad \text{and} \quad -\overset{\overset{\displaystyle O}{\|}}{\text{C}}\text{CH}_2\overset{\overset{\displaystyle O}{\|}}{\text{C}}-$$

Perform Step 2

Connecting the N atoms with the $C{=}O$, we get the polymer:

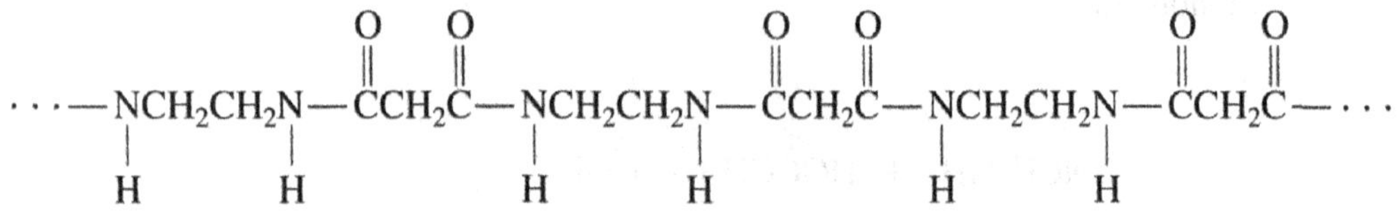

$$\cdots-\underset{\underset{\displaystyle H}{|}}{\text{NCH}_2\text{CH}_2}\underset{\underset{\displaystyle H}{|}}{\text{N}}-\overset{\overset{\displaystyle O}{\|}}{\text{C}}\text{CH}_2\overset{\overset{\displaystyle O}{\|}}{\text{C}}-\underset{\underset{\displaystyle H}{|}}{\text{NCH}_2\text{CH}_2}\underset{\underset{\displaystyle H}{|}}{\text{N}}-\overset{\overset{\displaystyle O}{\|}}{\text{C}}\text{CH}_2\overset{\overset{\displaystyle O}{\|}}{\text{C}}-\underset{\underset{\displaystyle H}{|}}{\text{NCH}_2\text{CH}_2}\underset{\underset{\displaystyle H}{|}}{\text{N}}-\overset{\overset{\displaystyle O}{\|}}{\text{C}}\text{CH}_2\overset{\overset{\displaystyle O}{\|}}{\text{C}}-\cdots$$

Example 3

Draw three repeating units of the polymer formed from the combination of units of the following monomer:

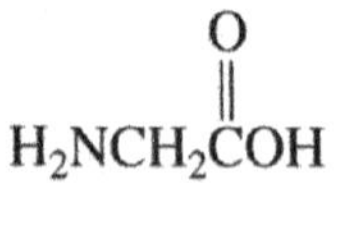

$$\text{H}_2\text{NCH}_2\overset{\overset{\displaystyle O}{\|}}{\text{C}}\text{OH}$$

Solution

Perform Step 1

Removing the $-\text{OH}$ groups from the $C{=}O$ and an H from the NH_2 we obtain:

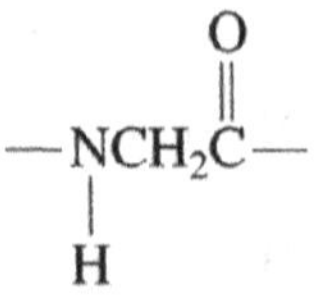

$$-\underset{\underset{\displaystyle H}{|}}{\text{NCH}_2}\overset{\overset{\displaystyle O}{\|}}{\text{C}}-$$

Perform Step 2

Connecting the N atoms with the C=O, we get the polymer:

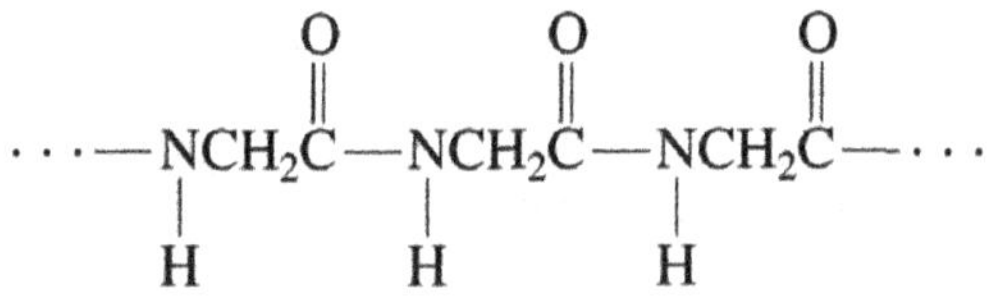

$$\cdots-\underset{\underset{H}{|}}{N}CH_2\overset{\overset{O}{\|}}{C}-\underset{\underset{H}{|}}{N}CH_2\overset{\overset{O}{\|}}{C}-\underset{\underset{H}{|}}{N}CH_2\overset{\overset{O}{\|}}{C}-\cdots$$

Practice Problems

18.3 Draw three repeating units of the polymer formed from the combination of units of the following monomer:

$$H_2NCH_2CH_2CH_2\overset{\overset{O}{\|}}{C}OH$$

18.4 Draw three repeating units of the polymer formed from the combination of the following monomers:

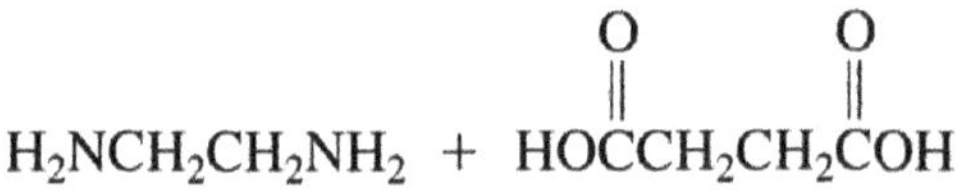

$$H_2NCH_2CH_2NH_2 \ + \ HO\overset{\overset{O}{\|}}{C}CH_2CH_2\overset{\overset{O}{\|}}{C}OH$$

Flowchart 18.3 Drawing the monomer used to synthesize a given alkene polymer

Flowchart 18.3

Summary

Method for drawing the monomer used to synthesize a given alkene polymer

Step 1
• **Identify the repeating units in the polymer.**

↓

Step 2
• **Draw one of these repeating units placing a double bond between the carbon atoms.**

Example 1

Write the monomer used to produce the following polymer:

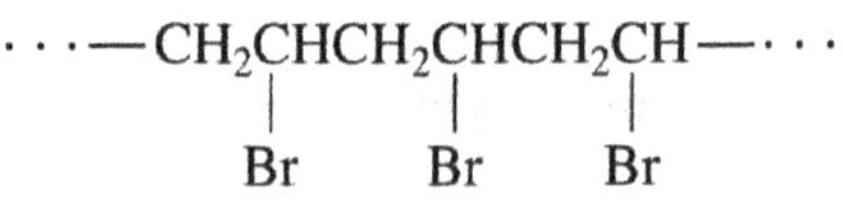

$$\cdots-CH_2\underset{\underset{Br}{|}}{C}HCH_2\underset{\underset{Br}{|}}{C}HCH_2\underset{\underset{Br}{|}}{C}H-\cdots$$

Solution

Perform Step 1

The repeating unit is

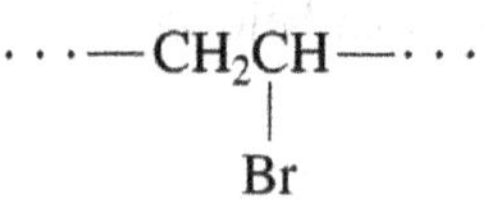

$$\cdots-CH_2\underset{\underset{Br}{|}}{C}H-\cdots$$

Perform Step 2

Putting a double bond between the two carbons, we obtain $CH_2\!\!=\!\!CHBr$ as the monomer used to produce the polymer.

Example 2

Write the monomer used to produce the following polymer.

$$\cdots-CF_2CHCF_2CHCF_2CH-\cdots$$
$$\quad\quad\;\; | \quad\quad | \quad\quad |$$
$$\quad\quad\;\; F \quad\quad F \quad\quad F$$

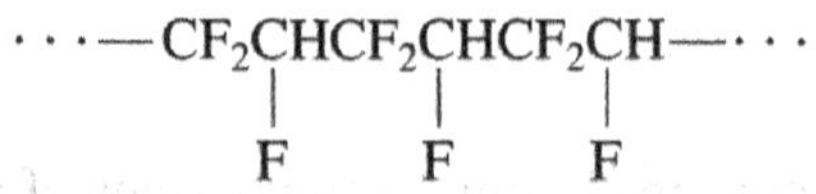

Solution

Perform Step 1

The repeating unit is:

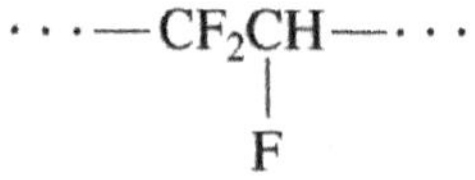

$$\cdots-CF_2CH-\cdots$$
$$\quad\quad\; |$$
$$\quad\quad\; F$$

Perform Step 2

Putting a double bond between the two carbons, we obtain $CF_2\!\!=\!\!CHF$ as the monomer used to produce the polymer.

Practice Problems

18.5 Write the monomer used to produce the following polymer:

$$\cdots-CF_2CHCF_2CHCF_2CH-\cdots$$
$$\quad\quad\;\; | \quad\quad\; | \quad\quad\; |$$
$$\quad\quad\;\; Br \quad\;\; Br \quad\;\; Br$$

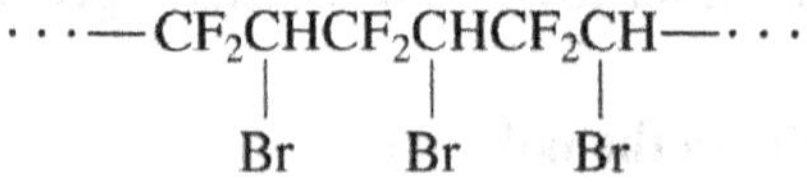

18.6 Write the monomer used to produce the following polymer:

$$\cdots-CHCHCHCHCHCH-\cdots$$
$$\quad\;\; | \; | \; | \; | \; | \; |$$
$$\quad\; Br\; Br\; Br\; Br\; Br\; Br$$

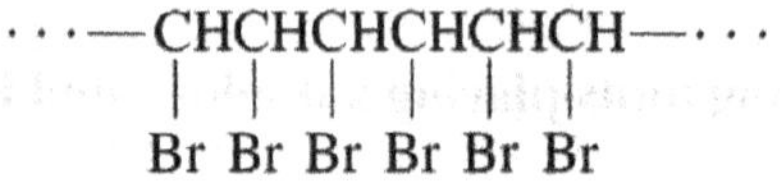

18.7 Write the monomer used to produce the following polymer:

$$\cdots-CH_2CHCH_2CHCH_2CH-\cdots$$
$$\quad\quad\quad | \quad\quad\; | \quad\quad\; |$$
$$\quad\quad\quad CH_3 \quad CH_3 \quad CH_3$$

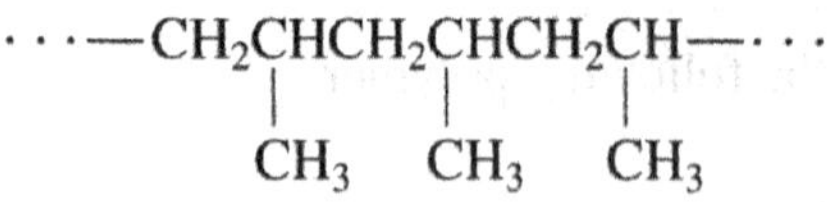

18.8 Write the monomer used to produce the following polymer:

$$\cdots-CH_2CHCH_2CHCH_2CH-\cdots$$

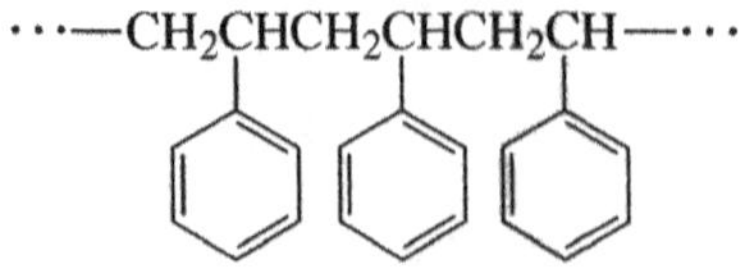

Quiz for Chapter 18 Problems

1. Draw two repeating units of the polymer formed from the monomer below:

$$HC\!\!=\!\!CH$$
$$\;\; | \quad\; |$$
$$\;\; I \quad\; I$$

2. Draw the monomer that was used to make the following polymer:

$$\cdots - \underset{\underset{Br}{|}}{\overset{\overset{H}{|}}{C}} - \underset{\underset{F}{|}}{\overset{\overset{H}{|}}{C}} - \underset{\underset{Br}{|}}{\overset{\overset{H}{|}}{C}} - \underset{\underset{F}{|}}{\overset{\overset{H}{|}}{C}} - \underset{\underset{Br}{|}}{\overset{\overset{H}{|}}{C}} - \underset{\underset{F}{|}}{\overset{\overset{H}{|}}{C}} - \cdots$$

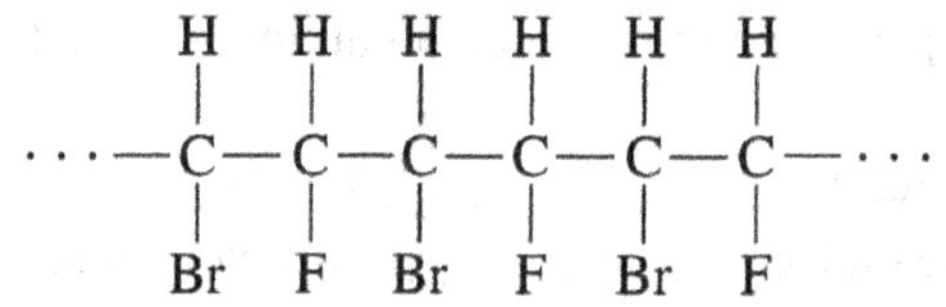

3. Fill in the blank in the following sentence:

The bond between the carbon from a carbonyl group and the nitrogen from an amine is called a(*n*) _____ bond.

4. Draw a monomer that can be used to make the following polymer:

$$\cdots - NH - \underset{\underset{CH_3}{|}}{CH} - \overset{\overset{O}{\|}}{C} - NH - \underset{\underset{CH_3}{|}}{CH} - \overset{\overset{O}{\|}}{C} - NH - \underset{\underset{CH_3}{|}}{CH} - \overset{\overset{O}{\|}}{C} - \cdots$$

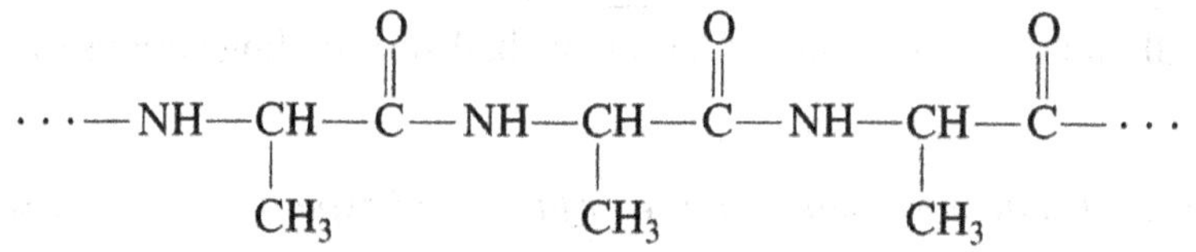

5. Fill in the blank in the following sentence:

When a carboxylic acid combines with an amine to form a polymer, the small molecule that is eliminated in the process is _____.

6. Alanine is one of 20 amino acids that are found in proteins. The structure of alanine is shown below. Write the structure of the polymer that would be formed from a combination of alanine units.

$$\underset{H_2NCHCOH}{\overset{\overset{H_3C \quad O}{| \quad \|}}{}}$$

7. Write the structure of the monomer that would be used to produce the following polymer:

$$\cdots - CH_2Cl_2CH_2Cl_2CH_2Cl_2 - \cdots$$

8. Fill in the blanks in the sentence below.

When an alkene polymer is formed from a monomer unit, the _____ in each monomer is converted to a _____ in the polymer.

9. Draw two repeating units of the polymer that is formed from the combination of units of the monomer shown below:

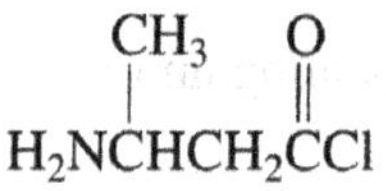

$$\underset{H_2NCHCH_2CCl}{\overset{\overset{CH_3 \quad O}{| \quad \|}}{}}$$

10. Complete the following sentence:

When the polymer is formed from the monomer in Question 9, the small molecule that is eliminated when the bond is formed is _____.

Exercises for Self-Testing

A. Completion. *Write the correct word(s) to complete each statement below.*

1. The polymer produced from chloroethene (vinyl chloride) is _____ _____, abbreviated as _____.

2. The monomer used to produce teflon is _____.

3. Glucose is a monosaccharide that exists in both _____ and _____ forms.
4. The bond between the carbon of a carbonyl group and the nitrogen on an amine is called a(n) _____ bond.
5. The amino acids in protein molecules are connected by _____ bonds.
6. Amino acids that must be supplied by the diet are called _____ amino acids.
7. The four bases in the DNA molecule are _____, _____, _____, and _____.
8. A sequence of three DNA bases that is used to specify a particular amino acid in protein synthesis is called a _____.
9. The two strands of the DNA molecule are known as the _____ _____ and are held together by _____ _____.
10. _____ are small parasites consisting of a protein shell surrounding strands of genetic material.

B. True/False. *Indicate whether each statement below is true or false, and explain why the false statements are false.*

_____ 1.   When polyethylene is formed from ethylene, all carbon–carbon double bonds are broken, and the polymer contains only single bonds.
_____ 2.   Ethane can be used as a monomer in polymerization.
_____ 3.   Amino acids that the body manufactures are called essential amino acids.
_____ 4.   Amide linkages and peptide linkages are structurally the same.
_____ 5.   Guanine and thymine are paired in the DNA double helix.
_____ 6.   Protein molecules are generally smaller than polypeptides.
_____ 7.   The monomer units for starch and cellulose are different monosaccharides.
_____ 8.   The components of the DNA backbone are phosphates and bases.
_____ 9.   Viruses are small parasitic living organisms that cause disease.
_____ 10. The bases in DNA are attached to the DNA backbone by covalent bonds.

C. Questions and Problems

1. Using polyethylene as an example, explain the difference between a monomer and a repeating unit.
2. Describe the mechanism by which the DNA molecule serves as the genetic blueprint for protein synthesis.
3. Draw the structure of the glucose molecule in its linear and cyclic forms.
4. Draw the monomer used to make the following polymer. Identify the repeating unit in the polymer, and indicate the name for the linkages.

$$\cdots - NH - \underset{\underset{CH_3}{|}}{CH} - \overset{\overset{O}{\|}}{C} - NH - \underset{\underset{CH_3}{|}}{CH} - \overset{\overset{O}{\|}}{C} - NH - \underset{\underset{CH_3}{|}}{CH} - \overset{\overset{O}{\|}}{C} - \cdots$$

5. Describe the mechanism by which viruses kill cells of living organisms.
6. Which pair of complementary bases, C & G, or A & T, would you expect to bind more strongly? Explain your answer.

Answers to Workbook Exercises and Problems

Chapter 1: What is Chemistry?

Practice Exercises

1.1 P_4 and H_2 are elements. SO_3, $C_6H_{12}O_6$, and NaCl are compounds.

1.2 (a) the Atlantic Ocean—heterogeneous mixture
(b) a cup of salt water—homogeneous mixture
(c) tincture of iodine—homogeneous mixture
(d) a bowl of cereal and milk—heterogeneous mixture
(e) a cloud in the sky—heterogeneous mixture

1.3 (b) explosion of a firecracker and (d) burning a piece of paper are not physical changes because new substance are produced as a result of the change

1.4 (a) reactivity with oxygen is the only chemical property listed

Practice Problems

1.1 Heterogeneous mixture

1.2 Elemental substance

1.3 Compound

1.4 Homogeneous mixture

1.5 Chemical

1.6 Physical

1.7 Physical

1.8 Chemical

Quiz

1. (a) Heterogeneous mixture
(b) Homogeneous mixture
(c) Heterogeneous mixture
(d) Elemental substance

2. (a) Physical
(b) Chemical
(c) Physical

3. New substance

4. (a) Compound
    (b) Elemental substance
    (c) Compound
    (d) Elemental substance

5. (a) Physical
    (b) Chemical
    (c) Chemical
    (d) Physical

6. Homogeneous mixtures are mixtures whose composition is constant throughout. Heterogeneous mixtures have variable composition throughout the mixture.

7. The milk would begin to develop a strong sour odor. It would also appear to be very clumpy. These changes indicate that a new substance has been produced; thus a chemical transformation must have taken place.

8. (Answers may vary.)
    (a) Chlorine gas
    (b) Carbon dioxide
    (c) The air we breathe
    (d) Chocolate-chip ice cream

9. False. Fizzing seltzer water contains not only water but also carbon dioxide, which produces the fizz.

10. Chemical

Exercises for Self-Testing

A. Completion

    1. homogeneous mixture   **2.** pure substances   **3.** matter, changes

    4. atom   **5.** element, compound   **6.** condensation, freezing

    7. sublimation   **8.** physical   **9.** products, reactants   **10.** bias

B. True/False

1, 3, 7, 8, and 10 are true.

2 is false because a fizzing soft drink contains a liquid and a gas. The composition of the mixture is not constant throughout.

4 is false because P_4 contains only one kind of atom (P) and is therefore an element.

5 is false because chemical properties can be observed only by causing the substance to undergo a chemical change.

6 is false because S is the symbol for sulfur. Na is the symbol for sodium.

9 is false. Because scientists are human beings, bias sometimes enters into the execution of the scientific method, affecting the results obtained.

C. Questions

    1. Elemental substances are made up of only one type of atom, whereas compounds are made up of two or more types of atoms. Examples of elemental substances are hydrogen (H_2), carbon (C), and silver (Ag). Examples of compounds are calcium fluoride (CaF_2), sodium hydroxide (NaOH), and glucose ($C_6H_{12}O_6$).

2. When the two metals are melted, mixed together, and then allowed to solidify, a solid solution is produced. The composition is constant throughout the mixture, making it a homogeneous mixture. If the two solids are ground into powders and mixed together, a tiny sample taken from one section of the mixture could differ from a tiny sample taken from another section. The mixture does not have constant composition throughout and is therefore a heterogeneous mixture.

3. Sublimation is the change of state from a solid to a gas (or vapor). Carbon dioxide (dry ice) sublimes under normal atmospheric conditions.

4. Akin to condensation being a reverse change of state to vaporization, sublimation is expected to have a reverse change of state. In sublimation, a solid becomes vapor, while in the reverse process called deposition, the vapor returns to the solid state.

5. Experiments are used to obtain data that can be summarized to postulate new laws.

6. Many correct answers are possible here. Biases might cause a scientist to selectively report only those data that support his or her biases, or to ignore the results of other scientists with whom he or she disagrees. Individual examples will vary.

Chapter 2: The Numerical Side of Chemistry

Practice Exercises

2.1 (a) 3.23   (b) 0.03   (c) both

(d) Many correct answers are possible. One possible set of measurements is 5.01, 5.02, and 5.01.

(e) Many correct answers are possible. One possible set of measurements is 3.23, 4.23, and 2.26.

2.2 (a) ± 0.001 (b) ± 0.01 (c) ± 0.1 (d) ± 1

2.3 (a) 2, the leading zeros (zeros before the 6) are not significant.
(b) 6, all digits are significant since the zeros are between nonzero digits.
(c) 3, 4, or 5 depending on where the uncertainty lies.
(d) 6, the trailing zeros are significant since they are after the decimal point.

2.4 (a) 6.78×10^{-1} (b) 5.7×10^3 (c) 7.9345×10^7 (d) 9.2×10^{-6}

2.5 (a) 0.694   (b) $0.000\,000\,461 \times 10^{-7}$ (c) 57,800   (d) 654,200,000

2.6 (a) 3.87   (b) 0.3   (c) 0.0338

2.7 (a) 473 mL = 0.473 L = 0.000 473 kL = 4.73×10^{-4} kL   (b) smaller

2.8 (a) 27 °C   (b) 250. °F

2.9 2.85 mL

2.10 (a) 955 cm   (b) 87 m   (c) 3,000,000 mL

2.11 $T = \dfrac{-G - H}{S}$

2.12 $k = \dfrac{1}{tA}$

2.13 7.31×10^3 cal

2.14 0.178 J/g·°C

Practice Problems

2.1 2

2.2 8

2.3 5

2.4 4

2.5 4

2.6 ambiguous, likely only 2

2.7 4

2.8 ambiguous, likely only 1.

2.9 3

2.10 0.006 74

2.11 6740

2.12 8,342,100,000

2.13 0.000 035 8

2.14 8.3×10^7

2.15 7.8200×10^{-4}

2.16 5.03×10^2

2.17 4.67×10^{-1}

2.18 2.9×10^1

2.19 6×10^{-4}

2.20 5.86×10^{-2}

2.21 3.9×10^{22}

2.22 937

2.23 5.175×10^{-2}

2.24 63

2.25 13.375

2.26 1.9

2.27 34

2.28 36.3

2.29 357

2.30 3.1

2.31 7.7×10^{-2}

2.32 31.3

2.33 0.0127 L of iron

2.34 1.2 mi/min

2.35 0.000 536 kg or 5.36×10^{-4} kg

2.36 250,000 cm^2 or 2.5×10^5 cm^2

2.37 \$0.17/in.^{2}

2.38 4.4×10^4 J

2.39 136 °C

Quiz

1. (a) 3
   (b) 5
   (c) 3
   (d) ambiguous, likely only 2
   (e) 7

2. (a) 2 zeros
   (b) 4 zeros
   (c) 3 zeros
   (d) 2 zeros
   (e) none

3. (a) 0.008 59
   (b) 276
   (c) 0.0276
   (d) 7,200,000,000

4. (a) 8.304×10^{-10}
   (b) 9.5×10^6
   (c) 1.3×10^{-2}
   (d) 5.83×10^1
   (e) 5.83×10^{-1}

5. (a) 51.8
   (b) 5.32
   (c) 244.8
   (d) 1583

6. (a) 5.0
   (b) 2.0
   (c) 3.0×10^4
   (d) 2.2×10^2

7. (a) 46
   (b) 2.2×10^5
   (c) 5.6
   (d) 6

8. (a) 0.0583 L or 5.83×10^{-2} L
   (b) 4.63×10^{-5} m
   (c) 8.39×10^5 g
   (d) 594 cm^2

9. 14.0 mL

10. 1.02×10^3 g

11. 305 in^3

12. \$0.21/in^2

13. 2.25×10^4 J

14. 375 g

15. 120 km/h

16. 156,000. indicates that there are six significant figures. 156,000 has only three significant figures. 156,000. assumes a more accurate measuring device.

17. 99 °F

18. (a) Less than
    (b) Equal to
    (c) Less than
    (d) Equal to
    (e) Less than

19. 19.3 g/mL

20. 356.3 K

Exercises for Self-Testing

A. Completion

1. accuracy; precision   2. last place (or last digit)   3. extensive; intensive
4. centimeter   5. nonzero digits   6. mega   7. density
8. 100 °; 212 °; 373   9. 854   10. $\dfrac{a - y}{b}$ 11. specific heat

B. True/False

1, 2, 3, 5, 7, 10, and 11 are true.

4 is false. The number has three significant digits.

6 is false. There are 5.6×10^3 mm in 5.6 m.

8 is false. A kelvin is the same size as a °C, which is larger than a °F.

9 is false. Precision means that there is good agreement among all values; more than one measurement must be taken.

C. Problems and Questions

1. (a) 5.43×10^5 mL   (b) 0.30 ft   (c) 516 months

2. (a) 21 °C   (b) 294 K

3. 1.17 g/mL

4. 2.4×10^7 mg

5. An intensive property is a property whose value does not depend on the amount of substance present. Color is an example of an intensive property. An extensive property is a property whose value depends on the amount of substance present. Length is an example of an extensive property.

6. $\dfrac{1000\ \text{L}}{1\ \text{kL}} \qquad \dfrac{1\ \text{kL}}{1000\ \text{L}}$

7. $n = PV/RT$

8. 9.06 mL

9. (a) 60.11×10^2 or 601.1×10^1 or 6.011×10^3 (b) 1.8×10^3

10. An exact value has an unlimited number of significant digits, whereas a measured value has a specific number of significant digits because there is some uncertainty in the measurement.

11. 2.69×10^4 J, or 2.69×10^1 kJ

Chapter 3: The Evolution of Atomic Theory

Practice Exercises

3.1 The atomic number is 18; the mass number is 40.

3.2 Isotope A: 27 protons, 27 electrons, and 32 neutrons.

Isotope B: 27 protons, 27 electrons, and 33 neutrons.

3.3 92 protons, 92 electrons, and 146 neutrons

3.4 107.9

3.5 $_{16}^{32}S^{2-}$

Practice Problems

3.1 15.79% C; 84.21% S

3.2 41.03% O; 58.97% Na

3.3 52.9% Al; 47.1% O

3.4 80.1%

3.5 39.948 amu

3.6 I < Br < Cl < F

3.7 I > Br > Cl > F

3.8 Sn, because it is lower in the periodic table.

Quiz

1. 3.25% H; 19.37% C; 77.38% O

2. 107.87 amu

3. Po < Pb < Tl < Cs

4. ^{39}K (38.9637 amu) will be present in the greatest amount because its atomic mass is the closest to the weighted average atomic mass of 39.098 amu.

5. Be < C < O < F < Ne

6. Tl

7. 95.71%

8. Cs > Rb > K > Na > Li

9. Xe

10. Ne > F > O > C > Be

Exercises for Self-Testing

A. Completion

1. atoms   2. plum pudding   3. alpha particles

4. protons, electrons, and neutrons   5. protons   6. nucleus, protons, neutrons

7. isotopes   8. 6, 6, 7   9. atomic, mass   10. atomic mass

11. groups (or families), periods   12. alkaline earth   13. decrease, increase

14. potassium   15. ionization potential   16. 3+   17. cations, anions

B. True/False

2, 3, 9, and 10 are true.

1 is false. Isotopes are atoms of the same element that have different numbers of neutrons.

4 is false. There are isotopes of nitrogen that contain a different number of neutrons (e.g., ^{15}N).

5 is false. Isotopes have the same number of protons but different numbers of neutrons.

6 is false. The elements are arranged according to increasing atomic number.

7 is false. Anions are negative ions, which contain more electrons than protons.

8 is false. Both calcium atom and its 2+ cation contain 20 protons in the nucleus. The calcium ion would contain 18 electrons.

C. Questions and Problems

1. The statement is false because isotopes are atoms of the same element that have different numbers of neutrons.

2. (a) He (b) K (c) Na (d) Fe (e) Ag (f) Au

3. (a) 85 protons, 85 electrons, 115 neutrons
   (b) 47 protons, 47 electrons, 61 neutrons
   (c) 7 protons, 7 electrons, 8 neutrons
   (d) 11 protons, 11 electrons, 12 neutrons

4. (a) $^{14}_{8}C$ (b) $^{3}_{1}H$

5. (a) electron (b) neutron (c) proton

6. Atomic mass of bromine = 79.9

7. F < O < N < P < As

   The atomic radii decrease going from left to right across a period, and increase going from top to bottom down a group in the periodic table.

8. F > O > N > P > As

9. (a) 26 protons, 30 neutrons, 24 electrons
   (b) 26 protons, 30 neutrons, 23 electrons
   (c) 47 protons, 60 neutrons, 46 electrons
   (d) 47 protons, 62 neutrons, 46 electrons

10. (a) 7 protons, 7 neutrons, 10 electrons
    (b) 34 protons, 46 neutrons, 36 electrons
    (c) 53 protons, 74 neutrons, 54 electrons
    (d) 35 protons, 45 neutrons, 36 electrons

Chapter 4: The Modern Model of the Atom

Practice Exercises

4.1 450 nm

4.2 The volume of water from the store can be obtained only in volumes of 16 oz., 32 oz., 48 oz., and so on. You cannot obtain volumes that are in between these values. However, you can obtain any volume you want from a water fountain. The volume of water obtainable from the store is quantized.

4.3 The repulsion between the negatively charged electrons would prevent smaller shells from holding as many electrons as larger shells.

4.4 12.8 eV

4.5 Red

4.6 The p subshell holds four more electrons than the s subshell. The f subshell holds four more electrons than the d subshell. Based on this pattern, an h subshell should hold 22 electrons.

4.7 6; VI

4.8 (a) $1s^2 2s^2 2p^6 3s^2 3p^6 4s^2 3d^{10} 4p^6 5s^2 4d^{10} 5p^5$ (b) $[\text{Kr}]5s^2 4d^{10} 5p^5$

4.9 (a) K^+ (b) P^{3-} (c) O^{2-} (d) Al^{3+}

4.10 (a) Ca_3P_2 (b) MgS

4.11 2 electrons; 2 electrons

Practice Problems

4.1 5.68×10^{-19} J

4.2 4.86×10^{-7} m or 486 nm

4.3 4.09×10^{-19} J

4.4 3.0 eV

4.5 1.0 eV

4.6 13.2 eV

4.7 20 electrons: $1s^2 2s^2 2p^6 3s^2 3p^6 4s^2$

4.8 34 electrons: $1s^2 2s^2 2p^6 3s^2 3p^6 4s^2 3d^{10} 4p^4$

4.9 82 electrons: $1s^2 2s^2 2p^6 3s^2 3p^6 4s^2 3d^{10} 4p^6 5s^2 4d^{10} 5p^6 6s^2 4f^{14} 5d^{10} 6p^2$

4.10 $[\text{Ne}]3s^2$

4.11 $[\text{Xe}]6s^2 4f^{14} 5d^{10} 6p^3$

4.12 $[\text{Ar}]4s^2 3d^3$

Quiz

1. 1.1 eV

2. 4.97×10^{-19} J

3. 7 electrons: $1s^2 2s^2 2p^6 3s^2 3p^6 4s^1$

4. Krypton: $[\text{Kr}]$

5. 12.8 eV

6. 2.7×10^{-19} J

7. 15 electrons: $1s^2 2s^2 2p^6 3s^2 3p^3$

8. The scaled energies are equivalent. The amount of energy released and absorbed would be 0.1 eV.

9. 48 electrons: $[Kr] 5s^2 4d^{10}$

10. 5.68×10^{-8} m or 56.8 nm

Exercises for Self-Testing

A. Completion

    1. wavelength   **2.** lower   **3.** line, continuous   **4.** quantized   **5.** Bohr

    6. primary quantum number   **7.** $2n^2$ **8.** valence shell   **9.** group number

    10. octet rule   **11.** ground-state   **12.** 2   **13.** Heisenberg uncertainty principle

    14. decreases, increases   **15.** orbitals   **16.** Rutherford   **17.** 1, 6   **18.** 3

B. True/False

1, 2, 3, 7, 10, and 12 are true.

4 is false. Only representative or main group elements can be determined from the group number. Not so with the transition elements.

5 is false. Energy is *absorbed* to move an electron to a higher energy level.

6 is false. Blue light has the shorter wavelength.

8 is false. Atoms that gain electrons become negatively charged anions.

9 is false. Atomic size decreases going from left to right.

11 is false. Ga has three valence electrons in the $n = 4$ shell.

C. Questions and Problems

1. $E = \dfrac{hc}{\lambda} = \dfrac{(6.63 \times 10^{-34}\,\text{J}\cdot\text{s})(3.00 \times 10^8\,\text{m/s})}{6.50 \times 10^{-7}\,\text{m}} = 3.06 \times 10^{-19}\,\text{J}$

   Blue light would have a shorter wavelength than orange light because it is higher-energy light.

2. See the electromagnetic spectrum in Section 4.1 of the textbook.

3. See the line spectrum of hydrogen in Section 4.1 of the textbook.

4. The Bohr diagram for a silicon atom is shown in the figure below.

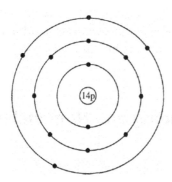

5. Energy is emitted when an electron falls from a higher to a lower state.

6. The Bohr diagrams are shown in the figure below.

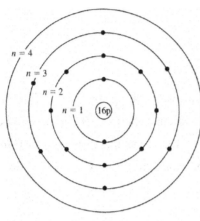

Ground state

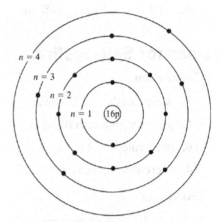

Excited state

7. (a) 1 (b) 2 (c) 3 (d) 4 (e) 5 (f) 6 (g) 7

8. $1s^2 2s^2 2p^6 3s^2 3p^6 4s^2 3d^{10} 4p^6 5s^2 4d^{10} 5p^6 6s^2 4f^{14} 5d^{10} 6p^4$ or $[Xe]6s^2 4f^{14} 5d^{10} 6p^4$ Po has six valence electrons.

9. (a) $BaCl_2$ (b) Al_2O_3 (c) Li_2O

10. (a) Sr: $1s^2 2s^2 2p^6 3s^2 3p^6 4s^2 3d^{10} 4p^6 5s^2$ or $[Kr]5s^2$
(b) Ag: $1s^2 2s^2 2p^6 3s^2 3p^6 4s^2 3d^{10} 4p^6 5s^1 4d^{10}$ or $[Kr]5s^1 4d^{10}$
(c) V: $1s^2 2s^2 2p^6 3s^2 3p^6 4s^2 3d^3$ or $[Ar]4s^2 3d^3$
(d) U: $1s^2 2s^2 2p^6 3s^2 3p^6 4s^2 3d^{10} 4p^6 5s^2 4d^{10} 5p^6 6s^2 4f^{14} 5d^{10} 6p^6 7s^2 5f^3$ or $[Rn]5f^3$

11. See the figure in Section 4.5 of the textbook.

12. In general, metals have fewer than four valence electrons, whereas nonmetals have four or more valence electrons, up to a maximum of 8.

13. (a) group number VIA (or 16); period number 6
(b) group number IIA (or 2); period number 4
(c) group number VIIA (or 7); period number 4

14. The Bohr model places the electrons in well-defined orbits around the nucleus. The quantum mechanical model places electrons in orbitals, which are regions of space around the nucleus in which it is reasonably probable that the electron would be found.

Chapter 5: Chemical Bonding and Nomenclature

Practice Exercises

5.1 (a) and (b) exist as individual molecules, whereas (c), (d), and (e) form ionic lattices.

5.2 Si atom has eight valence electrons; it has gained four electrons, one from each H atom.

5.3 (a) H_2S (b) As_2O_3

5.4 $\cdot \overset{\cdot\cdot}{\underset{\cdot}{As}} \cdot$ As has three unpaired electrons and two paired electrons. As will form bonds with three atoms.

5.5 $:\overset{\cdot\cdot}{\underset{\cdot\cdot}{Br}}:\overset{\cdot\cdot}{\underset{\cdot\cdot}{S}}:\overset{\cdot\cdot}{\underset{\cdot\cdot}{Br}}:$

5.6 $H:\overset{\cdot\cdot}{N}:\overset{\cdot\cdot}{\underset{\cdot\cdot}{F}}:$
$\qquad\ \ :\overset{}{\underset{\cdot\cdot}{F}}:$

5.7
$$H-C=C-H$$
$$\quad\ |\quad\ |$$
$$\quad\ H\quad H$$

5.8 $H-\overset{\cdot\cdot}{N}=\overset{\cdot\cdot}{N}-H$

5.9 These dot diagrams *are* resonance structures, because only the positions of the electrons have changed.

5.10 $\left[\ \begin{array}{c} :\overset{\cdot\cdot}{O}: \\ :\overset{\cdot\cdot}{\underset{\cdot\cdot}{O}}:\overset{\cdot\cdot}{\underset{\cdot\cdot}{S}}:\overset{\cdot\cdot}{\underset{\cdot\cdot}{O}}: \end{array}\ \right]^{2-}$

5.11 (a) and (d) are ionic, whereas (b), (c), and (e) are polar covalent.

For the polar covalent bonds:

(b) $\overset{\delta+\ \ \delta-}{H-Br}$ (c) $\overset{\delta+\ \ \delta-}{Si-C}$ (e) $\overset{\delta+\ \ \delta-}{B-I}$

5.12 (a) barium oxide   (b) calcium nitride
    (c) potassium oxide   (d) magnesium fluoride

5.13 (a) chromium(III) chloride   (b) copper(II) sulfide
    (c) iron(III) bromide   (d) manganese(II) fluoride

5.14 (a) magnesium nitrate   (b) aluminum nitrite
    (c) sodium bicarbonate   (d) iron(II) hydroxide

5.15 (a) silicon dioxide   (b) phosphorus triiodide
    (c) carbon tetrabromide   (d) boron trichloride

Practice Problems

5.1 $H:\overset{\cdot\cdot}{\underset{\cdot\cdot}{O}}:\overset{\cdot\cdot}{\underset{\cdot\cdot}{O}}:H$

5.2 $H:\overset{\cdot\cdot}{\underset{}{C}}::\overset{\cdot\cdot}{\underset{}{C}}:H$
(with H above each C)

5.3 $\left[\ :\overset{\cdot\cdot}{\underset{\cdot\cdot}{O}}:\overset{\cdot\cdot}{\underset{}{Cl}}:\overset{\cdot\cdot}{\underset{\cdot\cdot}{O}}:\ \right]^{-}$
$\qquad\quad :\overset{}{\underset{\cdot\cdot}{O}}:$

5.4 Barium chloride

5.5 Cesium nitride

5.6 Potassium iodide

5.7 Cobalt(II) iodide

5.8 Cuprous fluoride

5.9 Mercury(II) bromide

5.10 Diboron trioxide

5.11 Nitrogen trifluoride

5.12 Xenon tetrafluoride

5.13 Titanium(III) fluoride

5.14 Nickelous oxide

5.15 Tin(IV) chloride

5.16 Chlorous acid

5.17 Perbromic acid

5.18 Hydrofluoric acid

Quiz

1. (a) $:\overset{\cdot\cdot}{\underset{\cdot\cdot}{Cl}}:\overset{\cdot\cdot}{P}:\overset{\cdot\cdot}{\underset{\cdot\cdot}{Cl}}:$
   $\quad\quad:\overset{\cdot\cdot}{\underset{\cdot\cdot}{Cl}}:$

   (b) $:\overset{\cdot\cdot}{\underset{\cdot\cdot}{O}}::\overset{\cdot\cdot}{S}:\overset{\cdot\cdot}{\underset{\cdot\cdot}{O}}:$

2. (a) $\left[\begin{array}{c} H \\ H:\overset{\cdot\cdot}{N}:H \\ H \end{array}\right]^{+}$

   (b) $\left[:\overset{\cdot\cdot}{\underset{\cdot\cdot}{O}}::\overset{\cdot\cdot}{N}:\overset{\cdot\cdot}{\underset{\cdot\cdot}{O}}:\right]^{-}$

3. (a) Calcium phosphide
   (b) Iron(III) oxide

4. (a) Sulfuric acid (*Note:* this was given in a table in the text.)
   (b) Nitrous acid (*Note:* this was given in a table in the text.)

5. 2+, 1+

6. (a) Magnesium nitride
   (b) Nitrous acid
   (c) Hypochlorous acid

7. (a) 32 valence electrons
   (b) 32 valence electrons
   (c) 8 valence electrons
   (d) 18 valence electrons

8. (d) Hydrogen

9. (b) $CoCl_2$, (d) $HgCl_2$, and (e) $CrCl_2$

10. Nitrogen and oxygen form more than one binary molecular compound. If there are no prefixes in the name to indicate the number of nitrogens and oxygens in the formula, it is impossible to tell whether the compound's formula is NO, N_2O, NO_2, N_2O_5, or something else.

Exercises for Self-Testing

A. Completion

1. molecule   2. valence electrons   3. 2, 3   4. double bond

5. resonance structures   6. 2, 2   7. increases, decreases   8. fluorine

9. B, A   10. ionic*   11. CO_3^{2-} 12. molecules, ionic lattice

13. polar covalent   14. iron(II) oxide   15. binary covalent, ionic

*If the two elements are nonmetals, the bond will be polar covalent.

B. True/False

3, 5, 6, and 10* are true.

1 is false. Compounds are very different substances from the elements from which they are made, and will generally have properties that are very different.

2 is false. Polar covalent compounds (which are covalent since they involve shared electrons) are made of elements that share electrons unequally.

4 is false. The Lewis dot diagram for N will have two paired electrons and three unpaired electrons.

7 is false. Hydrogen reacts chemically to obtain two valence electrons. Noble gases don't react at all, for the most part.

8 is false. $MgCl_2$ is an ionic compound and will therefore exist in an ionic lattice.

9 is false. The correct name is calcium chloride.

*Even bonds with $\Delta EN < 0.4$ have some ionic character and are therefore somewhat polar; their polarity, however, is not significant. Many chemists consider them nonpolar covalent (e.g., C—H bonds in organic molecules).

C. Questions and Problems

1. (a) is ionic; (b) is (nonpolar) covalent, whereas (c) and (d) are polar covalent. The electronegativity difference is greater than 1.8 for CaO, and between 0.4 and 1.8 for the compounds in (b), (c), and (d). What makes (b) nonpolar even though it contains polar covalent bonds, is its tetrahedral symmetry (see Chapter 6).

2. (a) magnesium nitride   (b) diphosphorus pentoxide
   (c) ammonium bromide   (d) iron(III) oxide

3. H_2 < HI < HBr < HCl < HF

4. The charge on the ions determines the number of each ion present. Because the charges on the ions are fixed by the group number (or the formula of the polyatomic ion), there is no need to specify the number of each ion with Greek prefixes.

5. $\left[H \!:\! \overset{..}{\underset{H}{O}} \!:\! H \right]^{+}$

6. $:\!\overset{..}{\underset{..}{Cl}}\!:\!B\!:\!\overset{..}{\underset{..}{Cl}}\!:$ $:\!\overset{..}{\underset{..}{Cl}}\!:\!P\!:\!\overset{..}{\underset{..}{Cl}}\!:$
   $\quad\;:\!\overset{..}{\underset{..}{Cl}}\!:$ $\quad\;:\!\overset{..}{\underset{..}{Cl}}\!:$

   They would not have similar chemical properties. B is electron deficient; P has a lone pair.

7. (a) Li_2CO_3 (b) FeI_3 (c) $(NH_4)_2SO_4$ (d) $NaC_2H_3O_2$ (e) BCl_3

8. $H - \overset{\displaystyle H}{\underset{\displaystyle H}{\overset{|}{\underset{|}{C}}}} - C \equiv N :$

9. (a) diphosphorus trisulfide   (b) potassium phosphate
   (c) cobalt(III) sulfate   (d) xenon tetrafluoride

10. Sodium chloride is an ionic compound. It exists in an ionic lattice and is not composed of discrete molecules.

Chapter 6: The Shape of Molecules

Practice Exercises

6.1 V-shaped or bent, with bond angles of approximately 118°.

6.2 (a) $:\!\overset{..}{O}\!=\!C\!=\!\overset{..}{O}\!:$ (b) PH$_3$ structure   (c) $:\!\overset{..}{\underset{..}{F}}\!-\!\overset{..}{\underset{..}{F}}\!:$ (d) CH$_3$I structure

Nonpolar   Polar   Nonpolar   Polar

6.3 $SiHCl_3$, SO_2, and NH_3 experience dipole–dipole intermolecular attractive forces.

6.4 (a) ClF_3 is polar and will experience dipole–dipole intermolecular attractions
   (b) SeF_6 is nonpolar
   (c) $SeCl_4$ is polar and will experience dipole–dipole intermolecular attractions
   (d) SbF_5 is nonpolar

Practice Problems

6.1 Bent; $105°$ (two lone pairs)

6.2 Trigonal planar; $120°$

6.3 Tetrahedral; $109.5°$

6.4 Linear; $180°$

6.5 Linear; $180°$

6.6 PI_3 is a polar molecule; the net dipole moment is in the direction of the iodine atoms.

6.7 ONCl is a polar molecule. The shape is bent with the net dipole in the direction of the oxygen atom.

6.8 No net dipole exists. BCl_3 is a trigonal planar molecule; the bond dipole moments cancel each other.

Quiz

1. Trigonal planar; $120°$

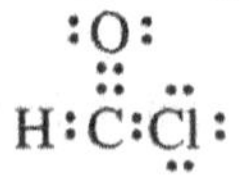

2. (a) Linear

(b) Trigonal planar

(c) Tetrahedral

3. Pyramidal

4. (a) Nonpolar

(b) Polar; the molecular dipole moment is in the direction of the fluorine atom.

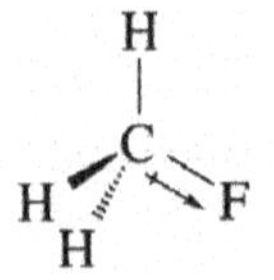

(c) Polar; the molecular dipole moment is in between the fluorine atoms.

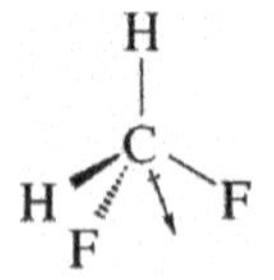

(d) Polar; the molecular dipole moment is in the direction of the fluorine atoms.

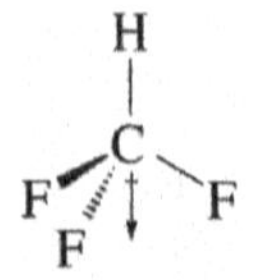

(e) Nonpolar

5. $C-Br < C-H < C-Cl < C-O < C-F$

6. False. When one or more lone pairs are present, the molecule will be pyramidal or bent.

7. Linear

8. BF_3, SO_3

9. Tetrahedral, bent

10. For the molecular shape with bond angles of ~105°:

(a) Four electron groups

(b) Two bonding pairs

(c) Two lone pairs

For the molecular shape with bond angles of ~118°:

(a) Three electron groups

(b) Two bonding pairs

(c) One lone pair

Exercises for Self-Testing

A. Completion

1. valence shell electron pair repulsion   **2.** V-shaped (or bent)   **3.** polar covalent

4. ionic   **5.** three   **6.** dipole–dipole   **7.** trigonal planar

8. three, one   **9.** lone pairs   **10.** polar

B. True/False

3, 4, 7, 8, and 9 are true.

1 is false. Molecules with polar bonds may be nonpolar if the geometry of the molecule is such that all of the individual bond dipoles cancel each other.

2 is false. A pyramidal molecule has four electron groups around the central atom, one of which is a lone pair.

5 is false. Hydrogen can form only single bonds with its one valence electron.

6 is false. The CH_2Cl_2 molecule is tetrahedral.

10 is false. *Intra*molecular forces (covalent bonds) are usually much stronger than *inter*molecular attractive forces.

C. Questions and Problems

1. (a) tetrahedral   (b) V-shaped (or bent)   (c) linear   (d) V-shaped (or bent)

2. (a) polar   (b) polar   (c) nonpolar   (d) nonpolar

3. (a) $\left[:\ddot{O} - \underset{\displaystyle \cdots}{N} \diagup\kern-1.2em\diagdown \ddot{O}: \right]^{-}$ (b) $:\ddot{O} = C = \ddot{O}:$ (c) $:\ddot{F} - \overset{\displaystyle \cdots}{\underset{\displaystyle |}{As}} - \ddot{F}:$
    $\qquad\qquad\qquad\qquad\qquad\qquad\qquad\qquad\qquad\qquad :\ddot{F}:$

4. The substance that is a liquid at room temperature will have larger intermolecular attractive forces than the substance that is a gas at room temperature because more energy (a higher temperature) is required to break the molecules apart.

5. The OCS molecule has a larger dipole moment than CO_2. The individual bond dipoles in OCS do not cancel each other because S and O have different electronegativities. The individual bond dipoles cancel in CO_2, making it a nonpolar molecule.

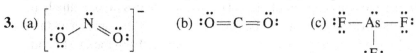

$\qquad\quad$ Net dipole$\qquad\qquad$ No net dipole

6. Linear; trigonal bipyramidal (2 bonds, 3 lone pairs); octahedral (2 bonds, 4 lone pairs).

Chapter 7: Intermolecular Forces and the Phases of Matter

Practice Exercises

7.1 (a) Added. Water droplets (fog) are being evaporated.
    (b) Removed. Liquid water is being converted to a solid.
    (c) Added. Liquid is being converted to a gas.

7.2 (a) London forces and dipole–dipole forces
    (b) London forces and dipole–dipole forces
    (c) London forces only
    (d) London forces only
    (e) London forces and dipole–dipole forces

7.3 (a), (b), and (c) will form hydrogen bonds.

7.4 $F_2 < CH_3CH_2CH_3 < CH_3F < H_2O < C_9H_{19}OH$

7.5 (a) molecular solid   (b) metal   (c) network solid   (d) ionic solid

Practice Problems

7.1 Dipole–dipole forces and London forces

7.2 Hydrogen bonding, dipole–dipole forces, and London forces

7.3 London forces only

7.4 CH_3OH has the higher boiling point because in addition to dipolar forces and London forces, it contains hydrogen bonding.

7.5 $CH_3CH_2CH_2CH_3 < CH_3CH_2OCH_3 < CH_3CH_2CH_2NH_2$

7.6 Even though HF forms smaller molecules than HCl, HF has the higher boiling point because of the presence of hydrogen bonds.

Quiz

1. Dipole–dipole forces and London forces

2. CH_3OCH_3 has the boiling point of 30 °C because it contains dipole–dipole forces and London forces. CH_3CH_2OH has the boiling point of 78 °C because in addition to dipolar forces and London forces, it contains hydrogen bonding from the O–H bond.

3. (a) CH_4
    (b) CH_3CH_2Cl

4. (a) London forces
    (b) Dipole–dipole forces and London forces
    (c) Hydrogen bonding, dipolar forces, and London forces

5. Higher

6. (a) $CH_3CH_2CH_2CH_3$ because it's larger and therefore has greater London forces.
    (b) CH_3CH_2OH because it has greater London forces.

7. CH_3NH_2

8. (b) CH_2Cl_2 and (c) CH_3OH

      STUDENT WORKBOOK AND SELECTED SOLUTIONS

9. $CH_4 < CH_3CH_3 < CH_3F < CH_3CH_2Cl < CH_3OH$

10. CH_3OH. Although both involve hydrogen bonding, oxygen is more electronegative than nitrogen. Therefore, the O$\cdots$H hydrogen bond is stronger than the N$\cdots$H hydrogen bond.

Exercises for Self-Testing

A. Completion

1. liquid, solid   2. London forces   3. nitrogen, oxygen, fluorine

4. sublimation   5. hydrogen bonding   6. stronger   7. O, N, F

8. stronger   9. nonmolecular   10. metallic

B. True/False

1, 2, 5, 7, and 9 are true.

3 is false. London forces are present in all molecules.

4 is false. Molecules with polar bonds will be nonpolar if the individual dipole moments cancel due to the geometry of the molecule.

6 is false. A very large molecule involving only London forces will have a higher boiling point than a very small molecule involving hydrogen bonding.

8 is false. The covalent bonds (*intra*molecular forces) holding network solids together are stronger than the *inter*molecular attractive forces holding the molecules of molecular solids together.

10 is false. All compounds that involve dipole–dipole forces also involve London forces.

C. Questions and Problems

1. Ethyl alcohol; in addition to being more polar than diethyl ether, it can participate in hydrogen bonding.

2. (a) The molecules of a liquid are not held in fixed positions but are free to move about the liquid. Hence, the molecules of a liquid will move to fit the shape of its container.
   (b) The molecules in a gas are spaced far apart and can be pushed closer together. The molecules in a liquid are in contact with each other and cannot be forced closer together.
   (c) The molecules of a gas feel little attraction for each other and will move rapidly throughout the volume of their container. The molecules of liquids and solids are held in contact with each other by attractive forces, and are not free to escape to fill the volume of their container.

3. The molecules of *n*-pentane, C_5H_{12}, will have the higher boiling point. Linear molecules have stronger London forces than more spherically shaped molecules because the linear molecules have greater contact with each other and thus will experience the intermolecular attractive forces to a greater extent.

4. (a) London forces and dipole–dipole forces
   (b) London forces
   (c) London forces, dipole–dipole forces, and hydrogen bonding
   (d) London forces
   (e) London forces and dipole–dipole forces

5. CCl_4 is a tetrahedral molecule in which the individual bond dipoles cancel each other, resulting in a net dipole of 0.

6. The strongest intermolecular attractive force operating between H_2O molecules is hydrogen bonding, whereas the strongest intermolecular attractive force operating in H_2S is dipole–dipole forces. Hydrogen bonding is a particularly strong form of dipole–dipole forces. Therefore, hydrogen bonding is a stronger force than dipole–dipole interactions that do not involve hydrogen bonding.

7. Liquid water must absorb energy to be converted to the gas state. When perspiration evaporates from the skin, the liquid perspiration is removing heat from the skin in order for the sweat to evaporate. The removal of heat from the skin causes the skin to feel cooler.

8. Atoms of helium can only interact with one another via London forces. At very low temperatures (4 K and below) these very weak forces among the tiny helium atoms cause the helium gas to condense.

9. F_2, Cl_2, Br_2, and I_2 are all diatomic, nonpolar molecules. The only attractive forces in which they can participate are London forces. The strength of London forces increases with the size of the molecules (which correlates with the total number of electrons). I_2 is the largest and, therefore, has the strongest London forces, to the extent that, even at room temperature, I_2 occurs as a solid.

10. CO_2 is a nonpolar molecule. Molecules of CO_2 can interact with one another using London forces only. When the temperature of solid CO_2 (dry ice) increases, the molecules of CO_2 attain enough kinetic energy for these forces to suddenly break and release CO_2 molecules into the gas phase, skipping the liquid state.

Chapter 8: Chemical Reactions

Practice Exercises

8.1 $Mg + I_2 \rightarrow MgI$

8.2 $4PH_3 + 9O_2 \rightarrow P_4O_{12} + 6H_2O$

8.3 (a) double replacement
    (b) decomposition
    (c) single replacement
    (d) combination

8.4 (a) is not a precipitation reaction.
    (b) is a precipitation reaction. $Mg(OH)_2$ is the precipitate.
    (c) is a precipitation reaction. $CaCO_3$ is the precipitate.

8.5 Balanced molecular equation: $Pb(NO_3)_2 + 2NaCl \rightarrow PbCl_2 + 2NaNO_3$
Ionic equation: $Pb^{2+} + 2NO_3^- + 2Na^+ + 2Cl^- \rightarrow PbCl_2(s) + 2Na^+ + 2NO_3^-$
Net ionic equation: $Pb^{2+} + 2Cl^- \rightarrow PbCl_2(s)$

8.6 Balanced molecular equation: $2HCl + Mg(OH)_2 \rightarrow MgCl_2 + 2H_2O$
Ionic equation: $2H^+ + 2Cl^- + Mg^{2+} + 2OH^- \rightarrow Mg^{2+} + 2Cl^- + 2H_2O$
Net ionic equation: $2H^+ + 2OH^- \rightarrow 2H_2O$ which reduces to $H^+ + OH^- \rightarrow H_2O$

Practice Problems

8.1 $NCl_3 + 3H_2O \rightarrow NH_3 + 3HOCl$

8.2 $2Al_2O_3 \rightarrow 4Al + 3O_2$

8.3 $PCl_5 + 4H_2O \rightarrow H_3PO_4 + 5HCl$

8.4 Balanced equation: $2NaOH(aq) + MgCl_2(aq) \rightarrow Mg(OH)_2(s) + 2NaCl(aq)$

Ionic equation:
$2Na^+(aq) + 2OH^-(aq) + Mg^{2+}(aq) + 2Cl^-(aq) \rightarrow Mg(OH)_2(s) + 2Na^+(aq)$
$+ 2Cl^-(aq)$

Net ionic equation:
$2OH^-(aq) + Mg^{2+}(aq) \rightarrow Mg(OH)_2(s)$

8.5 Balanced equation: $K_3PO_4(aq) + Al(NO_3)_3(aq) \rightarrow AlPO_4(s) + 3KNO_3(aq)$

Ionic equation:
$3K^+(aq) + PO_4^{3-}(aq) + Al^{3+}(aq) + 3NO_3^-(aq) \rightarrow AlPO_4(s) + 3K^+(aq) + 3NO_3^-(aq)$

Net ionic equation:
$PO_4^{3-}(aq) + Al^{3+}(aq) \rightarrow AlPO_4(s)$

8.6 Molecular neutralization reaction equation: $HBr(aq) + KOH(aq) \rightarrow KBr(aq) + H_2O(l)$

Ionic neutralization reaction equation:
$H^+(aq) + Br^-(aq) + K^+(aq) + OH^-(aq) \rightarrow K^+(aq) + Br^-(aq) + H_2O(l)$

Net ionic equation: $H^+(aq) + OH^-(aq) \rightarrow H_2O(l)$

8.7 Molecular neutralization reaction equation:
$2HNO_3(aq) + Ba(OH)_2(aq) \rightarrow Ba(NO_3)_2(aq) + 2H_2O(l)$

Ionic neutralization reaction equation:
$2H^+(aq) + 2NO_3^-(aq) + Ba^{2+}(aq) + 2OH^-(aq) \rightarrow Ba^{2+}(aq) + 2NO_3^-(aq) + 2H_2O(l)$

Net ionic equation:
$2H^+(aq) + 2OH^-(aq) \rightarrow 2H_2O(l)$, which reduces to $H^+(aq) + OH^-(aq) \rightarrow H_2O(l)$

8.8 Molecular neutralization reaction equation:
$H_2SO_4(aq) + 2NaOH(aq) \rightarrow Na_2SO_4(aq) + 2H_2O(l)$

Ionic neutralization reaction equation:
$2H^+(aq) + SO_4^{2-}(aq) + 2Na^+(aq) + 2OH^-(aq) \rightarrow 2Na^+(aq) + SO_4^{2-}(aq) + 2H_2O(l)$

Net ionic equation:
$2H^+(aq) + 2OH^-(aq) \rightarrow 2H_2O(l)$, which reduces to $H^+(aq) + OH^-(aq) \rightarrow H_2O(l)$

Quiz

1. $C_{10}H_{20} + 15O_2 \rightarrow 10CO_2 + 10H_2O$
2. $4PH_3 + 9O_2 \rightarrow P_4O_{12} + 6H_2O$; the coefficient is 4.
3. $MgCO_3$ and $CaSO_4$ are insoluble in water.
4. $CaCO_3$
5. Li^+, Cl^-, Ba^{2+}, NO_3^-

6. Molecular equation:

$$Na_2SO_4(aq) + Pb(NO_3)_2(aq) \rightarrow PbSO_4(s) + 2NaNO_3(aq)$$

Ionic equation:

$$2Na^+(aq) + SO_4^{2-}(aq) + Pb^{2+}(aq) + 2NO_3^-(aq) \rightarrow PbSO_4(s) + 2Na^+(aq) + 2NO_3^-(a$$

Net ionic equation:

$$Pb^{2+}(aq) + SO_4^{2-}(aq) \rightarrow PbSO_4(s)$$

7. Salt, water

8. Molecular equation: $2NH_4I(aq) + Pb(NO_3)_2(aq) \rightarrow PbI_2(s) + 2NH_4NO_3(aq)$

Ionic equation:

$$2NH_4^+(aq) + 2I^-(aq) + Pb^{2+}(aq) + 2NO_3^-(aq) \rightarrow PbI_2(s) + 2NH_4^+(aq) + 2NO_3^-(aq)$$

Net ionic equation:

$$Pb^{2+}(aq) + 2I^-(aq) \rightarrow PbI_2(s)$$

9. $CaSO_4$ is insoluble in water.

10. Neutralization

11. Coefficient, subscripts

Exercises for Self-Testing

A. Completion

1. chemical reactions   2. energy, energy

3. hydrogen, nitrogen, oxygen, fluorine, chlorine, bromine, iodine   4. new substances

5. $H^+ + OH^- \rightarrow H_2O$ 6. single replacement   7. precipitate   8. spectator ions

9. 8   10. phosphorus

B. True/False

2, 3, 5, 8, and 10 are true.

1 is false. All halides except halides of lead(II), silver(I), and mercury(I) are soluble in water.

4 is false. Only the coefficients can be changed to balance an equation. Changing subscripts will change the chemical identity of a substance.

6 is false. The number and type of atoms on both sides of the equation must be equal, but the coefficients are usually different.

7 is false. If the element is diatomic the number of atoms of the element will be twice the value of its coefficient.

9 is false. Although many reactions fit one of these categories, there are many that do not.

C. Questions and Problems

1. $MgCO_3$

2. $3Cl_2 + NH_3 \rightarrow NCl_3 + 3HCl$

3. (b) and (c)

4. Balanced molecular equation: $NH_4I + AgNO_3 \rightarrow AgI(s) + NH_4NO_3$

Ionic equation: $\cancel{NH_4^+} + I^- + Ag^+ + \cancel{NO_3^-} \rightarrow AgI(s) + \cancel{NH_4^+} + \cancel{NO_3^-}$

Net ionic equation: $Ag^+ + I^- \rightarrow AgI(s)$

5. Balanced molecular equation: $HBr + LiOH \rightarrow LiBr + H_2O$

   Ionic equation: $H^+ + Br^- + Li^+ + OH^- \rightarrow Li^+ + Br^- + H_2O$

   Net ionic equation: $H^+ + OH^- \rightarrow H_2O$

6. (b), (c), and (d)

7. (a), (b), and (c)

8. The moment the two clear solutions are added together, a yellow plume of solid PbI_2 will form and start settling at the bottom of the container. The net ionic equation is: $Pb^{2+}(aq) + 2I^-(aq) \rightarrow PbI_2(s)$.

Chapter 9: Stoichiometry and the Mole

Practice Exercises

9.1 342.15 g/mol

9.2 3.0 moles H_2O

9.3 2.8 moles NH_4Cl

9.4 7.5×10^4 g H_2O

9.5 72.15 g of C_5H_{12}, 255.99 g of O_2, 220.05 g of CO_2, and 108.09 g of H_2O

9.6 517 g NH_3

9.7 70.9%

9.8 226 g PBr_3

9.9 $C_3H_6O_2$

9.10 $C_8H_{10}N_4O_2$

9.11 46.646% N, 6.713% H, 20.000% C, and 26.641% O

9.12 C_2H_4O

Practice Problems

9.1 70.56 g NH_3

9.2 10.20 mol NaOH

9.3 225.2 g H_2O

9.4 61.8 g CO_2

9.5 8.15 g HCl

9.6 $C_8H_{20}Pb$

9.7 CH

9.8 $C_6H_6O_2$

9.9 $C_{23}H_{46}$

9.10 C_6H_{12}

9.11 Sn: 61%; F: 39%

9.12 Mg: 16.4%; N: 18.9%; O: 64.7%

Quiz

1. 2243.1 g CaO

2. 393.8 g H_2O

3. C_3H_6O

4. AlF_3

5. $C_6H_{12}O_6$

6. 969.6 g N_2

7. 3598 g S

8. Molecular, empirical

9. C: 72.35%; H: 13.88%; O: 13.77%

10. No, she is not correct. The limiting reactant is the reactant that has the smallest mole-to-coefficient ratio.

11. 2357 g $SiCl_4$

12. 833.5 g SiO_2

13. C_3H_3O

14. 6

15. Cl_2

16. 5.64 moles of HCl; 205.64 g of HCl.

Exercises for Self-Testing

A. Completion

1. stoichiometry　　2. Avogadro's number, 6.022×10^{23}　　3. 149.08 g/mol

4. 862 g　　5. 1.1×10^4　　6. theoretical yield　　7. limiting reactant　　8. 16

9. actual yield, theoretical yield, 100　　10. combustion analysis

11. empirical formula　　12. 36.11, 63.89　　13. 8　　14. 2.912 g

15. molar mass

B. True/False

1, 3, 5, 8, and 10 are true.

2 is false. Carbon must form four bonds to satisfy the octet rule. In CH_2, C atom would only have four valence electrons around it (review Chapter 6).

4 is false. The number of molecules in 1 mole of *any* substance is 6.022×10^{23}.

6 is false. The grams must be converted to moles, and the coefficients taken into account before the limiting reactant can be identified. It is often the case that there are more grams of the limiting reactant present than there are of the excess reactant(s).

7 is false. There are 2×10^{22} molecules in 10 g of O_2, and 6×10^{23} molecules in 2 g of H_2.

9 is false. The empirical formula can be determined from combustion analysis data. The molar mass is needed to determine the actual formula.

11 is false. The maximum yield of NH_3 would be 10.7 g.

12 is false. The maximum % yield is 100%.

C. Questions and Problems

1. (a) 92.7 g of CH_4 (b) 480 g of CH_4

2. (a) $3Cl_2 + NH_3 \rightarrow NCl_3 + 3HCl$ (b) 235 g of NCl_3
   (c) 17 g of NH_3 will be left over.

3. (a) NH_3O (b) NH_3O

4. (a) Yes, 0.48 g of oxygen   (b) C_3H_3O

5. (a) 9×10^{25} molecules H_2 (b) 6×10^{24} molecules N_2
   (c) 7×10^{23} molecules I_2

6. (a) 1.8×10^{24} molecules H_2 (b) 1.8×10^{24} molecules N_2
   (c) 1.8×10^{24} molecules I_2

7. 55.97%

8. The theoretical yield is usually larger than the actual yield because some of the product may be lost in the isolation of the product from the rest of the reaction mixture. Also, sometimes competing reactions convert some of the reactants to other products.

9. The empirical formula is the simplest formula and shows the *ratio* of the elements present in the compound. The actual formula shows the actual composition of the compound. Sometimes the actual composition of the compound is the same as the simplest formula, in which case the empirical formula and the actual formula are the same.

10. 56.34% carbon, 10.40% hydrogen, and 33.26% chlorine

11. (b) and (c). In both cases, the subscripts can be further simplified.

12. CH. Other compounds: C_2H_2 (acetylene), C_4H_4 (1, 2, 3-butatriene).

Chapter 10: Electron Transfer in Chemical Reactions

Practice Exercises

10.1 In NO_3^-: N owns zero electrons. There are five valence electrons on the free nitrogen atom. Each O owns eight electrons. There are six valence electrons on the free oxygen atom.

In PF_3: P owns two electrons. There are five valence electrons on the free phosphorus atom. Each F owns eight electrons. There are seven valence electrons on the free fluorine atom.

10.2 P has an oxidation state of +3; H has an oxidation state of +1; Cl has an oxidation state of −1.

10.3 (a) $C = -3; H = +1$ (b) $N = +5; O = -2$
   (c) $K = +1; Mn = +7; O = -2$ (d) $H = +1; O = -1$
   (e) $Ga = +3; O = -2$

10.4 (a) and (b) are both redox reactions.

10.5 Si is oxidized, and Cl is reduced.

10.6 Na is oxidized; Cl is reduced. Na is the reducing agent; Cl_2 is the oxidizing agent.

10.7

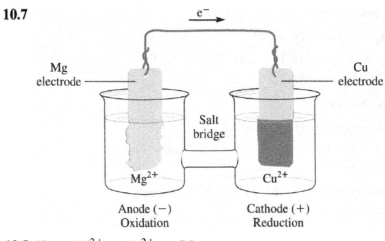

Anode (−)   Cathode (+)
Oxidation   Reduction

10.8 $Zn + Cd^{2+} \rightarrow Zn^{2+} + Cd$

10.9 (a) Pb   (b) Zn   (c) Cu

Practice Problems

10.1

Molecule	Dot diagram
CH_3F	H H:C:H :F:

Valence electrons on free atom	Electrons owned by each atom
H: 1 electron	H: 0 electrons
C: 4 electrons	C: 6 electrons
F: 7 electrons	F: 8 electrons

Oxidation state	Sum
H: $1 - 0 = +1$	H: $+1 \times 3 = +3$
C: $4 - 6 = -2$	C: $-2 \times 1 = -2$
F: $7 - 8 = -1$	F: $-1 \times 1 = -1$
	$= 0$

10.2

Molecule	Dot diagram
H_2O	H:O:H

Valence electrons on free atom	Electrons owned by each atom
H: 1 electron	H: 0 electrons
O: 6 electrons	O: 8 electrons

Oxidation state	Sum
H: $1 - 0 = +1$	H: $+1 \times 2 = +2$
O: $6 - 8 = -2$	O: $-2 \times 1 = -2$
	$= 0$

10.3

Molecule	Dot diagram
ClO_4^-	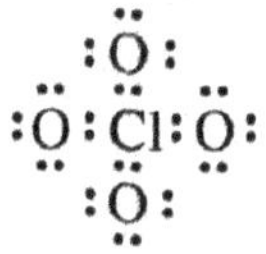

Valence electrons on free atom	Electrons owned by each atom
Cl: 7 electrons	Cl: 0 electrons
O: 6 electrons	O: 8 electrons

Oxidation state	Sum
Cl: $7 - 0 = +7$	Cl: $+7 \times 1 = +7$
O: $6 - 8 = -2$	O: $-2 \times 4 = -8$
	$= -1$

10.4

Molecule	Dot diagram
CO	$:C:::O:$

Valence electrons on free atom	Electrons owned by each atom
C: 4 electrons	C: 2 electrons
O: 6 electrons	O: 8 electrons

Oxidation state	Sum
C: $4 - 2 = +2$	C: $+2 \times 1 = +2$
O: $6 - 8 = -2$	O: $-2 \times 1 = -2$
	$= 0$

10.5 H: $+1$ (rule 3)

F: -1 (halide rule)

C: -2 (rule 7)

Sum of oxidation states:

H: $+1 \times 3 = +3$
F: $-1 \times 1 = -1$
C: $-2 \times 1 = -2$
$= 0$

10.6 H: +1 (rule 3)

S: −2 (rule 7)

Sum of oxidation states:

$$
\begin{array}{r}
\text{H: } +1 \times 2 = +2 \\
\underline{\text{S: } -2 \times 1 = -2} \\
= \ 0
\end{array}
$$

10.7 O: −2 (rule 2)

Cl: +7 (rule 7)

Sum of oxidation states:

$$
\begin{array}{r}
\text{O: } -2 \times 4 = -8 \\
\underline{\text{Cl: } +7 \times 1 = +7} \\
= -1
\end{array}
$$

10.8 O: −2 (rule 2)

C: +2 (rule 7)

Sum of oxidation states:

$$
\begin{array}{r}
\text{O: } -2 \times 1 = -2 \\
\underline{\text{C: } +2 \times 1 = +2} \\
= \ 0
\end{array}
$$

10.9 O: −2 (rule 2)

C: +3 (rule 7)

Sum of oxidation states:

$$
\begin{array}{r}
\text{O: } -2 \times 4 = -8 \\
\underline{\text{C: } +3 \times 2 = +6} \\
= -2
\end{array}
$$

10.10 $CaCl_2 + Na_2O \rightarrow NaCl + CaO$ is not a redox reaction:

Total charge:	$+2 - 2 = 0$	$+2 - 2 = 0$	$+1 - 1 = 0$	$+2 - 2 = 0$
Oxidation no.:	$+2 - 1$	$+1 - 2$	$+1 - 1$	$+2 - 2$

$$CaCl_2 \ + \ Na_2O \ \rightarrow \ NaCl \ + \ CaO$$

10.11 $2Al_2O_3 \rightarrow 4Al + 3O_2$ is a redox reaction:

Total charge:	$+6 - 6 = 0$	0	0
Oxidation no.:	$+3 - 2$	0	0

$$2Al_2O_3 \ \rightarrow \ 4Al \ + \ 3O_2$$

10.12 $Co + Cu^{2+} \rightarrow Co^{2+} + Cu$

10.13 $Zn + Sn^{2+} \rightarrow Zn^{2+} + Sn$

Quiz

1. (a)

Molecule	Dot diagram
HI	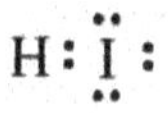 H : Ï :

Valence electrons on free atom	Electrons owned by each atom
H: 1 electron	H: 0 electrons
I: 7 electrons	I: 8 electrons

Oxidation state	Sum
H: $1 - 0 = +1$	H: $+1 \times 1 = +1$
I: $7 - 8 = -1$	I: $-1 \times 1 = -1$
	$= 0$

(b)

Molecule	Dot diagram
CH_2Cl_2	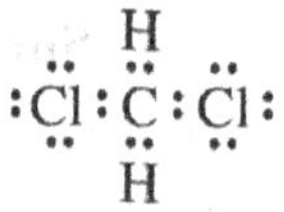 H :Cl : C : Cl: H

Valence electrons on free atom	Electrons owned by each atom
H: 1 electron	H: 0 electrons
C: 4 electrons	C: 4 electrons
Cl: 7 electrons	Cl: 8 electrons

Oxidation state	Sum
H: $1 - 0 = +1$	H: $+1 \times 2 = +2$
C: $4 - 4 = 0$	C: $0 \times 1 = 0$
Cl: $7 - 8 = -1$	Cl: $-1 \times 2 = -2$
	$= 0$

(c)

Molecule	Dot diagram
SO_3	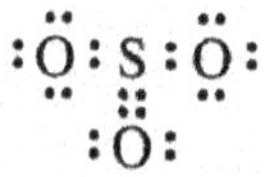 :Ö : S : Ö: :Ö:

Valence electrons on free atom	Electrons owned by each atom
S: 6 electrons	S: 0 electrons
O: 6 electrons	O: 8 electrons

Oxidation state	Sum
S: $6 - 0 = +6$	S: $+6 \times 1 = +6$
O: $6 - 8 = -2$	O: $-2 \times 3 = -6$
	$= \ 0$

(d)

Molecule	Dot diagram
C_2H_6	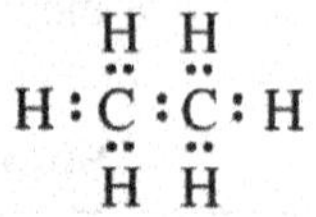 H H H : C : C : H H H

Valence electrons on free atom	Electrons owned by each atom
H: 1 electron	H: 0 electrons
C: 4 electrons	C: 7 electrons

Oxidation state	Sum
H: $1 - 0 = +1$	H: $+1 \times 6 = +6$
C: $4 - 7 = -3$	C: $-3 \times 2 = -6$
	$= \ 0$

2. (a) H: $+1$ (rule 3)
   O: -2 (rule 2)
   N: $+5$ (rule 7)
   Sum of oxidation states:
   $$H: -1 \times 1 = +1$$
   $$O: -2 \times 3 = -6$$
   $$N: +5 \times 1 = +5$$
   $$= \ 0$$

   (b) O: -2 (rule 2)
   S: $+2$ (rule 7)
   Sum of oxidation states:
   $$O: -2 \times 3 = -6$$
   $$S: +2 \times 2 = +4$$
   $$= -2$$

(c) O: −2 (rule 2)

Ca: +2 (rule 7)

S: +6 (rule 7)

Sum of oxidation states:

$$O: -2 \times 4 = -8$$
$$Ca: +2 \times 1 = +2$$
$$\underline{S: +6 \times 1 = +6}$$
$$= 0$$

3. (a) $B_2O_3 + 3Mg \rightarrow 2B + 3MgO$ is a redox reaction:

Total charge:	+6 − 6 = 0	0	0	+2 − 2 = 0
Oxidation no.:	+3 − 2	0	0	+2 − 2

$$B_2O_3 \;+\; 3Mg \;\rightarrow\; 2B + 3MgO$$

(b) $HF + KOH \rightarrow KF + H_2O$ is not a redox reaction:

Total charge:	+1 − 1 = 0	+1 − 2 + 1 = 0	+1 − 1 = 0	+2 − 2 = 0
Oxidation no.:	+1 − 1	+1 − 2 + 1	+1 − 1	+1 − 2

$$HF \;+\; KOH \;\rightarrow\; KF \;+\; H_2O$$

(c) $Zn + 2HCl \rightarrow ZnCl_2 + H_2$ is a redox reaction:

Total charge:	0	+1 − 1 = 0	+2 − 2 = 0	0
Oxidation no.:	0	+1 − 1	+2 − 1	0

$$Zn \;+\; 2HCl \;\rightarrow\; ZnCl_2 \;+\; H_2$$

4. (a) Zn is more active. The reaction is spontaneous as written.
   (b) Li is more active. The reaction is spontaneous as written.
   (c) Fe is more active. The reaction is spontaneous as written.
   (d) Fe is more active. The reaction is spontaneous in the opposite direction.

5. $Mn + Ni^{2+} \rightarrow Mn^{2+} + Ni$

6. more

7. (a) +4   (b) −3   (c) −2   (d) −1

8.

Total charge:	0	+2 + 6 − 8 = 0	+2 + 6 − 8 = 0	0
Oxidation no.:	0	+1 + 6 − 2	+2 + 6 − 2	0

$$Mg \;+\; H_2SO_4 \;\rightarrow\; MgSO_4 \;+\; H_2$$

H: $+1 \rightarrow 0$ a change

O: $-2 \rightarrow -2$ no change

Mg: $0 \rightarrow +2$ a change

S: $+6 \rightarrow +6$ no change

9. The ring must be made of gold because mercury is more active than gold. If the ring were silver, it would react with the mercury, losing electrons to the mercury ions and forming silver ions. Because gold is less active than mercury, no reaction will occur.

10. No, Cu is less active than Ni.

11. *Bookkeeping method:*

Polyatomic ion	Dot diagram
$PO_4{}^{3-}$	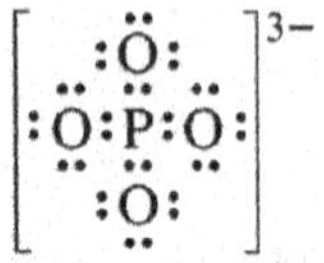 $\begin{bmatrix} :\ddot{O}: \\ :\ddot{O}:P:\ddot{O}: \\ :\ddot{O}: \end{bmatrix}^{3-}$

Valence electrons on free atom	Electrons owned by each atom
P: 5 electrons	P: 0 electrons
O: 6 electrons	O: 8 electrons

Oxidation state	Sum
P: $5 - 0 = +5$	P: $+5 \times 1 = +5$
O: $6 - 8 = -2$	O: $-2 \times 4 = -8$
	$= -3$

Shortcut method:

O: -2 (rule 2)

P: $+5$ (rule 7)

Sum of oxidation states:

O: $-2 \times 4 = -8$

P: $+5 \times 1 = +5$

$= -3$

12. (a) Mg, (b) Zn, and (d) Cr are all more active metals compared to iron and will be able to serve as sacrificial electrodes (anodes) when connected to iron.

Exercises for Self-Testing

A. Completion

1. electronegative   2. group number   3. 0   4. oxygen, more electronegative halogen

5. oxidation, reduction   6. Mg, Ag   7. salt bridge   8. gained, lost

9. more, less   10. corrosion

B. True/False

2, 4, 8, 9, 10, and 11 are true.

1 is false. This is not an electron-transfer (redox) reaction because none of the elements undergoes a change in its oxidation state when going from reactants to products.

3 is false. The oxidation number of Mn in MnO_4^- is +7.

5 is false. Oxidation occurs at the anode, and the anode is negative.

6 is false. Ag is above Au on the EMF series. Therefore, Ag will be preferentially oxidized over Au, and the reaction will be spontaneous.

7 is false. The more active metals above will reduce the ions of the less active metals below.

12 is false. The oxidation number of Fe increases from 0 to +2, so iron gets oxidized (loses electrons) and is, therefore, the reducing agent.

C. Questions and Problems

1. (a) $\begin{bmatrix} \overset{+1}{H} \\ \overset{+1}{H} \overset{..}{:} \overset{-3}{N} \overset{..}{:} H \\ \underset{+1}{H} \overset{..}{H} \underset{+1}{} \end{bmatrix}$ (b) $\overset{..}{:}\overset{0}{O}\overset{..}{=}\overset{+1}{O}-\overset{..}{\underset{-1}{O}}\overset{..}{:}$ (c) $\begin{bmatrix} \overset{-2}{:}\overset{..}{O}=\overset{+3}{\overset{..}{N}}-\overset{-2}{\overset{..}{O}}\overset{..}{:} \end{bmatrix}$ (d) $\overset{-2}{:}\overset{..}{O}:$ ... C ... $\overset{}{:}\overset{..}{\underset{-1}{Cl}}\overset{+4}{}\overset{}{\underset{-1}{Cl}}\overset{..}{:}$

2. (a) +4   (b) −3   (c) −2   (d) −1

3. Fe_2O_3 is the oxidizing agent; C is the reducing agent.

4. Magnesium is high in the electromotive force series, meaning that it is very easily oxidized to Mg^{2+} ions by other metals. Its metallic form therefore will not occur naturally. Gold, on the other hand, is low in the electromotive force series. It is not easily oxidized to its ionic form and is therefore found in the Earth's crust in its metallic form.

5.

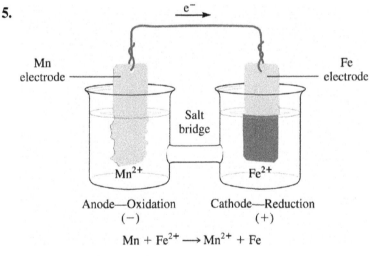

Mn electrode

Fe electrode

Salt bridge

Mn^{2+}

Fe^{2+}

Anode—Oxidation (−) Cathode—Reduction (+)

$Mn + Fe^{2+} \longrightarrow Mn^{2+} + Fe$

6. (a) $Ca + Co^{2+} \rightarrow Ca^{2+} + Co$
   (b) $Ag^+ + Ni \rightarrow Ni^{2+} + Ag$
   (c) $Mg + Mn^{2+} \rightarrow Mg^{2+} + Mn$

7. Copper is above silver in the EMF series and will be preferentially oxidized instead of silver, thereby preserving the silver metal. Copper, however, is below iron in the EMF series, so the iron will be oxidized before the copper.

8. Electrons must be produced on the anode and sent over the wire to the cathode, as it is on the cathode where they are needed for the reduction.

9. The salt bridge allows for the spectator ions, with which it is saturated, to travel to the anode solution (the anions) or to the cathode solution (the cations) and keep both solutions electrically neutral. It also prevents wholesale mixing of the two solutions.

10. Many different examples are possible. For instance:
$NaOH(aq) + HCl(aq) \rightarrow NaCl(aq) + H_2O(l)$.

All atoms in the above reaction equation maintain their oxidation numbers (Na:$+1$, O: -2, H:$+1$, Cl: -1). Since there are no changes in the oxidation numbers, no element is getting reduced or oxidized, and hence, this is not a redox reaction.

Chapter 11: What If There Were No Intermolecular Forces? The Ideal Gas

Practice Exercises

11.1 6.0 L

11.2 0.30 atm

11.3 18.7 L

11.4 6.34 atm

Practice Problems

11.1 4.19 atm

11.2 1047 K = 773 °C

11.3 0.387 mol

11.4 131 g/mol

11.5 34.0 g/mol

11.6 1.73 atm

11.7 0.794 L = 794 mL

11.8 78.5 L

11.9 1.0 mole of P_4

11.10 4.67×10^4 g of CO_2

11.11 12.6 moles of O_2

Quiz

1. 20.6 L

2. 23 mol

3. 4.17×10^3 g $CaCO_3$

4. (a) 3.40 L; they have an inversely proportional relationship.
   (b) 13.6 L; they have a directly proportional relationship.

5. 10.4 L

6. 24 °C

7. 4.86 L

8. 7.88 atm

9. (a) $2H_2O_2 \rightarrow 2H_2O + O_2$
   (b) 1.21 moles of H_2O_2

10. 5.00 L of He contains 0.41 moles of gas (a); 2.50 g of He contains 0.63 moles of gas (b); thus 2.50 g of He contains more moles than 5.00 L of He.

11. 3.14×10^3 g $KClO_3$

12. 30.1 g/mol

Exercises for Self-Testing

A. Completion

1. ideal gas   2. gases, liquids, solids   3. double   4. moles

5. 400   6. temperatures, pressures   7. 760   8. barometer

9. increases   10. increases

B. True/False

1, 2, 4, 6, and 8 are true.

3 is false. The pressure will decrease if the volume is increased.

5 is false. The number of moles of gas will not increase simply because the temperature has increased. More gas must be added to the container to increase the number of moles present.

7 is false. V and P are inversely proportional. Doubling volume would cut the pressure in half.

C. Questions and Problems

1. Air must be removed from the tire. The increase in temperature will cause the pressure to increase. The number of moles must be decreased to restore the pressure to the original value. Air must be let out of the tire to decrease the number of moles of gas present.

2. (a) The molecules of a liquid are not held in fixed positions but are free to move about the liquid. Hence, the molecules of a liquid will move to fit the shape of its container.
   (b) The molecules in a gas are spaced far apart and can be pushed closer together. The molecules in a liquid are in contact with each other and cannot be forced closer together.
   (c) The molecules of a gas feel little attraction for each other and will move rapidly throughout the volume of their container. The molecules of liquids and solids are held in contact with each other by attractive forces, and are not free to escape to fill the volume of their container.

3. 1.0×10^{23} molecules

4. (a) Increasing the temperature causes the molecules of gas to move faster. The faster the molecules move, the harder they hit the walls of the container, thereby causing the pressure to increase.

   (b) Increasing the volume means that the molecules have to travel farther to hit the walls of the container. The collisions with the walls will be less frequent, and the pressure will thus decrease.

   (c) Increasing the number of moles will result in more gas molecules hitting the walls of the container, thereby causing the pressure to increase.

5. At very low temperature, the gas molecules are moving relatively slowly and will feel the attractive forces between them more strongly. At very high pressure, the gas molecules are in relatively close proximity to each other and will feel the intermolecular attractive forces more strongly. In addition, the relatively small volumes necessary to cause very high pressures means that the volume occupied by the molecules of gas will not be negligible in relation to the volume of the container.

6. (c) Cl_2 is expected to deviate the most from the ideal gas behavior because chlorine gas is composed of the largest molecules, which are expected to exhibit the strongest intermolecular forces. (a) H_2 is expected to deviate the least from the ideal gas behavior because hydrogen gas is composed of the smallest molecules.

Chapter 12: Solutions

Practice Exercises

12.1 (a), (c), (d), and (e) are solutions.

12.2 The solute is soluble because the value of ΔE_{total} is negative.

12.3 (a), (d), and (e) are soluble in water. Water is a polar solvent and will dissolve polar solutes; only (a), (d), and (e) are polar solutes.

12.4 (a) and (c) will probably allow more solute to be dissolved at higher temperature. The solubility of solid solutes in liquid solvents generally increases with an increase in temperature.

12.5 0.47 M

12.6 3.6 L = 3600 mL

12.7 Mix 4.25 g of $NaNO_3$ with enough water to completely dissolve the $NaNO_3$. After all of the $NaNO_3$ is dissolved, add enough water to bring the total volume of solution to 500.0 mL.

12.8 20.8 mL

12.9 13.8%

12.10 25.00%

12.11 12.9%

12.12 0.502 g

12.13 6.00×10^3 mL = 6.00 L

12.14 102.8 °C

12.15 −0.42 °C

12.16 92 g/mol

Practice Problems

12.1 Place 293 g of NaCl in a 1.00-L volumetric flask, and add enough water to bring the level of the solution to 1.00 L.

12.2 Place 21.4 g of $C_{12}H_{22}O_{11}$ in a 250-mL volumetric flask, and add enough water to bring the level of the solution to 250 mL.

12.3 Place 556 mL of stock solution in a 1.00-L volumetric flask, and bring the solution level up to the desired volume.

12.4 Place 167 mL of stock solution in a 250-mL volumetric flask, and bring the solution level up to the desired volume.

12.5 12.5% CH_3OH

12.6 11.8% $C_{12}H_{22}O_{11}$

12.7 0.458 g of SiH_4

12.8 To prepare 45g of PbC_2O_4: Combine 608 mL of 0.250 M $Pb(NO_3)_2$ and 1.22 L of 0.125 M $K_2C_2O_4$.

12.9 0.0647 M HCl

12.10 0.0984 M H_2SO_4

12.11 176 g/mol

Quiz

1. 1.5 g of $BaCl_2$

2. Place 8.4 mL of $C_2H_2O_6$ in a 100-mL volumetric flask and bring the solution level up to the desired volume.

3. 4.7% ethanol

4. 81 mL of HCl

5. 0.0268 M H_2SO_4

6. 24.2% KNO_3

7. 20.1 g/mol

8. 40% KBr

9. 6.8% C_3H_6O

10. Place 44.81 g of Na_2SO_4 in a 500 mL volumetric flask, and add enough water to bring the level of the solution to 500.0 mL.

Exercises for Self-Testing

A. Completion

1. homogeneous   2. solute, solvent   3. solubility   4. like, like

5. solid, gaseous   6. gaseous   7. entropy   8. saturated   9. three

10. 3   11. hydrophobic, hydrophilic   12. micelles   13. moles, liter

14. concentrated   15. 20.0%   16. boiling-point elevation   17. less

18. molarity   19. lower   20. vapor pressure

B. True/False

5, 6, 8, 11, and 14 are true.

1 is false. Ethanol is the solvent because it is present in the greater amount.

2 is false. The ΔE value for step 1 of the solution process is always positive.

3 is false. The solute will be soluble if the entropy change is favorable for solution.

4 is false. The BF_3 molecule is nonpolar (because the individual bond dipoles cancel) and is soluble in nonpolar solvents.

7 is false. In order to determine the percent by volume, we need to know the volume of the *solution*. The percent by volume would be 15.0% if the volume of the solution were 100.0 mL, but that is just the volume of the *solvent*. The volume of the solution will be at least as great as the volume of the solvent but probably somewhat less than the sum of the volumes of the solvent and the solute.

9 is false. The solubility of gases in liquid solvents decreases with increasing temperature. This is because at higher temperatures thermal motions of gas molecules increase more than those of liquid molecules. These greater thermal motions of gas molecules result in the tendency of gas molecules to escape the mixture.

10 is false. Additional solute can usually be dissolved by heating the solution.

12 is false. The boiling-point elevation will be essentially the same because NaCl dissociates into two particles when it is dissolved in water.

13 is false. The vapor pressure of a solution is lower than the vapor pressure of the pure solvent.

C. Questions and Problems

1. As the temperature of the water increases, less gas is soluble. The bubbles form as the dissolved gas comes out of solution.

2. 15%

3. 0.833 M

4. This solution, with a molarity of 0.833 M, could not be used as a stock solution to make a solution with a molarity of 1.5 M. A stock solution must be *more* concentrated than the solution that will be prepared from it.

5. Place 60.0 mL of the solution (0.833 M) in a container. Add enough water to bring the total volume of the solution up to 500.0 mL.

6. 667 mL

7. Steps 1 and 2 of the solution process involve *breaking* intermolecular attractive forces, whereas step 3 involves *forming* intermolecular attractive forces.

8. If the solute and solvent molecules have similar intermolecular attractive forces (which is the case if they are "like" substances), the strength of the intermolecular attractive forces in step 3 of the solution process will be similar to the combined strength of the forces in steps 1 and 2. The solute molecules will have no problem intermingling with the solvent molecules, because the types of intermolecular forces are the same in both compounds.

9. The percent by mass and the percent by volume will be equal. If the densities of both the solute and the solvent are 1.0 g/mL, the masses in grams will be equal to the volumes in milliliters.

10.

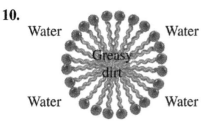

Water   Water

Greasy
dirt

Water   Water

The greasy dirt dissolves in the nonpolar interior section of the micelle. The polar outer section of the micelle dissolves in the water. Because the grease is dissolved in the water, and the micelle is dissolved in the water, the result is that the greasy dirt is dissolved in the soapy water.

11. 164 mL

12. (a) $-3.72\,°C$ (b) $-5.58\,°C$ (c) $-1.86\,°C$

13. 62.0 g/mol

14. Assuming complete dissociation of the ionic compounds $(Al_2(SO_4)_3, K_2CO_3,$ and LiBr) into ions and no such process possible for the molecular $C_6H_{12}O_6$, there will be 5 moles of ions in 1 L of $Al_2(SO_4)_3$ solution, 3 moles of ions in the solution of K_2CO_3 and 2 moles of ions in the solution of LiBr. There will be only 1 mole of molecules in 1 L of $C_6H_{12}O_6$ solution. The more particles in solution, the lower the vapor pressure. Therefore the order in terms of increasing vapor pressure above the solutions is: $Al_2(SO_4)_3 < K_2CO_3 < LiBr < C_6H_{12}O_6$.

15. 8.33×10^{-3} moles; 167 mL

Chapter 13: When Reactants Turn into Products

Practice Exercises

13.1 (a) endothermic

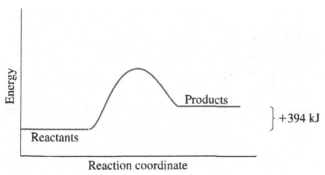

(b) endothermic

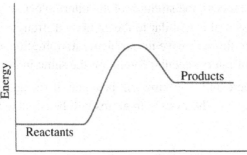

(c) exothermic

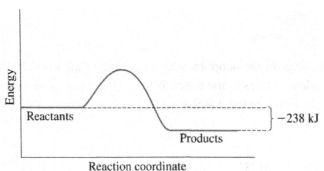

13.2 (a) The ΔE_{rxn} for the reaction is negative.

(b) The reactants have more energy.

(c) The ΔE_{rxn} for the reverse reaction is +560 kJ.

(d) The reaction container feels hot to the touch.

13.3 (a) The reaction is exothermic because it releases energy.

(b) The ΔE_{rxn} for the reaction is 180 kJ.

(c) The ΔE_{rxn} for the reverse reaction is +180 kJ.

(d) The reaction container feels hot to the touch because heat is being given off by the reaction.

13.4 (a) The reaction is endothermic.

(b) $\Delta E_{\text{rxn}} = +30 \text{ kJ/mol}$

(c) $E_{\text{a}} = 70 \text{ kJ/mol}$

13.5 The reaction that has 400 collisions per second and an orientation factor of 0.4 will occur faster. Forty percent of 400 is greater than 30% of 500.

13.6 The order with respect to NO is 2. The order with respect to O_2 is 1. The overall order of the reaction is 3.

13.7 (a) Reaction rate $= k[\text{R}]^x[\text{S}]^y$

(b) Experiments 1 and 2; experiments 1 and 3

(c) The exponent on R is 1; the exponent on S is 1.

(d) Reaction rate $= k[\text{R}]^1[\text{S}]^1$

(e) The order with respect to R is 1. The order with respect to S is 1. The overall order of the reaction is 2.

13.8 Reaction rate $= k[A][B]^2$

13.9 (a) $2H_2O_2 \rightarrow 2H_2O + O_2$

  (b) Rate $= k[H_2O_2][I^-]$

  (c) The catalyst is I^-.

  (d) The reaction intermediate is IO^-.

Practice Problems

13.1 The reaction is exothermic.

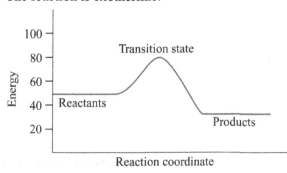

13.2 The reaction is endothermic.

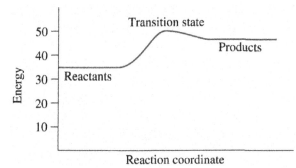

13.3 Rate $= k[R]^x[S]^y$

  (a) Use experiments 1 and 2 to find x: When [R] triples, the rate triples: $3^1 = 3$, $x = 1$

  (b) Use experiments 1 and 3 to find y: When [S] doubles, the rate doubles: $2^1 = 2$ $y = 1$

  Thus the rate law is:

  Reaction rate $= k[R]^1[S]^1$

  so, the overall order is $x + y = 2$.

13.4 $Br_2 \rightarrow 2Br$ (slow)

  $\dfrac{2Br + CHBr_3 \rightarrow HBr + CBr_4}{Br_2 + CHBr_3 \rightarrow HBr + CBr_4}$

  Note: Step 2 is termolecular (involves three molecules); possible, but very unlikely.

Quiz

1. Reactants; products
   Products; reactants

2. (a) Rate $= k[CO]^1[NO]^1$ (b) 2   (c) The rate will be doubled.

3. The reaction is endothermic.

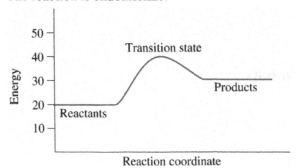

Reaction coordinate

4. Rate $= k[\text{NO}]^x[\text{O}_2]^y$

5. (a) The rate will increase by a factor of 4.

(b) The rate will be tripled.

(c) The rate will increase by a factor of 9.

(d) The rate will increase by a factor of 12.

6. (a) Exothermic

(b) The reactants are higher. Because the reaction releases energy, the products are 700 kJ lower in energy than the reactants.

7. NO + Cl$_2$ → NOCl + Cl (slow)

NO + Cl → NOCl (fast)
_____

2NO + Cl$_2$ → 2NOCl

8. 3

9. Rate $= k[\text{H}_2][\text{ICl}]$

10. Rate $= k[\text{A}]^2[\text{B}]^1$

Exercises for Self-Testing

A. Completion

1. reaction mechanism   2. endothermic   3. catalyst   4. transition state
5. 10 °C   6. enzyme   7. rate-determining step   8. activation energy   9. 16
10. transition state   11. reaction intermediate   12. positive   13. hot, cold
14. increase   15. activation energy

B. True/False

2, 3, 4, 6, and 7 are true.

1 is false. The size of the activation energy determines the speed of a reaction, not whether the reaction is endothermic or exothermic.

5 is false. The energy factor gives the proportion of collisions that will be of sufficient energy to cause a reaction to occur. The orientation factor gives the proportion of collisions that will be properly oriented to cause a reaction to occur.

8 is false. The rate law cannot prove a mechanism correct; it can only show that the mechanism is plausible.

9 is false. Tripling the concentration will increase the rate by a factor of 9.

10 is false. The rate law cannot be determined by the balanced equation. It must be experimentally determined by kinetics experiments.

C. Questions and Problems

1. The reaction cannot proceed any faster than its slowest step. Therefore, the overall reaction rate will be limited (determined) by the speed of this step.

2. (a) Reaction rate $= k[A]^x[B]^y$
   (b) Reaction rate $= k[A]^2[B]$
   (c) The order with respect to A is 2. The order with respect to B is 1. The overall order of the reaction is 3.

3. A transition state is an extremely short-lived, highly reactive species that appears at the top of the energy profile. It bares remnants of old bonds that will ultimately be broken and hints of new bonds that are in the process of being formed. A transition step does not appear in the proposed reaction mechanism. A reaction intermediate, on the other hand, is a little longer-lived but still a relatively unstable species that is produced in an earlier step of a reaction and then is quickly used up in a later step. An intermediate appears in the proposed reaction.

4. (a) The reaction is exothermic.

   (b)

   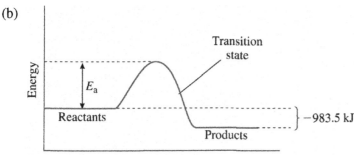

   (c) The ΔE_{rxn} for the reverse reaction is +983.5 kJ.
   (d) The reaction container will feel hot to the touch.

5. (a) Reaction rate $= k[CO][NO_2]$
   (b) The overall order of the reaction is 2.
   (c) The reaction rate will triple if the concentration of CO is tripled.

6. Enzymes are valuable because the other factors that are known to speed up reactions, such as higher temperature or increased reactant concentrations, typically cannot be used in biological systems. In particular, heating living organisms beyond certain temperatures results in its death. Enzymes catalyze biological processes without causing damage or death to the organisms.

7. Yes, the height of the reaction energy profile decreases when a catalyst is used, lowering the activation energy in either direction. The lower the activation energy, the faster the reaction.

Chapter 14: Chemical Equilibrium

Practice Exercises

14.1 A single arrow with an "x" through it.

14.2 The reverse reaction will occur to a large extent.

14.3 0.048

14.4 (a) no reaction (b) to the right (c) near the middle
  (d) to the right (e) to the left

14.5 (a) to the right (b) to the right (c) to the left (d) to the right

14.6 Heat is a product.

14.7 The equilibrium will shift to the left.

14.8 (a) $K_{eq} = \dfrac{[H_2]}{[H_2O]}$

  (b) $K_{eq} = [NH_3][H_2S]$

14.9 $K_{sp} = 5.11 \times 10^{-12}$

14.10 $s = 2.83 \times 10^{-14}$ M

14.11 $[N_2O_4] = 5.24$ M

Practice Problems

14.1 $K_{eq} = 0.207$

14.2 $K_{eq} = 9.5 \times 10^3$

14.3 $K_{sp} = 1.7 \times 10^{-6}$

14.4 $K_{sp} = 7.9 \times 10^{-9}$

14.5 $s = 8.8 \times 10^{-7}$ M

14.6 $[O_2] = 0.20$ M

Quiz

1. $K_{eq} = 1.8 \times 10^{-11}$

2. Products, reactants

3. $K_{sp} = 2.83 \times 10^{-11}$

4. $K_{eq} = \dfrac{[SO_3]^2}{[SO_2]^2[O_2]}$

5. $[X] = 0.314$ M

6. $K_{sp} = 4.0 \times 10^{-8}$

7. $K_{eq} = 4.0 \times 10^7$

8. $[O_2] = 8.0 \times 10^{-2}$ M

9. $s = 1.22 \times 10^{-8}$ M

10. AgCl is more soluble in water than AgBr because it has a higher concentration of Ag^+ ions. $[Ag^+]$ equals the moles of salt (AgCl or AgBr) that are dissolved per liter of solution. Because the $[Ag^+]$ is higher in the AgCl solution, AgCl is more soluble than AgBr.

11. The reaction $E \rightleftarrows F$ with a $K_{eq} = 1 \times 10^{-18}$ will contain more reactants in its equilibrium state because the $K_{eq} < 10^{-3}$, which means that the equilibrium lies far to the left. Therefore the reaction barely occurs, so mostly reactants are present.

12. $K_{eq} = [NH_3][H_2S]$

Exercises for Self-Testing

A. Completion

1. dynamic equilibrium   **2.** no reaction   **3.** gone to completion   **4.** rate, rate

5. left   **6.** away from, toward   **7.** exothermic   **8.** catalyst

9. products, reactants   **10.** forward rate, reverse rate   **11.** s

B. True/False

3, 4, 6, 10, 12, and 13 are true.

1 is false. The value of K_{eq} changes only if the temperature changes.

2 is false. There will be many more products in the mixture if K_{eq} is very large.

5 is false. When heat is added to an exothermic reaction, the equilibrium shifts to the left, and fewer products are produced.

7 is false. Adding a catalyst will decrease the time required for the system to reach the equilibrium state.

8 is false. Increasing the temperature of an exothermic reaction will shift the equilibrium toward the reactants.

9 is false. The value of K_{eq} is dependent on temperature and will change when the temperature changes.

11 is false. The value of K_{sp} is 1.59×10^{-10} if the solubility of AgCl is 1.26×10^{-5} M.

14 is false. The relationship between K_{sp} and s is: $K_{sp} = 4s^3$.

C. Questions and Problems

1. (a) $K_{eq} = \dfrac{[SO_3]^2}{[SO_2]^2[O_2]}$

   (b) $K_{eq} = \dfrac{[NO_2]^2}{[N_2O_4]}$

   (c) $K_{eq} = \dfrac{[N_2][O_2]}{[NO]^2}$

   (d) $K_{eq} = \dfrac{[CO][H_2]^3}{[CH_4][H_2O]}$

2. $K_{eq} = 0.58$

3. (a) to the left   (b) to the left   (c) to the left   (d) to the right
   (e) There will be no shift.

4.

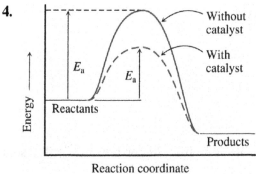

5. The temperature should be decreased. Because the reaction is exothermic, heat is a product, and a decrease in temperature will shift the equilibrium to the right. However, recall from Chapter 13, that temperature decrease will slow the reaction down!

6. $K_{eq} = \dfrac{[H_2O]^3}{[HCl]^6}$

7. $[B_2] = 3.14\ M$

8. (a) $K_{sp} = [Ca^{2+}][OH^-]^2$
   (b) $K_{sp} = [Mg^{2+}][CO_3{}^{2-}]$
   (c) $K_{sp} = [Pb^{2+}][F^-]^2$

9. For reactions that run to completion (i.e., all reactants are converted into products), simple rules of stoichiometry, as outlined in Chapter 9, apply. For reactions that reach an equilibrium significantly short of completion, a different approach is needed, in which the equilibrium constant, K_{eq}, is used in conjunction with the equilibrium constant expression, to determine the amount of product(s) that will be produced and the amount of reactant(s) that will remain in the mixture. It is a more challenging task to determine the yield of a reaction that reaches equilibrium instead of completion.

10. H_2O in reaction (i) is in the gaseous phase and, therefore, its concentration directly factors into the equilibrium calculation:

$$K_{eq} = \dfrac{[H_2O]^2}{[H_2]^2[O_2]}$$

In reaction (ii), H_2O is in the liquid phase and, as such, its concentration is already "built" into the K_{eq}. It is therefore skipped from the equilibrium constant expression:

$$K_{eq} = \dfrac{[F^-][H_3O^+]}{[HF]}$$

Chapter 15: Electrolytes, Acids, and Bases

Practice Exercises

15.1 (c), (d), and (e) are electrolytes; (a) and (b) are nonelectrolytes.

15.2 (a) weak acid   (b) salt   (c) strong acid   (d) salt   (e) strong acid

15.3 HCO_3^- will have a smaller K_a than H_2CO_3.

15.4 $HF + LiOH \rightarrow LiF + H_2O$

15.5 $HClO + NH_3 \rightarrow ClO^- + NH_4^+$

15.6 $[H_3O^+] = 1 \times 10^{-12}\ M$

15.7 The solution is basic.

15.8 pH $= 9$. The solution is basic.

15.9 $[H_3O^+] = 10^{-4}\ M$ (or 1×10^{-4}); $[OH^-] = 10^{-10}\ M$ (or 1×10^{-10}). The solution is acidic because the pH < 7.

15.10 (a) NH_3 (b) HBr   (c) H_3PO_4 (d) HSO_4^-

Br^- could not be used in a buffered solution because the base in a buffered solution must be a weak base or the conjugate base of a weak acid. HBr is a strong acid.

15.11 NH_4^+ will react with added base; NH_3 will react with added acid.

Practice Problems

15.1 (a) Strong acid; strong electrolyte
(b) Weak acid; weak electrolyte
(c) Weak acid; weak electrolyte

15.2 (a) Strong base; strong electrolyte
(b) Weak base; weak electrolyte
(c) Strong acid; strong electrolyte

15.3 K_2SO_4

15.4 HBr: acid

Br^-: base

NH_3: base

NH_4^+: acid

15.5 CH_3NH_2: base

$CH_3NH_3^+$: acid

H_2O: acid

OH^-: base

15.6 7.7×10^{-13} M

15.7 1.7×10^{-7} M

15.8 $H_2BO_3^- + H_3O^+ \rightarrow H_3BO_3 + H_2O$

15.9 $HCN + OH^- \rightarrow CN^- + H_2O$

Quiz

1. (a) Electrolyte (strong)
   (b) Nonelectrolyte
   (c) Electrolyte (weak)

2. (a) Na_3PO_4
   (b) $Mg(NO_3)_2$

3. (a) H_2BO_3
   (b) $C_2H_3O_2^-$
   (c) NH_3

4. (a), (c) and (d) are strong acids
   (b) is a weak acid

5. NaF

6. 1.0×10^{-7} M

7. (a) and (b) are weak bases
   (c) and (d) are strong bases

8. (a) NH_4^+

(b) HNO_3

(c) $H_2PO_4^-$

9. 1.1×10^{-10} M

10. (a) and (d)

(e) and (g)

(f) and (b)

(c) and (h)

11. (a) $HS^- + H_3O^+ \rightarrow H_2S + H_2O$

(b) $H_2S + OH^- \rightarrow HS^- + H_2O$

12. 1.0×10^{-9} M

Exercises for Self-Testing

A. Completion

1. hydrogen   **2.** Strong, weak   **3.** red, blue   **4.** nonelectrolytes

5. proton donors, proton acceptors   **6.** diprotic   **7.** salt, water   **8.** 5.0

9. weaker   **10.** buffered solution   **11.** 7   **12.** strong   **13.** H_2S, S^{2-}

14. 10.0   **15.** IA, IIA

B. True/False

1, 3, 6, and 9 are true.

2 is false. HCl is a strong electrolyte.

4 is false. NH_3 is a base that accepts an H^+ to produce NH_4^+.

5 is false. The pH of the solution is 2.

7 is false. The pH changes slightly when acid or base is added to a buffer.

8 is false. HSO_4^- is the conjugate base of H_2SO_4.

10 is false. The water acts as a Brønsted–Lowry base.

C. Questions and Problems

1. There are some substances, like NH_3, that do not dissociate into OH^- ions in water but still produce a basic solution. The definition of bases has to be broad enough to include them.

2. KNO_3

3. pH = 4.0

4. The value of K_a is a measure of the strength of an acid. The larger the value of K_a, the stronger the acid, and the more dissociated it will be in aqueous solution.

5. She needs a source of CN^- ions. She should choose KCN because KCN is a salt that will dissociate in water to produce the CN^- ions needed for the buffered solution. CH_3CN is not a salt, a strong acid, or a strong base and will not dissociate in water.

6. NH_3 is a Brønsted–Lowry because it has an ability to accept a proton from another substance. It is not an Arrhenius base because it is not a metal hydroxide capable of producing an OH^- ion when dissolved in water.

7. In endothermic processes heat is a "reactant". As heat is added to the water autodissociation reaction at equilibrium by increasing the temperature of water, the reaction will proceed forward (toward products) and the equilibrium will shift right. This means that when the new equilibrium is reached at a higher temperature, there will be more products present. In other words, while still equal to each other, $[H_3O^+]$ and $[OH^-]$ will both be greater than 1.0×10^{-7} M. Since $K_w = [H_3O^+][OH^-]$, K_w will be greater than 1.0×10^{-14}.

8. Since water autodissociation is an endothermic process, K_w increases with increasing temperature. At temperatures above 25 °C, $K_w > 1.0 \times 10^{-14}$, and since $[H_3O^+] = \sqrt{K_w}$, the square root of a number that is greater than 1.0×10^{-14} is greater than 1.0×10^{-7}. Hence, $[H_3O^+] > 1.0 \times 10^{-7}$ M, and pH $= -\log[1.0 \times 10^{-7}]$ is less than 7.

Chapter 16: Nuclear Chemistry

Practice Exercises

16.1 0.5838 g/mol

16.2 5.25×10^{10} kJ/mol

16.3 The binding energy per nucleon for $^{40}_{20}$Ca is 8.59 MeV/nucleon. The binding energy per nucleon for $^{24}_{12}$Mg is 8.30 MeV/nucleon. The $^{40}_{20}$Ca is the more stable nucleus.

16.4 $^{54}_{24}$Cr $> ^{108}_{47}$Ag $> ^{195}_{78}$Pt $> ^{200}_{79}$Au

16.5 4.70×10^{21} atoms (0.0781 moles). One hundred days is seven half-lives.

16.6 (a) $^{12}_{6}$C will lie within the band of stability. The n/p ratio is 1.
  (b) $^{12}_{7}$N will lie below the band of stability. There are too few neutrons. The n/p ratio is 0.71.
  (c) $^{30}_{12}$Mg will lie above the band of stability. There are too many neutrons. The n/p ratio is 1.5.
  (d) $^{235}_{92}$U will lie beyond the band of stability. There are too many protons (>84).

16.7 $^{198}_{80}$Hg will be produced.

16.8 $^{131}_{54}$Xe

16.9 22,860 years (four half-lives have passed)

16.10 2.4×10^4 years

Practice Problems

16.1 0.5444 g/mol

16.2 The mass defect for $^{12}_{6}$C is 0.0993 g/mol. The mass defect for $^{10}_{5}$B is 0.069 85 g/mol. Therefore, $^{12}_{6}$C has the greater mass defect.

16.3 7.58×10^8 kJ/mol of nucleons

16.4 7.73×10^8 kJ/mol of nucleons

16.5 12.5 g will remain after 45.8 years.

16.6 It will take approximately 162 seconds for the isotope to decay to 4 g.

16.7 1.05×10^9 years

16.8 2.47×10^4 years

Quiz

1. 1.89×10^4 years
2. 6.76×10^9 years
3. 0.0627 g/mol
4. 6.27×10^8 kJ/mol of nucleons
5. 8.29×10^8 kJ/mol of nucleons
6. 15.63 g will remain after 175 years.
7. It will take approximately 30 hours for the isotope to decay to 3 g.
8. The mass defect for $^{40}_{18}\text{Ar}$ is 0.009 31 g/mol of nucleons. The mass defect for $^{27}_{13}\text{Al}$ is 0.008 99 g/mol of nucleons. Therefore $^{40}_{18}\text{Ar}$ has the greater mass defect per nucleon.
9. Mass defect
10. 1/256 of a radioactive isotope with a 25-year half-life will remain after 200 years.

Exercises for Self-Testing

A. Completion

1. mass defect   2. $^{56}_{26}\text{Fe}$ 3. neutron   4. positron   5. above   6. 2, 4
7. above   8. gamma radiation   9. fission, fusion   10. critical mass

B. True/False

2, 3, 6, 9, and 10 are true.

1 is false. The protons and neutrons are the nucleons.

4 is false. The binding energy per mole of nucleons allows you to determine the more stable isotope.

5 is false. An isotope below the band of stability must emit a positron to become more stable.

7 is false. Alpha particles are slower and require less shielding to stop them.

8 is false. Magnetic resonance imaging is another name for nuclear magnetic resonance.

C. Questions and Problems

1. The missing mass is due to the mass defect. The missing mass is converted to energy and is a measure of the binding energy holding the nucleons together.
2. The isotope with a mass number of 57 is below the band of stability and decays by positron emission. The isotope with a mass number of 66 is above the band of stability and decays by beta emission.
3. The half-life of the substance is 6 years.
4. The other element would be lead (Pb).
5. Various answers are possible, such as the use of radioactive isotopes in radiation treatment for cancer patients, and the use of radioactive barium to monitor the digestive tract.

Chapter 17: The Chemistry of Carbon

Practice Exercises

17.1 (a) isomers   (b) the same   (c) different

17.2 (a) 7   (b) 5   (c) 6

17.3 (a) 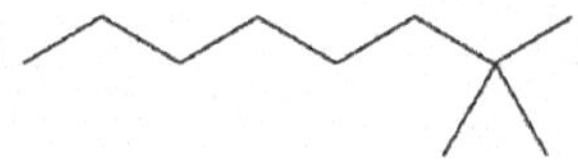   (b) 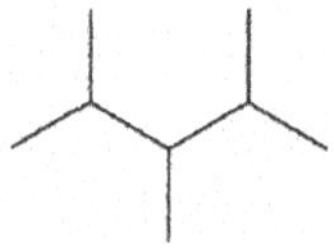   (c)

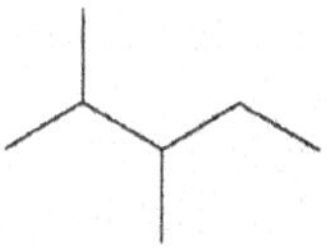

17.4 (a) saturated   (b) unsaturated; C_3H_8 (c) unsaturated; C_9H_{20} (d) saturated

17.5 (a) 3-ethylhexane   (b) 2-methyloctane   (c) 2,3-dimethlypentane
     (d) 1-heptyne   (e) 1-butene

17.6 (a) $CH_3 - CH_2 - \underset{\underset{\displaystyle CH_3}{\overset{\displaystyle |}{CH_2}}}{\overset{\displaystyle |}{CH}} - CH_2 - CH_2 - CH_2 - CH_2 - CH_2 - CH_3$

     (b) $CH_3 - \overset{\overset{\displaystyle CH_3}{\displaystyle |}}{CH} - CH_2 - CH_2 - \overset{\overset{\displaystyle CH_3}{\displaystyle |}}{CH} - CH_2 - CH_2 - CH_2$

     (c) $CH_3 - C \equiv C - CH_3$

     (d) $CH_3 - \underset{\underset{\displaystyle CH_3}{\displaystyle |}}{C} = CH - CH_2 - CH_3$

17.7 (a) amine   (b) alcohol   (c) alkyl halide   (d) carboxylic acid

17.8 (a) fluoroethane   (b) 3-bromohexane   (c) 3-iodopentane

17.9 $CH_3CH_3 < CH_3Br < CH_3CH_2Br < CH_3CH_2OH$

17.10 (a) dibutyl ether   (b) ethyl propyl ether

17.11 $CH_3CH_3 < CH_3Br < CH_3CH_2CH_2CH_2CH_2COOH < CH_3COOH$

17.12 (a) propanone   (b) pentanal   (c) 3-methylbutanal

Practice Problems

17.1 Saturated: $C_nH_{2n+2} = C_8H_{18}$

17.2 Unsaturated: $C_nH_{2n} = C_6H_{12}$

17.3 4-methyl-2-hexene

17.4 4-methyl-2-hexene

17.5 Ketone

17.6 Carboxylic acid

17.7 Alkyl halide (or haloalkane)

Quiz

1. 2-methyl-pentane

2. Alkane, alkene, alkyne

3. (a) Alcohol
   (b) Ether
   (c) Alkyl halide

4. (a) Saturated: $C_nH_{2n+2} = C_5H_{12}$
   (b) Unsaturated: $C_nH_{2n-2} = C_7H_{14}$

5. (a) R — OH
   (b) R — O — R

   (c) 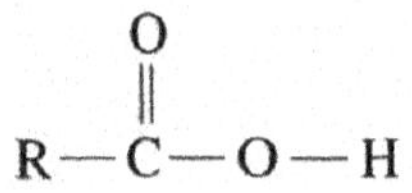
   $$\underset{}{R} - \overset{\displaystyle O}{\underset{\displaystyle \|}{C}} - O - H$$

6. (a) Propane   (b) Heptane

7. 4-methyl-1-hexene

8. Carboxylic acids, ketones, aldehydes

9. 26

10. The correct name should be 3-methyl-1-butene.

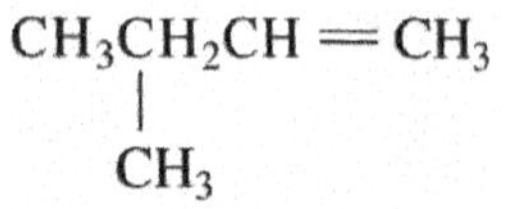

$$\underset{\underset{\displaystyle CH_3}{\displaystyle |}}{CH_3CH_2CH} = CH_3$$

Exercises for Self-Testing

A. Completion

1. graphite, diamond, fullerene   2. four   3. alkanes, alkenes, alkynes   4. 20
5. isomers   6. decane   7. alcohols   8. proof   9. amines   10. ethers
11. carboxylic acids, aldehydes, ketones   12. -one   13. carboxylic acids
14. hydrogen bonds   15. petroleum

B. True/False

1, 4, 5, and 8 are true.

2 is false. Alkynes contain at least one triple bond.

3 is false. The electronegativity difference is too small to result in polar molecules.

6 is false. The alkane would be C_6H_{14}.

7 is false. The larger the nonpolar hydrocarbon part of the molecule, the less soluble the carboxylic acid.

9 is false. Aldehydes and ketones are more soluble than alkanes. The lone pair of electrons on the oxygen atom of an aldehyde or a ketone can form hydrogen bonds with the hydrogen atoms in water.

10 is false. There is no hydrogen bonded to an N, O, or F atom, which is necessary for hydrogen bonding to occur between molecules.

C. Questions and Problems

1. (a) 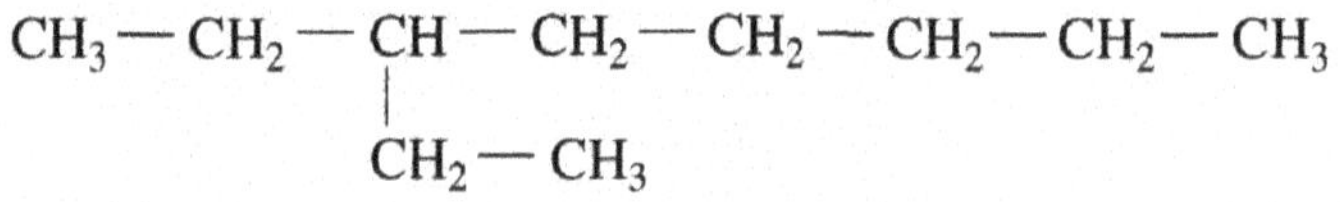 $CH_3-CH_2-CH-CH_2-CH_2-CH_2-CH_2-CH_3$
$|$
CH_2-CH_3

(b) 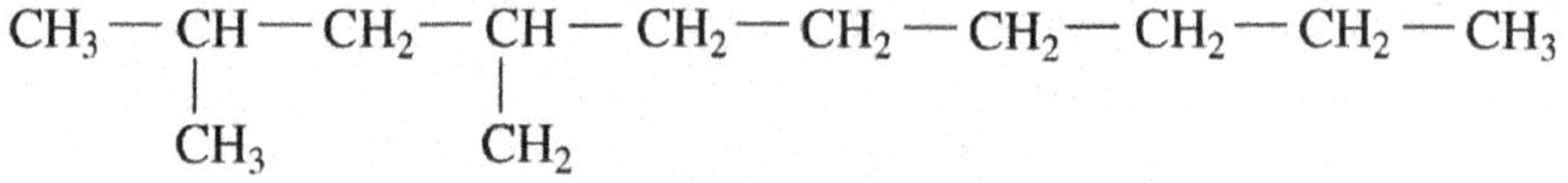 $CH_3-CH-CH_2-CH-CH_2-CH_2-CH_2-CH_2-CH_2-CH_3$
$||$
CH_3CH_2

(c) 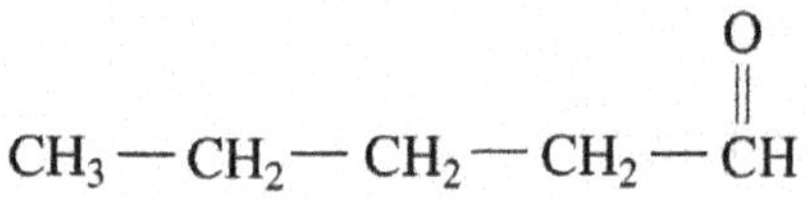 $CH_3-CH_2-CH_2-CH_2-\overset{\displaystyle O}{\overset{\|}{C}}H$

(d) 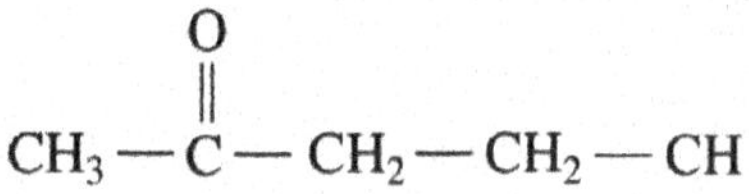 $CH_3-\overset{\displaystyle O}{\overset{\|}{C}}-CH_2-CH_2-CH$

(e) $CH_3-NH-CH_3$

(f) 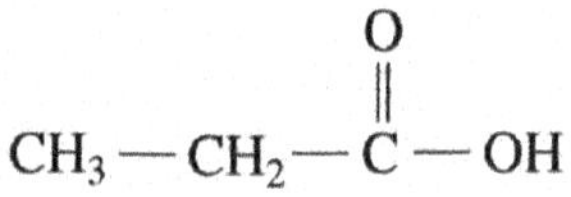 $CH_3-CH_2-\overset{\displaystyle O}{\overset{\|}{C}}-OH$

2. Examples will vary. The functional groups that must be in each are shown below.
   (a) amine: $-NH_2$, $-NHR$, $-NR_2$
   (b) haloalkane: $R-X$, where X is F, Cl, Br, or I
   (c) alkyne: $-C\equiv C-$
   (d) ether: $R-O-R$
   (e) carboxylic acid: $R-\overset{\displaystyle O}{\overset{\|}{C}}-OH$

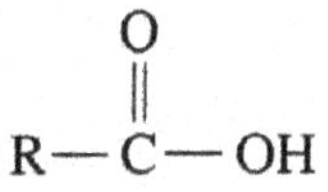

3. The molecules that exhibit hydrogen bonding (carboxylic acids) boil at a higher temperature than those that do not. The molecule that has dipole–dipole interactions (propyl chloride) boils at a higher temperature than the molecules that have only London forces holding them together (alkanes). The larger alkane boils at a higher temperature than the smaller one due to stronger London forces in the larger molecule.

4. $CH_3COOH + H_2O \rightleftharpoons CH_3COO^- + H_3O^+$

5. The product of ethanol metabolism is acetaldehyde, a compound that is harmless to humans if present in reasonable quantities. Methanol, on the other hand, is toxic to humans because the body metabolizes it to formaldehyde, a poison that results in blindness and death.

6. (b), (c) and (d) because they all contain polar carbon–oxygen and/or oxygen–hydrogen bonds. In addition, (b) and (d) have the capability of forming strong hydrogen bonds with water molecules.

7. Amines; just like ammonia, amines contain a nitrogen atom with a lone pair of electrons, which makes them good acceptors of electrons and the pyramidal molecular shape endows them with polarity.

Chapter 18: Synthetic and Biological Polymers

Practice Exercises

18.1 (a) The structure of 1-propene is:

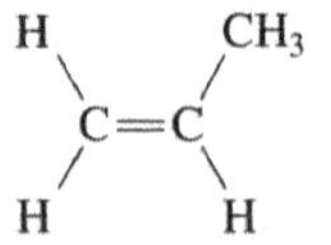

The structure of the polymer is, therefore:

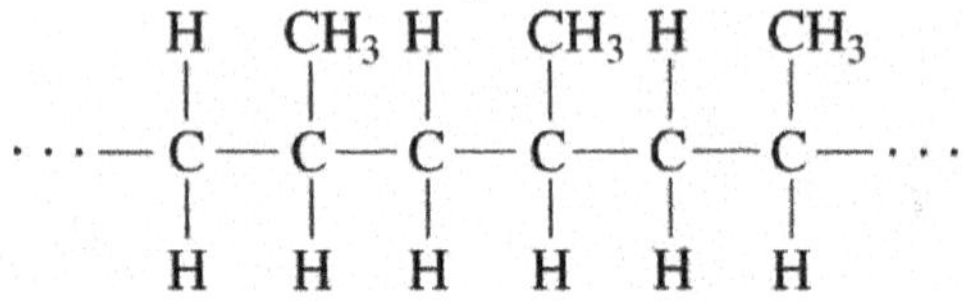

(b) The structure of 1,1-difluoroethene is:

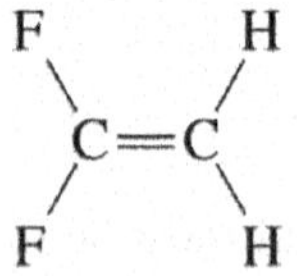

The structure of the polymer is, therefore:

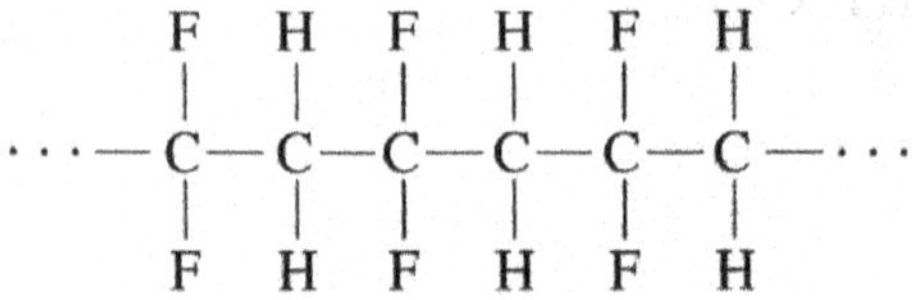

18.2 (a)

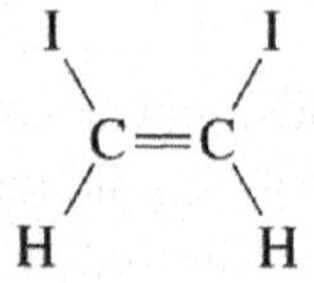

(b)

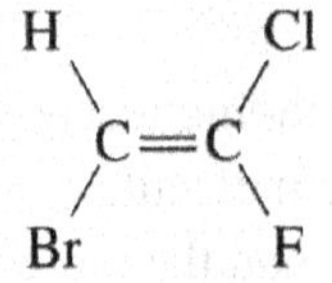

Practice Problems

18.1

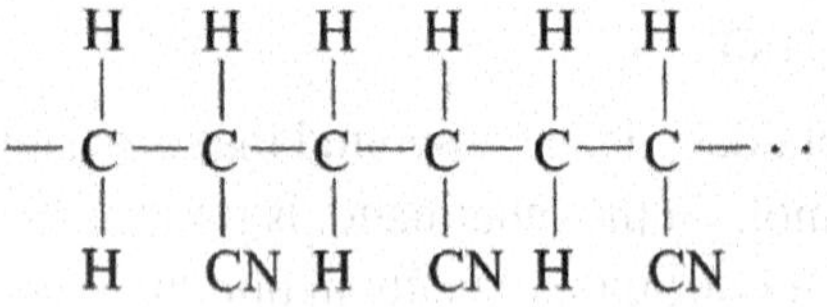

18.2

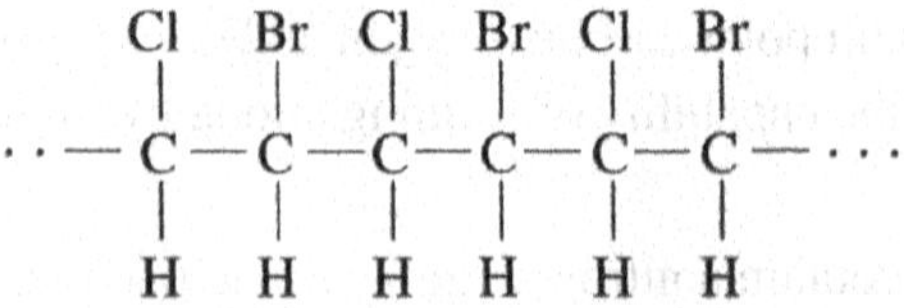

18.3

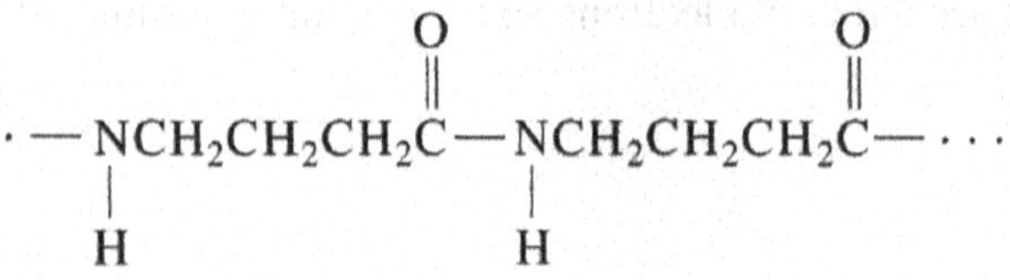

$$\cdots - \text{NCH}_2\text{CH}_2\text{CH}_2\overset{\displaystyle\overset{\text{O}}{\|}}{\text{C}} - \text{NCH}_2\text{CH}_2\text{CH}_2\overset{\displaystyle\overset{\text{O}}{\|}}{\text{C}} - \cdots$$

18.4

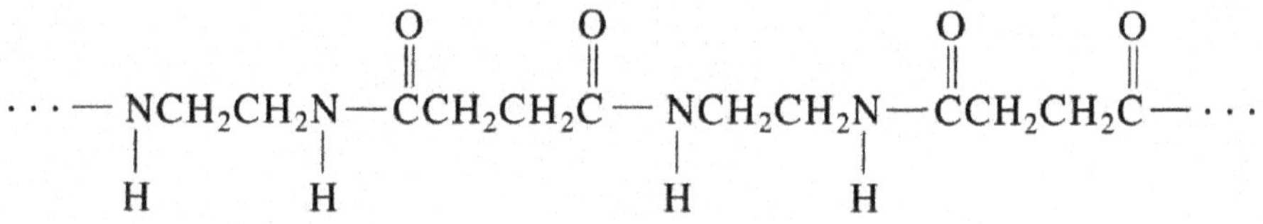

18.5

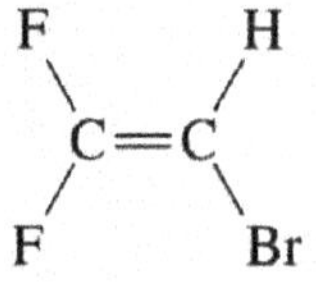

18.6

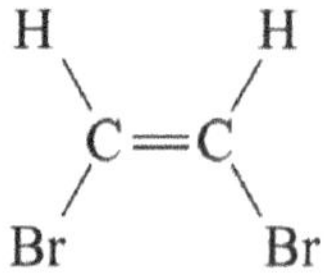

18.7

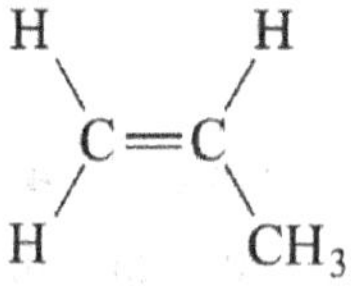

18.8

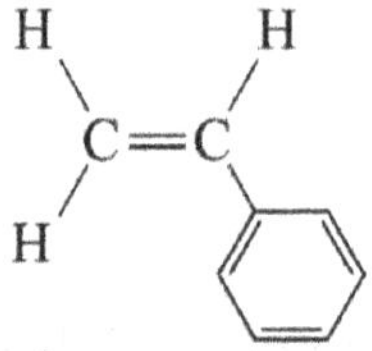

Quiz

1.

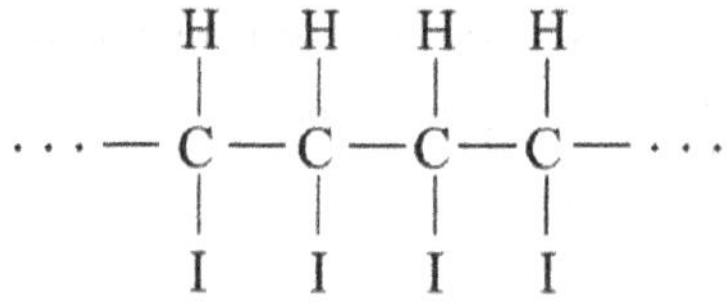

2.

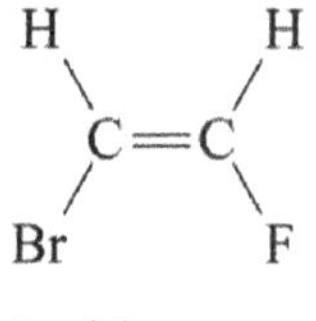

3. Amide

4.

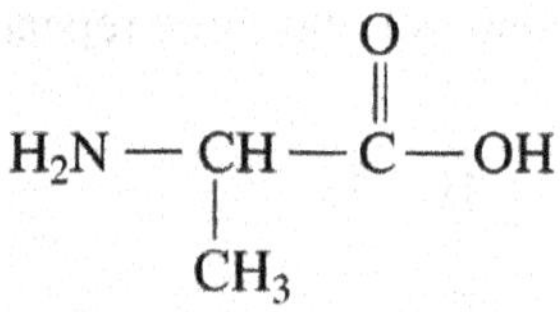

5. Water (H_2O)

6.

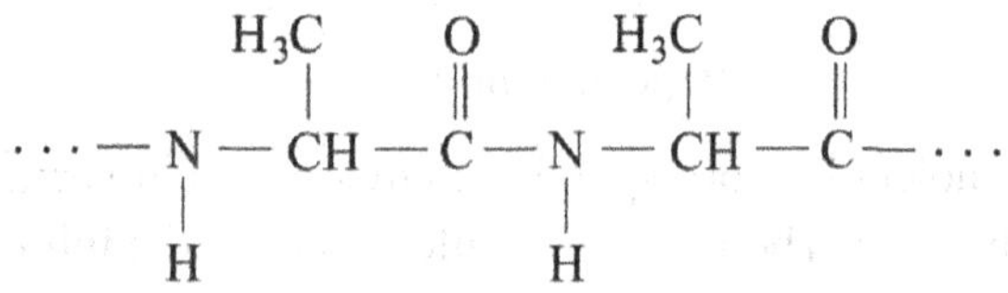

7.

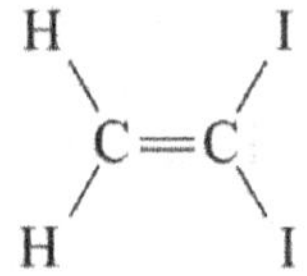

$$\begin{array}{ccc} H & & I \\ \diagdown & & \diagup \\ & C = C & \\ \diagup & & \diagdown \\ H & & I \end{array}$$

8. Double bond, single bond

9.

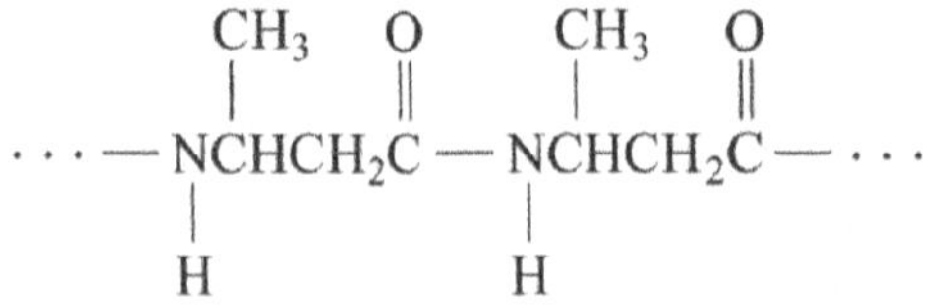

$$\cdots - \underset{\underset{H}{|}}{\overset{\overset{CH_3}{|}}{N}}CHCH_2\overset{\overset{O}{\|}}{C} - \underset{\underset{H}{|}}{\overset{\overset{CH_3}{|}}{N}}CHCH_2\overset{\overset{O}{\|}}{C} - \cdots$$

10. HCl

Exercises for Self-Testing

A. Completion

1. polyvinyl chloride, PVC   **2.** tetrafluoroethene   **3.** linear, cyclic   **4.** amide

5. peptide (or amide)   **6.** essential   **7.** adenine, cytosine, guanine, thymine

8. codon   **9.** double helix, hydrogen bonds   **10.** viruses

B. True/False

1, 4, and 10 are true.

2 is false. Ethane has no double or triple bonds that would break to form C—C bonds in a polymer.

3 is false. The ones that must be supplied by the diet are the essential amino acids.

5 is false. The pairs are cytosine and guanine; and adenine and thymine.

6 is false. Polypeptides contain fewer amino acids than proteins.

7 is false. The monomer units are glucose in each case.

8 is false. The components are sugars, phosphates, and bases.

9 is false. Viruses are nonliving organisms.

C. Questions and Problems

1. A monomer is the compound that is used to prepare the polymer. A repeating unit is the repeating structure that appears in the polymer. The monomer and three repeating units for polyethylene are shown below.

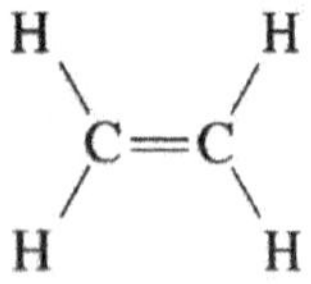

$$\begin{array}{ccc} H & & H \\ \diagdown & & \diagup \\ & C = C & \\ \diagup & & \diagdown \\ H & & H \end{array}$$

$$\cdots - \underset{\underset{H}{|}}{\overset{\overset{H}{|}}{C}} - \underset{\underset{H}{|}}{\overset{\overset{H}{|}}{C}} - \underset{\underset{H}{|}}{\overset{\overset{H}{|}}{C}} - \underset{\underset{H}{|}}{\overset{\overset{H}{|}}{C}} - \underset{\underset{H}{|}}{\overset{\overset{H}{|}}{C}} - \underset{\underset{H}{|}}{\overset{\overset{H}{|}}{C}} - \cdots$$

    Monomer 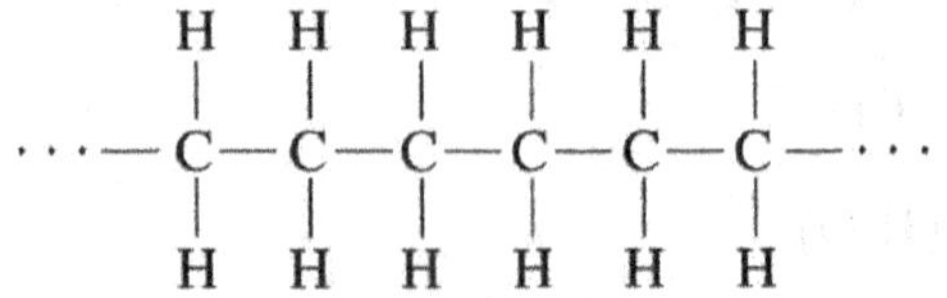   Repeating unit

2. Three-base sequences in DNA molecules correspond to individual amino acids that will make up the protein to be synthesized. The DNA molecule transmits the information to RNA molecules, which serve as the template for protein synthesis. The original sequence of bases in the DNA therefore determines the order in which the amino acids will be joined to form a protein. Hence, the DNA is the genetic blueprint for protein synthesis.

3.

4. The monomer is:

The repeating unit is:

The linkages are peptide linkages.

5. Once inside a living organism, the virus releases its DNA into the cells of the host, taking over the protein-making apparatus of the host cells and forcing them to reproduce more viruses. The host cells soon become overwhelmed by the viruses and die.

6. G & C; guanine (G) and cytosine (C) form three hydrogen bonds when they bind to each other, whereas adenine (A) and thymine (T) form only two hydrogen bonds.

Selected Solutions

Chapter 1

1   See solution in textbook.

3   Fog is a heterogeneous mixture. A microscopic spoon used to take samples would pick up water in some spots and air in others.

5   (d) and (e) are compounds. (a) is an elemental substance, (b) is a heterogeneous mixture, and (c) is a homogeneous mixture.

7   See solution in textbook.

9   False. It is ethanol in the gas state but still ethanol. Changes in state do not cause chemical change.

11 See solution in textbook.

13 (b) Both products contain hydrogen, H, an element not present in either reactant.

15 A law summarizes experimental data and states an experimentally proved relationship between natural phenomena.

17 Science is the experimental investigation and explanation of natural phenomena. Technology is the application of scientific knowledge. Here, the discovery of electromagnetic induction is an example of science and the use of magnets to produce electricity is an example of technology.

19 Numerous answers possible. Example: insecticides were developed in chemical laboratories to help control agricultural pests (positive result), but they pollute lakes and streams (negative result).

21 Numerous answers possible.

23 Yes, because every compound is a pure substance and a mixture is defined as two or more pure substances intermingled. Example: salt water is a mixture of the two compounds $NaCl$ and H_2O.

25 It is a heterogeneous mixture. A microscopic spoon would be able to extract either flour or sugar, depending on where in the mixture the spoon was placed. No amount of mechanical grinding of two solids can ever produce a solution (which, remember, is just another name for a homogeneous mixture).

1.27 Air (assuming it is free of all the pollution particles!) is a solution because the atoms and molecules of which it is composed are evenly dispersed throughout, resulting in a uniform composition throughout.

1.29 Na comes from *natrium,* the Latin word meaning "sodium," and Fe comes from *ferrum,* the Latin word meaning "iron." With only 26 letters in the English alphabet and 118 known elements, there are not enough letters for each element to have a one-letter symbol. Some names and symbols of atoms may be also influenced by other foreign languages (e.g., Ag from Greek *argentum*), names of scientists (e.g., Rf for Ernest *Rutherford*) or a country (e.g., Po for *Pologne–Poland* in French).

1.31 Lead, Pb; molybdenum, Mo; tungsten, W; chromium, Cr; mercury, Hg.

1.33 Ti, titanium; Zn, zinc; Sn, tin; He, helium; Xe, xenon; Li, lithium.

1.35 An elemental substance contains only one type of atom, while a compound contains two or more different types of atoms. Here, both S_8 and S_6 are elemental substances, while SO_2 is a compound.

1.37 Numerous answers possible. Example: the elements hydrogen and oxygen, both gases at room temperature, combine to form the compound water, a liquid at room temperature.

1.39 Chlorine, Cl_2, sulfur, S_8, and neon, Ne, are elemental substances (atoms of only one element); octane, C_8H_{18}, is a compound (atoms of more than one element).

1.41 The chemical formula for nonane is C_9H_{20}.

1.43 (a) Knowing the chemical formula N_2, you can say that the smallest possible piece of nitrogen gas contains two nitrogen atoms.

(b) Nitrogen gas, N_2, is not a compound. It is an elemental substance because it contains only one type of atom, N.

1.45 Sublimation is the process whereby matter changes directly from the solid state to the gas state.

1.47 The propane would be a gas, and the hexane would be a liquid.

1.49 Sublimation is occurring. The compound the mothballs are made of is going directly from the solid state to the gas state.

1.51 O_2 and O_3 are completely different substances and therefore have completely different physical and chemical properties. When determining chemical and physical properties, the *number* of atoms in a molecule is as important as *which* atoms are present.

1.53 (a) Sugar has undergone a chemical change because its chemical composition has been changed. New compounds, CO_2 and H_2O, have been formed.

(b) Condensation: $H_2O(g)$ condenses into $H_2O(l)$.

(c) The fog cannot be due to the condensation of $CO_2(g)$ as the temperature of the window is well above the condensation point of CO_2 of -78 °C.

1.55 Gasoline (C_8H_{18}) reacts with oxygen (O_2) in a car's engine, forming new compounds: $CO_2(g)$ and $H_2O(g)$. When water vapor condenses at the end of the exhaust pipe, it becomes $H_2O(l)$ and drips out.

1.57 This is a chemical change because the pure substance Ag is transformed into the different pure substance Ag_2S.

1.59 A law is a statement that summarizes experimental data. A theory is a statement that proposes an explanation of why a law is true.

1.61 Numerous answers possible. For instance, you may have noticed that some plants thrive in full sun. So, formulate a *prediction* or *hypothesis,* such as "the longer the exposure to sunlight, the faster the plant grows," then conduct a series of experiments, in which you measure the growth of several plants of the same kind when allowed different number of hours in the sun. Ensure that all other experimental conditions are the same (e.g., type of soil, temperature, moisture). Collect *data* for each exposure time and analyze them. Formulate a *theory* based on the collected data by attempting to explain the results. Then repeat the cycle by selecting a different plant type. If new experiments provide results that do not support your theory, you need to revise or replace the theory. If new data obtained with many more different plant types support your theory, you can attempt to formulate a *law.*

1.63 C, Na, and Hg are elemental substances; $NaHCO_3$ and CO_2 are compounds.

1.65 (a) Science. (b) Science. (c) Technology. (d) Science. (e) Technology.

1.67 This process is a chemical reaction because a new substance is formed.

1.69 Evaporation.

1.71 (a) Gold. (b) Mercury. (c) Potassium. (d) Phosphorus. (e) Silver.

1.73 This is the definition of a scientific law.

1.75 The sugar dissolved in water is a liquid that is a mixture, and the melted sugar is a liquid that is a pure substance.

1.77 Numerous answers possible. One is ammonia gas dissolved in water.

1.79 Numerous answers possible. One is oxygen gas dissolved in nitrogen gas, which is a homogeneous mixture because one substance *dissolved* in another forms a *solution*, which is another name for homogeneous mixture.

1.81 A chemical change; a change in color generally indicates that a new substance is formed.

1.83 You could use a magnet to separate the iron filings from the sand.

1.85 This separation cannot be done physically because pure water is a compound and the components of a compound can be separated only by *chemical* means.

1.87 This is a scientific theory because it offers an explanation ("because they all have a single electron . . .") for some experimentally observed behavior (that these metals react with water).

1.89 The 14-karat gold is the homogeneous mixture made by melting gold and other metals together and then letting the solution solidify.

1.91 True.

1.93 Bronze, which is a solid solution of copper metal and tin metal.

1.95 NaCl is the only compound in the list, being made of two types of atoms. Ozone and liquid nitrogen are elemental substances, 18-karat gold is a homogeneous mixture of gold and other metals, and iced tea is a heterogeneous mixture of a tea solution and solid H_2O.

1.97 The hydrogen peroxide (H_2O_2) in beaker (a) is undergoing a chemical change because it is converted into water (H_2O) and oxygen (O_2), both different substances. The water in beaker (b) is undergoing a physical change because it is still water after the change.

1.99 A degree Celsius, °C, is larger than a degree Fahrenheit, °F. It is about twice as large because a change of 100 °C represents the same temperature change as a difference of 180 °F. Note that on the Celsius scale the melting point of ice is 0 °C and the boiling point of water is 100 °C (a change of 100 °C). However, on the Fahrenheit scale, the melting point of ice is 32 °F and the boiling point of water is 212 °F (a change of 180 °F).

1.101 No. There is also a change in the substances, making this a chemical change.

Chapter 2

2.1 See solution in textbook.

2.3 Jack will be more accurate. If he completely fills the half-quart container twice, the total volume will be very close to 1 quart. However, Jill needs to estimate 1/40 of the 10-gallon container, which is difficult to do with much accuracy (1/40 because 1 gallon = 4 quarts).

2.5 The uncertainty is ± 0.1 gallon because the last digit in the measured volume, 16.0 gallons, is in the tenths column.

2.7 See solution in textbook.

2.9

	Number of significant figures	Uncertainty
10.0	3	± 0.1
0.004 60	3	± 0.00001
123	3	± 1

2.11 0.473 (the negative exponent means the number gets smaller).

2.13 See solution in textbook.

2.15 6000

2.17 $4.710\,000\,0 \times 10^{13}$. The fact that the uncertainty is ± 1 million tells you the final significant digit is in the 1-million column, which in this number is the fifth zero from the left.

2.19 See solution in textbook.

2.21 660. hours. The exact 3 has an infinite number of significant figures, meaning the number of significant figures in the answer is determined by the value 220. hours. The decimal point following the zero tells you this number has three significant figures, and that is how many the answer must have.

2.23 See solution in textbook.

2.25 See solution in textbook.

2.27 142 cm

-0.48 cm

141.52 cm, which rounded off to the correct number of significant figures is 142 cm.

2.29 4.736 km. The fact that 1 km is the same as 1000 m means that 4.736 km is the same as 4.736×1000 m = 4736 m.

2.31 See solution in textbook.

2.33 $1\ cm^3 = 1$ mL, which means that $246.7\ cm^3 = 246.7$ mL.

2.35 See solution in textbook.

2.37 $\dfrac{500.0\ \text{g}}{150.5\ \text{mL}} = 3.322\ \text{g/mL}$

2.39 $\dfrac{1\ \text{day}}{24\ \text{h}}\quad \dfrac{24\ \text{h}}{1\ \text{day}}$

2.41 $\dfrac{600.0\ \text{miles}}{1\ \cancel{\text{h}}} \times 50.0\ \cancel{\text{h}} = 3.00 \times 10^4\ \text{miles}$

2.43 $500.0 \, \cancel{L} \times \dfrac{1000 \, \cancel{mL}}{1 \, \cancel{L}} \times \dfrac{0.001 \, 30 \, g}{1 \, \cancel{mL}} = 650. \, g = 6.50 \times 10^2 \, g$

$650. \, \cancel{g} \times \dfrac{1 \, kg}{1000 \, \cancel{g}} = 0.650 \, kg$

2.45 Conversion factors: $\dfrac{6 \, cups \, flour}{1 \, cake} \qquad \dfrac{1 \, cup \, flour}{120.0 \, g \, flour}$

$6955 \, \cancel{g \, flour} \times \dfrac{1 \, \cancel{cup \, flour}}{120.0 \, \cancel{g \, flour}} \times \dfrac{1 \, cake}{6 \, \cancel{cups \, flour}} = 9.660 \, cakes$

You can bake nine cakes (it's not possible to bake a partial cake).

2.47 See solution in textbook.

2.49 See solution in textbook.

2.51 See solution in textbook.

2.53 $3.63 \, \cancel{kJ} \times \dfrac{1000 \, J}{1 \, \cancel{kJ}} = 3630 \, J$

$3.63 \, \cancel{kJ} \times \dfrac{1 \, Cal}{4.184 \, \cancel{kJ}} = 0.868 \, Cal$

$3.63 \, \cancel{kJ} \times \dfrac{1 \, cal}{4.184 \, \cancel{J}} = \dfrac{1000 \, \cancel{J}}{1 \, \cancel{kJ}} = 868 \, cal$

2.55 The 3 in "3 ft in a yard" is an exact number and therefore is really 3.0000... , with an unlimited number of significant figures. The 3 in "a certain piece of wood is 3 ft long" comes from a measurement and therefore has some uncertainty associated with it.

2.57 You should choose the accurate result because a precise value that is not accurate is useless. An average of accurate results that were not precise usually gets you closer to the true value. On the other hand, an average of inaccurate but precise results may be far off the true value. 1 in. above the average height measured accurately will be safer than 1 in. above the average height measured precisely, but inaccurately.

2.59 The person with the tape measure. He or she needs to make only one measurement, but the person with the ruler has to make at least 200 measurements and add them to get the length. There would be uncertainty associated with each measurement, resulting in a significant loss of accuracy in the result.

2.61 The uncertainty lies in the last digit written in the number. We often assume an uncertainty of ± 1 in the position of the uncertain digit. Some typical examples:

15.2 cm 15.2 ± 0.1 cm (if measured using a ruler marked with centimeters only)

1534 cm³ 1534 ± 1 cm³ (if measured using a cylinder marked every 10 cm³)

0.00987 g 0.00987 ± 0.00001 g (if measured using a so-called analytical balance)

2.63 (a) 12.60 ± 0.01 cm

(b) 12.6 ± 0.1 cm

(c) $0.000 \, 000 \, 03 \pm 0.000 \, 000 \, 01$ inch

(d) 125 ± 1 foot

2.65 Replacing the uncertain digit, 5, by 1 gives an uncertainty of 0.1 million years (or 100,000 years).

2.67 (a) 12.202   (b) No significant zeros.   (c) 2̲05   (d) 0.010̲

2.69 It is not clear whether 30 has one or two significant figures because the zero may or may not be significant. Adding the decimal point at the end of the number indicates that the trailing zero is significant, meaning 30. has two significant digits.

2.71 (a) 56.0 kg (three significant figures).

(b) 0.000 25 m (two significant figures).

(c) 5,600,000 miles (four significant figures, but you cannot tell that by looking at this standard notation).

(d) 2 ft (one significant figure).

2.73 (a) 3×10^1 ft

(b) 3.0×10^1 ft

(c) 3.00×10^1 ft

2.75 (a) 2.26×10^2 (b) 2.260×10^2 (c) 5.0×10^{-10} (d) 3×10^{-1}

(e) 3.0×10^{-1} (f) 9.00×10^8 (g) 9.000006×10^8

2.77 102 inches because the least certain measured value, either 100. or 2, has its uncertain digit in the ones position, which means the answer has its uncertain digit in the ones position. The uncertainty is ± 1 in.

2.79 The answer of 3.873143939 miles has too many reported digits. The result of the division of 20,450.2 ft by 5280 ft per mile should be reported as 3.87314 miles (six significant figures because 20,450.2 has six significant digits and 5280 is an exact number).

2.81 Length, meter; volume, cubic meter.

2.83 It is always correct to use cm^3 instead of mL. The two units are exactly equivalent.

2.85 (a) 2.31×10^9 m (b) 5.00×10^{-6} m (c) 1.004×10^0 m (d) 5.00×10^{-12} m
(e) 2.5×10^2 m

2.87 The Celsius and Fahrenheit scales can have negative temperature values. The Kelvin scale cannot because the zero point on the Kelvin scale is absolute zero. There is no colder temperature possible than absolute zero, 0 K.

2.89 32 °F; 0 °C; 273 K

2.91 (a) In the left cylinder, each shorter mark is 0.1 mL, which means the uncertain digit in a volume measurement must be in the hundredths position. The uncertainty is thus ± 0.01 mL. In the right cylinder, each shorter mark is 10 mL, which means the uncertain digit in a volume measurement is in the ones position and the uncertainty is ± 1 mL.

(b) The left cylinder contains 1.18 ± 0.01 mL. The right cylinder contains 98 ± 1 mL. Adding the two numbers yields 98 mL + 1.18 mL = 99.18 mL, which must be reported as 99 mL because the 98 value restricts your answer to being uncertain in the ones position. The uncertainty in this value is ± 1 mL.

2.93 The student who reports 1.5 cm used the ruler incorrectly. The ruler is marked in millimeters, which is tenths of centimeters. The uncertainty therefore lies in the hundredths place, and the measurement should be reported to the hundredths place—1.50 cm.

2.95 Density is the amount of mass in a given volume of a material. It is called a derived unit because it is a combination of one SI base unit, mass, and one SI derived unit, volume.

2.97 From Table 2.4, you know that the density of mercury at 25 °C is 13.6 g/mL. Therefore,

$$2.0 \, \text{L} \times \frac{1000 \, \text{mL}}{1 \, \text{L}} \times 13.6 \, \frac{\text{g}}{\text{mL}} = 27{,}200 \, \text{g} = 2.7 \times 10^4 \, \text{g}$$

2.99 First determine your own mass using a bathroom scale. One possible way would be then to enter the tub and fill the tub all the way to the top with water, making sure you are completely immersed. Next, carefully come out of the tub, sponging any residual water off yourself back into the tub. Using a measuring cup, refill the tub to the previous level. Keep track of the added volume. Your own volume will be equal to the volume of water that had to be replaced. Calculate density by dividing your mass by the volume of water that had to be replaced.

2.101 Place a mixture of gold and fool's gold in a container filled with liquid mercury, which has a density of 13.6 g/mL. Fool's gold, with a density of 5.02 g/mL, is less dense than the mercury and therefore floats. Gold, with a density of 19.3 g/mL, is denser than the mercury and therefore sinks.

2.103 $100.0 \, \text{miles} \times \dfrac{1 \, \text{h}}{45.0 \, \text{miles}} \times \dfrac{60 \, \text{min}}{1 \, \text{h}} = 133 \, \text{min}$

2.105 $100.0 \, \text{glonkins} \times \dfrac{0.911 \, \text{ounce}}{1 \, \text{glonkin}} \times \dfrac{28.35 \, \text{g}}{1 \, \text{ounce}} \times \dfrac{1 \, \text{mL}}{19.3 \, \text{g}} \times \dfrac{1 \, \text{L}}{1000 \, \text{mL}} = 0.134 \, \text{L}$

2.107 Note that since $1.00 \, \text{cm} = 10.0 \, \text{mm}$, $9.56 \times 10^2 \, \text{mm} = 95.6 \, \text{cm}$. Therefore:

Volume $= 10.2 \, \text{cm} \times 43.7 \, \text{cm} \times 95.6 \, \text{cm} = 4.26 \times 10^4 \, \text{cm}^3 = 4.26 \times 10^4 \, \text{mL}$.

$$4.26 \times 10^4 \, \text{mL} \times \frac{1 \, \text{L}}{1000 \, \text{mL}} = 42.6 \, \text{L}$$

2.109 (a) The length of the edge $= 100.0 \, \text{cm} + 1.40 \, \text{cm} = 101.4 \, \text{cm}$. You must report the answer to the tenths place because a sum cannot be more certain than the least certain measurement, which in this case is the 100.0 cm.

(b) Volume $= (101.4 \, \text{cm})^3 = 1.043 \times 10^6 \, \text{cm}^3 = 1.043 \times 10^6 \, \text{mL}$

(c) Density $= \dfrac{111 \, \text{kg}}{1.043 \times 10^6 \, \text{ml}} \times \dfrac{1000 \, \text{g}}{1 \, \text{kg}} = 0.106 \, \text{g/mL}$

2.111 You must convert both units of the given speed, and that means many conversion factors. Just take things one step at a time. Start with the numerator, meters to miles; then continue with the denominators, seconds to hours:

$$80.0 \, \frac{\text{m}}{\text{s}} \times \frac{3.28 \, \text{ft}}{1 \, \text{m}} \times \frac{1 \, \text{mile}}{5280 \, \text{ft}} \times \frac{60 \, \text{s}}{1 \, \text{min}} \times \frac{60 \, \text{min}}{1 \, \text{h}} = 179 \, \text{miles/h}$$

2.113 To solve for x means to get x alone on one side of the equals sign—in other words, to *isolate x*. For $y = z/x$, a good first step is to get x out of the denominator and onto the left side, accomplished by multiplying both sides by x:

$$x \times y = \frac{z}{\cancel{x}} \times \cancel{x} \Rightarrow xy = z$$

Dividing both sides by y isolates x:

$$\frac{x\cancel{y}}{\cancel{y}} = \frac{z}{y} \Rightarrow x = \frac{z}{y}$$

(b) First multiply both sides of the equation by x to get it out of the denominator and onto the left side:

$$x \times y = \frac{z}{2\cancel{x}} \times \cancel{x} \Rightarrow xy = \frac{z}{2}$$

Dividing both sides by y isolates x:

$$\frac{x\cancel{y}}{\cancel{y}} = \frac{z}{2y} \Rightarrow x = \frac{z}{2y}$$

2.115 One idea would be to place all x-containing terms on one side of the equation and the plain numbers on the other side of the equation. One can do this by subtracting $3x$ and adding 6 to both sides of the equation. Dividing both sides of the equation by 2 isolates x:

$$5x - 3x - \cancel{6} + \cancel{6} = \cancel{3x} - \cancel{3x} - 8 + 6 \Rightarrow 2x = -2$$

$$\frac{\cancel{2}x}{\cancel{2}} = \frac{-\cancel{2}}{\cancel{2}} \Rightarrow x = -1$$

2.117 With unit analysis, start with the information given and multiply by the appropriate conversion factor:

$$\frac{1.15\,\text{g}}{\cancel{\text{mL}}} \times 50.00\,\cancel{\text{mL}} = 57.5\text{g}$$

The answer is the same as in Problem 2.116.

2.119 1 cal is the amount of heat energy necessary to warm 1 g of water from $25\,°\text{C}$ to $26\,°\text{C}$.

2.121 The specific heat for any substance is the amount of heat energy necessary to increase the temperature of 1 g of the substance by $1\,°\text{C}$.

2.123 Use specific heat of water of $4.184\,\text{J/g} \cdot °\text{C}$ from Table 2.5:

$$\underbrace{2.00\,\cancel{\text{L}} \times \frac{1000\,\cancel{\text{mL}}}{1\,\cancel{\text{L}}} \times \frac{1.00\,\cancel{\text{g}}}{\cancel{\text{mL}}}}_{\text{Water mass}} \times \underbrace{\frac{4.184\,\text{J}}{\cancel{\text{g}} \cdot \cancel{°\text{C}}}}_{\substack{\text{Specific} \\ \text{heat of water}}} \times \underbrace{18.0\,\cancel{°\text{C}}}_{\substack{\text{Temperature} \\ \text{increase}}} = 1.51 \times 10^5\,\text{J}$$

$$1.51 \times 10^5\,\text{J} \times \frac{1\,\text{kJ}}{1000\,\cancel{\text{J}}} = 151 \times 10^2\,\text{kJ}$$

$$1.51 \times 10^2\,\text{kJ} \times \frac{1\,\text{Cal}}{4.184\,\cancel{\text{kJ}}} = 36.0\,\text{Cal}$$

Alternatively, you could use the specific heat of water of 1.000 cal/g · °C from Table 2.5:

$$2.00 \text{ L} \times \underbrace{\frac{1000 \text{ mL}}{1 \text{ L}} \times \frac{1.00 \text{ g}}{\text{mL}}}_{\text{Water mass}} \times \underbrace{\frac{1.000 \text{ cal}}{\text{g} \cdot °\text{C}}}_{\substack{\text{Specific} \\ \text{heat of water}}} \times \underbrace{18.0 \,°\text{C}}_{\substack{\text{Temperature} \\ \text{increase}}} = 3.60 \times 10^4 \text{ cal}$$

$$3.60 \times 10^4 \text{ cal} \times \frac{1 \text{ Cal}}{1000 \text{ cal}} = 36.0 \text{ Cal}$$

2.125 First, determine the amount of heat released when 2.50 g of wood is burned:

$$0.200 \text{ kg water} \times \frac{1000 \text{ g}}{1 \text{ kg}} \times \frac{4.184 \text{ J}}{\text{g} \cdot °\text{C}} \times 6.6 \,°\text{C} = 5.5 \times 10^3 \text{ J}$$

$$\frac{5.5 \times 10^3 \text{ J}}{2.50 \text{ g}} = 2.2 \times 10^3 \text{ J per gram of wood}$$

2.127 You need 2.44×10^4 J of heat energy, and each gram of wood supplies 2.2×10^3 J. Therefore, the mass of wood you need is

$$2.44 \times 10^4 \text{ J} \times \frac{1 \text{ g wood}}{2.2 \times 10^3 \text{ J}} = 11 \text{ g wood}$$

2.129 (c) 1230.0 m has five significant digits, and the converted value must also have five:

$$1230.0 \text{ m} \times \frac{1 \text{ km}}{1000 \text{ m}} = 1.2300 \text{ km}$$

Answer (d) has the correct number of significant digits but the wrong prefix on the unit:

$$1230.0 \text{ m} \times \frac{1000 \text{ mm}}{1 \text{ m}} = 1.2300 \times 10^6 \text{ mm} \neq 1.2300 \text{ mm}$$

2.131 $V = \left(\dfrac{4}{3}\right)\pi r^3 = \left(\dfrac{4}{3}\right) \times 3.14159 \times (4.00 \text{ cm})^3 = 268 \text{ cm}^3$

2.133 (b) 0.000 0003 L has only one significant figure, meaning the converted value can have only one. Answer (a) has the correct number of significant figures but the wrong prefix on the unit:

$$3 \times 10^{-6} \text{ L} \times \frac{1000 \text{ mL}}{1 \text{ L}} = 0.003 \text{ mL} \neq 3 \text{ mL}$$

2.135 $\dfrac{11.0 \text{ km}}{\text{L}} \times \dfrac{1 \text{ mi}}{1.61 \text{ km}} \times \dfrac{3.79 \text{ L}}{1 \text{ gallon}} = 25.9 \text{ mi/gallon}$

2.137 1 Calorie = 1000 calories = 1 kilocalorie; 1 calorie = 0.001 Calorie = 1 milliCalorie.

2.139 (a) Subtract 32 from both sides and then multiply both sides by 5/9:

$$°\text{F} - 32 = \frac{9}{5}°\text{C} + 32 - 32 \Rightarrow °\text{F} - 32 = \frac{9}{5}°\text{C}$$

$$\frac{5}{9} \times (°\text{F} - 32) = \frac{5}{9} \times \left(\frac{9}{5}°\text{C}\right) \Rightarrow \frac{5}{9} \times (°\text{F} - 32) = °\text{C}$$

(b) Divide both sides by nR:

$$\frac{PV}{nR} = \frac{nRT}{nR} \Rightarrow \frac{PV}{nR} = T$$

(c) Multiply both sides by λ and divide both sides by E:

$$E \times \lambda = \frac{hc}{\lambda} \times \lambda \Rightarrow E \times \lambda = hc$$

$$\frac{E\lambda}{E} = \frac{hc}{E} \Rightarrow \lambda = \frac{hc}{E}$$

2.141 (a) $\dfrac{1.34\text{ g}}{\text{L}} \times \dfrac{1\text{ L}}{1000\text{ mL}} = 1.34 \times 10^{-3}\text{ g/mL}$

(b) $\dfrac{1.34\text{ g}}{\text{L}} \times \dfrac{1\text{ kg}}{1000\text{ g}} = 1.34 \times 10^{-3}\text{ kg/L}$

(c) $\dfrac{1.34\text{ g}}{\text{L}} \times \dfrac{1\text{ kg}}{1000\text{ g}} \times \dfrac{1\text{ L}}{1000\text{ mL}} = 1.34 \times 10^{-6}\text{ kg/mL}$

2.143 Multiplying or dividing: the number of significant figures in the answer is determined by which multiplied/divided number has the fewest significant figures. Examples: $725 \times 2.6352 = 1.91 \times 10^3$, $427.45 \div 3.0 = 1.4 \times 10^2$. Adding or subtracting: the number of significant figures in the answer is determined by which added/subtracted number is least certain. Examples: $725 + 2.6352 = 728$, $427.45 - 3.0 = 424.5$.

2.145 (a) Neither accurate nor precise. The large spread between the highest and lowest values means the set is not precise. The average value $(6.38\text{ g} + 9.23\text{ g} + 4.36\text{ g}) \div 3 = 6.66\text{ g}$ tells you the set is not accurate.

(b) Both accurate (the average value is 8.56 g) and precise (very small spread in the three values).

(c) Accurate (the average is 8.54 g) but not precise (large spread).

(d) Precise (very small spread) but not accurate (average 6.26 g).

2.147 Density $= \dfrac{\text{Mass}}{\text{Volume}}$

Volume $= 3.0\text{ cm} \times 4.0\text{ cm} \times 5.0\text{ cm} = 60\text{ cm}^3$

Density $= \dfrac{470.0\text{ g}}{60.\text{ cm}^3} = 7.8\text{ g/cm}^3$

2.149 Volume $= 28.10\text{ mL} - 25.00\text{ mL} = 3.10\text{ mL}$

Density $= \dfrac{\text{Mass}}{\text{Volume}} = \dfrac{8.34\text{ g}}{3.10\text{ mL}} = 2.69\text{ g/mL}$

2.151 The conversion equation is $°\text{F} = 32 + \left(\dfrac{9}{5}\right)°\text{C}$. Make approximations the quick way by adding 30 instead of 32 and multiplying by 2 instead of $\dfrac{9}{5}$. Because $\dfrac{9}{5}$ (1.8) is only a little less than 2 and 32 is only a little more than 30, the errors introduced by making these two approximations tend to cancel, giving a result close to the actual value. As an example, convert 80.5 °C:

Quick way $80.5 \times 2 = 161$

$161 - 16 = 145$

$145 + 30 = 175\,°\text{F}$

Equation $32 + \dfrac{9}{5}(80.5\,°\text{C}) = 177\,°\text{F}$

2.153 Although water contains no calories, the body must expend ("burn") energy to raise the temperature of the ice-cold water from approximately 0 °C to the body temperature of 37 °C.

2.155 Density = Mass/Volume

Density of A = 200.0 g/25.64 mL = 7.800 g/ml

Density of B = 200.0 g/10.36 mL = 19.31 g/ml

Density of C = 200.0 g/17.54 mL = 11.40 g/ml

Therefore A is iron, B is gold, and C is lead.

2.157 (a) The uncertain digit, 8, is in the tenths position, making the uncertainty ± 0.1 m.

(b) The uncertain digit, 6, is in the ten-thousandths position, making the uncertainty ± 0.0001 g.

(c) $\pm 0.001 \times 10^3 = \pm 1$ L

(d) ± 1 cm

(e) No uncertainty because 18 here is an exact number.

2.159 A Calorie is 1000 calories.

2.161 (a) When adding/subtracting numbers written in scientific notation, first change all numbers to the same power of 10:

$$
\begin{array}{c}
5.03 \times 10^2 \\
+\ 0.81 \times 10^2 \\
\hline
5.84 \times 10^2
\end{array}
\quad \text{or} \quad
\begin{array}{c}
50.3 \times 10^1 \\
+\ 8.1 \times 10^1 \\
\hline
58.4 \times 10^1 = 5.84 \times 10^2
\end{array}
$$

(b) 4.4×10^{-1}; only two significant digits because of the 0.53.

(c) 2.01×10^{23}; three significant digits because the 3 is exact, meaning the number of significant digits is determined by the 6.02.

(d) As in part (a), change all numbers to the same power of 10:

$$
\begin{array}{c}
3.960 \times 10^3 \\
-0.462 \times 10^3 \\
\hline
3.498 \times 10^3
\end{array}
\quad \text{or} \quad
\begin{array}{c}
39.60 \times 10^2 \\
-\ 4.62 \times 10^2 \\
\hline
34.98 \times 10^2 = 3.498 \times 10^3
\end{array}
$$

2.163 (a) $20 \text{ atoms} \times \dfrac{20.2 \text{ atomic mass units}}{1 \text{ atom}} = 404$ atomic mass units

(b) $20 \text{ atoms} \times \dfrac{20.2 \text{ atomic mass units}}{1 \text{ atom}} \times \dfrac{1.66 \times 10^{-24} \text{ g}}{1 \text{ atomic mass units}} = 6.71 \times 10^{-22}$ g

(c) $6.022 \times 10^{23} \text{ atoms} \times \dfrac{20.2 \text{ atomic mass units}}{1 \text{ atom}} \times \dfrac{1.66 \times 10^{-24} \text{ g}}{1 \text{ atomic mass units}} = 20.2$ g

or

$6.022 \times 10^{23} \text{ atoms} \times \dfrac{6.71 \times 10^{-22} \text{ g}}{20 \text{ atoms}} = 20.2$ g

2.165 Solve the density equation for volume and then insert the data:

$$\text{Density} = \frac{\text{Mass}}{\text{Volume}} \qquad \text{Volume} = \frac{\text{Mass}}{\text{Density}}$$

(a) $\dfrac{15.0\ \text{g}}{0.997\ \text{g/mL}} = 15.0\ \text{mL}$

(b) $\dfrac{15.0\ \text{g}}{0.917\ \text{g/mL}} = 16.4\ \text{mL}$

(c) $\dfrac{15.0\ \text{g}}{0.7\ \text{g/mL}} = 21.4\ \text{mL} = 2 \times 10^1\ \text{mL}$ (only one significant figure allowed)

(d) $\dfrac{15.0\ \text{g}}{11.4\ \text{g/mL}} = 1.32\ \text{mL}$

(e) $\dfrac{15.0\ \text{g}}{13.6\ \text{g/mL}} = 1.10\ \text{mL}$

(f) $\dfrac{15.0\ \text{g}}{0.00018\ \text{g/mL}} = 8.3 \times 10^4\ \text{mL}$

2.167 (a) Six. (b) Three. (c) Two. (d) Four. (e) Four.

2.169 (a) True.

(b) False. When adding or subtracting a series of measured values, the number of significant figures in the answer is limited by the least certain measured value. Example: $3724 + 3.2 = 3727$. The 3.2 has the fewest significant figures but is certain to the tenths position. The 3724 is certain only to the ones position, and therefore is the measured value that determines the number of significant figures in the answer—the answer must be uncertain in the ones position. In this example, that restriction means four significant figures are allowed in the answer despite the 3.2.

2.171 $15\ \text{weeks} \times \dfrac{3\ \text{h}}{\text{week}} \times \dfrac{60\ \text{min}}{1\ \text{h}} \times \dfrac{60\ \text{s}}{1\ \text{min}} \times \dfrac{1000\ \text{ms}}{1\ \text{s}} = 1.6 \times 10^8\ \text{ms}$

2.173 (a) 5.02×10^5 (b) 3.8402×10^{-5} (c) 4.36×10^8

(d) 8.47×10^3 (e) 5.91×10^{-3} (f) 6.58×10^{-1}

2.175 $350\ \text{Martian dollars} \times \dfrac{1\ \text{U.S. dollar}}{1.54\ \text{Martian dollars}} = 227\ \text{U.S. dollars}$

2.177 (a) Two. (b) Two. (c) None. (d) Cannot tell. (e) Three.

2.179 No. In 580, the decimal point indicates that the trailing zero is significant. In 580, the trailing zero may or may not be significant; there is no way to tell with the value written this way.

2.181 It is more likely that the student's laboratory technique is bad because his measurements vary widely.

2.183 (a) Density = Mass/Volume = $195\ \text{g}/25.0\ \text{cm}^3 = 7.80\ \text{g/cm}^3$

(b) Volume = Mass/Density = $500.0\ \text{g}/7.80\ \text{g/cm}^3 = 64.1\ \text{cm}^3$

(c) The substance floats in mercury because it is less dense than mercury.

2.185 (a) $\dfrac{1.3 \times 10^{-3}\,\cancel{g}}{\cancel{mL}} \times \dfrac{1\,kg}{1000\,\cancel{g}} \times \dfrac{1000\,\cancel{mL}}{1\,L} = 1.3 \times 10^{-3}\,kg/L$

(b) $\dfrac{1.3 \times 10^{-3}\,\cancel{g}}{\cancel{mL}} \times \dfrac{1\,\cancel{kg}}{1000\,\cancel{g}} \times \dfrac{2.204\,1b}{1\,\cancel{kg}} \times \dfrac{1000\,\cancel{mL}}{1\,\cancel{L}} \times \dfrac{1\,\cancel{L}}{1.057\,\cancel{qt}} \times \dfrac{4\,\cancel{qt}}{1\,gallon} = 1.1 \times 10^{-2}\,1b/gallon$

2.187 (a) 1.0726×10^4. Changing both numbers in the sum from scientific notation to standard notation shows that both are uncertain in the ones position. Therefore, their sum has its uncertainty in the ones position: $9865 + 861 = 10726$.

(b) 4.4×10^{-19}. Only two significant figures allowed because of the 4.5.

(c) 1.471×10^{-18}. The 3.821 restricts the answer to four significant figures.

(d) 9.0618×10^2. Because the 5 is exact, the number of significant figures in the answer is determined by the two numbers of the subtraction. Standard notation shows that both numbers are uncertain in the tenths position, and therefore their difference has five significant figures: $4560.0 - 29.1 = 4530.9$. Dividing this value by the exact number 5 gives an answer having five significant figures.

2.189 (a) $\dfrac{1\,kg}{1000\,g} \qquad \dfrac{1000\,g}{1\,kg}$

(b) $\dfrac{1\,g}{0.001\,kg} \qquad \dfrac{0.001\,kg}{1\,g}$

(c) $\dfrac{1\,yd}{3\,ft} \qquad \dfrac{3\,ft}{1\,yd}$

(d) $\dfrac{1\,m}{100\,cm} \qquad \dfrac{100\,cm}{1\,m}$

(e) $\dfrac{1\,cm}{0.01\,m} \qquad \dfrac{0.01\,m}{1\,cm}$

2.191 (a) 4×10^3 (b) 0.37

(c) 10.12. The product of 6.23 and 0.042 is 0.26, only two significant digits but certain to the hundredths position. Add this to 9.86 and you get a sum certain to the hundredths position—legitimately gaining a significant digit.

2.193 (a) Two, determined by the 0.0080.

(b) Three. The $22.1 \times 10^2 = 2210$ is uncertain in the tenths position, meaning the sum must also be: $530 + 2210 = 2740 = 274 \times 10^3$.

(c) Five. The $5.830 \times 10^2 = 583.0$ is uncertain in the tenths position, and the same is true for the $22.100 \times 10^2 = 2210.0$. Therefore, the sum is also uncertain in the tenths position: $583.0 + 2210.0 = 2793.0 = 2.7930 \times 10^3$.

(d) Four, determined by 100.0 and 0.1500.

(e) Two, determined by 0.15.

2.195 More likely to be precise. Her good laboratory technique will yield measurements that are close to one another (high precision), but the volumes she reads will all be 5 mL higher than the true volume (low accuracy) because of the incorrect markings.

2.197 (a) Four.   (b) Two.   (c) Three.   (d) Eight.   (e) Four.

2.199 To compress the gas means to squeeze a given mass of it into a smaller volume. Compressing therefore increases the density of the gas because a given mass is forced to occupy a smaller volume. The relationship density = mass/volume tells you that, when mass stays constant, density must go up when volume goes down.

2.201 Water undergoes the most gradual temperature change because it has the highest specific heat. Mercury undergoes the fastest temperature change because it has the lowest specific heat.

2.203 Bar A represents a joule and bar B a calorie. A joule is about one-fourth of a calorie $(4.184 \text{ J} = 1 \text{ cal})$.

2.205 40.2×10^{-6} (or 4.02×10^{-5}) m; 40.2×10^3 (4.02×10^4) nm

2.207 Heat = Specific heat of water $\times$ Mass of water in calorimeter $\times$ Change in water temperature

$$\text{Heat} = \frac{4.184 \text{ J}}{\text{g} \cdot {}^\circ\text{C}} \times 1.00 \times 10^4 \text{ g} \times 20.0\,{}^\circ\!C = 8.37 \times 10^5 \text{ J or } 8.37 \times 10^2 \text{ kJ of energy.}$$

$$\text{Therefore the kJ/g} = \frac{8.37 \times 10^2 \text{ kJ}}{10.6 \text{ g}} = 79.0 \text{ kJ/g}$$

2.209 The mercury would have the smallest volume. It is by far the most dense, meaning that 50 g would fit into a smaller volume than any of the other substances.

Chapter 3

3.1 See solution in textbook.

3.3 See solution in textbook.

3.5 The law of conservation of matter requires that the total mass of the substances produced equal the total mass of coal plus oxygen. The coal seems to disappear because it is converted to carbon dioxide, which is a colorless, odorless gas. If you were to capture the carbon dioxide and determine its mass, you would find that mass to be equal to the combined mass of reacting coal and oxygen.

3.7 See solution in textbook.

3.9 The atoms of both isotopes have 35 electrons. In all neutral atoms, the number of electrons equals the number of protons.

3.11 See solution in textbook.

3.13 (a) The abundance of the ${}^{37}_{17}\text{Cl}$ isotope is $100\% - 75.77\% = 24.23\%$. (The 100 is taken to be an exact number.)

(b) Weighted average =

$$\left(\text{Atomic mass isotope 1} \times \frac{\% \text{ isotope 1}}{100\%} \right) + \left(\text{Atomic mass isotope 2} \times \frac{\% \text{ isotope 2}}{100\%} \right) + \cdots$$

The weighted average for chlorine is therefore

$$\left(34.969 \text{ amu} \times \frac{75.77}{100\%} \right) + \left(36.966 \text{ amu} \times \frac{24.23}{100\%} \right) = 35.45 \text{ amu}$$

(c) ${}^{37}_{17}\text{Cl}$ is $\dfrac{36.966}{34.969} = 1.0571$ times heavier than ${}^{35}_{17}\text{Cl}$.

3.15   No. The metal–nonmetal boundary passes only through the representative-elements region of the periodic table.

3.17   See solution in textbook.

3.19   If these three classes of elements were removed, the periodic table would be eight columns wide. Except for the first period, which would keep its periodicity of 2, the periodicity would be 8 in every period instead of the 2, 8, 8, 18, 18, 32, 32, periodicity that actually exists.

3.21   Ionization energy increases from left to right in the periodic table and decreases from top to bottom. Therefore, the order (smallest to largest) is Rb < Zr < Ag < I.

3.23   Eight protons means the atomic number is 8, making the element oxygen, O; eight neutrons means the mass number is $8 + 8 = 16$. Ten electrons means an unbalanced electrical charge of $10 - 8 = 2$, which is negative because electrons outnumber protons. The symbol is $^{16}_{8}O^{2-}$, and the group name is chalcogen.

3.25   (a) $\dfrac{94.8 \text{ g S}}{100.00 \text{ g sample}} \times 100\% = 94.08\% \text{ S}$

$\dfrac{(100.00 \text{ g} - 94.08 \text{ g}) \text{H}}{100.00 \text{ g sample}} \times 100\% = 5.92\% \text{ H}$

Because the sample contains only sulfur and hydrogen, the sum of their percentages must add up to 100%. Therefore, another way to calculate the percent H is to subtract the percent S from $100{:}100\% - 94.08\% = 5.92\% \text{ H}$ (the 100% is an exact number).

(b) The law of constant composition.

3.27   (a) The mass of A is $126.9 \text{ g} + 35.45 \text{ g} = 162.4 \text{ g}$. The percentages are therefore

$\dfrac{126.9 \text{ g I}}{162.4 \text{ g A}} \times 100\% = 78.14\% \text{ I in A}$

$100\% - 78.14\% = 21.86\% \text{ Cl in A}$

(b) The mass of B is $126.9 \text{ g} + 106.4 \text{ g} = 233.3 \text{ g}$.

$\dfrac{126.9 \text{ g I}}{233.3 \text{ g B}} \times 100\% = 54.39\% \text{ I in B}$

$100\% - 54.39\% = 45.61\% \text{ Cl in B}$

3.29   The carbon atom has four hooks, one for each H atom and, therefore, one more hook than the nitrogen atom in NH_3 would have three hooks:

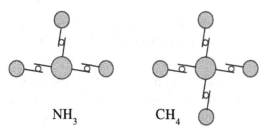

3.31 The three subatomic particles in an atom are: proton, neutron and electron. Their charges are: $1+$, 0, and $1-$, respectively. Their relative masses are: 1, 1, and $1/1836 = 5.486 \times 10^{-4}$, respectively.

3.33 (a) Rutherford reasoned that an atom must be mostly empty space (which is why most α particles passed straight through) but must have somewhere inside it an extremely tiny but massive part that carried a positive electrical charge (which is why a few α particles are deflected).

   (b) The physics of Rutherford's time predicted that any charged particle (such as an electron) moving in a circular path (as it was in Rutherford's model) must continuously radiate energy. This meant that Rutherford's moving electrons would continuously lose energy, slow down, and eventually spiral into the nucleus. Thus all atoms would collapse.

3.35 Elemental identity depends solely on the number of protons in an atom, which is called an atomic number. On the other hand, an atom's mass number is the sum of its protons and neutrons and identifies the isotope of an element. The mass number alone is not sufficient in determining elemental identity as there can be atoms of different elements with the same mass number, for instance: $^{15}_{8}O$ and $^{15}_{7}N$. Here, the atomic numbers of 8 and 7 identify atoms of oxygen and nitrogen, respectively, yet both have the same mass number of 15.

3.37 The diameter of an atom's nucleus is about 1/10,000 the diameter of the atom. Therefore, most of the atom is empty space occupied by the extremely tiny electrons.

3.39 $^{235}_{92}U$ and $^{238}_{92}U$.

3.41 The number of protons tells you the atomic number is 79. Add the proton number and neutron number to get the mass number: $79 + 118 = 197$. The symbol therefore is $^{197}_{79}Au$.

3.43 Because gold, Au, has 79 protons and lead, Pb, has 82, you would have to remove three protons from the lead nucleus.

3.45 Because all masses are given relative to the mass of the isotope $^{12}_{6}C$.

3.47 The atomic mass given for oxygen in the periodic table, 15.999 amu, is the weighted average of all naturally occurring oxygen isotopes. The atomic mass of carbon-12 is exactly 12 amu. Therefore an "average" oxygen atom is:

$$\frac{15.999 \text{ amu}}{12.0000 \text{ amu}} = 1.3333 \text{ times more massive than a carbon-12 atom.}$$

3.49 The atomic mass of oxygen, O, is 15.999 amu, and that of helium, He, is 4.003 amu, making an "average" oxygen atom 15.999 amu $/4.003$ amu $= 3.997$ times more massive than a helium atom.

3.51 (a) Exactly 235 amu.

   (b) With uranium-235 as the standard reference, the atomic masses of all the lighter elements would be a very small fraction of 235. Dealing with such small numbers would be an unnecessary nuisance.

3.53 Atomic mass is the weighted average of the atomic masses of all isotopes of an element. An atomic mass of 1.2000 amu, because it is higher than the Earth value of 1.0079 amu, means that the percentage of the higher-mass ^{2}H must be greater on the other planet than it is on Earth.

3.55 (a) Mendeleev ordered the elements according to atomic mass, and he stacked all elements having similar properties in the same vertical column.

(b) He discovered that his ordering gave eight columns; the properties repeated themselves every eight elements. This repeating behavior is called either chemical periodicity or periodic behavior.

(c) Mendeleev's ordering was by atomic mass; the modern ordering is by atomic number.

3.57 Chemical periodicity refers to the fact that the chemical properties repeat themselves every 8, 18, or 32 elements.

3.59 Considering just the 44 representative elements:

$$\% \text{ Metals} = \left(\frac{20}{44}\right) \times 100\% = 45.45\%$$

$$\% \text{ Nonmetals} = \left(\frac{17}{44}\right) \times 100\% = 38.64\%$$

$$\% \text{ Metalloids} = \left(\frac{7}{44}\right) \times 100\% = 15.91\%$$

Using the entire periodic table:

$$\% \text{ Metals} = \left(\frac{88}{112}\right) \times 100\% = 78.57\%$$

$$\% \text{ Nonmetals} = \left(\frac{17}{112}\right) \times 100\% = 15.18\%$$

$$\% \text{ Metalloids} = \left(\frac{7}{112}\right) \times 100\% = 6.25\%$$

Notice that the percentage of metals is much higher when you use the entire periodic table.

3.61 Hydrogen, because it is the only gas in the group and is not a metal. All of the other elements in group 1 are metals.

3.63 Atoms a period are arranged according to the increasing atomic number, which equals to the number of protons in a nucleus (and the number of electrons outside of a nucleus). This number increases by exactly one as you move to the right. So, adding one proton (and one electron) to an atom moves you exactly one spot to the right. For instance, adding one proton and one electron to an atom of carbon would result in one of lighter isotopes of nitrogen. Note that the mass number would remain the same as no neutron(s) is(are) added.

3.65 They are among the most unreactive substances known, and they are the only group in the periodic table that contains only gases.

3.67 The transition metal portion is 10 elements wide; the lanthanide/actinide portion is 14 elements wide.

3.69 Metalloids (or semimetals) have properties, including electrical conductivity, which are intermediate between those of metals and those of nonmetals. The elements silicon (Si), germanium (Ge), and arsenic (As) are examples of metalloids. Electrical conductivity of metalloids is between that of metals (conducting) and nonmetals (insulating), placing them in the category of materials called semiconductors.

3.71 (a) Removing a proton from an atom of chlorine would produce a 1– anion and decrease the atomic number by one. However, it is worthwhile to note that the energy required to remove a proton from a nucleus of an atom is extremely high!

(b) Removing a proton from an atom of chlorine would result in formation of an ion of an element with a different atomic number, and therefore it would no longer be that of chlorine. It would be a 1– anion of sulfur (which is unlikely to exist!).

(c) The student should have proposed to add one electron to a neutral atom of chlorine without changing the number of protons. Since electrons are negatively charged, an addition of one electron would result in a formation of a Cl^- anion.

3.73 No, student X correctly constructed the model, but student Y did not. Only the number of electrons can be changed to produce an ion. If the number of protons is changed, the elemental identity of the substance changes. Student X made $^{14}_{6}C^{2+}$, by adding two protons, student Y made $^{14}_{8}O^{2+}$.

3.75 "Adding an electron to a 1+ cation yields a neutral atom, but adding an electron to a neutral atom yields an anion."

3.77 First ionization energy decreases as you go down a group and increases as you go across a period.

3.79 Aluminum, Al, should be the most difficult to ionize because it is the smallest of the three atoms. Sodium, Na, should have the smallest first ionization energy because it is the largest of the three atoms. The smallest atom is the hardest to ionize because its electrons are closest to the positively charged nucleus.

3.81f Yes, there are minor exceptions. Ga has a lower first ionization energy than the elements to the left of it.

3.83 (a) The nitrogen gains three electrons. Nitrogen is to the right of lithium on the periodic table and therefore has a higher first ionization energy than lithium. The lithium, with a smaller first ionization energy, has a tendency to lose electrons and loses them to nitrogen.

(b) An anion because it acquires a negative charge after gaining electrons.

(c) $^{14}_{7}N^{3-}$

3.85 A line spectrum is made up of specific discrete colors separated by regions of darkness. A continuous spectrum consists of all colors, with adjacent colors blending smoothly into each other and no regions of darkness separating the colors.

3.87 Both compounds contain only Sn and O. Because the mass percents of all parts of a whole must add up to 100%, the mass percents of oxygen are:

Oxide A: 100% − 78.77% Sn = 21.23% O
Oxide B: 100% − 88.12% Sn = 11.88% O

3.89 The results from Rutherford's alpha-particle experiment. That some of the particles were deflected instead of passing right through the gold atoms is evidence that there must be something massive in each atom. That only a very few of the particles were deflected is evidence that this massive something must occupy a relatively tiny part of each atom.

3.91 She is correct. Isotopes of an element differ only in the number of neutrons in their nuclei.

3.93 By adding two electrons to the neutral atom.

3.95 Number of protons = atomic number = 9 protons.
Number of electrons = number of protons = 9 electrons.
Number of neutrons = mass number − atomic number = $19 - 9 = 10$ neutrons.

3.97 (a) Protons = atomic number = 35 protons.
Electrons = number of protons = 35 electrons.
Neutrons = mass number − atomic number = $79 - 35 = 44$ neutrons.

(b) Protons = atomic number = 35 protons.
Electrons = number of protons + one electron for each negative charge on ion = $35 + 1 = 36$ electrons.
Neutrons = mass number − atomic number = $81 - 35 = 46$ neutrons.

(c) Protons = atomic number = 11 protons.
Electrons = number of protons − one electron for each positive charge on ion = $11 - 1 = 10$ electrons.
Neutrons = mass number − atomic number = $23 - 11 = 12$ neutrons.

(d) Protons = atomic number = 1 proton.
Electrons = number of protons − one electron for each positive charge on ion = $1 - 1 = 0$ electrons.
Neutrons = mass number − atomic number = $3 - 1 = 2$ neutrons.

3.99 Because there are only two isotopes, the abundance of ^{109}Ag must be:
$100\% - 51.84\% = 48.16\%$.

$$\left(106.90509 \text{ amu} \times \frac{51.84\%}{100\%} \right) + \left(108.9047 \text{ amu} \times \frac{48.16\%}{100\%} \right) = 107.9 \text{ amu}$$

3.101 There are 44 known main-group elements. All of them are representative elements because *main group* and *representative* are different names for the same group of elements.

3.103 As is the case with the main-group metals, the transition metal groups are numbered according to the number of electrons they tend to lose.

3.105 Bromine, Br, has atomic number 35, meaning a neutral Br atom contains 35 electrons. Therefore the Br^- anion must contain $35 + 1 = 36$ electrons. To be a 1+ cation and contain 36 electrons, an atom must contain 37 protons, making it the element whose atomic number is 37—rubidium, Rb. The cation is Rb^+.

3.107 The size of magnesium is smaller than that of sodium. Therefore, the outer electrons are closer to the positively charged nucleus, making it harder to remove one of them.

3.109 (a) g Cl = $110.99 \text{ g} - 40.98 \text{ g} = 70.01 \text{ g Cl}$

(b) % Ca = $\dfrac{40.98 \text{ g}}{110.99 \text{ g}} \times 100\% = 36.92\% \text{ Ca}$

(c) % Cl = $\dfrac{70.01 \text{ g}}{110.99 \text{ g}} \times 100\% = 63.08\% \text{ Cl}$

3.111 The first ionization energy decreases when going down any group.

3.113 N, Si, P, K, Au

3.115 Properties of metals: shiny, good conductors of electricity and heat, malleable. Properties of nonmetals: brittle, poor conductors of electricity and heat. Metalloids can behave as metals or as nonmetals and generally have properties intermediate between those of metals and nonmetals.

3.117 (a) Alkali metals. (b) Alkaline earth metals. (c) Halogens. (d) Noble gases or rare gases.

3.119 (a) $^{17}_{8}O$ (b) $^{119}_{50}Sn$ (c) $^{23}_{11}Na$ (d) $^{58}_{28}Ni$ (e) $^{137}_{56}Ba$

3.121 Metals tend to lose electrons, whereas nonmetals tend to gain electrons. Metals tend to form cations and nonmetals tend to form anions.

3.123 Chemical periodicity refers to the regular way in which the chemical properties of the elements repeat as you move with atomic number through the periodic table. The properties of lithium repeat in sodium, and you get from lithium to sodium by moving across period 2. The same properties repeat in potassium, and you get from sodium to potassium by moving across period 3.

3.125 The tendency to lose electrons is measured by ionization energy. The element with the lowest first ionization energy has the most metallic character, therefore, and the element with the highest first ionization energy has the least metallic character: Ne < Be < Li < Na < Cs.

3.127 Mass number is the number of protons plus neutrons in an atom; it is always a whole number. Atomic mass is the weighted average of the masses of all of the isotopes of an element; it is usually not a whole number but is numerically very close to the mass number.

3.129 The elements in a group have similar characteristics, much as members of the same family do.

3.131 (a) Beryllium. (b) Magnesium. (c) Iron. (d) Sulfur. (e) Argon. (f) Copper.

3.133 (a) $^{16}_{8}O$

(b) $^{16}_{8}O^{2-}$

(c) The product is an anion because it is negatively charged after gaining two electrons.

3.135 $100{,}000 \times 24.25 \text{ mm} \left(\dfrac{\text{cm}}{10 \text{ mm}} \right) \times \left(\dfrac{\text{m}}{100 \text{ mm}} \right) \times \left(\dfrac{\text{mile}}{1609.344 \text{ mm}} \right) = 1.507 \text{ miles}$

Recall that an atom is 100,000 times larger than its nucleus.

3.137 Removing an electron from the ion formed after the first electron has been removed, forming a 2+ cation.

3.139 The size of the atom decreases when moving from left to right, making the electron closer to the positively charged nucleus; thus it is attracted more strongly. It takes more energy to remove an electron that is more strongly attracted to the nucleus.

3.141 We should expect similar chemical properties because these atoms both have 35 electrons. The chemical properties of atoms are determined by the number of electrons.

3.143 (a) 56.4 g of oxygen is unused

(b) $\% \text{ Si} = \dfrac{12.0 \text{ g Si}}{25.6 \text{ g SiO}_2} \times 100\% = 46.9\% \text{ Si}$

(c) $\% \text{ O} = \dfrac{13.6 \text{ g O}}{25.6 \text{ g SiO}_2} \times 100\% = 53.1\% \text{ O}$

3.145 Assuming the same charge (typically 1+), the extent to which a charged particle, such as an ionized atom, is deflected depends on its mass. The heavier the particle, the less deflected it is when it reaches the detector.

3.147 Terrestrial and extraterrestrial materials, such as rocks, have different isotopic make-ups. Mass spectrometer is a sensitive detector of isotopes based on their masses and it also allows to determine their relative abundances. Since different isotopes of one and the same element differ in mass, mass spectrometer will detect ions of different masses and/or different abundances, depending on the origin of the material.

Chapter 4

4.1 See solution in textbook.

4.3 $E = \dfrac{hc}{\lambda}$, where $h = 6.626 \times 10^{-34}\,\text{J}\cdot\text{s}$, $c = 3.00 \times 10^{8}\,\text{m/s}$, and λ is wavelength:

$$\lambda = \frac{hc}{E} = \frac{(6.626 \times 10^{-34}\,\text{J}\cdot\text{s}) \times (3.00 \times 10^{8}\,\text{m/s})}{3.50 \times 10^{-19}\,\text{J}} = 5.68 \times 10^{-7}\,\text{m} = 568\,\text{nm}$$

This wavelength corresponds to yellow light.

4.5 See solution in textbook.

4.7 Because the electron in the low-n shell is closer to the nucleus than the electron in the high-n shell is and feels a stronger attraction to the nucleus.

4.9 See solution in textbook.

4.11 A Li^{+} ion has lost one electron, and for the ion to be in the ground state, the two remaining electrons must be in the lowest shell:

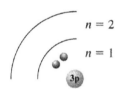

4.13 The $n = 1$ shell has three electrons in this diagram; the maximum is two electrons for this shell. The $n = 2$ shell has only six electrons; the maximum is eight for this shell, and it should be filled before an electron is placed in the $n = 3$ shell. The correct Bohr model for the F^{-} anion is

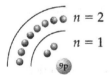

4.15 Arsenic, As, has 33 electrons, making the electron configuration $1s^2 2s^2 2p^6 3s^2 3p^6 3d^{10} 4s^2 4p^3$. There are five valence electrons (the $4s$ and $4p$ electrons), and the roman-numeral group number is VA. The number of valence electrons does agree with the group number.

4.17 See solution in textbook.

4.19 Pd has 46 electrons. When you subtract 1 from the period number each time you enter the d block, you get the configuration $1s^2 2s^2 2p^6 3s^2 3p^6 4s^2 3d^{10} 4p^6 5s^2 4d^8$.

4.21 Ra has 88 electrons. The first 54 go in the usual positions, giving the configuration $1s^2 2s^2 2p^6 3s^2 3p^6 4s^2 3d^{10} 4p^6 5s^2 4d^{10} 5p^6$. Then begin period 6 with $6s^2$, go one deep into $6d$ for $6 - 1 = 5d^1$, then go through the whole $6f$ row for $6 - 2 = 4f^{14}$. Now go back to $6 - 1 = 5d^1$ for nine more. Then $6p^6$ and end with $7s^2$. The configuration is therefore $1s^2 2s^2 2p^6 3s^2 3p^6 4s^2 3d^{10} 4p^6 5s^2 4d^{10} 5p^6 6s^2 4f^{14} 5d^{10} 6p^6 7s^2$. The first noble-gas element preceding radium is radon, Rn, making the abbreviated notation $[\text{Rn}]7s^2$.

4.23 See solution in textbook.

4.25 In $[\text{Xe}]6s^1$ the highest value of n is 6, making this a period 6 element. Because the element has only one valence electron, it is in group IA. The element is cesium, Cs. To be sure this is not a group B element masquerading as a group A element (see Solution 4.24), count the electrons: 54 to Xe + 1 = 55. Cesium is correct.

4.27 Atomic size decreases from left to right in the periodic table and increases from top to bottom. Therefore, the order (smallest to largest) is O < S < Mg < Sr < Rb.

4.29 Eight protons means the atomic number is 8, making the element oxygen, O; eight neutrons means the mass number is $8 + 8 = 16$. Ten electrons means an unbalanced electrical charge of $10 - 8 = 2$, which is negative because electrons outnumber protons. The symbol is $^{16}_{8}\text{O}^{2-}$, and the group name is chalcogen.

4.31

2 Na atoms

O atom

Two Na$^+$ ions, each with 11 protons and 10 electrons

O^{2-} ion with 8 protons and 10 electrons

4.33 Al is a group IIIA metal and thus loses three electrons to become Al^{3+}; O is a group VIA nonmetal and so gains two electrons to become O^{2-}. To have charge neutrality, the formula must be Al_2O_3.

4.35 $E = \dfrac{hc}{\lambda} \Rightarrow \dfrac{E\lambda}{h} = c = 3.00 \times 10^8 \, \text{m/s}$. Since $\dfrac{E\lambda}{h}$ is a constant, its value does not depend on the wavelength of light. In other words, since the energy and the wavelength are inversely proportional, when λ goes up, E goes down and $E\lambda$ remains unchanged.

4.37 $30 \text{ miles} \times \dfrac{1.61 \text{ km}}{1 \text{ mile}} \times \dfrac{1000 \text{ m}}{1 \text{ km}} \times \dfrac{1 \text{ s}}{3.00 \times 18^8 \text{ m}} = 1.6 \times 10^{-4} \text{ s}$

4.39 (2) is true. According to the equation $E = hc/\lambda$, the energy of light decreases as its wavelength increases. The equation indicates that the energy and wavelength are *inversely* related to each other because λ is in the denominator.

4.41 Gamma rays are much higher in energy and will damage cells and tissues. Radio waves do not have enough energy to cause damage to cells and tissues.

4.43 $E = \dfrac{(6.63 \times 10^{-34} \text{ J} \cdot \text{s}) \times (3.00 \times 10^8 \text{ m/s})}{10 \text{ pm}} \times \dfrac{1 \times 10^{12} \text{ pm}}{1 \text{ m}} = 1.99 \times 10^{-14} \text{ J}$

The X-Ray is

$$\dfrac{1.99 \times 10^{-14}}{1.99 \times 10^{-26}} = 1 \times 10^{12}$$

times more energetic than the radio wave of Problem 4.42.

4.45 You would appear first at the starting line, then a certain time later you would instantaneously appear some distance ahead, and later still some farther distance ahead, until finally you would appear instantaneously at the finish line.

4.47 *Quantized energy* means that the energy any object possesses can have only certain allowable values and no in-between values.

4.49 That something inside the atom is allowed to possess only certain energies, as opposed to any energy. Thus the atom can release only light energy of specific (quantized) energies rather than a continuous range of energies.

4.51 (a) is more stable because unlike charges—such as an electron and an atomic nucleus—attract each other and are stable when close to each other; (b) is less stable because like charges repel each other and are less stable when close to each other.

4.53 The n value for an electron dictates which shell the electron occupies. As n increases for an electron, both its energy and its distance from the nucleus increase.

4.55 Electron shell.

4.57 The n value for an electron dictates which shell the electron occupies. An increase in n value means the electron moves farther from the nucleus, and as a result of the move the electron's energy increases.

4.59 The lower shells are of lower energy; the electrons go into lower-energy positions before going into higher-energy (less stable) positions.

4.61 Bohr's rules say that inner shells are filled first. The atom of Problem 4.60 would have its lowest-energy (most stable) configuration if the electron in the $n = 3$ shell were moved to the $n = 2$ shell:

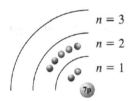

4.63 The energy of an electron depends on its distance from the nucleus. The Bohr model says each allowed distance from the nucleus has its own discrete, characteristic energy value. Therefore, saying an electron can be only at certain distances from the nucleus means the electron can have only certain energy values.

4.65 Because atoms in the same group have identical valence-shell configurations.

4.67 A ground-state configuration is the lowest-energy configuration possible, in which all electrons are in the shells having the lowest-energy values consistent with the restriction that $2n^2$ is the maximum number of electrons allowed in any shell. An excited-state configuration is one in which one or more electrons are boosted into higher-energy shells when the atom absorbs energy.

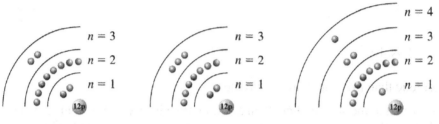

Mg ground state   Mg excited states

4.69

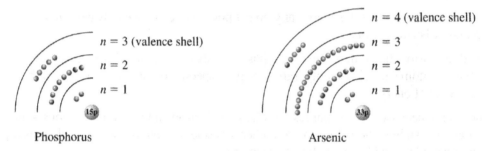

Phosphorus   Arsenic

These two elements have similar chemical properties because both have five electrons in the valence shell.

4.71 Nothing. The electron would remain in the ground state because it needs more than 5.1 eV to be excited into the $n = 2$ shell. It needs 10.2 eV to be excited from $n = 1$ (1.0 eV) to $n = 2$ (11.2 eV).

4.73 (a) 1: Nothing wrong. Seven protons but only six electrons means it is a cation. No shell filled beyond maximum capacity.
2: Nothing wrong. Same number of protons and electrons. No shell filled beyond maximum capacity.
3: Too many electrons in $n = 1$ shell, which holds a maximum of $2 \times 1^2 = 2$.

(b) 1: $^{14}_{7}N^+$ 2: $^{14}_{7}N$

(c) 1: Ground state because every electron has the lowest energy possible consistent with the $2n^2$ limit.
2: Excited state because one electron is in the $n = 3$ shell when the $n = 2$ shell is not filled to capacity.

4.75 ∿∿∿∿∿∿∿ Red light

∿∿∿∿∿∿∿∿∿∿∿∿ Blue light

Because it has the shorter wavelength, blue light has the higher energy. The blue light would therefore be associated with relaxation from a higher state, meaning in this case the relaxation from $n = 4$ to $n = 2$.

4.77 Because Planck's constant h is given in joule-seconds in the textbook, convert the given wavelength to meters before using the energy equation:

$$1772.6 \text{ nm} \times \frac{1 \text{ m}}{1 \times 10^9 \text{ nm}} = 1.7726 \times 10^{-6} \text{ m}$$

$$E = \frac{(6.626 \times 10^{-34} \text{ J} \cdot \text{s}) \times (3.00 \times 10^8 \text{ m/s})}{1.7726 \times 10^{-6} \text{ m}} = 1.12 \times 10^{-19} \text{ J}$$

$$1.12 \times 10^{-19} \text{ J} \times \frac{1 \text{ eV}}{1.602 \times 10^{-19} \text{ J}} = 0.700 \text{ eV}$$

The energy-level diagram in Section 4.4 of the textbook tells you the energy of the $n = 4$ shell is 13.8 eV. Subtract the emitted 0.700 eV from this value and you get 13.1 eV, which is the energy of the $n = 3$ shell. Therefore, the electron relaxed to the $n = 3$ shell.

4.79 The experimental evidence was that line spectra contained a number of closely spaced lines of nearly identical energy. Because the lines were of nearly identical energy, they must be in the same shell. Therefore, the small differences in energy suggested that the lines were the result of separations within a shell.

4.81 The s subshell holds a maximum of two electrons, the p holds 6, the d holds 10, and the f holds 14.

4.83 Drawing the diagram this way implies that the $3d$ subshell is larger than the $4s$ subshell. That is *not* so! An electron in the $4s$ subshell of an atom is, on average, farther from the nucleus than an electron in the $3d$ subshell is.

4.85 (a) $[\text{He}]2s^2 2p^1$

(b) $[\text{Ar}]4s^2 3d^1$

(c) $[\text{Ar}]4s^2 3d^7$

(d) $[\text{Ar}]4s^2 3d^{10} 4p^4$

(e) $[\text{Kr}]5s^2 4d^6$

4.87 (a) $[\text{Xe}]6s^2$

(b) $[\text{Xe}]6s^2 4f^{14} 5d^4$

(c) $[\text{Xe}]6s^2 4f^{14} 5d^{10} 6p^2$

(d) $[\text{Xe}]6s^2 5d^1 4f^2$

(e) $[\text{Rn}]7s^2 6d^1 5f^2$

4.89 (a) Incorrect; 2s should come after 1s.

(b) Incorrect; 2s subshell holds a maximum of two electrons and should be followed by 2p subshell.

(c) Incorrect; 1s subshell must come first in a ground-state configuration.

(d) Incorrect; 2p subshell holds a maximum of six electrons.

(e) Incorrect; 2p subshell should have six electrons before any electrons go into 3s subshell.

(f) Correct.

4.91 Lithium is in the second period and so its valence electrons have $n = 2$. Representative elements with an n value of $3 \times 2 = 6$, are in the sixth period: Cs, Ba, Tl, Pb, Bi, Po, At, and Rn.

4.93 In the d block, the n level that is filling is 1 less than the period. For example, the 3d electrons are being added in the fourth period. In the f block, the n level that is filling is 2 less than the period. For example, the 4f electrons are being added in the sixth period.

4.95 (a) Because the one additional electron for each element is placed in the same valence shell across an entire period, one might expect the atom size to stay the same.

(b) Atomic size decreases from left to right across a period because one electron is added to the valence shell *and* one proton is added to the nucleus. The added proton makes the nucleus more positive, which causes it to more strongly attract the surrounding electrons. The increased pull on the electrons shrinks the atom.

4.97 Beryllium has the smaller 1s subshell. The nuclear charge in Be is +4, and the nuclear charge in Li is +3. The greater charge in Be pulls the electrons in closer to the nucleus, shrinking the subshells.

4.99 Lithium is larger because atomic size decreases as you go from left to right in a period.

4.101 Si < Mg < Na < Rb

4.103 Fe < Ti < Hf < Cs.

4.105 As you go down a group, the size of the atom increases and the outermost electrons are farther away from the nucleus. Because these outer electrons are farther away, each one feels less pull from the nucleus and is easier to remove. Therefore, its separation requires less energy. As you go across a period, the atomic size decreases, and an outermost electron is closer to the positively charged nucleus. More energy is therefore needed to remove the electron.

4.107 By losing their valence electron(s).

4.109

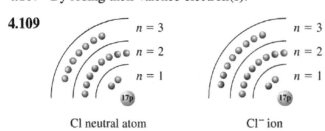

Cl neutral atom   Cl⁻ ion

There are seven electrons in the valence shell in Cl and eight in Cl⁻.

4.111 Each sodium atom has one valence electron, which it loses to obtain an octet. Each sulfur atom has six valence electrons and gains two to have an octet. Thus, each sulfur atom needs two sodium atoms, and the formula of the compound is Na_2S.

4.113 Because it is the same as the number of valence electrons, the roman-numeral group number allows us to predict how many electrons are gained or lost to form an octet. Elements with fewer than four valence electrons lose electrons to form an octet. Elements with more valence electrons gain electrons to form an octet. For example, elements in group IIA lose two electrons; elements in group VA gain three electrons.

4.115 XY_2. The group IIA element loses two electrons, and the group VIIA element gains one electron. Therefore, two Y atoms are needed for each X atom.

4.117 Aluminum has three valence electrons, which it loses to form a 3+ cation. Both oxygen and sulfur have six valence electrons and gain two to form a 2− anion. Because Al needs to lose three electrons and oxygen and sulfur need to gain two, each compound contains two Al^{3+} cations and three anions. The formulas are similar because both oxygen and sulfur have the same number of valence electrons.

4.119 It is true that Mg^{2+} and Na^+ have identical electron configurations; it is false that they have similar properties. The two ions have different charges, different sizes, and different chemical reactivities.

4.121 A representative nonmetal will gain enough electrons to have a total of eight in its valence shell. Because the roman-numeral group number is equal to the number of valence electrons, you can tell how may electrons it will gain by subtracting the number of valence electrons from 8. The charge on the nonmetal anion will be a negative number equal to 8 minus the roman-numeral group number.

4.123 By placing an atom's electrons in various subshells, Bohr's model places each electron in a particular location that we can, in theory, describe exactly. According to modern quantum mechanical theory, you cannot know precisely both (1) where an electron is at a given moment and (2) where it will be at any future moment. The more precisely you know one, the less precisely you know the other. Therefore, the Bohr concept of electrons traveling in exact orbits around the nucleus is incorrect.

4.125 An orbital is a region of space around the nucleus of an atom in which it is most probable that an electron will be found.

4.127 You would certainly *not* see any individual electrons. Orbitals would be seen as fuzzy clouds with different degrees of transparency. Fuzziness would signify an inability to see exact positions of electrons at any moment. The different degrees of transparency would signify different probabilities of finding electrons in different regions of the orbital. The more transparent the cloud, the less time an electron spends in an area; the more opaque, the more time. A 1s orbital would be spherical in shape and opaque near the middle; however, as you looked further away from the nucleus in any direction, the cloud would become more and more transparent. A 2p orbital would have an elongated dumbbell shape with the most opaque areas along its axis at a certain distance from the nucleus, becoming completely transparent at the nucleus and then again further away from the nucleus.

4.129 The speed at which light travels is 3.00×10^8 m/s. The time needed to travel the distance is

$$3.00 \times 10^3 \text{ miles} \times \frac{1.61 \text{ km}}{1 \text{ mile}} \times \frac{1000 \text{ m}}{1 \text{ km}} \times \frac{1 \text{ s}}{3.00 \times 10^8 \text{ m}} = 1.61 \times 10^{-2} \text{ s}$$

4.131 Because λ stands for wavelength, this is a simple conversion:

$$2.50 \times 10^{-5} \text{ m} \times \frac{1 \times 10^9 \text{ nm}}{1 \text{ m}} = 2.50 \times 10^4 \text{ nm}$$

$$\frac{2.50 \times 10^4 \text{ nm infrared}}{750 \text{ nm red}} = 33.3$$

The infrared wavelength is 33.3 times longer than the red wavelength.

4.133 Use the energy equation, in the form $\lambda = hc/E$ for the first row and in the form $E = hc/\lambda$ for the second and third rows.

Wavelength (m)	Wavelength (nm)	Energy (J)
3.52×10^{-8} m	35.2 nm	5.65×10^{-18} J
2.6020×10^{-6} m	2602.0 nm	7.64×10^{-20} J
7.85×10^{-12} m	7.85×10^{-3} nm	2.53×10^{-14} J

4.135 Energy is released. An electron initially in a shell having a higher n value needs less energy once it is in a shell having a lower n value. The electron releases the energy it no longer needs.

4.137 The ground state has a filled $n = 1$ shell and the rest of the electrons are in the $n = 2$ shell. Thus, relaxing only one of the electrons in the $n = 3$ shell gives a configuration that has lower energy than the original configuration but still is not the ground state:

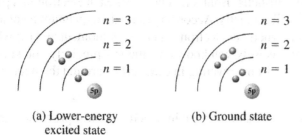

(a) Lower-energy excited state   (b) Ground state

4.139 You can get the energy of the emitted light from the energy-level diagram in Section 4.4 of the textbook:

$$(14.2 \text{ eV in } n = 6) - (1.0 \text{ eV in } n = 1) = 13.2 \text{ eV emitted}$$

$$13.2 \text{ eV} \times \frac{1.602 \times 10^{-19} \text{ J}}{1 \text{ eV}} = 2.11 \times 10^{-18} \text{ J}$$

The wavelength of light having this energy is

$$\lambda = \frac{hc}{E} = \frac{(6.626 \times 10^{-34} \text{ J} \cdot \text{s}) \times (3.00 \times 10^8 \text{ m/s})}{2.11 \times 10^{-18} \text{ J}} = 9.42 \times 10^{-8} \text{ m} \times \frac{1 \times 10^9 \text{ nm}}{1 \text{ m}} = 94.2$$

4.141 (a) Ground state. All 26 electrons are in the subshells assigned to them by the four-block periodic table guide.

(b) Incorrect. The $4s$ subshell, because it is lower in energy than the $3d$ subshell, should be filled before any electrons are put in the $3d$ subshell.

(c) Incorrect. The $2p$ subshell has more than its maximum allowable number of electrons.

4.143 The d subshell is being filled. It fills *after* the s subshell having the next highest principal quantum number n has been filled.

4.145 They have the same electron configuration: $1s^2\,2s^2\,2p^6$ and hence, the same number of electrons.

4.147 (a) N. Nitrogen has a nuclear charge of $+7$; fluorine has a nuclear charge of $+9$; and these two elements are in the same period and so they have the same number of electron shells. The more positive F nucleus pulls more on the electrons, making N electrons easier to break away.

(b) Ba. The outermost electrons in magnesium are in the $n = 3$ shell. Those in barium are in the $n = 6$ shell. The farther electrons feel less nuclear pull and are thus easier to break away.

(c) Ca. The outermost electrons in nitrogen are in the $n = 2$ shell. Those in calcium are in the $n = 3$ shell. Being farther from the nucleus, calcium valence electrons break away more easily.

4.149 (b) Cl < Br < As < Ca < Sr (b) Sr < Ca < As < Br < Cl

4.151 Rb < Ca < Se < S < F

4.153 The $4s$ electron, being a larger electron cloud, spends more time, on average, farther from the nucleus. This increases its energy.

4.155 The waves are more energetic than visible light because their wavelength is shorter.

4.157 Valence-shell electrons are shown in **bold font**.

(a) Ca^+: $1s^2\,2s^2\,2p^6\,3s^2\,3p^6\,\mathbf{4s^1}$

(b) Li: $1s^2\,\mathbf{2s^1}$

(c) P^{3-}: $1s^2\,2s^2\,2p^6\,\mathbf{3s^2\,3p^6}$

(d) Ar^+: $1s^2\,2s^2\,2p^6\,\mathbf{3s^2\,3p^5}$

(e) Si^{2+}: $1s^2\,2s^2\,2p^6\,\mathbf{3s^2}$

P^{3-} is the only one that has a valence-shell octet.

4.159 Valence-shell electrons are shown in **bold font**.

(a) Ar: $1s^2\,2s^2\,2p^6\,\mathbf{3s^2\,3p^6}$

(b) Na^+: $1s^2\,\mathbf{2s^2\,2p^6}$

(c) C^{2-}: $1s^2\,\mathbf{2s^2\,2p^4}$

(d) O^{2-}: $1s^2\,\mathbf{2s^2\,2p^6}$

(e) Ca^{2+}: $1s^2\,2s^2\,2p^6\,\mathbf{3s^2\,3p^6}$

Every atom/ion except C^{2-} has a valence shell octet.

4.161 (b) Antimony, atomic number 51, has 51 electrons in the neutral atom, and therefore (c), (d), and (e) are obviously wrong because they show 52, 41, and 41, respectively. (a) is wrong because it has electrons going into the $5d$ subshell while leaving the lower-energy $5p$ subshell empty.

4.163 To be quantized means that only a certain set amount of something is available. Thus (a) is quantized because if you want to, say, move the water from one place to another, you can move 16 oz or 32 oz or 48 oz and so forth but never any volumes between these. (b) is not quantized because there is a continuous flow of water from the fountain and you can get any volume you want. By the same reasoning, (c) is not quantized and (d) is.

4.165 Be < Mg < Ca < Sr < Ba.

4.167 (a) Quantized.   (b) Classical, quantum.

4.169 Lose, gain.

4.171 $E = \dfrac{(6.63 \times 10^{-34}\,\text{J}\cdot\text{s}) \times (3.00 \times 10^{8}\,\text{m/s})}{1.00 \times 10^{-12}\,\text{m}} = 1.99 \times 10^{-13}\,\text{J}$

4.173 That there are 32 electrons in this neutral atom tells you the atomic number is 32, making the element germanium, Ge. It is in group IVA (14) and period 4. Another way to see period and group number is from the way the electrons are configured. The highest occupied shell is $n = 4$, making this a period 4 element. The four valence electrons in this shell ($4s^2$ and $4p^2$) make it a group IVA element.

4.175 It is impossible to know with certainty both the position of an electron and where the electron is going. The more accurately one of these two details is known, the less accurately the other is known. This means that electrons must be viewed not as particles, but rather as probability clouds of negatively charged matter.

4.177 Lower in energy. The negative electrons are attracted to the positive nucleus, and you can model the attraction as a spring connecting the two. When the electron is in a close shell, the spring is relaxed and the electron has low energy. Just as you must put energy into a real-life spring to stretch it, you must put energy into the electron to "stretch" it to some shell farther from the nucleus.

4.179 The configuration is $1s^2 2s^2 2p^6 3s^2 3p^6$ for all four species. Ca^{2+} has lost two electrons to achieve the Ar noble gas configuration. K^+ has lost one electron to achieve it, and S^{2-} has gained two electrons.

4.181

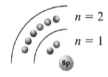

$n = 2$
$n = 1$

There are six electrons in the valence shell, and this shell can accept two more electrons to achieve a valence-shell octet.

4.183 The orbits in the Bohr atom are fixed paths in which electrons move. Orbitals are regions of space within which a negatively charged cloud of electrons has a certain probability of being found.

4.185 (a) Relaxation occurs from the $n = 4$; energy level to a point between allowed energy levels $n = 1$ and $n = 2$. This is strictly not allowed. Relaxation must be to the allowed energy levels only.

(b) 12.8 eV

(c)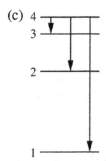

4.187 (b), (d), and (e) are correct statements.

4.189 Lithium's valence electron is most likely to be found in the volume described by a spherical shaped cloud about its nucleus.

4.191 The helium atom has lost one electron and is technically a helium cation, He^+.

4.193 AlN. Each Al atom will lose three electrons to form Al^{3+}, and each N atom will gain three electrons to form N^{3-}. Therefore, one Al will combine with one N to form the compound AlN.

4.195 (C) The $n = 1$ to $n = 4$ transition would absorb the largest amount of energy. Energy is absorbed when the transition is from lower to higher energy level, and the greatest amount of energy will be required by the transition involving the greatest distance between the energy levels involved.

4.197 Atomic size decreases from left to right, and increases from top to bottom. Therefore the orders are: (a) F < Cl < Br < I (b) S < Al < Mg < Na (c) Si < Al < Ge < Te

4.199 Nitrogen. There is a great increase in IE(6) compared to IE(5), indicating that the atom has five valence electrons.

4.201 (e) is the incorrect statement. Because the fourth ionization energy is very much greater than the third, we know that the atom has three valence electrons. The element could therefore not belong to group VA.

4.203 Group VIIA and period 2. Non-metallic character increases from left to right and decreases from top to bottom in the periodic table. (The elements in Group VIIIA are noble gases and do not fit our definition of nonmetals as elements that tend to gain electrons to become anions.)

4.205 (b) The $n = 4 \rightarrow n = 1$ transition. The emission of photons results from higher to lower energy transitions. The shortest wavelength (highest energy) will come from electrons falling the greatest distance, in this case from $n = 4$ to $n = 1$.

4.207 The $n = 2$ to $n = 1$ transition gives rise to the line labeled B. The greatest energy transition $(n = 3 \rightarrow n = 1)$ results in line A (shortest wavelength), and the lowest energy transition $(n = 3 \rightarrow n = 2)$ results in line C (longest wavelength).

Chapter 5

5.1 The force of attraction between the ions in MgO would be greater than in NaCl, by a factor of approximately 4. This estimate is based on the charges of the ions $(+2 \times -2 = -4$ for MgO; $+1 \times -1 = -1$ for NaCl$)$.

5.3 Table 5.1 of the textbook tells you silicon usually forms four bonds and bromine usually forms one bond. The formula is therefore $SiBr_4$.

5.5 F, group VIIA, has seven valence electrons and therefore needs to form one bond to achieve an octet. H, group IA, has one valence electron and needs to form one bond to obtain its duet. (Remember that hydrogen is stable with two valence electrons because its valence shell is $n = 1$.) The formula of the compound formed is therefore HF. The dot structure is $H\!:\!\ddot{\underset{\cdot\cdot}{F}}\!:$

5.7 See solution in textbook.

5.9 Each carbon atom has four valence electrons and therefore forms four bonds, either with H or with another C. Each hydrogen atom forms one bond:

$$\begin{array}{ccccccc}
 & H & & H & & H & \\
 & | & & | & & | & \\
H & - & C & - & C & - & C & - & H \\
 & | & & | & & | & \\
 & H & & H & & H &
\end{array}$$

5.11 Begin by drawing the individual dot diagrams with the atoms positioned as described in the problem. As a general rule, any C that has no H bonded to it is in the middle of the molecule:

$$\begin{array}{l}
H\cdot \\
H\cdot\ \cdot\dot{C}\cdot\ \cdot\dot{C}\cdot\ \cdot\dot{C}\cdot\ \cdot H \\
H\cdot
\end{array}$$

Pairing electrons to form single bonds gives

$$\begin{array}{ccccccc}
 & H & & & & & \\
 & | & & & & & \\
H & - & C & - & \dot{C} & - & \dot{C} & - & H \\
 & | & & & & & \\
 & H & & & & &
\end{array}$$

Let the two carbons that still have unpaired electrons form a multiple bond and you are done:

$$\begin{array}{ccccccc}
 & H & & & & & \\
 & | & & & & & \\
H & - & C & - & C & \equiv & C & - & H \\
 & | & & & & & \\
 & H & & & & &
\end{array}$$

5.13 Draw the atoms positioned as described, following the general rule that C bonded to no H is usually in the middle of the molecule:

$$\begin{array}{l}
\cdot H\quad :\ddot{O}\cdot\quad H\cdot \\
H\cdot\ \cdot\dot{C}\cdot\ \cdot\dot{C}\cdot\ \cdot\dot{C}\cdot\ \cdot H \\
\quad H\cdot\qquad\quad H\cdot
\end{array}$$

Form single bonds:

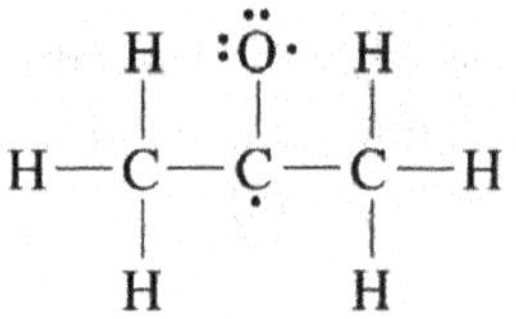

H :Ö· H
| | |
H—C—C—C—H
| |
H   H

Form a multiple bond with the remaining *unpaired* electrons:

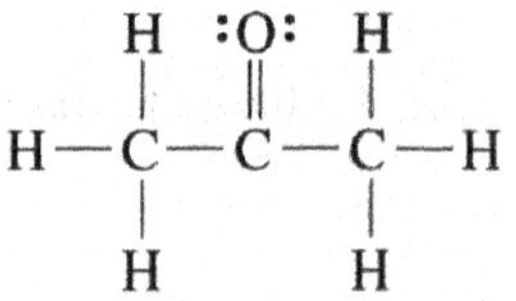

H :O: H
| ‖ |
H—C—C—C—H
| |
H   H

5.15 Step 1: SO₃ has 24 valence electrons: 6 from S and 3 × 6 = 18 from O.

Step 2: Make first atom in formula central and connect atoms via single bonds:

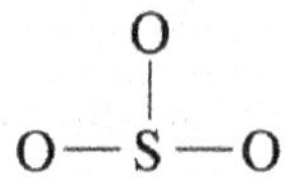

```
       O
       |
   O — S — O      6 electrons assigned
```

Step 3: Add remaining electrons (24 − 6 = 18) as lone pairs. Begin with the terminal atoms and complete the octet on one atom before going on to the next:

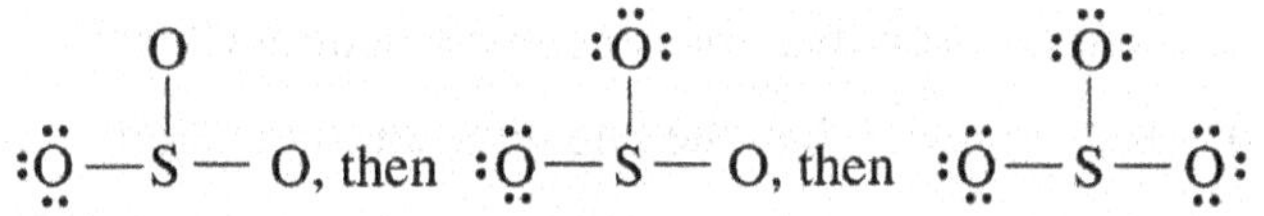

```
       O              :Ö:             :Ö:
       |               |               |
  :Ö — S — O, then :Ö — S — O, then :Ö — S — Ö:
```

One octet done,   Two octets done,   One octet done,
12 electrons   18 electrons   all 24 electrons
assigned   assigned   assigned

Step 4: You are not done because all 24 electrons have been assigned but S does not have an octet. To fix things, form an O=S double bond with any O atom. The three possible structures are all equally good because they all obey the octet rule:

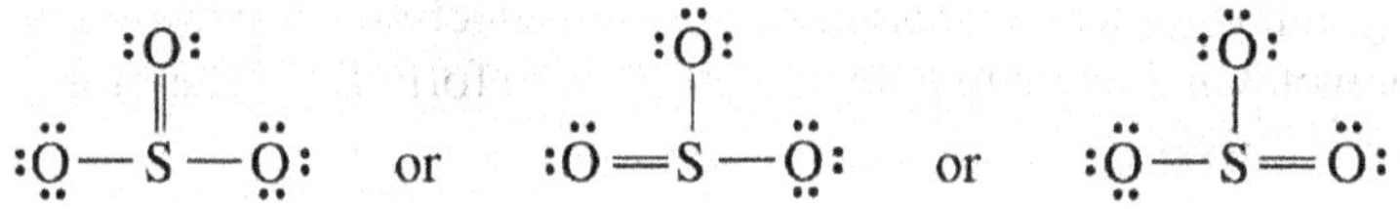

```
     :O:              :Ö:              :Ö:
      ‖                |                |
 :Ö — S — Ö:  or  :Ö = S — Ö:  or  :Ö — S = Ö:
```

5.17 Step 1: NO⁺ has 10 valence electrons: = 5 from N, 6 from O, and subtract 1 for the 1+ charge.

Step 2: N—O   2 electrons assigned

Step 3:   :N̈ — O, then   :N̈ — Ö

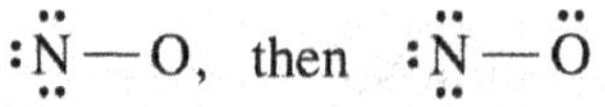

One octet done,   Still only one
8 electrons   octet done, but
assigned   all 10 electrons
                      assigned

Step 4: Use lone pairs to complete O octet (moving the O lone pair doesn't work):

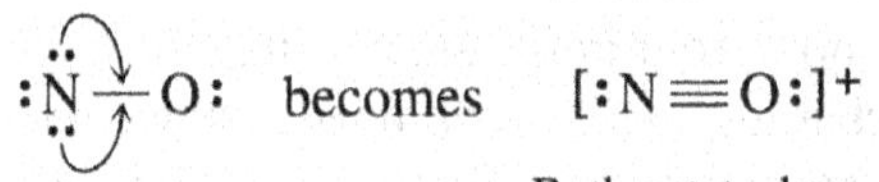

:N̈—O: becomes   [:N≡O:]⁺

Both octets done,
all 10 electrons
assigned

5.19 See solution in textbook.

5.21 EN for Si = 1.8 and EN for O = 3.5; therefore ΔEN = 1.7. Because ΔEN > 0 and the two atoms are both nonmetals, the bond is polar covalent. (At ΔEN = 1.7, this same bond would, by convention, be called ionic if the molecule were a metal/nonmetal combination.)

5.23 See solution in textbook.

5.25 They all contain two bromide ions because group IIA metals form 2+ cations and Br forms a 1− anion.

5.27 Because there are three Cl^- anions in the formula, titanium must have a 3+ charge to maintain electrical neutrality. $TiCl_3$ is therefore titanium(III) chloride.

5.29 See solution in textbook.

5.31 Phosphorus pentachloride.

5.33 See solution in textbook.

5.35 Calcium hypochlorite

5.37 (a) The 1− charge means this ion has one more valence electron than the seven found on a neutral I atom: $\left[:\ddot{\underset{..}{I}}:\right]^-$

(b) The 2− charge means two more electrons than the six on a neutral O atom: $\left[:\ddot{\underset{..}{O}}:\right]^{2-}$

(c) The 1− charge means one more electron than the seven on a neutral Cl atom: $\left[:\ddot{\underset{..}{Cl}}:\right]^-$

(d) The 1+ charge means one electron fewer than the one on a neutral H atom—in other words, no electrons: H^+

5.39 First, ions must be formed. This happens when valence electrons are transferred from one atom (most likely a metal) to another (most likely a nonmetal). The atom of a metal loses its valence electron(s) and becomes a positively charged ion, while the atom of a nonmetal gains valence electron(s) and becomes a negatively charged ion. When a negative ion comes into contact with a positive ion, an ionic bond forms.

5.41 (a) Magnesium is a group IIA metal and tends to lose two electrons to form Mg^{2+}. Bromine is a group VIIA nonmetal and tends to gain one electron to form Br^-. Therefore, two Br^- must react with one Mg^{2+}: $MgBr_2$.

(b) Beryllium is a group IIA metal and tends to lose two electrons to form Be^{2+}. Oxygen is a group VIA nonmetal and tends to gain two electrons to form O^{2-}. Therefore, these two substances react in a one-to-one ratio: BeO.

(c) Sodium is a group IA metal and loses one electron to form Na^+. Iodine is a group VIIA nonmetal and gains one electron to form I^-. These two substances therefore react in a one-to-one ratio: NaI.

5.43 Li_3N.

5.45 Coulomb's law explains how the value of the lattice energy is related to the distance between the ions and the charges on the ions in the lattice. For example, lattice energies for ionic compounds made of ions with charges of 2+/2− will be up to four items greater than those made of 1+/1− ions. And compounds made up of ions with charges 3+/3− will have a lattice energy nine times greater. Of course, the bigger the ion, the bigger the inter-ion distance, so that's why we can't predict the exact difference between the lattice energies.

5.47 Lattice energy is the energy required to break the ionic bond in an ionic compound and turn the solid into gaseous ions. These bonds must be broken in order for an ionic solid to melt. When the lattice is broken the ions are free to move about, just as when a solid melts.

5.49 A molecule is a stable collection of atoms that are bound together.

5.51 There is no such thing as an NaCl "molecule." The NaCl lattice has no discrete NaCl pairs that could be considered separate molecules. If you had an ability of looking into a sample of solid NaCl, you would see ions of Na^+ and Cl^- in the alternating points of its crystal lattice. On the other hand, H_2O is a molecular substance, in which there are distinct H_2O molecules. In solid H_2O you would see entire molecules of H_2O situated in the points of its crystal lattice.

5.53 Diatomic molecules in the atmosphere are nitrogen, N_2, and oxygen, O_2. Triatomic molecules in the atmosphere are ozone, O_3, water vapor, H_2O, carbon dioxide, CO_2, and nitrogen dioxide, NO_2.

5.55 Yes, a molecule is an elemental substance if all atoms it contains are of the same element. Examples are Cl_2, O_3, and S_8.

5.57 A covalent bond is *the additional force of attraction* that results from *valence electrons being shared between two nuclei.*

5.59 For any two systems that contain different amounts of energy, the system with less energy is more stable. The H_2 molecule is more stable than two separate H atoms because the energy of the H_2 molecule is less than the combined energy of the two separate atoms.

5.61 The H_2 bond distance is not shorter than 0.74 Å because if the atoms were closer than 0.74 Å to each other, the repulsion between the two positive nuclei and the repulsion between the two negative electrons would push the atoms apart. The bond distance is not longer than 0.74 Å because the atoms feel the greatest attractive force (and are therefore most stable) when they are 0.74 Å from each other.

5.63 Because the inner electrons are too close to the atom's nucleus to be shared with other atoms.

5.65 Boiling water will not produce hydrogen gas and oxygen gas because the energy required to break the covalent bonds in water is much more than the energy required to boil water.

5.67 A straight line.

5.69 The nitrogen atom is correct. The oxygen atom should be $\cdot \ddot{O} \colon$ because one electron must be placed on each of the four sides before any electrons are paired. Fluorine should have seven valence electrons, not six, and should be $\cdot \ddot{\ddot{F}} \colon$

5.71 (a) $\cdot \ddot{\ddot{S}} \colon$ (b) $\cdot \ddot{I} \colon$ (c) He: (d) $\cdot \dot{B} \cdot$ (e) $\colon \ddot{Ne} \colon$

5.73 As paired electrons or as unpaired electrons.

5.75 The diagram $\cdot \ddot{S} \colon$ tells you sulfur needs to form two bonds to have an octet, and the diagram $\cdot$ H tells you hydrogen needs to form one bond to have a duet. Therefore, the compound formed is H_2S, H—$\ddot{S}$—H

5.77 Three Br atoms combine with N, just as three Br atoms combine with P. Because N and P are both in group VA, they have the same number of valence electrons (five), and need to form the same number of bonds (three) to achieve an octet.

5.79 Step 1: The dot diagrams are $\cdot \dot{C} \cdot$ $\cdot \ddot{O} \cdot$ H $\cdot$. Because the molecule contains two C, six H, and one O, there are 20 electrons to be assigned.

Step 2: With only one C bonded to O, you can guess that O must also bond to at least one H:

$$
\begin{array}{c}
\text{H} \quad \text{H} \\
| \quad\quad | \\
\text{H—C—C—O—H} \\
| \quad\quad | \\
\text{H} \quad \text{H}
\end{array}
\qquad \text{16 electrons assigned}
$$

Step 3:

$$
\begin{array}{c}
\text{H} \quad \text{H} \\
| \quad\quad | \\
\text{H—C—C—\ddot{O}—H} \\
| \quad\quad | \\
\text{H} \quad \text{H}
\end{array}
$$

All octets (or duets) formed,
all 20 electrons assigned

There are eight bonding pairs and two lone pairs in the molecule. Because the beverage ethanol and the solvent ether have very different properties, it is evident that the arrangement of the atoms in a molecule is an important determinant of chemical properties.

5.81 Drawing the two electrons as unshared electrons would imply that a helium atom could either form two covalent bonds or form a 2+ ion. The first option is impossible, as it would have meant the presence of four valence electrons near helium (two more than allowed). If oxygen reacted with helium, one could envision an ionic bond forming between a He^{2+} ion and an O^{2-} ion.

5.83 Three bonds. The atom has five valence electrons and needs three more to achieve an octet.

5.85 The ethylene bond is weaker because it is only a double bond, whereas the bond in acetylene is a triple bond. The more shared electrons, the stronger the bond.

5.87 Step 1: SO_2 contains 18 valence electrons: 6 from $\cdot\ddot{S}:$ and $(2 \times 6) = 12$ from $\cdot\ddot{O}:$

Step 2: $O—S—O$ 4 electrons assigned

Step 3: $:\ddot{O}—S—O$, then $:\ddot{O}—S—\ddot{O}:$, then $:\ddot{O}—\ddot{S}—\ddot{O}:$

One octet done,	Two octets done,	Still only two
10 electrons	16 electrons	octets done, all
assigned	assigned	18 electrons assigned

Step 4: Convert lone pairs to double bonds to complete the octet for S. There are two resonance forms here:

$:\ddot{O}—\ddot{O}=\ddot{O}:$ and $:\ddot{O}=\ddot{O}—\ddot{O}:$

5.89 Yes, there should be similarities because both compounds have the same number of valence electrons and both contain only atoms from group VIA.

5.91 Because the bond between oxygen atoms in O_2 is a full double bond and the bonds in O_3 are 1.5 bonds, the O_2 bond is stronger and therefore more energy input is required to break it.

5.93 C_3H_6O has 24 valence electrons, $3 \times 4 = 12$ from $\cdot\dot{C}\cdot$, $1 \times 6 = 6$ from $\cdot\ddot{O}:$, and $6 \times 1 = 6$ from $\cdot H$, which is the number shown in the drawing. The error in the drawing is that the central carbon has only six electrons. Complete its octet by changing one of the O lone pairs to a double bond.

$$
\begin{array}{c}
\text{H} \quad :\ddot{O}: \quad \text{H} \\
| \quad\quad \| \quad\quad | \\
\text{H—C—C—C—H} \\
| \quad\quad\quad\quad | \\
\text{H} \quad\quad\quad \text{H}
\end{array}
$$

5.95 Step 1: NO_3^- contains 24 valence electrons: = from $\cdot\ddot{N}\cdot$, $3 \times 6 = 18$ from $\cdot\ddot{O}\!:$, and one to give the ion its $1-$ charge.

Step 2: O—N—O   6 electrons assigned
           |
          O

Step 3: $:\ddot{O}$—N—O , then $:\ddot{O}$—N—O , then $:\ddot{O}$—N—$\ddot{O}:$
         | | |
         O   :$\ddot{O}$:   :$\ddot{O}$:

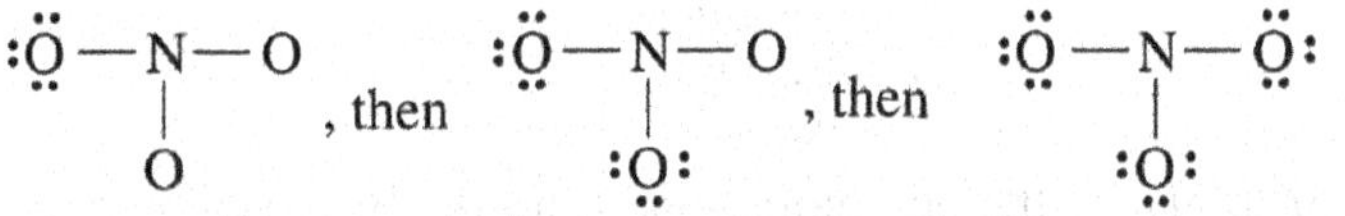

  One octet done,   Two octets done,   Three octets done,
   12 electrons   18 electrons   all 24 electrons
     assigned   assigned   assigned

Step 4: The N needs one more bond to complete its octet. Because any one of the O atoms could supply the needed lone pair, there are three resonance structures:

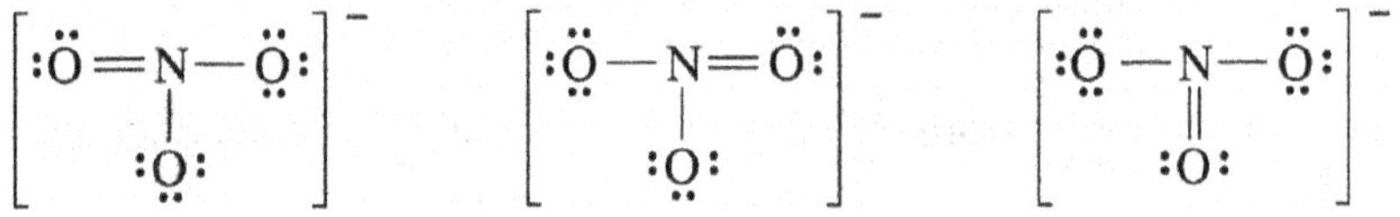

$$\left[:\ddot{O}\!=\!N\!-\!\ddot{O}: \right]^- \quad \left[:\ddot{O}\!-\!N\!=\!\ddot{O}: \right]^- \quad \left[:\ddot{O}\!-\!N\!-\!\ddot{O}: \right]^-$$
with $:\ddot{O}:$ below N in structures 1 and 2, and $:\ddot{O}:$ (double bonded) below N in structure 3.

5.97 Step 1: C_2H_4O has 18 valence electrons: $2 \times 4 = 8$ from $\cdot\ddot{C}\cdot$, $4 \times 1 = 4$ from $\cdot$H and 6 from $\cdot\ddot{O}\!:$

Step 2: Already done!

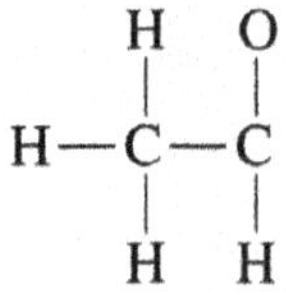

      H   O
      | |
  H—C—C   12 electrons assigned
      | |
      H   H

Step 3:
         H :$\ddot{O}$:
         | |
    H—C—C
         | |
         H   H

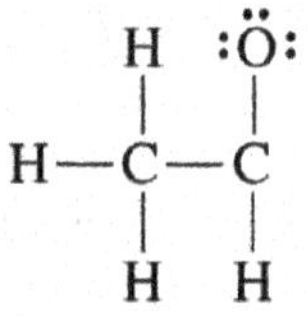

    One octet done,
  all 18 electrons assigned

Step 4: The C attached to O needs one more bond, meaning you must move one lone pair off the O:

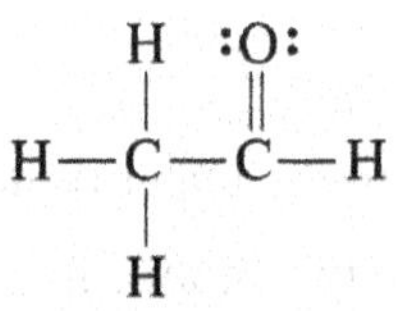

        H :O:
        | ‖
   H—C—C—H
        |
        H

5.99 Step 1: 24 valence electrons: $2 \times 4 = 8$ from $\cdot\ddot{C}\cdot$ $3 \times 1 = 3$ from $\cdot$H, and $2 \times 6 = 12$ from $\cdot\ddot{O}\!:$, and one more to account for the $1-$ charge.

Step 2: 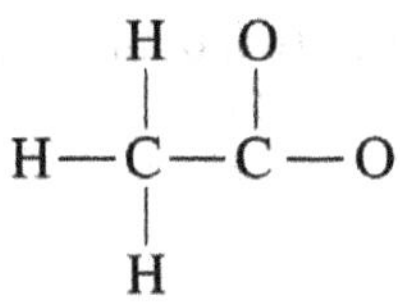
      H   O
      | |
  H—C—C—O   12 electrons assigned
      |
      H

Step 3:

H :Ö:
| |
H—C—C—O—H, then
|
H

One octet done,
20 electrons
assigned

H :Ö:
| |
H—C—C—Ö—H
|
H

Two octets done,
all 24 electrons
assigned

Step 4: Use an O lone pair to give C the one more bond it needs. Because the two O have the same number of lone pairs, either one can contribute and you get resonance forms:

$$\left[\begin{array}{c} H\ :O: \\ |\ \ || \\ H—C—C—\ddot{O}: \\ | \\ H \end{array} \right]^{-} \quad \text{and} \quad \left[\begin{array}{c} H\ :\ddot{O}: \\ |\ \ | \\ H—C—C=\ddot{O} \\ | \\ H \end{array} \right]^{-}$$

5.101 Step 1: 16 valence electrons: = from $\cdot\ddot{N}\cdot$, $2 \times 6 = 12$ from $\cdot\ddot{O}\colon$ and subtract one to account for the 1+ charge.

Step 2: O—N—O   4 electrons assigned

Step 3: $:\ddot{O}—N—\ddot{O}:$

One octet done,
all 16 electrons assigned

Two octets done,
all 16 electrons assigned

Step 4: N needs two more bonds. Take one lone pair from each O:

$$\left[:\ddot{O}=N=\ddot{O}: \right]^{+}$$

5.103 Eight valence electrons on each carbon atom and two valence electrons on each hydrogen atom, as there must be to satisfy the octet rule.

5.105 The important missing word is *shared*. Electronegativity is an indication of an atom's ability to attract *shared* electrons to itself.

5.107 Fluorine is the most electronegative element; francium is the least electronegative. If these two elements were brought together, the one valence electron of Fr would be transferred to F to form the ionic compound FrF.

5.109 (a) Electronegativity decreases going down a group.

(b) Electronegativity increases going across a period from left to right.

(c) Electronegativity increases going from the bottom left corner to the upper right corner.

5.111 In a covalent bond, the electrons are shared roughly equally by the two atoms. In a polar covalent bond, the electrons spend more time near the atom with the higher electronegativity than near the other atom. Diatomic covalent molecules are F_2, O_2, N_2, or any other diatomic molecule made up of atoms of the same element, for in all these cases $\Delta EN = 0$. Diatomic polar covalent molecules are CO, NO, LiAt, or any other diatomic molecule made up either of two nonidentical nonmetal atoms or of a metal/nonmetal combination with $\Delta EN < 1.7$.

5.113 The bonding would be covalent because $\Delta EN = 0$.

5.115 Anytime electrons are shared between atoms, the bond can be considered covalent. However, if the shared electrons spend more time nearer one atom than the other, it is as though the electrons have been partially transferred to one atom from the other—in other words, that an ionic bond has formed. Current thinking about covalent and ionic bonds is that they are not completely different from each other but rather exist along a continuum where one type blends into the other.

5.117 A is more electronegative because it carries the partial negative charge (δ^-), indicating that the electrons spend more time closer to A.

5.119 (a) MgO and NaCl are ionic (ΔEN $=$ 2.3 and 2.1, respectively).

(b) Cl_2 has nonpolar bonds (ΔEN $=$ 0).

(c) CF_4, PH_3, and SCl_2 have polar-covalent bonds (ΔEN $=$ 2.0, 1.0, and 0.5, respectively).

5.121 (a) Br_2 covalent because ΔEN $=$ 0.

(b) PCl_3 polar covalent because ΔEN $>$ 0 $(3.0 - 2.1 = 0.9)$ and all atoms are nonmetals.

(c) LiCl ionic because ΔEN $>$ 1.7 $(3.0 - 1.0 = 2.0)$ and one atom (Li) is a metal.

(d) ClF polar covalent because ΔEN $>$ 0 $(4.0 - 3.0 = 1.0)$ and both atoms are nonmetals.

(e) $MgCl_2$ ionic because ΔEN $>$ 1.7 $(3.0 - 1.2 = 1.8)$ and one atom (Mg) is a metal.

5.123 The suffix *-ide* indicates the negative part in a binary ionic compound; the negative ion gets this suffix.

5.125 Because the overall charge on the compound must be zero, the number of positive ions and negative ions must result in a total charge of zero. Examples are MgO (2+ and 2−) and $CaCl_2$ (2+ and two 1−).

5.127 (a) Calcium nitride

(b) Aluminum fluoride

(c) Sodium oxide

(d) Calcium sulfide

5.129 The transition metals can form more than one kind of cation. A roman numeral in parentheses is used to indicate the magnitude of the positive charge.

5.131 (a) Sodium sulfate   (b) Ammonium phosphate   (c) Potassium hypochlorite

(d) Calcium carbonate   (e) Aluminum nitrate

5.133 (a) Phosphorus trichloride   (b) Sulfur dioxide   (c) Dinitrogen tetroxide

(d) Pentaphosphorus decoxide

5.135 The oxide ion is a monoatomic anion of oxygen, whereas the peroxide ion is a diatomic anion of oxygen (both have an overall 2− charge).

$$:\ddot{\underset{\cdot\cdot}{O}}:^{2-} \qquad \left[:\ddot{\underset{\cdot\cdot}{O}} - \ddot{\underset{\cdot\cdot}{O}}:\right]^{2-}$$

Oxide ion   Peroxide ion

5.137 Ammonia is the electrically neutral molecule, NH_3. Ammonium is the cation, NH_4^+.

$$\text{H—}\overset{\displaystyle ..}{\underset{\displaystyle |}{\text{N}}}\text{—H}\qquad \left[\text{H—}\overset{\displaystyle \overset{|}{\text{H}}}{\underset{\displaystyle \underset{|}{\text{H}}}{\text{N}}}\text{—H}\right]^+$$

Ammonia   Ammonia ion

By analogy, phosphine is the electrically neutral molecule, PH_3, while phosphonium is the cation, PH_4^+.

$$\text{H—}\overset{\displaystyle ..}{\underset{\displaystyle |}{\text{P}}}\text{—H}\qquad \left[\text{H—}\overset{\displaystyle \overset{|}{\text{H}}}{\underset{\displaystyle \underset{|}{\text{H}}}{\text{P}}}\text{—H}\right]^+$$

Phosphine   Phosphonium

5.139 In the formula for a binary covalent compound, the less electronegative atom (N in this case) appears first, making N_2O correct and ON_2 incorrect. In the name, the more electronegative atom (O in this case) is given the *-ide* suffix because it is this atom that has the partial negative charge, and *-ide* means "negative."

5.141 You know the anion names from the *-ide* rule for single-atom ions and from having memorized the polyatomic ions in Table 5.5. (If you haven't memorized them yet, go do it now!)

Anion	Anion name	Acid formula	Acid name
F^-	Fluoride	HF	Hydrofluoric acid
NO_3^-	Nitrate	HNO_3	Nitric acid
Cl^-	Chloride	HCl	Hydrochloric acid
$C_2H_3O_2^-$	Acetate	$HC_2H_3O_2$	Acetic acid
NO_2^-	Nitrite	HNO_2	Nitrous acid

Two rules govern acid names: (1) with acids containing no O, add the prefix *hydro-* and the suffix *-ic acid* to the anion name; (2) with oxyacids, change the ending *-ate* in the name of a polyatomic ion to *-ic* and add the word *acid*, and change the ending *-ite* in the name of a polyatoic ion to *-ous* and add *acid*.

5.143 Table 5.5 in the textbook tells you NO_3^- is the nitrate ion. Because the oxyacid HNO_3 contains this ion, change the *-ate* in *nitrate* to *-ic acid*; this is nitric acid. The anion NO_2^- has fewer O than the nitrate ion and must therefore be the nit*rite* ion, which means HNO_2 is nitrous acid.

5.145 Hypochlor*ous* acid is formed from the hypochlor*ite* anion, ClO^-; the molecular formula for the acid is HClO. Perchlor*ic* acid is formed from the perchlor*ate* anion, ClO_4^-; the acid formula is $HClO_4$.

5.147 An ionic bond is the attractive force between oppositely charged ions that were formed by a transfer of electrons. It is similar to a covalent bond in that both types of bonds have approximately the same strength and both are the result of the attraction between positive and negative charges (attractions of electrons for nuclei in covalent bonds; attractions of ions for oppositely charged ions in ionic bonds). An ionic bond differs from a covalent bond in that an ionic bond is the result of a transfer of electrons, whereas a covalent bond is the result of a sharing of electrons.

5.149 Subtract the roman-numeral group number of element from 8.

5.151 Because they already have a valence octet of electrons.

5.153 You must look at the dot diagrams to answer this question:

$$\cdot \ddot{\text{F}} \text{:} + \cdot \ddot{\text{F}} \text{:} \longrightarrow \text{:} \ddot{\text{F}} \text{—} \ddot{\text{F}} \text{:}$$

$$\cdot \ddot{\text{O}} \text{:} + \cdot \ddot{\text{O}} \text{:} + \cdot \ddot{\text{O}} \text{:} \longrightarrow \text{:} \ddot{\text{O}} \text{—} \text{O} \text{—} \ddot{\text{O}} \text{:} \longrightarrow \text{:} \underset{\cdot\cdot}{\text{O}} {=} \text{O} {=} \underset{\cdot\cdot}{\text{O}} \text{:}$$

$$\text{H} \cdot + \cdot \dot{\underset{\cdot}{\text{C}}} \cdot + \cdot \ddot{\text{N}} \cdot \longrightarrow \text{H} \text{—} \text{C} \text{—} \ddot{\text{N}} \text{:} \longrightarrow \text{H} \text{—} \text{C} {\equiv} \ddot{\text{N}}$$

$$\text{H} \cdot + \text{H} \cdot + \cdot \dot{\underset{\cdot}{\text{C}}} \cdot + \cdot \ddot{\text{O}} \text{:} \longrightarrow \underset{\underset{\text{H}}{|}}{\text{H} \text{—} \text{C} \text{—} \ddot{\text{O}} \text{:}} \longrightarrow \underset{\underset{\text{H}}{|}}{\text{H} \text{—} \text{C} {=} \underset{\cdot\cdot}{\text{O}} \text{:}}$$

5.155 (b) is correct. (a) and (c) have all octets satisfied but incorrect numbers of electrons. There are 12 electrons to be assigned in this molecule: $2 \times 1 = 2$ from H and $2 \times 5 = 10$ from N. (a) has 14 electrons, and (c) has 10.

5.157 Steps 1 and 2: 28 electrons: $2 \times 1 = 2$ H, $3 \times 6 = 18$ O, $2 \times 4 = 8$ C; 12 already assigned.

Step 3:
$$\underset{\text{H} \text{—} \ddot{\text{O}} \text{—} \text{C} \text{—} \text{C} \text{—} \text{H}}{\overset{\overset{\cdot\cdot}{\text{:} \ddot{\text{O}} \text{:}} \quad \overset{\cdot\cdot}{\text{:} \ddot{\text{O}} \text{:}}}{\underset{|}{} \quad \underset{|}{}}}$$

Step 4:
$$\underset{\text{H} \text{—} \ddot{\text{O}} \text{—} \text{C} \text{—} \text{C} \text{—} \text{H}}{\overset{\text{:O:} \quad \text{:O:}}{\underset{\|}{} \quad \underset{\|}{}}}$$

(a) Two double bonds.   (b) Six lone pairs.

5.159 Two: $\left[\text{:} \ddot{\text{F}} \text{—} \ddot{\text{Br}} \text{—} \ddot{\text{F}} \text{:} \right]^{+}$

5.161 Calcium. You can determine relative electronegativities from the guideline that the lowest values are at the lower left of the periodic table and highest values are at the upper right.

5.163 Ionic.

5.165 (d) because the two I atoms have the same electronegativity.

5.167

Name	Formula
Silver nitrate	$AgNO_3$
Aluminum selenide	Al_2Se_3
Lithium oxide	Li_2O
Ammonium iodide	NH_4I
Copper(II) sulfate or cupric sulfate	$CuSO_4$
Potassium permanganate	$KMnO_4$
Calcium chlorate	$Ca(ClO_3)_2$

5.169

Name	Formula
Cadmium telluride	CdTe
Nitrogen triiodide	NI_3
Silicon tetriodide	SiI_4
Bromine trifluoride	BrF_3
Hydroiodic acid	HI (dissolved in water)
Hydrogen iodide	HI (as a pure gas)
Tetrasulfur tetranitride	S_4N_4

5.171 You need a metal/nonmetal combination or polyatomic ions to have an ionic compound, meaning you know (a), (b), and (e) are molecular compounds. For the other four, you should recognize Cl^- plus the anions from Table 5.5 of the textbook and therefore classify them as ionic compounds.

(a) Molecular; chlorine; 14 electrons: $:\ddot{Cl}—\ddot{Cl}:$

(b) Molecular (because this is *gaseous* HCl); hydrogen chloride gas; 8 electrons: $H—\ddot{Cl}:$

(c) Ionic; sodium chloride; anion, 8 electrons: $\left[:\ddot{Cl}:\right]^-$

(d) Ionic; magnesium chlorite; anion, 20 electrons: $\left[:\ddot{O}—\ddot{Cl}—\ddot{O}:\right]^-$

(e) Molecular; methanol; 14 electrons:
$$H—\overset{\overset{\displaystyle H}{|}}{\underset{\underset{\displaystyle H}{|}}{C}}—\ddot{O}—H$$

(f) Ionic; iron(III); nitrate or ferric nitrate; anion, 24 electrons:
$$\left[:\ddot{O}=N—\ddot{O}:\atop \qquad |\atop \qquad :O:\right]^-$$

(g) Ionic; lead acetate; anion, 24 electrons:
$$\left[H—\overset{\overset{\displaystyle H}{|}}{\underset{\underset{\displaystyle H}{|}}{C}}—\overset{\overset{\displaystyle :O:}{||}}{C}—\ddot{O}:\right]^-$$

5.173 Because in binary covalent compounds, the less electronegative element is listed first.

5.175 Because the term *ionic compound* is generally reserved for a metal bonded to a nonmetal. H and F are both nonmetals.

5.177 (a) Iron(III) sulfite or ferric sulfite   (b) Gold(III) nitrate

(c) Sodium dihydrogen phosphate   (d) Lead acetate or lead(II) acetate

5.179 (a) K, Li, Be   (b) Si, S, O   (c) Te, I, Br

5.181 (a) Cu_2S, polar covalent (metal/nonmetal, $\Delta EN < 1.7$)

(b) Al_4C_3, polar covalent (metal/nonmetal, $\Delta EN < 1.7$)

(c) I_2O_5, polar covalent (two nonmetals)

(d) ClF_3, polar covalent (two nonmetals)

5.183 Hydrogen sulfide refers to the covalent compound in the gas phase, and hydrosulfuric acid refers to the aqueous solution (hydrogen sulfide gas dissolved in water), which acts as an acid by dissociating to give H^+ ions in solution.

5.185 Manganate ion. The *per-* in *permanganate* tells you there are more than two oxyanions in this family, meaning a simple change from *-ate* to *-ite* is not correct. Following the pattern of the Cl oxyanions—perchlorate, chlorate, chlorite, hypochlorite—a loss of one O in the ion, from MnO_4^- to MnO_3^-, changes the name from *permanganate* to *manganate*.

5.187 There should be one lone pair of electrons.

5.189 (a) Atom B is more electronegative as the electrons spend more time near atom B in the AB molecule.

(b) Atom A has the δ^+ partial charge since atom B "hogs" the shared electrons, partially stripping negative charge from atom A.

(c) Atom A is most likely the metal, as metals tend to be less electronegative than nonmetals.

(d) ΔEN

5.191 (a) Equal sharing of two valence electrons—covalent bonding.

(b) Unequal sharing of two valence electrons—polar covalent bonding.

(c) Transfer of a valence electron from one atom to another—ionic bonding.

5.193

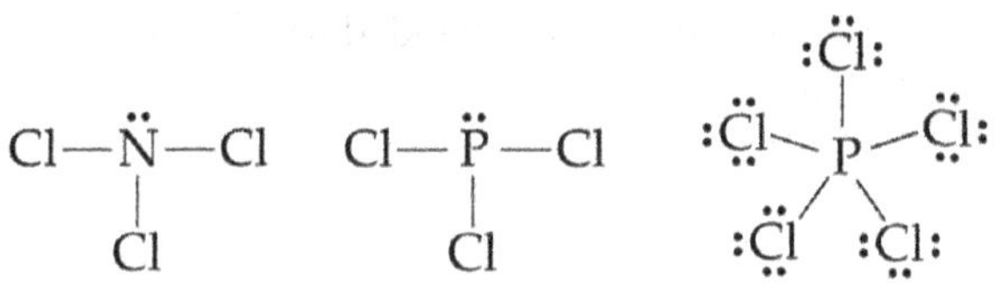

Phosphorus has $3s$, $3p$, and $3d$ orbitals in its outer shell, which can hold a total of 18 electrons. Atoms in periods three and above might be expected to exceed the octet rule.

5.195

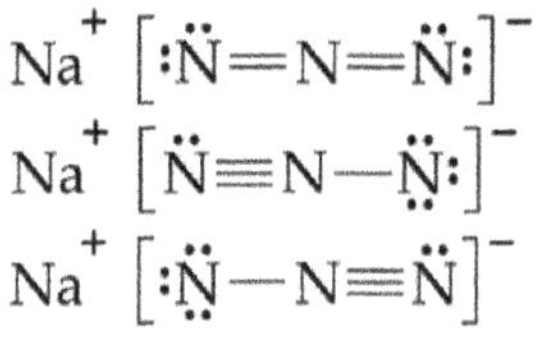

5.197 In order to form six bonds with fluorine atoms in SeF_6, selenium needs to expand its octet to 12 electrons. By doing so, it violates the octet rule. However, exceeding octet is possible for atoms in periods three and beyond. Since Se is in the fourth period, it can hold up to 18 electrons in its valence.

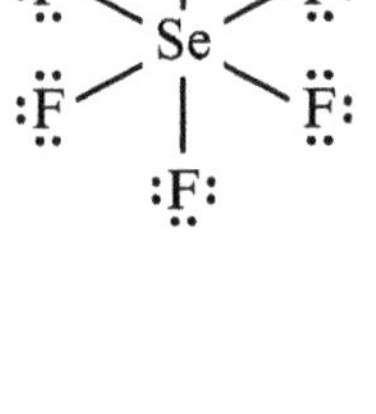

5.199 In addition to four sulfur–fluorine bonds, there is a lone pair of electrons on the S atom:

5.201 BeH_2 has four valence electrons and its Lewis dot diagram is:

H : Be : H

Be is electron deficient in BeH_2 because it has only four valence electrons and, therefore, it violates the octet rule. Hydrogen has access to two valence electrons, and while it does not satisfy the octet rule, it is not a violation since hydrogen satisfies a "duet" rule, that is the electron configuration of its closest noble gas, helium, which has only two valence electrons.

5.203 PF_5 is a 40 valence electron system, in which P has its octet expanded to 10 valence electrons.

5.205 XeF_4 is a 36 valence electron system, in which Xe has its octet expanded to 12 valence electrons. There are two lone pairs of electrons on the Xe atom.

Chapter 6

6.1 See solution in textbook.

6.3 Dot diagram: $:\ddot{C}l - C \equiv C - \ddot{C}l:$

Two groups of electrons around each carbon atom (remember that we count multiple bonds as single bonds when determining how the electrons are arranged) means the electrons are 180° apart; the molecular shape is therefore linear, and all bond angles are 180°:

6.5 See solution in textbook.

6.7 Dot diagram:

H—C≡N:

There are two bonding groups around the central C and no lone pairs (remember that lone pairs on peripheral atoms do not influence shape). Therefore the electron-group geometry and molecular shape are both linear:

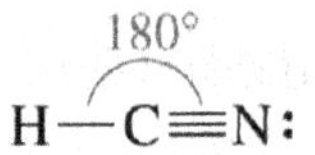

6.9 Dot diagram:

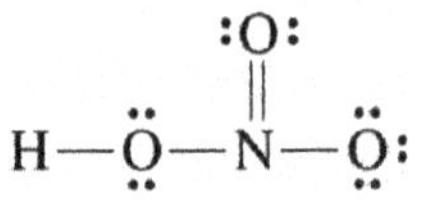

There are three bonding groups of electrons around the nitrogen and no lone pairs. Therefore, the electron-group geometry is trigonal planar, and the molecular shape is also trigonal planar, with bond angles of 120° around the N:

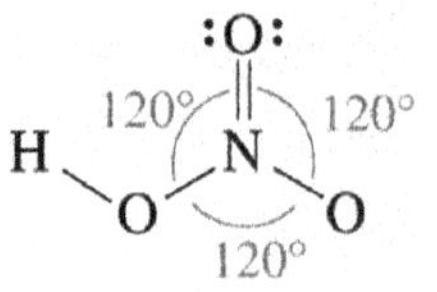

6.11 $CHCl_3$ is polar. The dot diagram

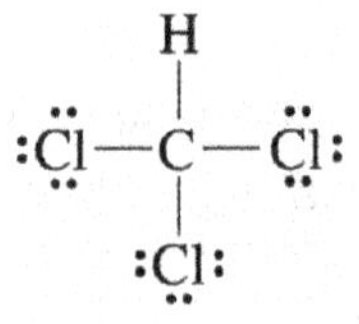

with its four electron groups around the central C tells you the molecule is tetrahedral. The C–H bond is treated as nonpolar, and the electronegativity difference between C and Cl tells you all three C–Cl bonds are polar. In a tetrahedral shape, there is no way three polar bonds can cancel one another, meaning the molecular is polar:

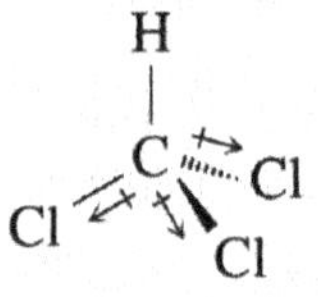

Individual bond
dipole moments

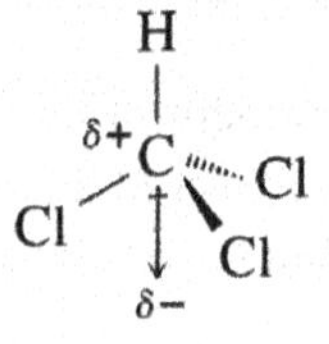

Overall molecule
dipole moments

6.13 The fact that NH_3 molecules are arranged away from the DNA strands so as not to interfere with the "bracing" of the cancer DNA strands together, and the flat arrangement and the cisplatin, allowing for the Pt to bind to the corresponding areas of the DNA strands, collectively prevent the cancer DNA strands from separating and, subsequently, replicating. Should the NH_3 point toward the strands, they would not bind to the DNA strands, which would then allow the strands to separate and then, replicate without impediment.

6.15 Because it is only the valence shell electrons that are involved in bonding, and that determines the shape of molecules.

6.17 $SOCl_2$ is sulfonyl chloride, also known as thionyl chloride. Assuming that bond angles are ideal for this molecule (they are not!), they are all $109.5°$.

6.19 (a)

(b) (1) Tetrahedral electron geometry, tetrahedral shape. (2) Trigonal planar electron geometry, trigonal planar shape. (3) Tetrahedral electron geometry, pyramidal shape. (4) Tetrahedral electron geometry, bent shape.

6.21 It is possible because the extent of bending depends on how many lone pairs surround the central atom. In a $118°$ bent molecule, a central atom is surrounded by two bonding groups and one lone pair. Think of this as a modification of a central atom having three bonding groups spaced $120°$ apart. Because a lone pair takes up more space than a bonding pair, however, the lone pair repels the two bonding groups slightly, and the bond angle is compressed by about $2°$, from $120°$ to $118°$. In a $105°$ bent molecule, a central atom is surrounded by two bonding groups and two lone pairs. Think of this as a modification of a central atom having four bonding groups spaced $109.5°$ apart. The two lone pairs repel the bonding groups even more than a single long pair does, and so the bond angle is compressed by about $4°$, from $109.5°$ to $105°$.

6.23 The shape of a molecule is determined by the arrangement of the atoms in the molecule, ignoring any lone pairs around the central atom. We do not call ammonia a tetrahedral molecule because the one N and three H atoms form a pyramid, as Table 6.2 of the textbook shows. The NH_3 molecule thus has a pyramidal shape.

6.25 (a)

(b) and (c) Call the carbons C1, C2, C3 from left to right. C1 and C2 have three bonding groups each, no lone pairs, meaning the shape in this part of the molecule is trigonal planar with 120° angles. C3 has four bonding groups, no lone pairs, and therefore the shape around this C is tetrahedral with 109.5° angles. The O has two bonding groups, two lone pairs, meaning the shape in this part of the molecule is bent with a 105° angle:

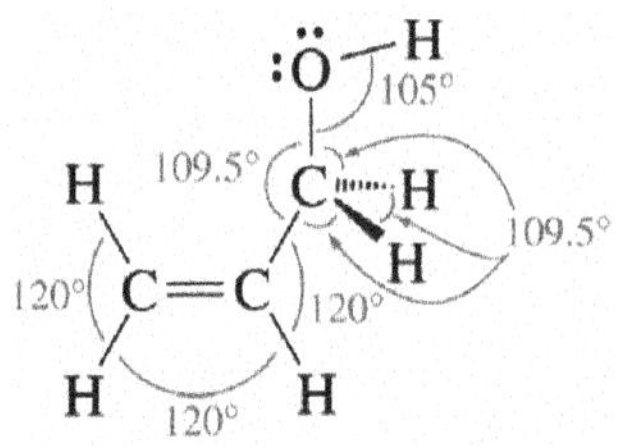

6.27 The first molecule is the flat molecule because the bond angles around each carbon are 120° (each carbon has three bonding groups around it). In the second molecule, each carbon has four bonding groups around it, resulting in bond angles of 109.5°. Therefore, it is not a flat molecule.

6.29 When it lines up in a specific orientation when placed between two metal plates of opposite electrical charge.

6.31 False. A molecule containing polar bonds is nonpolar if the individual dipole moments cancel each other, as in CO_2.

6.33 The degree of the unequal sharing of the electrons in the bond. The longer the arrow, the larger the magnitude of the dipole moment and the more polar the bond.

6.35 (a) The difference in the electronegativity value of the two bonded atoms determines the bond dipole moment; the greater the electronegativity difference, the more negative one atom is relative to the other and the greater the bond dipole moment.

(b) With an arrow pointing from the positive end of the bond to the negative end. The head of the arrow is at the negative end of the dipole, and the tail of the arrow carries a small perpendicular line so that this end looks like a plus sign: $\longmapsto$.

6.37 SO_2 is polar, while CO_2 is nonpolar. Because the molecule of SO_2 is bent (two bonding groups, one lone pair), the two bond dipole moments do not cancel each other. Instead, they add together to give a nonzero molecular dipole moment.

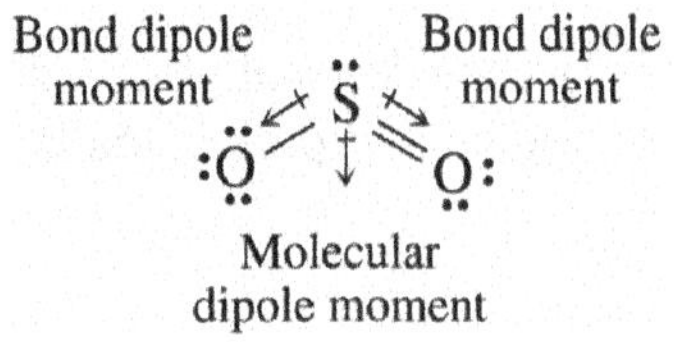

On the other hand, the molecule of CO_2 is linear and the two bond dipole moments cancel each other:

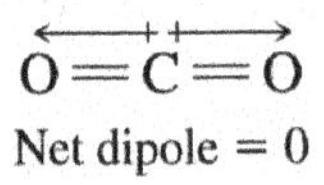

Net dipole = 0

6.39 (a) HF, because the electronegativity difference is greatest: $4.0 - 2.1 = 1.9$.

(b) HAt, because the electronegativity difference is least: $2.1 - 2.1 = 0$.

(c) $\overrightarrow{H-F}$ $\overleftrightarrow{H-Cl}$ $\overleftrightarrow{H-Br}$ $\overleftrightarrow{H-I}$ $H-At$

6.41 (a) Individual bond dipole moments

lead to molecular dipole moment

(b) Because there is only one polar bond in the molecule, the bond dipole moment and molecular dipole moment are identical:

(c) Individual bond dipole moments

lead to molecular dipole moment

(d) Because N and Cl have the same electronegativity value (3.0 to one decimal place), the N–Cl bond is nonpolar. With only one polar bond in the molecule, the bond dipole moment and molecular dipole moment are identical:

(e) This is a nonpolar molecule. The two bond dipole moments cancel so that there is no molecular dipole moment:

6.43 No. Dipole–dipole forces have only a fraction of the strength of the forces between ions because ions carry full electrical charges but dipole charges are partial.

6.45

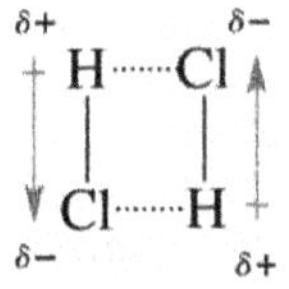

6.47 The lone pairs are important in determining the electron-groups geometry around an atom and thus in determining the shape of the molecule.

6.49

(a)

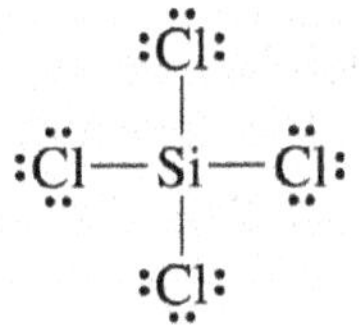

(b) and (c) Four bonding groups, no lone pairs around Si makes this a tetrahedral molecule:

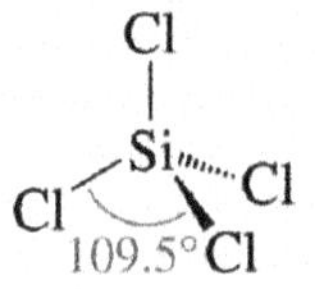

All bond angles 109.5°

(d) Because Cl is more electronegative than Si, the Cl atoms take the lion's share of the electrons:

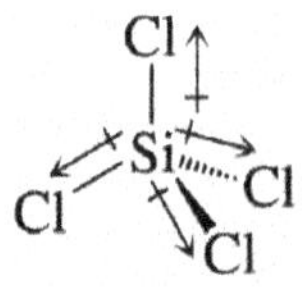

(e) The molecule is nonpolar because the four bond dipole moments cancel one another.

6.51 (a)

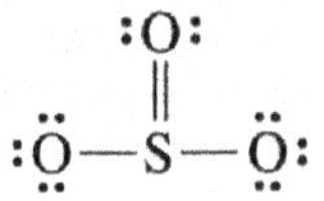

(b) and (c) Three bonding groups, no lone pairs, trigonal planar molecule:

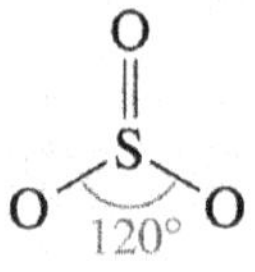

(d) O is more electronegative than S:

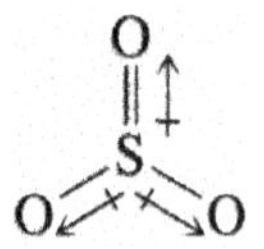

(e) The molecule is nonpolar because the three bond dipole moments, being 120° apart, all cancel.

6.53 (a) 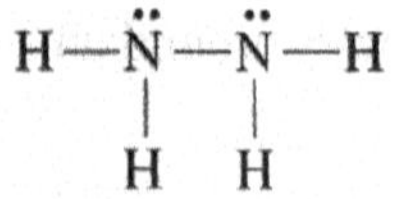 H—N̈—N̈—H
        | |
        H   H

(b) and (c) There are three bonding groups and one lone pair around each N, meaning the shape is pyramidal around each N. The electron-group geometry is tetrahedral, which means the bond angles should be 109.5°. However, the lone pair on each N compresses the angles to about 107°:

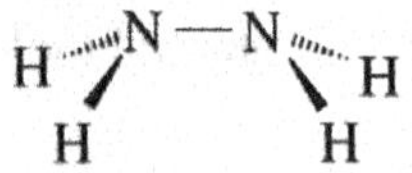

All angles 107°
(2° compression rule for period 2 atoms)

(d) N is more electronegative than H:

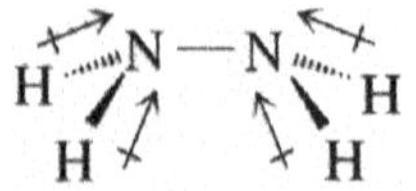

(e) The molecule is polar because the individual bond dipole moments do not cancel:

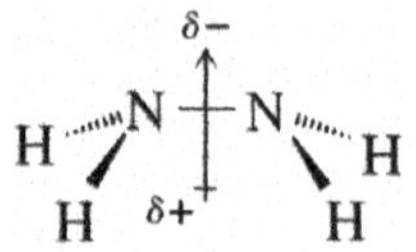

6.55 (a) :N≡N—Ö:

(b) and (c) Two bonding groups, no lone pairs around the central N makes this a linear molecule with 180° bond angles:

N≡N—O
  ⏝
 180°

(d) O is more electronegative than N:

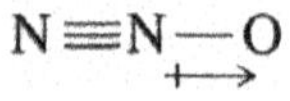

N≡N—O
    ⟼

(e) The molecule is polar because there is only one bond dipole moment. The bond dipole moment and molecular dipole moment are identical.

6.57 (a)
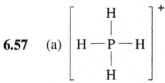

(b) and (c) Four bonding groups, no lone pairs, tetrahedral shape:

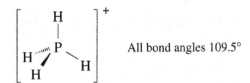   All bond angles 109.5°

(d) Because P and H have the same electronegativity value, there are no bond dipole moments in this molecule.

6.59 (a) Because the P–H bonds are nonpolar (the electronegativity difference is 0 for these two atoms: P 2.1, H 2.1), even though the PH_3 molecule is pyramidal, the zero bond dipole moments are adding up to ... zero.

(b) The P–F bonds are polar and they add up to a nonzero molecular dipole moment in the pyramidal molecular shape of PF_3.

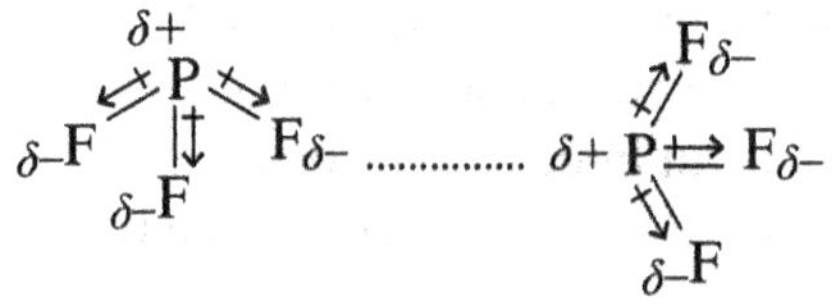

When two (or more) molecules of PF_3 are nearby, the partially negative portion of one polar molecule and the partially positive end of another molecule attract each other.

6.61 Consider each C atom separately. Each has four bonding groups and no lone electrons, meaning the molecular shape is tetrahedral about each C:

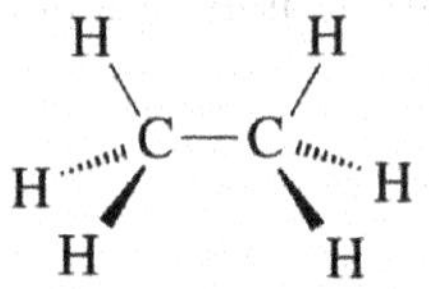

All bond angles 109.5°

6.63 Consider each interior atom separately. Left C: four bonding groups, no lone pairs, tetrahedral shape with 109.5° bond angles. O: two bonding groups, two lone pairs, bent shape with 105° bond angles. Right C: same as left C.

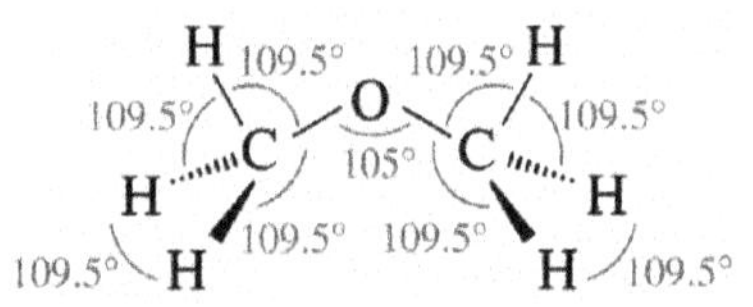

6.65   CH₄

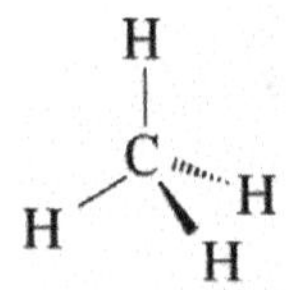

Because the C–H bond is taken to be nonpolar in this course, there are no dipole moments, and the molecule is nonpolar.

CH₃Cl

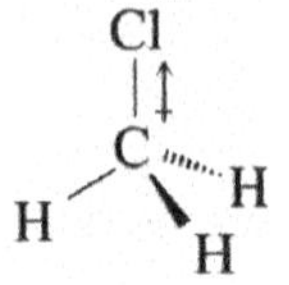

The C–Cl bond is the only polar bond here, making the molecule polar.

CH₂Cl₂

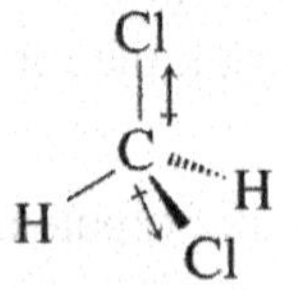

The two C–Cl bonds are polar, and the geometry of the molecule is such that the bond dipole moments do not cancel each other. Therefore the molecule is polar. Do not be misled by the seeming symmetry in the drawing. Remember, the molecule in three dimensions is a tetrahedron with the C buried in the interior. No matter which two corners hold the Cl atoms, the two dipole moments do not cancel each other.

CHCl₃

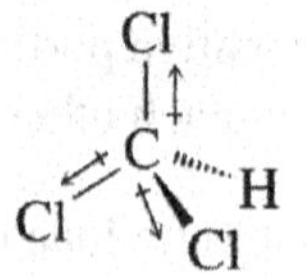

The three C–Cl dipole moments pull electrons to one side of the molecule, making it a polar molecule.

CCl₄

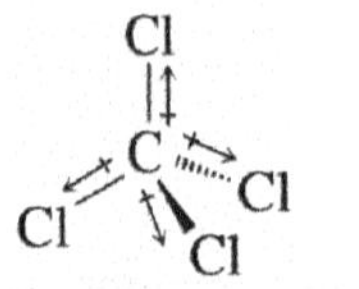

The symmetry in this molecule means all the bond dipole moments cancel, with the result that the molecule is nonpolar.

6.67 (a)

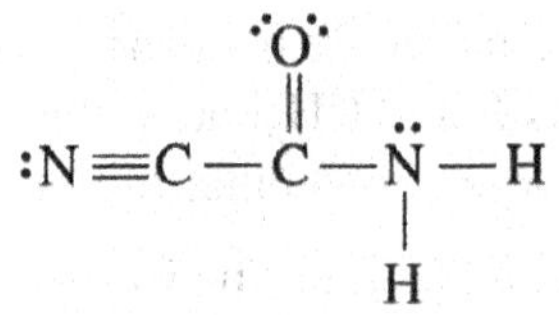

(b) Consider each interior atom separately. Left C: two bonding groups, no lone pairs, linear shape with 180° bond angles. Middle C: three bonding groups, no lone pairs, trigonal planar shape with 120° bond angles. N: three bonding groups, one lone pair, pyramidal shape with 107° bond angles.

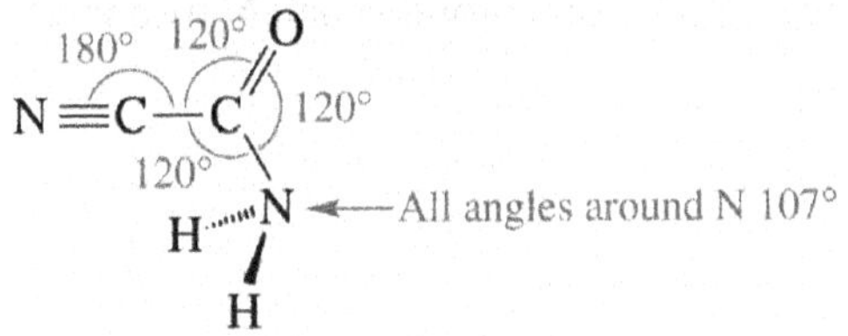

6.69

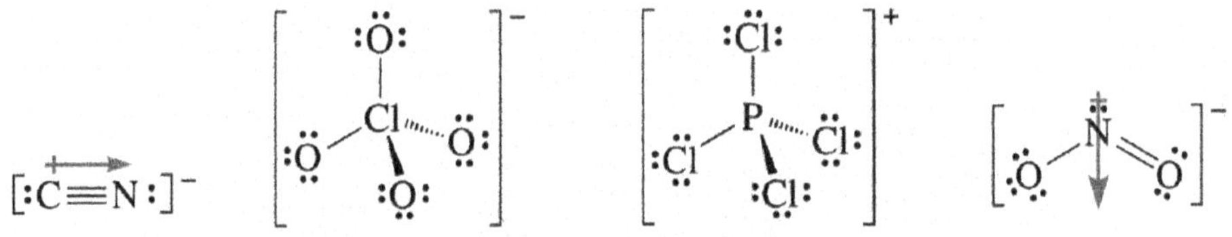

[:C≡N:]⁻			
Linear	Tetrahedral, no dipole moment	Tetrahedral, no dipole moment	Bent, 118° angle
(a)	(b)	(c)	(d)

6.71

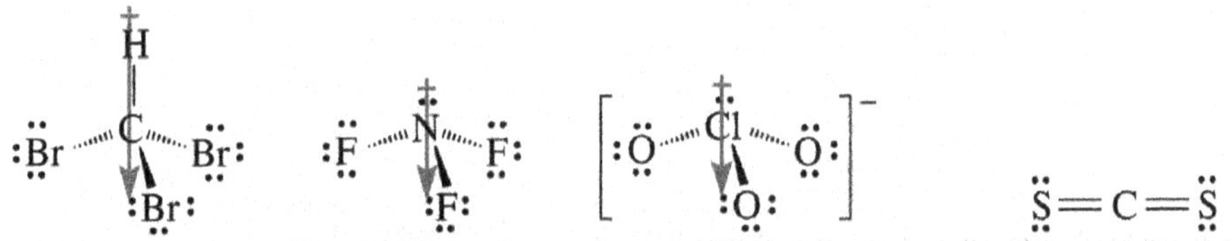

Tetrahedral	Pyramidal	Pyramidal	Linear, no dipole moment
(a)	(b)	(c)	(d)

6.73 All three are linear molecules, but only CO_2 and CSO have bond dipole moments:

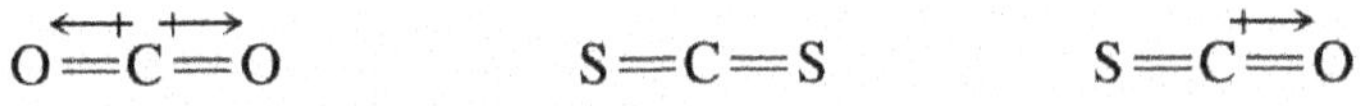

C–O ΔEN = 1.0   C–S ΔEN = 0
                   no bond dipole moments

CO_2 is a nonpolar molecule because the two bond dipole moments cancel each other. CS_2 is a nonpolar molecule because there are no bond dipole moments anywhere in the molecule. Only CSO is a polar molecule, because of its one bond dipole moment. The molecular dipole moment is identical to the bond dipole moment.

6.75 All three molecules are polar.

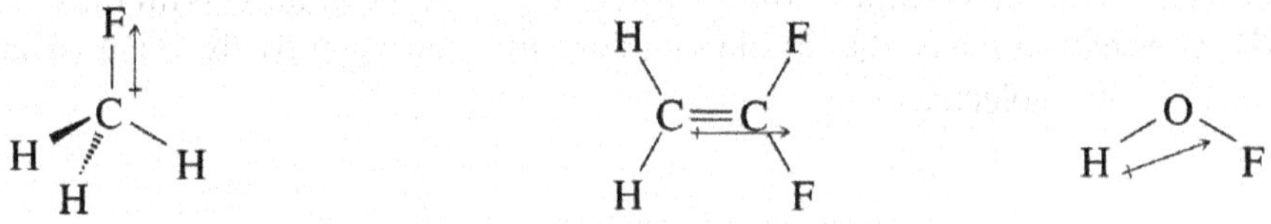

Tetrahdedral, all angles 109.5°   Trigonal planar, all angles 120°   Bent, 105° angle

6.77 H_2 is nonpolar because $\Delta EN = 0$. CO_2 is nonpolar because the molecule is linear and as a result the individual C–O dipole moments cancel each other. Both CH_3F and CH_3I are polar, but a C–F bond is more polar than a C–I bond. Therefore, CH_3F has the largest molecular dipole moment.

6.79 The student did not consider the three-dimensional shape of the molecule. The four electron groups have a tetrahedral arrangement, not the two-dimensional orientation shown in the dot diagram. Because there are two bonding groups and two lone pairs, the molecule has a tetrahedral electron-group geometry and a bent shape, no matter how it is drawn on paper.

6.81 In S_2F_2, two bonding groups and two lone pairs around each S means a bent shape with 105° bond angles. In N_2F_2, two bonding groups and one lone pair around each N means a bent shape with 118° bond angles.

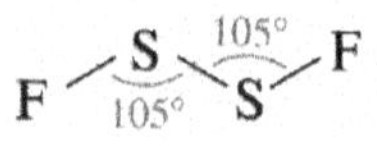

6.83 The bonding is

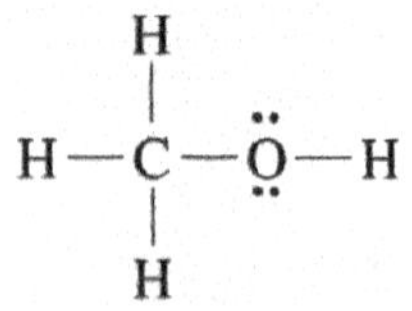

Four bonding groups around C means tetrahedral shape with 109.5° angles. Two bonding groups and two lone pairs around O means bent shape with a 105° angle:

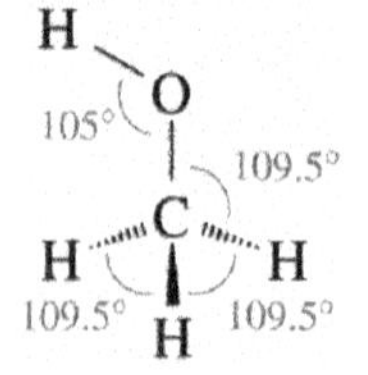

The molecule is polar because O is more electronegative than either H or C:

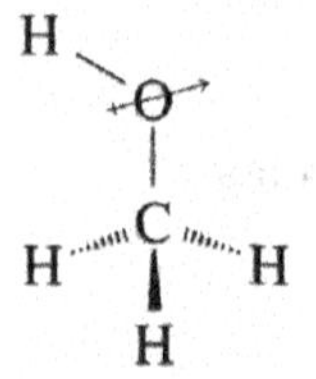

6.85 The electron-group geometry around an atom A in a molecule tells you the spatial arrangement of all the atoms bonded to A and any lone pairs associated with A. The molecular shape around A tells you the spatial arrangement of only the atoms bonded to A. When determining molecular shape, you use all the bonding groups shown in the electron-group geometry but none of the lone pairs.

6.87 If it had a trigonal planar shape, the NH_3 molecule would have no molecular dipole moment, and as a result there would be no dipole–dipole forces between molecules. With the correct pyramidal shape, NH_3 does have a molecular dipole moment, and consequently there are strong dipole–dipole forces between molecules.

6.89 The dot diagram,

$$\left[\ddot{N}=N=\ddot{N}\right]^{-}$$

shows two bonding groups and no lone pairs around the central atom, meaning the molecular shape is linear and the bond angle is 180°.

6.91 Oxygen, nitrogen, and fluorine are electron "hogs," meaning they have a high electronegativity. This makes the O–H, N–H, and F–H bonds very polar, with the result that molecules containing these bonds have large molecular dipole moments and relatively strong intermolecular forces.

6.93 (a) $:N\equiv C-C\equiv N:$

Linear electron-group geometry and linear molecular shape around each C. The two N–C bond dipole moments cancel so that there is no molecular dipole moment.

(b)

$$\left[\begin{array}{c} :\ddot{F}: \\ | \\ :\ddot{F}-B\overset{}{\diagdown}\ddot{F}: \\ :\ddot{F}: \end{array}\right]^{-}$$

Tetrahedral electron-group geometry and tetrahedral molecular shape around B. The four B–F bond dipole moments cancel so that there is no molecular dipole moment.

(c)

$$:\ddot{C}l:$$

Trigonal planar electron-group geometry and trigonal planar molecular shape around N. Because Cl and N have the same electronegativity value, there is no Cl–N bond dipole moment. Because O is more electronegative than N, each N–O bond has a dipole moment, and these two moments add vectorially to give the molecular dipole moment shown.

(d)

Tetrahedral electron-group geometry and bent molecular shape around Se. The Se–H electronegativity difference is 0.3, a bit smaller than the C–H difference of 0.4. Therefore because we consider the C–H bond to be nonpolar, we can assume the same about the Se–H bond. Therefore there are no bond dipole moments and no molecular dipole moment.

6.95 The dot diagrams,

$$H-\underset{|}{\overset{}{B}}-H \qquad H-\underset{|}{\overset{\cdot\cdot}{P}}-H$$
$$\;\;\;\;H \qquad\qquad\;\;\; H$$

tell you the electron-group geometry is trigonal planar in BH_3 but tetrahedral in PH_3. The molecular shapes are different, too: trigonal planar in BH_3 and pyramidal in PH_3. BH_3 is nonpolar because the B–H bond ($\Delta EN = 0.1$) is essentially nonpolar.

6.97 The student is incorrect. All the bonds in this molecule are nonpolar since the atoms at the ends of the bonds are identical. She should erase the bond dipole moment vectors.

6.99 (a) Molecule (a) contains the fluorine atoms as it has the larger bond dipole moment vectors (and F is more electronegative than Cl).

(b) No. Both are linear, and thus both would be nonpolar (neither would line up between charged plates).

6.101 (a) H—Ċ—H
‖
H

It is different because carbon does not have an octet of electrons. Also, not all of the electrons are in pairs.

(b)

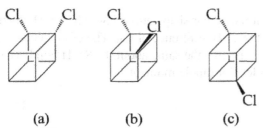

(c) Steric number is 3. The odd electron is not being counted as a separate electron domain.

(d)

The steric number is 4.

(e) Because the C–H bonds need more room so they won't bang into each other, they adopt a geometry where the bond angles can be 120° instead of 109.5°.

6.103 Drawing it as a puckered ring allows the carbons to have a tetrahedral geometry, with bond angles of 109.5°. Drawing it flat would make the bond angles 120°. The purple atoms are called *axial* because they are straight up and down, much like the axis of the earth. The blue atoms are called *equatorial* because they are horizontal and are similar to lines pointing out from the equator of the earth.

6.105 Cubane has bond angles of 90°, which are smaller than the preferred angle of 109.5° for carbon atoms with a steric number of 4. The cubane molecule is very unstable because of this.

The isomers of dichlorocubane are:

(a)   (b)   (c)

(a) and (b) are polar, with a net dipole toward the top of the molecule. (c) is nonpolar because the individual bond dipoles cancel each other.

6.107 In the trigonal bipyramidal electron group geometry, there are two distinct types of sites: two out of five groups of electrons can reside in the linear positions at the apexes of the structure, while the other three groups can reside in the trigonal part of the trigonal bipyramid. These latter positions allow electrons for more room and hence, less repulsions, due to the larger, 120° angles among these positions. On the other hand, electrons in the linear positions have only 90° angles separating them from the trigonal positions (the 180° angle between the two linear positions is cut in half by the trigonal plane) and are, therefore, smaller. Since lone pairs command more space than bonding electrons, when placing them in a trigonal bipyramidal electron group geometry, we do so in the trigonal part.

6.109 SeF$_4$ (34 valence electrons, SN = 5) has trigonal bipyramidal electron geometry and see-saw shape due to the presence of one lone pair (in the trigonal position), while SeF$_4{}^{2-}$ (36 valence electrons, SN = 6) has octahedral electron geometry and square planar shape due to the presence of two lone pairs of electrons.

selenium tetrafluoride   selenium tetrafluoride (2–) ion

6.111 ClF$_5$ (42 valence electrons, SN = 6) has octahedral electron geometry and square pyramidal shape due to the presence of one lone pair of electrons. Bond dipoles in the square planar positions cancel out (see previous problem), and the polarity of the Cl–F bond in the linear position is the sole contributor to the polarity of the ClF$_5$ molecule.

Chapter 7

7.1   See solution in textbook.

7.3

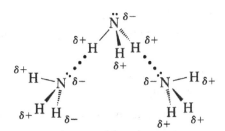

7.5   See solution in textbook.

7.7   (a) One electron from each H, five from N: H:N̈:H
          H

One electron from each H, five from P: H:P̈:H
          H

(b) NH$_3$ is polar because the H–N electronegativity difference causes the bonding electrons to be closer to the N. PH$_3$ is nonpolar because there is no H–P electronegativity difference.

(c) In PH$_3$, because this molecule has more electrons (18) than NH$_3$ (10 electrons).

(d) PH$_3$ because of stronger London forces. This is in fact not the case, however. See Practice Problem 7.8 for additional information.

7.9

7.11 See solution in textbook.

7.13 The intermolecular London forces holding the lattice together.

7.15 They move constantly in straight-line paths at very high speeds, changing direction only when they collide with one another or with the walls of their container.

7.17 When heat is added to a substance, the average kinetic energy of particles increases, and so does the speed with which they are moving. When enough heat is added, particles have too much kinetic energy to maintain significant intermolecular attractions with one another. That's when solids—in which particles are relatively still—melt, and liquids—in which particles are moving in a more or less "organized" fashion—boil. With an increase in the kinetic energy of particles, there is a progressive weakening of intermolecular attractive forces, which ultimately get completely broken, resulting in the formation of the least "organized" phase, namely, gas.

7.19 Heating causes the molecules of the liquid to move faster and faster until they have enough energy to overcome the attractive forces holding them in the liquid phase. When they overcome the attractive forces, they escape into the gas phase.

7.21 Because in both solids and liquids the molecules are very close to one another. They are *condensed* into a volume much smaller than what the gas phase occupies.

7.23 Gases have the least order, with the molecules moving everywhere in the volume of the container. Solids have the most order, with the molecules essentially fixed in position and unable to migrate. The order in liquids is between that in solids and that in gases; the molecules are confined to the liquid but free to migrate throughout the liquid.

7.25 The intermolecular forces holding the molecules together are dipole–dipole interactions.

7.27 CBr_4. Both molecules are nonpolar and therefore their liquid phase involves only London forces. Because CBr_4 is the larger molecule, it has more electrons and therefore stronger London forces. The stronger the London forces, the higher the boiling point.

7.29 Because propane molecules are much smaller than octane molecules, there are weaker London forces between propane molecules. Weaker London forces to overcome allow the propane molecules to separate more easily and enter the gas phase.

7.31 (a) HF is more polar because the H–F electronegativity difference is greater than the H–Cl electronegativity difference.

(b) For HF, because the partial charges in the HF molecules are greater than those in the HCl molecules, leading to a larger dipole moment in HF.

(c) HF. Because they have the larger dipole moment, HF molecules experience stronger inter-molecular forces, and stronger intermolecular forces mean less energy needs to be removed to change HF(g) to HF(l). That HF liquifies first is another way of saying HF has the higher condensation temperature. Because condensation temperature (gas $\rightarrow$ liquid) and boiling temperature (liquid $\rightarrow$ gas) are numerically equal, HF also has the higher boiling point.

(d) Yes; HCl experiences greater London forces because it is the larger molecule and so contains more electrons.

7.33 (a) CCl_4; both molecules are nonpolar, so the only intermolecular attractive forces at work here are the London forces, which correlate with the total number of electrons. CCl_4 has more electrons (74) than CS_2 (38).

(b) Cl_2; both molecules are nonpolar, so the only intermolecular attractive forces at work here are the London forces, which correlate with the total number of electrons. Cl_2 has more electrons (34) than F_2 (18).

(c) NH_3; both molecules have a total of 10 electrons, so the London forces will not be a differentiating factor. Dipole–dipole intermolecular attractions as well as hydrogen-bonding are both possible for the polar NH_3 but are absent for the nonpolar CH_4.

(d) CI_4; while CI_4 is nonpolar and can interact via London forces only, it has a much larger number of electrons (218) than the polar CH_3Cl (26). This example shows that, for large molecules, such as CI_4, London forces can provide stronger attractions than dipole–dipole interactions present between some polar molecules.

7.35 (b) will not form hydrogen bonds. Hydrogen bonds cannot form without the molecules containing O–H, N–H, and/or F–H bonds. In a molecule of CH_3OCH_3 (dimethyl ether), there is no hydrogen atom attached to the oxygen atom, as the Lewis dot diagram below shows:

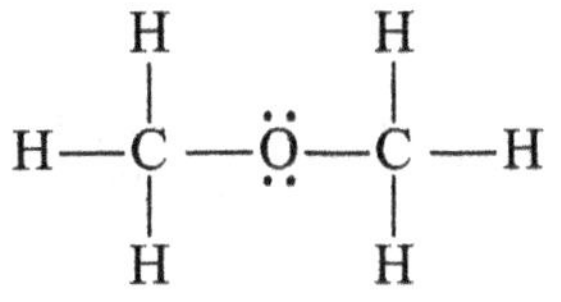

7.37 The compound with the higher boiling point must have hydrogen-bonding.

7.39 We mean the momentary dipole caused (induced) in one molecule by instantaneous electron imbalance in a neighboring molecule.

7.41 Because there are numerous N–H groups in the molecule. The H atom in an N–H can hydrogen bond either to some other N in the molecule or to an O.

7.43 If the nonpolar molecules are much larger than the polar molecules, the London forces in the non-polar liquid will be quite substantial and can be stronger than the dipole–dipole interactions in the polar liquid. The greater London forces make the boiling point of the nonpolar liquid higher than that of the polar liquid.

7.45 Sometimes. Compounds containing a group IA (1) or IIA (2) metal plus a nonmetal are usually ionic and therefore nonmolecular. Metals are nonmolecular compounds. It is generally not possible, however, to look at the formula of a covalent compound and tell whether its solid phase is molecular or nonmolecular.

7.47 Because nonmolecular solids are held together by much stronger forces—namely, covalent, ionic, or metallic bonds. Molecular solids are held together only by the much weaker intermolecular attractive forces—London forces, dipole–dipole interactions, and hydrogen bonding.

7.49 The high melting points of some metals, such as gold and iron, are evidence that metallic bonds can be as strong as ionic or covalent bonds.

7.51 (a) Metallic; potassium is classified as a metal.

   (b) Ionic; made of K^+ and Cl^- ions.

   (c) Network; the high melting point tells you this is a nonmolecular solid. The absence of ions tells you it is not ionic, that P is a nonmetal tells you it is not metallic, and so it must be a network nonmolecular solid.

   (d) Molecular solid, indicated by the relatively low melting point.

7.53 Because the attraction is the result of opposite partial charges attracting each other to form a hydrogen bond. If the molecules are oriented δ^+ to δ^+ or δ^- to δ^-, there can be no hydrogen bond formed:

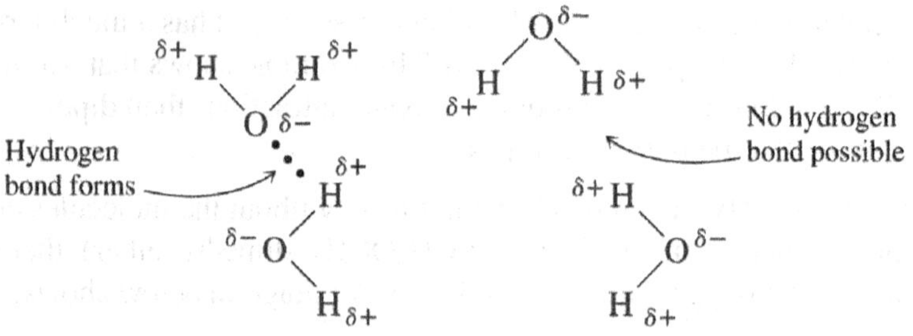

7.55 Because density defines the number of molecules per unit volume, there must be more water molecules per unit volume in the (denser) liquid phases than in the solid phase. (The molecules are packed more tightly in the liquid phase.)

7.57 The intermolecular forces between Br_2 molecules are greater than those between F_2 molecules or Cl_2 molecules.

7.59 CH_4 because it is both smaller and nonpolar. That CH_4 contains fewer electrons than CH_2Cl_2 means weaker London forces in CH_4 than in CH_2Cl_2; weaker London forces to overcome allows CH_4 to enter the gas phase at a lower temperature. That CH_4 is nonpolar means there are no intermolecular dipole–dipole attractions to be overcome. Because CH_2Cl_2 is slightly polar, there is also the possibility of intermolecular dipole–dipole attractions, which require more heat energy to overcome and therefore give CH_2Cl_2 a higher boiling point, making it more likely than CH_4 to be liquid at room temperature.

7.61 To determine relative boiling points, you must first determine the types of intermolecular forces in each substance. The weaker these forces, the easier it is to separate the molecules and consequently the lower the boiling point. (a) and (b) are nonpolar molecules, meaning only London forces. (c) has an –OH group, which means hydrogen bonds. (d) is polar, which means dipole–dipole attractions. Because hydrogen bonds are the strongest of the three, (c) should have the highest boiling point.

7.63 (a) CH_3OH hydrogen bonding; CH_3Cl dipole–dipole forces; CH_3CH_3 London forces; $CH_3CH_2CH_3$ London forces.

(b) $CH_3CH_3 < CH_3CH_2CH_3 < CH_3Cl < CH_3OH$. The reason nonpolar CH_3CH_3 has a lower boiling point than nonpolar $CH_3CH_2CH_3$ is that CH_3CH_3 is smaller and therefore fewer London forces are at work in a sample of CH_3CH_3.

7.65 In CH_2Cl_2, the δ^+ portion of one molecule attracts the δ^- portion of another:

7.67 (a) $H_2 < CH_4 < HCl < CH_3OH$

(b) H_2 and CH_4 are both nonpolar, meaning only the weakest intermolecular forces—London forces—are operating. Because H_2 has fewer electrons than CH_4, H_2 has weaker London forces and therefore the lower boiling point. In polar HCl, dipole–dipole attractions are present, and in polar CH_3OH, the –OH means hydrogen-bonding is present. Because hydrogen bonds are stronger than dipole–dipole interactions, HCl has a lower boiling point than CH_3OH.

7.69 (a) London forces. Each B–F bond is polar, but the symmetry of the trigonal planar molecule causes the three B–F dipole moments to cancel one another. The molecule is therefore nonpolar, and so only London forces are possible.

(b) Hydrogen-bonding, because of the –OH group.

(c) London forces. That Xe is a noble gas tells you there are no ionic or covalent bonds. It's not a molecule and therefore you can rule out dipole–dipole interactions. Hydrogen-bonding is impossible because there are no –OH, –NH, or –FH groups.

(d) Hydrogen-bonding, because of the HF "groups."

(e) Both dipole–dipole and London forces present; dipole–dipole is the stronger force.

7.71 SiO_2 is a network solid held together by covalent bonds. At room temperature, the atoms do not have anywhere near the energy required to break these bonds, which means the substance is a solid.

In a sample of CO_2, only London forces act between molecules. These forces are so weak that even at room temperature the CO_2 molecules have enough energy to overcome them, making CO_2 a gas at room temperature.

7.73   The substance having the stronger intermolecular forces has the higher boiling point, and hydrogen bonds are stronger than London forces. Thus ethylene glycol boils at the higher temperature because of the two –OH groups and the hydrogen-bonding they make possible. Pentane is essentially nonpolar because it contains only nonpolar C–C and C–H bonds. For this reason, the only intermolecular forces in pentane are London forces.

7.75   (a) Because C–C bonds are nonpolar, each graphite sheet must be nonpolar. This means the only attractive forces between sheets are London forces.

(b) Because the London forces between sheets in graphite are relatively weak, the sheets can slip past one another, making graphite slippery. The carbon atoms in diamond are all locked in position by extremely strong covalent bonds, preventing the atoms from moving and making diamond a very hard substance.

7.77   The valence electrons in a piece of metal are free to move about the entire lattice of metal atoms.

7.79   Because eicosane is so large—a string of 20 carbon atoms bonded to 42 hydrogen atoms—it contains a large number of electrons. The resulting London forces between molecules combine to produce an overall intermolecular attraction that is stronger than the hydrogen-bonding between water molecules. The substance with the greater intermolecular attraction has the higher melting and boiling points.

7.81 $CO_2 < SO_2 < CH_3CH_2OH < Al$. CO_2 is nonpolar (linear shape), meaning only London forces are possible. SO_2 is polar (bent shape) and experiences stronger dipolar attractive forces between molecules, giving it a higher boiling point than CO_2. Hydrogen-bonding is the strongest intermolecular force in CH_3CH_2OH, giving this substance a higher boiling point than either CO_2 or SO_2. Al is a metal, and the metallic bond is stronger than either London forces or hydrogen bonds, giving this substance the highest boiling point of the four.

7.83 NH_3 would appear to have too high a boiling point (it should be lower than NF_3, not higher). The reason it is higher is it can hydrogen bond. The hydrogen bonds between NH_3 molecules make it harder to boil.

7.85   (b) would have the higher boiling point because it is polar and has both London forces and dipolar forces of attraction between molecules. (a) is nonpolar and has just London forces of attraction between molecules.

7.87   Answers may vary. Examples are shown below.

(a) An ion–dipole interaction

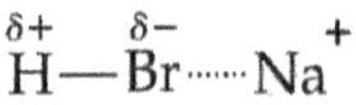

$$\overset{\delta+}{H}-\overset{\delta-}{Br}\cdots\overset{+}{Na}$$

(b) A dipole-induced dipole interaction

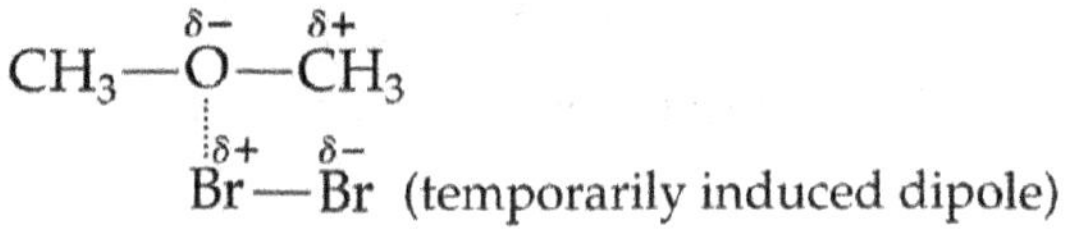

$$CH_3-\overset{\delta-}{O}-\overset{\delta+}{CH_3}$$
$$\overset{\delta+}{Br}-\overset{\delta-}{Br}\ \text{(temporarily induced dipole)}$$

(c) A dipole–dipole interaction

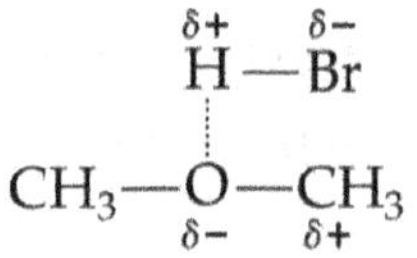

$$\overset{\delta+}{H}-\overset{\delta-}{Br}$$
$$CH_3-\underset{\delta-}{O}-\underset{\delta+}{CH_3}$$

(d) London forces

CH_4 & Br_2

Very weak due to temporarily induced dipoles

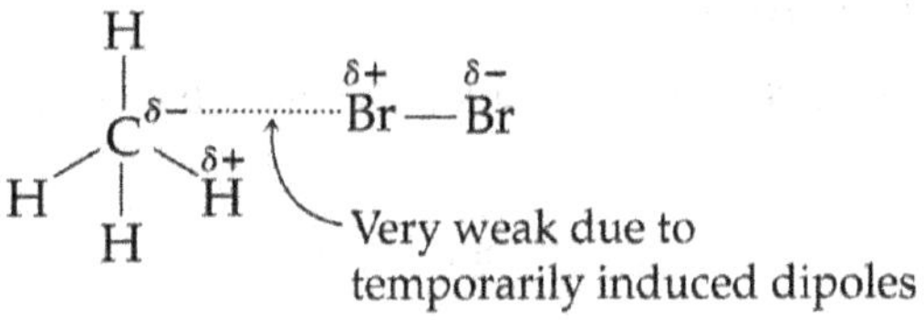

(e) Hydrogen-bonding
NA

7.89   (a) H-bond

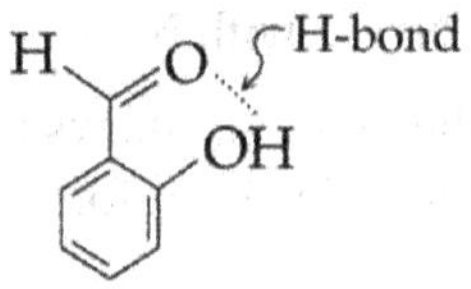

(b)

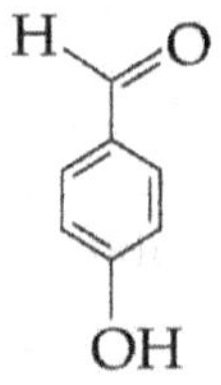

In compound A the O and H form an *intramolecular* hydrogen bond, as shown above. Therefore there is less *intermolecular* hydrogen-bonding than in compound B, where the hydrogen-bonding is holding the molecules more tightly to each other.

Chapter 8

8.1   See solution in textbook.

8.3   (1) Balance C by putting a 6 in front of CO_2: $C_6H_{12}O_6 + O_2 \rightarrow 6CO_2 + H_2O$.

(2) Balance H by putting a 6 in front of H_2O: $C_6H_{12}O_6 + O_2 \rightarrow 6CO_2 + 6H_2O$.

(3) Balance O by putting a 6 in front of O_2: $C_6H_{12}O_6 + 6O_2 \rightarrow 6CO_2 + 6H_2O$.

(4) The equation is balanced. Verify by counting atoms: 6C on each side, 12H on each side, 18O on each side.

8.5   See solution in textbook.

8.7   (1) You have five H left and two H right, which you can balance by putting a 2 in front of CH_3NH_2 and a 5 in front of H_2O. This step also balances N: $2CH_3NH_2 + O_2 \rightarrow CO_2 + 5H_2O + N_2$

(2) What you just did unbalanced C, which you can fix by putting a 2 in front of CO_2: $2CH_3NH_2 + O_2 \rightarrow 2CO_2 + 5H_2O + N_2$

(3) You now have everything balanced except O—two left, nine right. Balance by putting a 4.5 in front of O_2 to get $4.5 \times 2 = 9$O left: $2CH_3NH_2 + 4.5O_2 \rightarrow 2CO_2 + 5H_2O + N_2$

(4) The equation is balanced, but you should multiply through by 2 to get all whole-number coefficients: $4CH_3NH_2 + 9O_2 \rightarrow 4CO_2 + 10H_2O + 2N_2$

(5) Verify by counting atoms: 4C on each side, 20H on each side, 4N on each side, 18O on each side.

8.9 $2Pb(NO_3)_2(s) \xrightarrow{\text{Heat}} 2PbO(s) + 4NO(g) + 3O_2(g)$; decomposition.

8.11   See solution in textbook.

8.13   (a) In order to know what the spectator ions are, you must write the complete ionic equation.

Step 1: Identify the ions present initially: Ni^{2+}, NO_3^-, NH_4^+, PO_4^{3-}.

Step 2: Identify all possible new combinations: Ni^{2+} with PO_4^{3-}, NH_4^+ with NO_3^-.

Step 3: Determine whether any insoluble salts form. Table 8.1 of the textbook shows that $Ni_3(PO_4)_2$ is not soluble in water; all ammonium salts are soluble. Therefore, NH_4^+ and NO_3^- remain in solution. The complete ionic equation is:

$$3Ni^{2+}(aq) + 6NO_3^-(aq) + 6NH_4^+(aq) + 2PO_4^{3-}(aq) \rightarrow$$
$$Ni_3(PO_4)_2(s) + 6NO_3^-(aq) + 6NH_4^+(aq)$$

The spectator ions are NO_3^- and NH_4^+.

(b) $3Ni^{2+}(aq) + 2PO_4^{3-}(aq) \rightarrow Ni_3(PO_4)_2(s)$

8.15   It is true because neutralization requires one OH^- for every one H^+. Because 1 mole of HCl yields 1 mole of H^+, 1 mole of NaOH is sufficient for neutralization. Because 1 mole of H_2SO_4 yields 2 moles of H^+, 2 moles of NaOH are needed for neutralization.

8.17   Equation (a) represents a chemical reaction because new substances are produced. Equation (b) does not represent a chemical reaction because no new substances are produced; the products are the same substances as the reactants, but in different form.

8.19 The properties of a heterogeneous mixture of the reactants could be compared with those of the product. If electricity did not pass through the reactant mixture with zero resistance and/or if a magnet did not hover above the reactant mixture cooled to 90 K, that would be evidence that reactant and product are different substances.

8.21 Two Cl–Cl bonds and two C–H bonds (one from each carbon) must be broken. Two H–Cl bonds and two C–Cl bonds (one to each carbon) are formed.

8.23 (a) Cooling the reaction mixture slows down the molecules. When they are slowed down enough, the collisions will not have enough energy to break bonds in the reactants.

(b) Increasing the number of reactant molecules means more chances for collisions between molecules in a given time period. More collisions mean more chances for reactant bonds to break, which means more reactant molecules can be converted to product molecules in the given time period. In other words, the reaction goes faster.

8.25 No; in order to adhere to the law of conservation of matter, there must be the same number of atoms of *each* type on both sides of the equation. While the numbers of oxygen and carbon atoms are the same, 6 and 3, respectively, there are 4 atoms of iron on the left side and none on the right side. Also, there are 2 atoms of gold on the right side and none on the left side. A properly balanced reaction equation should have been: $2Fe_2O_3(s) + 3C(s) \rightarrow 3CO_2(g) + 4Fe(s)$.

8.27 Changing subscripts changes the chemical identity of a substance, which means the (altered) reactants or products shown in the equation do not represent the reaction.

8.29 Either "1 mole of methane and 2 moles of oxygen react to give 1 mole of carbon dioxide and 2 moles of water" or "1 molecule of methane and 2 molecules of oxygen react to give 1 molecule of carbon dioxide and 2 molecules of water."

8.31 If you balance in the order Al, Cl, H, the O is automatically balanced to give $Al_2Cl_6 + 6H_2O \rightarrow 2Al(OH)_3 + 6HCl$.

8.33 (a) H is balanced. Balancing C by putting a 2 in front of CO_2 gives you two O atoms on the left side and five O atoms on the right side, which you balance by giving O_2 a coefficient of 2.5: $C_2H_2 + 2.5O_2 \rightarrow H_2O + 2CO_2$. Multiplying through by 2 to obtain whole-number coefficients gives $2C_2H_2 + 5O_2 \rightarrow 2H_2O + 4CO_2$.

(b) Either "2 moles of acetylene and 5 moles of oxygen react to give 2 moles of water and 4 moles of carbon dioxide" or "2 molecules of acetylene and 5 molecules of oxygen react to give 2 molecules of water and 4 molecules of carbon dioxide."

8.35 Two approaches possible. (1) Writing $2NO_2$ to balance N gives five O left, six O right. Balance O by writing $\frac{1}{2}O_2$ to remove one O from right, then multiply through by 2 to get whole-number coefficients. (2) Writing $2NO_2$ to balance N gives five O left, six O right. To balance, you need more O left, and the only way to get more is to write $2N_2O_5$. Then rebalance N by writing $4NO_2$, a step that also balances O. Either way, the balanced equation is $2N_2O_5(g) \rightarrow 4NO_2(g) + O_2(g)$; decomposition.

8.37 If you balance first Fe by writing $2Fe(NO_3)_3$, then N by writing $6NaNO_3$, then Na by writing $3Na_2S$, S and O are automatically balanced: $2Fe(NO_3)_3(aq) + 3Na_2S(aq) \rightarrow Fe_2S_3(s) + 6NaNO_3(aq)$. This is both a double-replacement reaction and a precipitation reaction.

8.39 Two approaches possible. (1) First balance Li by writing 3Li, then balance N by writing $\frac{1}{2}N_2$, then multiply both sides of the equation by 2. (2) First balance N by writing $2Li_3N$, then balance Li by writing 6Li. Either way, you get the balanced equation: $6Li(s) + N_2(g) \rightarrow 2Li_3N(s)$; combination.

8.41 $2H_2O_2(l) \rightarrow 2H_2O(l) + O_2(g)$; decomposition.

8.43 $Ca^{2+}(aq) + B_4O_7^{2-}(aq) \rightarrow CaB_4O_7(s)$

$Mg^{2+}(aq) + B_4O_7^{2-}(aq) \rightarrow MgB_4O_7(s)$

$Na^+(aq)$ ions replace the $Ca^{2+}(aq)$ and $Mg^{2+}(aq)$ ions in the hard water.

8.45

Name	Formula	Soluble or insoluble in water?
Sodium phosphate	Na_3PO_4	Soluble
Barium acetate	$Ba(CH_3CO_2)_2$	Soluble
Ammonium sulfide	$(NH_4)_2S$	Soluble
Iron(II) carbonate	$FeCO_3$	Insoluble
Mercury(II) chloride	$HgCl_2$	Soluble
Cobalt(II) hydroxide	$Co(OH)_2$	Insoluble
Mercury(I) chloride	$HgCl$	Insoluble

8.47 (a) $Bi^{3+}(aq) + 3OH^-(aq) \rightarrow Bi(OH)_3(s)$

(b) $Sr^{2+}(aq) + SO_4^{2-}(aq) \rightarrow SrSO_4(s)$

(c) $3Cu^{2+}(aq) + 2PO_4^{3-}(aq) \rightarrow Cu_3(PO_4)_2(s)$

8.49 Iron(III) hydroxide and barium sulfate precipitate:

Balanced equation: $Fe_2(SO_4)_3(aq) + 3Ba(OH)_2(aq) \rightarrow 2Fe(OH)_3(s) + 3BaSO_4(s)$

Complete ionic equation:

$2Fe^{3+}(aq) + 3SO_4^{2-}(aq) + 3Ba^{2+}(aq) + 6OH^-(aq) \rightarrow 2Fe(OH)_3(s) + 3BaSO_4(s)$

Because there are no spectator ions, the net ionic equation is identical to the complete ionic equation.

8.51 (a) Numerous answers possible. You could mix an aqueous solution of calcium nitrate, $Ca(NO_3)_2$, with an aqueous solution of sodium sulfate, Na_2SO_4, and $CaSO_4(s)$ would precipitate. You could then filter the solution to isolate the $CaSO_4(s)$.

(b) $Ca^{2+}(aq) + 2NO_3^-(aq) + 2Na^+(aq) + SO_4^{2-}(aq) \rightarrow CaSO_4(s) + 2NO_3^-(aq) + 2Na^+(aq)$

(c) $Ca^{2+}(aq) + SO_4^{2-}(aq) \rightarrow CaSO_4(s)$

(d) In this double-replacement reaction, calcium cations (delivered with nitrate anions) react with sulfate anions (delivered with sodium cations). Calcium cations combine with sulfate anions to form an insoluble salt, calcium sulfate. Sodium cations and nitrate anions remain in solution as spectator ions.

8.53 The cloudiness means a precipitation reaction has taken place. According to Table 8.1 of the textbook, all nitrate salts are water-soluble, meaning the precipitate must be a silver salt (or hydroxide). From Table 8.1 you see that the possibilities are: $AgCl$, $AgBr$, AgI, Ag_3PO_4, Ag_2CO_3, Ag_2S, or $AgOH$.

8.55 (a) $H_2S(aq) + 2Na^+(aq) + 2OH^-(aq) \rightarrow 2Na^+(aq) + S^{2-}(aq) + 2H_2O(l)$

(b) Sodium sulfide, $Na_2S(s)$

8.57 Combine an aqueous solution of the oxalic acid with an aqueous solution of sodium hydroxide,

$H_2C_2O_4(aq) + 2NaOH(aq) \rightarrow Na_2C_2O_4(aq) + 2H_2O(l)$

and then evaporate off the $H_2O(l)$ to leave behind $Na_2C_2O_4(s)$.

8.59 (a) Lead(II) sulfate.

(b) $Pb^{2+}(aq) + SO_4^{2-}(aq) \rightarrow PbSO_4(s)$

(c) No, because $Pb(NO_3)_2$ is not a base. After the precipitation reaction, there is still $H^+(aq)$ in solution, making the solution acidic.

8.61 (a) $Ca(OH)_2(aq) + 2HNO_3(aq) \rightarrow Ca(NO_3)_2(aq) + 2H_2O(l)$

(b) $H^+(aq) + OH^-(aq) \rightarrow H_2O(l)$

8.63 (a) The hydrogen atom written at the right end of the formula, CH_3COOH.

(b) In methyl acetate, all the hydrogen atoms are covalently bonded to carbon. In acetic acid, the acidic hydrogen is bonded to oxygen. Whether or not a hydrogen atom is acidic must have something to do with which element it bonds to in a compound.

(c) Resonance forms.

8.65 In any covalent bond A:B, the electron from A is attracted both to its own nucleus and to nucleus B, and the electron from B is attracted both to its own nucleus and to nucleus A. Thus, using H1 and H2 to differentiate the two atoms in H_2, you can say that the added energy is used to pull the H1 electron in the bond away from the H2 nucleus and to pull the H2 electron away from the H1 nucleus.

8.67 (a) $SiO_2(s) + 2C(s) \rightarrow Si(s) + 2CO_2(g)$; single-replacement (the compound containing O replaces Si with C).

(b) Equation balanced as written; double-replacement.

(c) $4Al(s) + 3O_2(g) \rightarrow 2Al_2O_3(s)$; combination.

(d) $(NH_4)_2Cr_2O_7(s) \rightarrow N_2(g) + Cr_2O_3(s) + 4H_2O(g)$; decomposition.

8.69 (a) $CaCl_2$ and K_2CO_3

(b) Yes, because calcium carbonate is insoluble in water.

(c) $Ca^{2+}(aq) + CO_3^{2-}(aq) \rightarrow CaCO_3(s)$

8.71 Any sodium salt whose anion forms a water-insoluble precipitate with Cu^{2+}: Na_2CO_3, Na_3PO_4, NaOH, or Na_2S.

8.73 (a) (b)

(c) Nitrate ion, because HNO_3 dissociates in water to form $H^+(aq)$ and $NO_3^-(aq)$.

8.75 First add $NaCl(aq)$, $NaBr(aq)$, or $NaI(aq)$ to precipitate the water-insoluble lead(II) halide, which can be isolated via filtration. Then to the filtered solution, add any sodium salt whose anion forms a water-insoluble compound with $Ba^{2+}(aq)$ — Na_2SO_4, Na_3PO_4, Na_2CO_3, or Na_2S.

This will precipitate the insoluble Ba^{2+} salt, which can be isolated by filtration. The solution now contains just $Na^+(aq)$ cations.

8.77 Balance first C by writing $2CO_2$, then H by writing $8H_2O$, and last O by writing $11O_2$: $C_7H_{16}(l) + 11O_2(g) \rightarrow 7CO_2(g) + 8H_2O(g)$.

8.79 $2NaCl + Br_2 \rightarrow 2NaBr + Cl_2$

8.81 Initially, Fe and Cd are balanced, but you have two Cl on the left and three Cl on the right. Being that the subscript on the left Cl is 2, writing 1.5 in front of $CdCl_2$ gives you $2 \times 1.5 = 3Cl$ left (Cl is now balanced), but you've unbalanced Cd. Rebalance by writing 1.5 in front of Cd: $Fe + 1.5CdCl_2 \rightarrow FeCl_3 + 1.5Cd$. Then multiply through by 2 to get $2FeCl_3 + 3CdCl_2 \rightarrow 2FeCl_3 + 3Cd$; single-replacement.

8.83 The reaction should go faster because the reactants are moving faster at the higher temperature, resulting in more collisions having enough energy to cause the reactant bonds to break.

8.85 (b) and (c) represent chemical reactions because new substances are formed. In (a) and (d), no new substances are formed.

8.87 An acid is an H^+ donor; a base is an OH^- donor. Thus if a compound contains neither H nor OH, you know the compound is a salt.

    (a) KCl is a salt, because it does not contain acidic H or OH^- groups.

    (b) CH_3COOH is an acid, but note that only H bonded to O is acidic. The three H bonded to C are not acidic.

    (c) $Al(OH)_3$ is a base because of the presence of the OH^- groups.

    (d) H_3BO_3, is an acid because of the three acidic H.

    (e) $LiC_2H_3O_2$ is a salt because all three H are attached to one of the C and are therefore not acidic. The structure is

$$\begin{array}{c} H \\ | \\ H-C-C-OLi \\ | \quad \| \\ H \quad O \end{array}$$

8.89 Calcium is balanced. Balance first N by writing $2HNO_3$, then H by writing $2H_2O$. Doing so also balances O: $2HNO_3 + Ca(OH)_2 \rightarrow Ca(NO_3)_2 + 2H_2O$

    Complete ionic: $2H^+ + 2NO_3^- + Ca^{2+} + 2OH^- \rightarrow Ca^{2+} + 2NO_3^- + 2H_2O$

    Net ionic: $H^+ + OH^- \rightarrow H_2O$

8.91 (a) $Al(OH)_3$ (b) $Ca_3(PO_4)_2$ (c) $MgCO_3$

8.93 Balance Ca by writing $3Ca(OH)_2$, P by writing $2H_3PO_4$, H by writing $6H_2O$. Doing all this balances O automatically: $2H_3PO_4 + 3Ca(OH)_2 \rightarrow Ca_3(PO_4)_2 + 6H_2O$; double-replacement, which you can see more clearly by writing HOH for water:

    $2H_3PO_4 + 3Ca(OH)_2 \rightarrow Ca_3(PO_4)_2 + 6HOH$.

8.95 (a) Decomposition.   (b) Decomposition.

    (c) Single-replacement.   (d) Double-replacement.

8.97 Translating the words gives $KClO_3 \rightarrow KCl + O_2$. With K and Cl already balanced, balance O by writing $1.5O_2$ to get $1.5 \times 2 = 3O$ right. Then multiply through by 2 to get whole-number coefficients: $2KClO_3 \rightarrow 2KCl + 3O_2$

8.99 $Ca(OH)_2 + 2HCl \rightarrow CaCl_2 + 2H_2O$

8.101 K and O are balanced. With nine S left and two S right, writing $4.5K_2S_2O_3$ balances S but unbalances K and O, telling you that is not the way to go. Because the K subscript is the same left and right and the O subscript is the same left and right, the only way to balance these atoms is to have the K_2SO_3 coefficient be the same as the $K_2S_2O_3$ coefficient. If you use brute force and try each numeric coefficient in turn, you will discover that the lowest number that balances S is $8K_2SO_3 + S_8 \rightarrow 8K_2S_2O_3$; combination.

STUDENT WORKBOOK AND SELECTED SOLUTIONS

8.103 Translating the words gives $ZnS + O_2 \rightarrow ZnO + 2SO_2$, with only O unbalanced. Write $1.5O_2$, and then multiply through by 2 to get $2ZnS + 3O_2 \rightarrow 2ZnO + 2SO_2$

8.105 Translating to chemical symbols gives $C_3H_8 + O_2 \rightarrow CO_2 + H_2O$. Balance first C by writing $3CO_2$, then H by writing $4H_2O$, then O by writing $5O_2$: $C_3H_8 + 5O_2 \rightarrow 3CO_2 + 4H_2O$.

8.107 Translating to chemical symbols gives $CO + H_2O \rightarrow CO_2 + H_2$, which is balanced.

8.109 $Mn + 2S \rightarrow MnS_2$; combination.

8.111 The words translate to $H_2S + O_2 \rightarrow SO_2 + H_2O$, which is balanced by writing $1.5O_2$ and then multiplying through by 2: $2H_2S + 3O_2 \rightarrow 2SO_2 + 2H_2O$

8.113 (a) Cl^-, Br^-, I^-, PO_4^{3-}, CO_3^{2-}, S^{2-}, or OH^- because the sodium salt of any of these anions is soluble in water but the silver(I) salt of any of them is insoluble.

(b) SO_4^{2-}, PO_4^{3-}, or CO_3^{2-} because the sodium salt of any of these anions is soluble in water but the calcium salt is insoluble.

8.115 In each case you must first write the balanced equation, then convert to a complete ionic equation and identify the spectator ions. Then you have the information needed to write the net ionic equation.

(a) $Al(NO_3)_3(aq) + 3NaOH(aq) \rightarrow Al(OH)_3(s) + 3NaNO_3(aq)$
$Al^{3+}(aq) + 3NO_3^-(aq) + 3Na^+(aq) + 3OH^-(aq) \rightarrow Al(OH)_3(s) + 3Na^+(aq) + 3NO_3^-(aq)$
$Al^{3+}(aq) + 3OH^-(aq) \rightarrow Al(OH)_3(s)$

(b) $2K_3PO_4(aq) + 3CaCl_2(aq) \rightarrow Ca_3(PO_4)_2(s) + 6KCl(aq)$
$6K^+(aq) + 2PO_4^{3-}(aq) + 3Ca^{2+}(aq) + 6Cl^-(aq) \rightarrow Ca_3(PO_4)_2(s) + 6K^+(aq) + 6Cl^-(aq)$
$2PO_4^{3-}(aq) + 3Ca^{2+}(aq) \rightarrow Ca_3(PO_4)_2(s)$

(c) $MgSO_4(aq) + Na_2CO_3(aq) \rightarrow MgCO_3(s) + Na_2SO_4(aq)$
$Mg^{2+}(aq) + SO_4^{2-}(aq) + 2Na^+(aq) + CO_3^{2-}(aq) \rightarrow MgCO_3(s) + 2Na^+(aq) + SO_4^{2-}(aq)$
$Mg^{2+}(aq) + CO_3^{2-}(aq) \rightarrow MgCO_3(s)$

8.117 Double-replacement, which you can see more easily by writing the water molecule in the form HOH.

8.119 That the compound exists in solution as ions; there are no intact units in the solution.

8.121 Balanced: $2KCl(aq) + Pb(NO_3)_2(aq) \rightarrow PbCl_2(s) + 2KNO_3(aq)$

Complete ionic:

$2K^+(aq) + 2Cl^-(aq) + Pb^{2+}(aq) + 2NO_3^-(aq) \rightarrow PbCl_2(s) + 2K^+(aq) + 2NO_3^-(aq)$

Net ionic: $2Cl^-(aq) + Pb^{2-}(aq) \rightarrow PbCl_2(s)$

8.123 Neutralization.

8.125 Neutralization because an acid (H_2SO_4) and a base [$Ca(OH)_2$] react to form a salt ($CaSO_4$) plus water; double-replacement because SO_4^{2-} replaces $(OH)_2$ in the calcium compound and $(OH)_2$ replaces SO_4^{2-} in the "hydrogen compound"; precipitation because the water-insoluble $CaSO_4$ forms.

8.127 Slower because there will be fewer collisions in the increased volume in which the molecules are moving.

8.129 You need the balanced equation before you can write the ionic equations, and so write that first.

(a) Balanced: $AgNO_3(aq) + KI(aq) \rightarrow AgI(s) + KNO_3(aq)$

Complete ionic:

$Ag^+(aq) + NO_3^-(aq) + K^+(aq) + I^-(aq) \rightarrow AgI(s) + K^+(aq) + NO_3^-(aq)$

Net ionic: $Ag^+(aq) + I^-(aq) \rightarrow AgI(s)$

(b) Balanced: $Li_2SO_4(aq) + 2AgC_2H_3O_2(aq) \rightarrow 2LiC_2H_3O_2(aq) + Ag_2SO_4(aq)$

Complete ionic: $2Li^+(aq) + SO_4^{2-}(aq) + 2Ag^+(aq) + 2C_2H_3O_2^-(aq) \rightarrow$
$$2Li^+(aq) + 2C_2H_3O_2^-(aq) + 2Ag^+(aq) + SO_4^{2-}(aq)$$

There is no net ionic equation because all the ions in the complete ionic equation are spectator ions. No reaction is occurring, and the balanced equation is more properly written:

$Li_2SO_4(aq) + 2AgC_2H_3O_2(aq) \rightarrow$ no reaction

8.131 (a) The unbalanced equation is $NH_3(g) + F_2(g) \rightarrow NH_4F(s) + N_2(g)$. Your first thought might be to balance N, but you'll just unbalance N when you balance H because you cannot change H numbers without simultaneously changing N numbers. So begin with H, which you balance by writing: $4NH_3$ and $3NH_4F$:

$4NH_3 + F_2 \rightarrow 3NH_4F + N_2$

Now balance F by writing $1.5F_2$ to get $1.5 \times 2 = 3F$ atoms on the left side:

$4NH_3 + 1.5F_2 \rightarrow 3NH_4F + N_2$

All that's left is N, which stands at four left, five right. To lose one N right, write $0.5N_2$:

$4NH_3 + 1.5F_2 \rightarrow 3NH_4F + 0.5N_2$

Multiply both sides of the equation by 2, bring back the states of matter, and you are finished:

$8NH_3(g) + 3F_2(g) \rightarrow 6NH_4F(s) + N_2(g)$

(b)

(c)

8.133 Beaker (a) is incorrect. CH_2Cl_2 is not an ionic compound, and as such, it does not dissociate into ions upon dissolving.

8.135 $CH_4(g) + 2Cl_2(g) \rightarrow CH_2Cl_2(l) + 2HCl(g)$

One mole of methane and two moles of chlorine react to give one mole of dichloromethane and two moles of hydrogen chloride.

8.137 As an acid since it gives rise to H^+ ions in solution.

8.139 $2Pb(NO_3)_2(aq) \rightarrow 2PbO(s) + O_2(g) + 4NO_2(g)$

This is a decomposition reaction because one substance is breaking into more than one.

8.141 $NH_3(g) + H_2O(l) \rightarrow NH_4^+(aq) + OH^-(aq)$

This is an acid/base reaction because a proton (H^+) is being transferred from H_2O to NH_3.

8.143 $BaSO_4$ is insoluble and will not break up to produce Ba^{2+} ions. Therefore, there is no danger to the person ingesting $BaSO_4$. However, because $BaSO_4$ is soluble, it will dissolve and break into Ba^{2+} and CO_3^{2-} ions. The Ba^{2+} ions will poison the person ingesting them.

Chapter 9

9.1 See solution in textbook.

9.3 The two possible conversion factors are

$$\frac{3 \text{ blocks cream cheese}}{5 \text{ eggs}} \text{ and } \frac{5 \text{ eggs}}{3 \text{ blocks cream cheese}}$$

Use the second one because you want your answer in eggs:

$$21 \text{ blocks cream cheese} \times \frac{5 \text{ eggs}}{3 \text{ blocks cream cheese}} = \frac{21 \times 5}{3} \text{ eggs} = 35 \text{ eggs}$$

9.5 (a) $0.10 \text{ mol U} \times \dfrac{6.022 \times 10^{23} \text{ atoms U}}{1 \text{ mol U}} = \dfrac{21.5}{3} 6.0 \times 10^{22} \text{ atoms U}$

(b) $0.10 \text{ mol U} \times \dfrac{238.029 \text{ g U}}{1 \text{ mol U}} = 23.803 \text{ g U}$, which you must report as 24 g because 0.10 has

only two significant digits.

9.7 See solution in textbook.

9.9 The molecular formula for propane is C_3H_8—3 moles of C atoms for every 1 mole of propane. Therefore:

$$2 \text{ mol propane} \times \frac{3 \text{ mol carbon atoms}}{1 \text{ mol propane}} = 6 \text{ mol carbon atoms}$$

9.11 See solution in textbook.

9.13 (a) $5.000 \times 10^{24} \text{ molecules CH}_4 \times \dfrac{1 \text{ mol CH}_4}{6.022 \times 10^{23} \text{ molecules CH}_4} = 8.303 \text{ mol CH}_4$

(b) The molar mass of CH_4 is $(1 \times 12.011 \text{ g C}) + (4 \times 1.0079 \text{ g H}) = 16.043 \text{ g/mol}$. Thus:

$$8.303 \text{ mol CH}_4 \times \frac{16.043 \text{ g CH}_4}{1 \text{ mol CH}_4} = 133.2 \text{ g CH}_4$$

9.15 (a) $C_6H_6 + 3H_2 \rightarrow C_6H_{12}$

(b) 1 mole of benzene and 3 moles of hydrogen react to give 1 mole of cyclohexane.

(c) The molar mass of benzene is $(6 \times 12.011 \text{ g C}) + (6 \times 1.0079 \text{ g H}) = 78.113 \text{ g/mol}$. The balanced equation tells you that you need 1 mole of C_6H_6 to form 1 mole of C_6H_{12}:

$$1 \text{ mol C}_6\text{H}_{12} \times \frac{1 \text{ mol C}_6\text{H}_6}{1 \text{ mol C}_6\text{H}_{12}} \times \frac{78.113 \text{ g C}_6\text{H}_6}{1 \text{ mol C}_6\text{H}_6} = 78.113 \text{ g C}_6\text{H}_6$$

The molar mass of hydrogen is $(2 \times 1.0079 \text{ g H}) = 2.0158 \text{ g/mol}$. Because the equation says that 1 mole of C_6H_{12} requires 3 moles of hydrogen,

$$1 \text{ mol C}_6\text{H}_{12} \times \frac{3 \text{ mol H}_2}{1 \text{ mol C}_6\text{H}_{12}} \times \frac{2.0158 \text{ g H}_2}{1 \text{ mol H}_2} = 6.0474 \text{ g H}_2$$

(d) The most you can form is 1 mole of cyclohexane. The molar mass of cyclohexane is $(6 \times 12.011 \text{ g C}) + (12 \times 1.0079 \text{ g H}) = 84.161 \text{ g/mol}$. This 84.161 g is the theoretical yield of C_6H_{12}.

(e) $\text{Percent yield} = \dfrac{\text{Actual yield}}{\text{Theoretical yield}} \times 100\% = \dfrac{24.0 \text{ g}}{84.161 \text{ g}} \times 100\% = 28.5\%$

9.17 See solution in textbook.

9.19 $10.0 \text{ g glucose} \times \dfrac{1 \text{ mol glucose}}{180.155 \text{ g glucose}} \times \dfrac{6 \text{ mol } CO_2}{1 \text{ mol glucose}} \times \dfrac{44.009 \text{ g } CO_2}{1 \text{ mol } C_2O} = 14.7 \text{ g } CO_2$

9.21 See solution in textbook.

9.23 $10.0 \text{ g } H_2O \times \dfrac{1 \text{ mol } H_2O}{18.015 \text{ g } H_2O} \times \dfrac{6.022 \times 10^{23} \text{ } H_2O \text{ molecules}}{1 \text{ mol } H_2O} = 3.34 \times 10^{23} \text{ } H_2O \text{ molecules}$

9.25 (a) Step 1: $C_3H_8 + 5O_2 \rightarrow 3CO_2 + 4H_2O$

Step 2: $100.0 \text{ g } C_3H_8 \times \dfrac{1 \text{ mol } C_3H_8}{44.096 \text{ g } C_3H_8} = 2.268 \text{ mol } C_3H_8$

Step 2a is not necessary because reactions run in balanced fashion.

Step 3: $2.268 \text{ mol } C_3H_8 \times \dfrac{5 \text{ mol } O_2}{1 \text{ mol } C_3H_8} = 11.34 \text{ mol } O_2$

Step 4: $11.34 \text{ mol } O_2 \times \dfrac{31.998 \text{ g } O_2}{1 \text{ mol } O_2} = 362.9 \text{ g } O_2$

(b) Because the reaction is run in a balanced fashion, you may use either reactant to calculate theoretical yield:

$2.268 \text{ mol } C_3H_8 \times \dfrac{4 \text{ mol } H_2O}{1 \text{ mol } C_3H_8} \times \dfrac{18.015 \text{ g } H_2O}{1 \text{ mol } H_2O} = 163.4 \text{ g } H_2O \text{ theoretical yield}$

or

$11.34 \text{ mol } O_2 \times \dfrac{4 \text{ mol } H_2O}{5 \text{ mol } O_2} \times \dfrac{18.015 \text{ g } H_2O}{1 \text{ mol } H_2O} = 163.4 \text{ g } H_2O \text{ theoretical yield}$

9.27 (a) Step 1: Convert grams of CO_2 to moles of C and grams of H_2O to moles of H:

$0.686 \text{ g } CO_2 \times \dfrac{1 \text{ mol } CO_2}{44.009 \text{ g } CO_2} \times \dfrac{1 \text{ mol C}}{1 \text{ mol } CO_2} = 0.0156 \text{ mol C}$

$0.561 \text{ g } H_2O \times \dfrac{1 \text{ mol } H_2O}{18.015 \text{ g } H_2O} \times \dfrac{2 \text{ mol H}}{1 \text{ mol } H_2O} = 0.0623 \text{ mol H}$

Step 2: Convert to grams of C and H:

$0.0156 \text{ mol C} \times \dfrac{12.11 \text{ g C}}{1 \text{ mol C}} = 0.187 \text{ g C}$

$0.0623 \text{ mol H} \times \dfrac{1.0079 \text{ g H}}{1 \text{ mol H}} = 0.0628 \text{ g H}$

Step 3: Determine whether there is any O in the compound and, if so, convert to moles:

$0.250 \text{ g sample} - (0.0628 \text{ g H} + 0.187 \text{ g C}) = 0 \text{ g O}.$

Step 4: Divide subscripts through by the smallest subscript:

$$C_{\frac{0.0156}{0.0156}} H_{\frac{0.0623}{0.0156}} \rightarrow C_{1.00} H_{3.99} \rightarrow CH_4 \text{ molecular formula}$$

(b) $CH_4 + 2O_2 \rightarrow 2H_2O + CO_2$

9.29 (a) Use the four-step process for determining empirical formula from combustion analysis data. The four steps are shown here:

Step 1: Check that the percent by mass elemental compositions add up to 100%. If they do not, calculate the percent by mass of the additional element(s) by subtracting the total of the given percentages from 100%.

Step 2: Assume a 100 g sample of the compound.

Step 3: Convert grams of each element to moles of each element. This gives you the subscript in the formula for that element.

Step 4: Divide all subscripts by the smallest one to get the subscripts in the empirical formula.

1. 54.56% C $+ 9.16\%$ H $= 63.81\%$, which is less than 100%. Therefore the percent O $= 100\% - 63.81\% = 36.19\%$

2. Assume 100 g of compound (percentages then become grams). 54.65 g of carbon, 9.16 g of hydrogen, 36.19 g of oxygen

3. Calculate the moles of each element (moles $=$ grams/atomic mass).

$$\text{Moles of C} = 54.65 \text{ g C} \times \frac{1 \text{ mol C}}{12.011 \text{ g C}} = 4.550 \text{ moles C}$$

$$\text{Moles of H} = 9.16 \text{ g H} \times \frac{1 \text{ mol H}}{1.0079 \text{ g H}} = 9.088 \text{ moles H}$$

$$\text{Moles of O} = 36.19 \text{ g O} \times \frac{1 \text{ mol}}{15.999 \text{ g O}} = 2.262 \text{ moles O}$$

4. Divide by the smallest number to obtain the subscripts in the empirical formula. The smallest number is 2.262.

$$\text{Subscript for C:} \frac{4.550}{2.262} = 2$$

$$\text{Subscript for H:} \frac{9.088}{2.262} = 4$$

$$\text{Subscript for O:} \frac{2.262}{2.262} = 1$$

The empirical formula is therefore C_2H_4O.

(b) Divide the given molar mass of the compound by the molar mass of the empirical formula:

$$\frac{132.159}{44.053} = 3$$

The molecular formula is therefore $C_{2 \times 3}H_{4 \times 3}O_{1 \times 3} \rightarrow C_6H_{12}O_3$.

(c) The balanced equation for the combustion reaction is $C_6H_{12}O_3 + 15O_2 \rightarrow 12CO_2 + 12H_2O$

9.31 Step 1: Begin by assuming you have 1 mole of H_2O_2, and calculate the mass in grams of each element in that 1 mole of H_2O_2:

$$1 \text{ mol } H_2O_2 \times \frac{2 \text{ mol } H}{1 \text{ mol } H_2O_2} \times \frac{1.0079 \text{ g } H}{1 \text{ mol } H} = 2.0158 \text{ g } H$$

$$1 \text{ mol } H_2O_2 \times \frac{2 \text{ mol } O}{1 \text{ mol } H_2O_2} \times \frac{15.999 \text{ g } O}{1 \text{ mol } O} = 31.998 \text{ g } O$$

Step 2: Divide each calculated mass by the mass of the 1 mole of H_2O_2 you are assumed to be working with and multiply by 100: The mass of 1 mol H_2O_2 is its molar mass—34.014 g.

$$\frac{2.0158 \text{ g } H}{34.014 \text{ g } H_2O_2} \times 100\% = 5.926\% \text{ H}$$

$$\frac{31.998 \text{ g } O}{34.014 \text{ g } H_2O_2} \times 100\% = 94.07\% \text{ O}$$

9.33 Step 1: To get the empirical formula from percent composition, assume 100 g of compound (that way, percent values become grams):

$$89.09\% \text{ Cl} \rightarrow 89.09 \text{ Cl} \times \frac{1 \text{ mol Cl}}{35.453 \text{ g Cl}} = 2.513 \text{ mol Cl}$$

$$10.06\% \text{ C} \rightarrow 10.06 \text{ g C} \times \frac{1 \text{ mol C}}{12.011 \text{ g C}} = 0.8376 \text{ mol C}$$

$$0.84\% \text{ H} \rightarrow 0.84 \text{ g H} \times \frac{1 \text{ mol H}}{1.0079 \text{ g H}} = 0.8334 \text{ mol H}$$

Step 2: Use the calculated number of moles as subscripts and divide through by the smallest:

$$Cl_{\frac{2.513}{0.8334}} C_{\frac{0.8376}{0.8334}} H_{\frac{0.8334}{0.8334}} \rightarrow Cl_{3.02}C_{1.01}H_{1.00} \rightarrow Cl_3CH \text{ empirical formula}$$

The molecular mass of this empirical formula is 119.4 g/mol. That the molar mass of the compound is given as 119.378 g/mol tells you the molecular formula is the same as the empirical formula.

9.35 A balanced equation is like a recipe because the equation specifies how much of each reactant must be used to give a certain amount of product.

9.37 (a) $3 \text{ cakes} \times \frac{5 \text{ eggs}}{1 \text{ cake}} = 15 \text{ eggs}$

(b) $63 \text{ blocks of cream cheese} \times \frac{5 \text{ eggs}}{3 \text{ blocks of cream cheese}} = 105 \text{ eggs}$

(c) $5 \text{ cakes} \times \frac{5 \text{ eggs}}{1 \text{ cake}} = 25 \text{ eggs}$; Then add 2 eggs: $25 + 2 = 27$.

(d) $3.011 \times 10^{24} \text{ eggs} \times \frac{1 \text{ cake}}{5 \text{ eggs}} \times \frac{1 \text{ mol of cakes}}{6.022 \times 10^{23} \text{ cakes}} = 1 \text{ mole of cakes!}$

9.39 Always.

9.41 (a) There are 6.022×10^{23} bicycles in 1 mole of bicycles.

(b) Because there are two tires on every bicycle, the number of tires in 1 mole of bicycles is two times $6.022 \times 10^{23} = 1.204 \times 10^{24}$.

(c) $1 \text{ mole of bicycles} \times \dfrac{6.022 \times 10^{23} \text{ bicycles}}{1 \text{ mole of bicycles}} \times \dfrac{2 \text{ tires}}{1 \text{ bicycle}} \times \dfrac{24 \text{ spokes}}{1 \text{ tire}} = 2.9 \times 10^{25} \text{ spokes}$

(d) $1 \text{ mole of spokes} \times \dfrac{6.022 \times 10^{23} \text{ spokes}}{1 \text{ mole of spokes}} \times \dfrac{1 \text{ tire}}{24 \text{ spokes}} \times \dfrac{1 \text{ bicycle}}{2 \text{ tires}} = 1.3 \times 10^{22} \text{ bicycles}$

9.43 $2.5 \text{ moles of pennies} \times \dfrac{6.022 \times 10^{23} \text{ pennies}}{1 \text{ mole of pennies}} = 1.5 \times 10^{24} \text{ pennies}$

$1.5 \times 10^{24} \text{ pennies} \times \dfrac{1 \text{ dollar}}{100 \text{ pennies}} = 1.5 \times 10^{22} \text{ dollars}$

9.45 Two molecules of sulfur dioxide react with 1 molecule of oxygen to give 2 molecules of sulfur trioxide. Two moles of sulfur dioxide react with 1 mole of oxygen to give 2 moles of sulfur trioxide.

9.47 The mass of 1 mole of atoms of an element is equal to the element's atomic mass expressed in grams.

9.49 A compound's molar mass is calculated by summing the molar masses of all the atoms in the compound.

9.51 The atomic mass of this isotope is exactly 12 amu, by definition. Therefore, the molar mass is exactly 12 g, meaning that 12 g of $^{12}_{6}C$ is 1 mole and must contain 6.022×10^{23} atoms.

9.53 The subscript on each atom in the chemical formula tells you the number of moles of the atom in 1 mole of the compound:

(a) $1 \text{ mol } C_6H_{12}O_6 \times \dfrac{6 \text{ mol C}}{1 \text{ mol } C_6H_{12}O_6} = 6 \text{ mol C}$

(b) $1 \text{ mol } C_6H_{12}O_6 \times \dfrac{12 \text{ mol H}}{1 \text{ mol } C_6H_{12}O_6} = 12 \text{ mol H}$

(c) $1 \text{ mol } C_6H_{12}O_6 \times \dfrac{6 \text{ mol O}}{1 \text{ mol } C_6H_{12}O_6} \times \dfrac{6.22 \times 10^{23} \text{ atoms O}}{1 \text{ mol O}} = 3.613 \times 10^{24} \text{ atoms O}$

9.55 The theoretical yield of a reaction is the maximum amount of product that can be made from a given amount of reactants.

9.57 Some reactions have competing side reactions that use up reactants to produce unwanted side products. Sometimes difficulties in collecting the desired product make it impossible to recover all of it.

9.59 Percent yield $= \dfrac{15.5 \text{ g}}{40.0 \text{ g}} \times 100\% = 38.8\%$

9.61 (a) $2HCl + Zn \rightarrow H_2 + ZnCl_2$

(b) 2 moles of hydrogen chloride and 1 mole of zinc react to give 1 mole of hydrogen and 1 mole of zinc chloride.

(c) You need 2 moles of HCl and 1 mole of Zn. The molar masses are

1.0079 + 35.453 = 36.461 g/mol for HCl and 65.39 g/mol for Zn:

$$2 \text{ mol HCl} \times \frac{36.461 \text{ g}}{1 \text{ mol HCl}} = 72.922 \text{ g HCl}$$

$$1 \text{ mol Zn} \times \frac{65.39 \text{ g}}{1 \text{ mol Zn}} = 65.39 \text{ Zn}$$

(d) The theoretical yield is 1 mole of H_2, and the molar mass of H_2 is 2.0158 g/mol:

$$1 \text{ mol } H_2 \times \frac{2.0158 \text{ g } H_2}{1 \text{ mol } H_2} = 2.0158 \text{ g } H_2$$

(e) Percent yield $= \dfrac{\text{Actual yield}}{\text{Theoretical yield}} \times 100\% = \dfrac{2.00 \text{ g}}{2.0158 \text{ g}} \times 100\% = 99.2\%$

9.63 (a) $\dfrac{6 \text{ mol C atoms}}{1 \text{ mol } C_6H_{12}O_6}, \dfrac{12 \text{ mol H atoms}}{1 \text{ mol } C_6H_{12}O_6}, \dfrac{1 \text{ mol } C_6H_{12}O_6}{6 \text{ mol O atoms}}, \dfrac{6 \text{ mol C atoms}}{12 \text{ mol H atoms}},$

$\dfrac{12 \text{ mol H atoms}}{6 \text{ mol O atoms}}, \dfrac{12 \text{ mol H atoms}}{6 \text{ mol C atoms}}$; others possible.

(b) $72{,}000 \text{ H atoms} \times \dfrac{6 \text{ C atoms}}{12 \text{ H atoms}} = 36{,}000 \text{ C atoms}$

(c) $72{,}000 \text{ H atoms} \times \dfrac{1 \text{ molecule of } C_6H_{12}O_6}{12 \text{ H atoms}} \times \dfrac{1 \text{ mole of } C_6H_{12}O_6}{6.022 \times 10^{23} \text{ molecules of } C_6H_{12}O_6} \times$

$\dfrac{180.157 \text{ g}}{1 \text{ mole of } C_6H_{12}O_6} = 1.8 \times 10^{-18} \text{ g}$

9.65 (a) $I_2 + 3Cl_2 \rightarrow 2ICl_3$

(b) $\dfrac{1 \text{ mol } I_2}{3 \text{ mol } Cl_2}, \dfrac{1 \text{ mol } I_2}{2 \text{ mol } ICl_3}, \dfrac{3 \text{ mol } Cl_2}{2 \text{ mol } ICl_3}, \dfrac{3 \text{ mol } Cl_2}{1 \text{ mol } I_2}, \dfrac{2 \text{ mol } ICl_3}{1 \text{ mol } I_2}, \dfrac{2 \text{ mol } ICl_3}{3 \text{ mol } Cl_2}$

(c) $\dfrac{2 \text{ mol I atoms}}{1 \text{ mol } I_2 \text{ molecules}}, \dfrac{1 \text{ mol } I_2 \text{ molecules}}{2 \text{ mol I atoms}}, \dfrac{2 \text{ mol Cl atoms}}{1 \text{ mol } Cl_2 \text{ molecules}}, \dfrac{1 \text{ mol } Cl_2 \text{ molecules}}{2 \text{ mol Cl atoms}},$

$\dfrac{1 \text{ mol I atoms}}{1 \text{ mol } ICl_3 \text{ molecules}}, \dfrac{1 \text{ mol } ICl_3 \text{ molecules}}{1 \text{ mol I atoms}}, \dfrac{3 \text{ mol Cl atoms}}{1 \text{ mol } ICl_3 \text{ molecules}},$

$\dfrac{1 \text{ mol } ICl_3 \text{ molecules}}{3 \text{ mol Cl atoms}}$

9.67 (a) $24.0 \text{ g } O_2 \times \dfrac{1 \text{ mol } O_2}{31.998 \text{ g } O_2} = 0.750 \text{ mol } O_2$

(b) $0.750 \text{ mol } O_2 \times \dfrac{2 \text{ moles O atoms}}{1 \text{ mol } O_2} = 1.50 \text{ moles O atoms}$

9.69 (a) The molar mass of H_2SO_4 is

$$\begin{array}{ll} \text{H} & 2 \times 1.0079 \text{ g/mol} = 2.0158 \text{ g/mol} \\ \text{S} & 1 \times 32.064 \text{ g/mol} = 32.64 \text{ g/mol} \\ \underline{\text{O} \quad 4 \times 15.999 \text{ g/mol} = 63.996 \text{ g/mol}} \\ \qquad\qquad\qquad\qquad\quad 98.076 \text{ g/mol} \end{array}$$

(b) 98.076 g

(c) $2.50 \text{ mol } H_2SO_4 \times \dfrac{98.076 \text{ g } H_2SO_4}{1 \text{ mol } H_2SO_4} = 245 \text{ g}$

(d) $1000 \text{ molecules } H_2SO_4 \times \dfrac{1 \text{ mol } H_2SO_4}{6.022 \times 10^{23} \text{ molecules } H_2SO_4} \times \dfrac{98.076 \text{ g } H_2SO_4}{1 \text{mol } H_2SO_4}$

$= 1.629 \times 10^{-19} \text{ g } H_2SO_4$

9.71 $1.00 \times 10^9 \text{ molecules } H_2O \times \dfrac{1 \text{ mol } H_2O}{6.022 \times 10^{23} \text{ molecules } H_2O} \times \dfrac{18.015 \text{ g } H_2O}{1 \text{ mol } H_2O} = 2.99 \times 10^{-14} \text{ g } H_2O$

9.73 (a) $20.0 \text{ g } H_2O_2 \times \dfrac{1 \text{ mol } H_2O_2}{34.014 \text{ g } H_2O_2} \times \dfrac{2 \text{ mol } H_2O}{2 \text{ mol } H_2O_2} \times \dfrac{18.015 \text{ g } H_2O}{1 \text{ mol } H_2O} = 10.6 \text{ g } H_2O$

(b) $20.0 \text{ g } H_2O \times \dfrac{1 \text{ mol } H_2O}{18.015 \text{ g } H_2O} \times \dfrac{2 \text{ mol } H_2O_2}{2 \text{ mol } H_2O} \times \dfrac{34.014 \text{ g } H_2O_2}{1 \text{ mol } H_2O_2} = 37.8 \text{ g } H_2O_2$

(c) $20.0 \text{ g } O_2 \times \dfrac{1 \text{ mol } O_2}{31.998 \text{ g } O_2} \times \dfrac{2 \text{ mol } H_2O_2}{2 \text{ mol } O_2} \times \dfrac{34.014 \text{ g } H_2O_2}{1 \text{ mol } H_2O_2} = 42.5 \text{ g } H_2O_2$

9.75 Since B was the limiting reactant, some of the excess reactant A will be left behind. At the end of the reaction, you will discover that the product C is contaminated with the leftover reactant A. Since both A and C are solids, it will be rather difficult, and hence expensive, to separate them from each other.

9.77 To determine which is the limiting ingredient, multiply each given amount by the appropriate conversion factor:

$25 \text{ eggs} \times \dfrac{1 \text{ cheesecake}}{5 \text{ eggs}} = 5 \text{ cheesecakes}$

$9 \text{ blocks cream cheese} \times \dfrac{1 \text{ cheesecake}}{3 \text{ blocks cream cheese}} = 3 \text{ cheesecakes}$

$4 \text{ cups sugar} \times \dfrac{1 \text{ cheesecake}}{1 \text{ cups sugar}} = 4 \text{ cheesecakes}$

Because it gives the smallest number, the cream cheese is the limiting ingredient. You can make three cheesecakes.

9.79 The first thing you must do is balance the equation: $4P + 5O_2 \rightarrow 2P_2O_5$

(a) $20.0 \text{ g } O_2 \times \dfrac{1 \text{ mol } O_2}{31.998 \text{ g } O_2} \times \dfrac{4 \text{ mol } P}{5 \text{ mol } O_2} \times \dfrac{30.973 \text{ g } P}{1 \text{ mol } P} = 15.5 \text{ g } P$

(b) Because part (a) was calculated for running the reaction in a balanced fashion, you may use either mass to calculate the theoretical yield:

$20.0 \text{ g } O_2 \times \dfrac{1 \text{ mol } O_2}{31.998 \text{ g } O_2} \times \dfrac{2 \text{ mol } P_2O_5}{5 \text{ mol } O_2} \times \dfrac{141.941 \text{ g } P_2O_5}{1 \text{ mol } P_2O_5} = 35.5 \text{ g } P_2O_5$

or

$15.5 \text{ g } P \times \dfrac{1 \text{ mol } P}{30.973 \text{ g } P} \times \dfrac{2 \text{ mol } P_2O_5}{4 \text{ mol } P} \times \dfrac{141.941 \text{ g } P_2O_5}{1 \text{ mol } P_2O_5} = 35.5 \text{ g } P_2O_5$

9.81 (a) $H_2 + Br_2 \rightarrow 2HBr$

(b) The balanced equation tells you the H_2 and Br_2 combine in a one-to-one ratio.

$$10.079 \text{ g } H_2 \times \frac{1 \text{ mol } H_2}{2.0158 \text{ g } H_2} = 5.0000 \text{ mol } H_2$$

Because you have 7.00 moles of Br_2 but only 5.0000 moles of H_2, the H_2 is the limiting reactant.

(c) Because H_2 is the limiting reactant, you must use it to calculate the theoretical yield:

$$5.0000 \text{ mol } H_2 \times \frac{2 \text{ mol } HBr}{1 \text{ mol } H_2} = 10.000 \text{ mol } HBr$$

(d) $10.000 \text{ mol } HBr \times \dfrac{80.912 \text{ g } HBr}{1 \text{ mol } HBr} = 809.00 \text{ g } HBr$

(e) You begin with 5.0000 moles of H_2 and 7.00 moles of Br_2. You know from the one-to-one reactant ratio in the balanced equation that 5.00 moles of Br_2 react with 5.00 moles of H_2. Therefore, the excess amount of Br_2 is $7.00 \text{ mol} - 5.00 \text{ mol} = 2.00 \text{ mol}$.

(f) $2.00 \text{ mol } Br_2 \times \dfrac{159.81 \text{ g } Br_2}{1 \text{ mol } Br_2} = 320 \text{ g } Br_2$

9.83 (a) Step 1: $2Na(s) + Br_2(l) \rightarrow 2NaBr(s)$

(b) Step 2: $5.00 \text{ g } Na \times \dfrac{1 \text{ mol } Na}{22.9898 \text{ g } Na} = 0.217 \text{ mol } Na$

$$30.0 \text{ g } Br_2 \times \frac{1 \text{ mol } Br_2}{159.8 \text{ g } Br_2} = 0.188 \text{ mol } Br_2$$

Step 2a: Na $\dfrac{0.217}{2} = 0.109$

Br$_2$ $\dfrac{0.188}{1} = 0.188$

Because 0.109 is the smaller number, Na is the limiting reactant.

(c) Steps 3 and 4: $0.217 \text{ mol } Na \times \dfrac{2 \text{ mol } NaBr}{2 \text{ mol } Na} \times \dfrac{102.89 \text{ g } NaBr}{1 \text{ mol } NaBr} = 22.3 \text{ g } NaBr$

(d) Because the problem deals with theoretical yields, all the $5.00 \text{ g} = 0.217$ moles of limiting reactant Na is used up. The amount of Br_2 it combined with is

$$0.217 \text{ mol } Na \times \frac{1 \text{ mol } Br_2}{2 \text{ mol } Na} \times \frac{159.82 \text{ g } Br_2}{1 \text{ mol } Br_2} = 17.3 \text{ g } Br_2$$

The leftover Br_2 is therefore: $30.0 \text{ g} - 17.3 \text{ g} = 12.7 \text{ g}$.

(e) $\dfrac{14.7 \text{ g } NaBr}{22.3 \text{ g } NaBr} \times 100\% = 65.9\%$

9.85 (a) Step 1: $2 \text{ Na} + H_2 \rightarrow 2 \text{ NaH}$

(b) Step 2: $10.00 \text{ g } Na \times \dfrac{1 \text{ mol } Na}{22.9898 \text{ g } Na} = 0.4350 \text{ mol } Na$

$$0.0235 \text{ g } H_2 \times \frac{1 \text{ mol } H_2}{2.0158 \text{ g } H_2} = 0.0117 \text{ mol } H_2$$

Step 2a: Na $\dfrac{0.4350}{2} = 0.2175$

$\quad$ H$_2$ $\dfrac{0.0117}{1} = 0.0117$ limiting reactant

(c) Steps 3 and 4: $0.0117 \text{ mol H}_2 \times \dfrac{2 \text{ mol NaH}}{1 \text{ mol H}_2} \times \dfrac{23.998 \text{ g NaH}}{1 \text{ mol NaH}} = 0.562 \text{ g NaH}$

(d) All of the H$_2$ is used up, consuming

$\quad 0.0117 \text{ mol H}_2 \times \dfrac{2 \text{ mol Na}}{1 \text{ mol H}_2} \times \dfrac{22.9898 \text{ g Na}}{1 \text{ mol Na}} = 0.538 \text{ g Na}$

The leftover Na is therefore: $10.00 \text{ g} - 0.538 \text{ g} = 9.46 \text{ g}$.

(e) $\dfrac{0.428 \text{ g NaH}}{0.562 \text{ g NaH}} \times 100\% = 76.2\%$

9.87 Step 1: $\dfrac{\text{Molar mass of compound}}{\text{Molar mass of empirical formula}} = \dfrac{90 \text{ g/mol}}{44.053 \text{ g/mol}} = 2$ approximately

$\quad$ Step 2: $C_{2\times2}H_{2\times4}O_{2\times1} \rightarrow C_4H_8O_2$

9.89 (a) Use the four-step process for determining empirical formula from combustion analysis data. The four steps are shown here:

1. $40.00\% \text{ C} + 6.71\% \text{ H} = 46.71\%$, which is less than 100%. Therefore the percent O $= 100\% - 46.71\% = 53.29\%$

2. Since the mass of the sample is provided, calculate the mass due to each element:
   Mass of C $= 0.4000 \times 1.540 \text{ g} = 0.6160 \text{ g C}$
   Mass of H $= 0.0671 \times 1.540 \text{ g} = 0.103 \text{ g H}$
   Mass of O $= 0.5329 \times 1.540 \text{ g} = 0.8207 \text{ g O}$

3. Calculate the moles of each element (moles = grams/atomic mass).

   Moles of C $= 0.6160 \text{ g C} \times \dfrac{1 \text{ mol C}}{12.011 \text{ g C}} = 0.05129$ moles C

   Moles of H $= 0.103 \text{ g H} \times \dfrac{1 \text{ mol H}}{1.0079 \text{ g H}} = 0.102$ moles H

   Moles of O $= 0.8207 \text{ g O} \times \dfrac{1 \text{ mol}}{15.999 \text{ g O}} = 0.05129$ moles O

4. Divide by the smallest number to obtain the subscripts in the empirical formula. The smallest number is 0.05129.

   Subscript for C: $\dfrac{0.05129}{0.05129} = 1$

   Subscript for H: $\dfrac{0.102}{0.05129} = 2$

   Subscript for O: $\dfrac{0.05129}{0.05129} = 1$

   The empirical formula is therefore CH$_2$O.

(b) Divide the given molar mass of the compound by the molar mass of the empirical formula:

$$\frac{30}{30.026} = 1$$

Because the molar mass is equal to the mass of the empirical formula. The molecular formula is also CH_2O.

9.91 (a) Use the four-step process for determining empirical formula from combustion analysis data. The four steps are shown here:

1. $92.3\% \, C + 7.7\% \, H = 100.0\%$. Therefore the sample does not contain any oxygen.

2. Since the mass of the sample is provided, calculate the mass due to each element:

Mass of $C = 0.923 \times 1.000 \, g = 0.923 \, g \, C$
Mass of $H = 0.077 \times 1.000 \, g = 0.077 \, g \, H$

3. Calculate the moles of each element (moles = grams/atomic mass).

$$\text{Moles of } C = 0.923 \, g \, C \times \frac{1 \, mol \, C}{12.011 \, g \, C} = 0.0768 \text{ moles C}$$

$$\text{Moles of } H = 0.077 \, g \, H \times \frac{1 \, mol \, H}{1.0079 \, g \, H} = 0.076 \text{ moles H}$$

4. Divide by the smallest number to obtain the subscripts in the empirical formula. The smallest number is 0.076.

Subscript for C: $\dfrac{0.0768}{0.076} = 1$

Subscript for H: $\dfrac{0.076}{0.076} = 1$

The empirical formula is therefore CH.

(b) Divide the given molar mass of the compound by the molar mass of the empirical formula:

$$\frac{78}{13.019} = 6$$

The molecular formula is $C_{1\times6} H_{1\times6} = C_6H_6$.

9.93 (a) Use the four-step process for determining empirical formula from combustion analysis data.

1. $55.55\% \, C + 3.15\% \, H = 58.70\%$, which is less than 100%. Therefore, the percent $Cl = 100\% - 58.70\% = 41.30\%$.

2. Assume 100 g of compound (percentages then become grams).
54.65 g of carbon, 3.15 g of hydrogen, 41.3 g of chlorine

3. Calculate the moles of each element (moles = grams/atomic mass).

$$\text{Moles of } C = 55.55 \, g \, C \times \frac{1 \, mol \, C}{12.011 \, g \, C} = 4.625 \text{ moles C}$$

$$\text{Moles of } H = 3.15 \, g \, H \times \frac{1 \, mol \, H}{1.0079 \, g \, H} = 3.125 \text{ moles H}$$

$$\text{Moles of } Cl = 41.3 \, g \, Cl \times \frac{1 \, mol}{35.453 \, g \, Cl} = 1.165 \text{ moles Cl}$$

4. Divide by the smallest number to obtain the subscripts in the empirical formula. The smallest number is 1.165.

Subscript for C: $\dfrac{4.550}{1.165} = 3.91$ (approximately 4)

Subscript for H: $\dfrac{3.125}{1.165} = 2.67$ (2 2/3 or 8/3)

Subscript for Cl: $\dfrac{1.165}{1.165} = 1$

The subscripts in the empirical formula are therefore $C_4H_{2.67}Cl$. However, note that the subscripts are not all integral (whole) numbers. To get all whole numbers, we must multiply each subscript by 3, to obtain the actual empirical formula of $C_{12}H_8Cl_3$.

(b) If the empirical formula is one-third the mass of its actual formula, we multiply the subscripts in the empirical formula by three to get the molecular formula, which is $C_{36}H_{24}Cl_9$.

9.95 Step 1: To get the empirical formula from percent composition, assume 100 g of compound (that way, percent values become grams):

$$66.63\% \ C \rightarrow 66.63 \ g\,C \times \frac{1 \ mol \ C}{12.011 \ g\,C} = 5.547 \ mol \ C$$

$$11.18\% \ H \rightarrow 11.18 \ g\,H \times \frac{1 \ mol \ H}{1.0079 \ g\,H} = 11.09 \ mol \ H$$

$$22.19\% \ O \rightarrow 22.19 \ g\,O \times \frac{1 \ mol \ O}{15.999 \ g\,O} = 1.387 \ mol \ O$$

Step 2: Use the calculated numbers of moles as subscripts and divide through by the smallest:

$$C_{\underset{1.387}{5.547}}H_{\underset{1.387}{11.09}}O_{\underset{1.387}{1.387}} \rightarrow C_{4.00}H_{8.00}O_{1.00} \rightarrow C_4H_8O$$

9.97 First determine the percent by mass oxygen:

$$100\% - (26.4\% \ Na + 36.8\% \ S) = 36.8\% \ O$$

Step 1: Assume 100 g of compound and convert masses to moles:

$$26.4\% \ Na \rightarrow 26.4 \ g\,Na \times \frac{1 \ mol \ Na}{22.9898 \ g\,Na} = 1.15 \ mol \ Na$$

$$36.8\% \ S \rightarrow 36.8 \ g\,S \times \frac{1 \ mol \ S}{32.066 \ g\,S} = 1.15 \ mol \ S$$

$$36.8\% \ O \rightarrow 36.8 \ g\,O \times \frac{1 \ mol \ O}{15.999 \ g\,O} = 2.30 \ mol \ O$$

Step 2: $Na_{\underset{1.15}{1.15}}S_{\underset{1.15}{1.15}}O_{\underset{1.15}{2.30}} \rightarrow Na_{1.00}S_{1.00}O_{2.00} \rightarrow NaSO_2$

9.99 Step 1: Assume you have 1 mole of C_2H_6O. That way, the subscripts tell you how many moles you have of each element in the compound. Use this information to convert element moles to element masses:

$$2 \text{ mol C} \times \frac{12.011 \text{ g C}}{1 \text{ mol C}} = 24.022 \text{ g C}$$

$$6 \text{ mol H} \times \frac{1.0079 \text{ g H}}{1 \text{ mol H}} = 6.474 \text{ g H}$$

$$1 \text{ mol O} \times \frac{15.999 \text{ g O}}{1 \text{ mol O}} = 15.999 \text{ g O}$$

Step 2: Divide each mass calculated in step 1 by the mass of the 1 mole of C_2H_6O you are assumed to have. The mass in grams of that 1 mole is numerically equal to the molar mass. The molar mass of C_2H_6 is 46.068 g/mol, which means you are working with 46.068 g of C_2H_6O.

Multiply each quotient by 100% to get the mass percent of each element:

$$\frac{24.022 \text{ g C}}{46.068 \text{ g C}_2\text{H}_6\text{O}} \times 100\% = 52.14\% \text{ C}$$

$$\frac{6.0474 \text{ g C}}{46.068 \text{ g C}_2\text{H}_6\text{O}} \times 100\% = 13.13\% \text{ H}$$

$$\frac{15.999 \text{ g C}}{46.068 \text{ g C}_2\text{H}_6\text{O}} \times 100\% = 34.73\% \text{ O}$$

9.101 Since P_2O_5 can be viewed as the empirical formula for P_4O_{10}, there is no difference in the % by mass composition for P and for O between these two formulas.

9.103 (a)

5 mol C	=	5×12.011 g/mol =	60.055 g/mol
10 mol H	=	10×1.0079 g/mol =	10.079 g/mol
5 mol O	=	5×15.999 g/mol =	79.995 g/mol

$$1 \text{ mol C}_5\text{H}_{10}\text{O}_5 \rightarrow 150.129 \text{ g/mol}$$

(b) $3.87 \text{ mol} \times \dfrac{150.129 \text{ g}}{1 \text{ mol}} = 581 \text{ g}$

(c) $3.87 \text{ mol} \times \dfrac{6.022 \times 10^{23} \text{ molecules}}{1 \text{ mol}} = 2.33 \times 10^{24} \text{ molecules}$

(d) $3.87 \text{ mol ribose} \times \dfrac{5 \text{ mol O}}{1 \text{ mol ribose}} \times \dfrac{6.022 \times 10^{23} \text{ atoms O}}{1 \text{ mol O}} = 1.16 \times 10^{25} \text{ atoms O}$

(e) $1.16 \times 10^{25} \text{ atoms O} \times \dfrac{15.999 \text{ g}}{6.022 \times 10^{23} \text{ atoms O}} = 309 \text{ g}$

9.105 No, because the theoretical yield is 100% of what you can possibly get.

9.107 C_2H_4 molar mass

2×12.011 g/mol	=	24.022 g/mol
4×1.0079 g/mol	=	4.032 g/mol
		28.054 g/mol

CO molar mass

1×12.011 g/mol	=	12.011 g/mol
1×15.999 g/mol	=	15.999 g/mol
		28.010 g/mol

N_2 molar mass $2 \times 14.007 \text{ g/mol} = 28.014 \text{ g/mol}$

The 1 mole of C_2H_4 has the greatest mass.

9.109 (a) $12 \times 12.011 \text{ g/mol C} = 144.13 \text{ g/mol}$

$22 \times 1.0079 \text{ g/mol H} = 22.17 \text{ g/mol}$

$11 \times 15.999 \text{ g/mol O} = 175.99 \text{ g/mol}$

$$\overline{ 342.29 \text{ g/mol } C_{12}H_{22}O_{11}}$$

(b) $1.25 \text{ mol} \times \dfrac{342.29 \text{ g}}{1 \text{ mol}} = 428 \text{ g}$

(c) $1.25 \text{ mol} \times \dfrac{6.022 \times 10^{23} \text{ molecules}}{1 \text{ mol}} = 7.53 \times 10^{23} \text{ molecules}$

(d) $1.25 \text{ mol sucrose} \times \dfrac{22 \text{ mol H}}{1 \text{ mol sucrose}} \times \dfrac{6.022 \times 10^{23} \text{ H atoms}}{1 \text{ mol H}} = 1.66 \times 10^{25} \text{ H atoms}$

(e) $1.66 \times 10^{25} \text{ H atoms} \times \dfrac{1.0079 \text{ amu}}{1 \text{ H atom}} \times \dfrac{1.661 \times 10^{-24} \text{ g}}{1 \text{ amu}} = 27.8 \text{ g}$

9.111 (a) $2Cu_2O(s) + C(s) \rightarrow 4Cu(s) + CO_2(g)$

(b) Begin by converting the Cu_2O mass from tons to grams:

$1.000 \text{ ton} \times \dfrac{2000 \text{ lb}}{1 \text{ ton}} \times \dfrac{1 \text{ kg}}{2.205 \text{ lb}} \times \dfrac{1000 \text{ g}}{1 \text{ kg}} = 9.070 \times 10^5 \text{ g}$

Convert to moles:

$9.070 \times 10^5 \text{ g Cu}_2\text{O} \times \dfrac{1 \text{ mol Cu}_2\text{O}}{143.08 \text{ g Cu}_2\text{O}} = 6339 \text{ mol Cu}_2\text{O}$

Use the equation coefficients to determine the number of moles of C that react with this much Cu_2O, then convert to mass:

$6339 \text{ mol Cu}_2\text{O} \times \dfrac{1 \text{ mol C}}{2 \text{ mol Cu}_2\text{O}} = 3170 \text{ mol C}$

$3170 \text{ mol C} \times \dfrac{12.011 \text{ g C}}{1 \text{ mol C}} = 3.807 \times 10^4 \text{ g C}$

This is the amount of C you need but you must account for the fact that the coke is only 95% by mass C:

$3.807 \times 10^4 \text{ g C} \times \dfrac{100 \text{ g coke}}{95 \text{ g C}} \times \dfrac{1 \text{ kg}}{1000 \text{ g}} = 40.07 \text{ kg coke}$

9.113 (a) $10.0 \text{ g PF}_3 \times \dfrac{1 \text{ mol PF}_3}{87.968 \text{ g PF}_3} \times \dfrac{3 \text{ mol CO}}{2 \text{ mol PF}_3} \times \dfrac{28.010 \text{ g CO}}{1 \text{ mol CO}} = 4.78 \text{ g CO}$

(b) Determine the limiting reactant:

$Fe(CO)_5 \dfrac{5.0}{1} = 5.0$

$PF_3 \dfrac{8.0}{2} = 4.0 \text{ smallest ratio, limiting reactant}$

$H_2 \dfrac{6.0}{1} = 6.0$

$$8.0 \text{ mol PF}_3 \times \frac{3 \text{ mol CO}}{2 \text{ mol PF}_3} \times \frac{28.010 \text{ g CO}}{1 \text{ mol CO}} = 3.4 \times 10^2 \text{ g CO}$$

(c) $25.0 \text{ g Fe(CO)}_5 \times \dfrac{1 \text{ mol Fe(CO)}_5}{195.9 \text{ g Fe(CO)}_5} = 0.128 \text{ mol Fe(CO)}_5$

$10.0 \text{ g PF}_3 \times \dfrac{1 \text{ mol PF}_3}{87.968 \text{ g PF}_3} = 0.114 \text{ mol PF}_3$

$\text{Fe(CO)}_5 \dfrac{0.128}{1} = 0.128$

$\text{PF}_3 \dfrac{0.114}{2} = 0.057 \text{ limiting reactant}$

$0.114 \text{ mol PF}_3 \times \dfrac{3 \text{ mol CO}}{2 \text{ mol PF}_3} = 0.171 \text{ mol CO}$

(d) $5.00 \text{ L H}_2 \times \dfrac{0.0820 \text{ g H}_2}{1 \text{ L H}_2} \times \dfrac{1 \text{ mol H}_2}{2.0158 \text{ g H}_2} = 0.203 \text{ mol H}_2$

$0.203 \text{ mol H}_2 \times \dfrac{3 \text{ mol CO}}{1 \text{ mol H}_2} \times \dfrac{28.010 \text{ g CO}}{1 \text{ mol CO}} = 17.1 \text{ g CO theoretical yield}$

$\text{Percent yield} = \dfrac{13.5 \text{ g}}{17.1 \text{ g}} \times 100\% = 78.9\%$

9.115 (a) $5.00 \text{ g AgNO}_3 \times \dfrac{1 \text{ mol AgNO}_3}{169.9 \text{ g AgNO}_3} = 0.0294 \text{ mol AgNO}_3$

(b) $0.0294 \text{ mol AgNO}_3 \times \dfrac{3 \text{ mol O}}{1 \text{ mol AgNO}_3} = 0.0882 \text{ mol O}$

(c) $0.0294 \text{ mol AgNO}_3 \times \dfrac{1 \text{ mol N}}{1 \text{ mol AgNO}_3} \times \dfrac{14.007 \text{ g N}}{1 \text{ mol N}} = 0.412 \text{ g N}$

(d) $0.0294 \text{ mol AgNO}_3 \times \dfrac{1 \text{ mol Ag}}{1 \text{ mol AgNO}_3} \times \dfrac{6.022 \times 10^{23} \text{ Ag atoms}}{1 \text{ mol Ag}} = 1.77 \times 10^{22} \text{ Ag atoms}$

9.117 (a) $0.50 \text{ mol F}_2 \times \dfrac{1 \text{ mol N}_2}{3 \text{ mol F}_2} = 0.17 \text{ mol N}_2$

$0.50 \text{ mol N}_2 = 0.17 \text{ mol N}_2 = 0.33 \text{ mol N}_2 \text{ left over}$

(b) $20.0 \text{ mol F}_2 \times \dfrac{1 \text{ mol N}_2}{3 \text{ mol F}_2} = 6.67 \text{ mol N}_2 \text{ used up}$

$12.0 \text{ mol N}_2 - 6.67 \text{ mol N}_2 = 5.3 \text{ mol N}_2 \text{ left over}$

(c) The two reactants are present in the needed 3:1 ratio, meaning both are used up completely.

(d) $100 \text{ molecules N}_2 \times \dfrac{3 \text{ molecules F}_2}{1 \text{ molecule N}_2} = 300 \text{ molecules F}_2 \text{ used up}$

$500 \text{ molecules F}_2 - 300 \text{ molecules F}_2 = 200 \text{ molecules F}_2 \text{ left over}$

(e) From the preceding problem, $15.0 \text{ g F}_2 = 0.395 \text{ mol F}_2$

$0.395 \text{ mol F}_2 \times \dfrac{1 \text{ mol N}_2}{3 \text{ mol F}_2} = 0.132 \text{ mol N}_2 \text{ used up}$

$$0.132 \ \text{mol N}_2 \times \frac{28.014 \ \text{g N}_2}{1 \ \text{mol N}_2} = 3.69 \ \text{g N}_2 \ \text{used up}$$

$$5.00 \ \text{g N}_2 - 3.69 \ \text{g N}_2 = 1.31 \ \text{g N}_2 \ \text{left over}$$

(f) From the preceding problem, $70.0 \ \text{mg F}_2 = 1.84 \times 10^{-3} \ \text{mol F}_2$

$$1.84 \times 10^{-3} \ \text{mol F}_2 \times \frac{1 \ \text{mol N}_2}{3 \ \text{mol F}_2} = 6.13 \times 10^{-4} \ \text{mol N}_2 \ \text{used up}$$

$$6.13 \times 10^{-4} \ \text{mol N}_2 \times \frac{28.018 \ \text{g N}_2}{1 \ \text{mol N}_2} = 0.0172 \ \text{g N}_2 = 17.2 \ \text{mg N}_2 \ \text{used up}$$

$$20.0 \ \text{mg N}_2 - 17.2 \ \text{mg N}_2 = 2.8 \ \text{mg N}_2 \ \text{left over}$$

9.119 (a) $Ca_3P_2 + 6H_2O \rightarrow 3Ca(OH)_2 + 2PH_3$

(b) $60.0 \ \text{g } Ca_3P_2 \times \dfrac{1 \ \text{mol } Ca_3P_2}{182.2 \ \text{g } Ca_3P_2} = 0.329 \ \text{mol } Ca_3P_2$

$$0.329 \ \text{mol } Ca_3P_2 \times \frac{6 \ \text{mol } H_2O}{1 \ \text{mol } Ca_3P_2} \times \frac{18.015 \ \text{g } H_2O}{1 \ \text{mol } H_2O} = 35.6 \ \text{g } H_2O$$

(c) Because you calculated the exact amount of H_2O needed to react completely with 60.0 g of Ca_3P_2, you may use either reactant to calculate theoretical yield:

$$60.0 \ \text{g } Ca_3P_2 = 0.329 \ \text{mol } Ca_3P_2 \times \frac{2 \ \text{mol } PH_3}{1 \ \text{mol } PH_3} \times \frac{33.998 \ \text{g } PH_3}{1 \ \text{mol } PH_3} = 22.4 \ \text{g } PH_3$$

$$35.6 \ \text{g } H_2O \times \frac{1 \ \text{mol } H_2O}{18.015 \ \text{g } H_2O} \times \frac{2 \ \text{mol } PH_3}{6 \ \text{mol } H_2O} \times \frac{33.998 \ \text{g } PH_3}{1 \ \text{mol } PH_3} = 22.4 \ \text{g } PH_3$$

9.121 Assume you have 1 mole of $C_8H_9NO_2$.

$$8 \ \text{mol C} \times \frac{12.011 \ \text{g C}}{1 \ \text{mol C}} = 96.088 \ \text{g C}$$

$$9 \ \text{mol H} \times \frac{1.0079 \ \text{g H}}{1 \ \text{mol H}} = 9.0711 \ \text{g H}$$

$$1 \ \text{mol N} \times \frac{14.007 \ \text{g N}}{1 \ \text{mol N}} = 14.007 \ \text{g N}$$

$$2 \ \text{mol O} \times \frac{15.999 \ \text{g O}}{1 \ \text{mol O}} = 31.998 \ \text{g O}$$

The molar mass of $C_8H_9NO_2$ is 151.16 g/mol, meaning your 1 mole has a mass of 151.16 g.

$$\frac{96.088 \ \text{g C}}{151.16 \ \text{mol } C_8H_9NO_2} \times 100\% = 63.57\% \ \text{C}$$

$$\frac{9.0711 \ \text{g H}}{151.16 \ \text{mol } C_8H_9NO_2} \times 100\% = 6.001\% \ \text{H}$$

$$\frac{14.007 \ \text{g N}}{151.16 \ \text{mol } C_8H_9NO_2} \times 100\% = 9.266\% \ \text{N}$$

$$\frac{31.998 \ \text{g O}}{151.16 \ \text{mol } C_8H_9NO_2} \times 100\% = 21.17\% \ \text{g O}$$

9.123 The balanced equation is $C_{12}H_{22}O_{11} + 12O_2 \rightarrow 12CO_2 + 11H_2O$.

$$2.00 \text{ g sucrose} \times \frac{1 \text{ mol sucrose}}{342.3 \text{ g sucrose}} = 5.84 \times 10^{-3} \text{ mol sucrose}$$

$$5.84 \times 10^{-3} \text{ mol sucrose} \times \frac{12 \text{ mol } CO_2}{1 \text{ mol sucrose}} \times \frac{44.009 \text{ g } CO_2}{1 \text{ mol } CO_2} = 3.08 \text{ g } CO_2$$

$$5.84 \times 10^{-3} \text{ mol sucrose} \times \frac{11 \text{ mol } H_2O}{1 \text{ mol sucrose}} \times \frac{18.015 \text{ g } H_2O}{1 \text{ mol } H_2O} = 1.16 \text{ g } H_2O$$

9.125 (a) $4BF_3 + 3H_2O \rightarrow H_3BO_3 + 3HBF_4$

(b) Theoretical yield is

$$24.2 \text{ g } BF_3 \times \frac{1 \text{ mol } BF_3}{67.81 \text{ g } BF_3} \times \frac{3 \text{ mol } HBF_4}{4 \text{ mol } BF_3} \times \frac{87.81 \text{ g } HBF_4}{1 \text{ mol } HBF_4} = 23.5 \text{ g } HBF_4$$

$$\% \text{ yield} = \frac{14.8 \text{ g}}{23.5 \text{ g}} \times 100\% = 63.0\%$$

9.127 Assume 100 g of chrome yellow.

$$64.11\% \text{ Pb} \rightarrow 64.11 \text{ g Pb} \times \frac{1 \text{ mol Pb}}{207.19 \text{ g pb}} = 0.3094 \text{ mol Pb}$$

$$16.09\% \text{ Cr} \rightarrow 16.09 \text{ g Cr} \times \frac{1 \text{ mol Cr}}{51.996 \text{ g Cr}} = 0.3094 \text{ mol Cr}$$

$$19.80\% \text{ O} \rightarrow 19.80 \text{ g O} \times \frac{1 \text{ mol O}}{15.999 \text{ g O}} = 1.238 \text{ mol O}$$

$$Pb_{\frac{0.3094}{0.3094}} Cr_{\frac{0.3094}{0.3094}} O_{\frac{1.238}{0.3094}} \rightarrow Pb_{1.00}Cr_{1.00}O_{4.00} \rightarrow PbCrO_4$$

9.129 The balanced equation is $3SiCl_4 + 4NH_3 \rightarrow Si_3N_4 + 12HCl$. Find out which reactant is the limiting one:

$$64.2 \text{ g } SiCl_4 \times \frac{1 \text{ mol } SiCl_4}{169.9 \text{ g } SiCl_4} = 0.378 \text{ mol } SiCl_4$$

$$20.0 \text{ g } NH_3 \times \frac{1 \text{ mol } NH_3}{17.03 \text{ g } NH_3} = 1.17 \text{ mol } NH_3$$

$$SiCl_4 \frac{0.378}{3} = 0.126 \text{ limiting reactant}$$

$$NH_3 \frac{1.17}{4} = 0.293$$

The theoretical yield is

$$0.378 \text{ mol } SiCl_4 \times \frac{1 \text{ mol } Si_3N_4}{3 \text{ mol } SiCl_4} \times \frac{140.3 \text{ g } Si_3N_4}{1 \text{ mol } Si_3N_4} = 17.7 \text{ g } Si_3N_4$$

Actual yield $= 0.960 \times 17.7 \text{ g} = 17.0 \text{ g } Si_3N_4$

9.131 The balanced equation is $Cr_2O_3 + 3H_2S \rightarrow Cr_2S_3 + 3H_2O$.

    (a) $13.6 \text{ g } Cr_2O_3 \times \dfrac{1 \text{ mol } Cr_2O_3}{151.99 \text{ g } Cr_2O_3} \times \dfrac{3 \text{ mol } H_2S}{1 \text{ mol } Cr_2O_3} \times \dfrac{34.080 \text{ g } H_2S}{1 \text{ mol } H_2S} = 9.15 \text{ g } H_2S$

    (b) $13.6 \text{ g } Cr_2O_3 \times \dfrac{1 \text{ mol } Cr_2O_3}{151.99 \text{ g } Cr_2O_3} \times \dfrac{1 \text{ mol } Cr_2S_3}{1 \text{ mol } Cr_2O_3} \times \dfrac{200.2 \text{ g } Cr_2S_3}{1 \text{ mol } Cr_2S_3} = 17.9 \text{ g } Cr_2S_3$

    or

    $9.15 \text{ g } H_2S \times \dfrac{1 \text{ mol } H_2S}{34.080 \text{ g } H_2S} \times \dfrac{1 \text{ mol } Cr_2S_3}{3 \text{ mol } H_2S} \times \dfrac{200.2 \text{ g } Cr_2S_3}{1 \text{ mol } Cr_2S_3} = 17.9 \text{ g } Cr_2S_3$

9.133 (a) $SiCl_4 + 2H_2O \rightarrow SiO_2 + 4HCl$

    (b) $120.0 \text{ g } HCl \times \dfrac{1 \text{ mol } HCl}{36.46 \text{ g } HCl} = 3.291 \text{ mol } HCl$

    $3.291 \text{ mol } HCl \times \dfrac{1 \text{ mol } SiCl_4}{4 \text{ mol } HCl} \times \dfrac{169.9 \text{ g } SiCl_4}{1 \text{ mol } SiCl_4} = 139.8 \text{ g } SiCl_4$

    $3.291 \text{ mol } HCl \times \dfrac{2 \text{ mol } H_2O}{4 \text{ mol } HCl} \times \dfrac{18.015 \text{ g } H_2O}{1 \text{ mol } H_2O} = 29.64 \text{ g } H_2O$

    (c) $3.291 \text{ mol } HCl \times \dfrac{1 \text{ mol } SiO_2}{4 \text{ mol } HCl} \times \dfrac{60.08 \text{ g } SiO_2}{1 \text{ mol } SiO_2} = 49.44 \text{ g } SiO_2$

    or

    $139.8 \text{ g } SiCl_4 \times \dfrac{1 \text{ mol } SiCl_4}{169.9 \text{ g } SiCl_4} \times \dfrac{1 \text{ mol } SiO_2}{1 \text{ mol } SiCl_4} \times \dfrac{60.08 \text{ g } SiO_2}{1 \text{ mol } SiO_2} = 49.43 \text{ g } SiO_2$

    or

    $29.64 \text{ g } H_2O \times \dfrac{1 \text{ mol } H_2O}{18.015 \text{ g } H_2O} \times \dfrac{1 \text{ mol } SiO_2}{2 \text{ mol } H_2O} \times \dfrac{60.08 \text{ g } SiO_2}{1 \text{ mol } SiO_2} = 49.42 \text{ g } SiO_2$

9.135 The balanced equation is $Fe_2O_3 + 3CO \rightarrow 2Fe + 3CO_2$.

    $24.0 \text{ g } Fe_2O_3 \times \dfrac{1 \text{ mol } Fe_2O_3}{159.687 \text{ g } Fe_2O_3} = 0.150 \text{ mol } Fe_2O_3$

    $34.0 \text{ g } CO \times \dfrac{1 \text{ mol } CO}{28.01 \text{ g } CO} = 1.21 \text{ mol } CO$

    $Fe_2O_3 \quad \dfrac{0.150}{1} = 0.150 \text{ limiting reactant}$

    $CO \quad \dfrac{1.21}{3} = 0.403$

    Theoretical yield:

    $0.150 \text{ mol } Fe_2O_3 \times \dfrac{3 \text{ mol } CO_2}{1 \text{ mol } Fe_2O_3} \times \dfrac{44.009 \text{ g } CO_2}{1 \text{ mol } CO_2} = 19.8 \text{ g } CO_2$

9.137 $200.0 \text{ g } N_2O_5 \times \dfrac{1 \text{ mol } N_2O_5}{108.0 \text{ g } N_2O_5} \times \dfrac{2 \text{ mol } N}{1 \text{ mol } N_2O_5} \times \dfrac{6.022 \times 10^{23} \text{ N atoms}}{1 \text{ mol } N} = 2.230 \times 10^{24} \text{ N atoms}$

    $200.0 \text{ g } N_2O_5 \times \dfrac{1 \text{ mol } N_2O_5}{108.0 \text{ g } N_2O_5} \times \dfrac{5 \text{ mol } O}{1 \text{ mol } N_2O_5} \times \dfrac{6.022 \times 10^{23} \text{ O atoms}}{1 \text{ mol } O} = 5.576 \times 10^{24} \text{ O atoms}$

9.139 The balanced equation is $P_4O_{10} + 6H_2O \rightarrow 4H_3PO_4$.

$$52.5 \text{ g } P_4O_{10} \times \frac{1 \text{ mol } P_4O_{10}}{283.9 \text{ g } P_4O_{10}} = 1.185 \text{ mol } P_4O_{10}$$

$$25.0 \text{ g } H_2O \times \frac{1 \text{ mol } H_2O}{18.015 \text{ g } H_2O} = 1.39 \text{ mol } H_2O$$

$$P_4O_{10} \frac{0.185}{1} = 0.185 \text{ limiting reactant}$$

$$H_2O \frac{1.39}{6} = 0.232$$

$$0.185 \text{ mol } P_4O_{10} \times \frac{4 \text{ mol } H_3PO_4}{1 \text{ mol } P_4O_{10}} \times \frac{97.9 \text{ g } H_3PO_4}{1 \text{ mol } H_3PO_4} = 72.4 \text{ g } H_3PO_4$$

9.141 The balanced equation is $Cl_2O_7 + H_2O \rightarrow 2HClO_4$.

(a) Rearrange the percent yield equation by solving for theoretical yield:

$$\text{Percent yield} = \frac{\text{Actual yield}}{\text{Theoretical yield}} \times 100\%$$

$$\text{Theoretical yield} = \frac{\text{Actual yield}}{\text{Percent yield}} \times 100\% = \frac{52.8 \text{ g } HClO_4}{82.0\%} \times 100\% = 64.4 \text{ g } HClO_4$$

(b) You must use the theoretical yield in calculating how much of each reactant was used up.

$$64.4 \text{ g } HClO_4 \times \frac{1 \text{ mol } HClO_4}{100.5 \text{ g } HClO_4} \times \frac{1 \text{ mol } Cl_2O_7}{2 \text{ mol } HClO_4} \times \frac{182.9 \text{ g } Cl_2O_7}{1 \text{ mol } Cl_2O_7} = 58.6 \text{ g } Cl_2O_7$$

$$64.4 \text{ g } HClO_4 \times \frac{1 \text{ mol } HClO_4}{100.5 \text{ g } HClO_4} \times \frac{1 \text{ mol } H_2O}{2 \text{ mol } HClO_4} \times \frac{18.015 \text{ g } H_2O}{1 \text{ mol } H_2O} = 5.77 \text{ g } H_2O$$

9.143 The balanced equation is $C_2H_4(g) + 6F_2(g) \rightarrow 2CF_4(g) + 4HF(g)$.

$$2.78 \text{ g } C_2H_4 \times \frac{1 \text{ mol } C_2H_4}{28.054 \text{ g } C_2H_4} = 0.0991 \text{ mol } C_2H_4$$

$$0.0991 \text{ mol } C_2H_4 \times \frac{2 \text{ mol } CF_4}{1 \text{ mol } C_2H_4} \times \frac{88.003 \text{ g } CF_4}{1 \text{ mol } CF_4} = 17.4 \text{ g } CF_4$$

$$0.0991 \text{ mol } C_2H_4 \times \frac{4 \text{ mol } HF}{1 \text{ mol } C_2H_4} \times \frac{20.006 \text{ g } HF}{1 \text{ mol } HF} = 7.93 \text{ g } HF$$

9.145 The balanced equation is $2HSbCl_4 + 3H_2S \rightarrow Sb_2S_3 + 8HCl$.

$$118.2 \text{ g } HSbCl_4 \times \frac{1 \text{ mol } HSbCl_4}{264.6 \text{ g } HSbCl_4} = 0.4467 \text{ mol } HSbCl_4$$

$$47.9 \text{ g } H_2S \times \frac{1 \text{ mol } H_2S}{34.080 \text{ g } H_2S} = 1.41 \text{ mol } H_2S$$

$$HSbCl_4 \frac{0.4467}{2} = 0.2234 \text{ limiting reactant}$$

$$H_2S \frac{1.41}{3} = 0.470$$

Theoretical yield:

$$0.4467 \text{ mol HSbCl}_4 \times \frac{8 \text{ mol HCl}}{2 \text{ mol HSbCl}_4} \times \frac{36.46 \text{ g HCl}}{1 \text{ mol HCl}} = 65.15 \text{ g HCl}$$

$$\text{Percent yield} = \frac{41.6 \text{ g}}{65.15 \text{ g}} \times 100\% = 63.9\%$$

9.147 (a) $4NH_3 + 6NO \rightarrow 5N_2 + 6H_2O$

(b) $15 \text{ g NH}_3 \times \dfrac{1 \text{ mol NH}_3}{17.03 \text{ g NH}_3} = 0.881 \text{ mol NH}_3$

$22.0 \text{ g NO} \times \dfrac{1 \text{ mol NO}}{30.01 \text{ g NO}} = 0.733 \text{ mol NO}$

$NH_3 \dfrac{0.881}{4} = 0.220$

$NO \dfrac{0.733}{3} = 0.122 \text{ limiting reactant}$

$0.733 \text{ mol NO} \times \dfrac{5 \text{ mol N}_2}{6 \text{ mol NO}} \times \dfrac{28.014 \text{ g N}_2}{1 \text{ mol N}_2} = 17.1 \text{ g N}_2 \text{ theoretical yield}$

$\dfrac{13.3 \text{ g}}{17.1 \text{ g}} \times 100\% = 77.8\% \text{ yield}$

9.149 The balanced equation is $CaC_2 + 3CO \rightarrow 4C + CaCO_3$. That some CaC_2 is left over tells you CO is the limiting reactant and therefore the one to use to calculate theoretical yield. Determine what mass of CO was used to get the $CaCO_3$:

$$135.4 \text{ g CaCO}_3 \times \frac{1 \text{ mol CaCO}_3}{100.1 \text{ g CaCO}_3} \times \frac{3 \text{ mol CO}}{1 \text{ mol CaCO}_3} \times$$

$$\frac{28.01 \text{ g CO}}{1 \text{ mol CO}} = 113.7 \text{ g CO present at beginning of reaction}$$

Now determine what mass of CaC_2 combined with the 113.7 g of CO:

$$113.7 \text{ g CO} \times \frac{1 \text{ mol CO}}{28.01 \text{ g CO}} \times \frac{1 \text{ mol CaC}_2}{3 \text{ mol CO}} \times \frac{64.10 \text{ g CaC}_2}{1 \text{ mol CaC}_2} = 86.73 \text{ g CaC}_2 \text{ used up}$$

 86.73 g CaC_2 used
 38.5 g CaC_2 left over
 125.2 g CaC_2 present at beginning of reaction

9.151 (a) $S + 2H_2SO_4 \rightarrow 3SO_2 + 2H_2O$

(b) $4.80 \text{ g S} \times \dfrac{1 \text{ mol S}}{32.06 \text{ g S}} = 0.150 \text{ mol S}$

$16.20 \text{ g H}_2SO_4 \times \dfrac{1 \text{ mol H}_2SO_4}{98.07 \text{ g H}_2SO_4} = 0.1652 \text{ mol H}_2SO_4$

$S \qquad \dfrac{0.150}{1} = 0.150$

$$\text{H}_2\text{SO}_4 \quad \frac{0.1652}{2} = 0.0826 \text{ limiting reactant}$$

$$0.1652 \text{ mol H}_2\text{SO}_4 \times \frac{3 \text{ mol SO}_2}{2 \text{ mol H}_2\text{SO}_4} \times \frac{64.06 \text{ g SO}_2}{1 \text{ mol SO}_2} = 15.87 \text{ g SO}_2$$

9.153 No. The number of moles is proportional to the number of product and reactant particles, which can change during a reaction. For example, in the reaction $2\text{H}_2 + \text{O}_2 \rightarrow 2\text{H}_2\text{O}$, you begin with 3 moles of reactants and end up with 2 moles of product.

9.155 The balanced equation is $2\text{NH}_4\text{F} + \text{Ca(NO}_3)_2 \rightarrow \text{CaF}_2 + 2\text{N}_2\text{O} + 4\text{H}_2\text{O}$.

$$22.8 \text{ g NH}_4\text{F} \times \frac{1 \text{ mol NH}_4\text{F}}{37.04 \text{ g NH}_4\text{F}} = 0.616 \text{ mol NH}_4\text{F}$$

$$38.2 \text{ g Ca(NO}_3)_2 \times \frac{1 \text{ mol Ca(NO}_3)_2}{164.1 \text{ g Ca(NO}_3)_2} = 0.0233 \text{ mol Ca(NO}_3)_2$$

$$\text{NH}_4\text{F} \quad \frac{0.616}{2} = 0.308$$

$$\text{Ca(NO}_3)_2 \quad \frac{0.233}{1} = 0.233 \text{ limiting reactant}$$

$$0.233 \text{ mol Ca(NO}_3)_2 \times \frac{1 \text{ mol CaF}_2}{1 \text{ mol Ca(NO}_3)_2} \times \frac{78.08 \text{ g CaF}_2}{1 \text{ mol CaF}_2} = 18.2 \text{ g CaF}_2$$

$$0.233 \text{ mol Ca(NO}_3)_2 \times \frac{2 \text{ mol N}_2\text{O}}{1 \text{ mol Ca(NO}_3)_2} \times \frac{44.01 \text{ g N}_2\text{O}}{1 \text{ mol N}_2\text{O}} = 20.5 \text{ g N}_2\text{O}$$

$$0.233 \text{ mol Ca(NO}_3)_2 \times \frac{4 \text{ mol H}_2\text{O}}{1 \text{ mol Ca(NO}_3)_2} \times \frac{18.015 \text{ g H}_2\text{O}}{1 \text{ mol H}_2\text{O}} = 16.8 \text{ g H}_2\text{O}$$

9.157 The balanced equation is $\text{PbS} + 4\text{H}_2\text{O}_2 \rightarrow \text{PbSO}_4 + 4\text{H}_2\text{O}$.

(a) $63.2 \text{ g PbS} \times \dfrac{1 \text{ mol PbS}}{239.3 \text{ g PbS}} = 0.264 \text{ mol PbS}$

$$48.0 \text{ g H}_2\text{O}_2 \times \frac{1 \text{ mol H}_2\text{O}_2}{34.01 \text{ g H}_2\text{O}_2} = 1.41 \text{ mol H}_2\text{O}_2$$

$$\text{PbS} \quad \frac{0.264}{1} = 0.264 \text{ limiting reactant}$$

$$\text{H}_2\text{O}_2 \quad \frac{1.41}{4} = 0.352$$

(b) $0.264 \text{ mol PbS} \times \dfrac{4 \text{ mol H}_2\text{O}_2}{1 \text{ mol PbS}} \times \dfrac{34.01 \text{ g H}_2\text{O}_2}{1 \text{ mol H}_2\text{O}_2} = 35.9 \text{ g H}_2\text{O}_2 \text{ used up}$

Because the reaction began with 48.0 g of H_2O_2, the amount remaining is

$$48.0 \text{ g} - 35.9 \text{ g} = 12.1 \text{ g}.$$

9.159 The balanced equation is $\text{CaCN}_2 + 3\text{H}_2\text{O} \rightarrow \text{CaCO}_3 + 2\text{NH}_3$.

$$5.65 \text{ g CaCN}_2 \times \frac{1 \text{ mol CaCN}_2}{80.11 \text{ g CaCN}_2} = 0.0705 \text{ mol CaCN}_2$$

$$12.2 \text{ g H}_2\text{O} \times \frac{1 \text{ mol H}_2\text{O}}{18.015 \text{ g H}_2\text{O}} = 0.677 \text{ mol H}_2\text{O}$$

$$\text{CaCN}_2 \quad \frac{0.0705}{1} = 0.0705 \text{ limiting reactant}$$

$$\text{H}_2\text{O} \quad \frac{0.667}{3} = 0.226$$

$$0.0705 \text{ mol CaCN}_2 \times \frac{2 \text{ mol NH}_3}{1 \text{ mol CaCN}_2} \times \frac{17.03 \text{ g NH}_3}{1 \text{ mol NH}_3} = 2.40 \text{ g NH}_3 \text{ theoretical yield}$$

Actual yield $= 0.860 \times 2.40 \text{ g} = 2.06 \text{ g NH}_3$

9.161 $5.00 \times 10^{11} \text{ Ti atoms} \times \dfrac{47.90 \text{ amu}}{1 \text{ Ti atom}} \times \dfrac{1.66 \times 10^{-24} \text{ g}}{1 \text{ amu}} = 3.98 \times 10^{-11} \text{ g}$

9.163 Assume 100 g of arginine.

$$41.37\% \text{ C} \rightarrow 41.37 \text{ g C} \times \frac{1 \text{ mol C}}{12.011 \text{ g C}} = 3.444 \text{ mol C}$$

$$8.10\% \text{ H} \rightarrow 8.10 \text{ g H} \times \frac{1 \text{ mol H}}{1.0079 \text{ g H}} = 8.037 \text{ mol H}$$

$$32.16\% \text{ N} \rightarrow 32.16 \text{ g N} \times \frac{1 \text{ mol N}}{14.007 \text{ g N}} = 2.296 \text{ mol N}$$

$$18.37\% \text{ O} \rightarrow 18.37 \text{ g O} \times \frac{1 \text{ mol O}}{15.999 \text{ g O}} = 1.148 \text{ mol O}$$

$$\text{C}_{\frac{3.444}{1.148}} \text{H}_{\frac{8.037}{1.148}} \text{N}_{\frac{2.296}{1.148}} \text{O}_{\frac{1.148}{1.148}} \rightarrow \text{C}_{3.00}\text{H}_{7.00}\text{N}_{2.00}\text{O}_{1.00} \rightarrow \text{C}_3\text{H}_7\text{O}_2\text{N empirical formula}$$

$$\frac{174 \text{ g/mol arginine}}{87.1 \text{ g/mol C}_3\text{H}_7\text{N}_2\text{O}} = 2 \text{ approximately}$$

$$\text{C}_{2 \times 3}\text{H}_{2 \times 7}\text{N}_{2 \times 2}\text{O}_{2 \times 1} \rightarrow \text{C}_6\text{H}_{14}\text{N}_4\text{O}_2 \text{ molecular formula}$$

9.165 $11.0 \text{ g Cu}_2\text{O} \times \dfrac{1 \text{ mol Cu}_2\text{O}}{143.09 \text{ g Cu}_2\text{O}} \times \dfrac{2 \text{ mol Cu}}{1 \text{ mol Cu}_2\text{O}} \times \dfrac{63.546 \text{ g Cu}}{1 \text{ mol Cu}} = 9.77 \text{ g Cu}$

$12.6 \text{ g Cu}_2\text{S} \times \dfrac{1 \text{ mol Cu}_2\text{S}}{159.16 \text{ g Cu}_2\text{S}} \times \dfrac{2 \text{ mol Cu}}{1 \text{ mol Cu}_2\text{S}} \times \dfrac{63.546 \text{ g Cu}}{1 \text{ mol Cu}} = 10.06 \text{ g Cu}$

The Cu_2S sample contains more Cu.

9.167 $4.25 \times 10^{22} \text{ atoms S} \times \dfrac{1 \text{ mol S}}{6.022 \times 10^{23} \text{ atoms S}} = 0.0706 \text{ mol S}$

$$0.0706 \text{ mol S} \times \frac{1 \text{ mol Al}_2\text{S}_3}{3 \text{ mol S}} = 0.0235 \text{ mol Al}_2\text{S}_3$$

$$0.0235 \text{ mol Al}_2\text{S}_3 \times \frac{150.2 \text{ mol Al}_2\text{S}_3}{1 \text{ mol Al}_2\text{S}_3} = 3.53 \text{ g Al}_2\text{S}_3$$

9.169 $250 \text{ mg} \times \dfrac{1 \text{ g}}{1000 \text{ mg}} \times \dfrac{1 \text{ mol C}_9\text{H}_8\text{O}_4}{180.2 \text{ g C}_9\text{H}_8\text{O}_4} \times \dfrac{6.022 \times 10^{23} \text{ molecules C}_9\text{H}_8\text{O}_4}{1 \text{ mol C}_9\text{H}_8\text{O}_4}$

$\qquad = 8.35 \times 10^{20} \text{ molecules C}_9\text{H}_8\text{O}_4$

There are nine C atoms in each molecule:

$$8.35 \times 10^{20} \text{ molecules C}_9\text{H}_8\text{O}_4 \times \frac{9 \text{ atoms C}}{1 \text{ molecule C}_9\text{H}_8\text{O}_4} = 7.52 \times 10^{21} \text{ atoms C}$$

9.171 Use the four-step procedure for determining empirical formula demonstrated in Problem 9.29 to get the empirical formula and the molecular formula.

(a) The empirical formula is $C_8H_8O_3$.

(b) Because the molar mass (152 g/mole) divided by the empirical formula mass (152 g/mole) is 1, the molecular formula is also $C_8H_8O_3$.

9.173 The balanced equation is $16Cr + 3S_8 \rightarrow 8Cr_2S_3$. First determine the theoretical yield:

$$\text{Theoretical yield} = \frac{\text{Actual yield}}{\text{Percent yield}} \times 100\% = \frac{235.0 \text{ g } Cr_2S_3}{63.80\%} \times 100\% = 368.3 \text{ g } Cr_2S_3$$

$$368.3 \text{ g } Cr_2S_3 \times \frac{1 \text{ mol } Cr_2S_3}{200.2 \text{ g } Cr_2S_3} \times \frac{16 \text{ mol } Cr}{8 \text{ mol } Cr_2S_3} \times \frac{52.00 \text{ g } Cr}{1 \text{ mol } Cr} = 191.3 \text{ g } Cr$$

$$368.3 \text{ g } Cr_2S_3 \times \frac{1 \text{ mol } Cr_2S_3}{200.2 \text{ g } Cr_2S_3} \times \frac{3 \text{ mol } S_8}{8 \text{ mol } Cr_2S_3} \times \frac{256.5 \text{ g } S_8}{1 \text{ mol } S_8} = 177.0 \text{ g } S_8$$

9.175 (a)

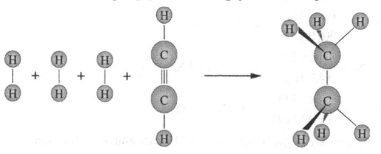

(b) $2H_2(g) + C_2H_2(g) \rightarrow C_2H_6(g)$

(c) Three moles of hydrogen gas and one mole of acetylene (C_2H_2) gas react to give one mole of ethane (C_2H_6) gas.

(d) $(10.0 \text{ g } C_2H_6)\left(\dfrac{\text{mole } C_2H_6}{30.0694 \text{ g } C_2H_6}\right) \times \left(\dfrac{2 \text{ mole } H_2}{\text{mole } C_2H_6}\right)\left(\dfrac{2.0158 \text{ g } H_2}{\text{mole } H_2}\right) = 1.34 \text{ g } H_2$

9.177 (a) 1 cup sugar + 3 pkgs cream cheese + 4 eggs $\rightarrow$ 12 dessert squares

(b) Still only 12 dessert squares

(c) The sugar

(d) 637 cups sugar/1 = 637

2498 pkgs cream cheese/3 = 832.7

4024 eggs/4 = 1006

The limiting reagent is sugar.

637 cup sugar $\times$ (12 dessert squares/1 cup sugar) = 7644 dessert squares.

637 cups sugar $\times$ (3 pkgs cream cheese/1 cup sugar) = 1911 pkgs cream cheese used, leaving 2498 − 1911 = 587 pkgs cream cheese left over.

637 cups sugar $\times$ (4 eggs/1 cup sugar)

= 2548 eggs used, leaving 4024 − 2548 = 1476 eggs left over.

9.179 (a) $CH_4(g) + 2O_2(g) \rightarrow 2H_2O(g) + CO_2(g)$

(b) 5 moles O_2 × (1 mole CO_2/2 moles O_2) = 2.5 moles CO_2

(c) 5 g O_2 × (1 mole O_2/31.998 g O_2) × (1 mole CO_2/2 mole O_2) × (6.022 × 10²³ molecules CO_2/mole CO_2) = 4.70 × 10²² molecules CO_2.

9.181 The percents of methane and ethane are 25.6% methane, 74.4% ethane

The % carbon in methane is

$$\frac{\text{mass of carbon}}{\text{Molar mass of } CH_4} \times 100\% = \frac{12.011}{16.043} \times 100\% = 74.868\% \text{ C}$$

The % carbon in ethane is

$$\frac{\text{mass of carbon}}{\text{Molar mass of } C_2H_6} \times 100\% = \frac{24.022}{30.069} \times 100\% = 79.889\% \text{ C}$$

The % carbon in carbon dioxide is

$$\frac{\text{mass of carbon}}{\text{Molar mass of } CO_2} \times 100\% = \frac{12.011}{44.009} \times 100\% = 27.292\% \text{ C}$$

To determine the grams of methane and ethane:

(a) 25.6% of 1.50 g = 0.384 g methane

(b) 1.50 − 0.384 = 1.12 g ethane

To determine the grams of C from methane and ethane

(a) 74.868% of 0.384 = 0.287 g C from methane

(b) 79.889% of 1.12 = 0.895 g C from ethane

Hence, there is 1.182 g C total.

If 27.292% of the total mass of CO_2 due to C is 1.182 g, then the total mass of CO_2 = 4.330 g

Alternatively, the stoichiometry of the combustion reactions can be used:

a) Methane combustion: $CH_4 + 2O_2 \rightarrow CO_2 + 2H_2O$

0.384 g methane = 0.0239 mol methane = 0.0239 mol CO_2 = 1.05 g CO_2

b) Ethane combustion: $2C_2H_6 + 7O_2 \rightarrow 4CO_2 + 6H_2O$

1.12 g ethane = 0.0372 mol ethane = 0.0745 mol CO_2 = 3.28 g CO_2

1.05 + 3.28 = 4.33 g CO_2

9.183 (a) Use the four-step process for determining empirical formula from combustion analysis data.

1. 40.92% C + 4.58% H + 54.40 g O = 100% = 100.00%

2. Assume 100 g of compound (percentages then become grams). 40.92 of carbon; 4.58 g of hydrogen, 54.40 g of oxygen.

3. Calculate the moles of each element (moles = grams/atomic mass).

$$\text{Moles of C} = 40.92 \text{ g C} \times \frac{1 \text{ mol C}}{12.011 \text{ g C}} = 3.407 \text{ moles C}$$

$$\text{Moles of H} = 4.58 \text{ g H} \times \frac{1 \text{ mol H}}{1.0079 \text{ g H}} = 4.544 \text{ moles H}$$

$$\text{Moles of O} = 54.40 \text{ g O} \times \frac{1 \text{ mol O}}{15.999 \text{ g O}} = 3.400 \text{ moles O}$$

4. Divide by the smallest number to obtain the subscripts in the empirical formula. The smallest number is 3.400.

$$\text{Subscript for C:} \frac{3.407}{3.400} = 1$$

$$\text{Subscript for H:} \frac{4.544}{3.400} = 1.34$$

$$\text{Subscript for O:} \frac{3.400}{3.400} = 1$$

Multiplying each calculated subscript by three to make the subscripts whole numbers, the empirical formula is determined to be $C_3H_4O_3$.

Upon examining the ball and stick picture, we see that the molecular formula is $C_6H_8O_6$.

9.185 (a) Determining the mass percents.

First calculate the moles of hydrogen from the moles of water:

$$\text{Moles of water} = \frac{1.284 \text{ g H}_2\text{O}}{18.01 \text{ g/mole}} = 0.0713 \text{ moles of H}_2\text{O}$$

Moles of hydrogen $= 2 \times$ moles of $H_2O = 2 \times 0.0713 = 0.1426$ moles of H

$$\text{Grams of hydrogen} = 0.1426 \text{ moles} \times \frac{1.0079 \text{ g H}}{\text{mole H}} = 0.1437 \text{ g H}$$

Grams of carbon $= 1.000$ g sample $- 0.1426$ g H $= 0.85714$ g C

To get mass percents:

$$\% \text{ H} = \frac{0.1437 \text{ g H}}{1.000 \text{ g compound}} \times 100\% = 14.37\% \text{ H}$$

$$\% \text{ C} = \frac{0.85714 \text{ g H}}{1.000 \text{ compound}} \times 100\% = 85.714\% \text{ C}$$

(b) Determining the empirical formula.

Using the steps to calculate empirical formula from percent composition:

1. 85.71% C + 14.37% H = 100.1% (100% considering inaccuracies of measurement.)

2. Assume 100 g of compound; percentages then become grams: 85.71 g C; 14.37 g H.

3. Calculate the mole of each element (moles = grams/atomic mass).

$$\text{Moles of C} = 85.71 \text{ g C} \times \frac{1 \text{ mol C}}{12.011 \text{ g C}} = 7.142 \text{ moles C}$$

$$\text{Moles of H} = 14.37 \text{ g H} \times \frac{1 \text{ mol H}}{1.0079 \text{ g}} = 14.26 \text{ moles H}$$

4. Divide by the smallest number to obtain the subscripts in the empirical formula. The smallest number is 7.142.

$$\text{Subscript for C:} \frac{7.142}{7.142} = 1.00$$

$$\text{Subscript for H:} \frac{14.26}{7.142} = 2.00$$

The empirical formula is therefore: CH_2.

(c) **To determine the molecular formula:**

Divide the given molar mass of the compound by the molar mass of the empirical formula to determine the number that the subscripts in the empirical formula should be multiplied by:

$$\frac{71 \text{ g/mol}}{14 \text{ g/mol}} = 5; \text{ The molecular formula is therefore } C_{1\times5} H_{2\times5} \rightarrow C_5H_{10}.$$

Chapter 10

10.1   See solution in textbook.

10.3   (a) Dot diagram:

$$\begin{array}{c} \text{H} \\ \text{H} \!:\! \overset{..}{\underset{..}{\text{C}}} \!:\! \text{H} \\ \text{H} \end{array}$$

We determine oxidation number by subtracting the number of valence electrons assigned by oxidation-state bookkeeping from the number of valence electrons each free atom normally owns. Because C is more electronegative than H, all the electrons in the dot diagram are assigned to C and none is assigned to H:

$$\begin{array}{c} \text{H} \\ \\ \text{H} \quad \overset{..}{\underset{..}{:\text{C}:}} \quad \text{H} \\ \\ \text{H} \end{array}$$

Atom	Valence electrons in free atom		Electrons assigned by oxidation-state bookkeeping		Oxidation state
C	4	−	8	=	−4
Each H	1	−	0	=	+1

(b) Eight

(c) Four more because a free C atom has four valence electrons.

10.5   See solution in textbook.

10.7   Each O: −2 (rule 2; O assigned first because no rule for Fe); each Fe: +3 (rule 7: three O of oxidation state −2 is $3 \times (-2) = -6$, which means combined oxidation state of the two Fe must be +6: $+6/2 = +3$ for each Fe).

10.9 Each O: -2 (rule 2); C $+4$ (rule 7: three O of oxidation state -2 is $3 \times (-2) = -6$; charge on ion is $2-$, meaning charge on C must be such that C $+$ 3 O sum is -2: $+4 + (-6) = -2$).

10.11 See solution in textbook.

10.13 This is a redox reaction. The oxidation number of Ca changes from 0 to $+2$ (Ca loses two electrons), and the oxidation number of H changes from $+1$ to 0 (each H atom gains one electron). The Ca is oxidized, and the H is reduced.

10.15 See solution in textbook.

10.17 Ca is oxidized, and H is reduced (see Practice Problem 10.13). Therefore, Ca is the reducing agent, and H^+ is the oxidizing agent.

10.19 (a) Pb^{2+} (reduced from $+2$ to 0).

(b) Ni (oxidized from 0 to $+2$).

(c) The Pb cathode because Pb metal plates out on the electrode.

(d) The Ni anode because Ni metal atoms from the electrode dissolve and become Ni^{2+} ions in solution.

10.21 See solution in textbook.

10.23 Draw two beakers, one containing a solid Al electrode in a solution of Al^{3+} ions and the other containing a solid Zn electrode in a solution of Zn^{2+} ions. Connect the electrodes by a wire and the beakers by a salt bridge. Because Al is higher than Zn on the EMF list, electrons flow from the Al electrode to the Zn electrode, making Al the anode $(-)$ and Zn the cathode $(+)$. Al atoms lose electrons, which means they are oxidized. Zn^{2+} ions gain electrons, which means they are reduced.

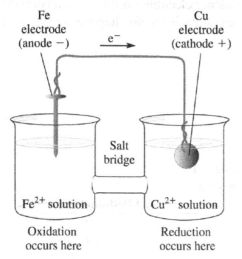

10.25 The electrical voltage is analogous to water pressure, and the electron flow is analogous to water flow.

10.27 One in which atoms of one of the reactants lose electrons, while atoms of some other reactant gain electrons.

10.29 Electron bookkeeping is a method of keeping track of all of the electrons owned by every atom on both sides of an equation representing a redox reaction.

10.31 Because the oxidation state of the atom is the difference between the number of valence electrons in the free atom and the number of valence electrons assigned to the atom by oxidation-state bookkeeping. The number of valence electrons in the free atom must be known in order to determine this difference.

10.33 Dot diagram:

$$:\ddot{F}-\overset{\textstyle\cdot\cdot}{N}-\ddot{F}:$$
$$\qquad\quad|$$
$$\qquad\quad H$$

Because F is more electronegative than N, each F owns eight electrons: all three of its lone pairs plus both electrons from the F–N bond. Because N is more electronegative than H, N owns four electrons: one lone pair plus both electrons from the N–H bond. Having given its one electron to the more electronegative N, H owns no electrons:

$$:\ddot{F}: \quad \overset{\textstyle\cdot\cdot}{N} \quad :\ddot{F}:$$
$$\qquad\quad H$$

10.35 Dot diagram:

$$\left[:\ddot{O}-N=\ddot{O}\right]^{-}$$
$$\qquad\ \ |$$
$$\qquad\ :\overset{\cdot\cdot}{O}:$$

Because O is more electronegative than N, each O owns eight electrons, all of its lone pairs plus all electrons from the N–O bonds. The N owns no electrons because it gave all its electrons from its N–O bonds to the more electronegative O:

$$\left[:\ddot{O}: \quad N \quad :\ddot{O}:\right]^{-}$$
$$\qquad\quad :\overset{\cdot\cdot}{O}:$$

10.37 (a) Dot diagram: H:H Each H owns one electron because the two atoms have the same electronegativity.

H: 1 free-atom valence e^- − 1 assigned e^- = 0 oxidation state.

(b) Dot diagram: $:\ddot{O}::\ddot{O}:$ Each O owns six electrons, four from its two lone pairs plus two from the double bond because the two atoms have the same electronegativity.

O: 6 free-atom valence e^- − 6 assigned e^- = 0 oxidation state.

(c) Dot diagram: $:\ddot{Cl}:\ddot{Cl}:$ Each Cl owns seven electrons, six from its three lone pairs plus one of the bond electrons.

Cl: 7 free-atom valence e^- − 7 assigned e^- = 0 oxidation state.

(d) Dot diagram: $:\ddot{I}-\ddot{F}:$ Fluorine, being more electronegative than iodine, owns eight electrons, six from its three lone pairs, plus two from the single bond. Iodine owns six electrons from its three lone pairs.

F: 7 free-atom valence e^- − 8 assigned e^- = −1 oxidation state.

I: 7 free-atom valence e^- − 6 assigned e^- = +1 oxidation state.

10.39 The sum must always equal the overall charge on the molecule.

10.41 (a) The equation for calculating oxidation state is "number of free-atom valence electrons-number of assigned electrons." In order for this quantity to be a negative number, as is the −2 oxidation state of this problem, the term "number of assigned electrons" must be greater than the term "number of free-atom valence electrons." Thus, any atom having a negative oxidation state has been assigned more electrons than the number of free-atom valence electrons.

(b) The −2 tells you the number of electrons assigned must be two more than the number of valence electrons in the free atom.

10.43 (a) Each Cl: -1 by the halide rule (applies because Cl is more electronegative than P); P: $+3$ by rule 7: $3 \times (-1) = -3, -3 + ? = 0, ? = +3$.

(b) Each H: $+1$ by rule 3; S: -2 by rule 7: $2 \times (+1) = +2, +2 + ? = 0, ? = -2$.

(c) Each O: -2 by rule 2; Mn: $+7$ by rule 7: $4 \times (-2) = -8, -8 + ? = -1, ? = +7$.

(d) H: $+1$ by rule 3; each O: -2 by rule 2; N: $+5$ by rule 7: $[1 \times (+1)] + [3 \times (-2)] = -5$, $-5 + ? = 0, ? = +5$.

(e) Each H: $+1$ by rule 3; each O: -2 by rule 2; C: $+2$ by rule 7: $[2 \times (+1)] + [2 \times -2)] = -2$, $-2 + ? = 0, ? = +2$.

(f) Each O: -2 by rule 2; each S: $+2$ by rule 7: $3 \times (-2) = -6, -6 + ? = -2, ? = +4$. This $+4$ oxidation state must be spread over the two S, giving each S a $+2$ state.

10.45 (a) O: -2 by rule 6.

(b) Each Li: $+1$ by rule 4; N: -3 by rule 7: $3 \times (+1) = +3, +3 + ? = 0, ? = -3$.

(c) Mg: $+2$ by rule 5; each O: -2 by rule 2; S: $+6$ by rule 7: $+2 + [4 \times (-2)] = -6$, $-6 + ? = 0, ? = +6$.

(d) Each O: -2 by rule 2; Mn: $+4$ by rule 7: $2 \times (-2) = -4; -4 + ? = 0, ? = +4$.

10.47 Recall from Chapter 5 that ozone has two resonance forms: $:\ddot{O}-\ddot{O}=\ddot{O}:$ and $:\ddot{O}=\ddot{O}-\ddot{O}:$

The oxidation numbers of the oxygen atoms are as follows: $\quad -1 \quad +1 \quad 0 \qquad 0 \quad +1 \quad -1$

The average oxidation number of the left terminal O is $(-1 + 0)/2 = -\frac{1}{2}$, on the central O is $(+1 + 1)/2 = +1$, and on the right terminal O is $(0 - 1)/2 = -\frac{1}{2}$. This is why the second student proposed the $-\frac{1}{2}$ oxidation states for the terminal O atoms.

10.49 Reduction is the gain of valence electrons by an atom or ion in an oxidation–reduction reaction. Oxidation is the loss of valence electrons by an atom or ion in an oxidation–reduction reaction.

10.51 If the oxidation number of any atom changes during a reaction, the reaction is a redox reaction.

10.53 Oxidation state increases when an atom is oxidized.

10.55 Because the oxidation and reduction that occur in the redox reaction involve a transfer of electrons, by definition. Electrons are transferred from the species being oxidized to the species being reduced.

10.57 (a) In reaction (a), the oxidation-state change $0 \rightarrow +1$ for Na means Na lost an electron and so was oxidized; the change $+1 \rightarrow 0$ for H means H gained an electron and so was reduced. In reaction (c), the change $+2 \rightarrow +4$ for C means C lost two electrons and so was oxidized; the change $0 \rightarrow -2$ for O means O gained two electrons and so was reduced.

(b) The oxidizing agent is the reactant that contains the atom that is reduced: H_2O in reaction (a) and O_2 in reaction (c).

(c) The reducing agent is the reactant that contains the atom that is oxidized: Na in reaction (a) and CO in reaction (c).

10.59 (a) In reaction (b), the oxidation-state change $0 \rightarrow +3$ for Fe means Fe lost electrons and so was oxidized; the change $+5 \rightarrow +2$ for N means N gained electrons and so was reduced.

In (c), the change $-3 \rightarrow +4$ for C means C lost electrons and so was oxidized; the change $0 \rightarrow -2$ for O means O gained electrons and so was reduced.

In (d), the change $+1 \rightarrow 0$ for Ag^+ means that Ag^+ gained an electron and so it was reduced; the change $-1 \rightarrow 0$ for Br^- means that Br^- lost an electron and so it was oxidized.

(b) The oxidizing agent is the reactant containing the atom that is reduced: NO_3^- in (b), O_2 in (c), and AgBr in (d).

(c) The reducing agent is the reactant containing the atom that is oxidized: Fe in (b), C_2H_6 in (c), and AgBr in (d). (AgBr acts as both reducing agent and oxidizing agent because it is the only reactant present. The Ag^+ is reduced and can therefore be thought of as the "true" oxidizing agent; the Br^- is oxidized and can be thought of as the "true" reducing agent.)

10.61 Yes, because the reaction is an electron-transfer (redox) reaction.

10.63 (a) In hydroquinone, the two C bonded to O are oxidized from $+1$ to $+2$. This can be seen when you assign oxidation numbers to the hydroquinone atoms and the quinone atoms:

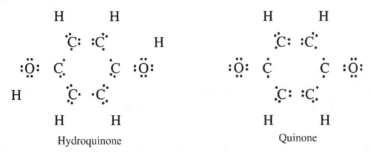

Hydroquinone   Quinone

In hydroquinone, the oxidation state of these two C is:

4 free-atom valence e^- − 3 assigned e^- = $+1$.

In quinone, their oxidation state is:

4 free-atom valence e^- − 2 assigned e^- = $+2$.

Each C has lost an electron and so has been oxidized.

(b) Ag is reduced, changing from Ag^+ as a reactant to Ag as a product. Each Ag^+ gained an electron.

(c) The oxidizing agent is the reactant containing the atom that is reduced, AgBr.

(d) The reducing agent is the reactant containing the atom that is oxidized, hydroquinone.

10.65

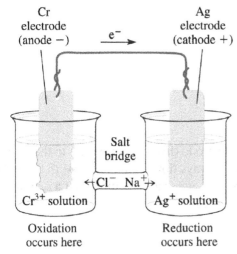

Cr is oxidized from 0 to $+3$. Ag^+ is reduced from $+1$ to 0. Cl^- ions from the salt bridge flow toward the anode solution to neutralize the newly formed Cr^{3+} anions. Na^+ ions, on the other hand, flow toward the cathode solution in order to replenish the positive charges, which have been diminished due to the reduction of Ag^+ ions to neutral $Ag(s)$.

10.67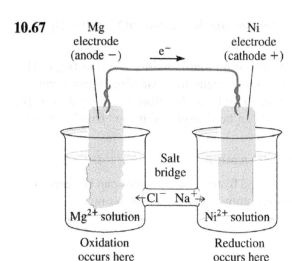

Mg is oxidized from 0 to +2. Ni^{2+} is reduced from +2 to 0. Cl^- ions from the salt bridge flow toward the anode solution to neutralize the newly formed Mg^{2+} anions. Na^+ ions, on the other hand, flow toward the cathode solution in order to replenish the positive charges, which have been diminished due to the reduction of Ni^{2+} to neutral $Ni(s)$.

10.69 (a) The increase in Sn^{2+} concentration tells you that the reaction $Sn \rightarrow Sn^{2+} + 2\,e^-$ is taking place, meaning that tin is being oxidized (from 0 to +2).

(b) The decrease in Cu^{2+} concentration tells you that the reaction $Cu^{2+} + 2\,e^- \rightarrow Cu$ is taking place, meaning that Cu^{2+} is being reduced (from +2 to 0).

(c)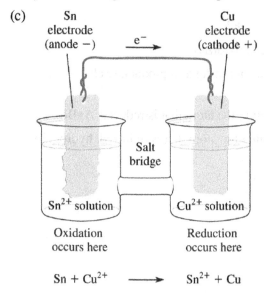

$$Sn + Cu^{2+} \longrightarrow Sn^{2+} + Cu$$

10.71 Electrons are negative and flow from the negative electrode (like charges repel) to the positive electrode (unlike charges attract). Because electrons always flow from the anode to the cathode, the anode must be negative and the cathode positive. Or, you can remember that a *cation* is positive (so is a *cathode*) and an *anion* is negative (so is an *anode*).

10.73 (a), (b), (d), (e), (f): See diagram below.

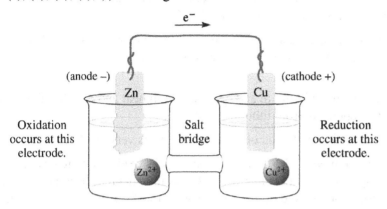

(c) $Zn + Cu^{2+} \rightarrow Zn^{2+} + Cu$.

(g) The Zn^{2+} concentration increases with time as Zn atoms are oxidized to Zn^{2+} ions.

(h) The Cu^{2+} concentration decreases with time as Cu^{2+} ions are reduced to Cu atoms.

(i) The Zn electrode is eaten away as Zn atoms leave the electrode and enter the solution as Zn^{2+} ions.

(j) The Cu electrode gains mass as Cu^{2+} plates out as Cu.

10.75 One could envision setting up a number of cells consisting of pairs of metals immersed in solutions of their respective ions and measuring electron flow. Spontaneous reactions—the only type that will take place—will always involve the more active metal getting oxidized (as evidenced by the electrode made from this metal getting pitted and losing its mass) and the less active metal getting reduced (as evidenced by the plating of its ions onto the surface of the electrode and the increasing mass of the electrode). Say, you have ten different metals (and their ion solutions + supplies); you can then construct as many as 45 different cells. Select one metal as a reference. Then construct nine cells, in which you utilize your reference electrode and the remaining metals. Keeping good records, note your nine observations. Then choose another metal and proceed to construct eight new cells (you do not need to construct the cell with the first reference metal, as it has already been observed before). When you have performed 45 measurements, you should be able to rank metals in terms of their activity and derive your version of the EMF series table. Of course, if you had a voltmeter and knew how to interpret the voltage, you could have saved yourself a lot of work, as the ranking would be based on voltmeter readings!

10.77 No reaction occurs.

10.79 (a) is spontaneous because Ag is more active than Au; (b) is not spontaneous for the same reason.

10.81

Ni electrode (anode −) e^- Cu electrode (cathode +)

Salt bridge

Ni^{2+} solution Cu^{2+} solution

Oxidation occurs here   Reduction occurs here

The reaction is $Cu^{2+} + Ni \rightarrow Cu + Ni^{2+}$ because Ni is higher than Cu on the EMF list. The Ni is oxidized from 0 to +2 and the Cu^{2+} is reduced from +2 to 0.

10.83 (a) Au is the lowest on the EMF list, meaning that Au^{3+} accepts electrons with the least resistance. Since Li has the greatest tendency to lose electrons, Li and Au should combine to give the highest-voltage battery.

   (b) Since water reacts with elemental lithium and generates both heat and a highly flammable $H_2(g)$, water should be kept away from lithium metal.

   (c) Anode; as represented by being at the top of the EMF list, Li is very easily oxidized to Li^+.

10.85 (a) Magnesium is relatively high on the EMF list and therefore is oxidized more easily than many other metals. When magnesium is in contact with a less active metal, it becomes an anode. The material of the anode is dissolved during the electron transfer from the anode to the cathode, and the mass of the anode decreases until there is none left, thus sacrificing itself to protect the metal it is attached to.

   (b) Iron is less active than magnesium so, when magnesium is in contact with iron, magnesium becomes an anode and iron becomes a cathode. First, this partially protects iron from getting oxidized, and second, when magnesium comes into contact with iron in its oxidized state (Fe^{2+}), the magnesium oxidizes and loses electrons to the iron, reducing it back to elemental iron.

   (c) The steel hull. Steel is in large part made of iron, which is less active than magnesium and will, therefore, become the positive electrode (i.e., the cathode), toward which the negatively charged electrons from magnesium flow.

   (d) To form a battery, the cathode and the anode must be separated from one another so that a wire can then connect them. In this case, the anode and cathode are in direct contact.

10.87 The statue is made of sheets of copper metal placed over a steel (iron) skeleton. Iron is more active than copper. When the copper sheets began to oxidize $(Cu \rightarrow Cu^{2+})$, the iron skeleton gave up its electrons to reduce the copper ions $(Cu^{2+} \rightarrow Cu)$, meaning that the iron skeleton was being eaten away as Fe atoms oxidized to Fe^{2+} ions. The solution to the problem was to replace the dissolving steel skeleton with a reinforced, corrosion-resistant, stainless steel skeleton incorporating high-tech plastics to isolate the steel from the copper.

10.89 +7 by rule 7: each H: +1; each O: −2; therefore because $[2 \times (+1)] + [6 \times (-2)] = -10$, I is $-10 + ? = -3, ? = +7$.

10.91 C in C_8H_{18} is oxidized; O in O_2 is reduced.

C_8H_{18}: The shortcut method, with +1 for each H, yields a fractional oxidation state for each C, making the electron-bookkeeping method a better choice. Assigning all electrons to the more electronegative C, the electron-assignment diagram is

H   H   H   H   H   H   H   H

H :C̈· ·C̈· ·C̈· ·C̈· ·C̈· ·C̈· ·C̈· ·C̈: H

H   H   H   H   H   H   H   H

Two terminal C: 4 free-atom valence e^- − 7 assigned $e^- = -3$.

Six interior C: 4 free-atom valence e^- − 6 assigned $e^- = -2$.

Each H: 1 free-atom valence e^- – 0 assigned $e^- = +1$.

CO_2: each O: –2 by rule 2; C: +4 by rule 7: $2 \times (-2) = -4, -4 + ? = 0, ? = +4$.

The C atoms in C_8H_{18} are oxidized either from –3 to +4 or from –2 to +4.

O_2: each O: 0 by rule 1.

H_2O: each H: +1 by rule 3; O: –2 by rule 2.

CO_2: each O: –2 by rule 2.

The O atoms in O_2 are reduced from 0 to –2.

10.93 Spontaneous reactions occur in (c) because Zn is more active than Ag and in (e) because Cu is more active than Hg. In (a), (b), and (d), the metal of which the strip is made is less active than the metal whose ions are in solution.

10.95 Li^+. To reduce an ion means to add electrons to it. The fact that Li metal is easier to oxidize than Mn, Hg, Fe, or Mg (Li highest in textbook EMF series) means the reduction $Li^+ + e^- \rightarrow Li$ is most difficult to accomplish.

10.97 Because copper is the anode, it is being oxidized to Cu^{2+} and losing electrons. At the cathode, Rh^{3+} ions gain electrons to form Rh metal. Because copper metal can give electrons to Rh^{3+} ions, rhodium must be lower in the EMF series than copper. (The series shown in the textbook is not exhaustive, which is why Rh is not shown.)

10.99 (a) and (c).

(a) SeO_3^{2-}: each O: –2 by rule 2; Se: +4 by rule 7: $3 \times (-2) = -6, -6 + ? = -2, ? = +4$.

Se: 0 by rule 1.

Se is reduced from +4 to 0, making SeO_3^{2-} the oxidizing agent (which is the reactant containing the atom that is reduced).

I^-: –1 by rule 6.

I_2: each I has oxidation number of 0 by rule 1.

I is oxidized from –1 to 0, making I^- the reducing agent (which is the reactant containing the atom that is oxidized).

(c) HCl: H: +1 by rule 3; Cl: –1 by either rule 6 or halide rule.

Cl_2: each Cl has oxidation number of 0 by rule 1.

Cl is oxidized from –1 to 0, making HCl the reducing agent (which is the reactant containing the atom that is oxidized).

O_2: each O has oxidation number of 0 by rule 1.

H_2O: each H: +1 by rule 3; O: –2 by rule 2.

O is reduced from 0 to –2, making O_2 the oxidizing agent (the reactant containing the atom that is reduced).

10.101 Electron assignment:

$$\begin{array}{ccc}
 & \ddot{O} & \\
 & \dot{C} & \\
H \quad \ddot{N} & \dot{C} & \ddot{F} \\
C & \ddot{N} & \ddot{C} \\
\ddot{O} & \ddot{N} & H \\
 & H &
\end{array}$$

Each O: 6 free-atom valence e^- − 8 assigned e^- = −2

Each H: 1 free-atom valence e^- − 0 assigned e^- = +1

F: 7 free-atom valence e^- − 8 assigned e^- = −1

Each N: 5 free-atom valence e^- − 8 assigned e^- = −3

12 o'clock C: 4 free-atom valence e^- − 1 assigned e^- = +3

2 o'clock C: 4 free-atom valence e^- − 3 assigned e^- = +1

4 o'clock C: 4 free-atom valence e^- − 4 assigned e^- = 0

8 o'clock C: 4 free-atom valence e^- − 0 assigned e^- = +4

10.103 (a) Each H: +1 by rule 3; each C: −3 by rule 7: $6 \times (+1) + ? = 0, ? = -6$, making each C $-6/2 = -3$.

(b) Each H: +1 by rule 3; each C: −2 by rule 7: $4 \times (+1) + ? = 0, ? = -4$, making each C $-4/2 = -2$.

(c) Each H: +1 by rule 3; each C: −1 by rule 7: $2 \times (+1) + ? = 0, ? = -2$, making each C $-2/2 = -1$.

10.105 Na_2SO_3:

Each Na: +1 by rule 4; each O: −2 by rule 2; S: +4 by rule 7: $[2 \times (+1)] + [3 \times (-2)] = -4$, $-4 + ? = 0, ? = +4$.

$KMnO_4$:

K: +1 by rule 4; each O: −2 by rule 2; Mn: +7 by rule 7: $[1 \times (+1)] + [4 \times (-2)] = -7$, $-7 + ? = 0, ? = +7$.

H_2O:

Each H: +1 by rule 3; O: −2 by rule 2.

Na_2SO_4:

Each Na: +1 by rule 4; each O: −2 by rule 2; S: +6 by rule 7: $[2 \times (+1)] + [4 \times (-2)] = -6$, $-6 + ? = 0, ? = +6$.

MnO_2:

Each O: −2 by rule 2; Mn: +4 by rule 7: $2 \times (-2) = -4, -4 + ? = 0, ? = +4$.

KOH:

K: +1 by rule 4; O: −2 by rule 2; H: +1 by rule 3.

Mn is reduced from +7 to +4, making $KMnO_4$ the oxidizing agent (which is the reactant containing the atom that is reduced).

S is oxidized from +4 to +6, making Na_2SO_3 the reducing agent (which is the reactant containing the atom that is oxidized).

10.107 (a) Each F: -1 by the halide rule, which applies because P is less electronegative than F; P: $+5$ by rule 7: $6 \times (-1) = -6, -6 + ? = -1, ? = +5$.

(b) Each O: -2 by rule 2; each Mo: $+6$ by rule 7: Mo_2 $7 \times (-2) = -14, -14 + ? = -2$, $? = +12$, making each Mo: $+12/2 = +6$.

(c) H: $+1$ by rule 3; each O: -2 by rule 2; Pb: $+2$ by rule 7: $[1 \times (+1)] + [2 \times (-2)] = -3, -3 + ? = -1, ? = +2$.

(d) H: $+1$ by rule 3; each O: -2 by rule 2; each C: $+3$ by rule 7: C_2 $[1 \times (+1)] + [4 \times (-2)] = -7, -7 + ? = -1, ? = +6$, making each C: $+6/2 = +3$.

10.109 (a) Because it is the cathode, the nickel electrode gains electrons converting Ni^{2+} to $Ni(s)$. This conversion lowers the amount of positive charge on the cathode side, and to compensate K^+ ions flow toward the cathode to neutralize the buildup of negative charge.

(b) Because it is the anode, the zinc electrode gives up electrons as Zn atoms are converted to Zn^{2+} ions in solution. This conversion raises the amount of positive charge on the anode side, and to compensate Cl^- ions flow toward the anode to neutralize the buildup of positive charge.

10.111 IO_4^-: each O: -2 by rule 2; I: $+7$ by rule 7: $4 \times (-2) = -8, -8 + ? = -1, ? = +7$

I^-: -1 by rule 6

H^+: $+1$ by rule 3

I_2: each I has the oxidation number of 0 by rule 1

H_2O: each H: $+1$ by rule 3, O: -2 by rule 2

The only changes in oxidation number are for I, which appears in both I^- and IO_4^-. The I in IO_4^- is reduced from $+7$ to 0, meaning IO_4^- is the oxidizing agent (which is the reactant containing the atom that is reduced). The I^- is oxidized from -1 to 0, meaning it is the reducing agent (which is the reactant containing the atom that is oxidized).

10.113 Only when the fluorine is in F_2, where, by rule 1, each F has an oxidation state of 0. Otherwise, fluorine will always get complete ownership of all bonding electrons because it is the most electronegative atom. Owning all bonding electrons (eight) always gives fluorine an oxidation state of -1 because the free atom has seven valence electrons.

10.115 Dot diagram: $H-C\equiv\overset{..}{N}$ Electron assignment: $H \quad :C \quad :\overset{..}{\underset{..}{N}}:$

H: 1 free-atom valence e^- $-$ 0 assigned e^- $= +1$

C: 4 free-atom valence e^- $-$ 2 assigned e^- $= +2$

N: 5 free-atom valence e^- $-$ 8 assigned e^- $= -3$

10.117 $+4$ to -4. In a compound such as CF_4, carbon has no bonding electrons assigned to it because fluorine is more electronegative: 4 free-atom valence e^- $-$ 0 assigned e^- $= +4$. In a compound such as CH_4, carbon is assigned all eight bonding electrons because hydrogen is less electronegative than carbon: 4 free-atom valence e^- $-$ 8 assigned e^- $= -4$.

10.119 Because zinc is a more active metal than iron. If any iron is oxidized to Fe^{2+}, the zinc metal transfers electrons to the Fe^{2+} and reduces it back to iron metal. In this way zinc acts as a sacrificial electrode. The zinc ions formed from the zinc metal form a protective coating of $Zn_2(OH)_2CO_3$.

10.121 (a) Not an electron-transfer reaction; P stays $+3$, F stays -1, H stays $+1$, O stays -2.

    (b) H_2: each H has the oxidation number of 0 by rule 1.

       Cl_2: each Cl has the oxidation number of 0 by rule 1.

       HCl: H: $+1$ by rule 1, Cl: -1 by rule 7 or halide rule.

       H_2 is oxidized as H changes from 0 in H_2 to $+1$ in HCl; thus, H_2 is the reducing agent. Cl_2 is reduced as Cl changes from 0 in Cl_2 to -1 in HCl; thus, Cl_2 is the oxidizing agent.

    (c) Cr_2O_3: each O: -2 by rule 2; each Cr: $+3$ by rule 7: Cr_2 $3 \times (-2) = -6, -6 + ? = 0$, $? = +6$, making each Cr $+6/2 = +3$.

       Si: 0 by rule 1.

       Cr: 0 by rule 1.

       SiO_2: each O: -2 by rule 2; Si: $+4$ by rule 7: $2 \times (-2) = -4, -4 + ? = 0, ? = +4$.

       Cr_2O_3 is reduced as Cr changes from $+3$ in Cr_2O_3 to 0 in Cr; thus, Cr_2O_3 is the oxidizing agent. Si is oxidized from 0 in Si to $+4$ in SiO_2; thus, Si is the reducing agent.

    (d) Not an electron-transfer reaction; H stays $+1$, Cl stays -1, Na stays $+1$, O stays -2.

10.123 (a) Na: $+1$ by rule 4; O: -2 by rule 2; P: $+5$ by rule 7: $[1 \times (+1)] + [3 \times (-2)] = -5$, $-5 + ? = 0, ? = +5$.

    (b) Each O: -2 by rule 2; each H: $+1$ by rule 3; B: $+3$ by rule 7: $[3 \times -2)] + [3 \times (+1)] =$ $-3, -3 + ? = 0, ? = +3$. (Alternatively, you can say that B, which must carry a $3+$ ionic charge to balance the 3 OH^-, is $+3$ by rule 6.)

    (c) Each O: -2 by rule 2; each V: $+5$ by rule 7: V_2 $5 \times (-2) = -10, -10 + ? = 0, ? = +10$, making each V $+ 10/2 = +5$.

    (d) Each K: $+1$ by rule 4; each F: -1 by the halide rule; Ti: $+4$ by rule 7: $[2 \times (+1)] + [6 \times (-1)] = -4, -4 + ? = 0, ? = +4$.

10.125 That the Zn^{2+} concentration decreases tells you Zn^{2+} is being reduced to Zn metal atoms. That the Ti^{2+} concentration increases tells you Ti metal atoms are being oxidized to Ti^{2+} ions.

    (a) The titanium electrode is the anode because the anode is where oxidation occurs.

    (b) The zinc electrode is the cathode because the cathode is where reduction occurs.

    (c) Because titanium gives up electrons to Zn^{2+}, titanium must be higher on the EMF scale than zinc.

    (d) $Ti + Zn^{2+} \rightarrow Ti^{2+} + Zn$

10.127 Yes, it is redox, because oxidation numbers are changing. In H_2O_2, each oxygen atom has an oxidation state of -1. On the product side, the oxygen atom in water has an oxidation state of -2, and each oxygen atom in O_2 has an oxidation state of zero. The vapor is most likely water vapor, rich in O_2 gas.

10.129 Not using the shortcut rules achieves the oxidation states shown below. You achieve the same result with the shortcut rules if you use the halide rule for the F only (logical, as it is more electronegative than I).

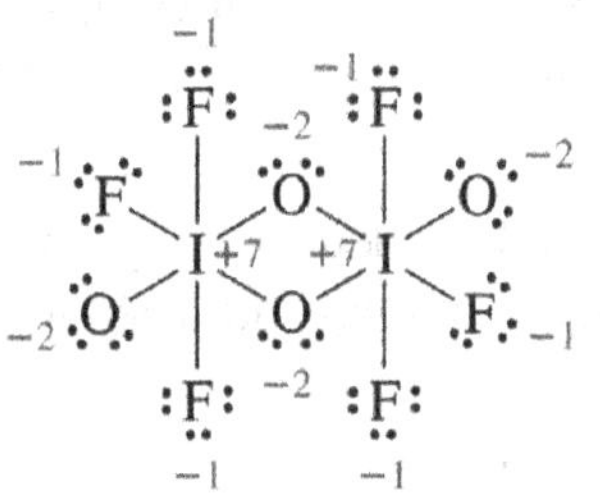

The red (central) carbon gets oxidized and the blue (outer) carbon atom gets reduced. Proof is supplied by the way the oxidation states of these atoms change.

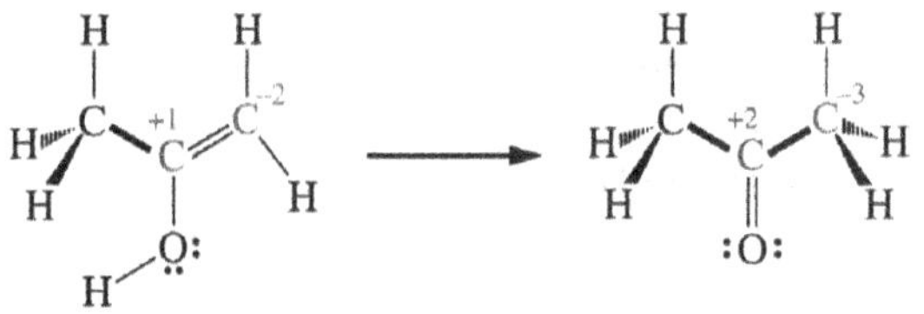

10.131 In a combustion reaction that involves reaction of a substance containing carbon with O_2, the oxidation number of carbon increases when CO_2 is produced. When methane reacts with Cl_2 to form chlorinated compounds, the oxidation number of carbon also increases. Therefore, the word "burn" can be used when talking about both of these combustion reactions.

10.133 (a) Oxidation occurs with the oxidation number increases when going from reactant to product. The oxidation state of C in ethane is -3. The oxidation state of C in the aldehyde is -1, and the oxidation state of C in the alcohol is -2. Both -1 and -2 are greater than -3, so oxidation is occurring. The C in ethane must lose two electrons to produce the aldehyde (to go from -3 to -1), and it must lose one electron to produce the alcohol (to go from -3 to -2).

(b) The aldehyde is the more oxidized form. It has lost more negative charge. The C in ethane has lost two electrons to produce the aldehyde (from -3 to -1), and it lost one electron to produce the alcohol (from -3 to -2).

10.135 The conclusion would be the same—both H and F would still be reduced. In H_2O_2, all electrons would be assigned to the O (more electronegative element), giving each H an oxidation state of $+1$ and each O an oxidation state of -1. Therefore, H would be going from a state of $+1$ to 0 during the reaction. In F_2O_2 all electrons would be assigned to the F (more electronegative element), giving each O an oxidation state of $+1$ and each F an oxidation state of -1. Therefore, F would be going from a state of 0 to -1 during the reaction.

10.137 (a) Based on the reactions presented, W reduces X, X reduces Z, Y reduces Z, X reduces Y, so the order is W, X, Y, and Z from most active to least active.

(b) Neither of the reactions would be expected to occur.

The first one is nonspontaneous because Y will not reduce W.

The second one is nonspontaneous because Z will not reduce W.

10.139 (a) ClO_4^-: each O: -2 by rule 2; Cl: $+7$ by rule 7: $4 \times (-2) = -8, -8 + ? = -1, ? = +7$

(b) Halide rule does not work here, as Cl is bonded to the more electronegative O atoms.

(c) With such a large positive oxidation number $(+7)$, it is expected that Cl will vigorously draw electrons toward itself, oxidizing other molecules. So, the perchlorate ion is expected to be a powerful oxidizing agent.

(d) Being in the VII A group, chlorine has seven valence electrons. It is therefore impossible for chlorine to lose more than seven electrons (core electrons are too strongly attracted to the nucleus to be removed).

Chapter 11

11.1 See solution in textbook.

11.3 (a) 5×760 mm Hg $= 3800$ mm Hg $= 3.80 \times 10^3$ mm Hg

(b) 3.80×10^3 mm Hg $\times \dfrac{1 \text{ cm}}{10 \text{ mm}} \times \dfrac{1 \text{ in.}}{2.54 \text{ cm}} = 150$ in Hg

11.5 See solution in textbook.

11.7 $P = \dfrac{nRT}{V}$

$n = \dfrac{PV}{RT}$

Pressure in atm $= 40.0 \text{ lb/in.}^2 \times \dfrac{760 \text{ mm Hg}}{14.696 \text{ lb/in.}^2} \times \dfrac{1 \text{ atm}}{760 \text{ mm Hg}} = 2.72$ atm

Because 760 mm Hg $= 1$ atm, you could also use the conversion factor
1 atm $= 14.696 \text{ lb/in.}^2$:

$40.0 \text{ lb/in.}^2 \times \dfrac{1 \text{ atm}}{14.696 \text{ lb/in.}^2} = 2.72$ atm

Volume in L $= 10.5 \text{ gal} \times \dfrac{3.785 \text{ L}}{1 \text{ gal}} = 39.7$ L

$22.5\,°C + 273.15 = 295.7$ K

$n = \dfrac{PV}{RT} = \dfrac{2.72 \text{ atm} \times 39.7 \text{ L}}{\dfrac{0.0821 \text{ L} \cdot \text{atm}}{\text{K} \cdot \text{mol}} \times 295.7 \text{ K}} = 4.45$ mol O_2

$4.45 \text{ mol } O_2 \times \dfrac{31.998 \text{ g } O_2}{1 \text{ mol } O_2} = 142 \text{ g } O_2$

11.9 See solution in textbook.

11.11 Mass divided by molar mass gives number of moles:

$\dfrac{m}{MM} = \dfrac{g}{g/\text{mol}} = g \div \dfrac{g}{\text{mol}} = g \times \dfrac{\text{mol}}{g} = \text{mol}$

11.13 See solution in textbook.

11.15 Because the molecules in the gas phase are moving rapidly throughout the container. There is no order to their movement; it is completely random. In the liquid phase, the molecules are confined to a smaller space and are moving throughout the liquid only rather than throughout the container. In the solid phase, the molecules remain in relatively fixed positions.

11.17 Yes, the gas will leak out even though the pressure inside the container is the same as the pressure outside. Because all the gas molecules in the container are moving along random straight-line paths, some will find themselves on paths that carry them through the tiny hole.

11.19 Because the air pressure above the liquid inside the straw has been decreased. (You have created a near-vacuum inside the straw.) The unchanged air pressure outside the straw pushes down on the liquid outside the straw, forcing some of it up into the straw.

11.21 760 mm Hg; 1 atm.

11.23 (a) $29.7 \text{ in. Hg} \times \dfrac{2.54 \text{ cm}}{1 \text{ in.}} \times \dfrac{10 \text{ mm}}{1 \text{ cm}} = 754 \text{ mm Hg}$

 (b) $754 \text{ mm Hg} \times \dfrac{1 \text{ atm}}{760 \text{ mm Hg}} = 0.992 \text{ atm}$

11.25 By adding 273.15 to the Celsius temperature: $22.0\,°C + 273.15 = 295.2 \text{ K}$.

11.27 $K = °C + 273.15 = -2.0\,°C + 273.15 = 271.2 \text{ K}$

$2.35 \text{ g CH}_4 \times \dfrac{1 \text{ mol CH}_4}{16.04 \text{ g CH}_4} = 0.147 \text{ mol CH}_4$

11.29 The pressure decreases. The pressure is due to collisions between the gas molecules and the walls of the container. If the container is bigger, the molecules have farther to go before they hit the walls and therefore hit the walls less frequently. Less-frequent collisions result in lower pressure.

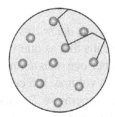

Large volume,
less frequent collisions with walls,
low pressure

Small volume,
frequent collisions with walls,
high pressure

11.31 The pressure increases. The pressure is due to collisions between the gas molecules and the walls of the container. If there are more molecules inside the container, there are more collisions per unit time and the pressure increases.

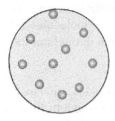

Few gas molecules,
few collisions with walls,
low pressure

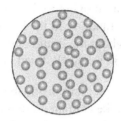

Many gas molecules,
many collisions with walls,
high pressure

11.33 Sample 2. In fact, the temperature of this sample has been lowered to the point where the gas has condensed to the liquid phase, which is why the molecules are clumped together at the bottom of the container. Because sample 2 is really a liquid and not a gas, we used quotation marks for "gas."

11.35 From the ideal gas law: $PV = nRT$, when n and V are constant:

$$P = \frac{nRT}{V} = \underbrace{\frac{nR}{V}}_{\text{constant}} T \Rightarrow P \propto T$$

The pressure of a gas is directly proportional to temperature: P is proportional to T.

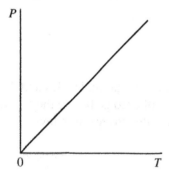

11.37 Inverse proportionality means that as one value *increases*, the other one *decreases*; or that as one value *decreases*, the other one *increases*. Direct proportionality means that as one value *increases*, the other one *also increases*; or that as one value *decreases*, the other one *also decreases*. An example of inverse proportionality is the way gas pressure varies with volume. If a certain gas sample exerts a pressure of 4 atm when it occupies a volume of 10 L at a temperature T, it will exert a pressure of 2 atm when it occupies a volume of 20 L at the same temperature T. An example of direct proportionality is the relationship between gas volume and temperature. If a gas occupies a volume of 3 L at 200 K and a pressure P, it will occupy a volume of 6 L at 400 K and the same pressure P.

11.39 Inverse proportion. As his age increases, his hair decreases.

11.41 Liters-atmospheres ÷ Kelvin-moles: $L \cdot atm / K \cdot mol$. The units are important because they indicate the units in which the volume, pressure, temperature, and quantity of gas must be expressed when using the ideal gas law. The volume must be in *liters*, the pressure in *atmospheres*, the temperature in *Kelvin*, and the quantity of gas in *moles*.

11.43 To derive the formula and the units of T, manipulate the ideal gas law, as shown below:

$$P = \frac{nRT}{V}$$

$$PV = \frac{nRT\cancel{V}}{\cancel{V}}$$

$$PV = nRT$$

$$\frac{PV}{nR} = \frac{\cancel{n}\cancel{R}T}{\cancel{n}\cancel{R}}$$

$$\frac{PV}{nR} = T, \text{ or, more conventionally, } T = \frac{PV}{nR}$$

$$\text{units of } T = \frac{\cancel{atm} \cdot \cancel{L}}{\cancel{mol} \cdot \dfrac{\cancel{L}\,\cancel{atm}}{\cancel{mol}\, K}} = \frac{1}{\dfrac{1}{K}} = K$$

11.45 (a) You should get the value of R, as you can see by solving the ideal gas equation for R:

$$P = \frac{nRT}{V}$$

$$PV = \frac{nRT\cancel{V}}{\cancel{V}}$$

$$PV = nRT$$

$$\frac{PV}{nT} = \frac{\cancel{n}R\cancel{T}}{\cancel{n}\cancel{T}}$$

$$\frac{PV}{nT} = R$$

For an ideal gas, the ratio PV/nT *always* gives the numeric value 0.0821 (and the units, of course, are $L \cdot atm/K \cdot mol$).

(b) No units; numerical value of 1.

11.47 The pressure would increase sixfold. According to the ideal gas law, $P = nRT/V$, doubling n doubles the pressure and tripling T triples the pressure. Making both changes at once results in a sixfold increase in pressure:

$$P = \frac{nRT}{V}$$

$$\frac{(2n)\,R(3T)}{V} = \frac{6nRT}{V} = 6P$$

11.49 You use the ideal gas law solved for n. First, though, you must change the pressure to atmospheres, the volume to liters, and the temperature to Kelvin.

$$745.5 \text{ mm Hg} \times \frac{1 \text{ atm}}{760.0 \text{ mm Hg}} = 0.9809 \text{ atm}$$

$$250.0 \text{ mL} \times \frac{1 \text{ L}}{1000 \text{ mL}} \times 0.2500 \text{ L}$$

$$25.5 \,°C + 273.15 = 298.7 \text{ K}$$

$$n = \frac{PV}{RT} = \frac{0.9809 \text{ atm} \times 0.2500 \text{ L}}{\dfrac{0.0821 \text{ L} \cdot \text{atm}}{\text{K} \cdot \text{mol}} \times 298.7 \text{ K}} = 0.0100 \text{ mol}$$

11.51 $T = \dfrac{PV}{nR} = \dfrac{100 \text{ atm} \times 4.0 \text{ L}}{2.0 \text{ mol} \times \dfrac{0.0821 \text{ L} \cdot \text{atm}}{\text{K} \cdot \text{mol}}} = 2.4 \times 10^3 \text{ K}$

$$°C = K - 273.15 = 2.4 \times 10^3 - 273.15 = 2.4 \times 10^3 - 0.27315 \times 10^3 = 2.1 \times 10^3 \,°C$$

11.53 (a) Before using the ideal gas law solved for moles, you must change the pressure to atmospheres and the temperature to Kelvin:

$$30 \text{ psi} \times \frac{1 \text{ atm}}{14.696 \text{ psi}} = 2.0 \text{ atm}$$

$$22 \,°C + 273.15 = 295 \text{ K}$$

$$n = \frac{PV}{RT} = \frac{2.0 \text{ atm} \times 20.0 \text{ L}}{\dfrac{0.0821 \text{ L} \cdot \text{atm}}{\text{K} \cdot \text{mol}} \times 295 \text{ K}} = 1.7 \text{ mol}$$

(b) $1.7 \text{ mol N}_2 \times \dfrac{28.01 \text{ g N}_2}{1 \text{ mol N}_2} = 48 \text{ g N}_2$

$$48 \text{ g N}_2 \times \frac{1 \text{ lb N}_2}{453.6 \text{ g N}_2} = 0.11 \text{ lb N}_2$$

11.55 (a) Use the ideal gas equation solved for V: $V = nRT/P$. All quantities are in the correct units except temperature.

$$22.5 \,°C + 273.15 = 295.7 \text{ K}$$

$$V = \frac{nRT}{P} = \frac{1 \text{ mol} \times \dfrac{0.0821 \text{ L} \cdot \text{atm}}{\text{K} \cdot \text{mol}} \times 295.7 \text{ K}}{1.00 \text{ atm}} = 24.3 \text{ L H}_2$$

(b) $1 \text{ mol H}_2 \times \dfrac{2 \text{ mol HCl}}{1 \text{ mol H}_2} = 2 \text{ mol HCl}$ or $1 \text{ mol Cl}_2 \times \dfrac{2 \text{ mol HCl}}{1 \text{ mol Cl}_2} = 2 \text{ mol HCl}$

$$V = \frac{nRT}{P} = \frac{2 \text{ mol} \times \dfrac{0.0821 \text{ L} \cdot \text{atm}}{\text{K} \cdot \text{mol}} 295.7 \text{ K}}{1.00 \text{ atm}} = 48.6 \text{ L HCl}$$

11.57 Use the ideal gas law expressed in terms of sample mass and molar mass. The given temperature and pressure must be converted to the proper units.

$25.0 \,^\circ\text{C} + 273.15 = 298.2 \text{ K}$

$$765.0 \text{ mm Hg} \times \frac{1 \text{ atm}}{760.0 \text{ mm Hg}} = 1.007 \text{ atm}$$

$$\frac{m}{MM} = \frac{PV}{RT}$$

$$mRT = (MM)PV$$

$$MM = \frac{mRT}{PV} = \frac{7.24 \text{ g} \times 0.0821\dfrac{\text{L} \cdot \text{atm}}{\text{K} \cdot \text{mol}} \times 298.2 \text{ K}}{1.007 \text{ atm} \times 4.00 \text{ L}} = 44.0 \text{ g/mol}$$

11.59 Use the ideal gas law expressed in terms of molar mass and sample mass. The given volume, pressure, and temperature must be converted to the proper units, and the molar mass of H_2 is 2.0158 g/mol.

$$1610.2 \text{ mL} \times \frac{1 \text{ L}}{1000 \text{ mL}} = 1.6102 \text{ L}$$

$$745.4 \text{ mm Hg} \times \frac{1 \text{ atm}}{760.0 \text{ mm Hg}} = 0.9808 \text{ atm}$$

$22.7 \,^\circ\text{C} + 273.15 = 295.9 \text{ K}$

$$\frac{m}{MM} = \frac{PV}{RT}$$

$$m = \frac{(MM)PV}{RT} = \frac{2.0158 \text{ g/mol} \times 0.9808 \text{ atm} \times 1.6102 \text{ L}}{0.0821\dfrac{\text{L} \cdot \text{atm}}{\text{K} \cdot \text{mol}} \times 295.9 \text{ K}} = 0.131 \text{ g}$$

11.61 (a) The definition of density is mass divided by volume. You know a form of the ideal gas law that contains m and V, and if you solve this equation for m/V, you'll have an expression that tells
you something about the density of these two gases.

$$\frac{m}{MM} = \frac{PV}{RT}$$

$$\frac{m}{V} = \frac{(MM)P}{RT} = \text{Density}$$

P and T are the same in the two balloons, and R is a constant. Therefore you can ignore them and focus on the relationship between MM and density. MM being in the numerator means density is directly proportional to MM. Because the molar masses are 2.0158 g/mol for H_2 and 4.003 g/mol for He, the density must be greater in the He balloon.

(b) H_2 density $= 2.0158 \text{ g/mol} \times \dfrac{P}{RT}$

He density $= 4.003 \text{ g/mol} \times \dfrac{P}{RT}$

$\dfrac{4.003}{2.0158} = 1.986$

The He density is 1.986 times greater than the H_2 density.

11.63 $10.0\,°C + 273.15 = 283.2 \text{ K}$

$20.0\,°C + 273.15 = 293.2 \text{ K}$

$50.5 \text{ lb/in.}^2 \times \dfrac{1 \text{ atm}}{14.70 \text{ lb/in.}^2} = 3.44 \text{ atm}$

Step 1: $P_i = 3.44 \text{ atm} \qquad P_f = ?$

$\qquad\qquad V_i = ? \qquad\qquad\quad V_f = ? = V_i$

$\qquad\qquad n_i = ? \qquad\qquad\quad n_f = ? = n_i$

$\qquad\qquad T_i = 283.2 \text{ K} \qquad T_f = 293.2 \text{ K}$

Step 2: $\dfrac{P_i V_i}{n_i T_i} = \dfrac{P_f V_f}{n_f T_f}$

$\qquad\qquad \dfrac{P_i V_i}{n_i T_i} = \dfrac{P_f V_i}{n_i T_f}$

$\qquad\qquad \dfrac{P_i}{T_i} = \dfrac{P_f}{T_f}$

Step 3: $P_f = \dfrac{T_f P_i}{T_I} = \dfrac{293.2 \text{ K} \times 3.44 \text{ atm}}{283.2 \text{ K}} = 3.56 \text{ atm}$

$3.56 \text{ atm} \times \dfrac{14.70 \text{ lb/in.}^2}{1 \text{ atm}} = 52.3 \text{ lb/in.}^2$

11.65 When the temperature in *Kelvin* doubles, the average energy of motion doubles as well. Since the mass of particles does not change upon heating, the average speed with which the gas particles move must increase (though it does not double!). Assuming there is no change in volume of the gas sample, this additional speed with which the molecules move results in (i) more collisions with the walls of the container in a given time period and (ii) more forceful collisions. Both effects explain the increase in pressure.

11.67 Step 1: $P_i = ? \quad P_f = ?$

$\qquad\qquad V_i = ? \quad V_f = ? = 0.5V_i$

$\qquad\qquad n_i = ? \quad n_f = ? = n_i$

$\qquad\qquad T_i = ? \quad T_f = ? = 2T_i$

Step 2: $\dfrac{P_iV_i}{n_iT_i} = \dfrac{P_fV_f}{n_fT_f}$

$$\dfrac{P_i\cancel{V_i}}{\cancel{n_i}\cancel{T_i}} = \dfrac{P_f \times 0.5\,\cancel{V_i}}{\cancel{n_i} \times 2\,\cancel{T_i}}$$

$$P_i = \dfrac{0.5P_f}{2}$$

$$\dfrac{2\,P_i}{0.5} = P_f = 4\,P_i$$

The final pressure is four times greater than the initial pressure.

11.69 (a) The reaction stoichiometry tells you that it takes 1 mole of $O_2(g)$ to produce 2 moles of $SO_3(g)$. At STP, 1 mole of O_2 gas has a volume of 22.4 L.

    (b) The required amount is still 1 mole of H_2O, but now conditions are not STP, which means you must use the ideal gas equation to calculate the required volume.

    $25.0\,^\circ C + 273.15 = 298.2\ K$

$$V = \dfrac{nRT}{P} = \dfrac{1\ \cancel{mol} \times 0.0821\dfrac{L\cdot\cancel{atm}}{\cancel{K}\cdot\cancel{mol}} \times 298.2\ \cancel{K}}{1\ \cancel{atm}} = 24.5\ L$$

11.71 First find out how much H_2 gas must be present to create the given pressure in the cylinder at the given temperature. Then use stoichiometry to find out the mass of Zn needed to produce that amount of H_2 gas.

    $25.0\,^\circ C + 273.15 = 298.2\ K$

$$n = \dfrac{PV}{RT} = \dfrac{10.0\ \cancel{atm} \times 50.0\ \cancel{L}}{0.0821\,\dfrac{\cancel{L}\cdot\cancel{atm}}{\cancel{K}\cdot mol} \times 298.2\ \cancel{K}} = 20.4\ mol\ H_2$$

    $20.4\ \cancel{mol\ H_2} \times \dfrac{1\ \cancel{mol\ Zn}}{1\ \cancel{mol\ H_2}} \times \dfrac{65.39\ g\ Zn}{1\ \cancel{mol\ Zn}} = 1.33 \times 10^3\ g\ Zn$

11.73 Because the model assumes that all gases consist of identical particles that have no attraction for one another and do not interact with one another in any way. Because there is no interaction, the chemical identity of the particles does not matter.

11.75 Convert the given data to match the units in R, use the ideal gas law to find the number of moles of H_2 produced, then convert moles to grams.

    $100.0\ \cancel{mL} \times \dfrac{1\ L}{1000\ \cancel{mL}} = 0.1000\ L$

    $745.5\ \cancel{mm\ Hg} \times \dfrac{1\ atm}{760.0\ \cancel{mm\ Hg}} = 0.9809\ atm$

    $24.0\,^\circ C + 273.15 = 297.2\ K$

$$n = \frac{PV}{RT} = \frac{0.9809 \text{ atm} \times 0.1000 \text{ L}}{0.0821 \dfrac{\text{L} \cdot \text{atm}}{\text{K} \cdot \text{mol}} \times 297.2 \text{ K}} = 4.02 \times 10^{-3} \text{ mol H}_2$$

$$4.02 \times 10^{-3} \text{ mol H}_2 \times \frac{2.0158 \text{ g H}_2}{1 \text{ mol H}_2} = 8.10 \times 10^{-3} \text{ g H}_2$$

11.77 (a) $MM = \dfrac{mRT}{PV} = \dfrac{2.678 \text{ g} \times 0.0821 \dfrac{\text{L} \cdot \text{atm}}{\text{K} \cdot \text{mol}} \times 273.15 \text{ K}}{1 \text{ atm} \times 2.00 \text{ L}} = 30.0 \text{ g/mol}$

(b) To determine the empirical formula, assume you have 100.0 g of the sample rather than the 2.678 g mentioned in the problem statement.

The percentages then equal the grams.

$$80.0\%\text{C} \rightarrow 80.0 \text{ g C} \times \frac{1 \text{ mol C}}{12.011 \text{ g C}} = 6.66 \text{ mol C}$$

$$20.0\% \text{ H} \rightarrow 20.0 \text{ g H} \times \frac{1 \text{ mol H}}{1.0079 \text{ g H}} = 19.8 \text{ mol H}$$

$$\text{C}_{\frac{6.66}{6.66}} \text{H}_{\frac{19.8}{6.66}} \rightarrow \text{C}_1 \text{H}_{2.97} \rightarrow \text{CH}_3$$

(c) The molar mass of the empirical formula CH_3 determined in (b) is 15.035 g/mol.

The molar mass of the compound is, from (a), 30.0 g/mol.

$$\frac{30.0 \text{ g/mol}}{15.035 \text{ g/mol}} = 2.00$$

$$\text{C}_{2 \times 1} \text{H}_{2 \times 3} \rightarrow \text{C}_2\text{H}_6$$

11.79 Step 1: $\quad P_i = ? \qquad P_f = 3P_i$

$\qquad\qquad\quad V_i = ? \qquad V_f = 2V_i$

$\qquad\qquad\quad n_i = ? \qquad n_f = n_i$

$\qquad\qquad\quad T_i = 293.2 \text{ K} \quad T_f = ?$

Step 2: $\quad \dfrac{P_i V_i}{n_i T_i} = \dfrac{P_f V_f}{n_f T_f}$

$\qquad\qquad \dfrac{P_i V_i}{n_i T_i} = \dfrac{3P_i 2V_i}{n_i T_f}$

$\qquad\qquad\quad \dfrac{1}{T_i} = \dfrac{6}{T_f}$

Step 3: $\quad T_f = 6T_i = 6 \times 293.2 \text{ K} = 1759 \text{ K}$

$\qquad\qquad 1759 \text{ K} - 273.15 = 1486\,°\text{C}$

11.81 Use the ideal gas law expressed in terms of molar mass and sample mass. Because the equation requires volume to be in liters, convert the density to grams per liter:

$$\frac{2.26 \times 10^{-3}\,g}{mL} \times \frac{1000\,mL}{1\,L} = 2.26\,g/L$$

$$655\,mm\,Hg \times \frac{1\,atm}{760\,mm\,Hg} = 0.862\,atm$$

You know density is mass/volume, meaning you can replace m/V by the density value.

$$25.0\,°C + 273.15 = 298.2\,K$$

$$MM = \frac{mRT}{PV} = \frac{m}{V} \times \frac{RT}{P} = \text{Density} \times \frac{RT}{P} = 2.26\frac{g}{L} \times \frac{0.0821\frac{L \cdot atm}{K \cdot mol} \times 298.2\,K}{0.862\,atm} = 64.2\,g/mol$$

11.83 $$230.0\,mL \times \frac{1\,L}{1000\,mL} = 0.2300\,L$$

$$745.0\,mm\,Hg \times \frac{1\,atm}{760.0\,mm\,Hg} = 0.9803\,atm$$

$$34.0\,°C + 273.15 = 307.2\,K$$

$$n = \frac{PV}{RT} = \frac{0.9803\,atm \times 0.2300\,L}{0.0821\frac{L \cdot atm}{K \cdot mol} \times 307.2\,K} = 8.940 \times 10^{-3}\,mol\,N_2O_4$$

$$8.940 \times 10^{-3}\,mol\,N_2O_4 \times \frac{2\,mol\,N\,atoms}{1\,mol\,N_2O_4} \times \frac{6.022 \times 10^{23}\,N\,atoms}{1\,mol\,N\,atoms} = 1.077 \times 10^{22}\,N\,atoms$$

11.85 The same in both containers because the identity of the gas does not matter. $P = nRT/V$, $n = 1$ mol in both containers, R is the same in both containers, $T = 300\,K$ in both containers, $V = 1\,L$ in both containers. Therefore P must be the same in the two containers.

11.87 $25.0\,°C + 273.15 = 298.2\,K$

$$0.200\,g\,O_2 \times \frac{1\,mol\,O_2}{31.998\,g\,O_2} \times \frac{1\,mol\,N_2O_4}{2\,mol\,O_2} = 3.13 \times 10^{-3}\,mol\,N_2O_4$$

$$V = \frac{nRT}{P} = \frac{3.13 \times 10^{-3}\,mol \times 0.0821\frac{L \cdot atm}{K \cdot mol} \times 298.2\,K}{0.996\,atm} = 0.0769\,L = 76.9\,mL\,N_2O_4$$

11.89 The ideal gas equation solved for V, $V = nRT/P$, tells you that the volume of a gas is *directly* proportional to the temperature. Thus, if the temperature decreases (as it would in a freezer), the volume must also decrease.

11.91 (a) n and V constant. That the bulb is sealed tells you n is constant. That the bulb is made of glass, a rigid material, tells you V is constant. That T changes is stated in the problem. P changes because T changes and, according to the ideal gas law, P is directly proportional to T. (Here P increases because T increases.)

(b) n and P constant. The balloon can be assumed to be tied tightly, making n constant because no He can escape. That T changes is stated in the problem. P is constant because the pressure inside the balloon must equal the exterior (atmospheric) pressure at all times. V changes because T changes and the ideal gas equation shows that V is directly proportional to T: $P = nRT/V$, $V = nRT/P$. (Here V decreases because T decreases.)

(c) T and V constant. That n changes is stated in the problem. Because the cylinder is made of steel and is sealed, V cannot change. T is constant because the CO_2 is assumed to behave ideally. P changes because it is directly proportional to n. (Here P increases because n increases.) [Note: For any real gas, T would increase slightly as n was increased at constant V.]

11.93 Density $= \dfrac{m}{V} = \dfrac{(MM)P}{RT}$

Because all the gases are at 0.0 °C and 1.0 atm, P, T, and R can be ignored so that this expression reduces to

Density $= \dfrac{m}{V}$ is proportional to MM

Thus, the greater the molar mass, the greater the density. The molar masses are CO_2, 44 g/mol; H_2, 2 g/mol; O_2, 32 g/mol; CH_4, 16 g/mol; He, 4 g/mol. Therefore the order of increasing density is $H_2 < He < CH_4 < O_2 < CO_2$.

11.95 Molar mass; all the others are needed in the ideal gas equation.

11.97 In each case, solve the ideal gas equation for the unknown quantity.

Gas A

$$757 \text{ mm Hg} \times \frac{1 \text{ atm}}{760 \text{ mm Hg}} = 0.996 \text{ atm}$$

$$952 \text{ mL} \times \frac{1 \text{ L}}{1000 \text{ mL}} = 0.952 \text{ L}$$

$$T = \frac{PV}{nR} = \frac{0.996 \text{ atm} \times 0.952 \text{ L}}{0.300 \text{ mol} \times 0.0821 \dfrac{\text{L} \cdot \text{atm}}{\text{K} \cdot \text{mol}}} = 38.5 \text{ K}$$

$$38.5 \text{ K} - 273.15 = 234.7 \text{ °C}$$

Gas B

$$27.0 \text{ °C} + 273.15 = 300.2 \text{ K}$$

$$V = \frac{nRT}{P} = \frac{5.00 \text{ mol} \times 0.0821 \dfrac{\text{L} \cdot \text{atm}}{\text{K} \cdot \text{mol}} \times 300.2 \text{ K}}{1.20 \text{ atm}} = 103 \text{ L}$$

Gas C

$$800. \text{ mm Hg} \times \frac{1 \text{ atm}}{760.0 \text{ mm Hg}} = 1.053 \text{ atm}$$

$$n = \frac{PV}{RT} = \frac{1.053 \text{ atm} \times 1.20 \text{ L}}{0.0821 \dfrac{\text{L} \cdot \text{atm}}{\text{K} \cdot \text{mol}} \times 298 \text{ K}} = 0.0516 \text{ mol}$$

Gas D

$$750 \ \text{mL} \times \frac{1 \ \text{L}}{1000 \ \text{mL}} = 0.750 \ \text{L}$$

$$0.0\,°\text{C} + 273.15 = 273.2 \ \text{K}$$

$$P = \frac{nRT}{V} = \frac{0.0875 \ \text{mol} \times 0.0821 \dfrac{\text{L} \cdot \text{atm}}{\text{K} \cdot \text{mol}} \times 273.2 \ \text{K}}{0.750 \ \text{L}} = 2.62 \ \text{atm}$$

11.99 Step 1: $P_i = 1 \ \text{atm}$ $P_f = 1 \ \text{atm} = P_i$

   $V_i = 2.40 \ \text{L}$ $V_f = 7.20 \ \text{L}$

   $n_i = ?$ $n_f = ? = n_i$

   $T_i = 22\,°\text{C} = 295 \ \text{K}$ $T_f = ?$

Step 2: $\dfrac{P_i V_i}{n_i T_i} = \dfrac{P_f V_f}{n_f T_f}$

   $\dfrac{\cancel{P_i} V_i}{\cancel{n_i} T_i} = \dfrac{\cancel{P_i} V_f}{\cancel{n_i} T_f}$

   $\dfrac{V_i}{T_i} = \dfrac{V_f}{T_f}$

Step 3: $T_f = \dfrac{T_i V_f}{V_i} = \dfrac{295 \ \text{K} \times 7.20 \ \text{L}}{2.40 \ \text{L}} = 885 \ \text{K}$

   $885 \ \text{K} - 273.15 = 612\,°\text{C}$

11.101 (a) $25.0\,°\text{C} + 273.15 = 298.2 \ \text{K}$

   (b) $5.00 \ \text{gallons} \times \dfrac{3.785 \ \text{L}}{1 \ \text{gallon}} = 18.9 \ \text{L}$

   (c) $755 \ \text{mm Hg} \times \dfrac{1 \ \text{atm}}{760 \ \text{mm Hg}} = 0.993 \ \text{atm}$

11.103 $\dfrac{5}{9} \times (32\,°\text{F} - 32) = 0\,°\text{C} + 273.15 = 273 \ \text{K}$

   $\dfrac{5}{9} \times (80\,°\text{F} - 32) = 27\,°\text{C} + 273.15 = 300 \ \text{K}$

   Step 1: $P_i = 2.20 \ \text{atm}$ $P_f = ?$

      $V_i = ?$ $V_f = ? = V_i$

      $n_i = ?$ $n_f = ? = n_i$

      $T_i = 273 \ \text{K}$ $T_f = 300 \ \text{K}$

Step 2: $\dfrac{P_i V_i}{n_i T_i} = \dfrac{P_f V_f}{n_f T_f}$

$\dfrac{P_i \cancel{V_i}}{\cancel{n_i} T_i} = \dfrac{P_i \cancel{V_i}}{\cancel{n_i} T_f}$

$\dfrac{P_i}{T_i} = \dfrac{P_f}{T_f}$

Step 3: $P_f = \dfrac{T_f P_i}{T_i} = \dfrac{300 \text{ K} \times 2.20 \text{ atm}}{273 \text{ K}} = 2.42 \text{ atm}$

11.105 False. Because the relationship between P, V, T, and n in an ideal gas is independent of the identity of the gas.

11.107 (a) For this stoichiometry problem, you need the molar mass of $KClO_3$, which is 122.5 g/mol.

$$500.0 \text{ g } \cancel{KClO_3} \times \dfrac{1 \text{ mol } \cancel{KClO_3}}{122.5 \text{ g } \cancel{KClO_3}} \times \dfrac{3 \text{ mol } O_2}{2 \text{ mol } \cancel{KClO_3}} = 6.123 \text{ mol } O_2$$

(b) $24.0\,°C + 273.15 = 297.2 \text{ K}$

$$750.0 \text{ } \cancel{mm\,Hg} \times \dfrac{1 \text{ atm}}{760.0 \text{ } \cancel{mm\,Hg}} = 0.9868 \text{ atm}$$

$$V = \dfrac{nRT}{P} = \dfrac{6.123 \text{ } \cancel{mol} \times 0.0821\dfrac{\text{L} \cdot \cancel{atm}}{\text{K} \cdot \cancel{mol}} \times 297.2 \text{ K}}{0.9868 \text{ } \cancel{atm}} = 151.4 \text{ L}$$

(c) STP means $0\,°C = 273.15$ K and 1 atm.

$$V = \dfrac{6.123 \text{ } \cancel{mol} \times 0.0821\dfrac{\text{L} \cdot \cancel{atm}}{\text{K} \cdot \cancel{mol}} \times 273.15 \text{ K}}{1 \text{ } \cancel{atm}} = 137.3 \text{ L}$$

11.109 (a) If the temperature of a gas is doubled while the pressure is kept constant, the volume of the gas *doubles*. (Volume is directly proportional to temperature.)

(b) If the pressure of a gas is halved while the temperature is kept constant, the volume of the gas *doubles*. (Volume is inversely proportional to pressure.)

11.111 (a) $MM = \dfrac{m_{sample}RT}{PV} = \dfrac{1.25 \text{ g} \times 0.0821\dfrac{\cancel{L} \cdot \cancel{atm}}{\text{K} \cdot \cancel{mol}} \times 273 \text{ K}}{0.400 \text{ } \cancel{atm} \times \cancel{2.50\,L}} = 28.0 \text{ g/mol}$

(b) The molar mass of CH_2 is 14.027 g/mol.

$$\dfrac{28.0 \text{ g/mol}}{14.027 \text{ g/mol}} = 2.00$$

$$C_{2\times1}H_{2\times1} = C_2H_4$$

11.113 (a) $30.2 \text{ in Hg} \times \dfrac{760 \text{ mm Hg}}{29.9 \text{ in Hg}} = 768 \text{ mm Hg}$

(b) $890.00 \text{ mm Hg} \times \dfrac{1 \text{ atm}}{760 \text{ mm Hg}} = 1.17 \text{ atm}$

(c) $300.0 \text{ lb/in.}^2 \text{ Hg} \times \dfrac{1 \text{ atm}}{14.696 \text{ lb/in.}^2} = 20.41 \text{ atm}$

11.115 $40\,^{\circ}\text{C} + 273.15 = 313 \text{ K}$

$$MM = \frac{m_{\text{sample}}RT}{PV} = \frac{48.3 \text{ g} \times 0.0821\dfrac{\text{L} \cdot \text{atm}}{\text{K} \cdot \text{mol}} \times 313 \text{ K}}{3.10 \text{ atm} \times 10.0 \text{ L}} = 40.0 \text{ g/mol}$$

11.117 Because you are given density, which is mass/volume, use the ideal gas law expressed in terms of molar mass and sample mass. That way you can use the density value for m/V in the equation.

$0\,^{\circ}\text{C} + 273.15 = 273 \text{ K}$

$$\frac{m}{MM} = \frac{PV}{RT}$$

$$MM = \frac{mRT}{PV} = \frac{m}{V} \times \frac{RT}{P} = 1.52\frac{\text{g}}{\text{L}} \times \frac{0.0821\dfrac{\text{L} \cdot \text{atm}}{\text{K} \cdot \text{mol}} \times 273 \text{ K}}{1 \text{ atm}} = 34.1 \text{ g/mol}$$

11.119 (a) The volume is halved because volume is inversely proportional to pressure.

(b) The volume is doubled because volume is directly proportional to temperature.

(c) The volume remains unchanged because the two changes cancel each other. Doubling P changes the volume to $0.5\ V_i$, and doubling T changes the volume to $2\ V_i$: $(0.5 \times 2)V_i = V_i$.

11.121 This can be accomplished by doubling the temperature (in Kelvin) or by doubling the number of moles of gas in the tank.

11.123 (a) It will eventually liquefy. Cooling slows the molecules down. Pushing down on the piston squeezes them closer together. Both allow the attractive forces between the molecules to eventually "take hold" and lead to a condensed phase.

(b) The ideal gas law will do a very poor job of describing a gas under such conditions. Remember, the ideal gas law was derived for a gas where there are no forces of attraction between the molecules. Under conditions close to condensation to a liquid, the intermolecular forces are beginning to "take hold" and are quite important (whereas they can be neglected at conditions far from liquefaction).

11.125 There is an assumption of *absolutely no attractive* intermolecular forces existing among the molecules of an ideal gas. However, molecules of all real gases do interact with one another to some extent. These interactions usually can be ignored, as we do when we apply the ideal gas equation to a real gas, when the temperatures are not too low, and/or when the pressures are not too high.

11.127 The pressure due to methane would be 1.25 atm, or half the total pressure, because half the moles of total gas is methane. The 1.25 atm is called the partial pressure due to methane.

11.129 Imagine a column of Hg 1″ × 1″ × 76 cm high.

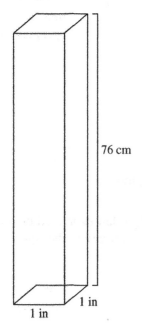

76 cm

1 in
1 in

Volume of Hg in the column is:

$$(1 \text{ in.}^2)\left(\frac{2.54 \text{ cm}}{\text{in.}}\right)^2 (76 \text{ cm}) = 490.3 \text{ cm}^3 = 490.3 \text{ mL}$$

Mass of this column of Hg is:

$$(490.3 \text{ mL})\left(\frac{13.6 \text{ g}}{\text{mL}}\right)\left(\frac{\text{lbs}}{453.6 \text{ g}}\right) = 14.7 \text{ lbs}$$

$$\text{Pressure} = \frac{\text{force}}{\text{area}} = \frac{14.7 \text{ lbs}}{1 \text{ in.}^2} = 14.7 \frac{\text{lbs}}{\text{in.}^2}$$

11.131 (a) Solve for pressure from the ideal gas equation, $PV = nRT$:

$$P = \frac{nRT}{V} = \frac{(1.00 \text{ mol})(0.082 \text{ L} \cdot \text{atm}/\text{K} \cdot \text{mol})(298 \text{ K})}{1.00 \text{ L}} = 24.5 \text{ atm } CO_2$$

(b) Use the van der Waals equation with the following constants for CO_2: a $= 3.59$ atm L^2/mol^2 a
b $= 0.0427$ L/mol.

$$\underbrace{P_{\text{real}} + 3.59 \frac{\text{atm L}^2}{\text{mol}^2}\left(\frac{1.00 \text{ mol}}{1.00 \text{ L}}\right)^2}_{3.59 \text{ atm}} \underbrace{\left(\frac{1.00 \text{ L}}{1.00 \text{ mol}} - 0.0427 \frac{\text{L}}{\text{mol}}\right)}_{0.957 \frac{\text{L}}{\text{mol}}} = \underbrace{\left(0.0821 \frac{\text{L atm}}{\text{mol K}}\right)(298 \text{ K})}_{24.5 \frac{\text{L atm}}{\text{mol}}}$$

$$[P_{\text{real}} + 3.59 \text{ atm}] \times 0.957 \frac{\text{L}}{\text{mol}} = 24.5 \frac{\text{L atm}}{\text{mol}}$$

$$P_{\text{real}} + 3.59 \text{ atm} = \frac{24.5 \frac{\text{L atm}}{\text{mol}}}{0.957 \frac{\text{L}}{\text{mol}}} = 25.6 \text{ atm}$$

$$P_{\text{real}} = 25.6 \text{ atm} - 3.59 \text{ atm} = 22.0 \text{ atm}$$

(c) The percent difference between the ideal pressure and that obtained from the van der Waals equation is 10%, as shown below:

$$\frac{(24.5 \text{ atm} - 22.0 \text{ atm})}{24.5 \text{ atm}} \times 100\% = 10\%$$

(d) As the temperature increases, the attractive intermolecular forces are expected to get weakened, making the real gas more and more as the ideal gas. Therefore, the expectation is that as the temperature increases, the percent difference between the real and ideal pressure should decrease. Indeed, P_{ideal} at 1000 K = 82.1 atm, P_{real} at 1000 K = 78.6 atm, and the percent difference calculated as in part (c) of this problem would be only 4.3%.

Chapter 12

12.1 See solution in textbook.

12.3 Because 135-proof vodka is $135/2 = 67.5\%$ alcohol and $100\% - 67.5\% = 32.5\%$ water, alcohol is the major component and therefore the solvent.

12.5 Less than in water. Because carbon tetrachloride is a nonpolar molecule, it does not attract ions. With little or no attraction between the Na^+ and Cl^- ions and the carbon tetrachloride molecules, there is little or no energy released in the solvation step.

12.7 See solution in textbook.

12.9 Because $\Delta E_{\text{solvation}}$ is the only negative energy value when a solute dissolves, a more negative ΔE_{total} usually means a more negative $\Delta E_{\text{solvation}}$, which measures how strongly solute and solvent particles attract one another. The stronger this attraction, the more likely the solute is to dissolve in the solvent.

12.11 See solution in textbook.

12.13 The biggest difference is in the solute-separation step. Gases do not exist in a lattice, which means the solute particles are already separated. Hence, $\Delta E_{\text{solute separation}}$ is essentially zero for gases. For both solvent separation and solvation, the energy needed in dissolution of a gas is roughly the same as the energy needed in dissolution of a solid.

12.15 $25.0 \text{ g glucose} \times \dfrac{1 \text{ mol glucose}}{180.155 \text{ g glucose}} \times \dfrac{1 \text{ L solution}}{2.55 \text{ mol glucose}} \times \dfrac{1000 \text{ mL solution}}{1 \text{ L solution}} = 54.4 \text{ mL solution}$

12.17 $400.0 \text{ mL solution} \times \dfrac{1 \text{ L solution}}{1000 \text{ mL solution}} \times \dfrac{2.00 \text{ mol NaCl}}{1 \text{ L solution}} \times \dfrac{58.443 \text{ g NaCl}}{1 \text{ mol NaCl}}$

$= 46.8 \text{ g NaCl required}$

After placing this amount of solid NaCl in a 400-mL volumetric flask, add some water, swirl to dissolve the solid, and then add enough water to raise the level to the 400.0-mL mark. Because the NaCl takes up some volume (even after it has dissolved), the amount of water added is slightly less than 400.0 mL.

12.19 See solution in textbook.

12.21 $65.0 \text{ g solution} \times \dfrac{12.5 \text{ g sucrose}}{100 \text{ g solution}} = 8.12 \text{ g sucrose}$

12.23 See solution in textbook.

12.25 Number of moles = Volume of solution in liters × Molarity of solution

$$\text{Volume} = \frac{\text{Number of moles}}{\text{Molarity}} = \frac{0.500 \text{ mol}}{0.350 \text{ mol/L}} = 1.43 \text{ L}$$

You can also solve this problem by using molarity as a conversion factor:

$$0.500 \text{ mol BaCl}_2 \times \frac{1 \text{ L solution}}{0.350 \text{ mol BaCl}_2} = 1.43 \text{ L solution}$$

12.27 See solution in textbook.

12.29 Step 1: $Fe^{3+}(aq) + 3OH^-(aq) \rightarrow Fe(OH)_3(s)$

Step 2: $20.0 \text{ g Fe(OH)}_3 \times \dfrac{1 \text{ mol Fe(OH)}_3}{106.9 \text{ g Fe(OH)}_3} = 0.187 \text{ mol Fe(OH)}_3$

Step 3: $0.187 \text{ mol Fe(OH)}_3 \times \dfrac{1 \text{ mol Fe}^{3+}}{1 \text{ mol Fe(OH)}_3} = 0.187 \text{ mol Fe}^{3+}$

$0.187 \text{ mol Fe(OH)}_3 \times \dfrac{3 \text{ mol OH}^-}{1 \text{ mol Fe(OH)}_3} = 0.561 \text{ mol OH}^-$

Step 4: $0.187 \text{ mol Fe}^{3+} \times \dfrac{1 \text{ mol Fe(NO}_3)_3}{1 \text{ mol Fe}^{3+}} \times \dfrac{1 \text{ L Fe(NO}_3)_3 \text{ solution}}{0.250 \text{ mol Fe(NO}_3)_3} = 0.748 \text{ L Fe(NO}_3)_3 \text{ solution}$

$0.561 \text{ mol OH}^- \times \dfrac{1 \text{ mol Ba(OH)}_2}{2 \text{ mol OH}^-} \times \dfrac{1 \text{ L Ba(OH)}_2 \text{ solution}}{0.150 \text{ mol Ba(OH)}_2} = 1.87 \text{ L Ba(OH)}_2 \text{ solution}$

Combine 0.748 L of the iron(III) nitrate solution with 1.87 L of the barium hydroxide solution, and then filter off the precipitated $Fe(OH)_3$. The amount to be collected for a 67.5% yield is

$$\text{Actual yield} = \frac{\% \text{ yield} \times \text{theoretical yield}}{100\%} = \frac{67.5\% \times 20.0 \text{ g}}{100\%} = 13.5 \text{ g Fe(OH)}_3$$

12.31 Step 1: $H^+(aq) + OH^-(aq) \rightarrow H_2O(l)$

Step 2: $0.03860 \text{ L NaOH solution} \times \dfrac{0.100 \text{ mol OH}^-}{1 \text{ L NaOH solution}} = 0.00386 \text{ mol OH}^-$

Step 3: $0.00386 \text{ mol OH}^- \times \dfrac{1 \text{ mol H}^+}{1 \text{ mol OH}^-} \times \dfrac{1 \text{ mol H}_3PO_4}{3 \text{ mol H}^+} = 1.29 \times 10^{-3} \text{ mol H}_3PO_4$

Step 4: $\dfrac{1.29 \times 10^{-3} \text{ mol H}_3PO_4}{0.05000 \text{ L}} = 0.0258 \text{ M H}_3PO_4$

12.33 See solution in textbook.

12.35 When the vapor pressure of an aqueous solution at sea level is 760 mm Hg (1 atm), the solution boils. Therefore, one way to determine the temperature at which the solution has a vapor pressure of 760 mm Hg is to determine the boiling point.

That $Ca(NO_3)_2$ is dissolved tells you the boiling point is higher than 100.0 °C. Therefore, calculate how much the dissolved salt elevates the boiling point and you have the temperature at which the vapor pressure is 760 mm Hg.

Each mole of calcium nitrate dissociates in water to produce 3 moles of solute particles: $Ca(NO_3)_2 \rightarrow Ca^{2+} + 2NO_3^-$. The number of moles of solute particles in solution is therefore

$$100.0 \text{ g } \cancel{Ca(NO_3)_2} \times \frac{1 \text{ mol } \cancel{Ca(NO_3)_2}}{164.1 \text{ g } \cancel{Ca(NO_3)_2}} \times \frac{3 \text{ mol solute particles}}{1 \text{ mol } \cancel{Ca(NO_3)_2}} = 1.828 \text{ mol solute particles}$$

This amount of solute particles causes a boiling-point elevation of

$$\Delta T_b = \frac{0.52 \dfrac{°C \cdot \cancel{\text{kg solvent}}}{\cancel{\text{mol solute particles}}} \times 1.828 \text{ } \cancel{\text{mol solute particles}}}{0.4500 \text{ } \cancel{\text{kg solvent}}} = 2.1 \text{ °C}$$

The elevated boiling point is therefore $100.0 \text{ °C} + 2.1 \text{ °C} = 102.1 \text{ °C}$, and it is at this temperature that the vapor pressure is 760 mm Hg.

12.37 A solute is any substance in a solution other than the solvent.

A solvent is the substance in a solution present in the greatest amount.

A solution is a homogeneous mixture of two or more substances.

12.39 A soft drink can be called a solution only if there are no visible bubbles of CO_2 gas. If such bubbles are visible, the CO_2 is in the gas phase and you have a heterogeneous mixture. If there are no bubbles visible, then the soft drink is indeed a solution of CO_2 gas, sugar, coloring agents, and so forth dissolved in water. The atmosphere you breathe, as long as it is clean, is a homogeneous mixture of $N_2(g)$, $O_2(g)$, $CO_2(g)$, $H_2O(g)$, and other gases.

12.41 As noted in Section 12.1 of the textbook, vinegar is mostly water. Therefore, the solvent is water because it is the component present in the greater amount. The solute is the acetic acid dissolved in the water.

12.43 (a) Solution; the two molten metals mix completely and uniformly, with the result that the cooled solid is uniform in composition and therefore a solution.

(b) Solution; the filtering removes anything not in solution.

(c) Heterogeneous mixture; visual inspection shows you the nonuniformity.

(d) Solution; what you exhale is a solution of various gases, just as the atmosphere you inhale is.

(e) Heterogeneous mixture; no matter how much you shake, sooner or later the two components separate, telling you a true solution never formed.

12.45 Heterogeneous mixture because each solid exists only in its own lattice and cannot mix into the other lattice.

12.47 Intermolecular attractive forces are responsible for gas molecules coming together and condensing to the liquid and solid phases. Therefore, any everyday instance of liquid H_2O—raindrops, the ocean—or solid H_2O—ice, snow—is evidence of these attractive forces.

12.49 When a piece of NaCl is melted, the lattice is broken and the Na^+ and Cl^- ions become mobile. The same thing happens when NaCl dissolves in water. The difference is that in melting, the energy provided by the heat breaks the bonds, and in dissolving, the solute–solvent interactions break the bonds.

12.51 Absorbs energy because attractive forces must be overcome. Overcoming an attractive force *always* requires an input of energy.

12.53 (a) The surrounding of solute particles by molecules of water.

(b) Because intermolecular forces are being *formed* between the solute particles and the water molecules during the hydration process, energy is released. Forming an attractive force *always* releases energy.

12.55 1. Solute separation—solute–solute attractive forces overcome (solute lattice broken apart):

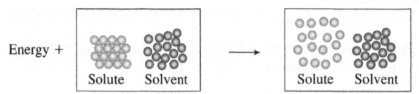

2. Solvent separation—solvent–solvent attractive forces overcome:

Energy + | Solute Solvent → Solute Solvent

3. Solvation—solute–solvent attractive forces formed:

Solute Solvent → Solution + Energy

12.57 The attraction between the ions and water molecules. The Na^+ ions are attracted to the δ^- end of the water molecules; the Cl^- ions are attracted to the δ^+ ends of the water molecules. Once the ions are surrounded by H_2O molecules, there is very little tendency to re-form the NaCl lattice.

12.59 The water molecule forms a hydrogen bond with the alcohol. The bond is either between the O of the alcohol $-OH$ group and one water H or between the H of the alcohol $-OH$ group and the water O:

This interaction is rightly called hydration because it is a solvation process with water as the solvent.

12.61 Probably not soluble at all. The step in which solute–solvent interactions take place is the solvation step. Because the energy released during this step is a lot less than the energy absorbed for solute separation and solvent separation, ΔE_{total} is positive, indicating the solid is insoluble in this solvent.

12.63 As in Problem 11.62(a), it is a combination of several factors. The fact that energy is released in such a large amount is evidence of much less energy being required for the solute-separation and solvent-separation steps compared to the energy being released in the solvation (here, hydration) step.

12.65 Because water molecules are polar and hexane molecules are nonpolar. The Na^+ and Cl^- ions are strongly attracted to the δ^+ and δ^- parts of the water molecules, and consequently the solvation step releases more energy than what is absorbed during solute separation and solvent separation. ΔE_{total} is negative, and the solid NaCl dissolves in water. The ions are not attracted to the nonpolar hexane molecules, and consequently the solvation step releases very little energy, less than what is absorbed during solute separation and solvent separation. ΔE_{total} is positive, and the solid NaCl does not dissolve in hexane.

12.67 $AlCl_3$, because hydration energy is a function of how strongly solute particles and solvent particles attract each other. Both $AlCl_3$ and NaCl are ionic compounds, and so the interaction with water molecules is an ion–dipole interaction. Because the attraction gets stronger with increasing charge on the ion, the $Al^{3+} \cdots H_2O$ attraction is stronger than the $Na^+ \cdots H_2O$ attraction.

12.69 Polar solutes dissolve in polar solvents; nonpolar solutes dissolve in nonpolar solvents.

12.71 (a) Because attractive forces between solute particles are being broken.

(b) Because attractive forces between solvent particles are being broken.

(c) Because attractive forces between solute particles and solvent particles are being formed.

(d) Soluble because ΔE_{total} is negative.

(e) Warmer because the energy released in the solvation step is more than the energy absorbed in the solute-separation and solvent-separation steps. The excess energy released is heat energy that warms the solution.

12.73 B. The described difference means that $\Delta E_{solute\ separation,\ A}$ is larger than $\Delta E_{solute\ separation,\ B}$. That this is the only significant difference tells you that $\Delta E_{solvent\ separation,\ A} = \Delta E_{solvent\ separation,\ B}$ and $\Delta E_{solvation,\ A} = \Delta E_{solvation,\ B}$. Mathematically, this means that $\Delta E_{total,\ B}$ is more negative than $\Delta E_{total,\ A}$. The more negative ΔE_{total}, the more soluble a compound. (Note that a number does not have to be negative in order to be "more negative." For instance, 3 is more negative than 4.)

12.75 The intermolecular attractive forces holding the liquid solute molecules are overcome. The step absorbs energy because attractive forces are being overcome.

12.77 Because the interactions between the nonpolar oil molecules and the solvent water molecules are not strong very little energy is released in the solvation step and, as a result, ΔE_{total} for the dissolution is likely to be positive, making the oil insoluble.

12.79 In order for anything to happen spontaneously, the entropy of the universe must increase.

12.81 Because the state of higher entropy is the one in which the dye molecules are dispersed. There are many more ways to have the dye molecules randomly dispersed than to have them grouped together in the original blue drop.

12.83 True. The second law of thermodynamics tells you that any time a process occurs spontaneously, the entropy of the universe increases.

12.85 Total chaos, in which everything has gone back to a totally random state.

12.87 Increasing pressure increases the solubility of gaseous solutes because the increased pressure above the solution forces more solute molecules into solution.

12.89 The trend is an increase in solubility as temperature increases.

12.91 Aquatic life can survive only if the water it lives in contains dissolved O_2 gas. Discharging hot water heats all the water in the river or lake, and at the higher temperatures less O_2 gas can dissolve in the water.

12.93 The sucrose solubility graph in the textbook tells you the solubility at 90 °C is 420.0 g/100.0 g of water.

$$100.0 \text{ lb water} \times \frac{453.6 \text{ g water}}{1 \text{ lb water}} \times \frac{420.0 \text{ g sucrose}}{100.0 \text{ g water}} = 1.905 \times 10^5 \text{ g sucrose}$$

12.95 At 10 °C:

$$250.0 \text{ g water} \times \frac{35.0 \text{ g Li}_2\text{SO}_4}{100 \text{ g water}} = 87.5 \text{ g Li}_2\text{SO}_4$$

At 50 °C:

$$250.0 \text{ g water} \times \frac{32.5 \text{ g Li}_2\text{SO}_4}{100 \text{ g water}} = 81.3 \text{ g Li}_2\text{SO}_4$$

$$87.5 \text{ g} - 81.3 \text{ g} = 6.2 \text{ g more at 10 °C}$$

12.97 Problem 12.94 tells you that at 50 °C the solubility of KCl is 42.6 g/100 g water.

$$2.00 \text{ L water} \times \frac{1000 \text{ mL water}}{1 \text{ L water}} \times \frac{1.00 \text{ g water}}{1 \text{ mL water}} \times \frac{42.6 \text{ g KCl}}{100 \text{ g water}} = 852 \text{ g KCl}$$

12.99 You know from the definition of molarity that 1.00 L of a 1.00 M $CaCl_2$ solution contains 1.00 mole of $CaCl_2$. Because your assistant needs to know what mass of $CaCl_2$ to use, your first step is to determine the number of grams in 1.00 mole:

$$1.00 \text{ L solution} \times \frac{1.00 \text{ mol CaCl}_2}{1 \text{ L solution}} \times \frac{110.984 \text{ g CaCl}_2}{1 \text{ mol CaCl}_2} = 111.0 \text{ g CaCl}_2$$

Your instructions should therefore be, "Place 111.0 g of $CaCl_2$ in a 1-L volumetric flask, dissolve in about 500 mL of water, and then add enough water to bring the volume to the 1-L mark."

12.101 You know from the definition of molarity that 1.00 L of a 0.250 M sucrose solution contains 0.250 mole or 0.250 mol of sucrose. Because your assistant needs to know what mass of sucrose to use, your first step is to determine the number of grams in 0.250 mole:

$$1.00 \text{ L solution} \times \frac{0.250 \text{ mol } C_{12}H_{22}O_{11}}{1 \text{ L solution}} \times \frac{342.3 \text{ g } C_{12}H_{22}O_{11}}{1 \text{ mol } C_{12}H_{22}O_{11}} = 85.6 \text{ g } C_{12}H_{22}O_{11}$$

Your instructions should therefore be, "Place 85.6 g of sucrose in a 1-L volumetric flask, dissolve in about 500 mL of water, and then add enough water to bring the volume to the 1-L mark."

12.103 (a) $2.50 \text{ mol NaCl} \times \dfrac{58.443 \text{ g NaCl}}{1 \text{ mol NaCl}} = 146 \text{ g NaCl}$

(b) Yes. Proof:

$$\frac{2.50 \text{ mol NaCl}}{500.0 \text{ mL}} \times \frac{1000 \text{ mL}}{1 \text{ L}} = \frac{5.00 \text{ mol NaCl}}{L} = 5.00 \text{ M}$$

12.105 (a) $2500.0 \text{ mL} \times \dfrac{1 \text{ L}}{1000 \text{ mL}} \times \dfrac{0.250 \text{ mol NaCl}}{1 \text{ L}} = 0.625 \text{ mol NaCl}$

(b) $0.625 \text{ mol NaCl} \times \dfrac{2 \text{ mol ions}}{\text{mol NaCl}} = 1.25 \text{ mol ions}$

(c) $0.625 \text{ mol NaCl} \times \dfrac{58.443 \text{ g}}{1 \text{ mol NaCl}} = 36.5 \text{ g NaCl}$

12.107 $5.00 \text{ g NaCl} \times \dfrac{1 \text{ mol NaCl}}{58.443 \text{ g NaCl}} \times \dfrac{1 \text{ L}}{1 \text{ mol NaCl}} \times \dfrac{1000 \text{ mL}}{1 \text{ L}} = 85.6 \text{ mL}$

12.109 You need to know what volume of stock solution goes into a 250-mL volumetric flask. You can determine this volume either by calculating numbers of moles or by using the dilution equation. Working with moles, you find

$$250.0 \text{ mL} \times \frac{1 \text{ L}}{1000 \text{ mL}} \times \frac{0.348 \text{ mol glucose}}{1 \text{ L}} = 0.0870 \text{ mol sucrose needed in 250-mL flask}$$

$$0.0870 \text{ mol sucrose} \times \frac{1 \text{ L stock solution}}{0.500 \text{ mol sucrose}} = 0.174 \text{ L stock solution} = 174 \text{ mL stock solution}$$

With the dilution equation, you get

$$V_{\text{stock solution}} = \frac{0.348 \text{ M} \times 250.0 \text{ mL}}{0.500 \text{ M}} = 174 \text{ mL}$$

Therefore, your instructions should be, "Place 174 mL of the stock solution in a 250-mL volumetric flask and add enough water to bring the total volume to 250.0 mL."

12.111 First, molarity represents moles of *solute* per volume of solution and, therefore, does not directly address the moles of *solvent*. Secondly, dividing molarity by the volume of solution does not result in moles, as shown below:

$$\frac{\left(\dfrac{\text{Moles}}{\text{Liter}}\right)}{\text{Liter}} = \frac{\text{Moles}}{\text{Liter}^2} \quad !$$

Instead, to obtain the moles of solute, the student should *multiply* the molarity of the solution by the volume of the solution:

$$\frac{\text{Moles}}{\text{Liter}} \times \text{Liter} = \text{Moles}$$

12.113 A concentration of 90 proof means $90/2 = 45\%$ alcohol. If this is volume percent, a 90-proof drink contains 45 mL of alcohol in every 100 mL of the drink.

12.115 (a) $100.0 \text{ g solution} \times \dfrac{25.0 \text{ g solute}}{100 \text{ g solution}} = 25.0 \text{ g solute}$

(b) $48.0 \text{ g solution} \times \dfrac{25.0 \text{ g solute}}{100 \text{ g solution}} = 12.0 \text{ g solute}$

(c) $56.5 \text{ g solute} \times \dfrac{100 \text{ g solution}}{25.0 \text{ g solute}} = 226 \text{ g solution}$

12.117 (a) $1.00 \text{ L solution} \times \dfrac{1000 \text{ mL solution}}{1 \text{ L solution}} \times \dfrac{5.00 \text{ mL alcohol}}{100.0 \text{ mL solution}} = 50.0 \text{ mL alcohol}$

You would place 50.0 mL of alcohol in a 1-L volumetric flask, add about 900 mL of water, mix thoroughly, and then add water to the 1-L mark.

(b) The plan of action is to find the mass of alcohol [using the provided density and the volume of alcohol calculated in part (a)], then the moles, and finally, the molarity of alcohol in the solution:

$$\text{Mass of alcohol} = \text{Density} \times \text{Volume of alcohol} = 0.789 \, \frac{\text{g}}{\text{mL}} \times 50.0 \text{ mL} = 39.4 \text{ g alcohol}$$

$$\text{Moles of alcohol} = \frac{\text{Mass of alcohol}}{\text{MM of alcohol}} = \frac{39.4 \text{ g alcohol}}{46.07 \text{ g/mol}} = 0.856 \text{ moles of alcohol}$$

$$\text{Molarity} = \frac{\text{Moles of alcohol}}{\text{Volume of solution}} = \frac{0.856 \text{ moles of alcohol}}{1.00 \text{ L}} = 0.856 \text{ M}$$

12.119 A concentration of 25.0 mass % means 25.0 g of NaCl per 100.0 g of solution. First use this information to get the number of moles of NaCl in 100.0 g of solution, then use density to get the volume of the 100.0 g of solution. You then have enough information to calculate molarity:

$$25.0 \text{ g NaCl} \times \frac{1 \text{ mol NaCl}}{58.443 \text{ g NaCl}} = 0.428 \text{ mol NaCl in } 100.0 \text{ g of solution}$$

$$100.0 \text{ g solution} \times \frac{1 \text{ mL solution}}{1.05 \text{ g solution}} = \frac{1 \text{ L solution}}{1000 \text{ mL solution}} = 0.0952 \text{ L solution}$$

$$\frac{0.428 \text{ mol NaCl}}{100.0 \text{ g solution}} \times \frac{100.0 \text{ g solution}}{0.0952 \text{ L solution}} = 4.50 \, \frac{\text{mol NaCl}}{\text{L solution}} = 4.50 \text{ M NaCl}$$

12.121 We mean the particles move in random, zigzag paths through the solution. They move this way because they are constantly buffeted by moving solvent molecules.

12.123 (a) $Ba^{2+}(aq) + 2OH^-(aq) \rightarrow Ba(OH)_2(s)$

(b) $5.00 \text{ g Ba(OH)}_2 \times \dfrac{1 \text{ mol Ba(OH)}_2}{171.3 \text{ g Ba(OH)}_2} = 0.0292 \text{ mol Ba(OH)}_2$

$$0.0292 \ \text{mol Ba(OH)}_2 \times \frac{1 \ \text{mol Ba}^{2+}}{1 \ \text{mol Ba(OH)}_2} \times \frac{1 \ \text{mol Ba(NO}_3)_2}{1 \ \text{mol Ba}^{2+}} \times \frac{1000 \ \text{mL Ba(NO}_3)_2 \ \text{solution}}{0.755 \ \text{mol Ba(NO}_3)_2}$$

$$= 38.7 \ \text{mL Ba(NO}_3)_2 \ \text{solution}$$

$$0.0292 \ \text{mol Ba(OH)}_2 \times \frac{2 \ \text{mol OH}^-}{1 \ \text{mol Ba(OH)}_2} \times \frac{1 \ \text{mol Ca(OH)}_2}{2 \ \text{mol OH}^-} \times \frac{1000 \ \text{mL Ca(OH)}_2 \ \text{solution}}{1.250 \ \text{mol Ca(OH)}_2}$$

$$= 23.4 \ \text{mL Ca(OH)}_2 \ \text{solution}$$

(c) From Section 8.3 of the textbook,

$$\% \ \text{yield} \times \frac{\text{Actual yield}}{\text{Theoretical yield}} \times 100\%$$

$$\text{Theoretical yield} = \frac{\text{Actual yield}}{\% \ \text{yield}} \ \text{yield} \times 100\% = \frac{5.00 \ \text{g}}{85.0\%} \times 100\% = 5.88 \ \text{g}$$

$$5.88 \ \text{g Ba(OH)}_2 \times \frac{1 \ \text{mol Ba(OH)}_2}{171.3 \ \text{g Ba(OH)}_2} = 0.0343 \ \text{mol Ba(OH)}_2$$

$$0.0343 \ \text{mol Ba(OH)}_2 \times \frac{1 \ \text{mol Ba}^{2+}}{1 \ \text{mol Ba(OH)}_2} \times \frac{1 \ \text{mol Ba(NO}_3)_2}{1 \ \text{mol Ba}^{2+}} \times \frac{1000 \ \text{mL Ba(NO}_3)_2 \ \text{solution}}{0.755 \ \text{mol Ba(NO}_3)_2}$$

$$= 45.4 \ \text{mL Ba(NO}_3)_2 \ \text{solution}$$

$$0.0343 \ \text{mol Ba(OH)}_2 \times \frac{2 \ \text{mol OH}^-}{1 \ \text{mol Ba(OH)}_2} \times \frac{1 \ \text{mol Ca(OH)}_2}{2 \ \text{mol OH}^-} \times \frac{1000 \ \text{mL Ca(OH)}_2 \ \text{solution}}{1.250 \ \text{mol Ca(OH)}_2}$$

$$= 27.4 \ \text{mL Ca(OH)}_2 \ \text{solution}$$

You could also use the shortcut method of dividing each volume calculated in (b) by the percent yield in decimal form:

$$\frac{38.7 \ \text{mL Ba(NO}_3)_2 \ \text{solution}}{0.850} = 45.5 \ \text{mL Ba(NO}_3)_2 \ \text{solution}$$

$$\frac{23.4 \ \text{mL Ca(OH)}_2 \ \text{solution}}{0.850} = 27.5 \ \text{mL Ca(OH)}_2 \ \text{solution}$$

12.125 (a) $3Pb^{2+}(aq) + 2PO_4{}^{3-}(aq) \rightarrow Pb_3(PO_4)_2(s)$

(b) The procedure for determining theoretical yield when one reactant is limiting is described in Section 8.4 of the textbook.

Step 1: See part (a).

Step 2: $50.0 \ \text{mL Pb(C}_2\text{H}_3\text{O}_2)_2 \ \text{solution} \times \dfrac{0.800 \ \text{mol Pb(C}_2\text{H}_3\text{O}_2)_2}{1000 \ \text{mL Pb(C}_2\text{H}_3\text{O}_2)_2 \ \text{solution}} \times$

$$\frac{1 \ \text{mol Pb}^{2+}}{1 \ \text{mol Pb(C}_2\text{H}_3\text{O}_2)_2} = 0.0400 \ \text{mol Pb}^{2+}$$

$$100.0 \ \text{mL Na}_3\text{PO}_4 \ \text{solution} \times \frac{0.800 \ \text{mol Na}_3\text{PO}_4}{1000 \ \text{mL Na}_3\text{PO}_4 \ \text{solution}} \times \frac{1 \ \text{mol PO}_4{}^{3-}}{1 \ \text{mol Na}_3\text{PO}_4}$$

$$= 0.0800 \ \text{mol PO}_4{}^{3-}$$

Step 2a:

$$\frac{0.0400 \text{ mol Pb}^{2+}}{3} = 0.0133 \leftarrow \text{Smaller number, limiting reactant}$$

$$\frac{0.0800 \text{ mol PO}_4{}^{3-}}{2} = 0.0400$$

Steps 3 and 4:

$$0.0400 \text{ mol Pb}^2 \times \frac{1 \text{ mol Pb}_3(\text{PO}_4)_2}{3 \text{ mol Pb}^{2+}} \times \frac{811.5 \text{ g Pb}_3(\text{PO}_4)_2}{1 \text{ mol Pb}_3(\text{PO}_4)_2} = 10.8 \text{ g Pb}_3(\text{PO}_4)_2$$

(c) You must first determine the number of moles of $PO_4{}^{3-}$ not used:

$$0.0400 \text{ mol Pb}^{2+} \times \frac{2 \text{ mol PO}_4{}^{3-}}{3 \text{ mol Pb}^{2+}} = 0.0267 \text{ mol PO}_4{}^{3-} \text{ consumed}$$

$$0.800 \text{ mol PO}_4{}^{3-} - 0.0267 \text{ mol PO}_4{}^{3-} = 0.0533 \text{ mol PO}_4{}^{3-} \text{ remaining}$$

This amount of $PO_4{}^{3-}$ is present in 100.0 mL + 50.0 mL = 150.0 mL of solution, making the molar concentration of $PO_4{}^{3-}$:

$$\frac{0.0533 \text{ mol}}{0.1500 \text{ L}} = 0.355 \text{ M}$$

12.127 (a) $H^+(aq) + OH^-(aq) \rightarrow H_2O(l)$

(b) The number of moles of OH^- used to neutralize the acid is:

$$0.02755 \text{ L NaOH} \times \frac{1.00012 \text{ mol NaOH}}{1 \text{ L NaOH}} \times \frac{1 \text{ mol OH}^-}{1 \text{ mol NaOH}} = 0.02756 \text{ mol OH}^-$$

(c) From the stoichiometry of the equation in part (a) and the fact that there are two H^+ in H_2SO_4:

$$0.02756 \text{ mol OH}^- \times \frac{1 \text{ mol H}^+}{1 \text{ mol OH}^-} \times \frac{1 \text{ mol H}_2\text{SO}_4}{2 \text{ mol H}^+} = 0.01378 \text{ mol H}_2\text{SO}_4$$

(d) This number of moles in the 25.00-mL acid sample means the molar concentration of HBr is

$$\frac{0.01378 \text{ mol}}{0.02500 \text{ L}} = 0.5512 \text{ M}$$

12.129 (a) A heating curve tells you (1) when a substance is changing phase (horizontal portions) and (2) when the temperature of the substance is changing (sloping portions). Melting is the phase change from solid to liquid, and therefore the lower horizontal portion tells you that the melting point for this substance is $-115.0\,°C$.

(b) Freezing is the phase change from liquid to solid, which means the same horizontal portion used in part (a) also tells you the freezing point, $-115.0\,°C$.

(c) Boiling is the phase change from liquid to gas, and therefore the upper horizontal portion of the heating curve tells you that the boiling point for this substance is $78.4\,°C$.

(d) Condensation is the phase change from gas to liquid, which means the upper horizontal line tells you the condensation point as well as the boiling point. Both are $78.4\,°C$.

(e) Because the temperature of a substance does not change whenever the substance is changing phase at its freezing/melting point or its boiling/condensing point.

(f) Ethanol.

12.131 To break the hydrogen bonds holding the water molecules together in the liquid phase.

12.133 The pressure exerted by the gas phase of a substance when the gas phase is in dynamic equilibrium with the liquid phase of the substance.

12.135 Water boils when its vapor pressure is equal to the downward pressure exerted by the air on the water surface. Because the vapor pressure of water at 20 °C is 17.54 mm Hg, you get water to boil at this temperature by using a vacuum pump to reduce the air pressure above the water surface to 17.54 mm Hg or less.

12.137 Any liquid boils when its vapor pressure is equal to atmospheric pressure, which is normally 760 mm Hg. Because 80.1 °C is the normal boiling point of benzene, the vapor pressure at that temperature must be atmospheric pressure, 760 mm Hg.

12.139 The two solutions have the same freezing point when they contain the same number of moles of solute particles. NaCl dissociates to 2 moles of solute particles for every 1 mole of NaCl, which means the number of moles of solute particles in the 28.7 g is

$$28.7 \text{ g NaCl} \times \frac{1 \text{ mol NaCl}}{58.443 \text{ g NaCl}} \times \frac{2 \text{ mol solute particles}}{1 \text{ mol NaCl}} = 0.982 \text{ mol solute particles}$$

Because sucrose does not dissociate when it dissolves, each mole of sucrose yields 1 mole of solute particles. You therefore need 0.982 mole of sucrose, which means a mass of

$$0.982 \text{ mol sucrose} \times \frac{342 \text{ g sucrose}}{1 \text{ mol sucrose}} = 336 \text{ g sucrose}$$

12.141 The two solutions have the same vapor pressure when they contain the same number of moles of solute particles. Because each mole of Na_2SO_4 dissociates to 3 moles of solute particles, the Na_2SO_4 solution contains

$$45.6 \text{ g Na}_2\text{SO}_4 \times \frac{1 \text{ mol Na}_2\text{SO}_4}{142 \text{ g Na}_2\text{SO}_4} \times \frac{3 \text{ mol solute particles}}{1 \text{ mol Na}_2\text{SO}_4} = 0.963 \text{ mol solute particles}$$

She therefore needs 0.963 mol of solute particles from NaCl, where 1 mole of the salt yields 2 moles of particles:

$$0.963 \text{ mol solute particles} \times \frac{1 \text{ mol NaCl}}{2 \text{ mol solute particles}} \times \frac{58.4 \text{ g NaCl}}{1 \text{ mol NaCl}} = 28.1 \text{ g NaCl}$$

12.143 The ΔT_b and ΔT_f equations require the solvent mass in kilograms:

$$100.0 \text{ mL benzene} \times \frac{0.874 \text{ g benzene}}{1 \text{ mL benzene}} \times \frac{1 \text{ kg benzene}}{1000 \text{ g benzene}} = 0.0874 \text{ kg benzene}$$

The equations require the solute particles in moles. The solute cetyl alcohol does not dissociate when it dissolves:

$$5.75 \text{ g C}_{16}\text{H}_{34}\text{O} \times \frac{1 \text{ mol C}_{16}\text{H}_{34}\text{O}}{242.4 \text{ g C}_{16}\text{H}_{34}\text{O}} \times \frac{1 \text{ mol solute particles}}{1 \text{ mol C}_{16}\text{H}_{34}\text{O}} = 0.0237 \text{ mol solute particles}$$

(a) $\Delta T_b = \dfrac{2.53\,\dfrac{°C \cdot kg\ benzene}{mol\ solute\ particles} \times 0.0237\ mol\ solute\ particles}{0.0874\ kg\ benzene} = 0.686\,°C$

Elevated boiling point $=$ Normal boiling point $+\ \Delta T_b = 80.1\,°C + 0.686\,°C = 80.8\,°C$

(b) $\Delta T_f = \dfrac{5.12\,\dfrac{°C \cdot kg\ benzene}{mol\ solute\ particles} \times 0.0237\ mol\ solute\ particles}{0.0874\ kg\ benzene} = 1.39\,°C$

Depressed freezing point $=$ Normal freezing point $-\ \Delta T_f = 5.53\,°C - 1.39\,°C = 4.14\,°C$

12.145 The ΔT_f equation solved for moles of solute particles tells you how many moles of cholesterol is represented by the 1.56 g. To use the equation, you'll need the solvent mass in kilograms and a value for ΔT_f

$50.0\ mL\ cyclohexane \times \dfrac{0.779\ g\ cyclohexane}{1\ mL\ cyclohexane} \times \dfrac{1\ kg\ cyclohexane}{1000\ g\ cyclohexane} = 0.0390\ kg\ cyclohexane$

$\Delta T_f =$ Normal freezing point $-$ Depressed freezing point $= 6.47\,°C - 4.40\,°C = 2.07\,°C$

$\Delta T_f = \dfrac{K_f \times Mol\ solute\ particles}{Kilograms\ solvent}$

Mol solute particles $= \dfrac{Kilograms\ solvent \times \Delta T_f}{K_f} = \dfrac{0.0390\ kg\ cyclohexane \times 2.07\,°C}{20.0\,\dfrac{°C \cdot kg\ cyclohexane}{mol\ solute\ particles}}$

$= 0.00404\ mol\ solute\ particles$

Assuming cholesterol does not dissociate when it dissolves, you get:

$0.00404\ mol\ solute\ particles \times \dfrac{1\ mol\ cholesterol}{1\ mol\ solute\ particles} = 0.00404\ mol\ cholesterol$

This is the number of moles of cholesterol in the 1.56 g, making the molar mass

$\dfrac{1.56\ g}{0.00404\ mol} = 386\ g/mol$

12.147 (a) The nonpolar tails of a soap are hydrophobic, meaning "water-fearing," while the polar heads of a soap are hydrophilic, meaning "water-loving." When a soap is dissolved in pool water, the hydrophilic heads maximize contact with water, while the hydrophobic tails try to stay as far away from water as possible, that is, point toward air above the water.

(b) Water is polar and so are the heads of a soap. Therefore, heads will orient themselves so as to maximize contact with water. Tails are nonpolar and so they will orient themselves *away* from water and toward the air (air can be viewed primarily as a mixture of nonpolar gases: N_2, O_2, CO_2).

(c) Soap disrupts the strong hydrogen bonds present in water, including those on the water's surface. This means that the "skin" on the water's surface is weakened to the extent that it gets easily pierced by the legs of water bugs.

12.149 By forming a micelle that keeps the hydrophobic portion of each soap molecule away from the water.

12.151 The hydrophobic portion is the hydrocarbon chain. The hydrophilic portion is

$$
O-\overset{\overset{\displaystyle O}{\|}}{\underset{\underset{\displaystyle O}{\|}}{S}}-O^- \ Na^+.
$$

12.153 The amount of hydration energy released is proportional to the strength of the intermolecular forces between solute molecule and water molecule. The stronger these forces, the greater the amount of energy released. Therefore $CH_3CH_2OH(l)$ releases the most energy because these molecules contain an $-OH$ group and can therefore form hydrogen bonds with water molecules. $C_8H_{18}(l)$ releases the least energy because the only forces possible between these nonpolar molecules and water molecules are London forces. Because hydrogen bonds are stronger than London forces, $CH_3CH_2OH(l)$ releases more energy.

An intermediate amount of energy is released by $CH_3Cl(l)$, which interacts with water molecules via dipole–dipole interactions. (Recall from Section 10.3 of the textbook that hydrogen bonds are considerably stronger than dipole–dipole interactions.)

12.155 Step (b) because a negative ΔE value means, by convention, that energy is released, and energy is released whenever two particles attract each other. Steps (a) and (c) both have a positive energy change (these steps absorb energy).

12.157 $0.02886 \ \text{L solution} \times \dfrac{5.20 \times 10^{-3} \ \text{mol KMnO}_4}{1 \ \text{L solution}} = 1.50 \times 10^{-4} \ \text{mol KMnO}_4$

12.159 The dilution equation tells you how much stock solution you need to dilute:

$$
V_{\text{stock solution}} = \frac{0.350 \ \text{M} \times 250.0 \ \text{ml}}{6.00 \ \text{M}} = 14.6 \ \text{mL}
$$

Transfer 14.6 mL of the stock solution to a 250-mL volumetric flask and dilute to the mark with water.

12.161 Because molarity is moles per liter, the first step is to convert the given volume to liters:

$$
200 \ \text{cm}^3 \times \frac{1 \ \text{mL}}{1 \ \text{cm}^3} \times \frac{1 \ \text{L}}{1000 \ \text{mL}} = 0.200 \ \text{L}
$$

(a) $0.200 \ \text{L} \times \dfrac{0.22 \ \text{mol NaCl}}{\text{L}} \times \dfrac{1 \ \text{mol Na}^+}{1 \ \text{mol NaCl}} = 0.0400 \ \text{mol Na}^+$

$0.22 \ \text{L} \dfrac{0.200 \ \text{mol NaCl}}{\text{L}} \times \dfrac{1 \ \text{mol Cl}^-}{1 \ \text{mol NaCl}} = 0.0400 \ \text{mol Cl}^-$

(b) $0.200 \ \text{L} \times \dfrac{0.350 \ \text{mol K}_3\text{PO}_4}{\text{L}} \times \dfrac{3 \ \text{mol K}^+}{1 \ \text{mol K}_3\text{PO}_4} = 0.210 \ \text{mol K}^+$

$0.200 \ \text{L} \times \dfrac{0.350 \ \text{mol K}_3\text{PO}_4}{\text{L}} \times \dfrac{1 \ \text{mol PO}_4{}^{3-}}{1 \ \text{mol K}_3\text{PO}_4} = 0.0700 \ \text{mol PO}_4{}^{3-}$

(c) $0.200 \cancel{L} \times \dfrac{1.44 \cancel{\text{mol Al(NO}_3)_3}}{\cancel{L}} \times \dfrac{1 \text{ mol Al}^{3+}}{1 \cancel{\text{mol Al(NO}_3)_3}} = 0.288 \text{ mol Al}^{3+}$

$0.200 \cancel{L} \times \dfrac{1.44 \cancel{\text{mol Al(NO}_3)_3}}{\cancel{L}} \times \dfrac{3 \text{ mol NO}_3^-}{1 \cancel{\text{mol Al(NO}_3)_3}} = 0.864 \text{ mol NO}_3^-$

12.163 (a) $4.70 \text{ g } \cancel{\text{CuSO}_4} \times \dfrac{1 \text{ mol CuSO}_4}{160 \text{ g } \cancel{\text{CuSO}_4}} = 0.0294 \text{ mol CuSO}_4$

$\dfrac{0.0294 \text{ mol}}{150.0 \cancel{\text{cm}^3}} \times \dfrac{1 \cancel{\text{cm}^3}}{1 \cancel{\text{mL}}} \times \dfrac{1000 \cancel{\text{mL}}}{1 \text{ L}} = 0.196 \text{ mol/L} = 0.196 \text{ M}$

(b) $1.00 \cancel{\text{mL}} \times \dfrac{1 \cancel{L}}{1000 \cancel{\text{mL}}} \times \dfrac{0.196 \text{ mol}}{\cancel{L}} = 1.96 \times 10^{-4} \text{ mol}$

(c) To calculate percent by mass, both solute amount and solution amount must be in mass units. You already have solute mass and therefore need to calculate only solution mass:

$150.0 \cancel{\text{cm}^3} \times \dfrac{1 \cancel{\text{mL}}}{1 \cancel{\text{cm}^3}} \times \dfrac{1.01 \text{ g}}{\cancel{\text{mL}}} = 152 \text{ g}$

$\dfrac{4.70 \text{ g solute}}{152 \text{ g solution}} \times 100\% = 3.09 \text{ mass \%}$

12.165 Get the empirical formula from the combustion data, assuming a 100.0 g sample so that percents can be directly converted to grams. The amount of O is determined by difference:

$100.0\% - 40.9\% \text{ C} - 4.48\% \text{ H} - 54.5\% \text{ O}.$

$40.9 \text{ g C} \times \dfrac{1 \text{ mol C}}{12.011 \text{ g C}} = 3.41 \text{ mol C}$

$4.58 \text{ g H} \times \dfrac{1 \text{ mol H}}{1.0079 \text{ g H}} = 4.54 \text{ mol H}$

$54.5 \text{ g O} \times \dfrac{1 \text{ mol O}}{15.999 \text{ g O}} = 3.41 \text{ mol O}$

$C_{\frac{3.41}{3.41}} H_{\frac{4.54}{3.41}} O_{\frac{3.41}{3.41}} \rightarrow C_{1.00} H_{1.33} O_{1.00} \rightarrow C_{1.00 \times 3} H_{1.33 \times 3} O_{1.00 \times 3} = C_3 H_4 O_3$

A compound having this empirical formula has a molar mass of 88.1 g/mol. To determine whether or not this empirical formula is the molecular formula, you need to know the vitamin C molar mass, which you get from the freezing point of the vitamin C solution:

$\text{Mol solute} = \dfrac{\Delta T_f \times \text{kg solvent}}{K_f} = \dfrac{2.05 \text{ }°C \times 0.100 \text{ } \cancel{\text{kg solvent}}}{1.86 \dfrac{°C \cdot \cancel{\text{kg solvent}}}{\text{mol solute}}} = 0.110 \text{ mol solute}$

The molar mass of the vitamin C is therefore

$\dfrac{19.40 \text{ g}}{0.110 \text{ mol}} = 176 \text{ g/mol}$

Because $(176 \text{ g/mol})/(88.1 \text{ g/mol}) = 2$, the molecular formula is $C_{2 \times 3} H_{2 \times 4} O_{2 \times 3} = C_6 H_8 O_6$.

12.167 Wine that is 24 proof is $24/2 = 12\%$ alcohol, by definition, which means each gallon of wine contains 0.12 gallon of alcohol, as you can see by rearranging the expression for percent composition by volume:

$$\text{Volume of solute} = \frac{12\% \times 1 \text{ gallon wine}}{100\%} = 0.12 \text{ gallon alcohol}$$

$$0.100 \text{ gallon alcohol} \times \frac{1 \text{ gallon wine}}{0.12 \text{ gallon alcohol}} = 0.83 \text{ gallon wine}$$

12.169 (a) Use the dilution equation with $V_{\text{diluted solution}} = 125.0 \text{ mL} + 50.0 \text{ mL} = 175.0 \text{ mL}$:

$$M_{\text{diluted solution}} = \frac{0.250 \text{ M} \times 50.0 \text{ mL}}{175.0 \text{ mL}} = 0.0714 \text{ M}$$

(b) $0.0714 \text{ M } (NH_4)_3PO_4 = \dfrac{3 \text{ mol NH}_4^+}{1 \text{ mol } (NH_4)_3PO_4} = 0.214 \text{ M NH}_4^+$

(c) $0.0714 \text{ M } (NH_4)_3PO_4 \times \dfrac{1 \text{ mol PO}_4^{3-}}{1 \text{ mol } (NH_4)_3PO_4} = 0.0714 \text{ M PO}_4^{3-}$

12.171 The net ionic equation is $Ba^{2+}(aq) + SO_4^{2-}(aq) \rightarrow BaSO_4(s)$. In order to determine how much sulfate ion you need, you must know how much barium ion you have:

$$0.2500 \text{ L} \times \frac{0.600 \text{ mol Ba(NO}_3)_2}{\text{L}} \times \frac{1 \text{ mol Ba}^{2+}}{1 \text{ mol Ba(NO}_3)_2} = 0.150 \text{ mol Ba}^{2+}$$

You therefore need 0.150 mole of SO_4^{2-}, and the volume of sodium sulfate solution containing that amount of SO_4^{2-} is

$$0.150 \text{ mol SO}_4^{2-} \times \frac{1 \text{ mol Na}_2SO_4}{1 \text{ mol SO}_4^{2-}} \times \frac{1 \text{ L}}{0.500 \text{ mol Na}_2SO_4} = 0.300 \text{ L}$$

12.173 The complete ionic equation is

$$2 \text{ Na}^+(aq) + 2 I^-(aq) + Pb^{2+}(aq) + 2 C_2H_3O_2^-(aq) \rightarrow PbI_2(s) + 2 \text{ Na}^+(aq) + 2 C_2H_3O_2^-(aq)$$

telling you the spectator ions are $Na^+(aq)$ and $C_2H_3O_2^-(aq)$. The amount of lead ion in the 100.0 mL of $Pb(C_2H_3O_2)_2$ solution is

$$0.1000 \text{ L} \times \frac{0.300 \text{ mol Pb(C}_2H_3O_2)_2}{\text{L}} \times \frac{1 \text{ mol Pb}^{2+}}{1 \text{ mol Pb(C}_2H_3O_2)_2} = 0.0300 \text{ mol Pb}^{2+}$$

Because 1 mole of Pb^{2+} reacts with 2 moles of I^-, the volume of NaI solution needed is

$$0.0300 \text{ mol Pb}^{2+} \times \frac{2 \text{ mol I}}{1 \text{ mol Pb}^{2+}} \times \frac{1 \text{ mol NaI}}{1 \text{ mol I}} \times \frac{1 \text{ L}}{0.245 \text{ mol NaI}} = 0.245 \text{ L}$$

12.175 Adding the 27.65 mL of water to the flask does not change the number of moles of acid in the flask. All the acid that reacts with the NaOH comes from the 25.00 mL sample. The number of moles of HCl in this sample is

$$0.02870 \text{ L NaOH} \times \frac{0.1004 \text{ mol OH}^-}{1 \text{ L NaOH}} \times \frac{1 \text{ mol H}^+}{1 \text{ mol OH}^-} \times \frac{1 \text{ mol HCl}}{1 \text{ mol H}^+} = 0.002881 \text{ mol HCl}$$

Because this number of moles is in 25.00 mL, the concentration of the sample solution is

$$\frac{0.002881 \text{ mol}}{0.02500 \text{ L}} = 0.1152 \text{ mol/L} = 0.1152 \text{ M}$$

Remember, in a titration, it's the number of moles in the flask that is important, not the volume of solution in the flask.

12.177 No. A compound is a single pure substance, but a solution must contain at least two substances, mixed homogeneously.

12.179 More soluble. I_2 and CCl_4 are both nonpolar molecules, and H_2O is a polar molecule. Because "like dissolves like," you expect I_2 to be more soluble in CCl_4 than in H_2O. In terms of energy, the attractive forces that develop between I_2 and CCl_4 are greater than the forces that develop between I_2 and H_2O. The I_2–CCl_4 attractions release more energy to drive the energy-absorbing solute-separation and solvent-separation steps.

12.181 Aluminum phosphate contains Al^{3+} ions, which carry three times the positive charge of the Na^+ ions in sodium phosphate. Because of this higher charge, the ionic bonds in the $AlPO_4$ lattice are stronger than those in the Na_3PO_4 lattice. The strongly held together $AlPO_4$ lattice makes it insoluble, while the much weaker ionic bonds in the Na_3PO_4 lattice make it very soluble.

12.183 Whether or not a solute dissolves in a solvent is decided not only by changes in energy but also by changes in *entropy*.

12.185 (a) An endothermic reaction is one that absorbs energy from its surroundings. According to the convention taught in the textbook, a reaction that absorbs energy has a positive ΔE_{total} value.

(b) The entropy for the universe increases when the solute dissolves because there is less order in the arrangement of the dissolved solute particles than in their arrangement in the undissolved solid.

12.187 The graph tells you that, at 20.0 °C, the solubility of NaCl in water is 38 g per 100.0 g of water. In 75 g of 20.0 °C water, therefore, you can dissolve

$$75 \text{ g water} \times \frac{38 \text{ g NaCl}}{100.0 \text{ g water}} = 29 \text{ g NaCl}$$

Only 29 g of NaCl dissolves, which means the answer is no.

12.189 The graph tells you that, at 20.0 °C the solubility of glucose in water is 86 g per 100.0 g of water. In 60.0 g of 20.0 °C water, therefore, a saturated solution contains:

$$60.0 \text{ g water} \times \frac{86 \text{ g glucose}}{100.0 \text{ g water}} = 52 \text{ g glucose}$$

12.191

All the portions carrying a charge, either partial or full are hydrophilic because they can form either hydrogen bonds or ion–dipole attractions with water molecules.

12.193 Fat molecule:

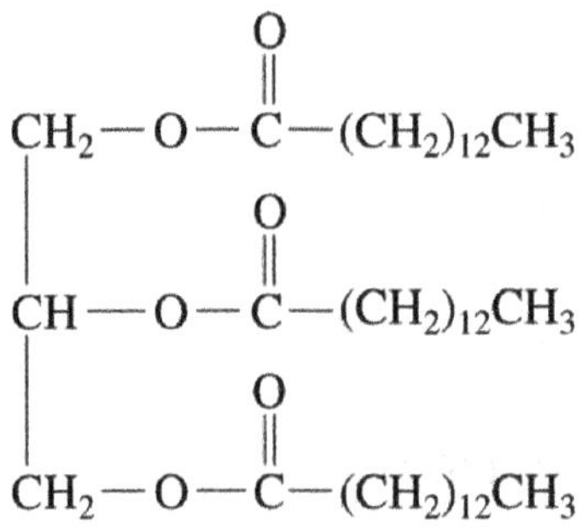

$$
\begin{array}{l}
\text{CH}_2-\text{O}-\overset{\displaystyle\text{O}}{\overset{\|}{\text{C}}}-(\text{CH}_2)_{12}\text{CH}_3 \\[2pt]
\;\;| \\[2pt]
\text{CH}-\text{O}-\overset{\displaystyle\text{O}}{\overset{\|}{\text{C}}}-(\text{CH}_2)_{12}\text{CH}_3 \\[2pt]
\;\;| \\[2pt]
\text{CH}_2-\text{O}-\overset{\displaystyle\text{O}}{\overset{\|}{\text{C}}}-(\text{CH}_2)_{12}\text{CH}_3
\end{array}
$$

Each soap molecule is:

$$
\text{K}^+\text{O}^- - \overset{\displaystyle\text{O}}{\overset{\|}{\text{C}}}-(\text{CH}_2)_{12}\text{CH}_3
$$

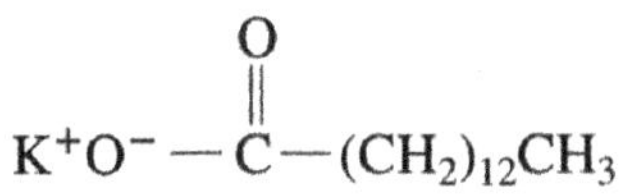

12.195 The soap molecules in a vesicle align themselves such that there are polar heads (gray circles in the textbook drawing) lining both the outside and the hollowed-out center of the structure. (The tails from the two layers intermingle and "dissolve" into one another.) Because there are polar heads at the center, polar water molecules migrate there to interact with the polar heads.

12.197 (a) $M_{\text{diluted solution}} = \dfrac{0.500\ \text{M} \times 0.0450\ \cancel{L}}{1.50\ \cancel{L}} = 0.0150\ \text{M}$

(b) The formula for aluminum sulfate is $Al_2(SO_4)_3$.

$$
\frac{0.015\ \cancel{\text{mol Al}_2(\text{SO}_4)_3}}{1\ \text{L}} \times \frac{2\ \text{mol Al}^{3+}}{1\ \cancel{\text{mol Al}_2(\text{SO}_4)_3}} = 0.0300\frac{\text{mol Al}^{3+}}{\text{L}} = 0.0300\ \text{M}
$$

(c) $\dfrac{0.0150\ \cancel{\text{mol Al}_2(\text{SO}_4)_3}}{1\ \text{L}} \times \dfrac{3\ \text{mol SO}_4^{2-}}{1\ \cancel{\text{mol Al}_2(\text{SO}_4)_3}} = 0.0450\dfrac{\text{mol SO}_4^{2-}}{\text{L}} = 0.0450\ \text{M}$

12.199 Your first step is to determine the number of moles of each base in the $60.0\ \text{mL} + 60.0\ \text{mL} = 120.0\ \text{mL}$ of solution. Because the product MV equals number of moles, you can say (remembering that V must be in liters)

$0.250\ \text{M} \times 0.0600\ \text{L} = 0.0150\ \text{mol NaOH}$

$0.125\ \text{M} \times 0.0600\ \text{L} = 0.00750\ \text{mol Ba(OH)}_2$

Now determine how much OH^- these quantities yield:

$0.0150\ \cancel{\text{mol NaOH}} \times \dfrac{1\ \text{mol OH}^-}{1\ \cancel{\text{mol NaOH}}} = 0.0150\ \text{mol OH}^-$

$0.00750\ \cancel{\text{mol Ba(OH)}_2} \times \dfrac{2\ \text{mol OH}^-}{1\ \cancel{\text{mol Ba(OH)}_2}} = 0.0150\ \text{mol OH}^-$

You therefore have $0.0150\ \text{mol} + 0.0150\ \text{mol} = 0.0300\ \text{mol OH}^-$ in the 120.0 mL, meaning the molarity is:

$\dfrac{0.0300\ \text{mol}}{0.1200\ \text{L}} = 0.250\dfrac{\text{mol}}{\text{L}} = 0.250\ \text{M}$

12.201 $0.5000\ \cancel{\text{L solution}} \times \dfrac{0.300\ \cancel{\text{mol NaOH}}}{1\ \cancel{\text{L solution}}} \times \dfrac{39.997\ \text{g NaOH}}{1\ \cancel{\text{mol NaOH}}} = 6.00\ \text{g NaOH}$

12.203 The product MV gives number of moles (V must be in liters).

$$5.20 \times 10^{-3} \text{ M} \times 0.02868 \text{ L} = 1.49 \times 10^{-4} \text{ mol}$$

12.205 The number of moles of H^+ to be neutralized is

$$0.0500 \text{ L} \times \frac{0.0100 \text{ mol } H_2SO_4}{1 \text{ L}} \times \frac{2 \text{ mol } H^+}{1 \text{ mol } H_2SO_4} = 0.00100 \text{ mol } H^+$$

The volume of base solution containing this many moles of OH^- is:

$$0.00100 \text{ mol } OH^- \times \frac{1 \text{ mol NaOH}}{1 \text{ mol } OH^-} \times \frac{1 \text{ L NaOH}}{0.0150 \text{ mol NaOH}} = 0.0667 \text{ L} = 66.7 \text{ mL}$$

12.207 Percent composition by mass is calculated with solute and solution masses. Therefore, your first step is to use the density to convert the given solvent volume to mass:

$$1.00 \text{ L} \times \frac{1000 \text{ mL}}{1 \text{ L}} \times \frac{1.00 \text{ g}}{\text{mL}} = 1.00 \times 10^3 \text{ g water}$$

$$\frac{5.00 \text{ g sucrose}}{5.00 \text{ g sucrose} + (1.00 \times 10^3 \text{ g water})} \times 100\%$$

$$= \frac{5.00 \text{ g sucrose}}{(0.00500 \times 10^3 \text{ g sucrose}) + (1.00 \times 10^3 \text{ g water})} \times 100\%$$

$$= \frac{5.00 \text{ g sucrose}}{1.01 \times 10^3 \text{ g solution}} \times 100\% = 0.495 \text{ mass \%}$$

12.209 Air being 21 vol % O_2 means each 100 mL of air contains 21 mL of O_2, information you use as a conversion factor:

$$200.0 \text{ L air} \times \frac{1000 \text{ mL air}}{1 \text{ L air}} \times \frac{21 \text{ mL } O_2}{100 \text{ mL air}} = 4.2 \times 10^4 \text{ mL } O_2 \text{ in 200.0 L of air}$$

Now use the ideal gas equation solved for n to determine moles of O_2 (remembering that you must convert O_2 volume to liters to match the units of R):

$$n = \frac{1.00 \text{ atm} \times 42 \text{ L}}{0.0821 \dfrac{\text{L} \cdot \text{atm}}{\text{K} \cdot \text{mol}} \times 298 \text{ K}} = 1.7 \text{ mol } O_2$$

$$1.7 \text{ mol } O_2 = \frac{31.998 \; O_2}{1 \text{ mol } O_2} = 54 \text{ g } O_2$$

12.211 11.5 mass % means each 100.0 g of steel contains 11.5 g of Cr.

$$250.0 \text{ lb steel} \times \frac{453.6 \text{ g steel}}{1 \text{ lb steel}} \times \frac{11.5 \text{ g Cr}}{100.0 \text{ g steel}} = 1.30 \times 10^4 \text{ g Cr}$$

12.213 (a) The procedure for determining theoretical yield is described in Section 8.4 of the textbook.

Step 1: $Ca^{2+}(aq) + 2F^-(aq) \rightarrow CaF_2(s)$

Step 2: $0.0265 \text{ L Ca(NO}_3)_2 \times \dfrac{0.100 \text{ mol Ca(NO}_3)_2}{1 \text{ L Ca(NO}_3)_2} \times \dfrac{1 \text{ mol } Ca^{2+}}{1 \text{ mol Ca(NO}_3)_2}$

$= 0.00265 \text{ mol } Ca^{2+}$

$$0.0498 \text{ L NaF solution} \times \frac{0.100 \text{ mol NaF}}{1 \text{ L NaF solution}} \times \frac{1 \text{ mol F}^-}{1 \text{ mol NaF}} = 0.00498 \text{ mol F}^-$$

Step 2a: $\dfrac{0.00265 \text{ mol Ca}^{2+}}{1} = 0.00265$

$\dfrac{0.00498 \text{ mol F}^-}{2} = 0.00294 \leftarrow$ Smaller number, limiting reactant

Steps 3 and 4: $0.00498 \text{ mol F}^- \times \dfrac{1 \text{ mol CaF}_2}{2 \text{ mol F}^-} \times \dfrac{78.074 \text{ g CaF}_2}{1 \text{ mol CaF}_2} = 0.194 \text{ g CaF}_2$ consumed

(b) You must first determine the number of moles of Ca^{2+} not used in the reaction. Use the equation coefficients to determine how much Ca^{2+} reacted with the F^-:

$$0.00498 \text{ mol F}^- \times \frac{1 \text{ mol CaF}_2}{2 \text{ mol F}^-} \times \frac{1 \text{ mol Ca}^{2+}}{1 \text{ mol CaF}_2} = 0.00249 \text{ mol Ca}^{2+} \text{ used}$$

$0.00265 \text{ mol Ca}^{2+} - 0.00249 \text{ mol Ca}^{2+} = 0.00016 \text{ mol Ca}^{2+}$ left over

This amount of Ca^{2+} is present in $26.5 \text{ mL} + 49.8 \text{ mL} = 76.3 \text{ mL}$ of solution, making the molar Ca^{2+} concentration

$$\frac{0.00016 \text{ mol}}{0.0763 \text{ L}} = 0.0021 \text{ M}$$

12.215 (a) $Ag^+(aq) + Br^-(aq) \rightarrow AgBr(s)$

(b) $20.0 \text{ g NaBr} \times \dfrac{1 \text{ mol NaBr}}{102.894 \text{ g NaBr}} \times \dfrac{1 \text{ mol Br}^-}{1 \text{ mol NaBr}} = \dfrac{0.194 \text{ mol Br}^-}{1} = 0.194$

$0.0500 \text{ L AgNO}_3 \text{ solution} \times \dfrac{2.00 \text{ mol AgNO}_3}{1 \text{ L solution}} \times \dfrac{1 \text{ mol Ag}^+}{1 \text{ mol AgNO}_3}$

$= \dfrac{0.100 \text{ mol Ag}^+}{1} = 0.100 \leftarrow$ Smaller number, limiting reactant

Use the limiting reactant to calculate theoretical yield:

$$0.100 \text{ mol Ag}^+ \times \frac{1 \text{ mol AgBr}}{1 \text{ mol Ag}^+} \times \frac{187.772 \text{ g AgBr}}{1 \text{ mol AgBr}} = 18.8 \text{ g AgBr}$$

(c) $\dfrac{15.0 \text{ g}}{18.8 \text{ g}} \times 100\% = 79.8\%$

(d) $0.100 \text{ mol Ag}^+ \times \dfrac{1 \text{ mol Br}^-}{1 \text{ mol Ag}^+} = 0.100 \text{ mol Br}^-$ consumed

The Br^- remaining in the 50.0 mL of solution is thus

$0.194 \text{ mol} - 0.100 \text{ mol} = 0.094 \text{ mol}$
making the Br^- molar concentration

$$\frac{0.094 \text{ mol}}{0.0500 \text{ L}} = 1.88 \text{ M}$$

12.217 The pressure does not remain at zero because some of the liquid water evaporates to create water vapor. The pressure stops at 23.76 mm Hg because at this point at room temperature, equilibrium is reached between evaporation and condensation.

12.219 (a) Situation (a) represents a soluble solute as solvation releases more than enough energy to accomplish the first two steps.

(b) Situation (b) represents an insoluble solute as there is a tremendous energy deficit (not nearly enough energy is released by solvation to accomplish the first two steps).

(c) Situation (c) has only a minor energy deficit and thus may be tipped toward solubility by an entropy increase.

12.221 $\left(\dfrac{1.0000 \text{ g } H_2O}{\text{mL } H_2O}\right)\left(\dfrac{\text{mole } H_2O}{18.015 \text{ g } H_2O}\right)\left(\dfrac{100 \text{ mL } H_2O}{\text{L } H_2O}\right) = 55.51\dfrac{\text{mole}}{\text{liter}} = 55.51 \text{ M}$

12.223 The freezing point of the cyclohexane is depressed by $6.47\,^\circ C - 3.97\,^\circ C = 2.50\,^\circ C = \Delta T_f$

So, $\Delta T_f = 2.50\,^\circ C = \left(\dfrac{\left(\dfrac{20.0\,^\circ C \text{ kg solvent}}{\text{mole solute}}\right)(\text{mole of solute})}{0.100 \text{ kg solvent}}\right)$

Yielding moles of solute $= 0.0125$ mole

Now, you were told that 0.100 moles of solute were actually dissolved. Yet according to the freezing-point depression, only 0.0125 moles of particles were in solution! This is one-eighth of the moles of solute that was put in and one-sixteenth of the number of individual ions we would expect if this compound dissociated into ions like NaCl does! This supports the notion that CH_3Li forms soluble neutral structures, as the one shown in the text of the problem, in which $\dfrac{0.100 \text{ moles}}{0.0125 \text{ moles}} = 8$ ions must cluster together to act as one particle.

12.225 When an ordered solid solute dissolves in a solvent, the resulting solution represents a state of increased disorder and increased entropy. The amount of entropy increase will depend on the number of dissolved solute particles, not on what they are. This is consistent with how the colligative properties behave, so the colligative properties must be tied to entropy increase when a solid solute is dissolved in a solvent.

12.227 (a) Well, how you answer depends on your point of view, but without a doubt it solves the solvent-separation step and the solution step. As for which of these two is most responsible, we will choose solvation (but the choice of the solvent-separation step would not be wrong). Read on.

(b) Solvent separation requires a lot of energy because of the strong hydrogen bonds between water molecules. Now consider the solute. Although it has an OH group capable of hydrogen-bonding to water, most of the molecule is organic (nonpolar) and cannot interact strongly with water molecules. Thus, to accommodate the large solute molecule, many solvent hydrogen bonds must be broken, but few solvent–solute hydrogen bonds can form to supply the energy for solvent separation. So, is it solvent separation or solvation that is the problem here? It is both.

Chapter 13

13.1   See solution in textbook.

13.3   (a) Because $\Delta E_{rxn} = E_{products} - E_{reactants}$ is negative, $E_{reactants}$ must be higher than $E_{products}$.

(b) $E_{reactants} = E_{products} - \Delta E_{rxn} = 20\,kJ/mol - (-450\,kJ/mol) = +470\,kJ/mol$.

(c) Exothermic, because ΔE_{rxn} is negative.

(d) Downhill, starting high ($E_{reactants} = 470\,kJ/mol$) and ending lower ($E_{products} = 20\,kJ/mol$).

(e)

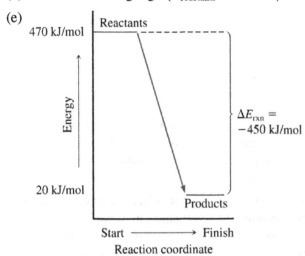

13.5   (a) That the container gets cold tells you energy (in the form of heat) is absorbed into the reaction mixture. This means the reaction is endothermic, and therefore ΔE_{rxn} must be a positive number: $\Delta E_{forward\ rxn} = +250\,kJ/mol$.

(b) A change of sign gives $\Delta E_{reverse\ rxn} = -250\,kJ/mol$.

13.7   See solution in textbook.

13.9   200 kJ. The negative value of ΔE_{rxn} means an exothermic reaction. Therefore, the amount of energy released as product forms is 100 kJ greater than the amount of energy absorbed as reactants break up. The 2 moles of AB formed releases $2\,mol \times 150\,kJ/1\,mol = 300\,kJ$, making the amount absorbed by the reactants $300\,kJ - 100\,kJ = 200\,kJ$.

13.11   See solution in textbook.

13.13   False. It is the size of the E_a barrier that determines the speed of any exothermic reaction, not how much heat energy is given off as the reaction proceeds.

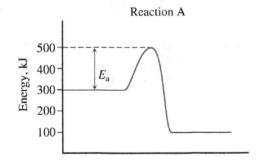

$\Delta E_{rxn} = 100\,kJ - 300\,kJ = -200\,kJ$; more exothermic than reaction B

$E_a = 500\,kJ - 300\,kJ = 200\,kJ$; slower than reaction B

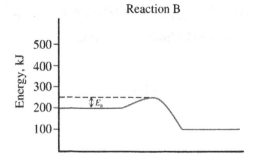

Reaction B

$\Delta E_{rxn} = 100\,kJ - 200\,kJ = -100\,kJ$; less exothermic than reaction A

$E_a = 250\,kJ - 200\,kJ = 50\,kJ$; faster than reaction A

13.15 (a) The rule of thumb is that reaction rate doubles with every 10 °C increase in temperature. Practice Problem 13.14(b) shows that reaction rate is expressed as number of effective collisions per second. Thus, the doubled reaction rate means the number of effective collisions increases from 20 per second to 40 per second.

    (b) Because increasing the temperature causes the molecules to have more kinetic energy, and consequently more of them have energy equal to or greater than E_a. In mathematical terms, the energy factor in Equation 13.1 of the textbook increases while the orientation factor remains unchanged. Therefore, the number of effective collisions increases.

    (c) (200 sufficiently energetic collisions/s) $\times$ 0.2 = 40 effective collisions/s, forming 40 CH_3OH molecules/s.

13.17 See solution in textbook.

13.19 (a) k for the NO reaction is much larger than k for the H_2O_2 reaction. Temperature increases affect reaction rates, as do reactant concentrations. Because both these factors are the same for the two reactions, the vast difference in rates must be due to different k values.

    (b) Because faster reactions have smaller E_a values, the NO E_a must be much smaller than the H_2O_2 E_a

    (c) Increase $[H_2O_2]$, $[I^-]$, and/or $[H^+]$; increase the temperature at which the reaction is run; add a catalyst to the reaction mixture.

13.21 See solution in textbook.

13.23 Rate $= k[NO]^2[O_2]$. To get the NO exponent, use experiments 1 and 2, which say that when $[NO]$ was doubled while $[O_2]$ was held constant, the rate quadrupled. This means the reaction is second order with respect to NO. (Note that you could have used experiments 3 and 4 to get this same result.)

    To get the O_2 exponent, use experiments 1 and 3, which say that when $[O_2]$ was doubled while $[NO]$ was held constant, the rate doubled. This means the reaction is first order with respect to O_2. (Note that you could have used experiments 2 and 4 to get this same result.)

13.25 (a) and (b). Numerous answers possible, such as

$$
\begin{aligned}
Y_2 + Y_2 &\xrightarrow{\text{Slow}} \; Y_3 + Y \\
Y + X_2 &\xrightarrow{\text{Fast}} XY + X \\
Y_3 + X &\xrightarrow{\text{Fast}} XY + Y_2 \\
\hline
X_2 + Y_2 &\longrightarrow XY + XY
\end{aligned}
$$

$$
\begin{aligned}
Y_2 + Y_2 &\xrightarrow{\text{Slow}} \; Y_3 + Y \\
X_2 + Y_3 &\xrightarrow{\text{Fast}} XY + XY_2 \\
XY_2 &\xrightarrow{\text{Fast}} X + Y_2 \\
X + Y &\xrightarrow{\text{Fast}} XY \\
\hline
X_2 + Y_2 &\longrightarrow XY + XY
\end{aligned}
$$

$$
\begin{aligned}
Y_2 + Y_2 &\xrightarrow{\text{Slow}} \; Y_3 + Y \\
X_2 + X_2 &\xrightarrow{\text{Fast}} X_3 + X \\
X + Y_3 &\xrightarrow{\text{Fast}} XY + Y_2 \\
X_3 + Y &\xrightarrow{\text{Fast}} XY + Y_2 \\
\hline
X_2 + Y_2 &\longrightarrow XY + XY
\end{aligned}
$$

$$
\begin{aligned}
Y_2 + Y_2 &\xrightarrow{\text{Slow}} \; Y_3 + Y \\
X_2 + X_2 &\xrightarrow{\text{Fast}} X_4 \\
X_4 + Y &\xrightarrow{\text{Fast}} XY + X_3 \\
Y_3 + X_3 &\xrightarrow{\text{Fast}} XY + X_2 + Y_2 \\
\hline
X_2 + Y_2 &\longrightarrow XY + XY
\end{aligned}
$$

(c) For any mechanism you postulate, design an experiment that detects one or more of the reaction intermediates.

13.27 There are no C–H bonds in either product, which means all four C–H bonds in $H_2C{=}CH_2$ were broken. There are no C–C bonds in either product, which means the $C{=}C$ bond was broken. There are no O–O bonds in either product, which means the three O–O bonds in 3 O_2 were broken.

There are no C–O bonds in either reactant, meaning four C–O bonds formed to create the 2 CO_2. There are no O–H bonds in either reactant, meaning four O–H bonds formed to create the 2 H_2O.

13.29 (a) Yes, because the Cl is substituted for the I.

(b) No, in this reaction two neutral reactants produce a pair of ions.

13.31 The sequence of steps molecules go through when changing from reactants to products during the chemical reaction.

13.33 (a) Reactions occur among a huge number of molecules (on the order of magnitude of Avogadro's constant), are too fast (on the order of magnitude of or less than 10^{-6} s) or too exothermic ($\Delta E_{rxn} \ll 0$) for us to see what is happening in the reaction mixture. Also, if a reaction does indeed occur as a sequence of two or more steps, it is possible that some molecules will be engaged in the earlier steps, while others will be already engaged in the later steps.

(b) They study the kinetics of the reaction by determining how the reaction rate changes as temperature and reactant concentrations are varied.

13.35 Because of the way ΔE_{rxn} is defined ($\Delta E_{rxn} = E_{products} - E_{reactants}$), a negative ΔE_{rxn} means the energy of the products is less than the energy of the reactants. Therefore, compound B, the product, is at a lower energy level than compound A, the reactant, by 100 kJ/mol.

13.37 Compound B, the product, because it has the energy that compound A originally had plus the energy absorbed from the surroundings.

13.39 The reaction in Problem 13.37 is endothermic because it absorbs energy from the surroundings; because it lowers the temperature of the surroundings as it absorbs energy, this reaction can "supply cold." The reaction in Problem 13.38 is exothermic because it releases energy into the surroundings; because it increases the temperature of the surroundings as it releases energy, this reaction can supply heat.

13.41 (a) No, $\Delta E_{rxn} = E_{products} - E_{reactants}$. In the context of a chemical reaction, products represent the final situation, while reactants represent the initial situation.

(b) Because the definition of ΔE_{rxn} is $E_{products} - E_{reactants}$, a negative ΔE_{rxn} means that the energy of the products is less than the energy of the reactants. The energy initially in the reactants but not needed by the products is released into the surroundings, making the reaction exothermic.

13.43 An energy-uphill reaction is one in which the products have more energy than the reactants—in other words, an energy-absorbing, endothermic reaction. An energy-downhill reaction is one in which the products have less energy than the reactants—in other words, an energy-releasing, exothermic reaction.

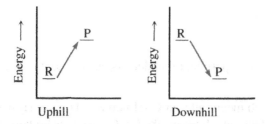

13.45 The reactants must absorb energy from the surroundings.

13.47 $\Delta E_{rxn} = E_{absorbed} - E_{released} = 800 \text{ kJ} - 400 \text{ kJ} = 400 \text{ kJ}$. The reaction is endothermic because ΔE_{rxn} is positive, which means the energy absorbed to break reactant bonds is greater than the energy released as product bonds form.

13.49 The minimum amount of energy reactant molecules must have in order for a reaction to occur.

13.51 False. The speed of any chemical reaction depends on the size of E_a, not on whether the reaction is exothermic (energy-downhill) or endothermic (energy-uphill). The smaller the E_a value, the faster the reaction.

13.53 Exothermic means the products must have less energy than the reactants. You can use any numeric values you like on the energy axis, just so long as there is a 20 kJ difference between E_a and $E_{reactants}$ and a 40 kJ difference between $E_{reactants}$ and $E_{products}$. Here are just two of the many possibilities for what your graph could look like. Note that no matter what values you use on the energy axis, the shape of the curve never changes.

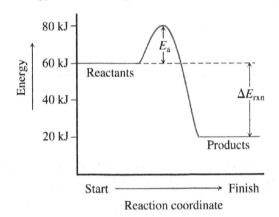

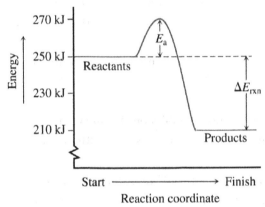

13.55

$\Delta E_{rxn} = E_{products} - E_{reactants} = 60\,kJ - 30\,kJ = +30\,kJ$

$E_a = E_{transition\ state} - E_{reactants} = 100\,kJ - 30\,kJ = +70\,kJ$

The positive ΔE_{rxn} value makes this an endothermic reaction.

13.57 Inversely. The larger the E_a is, the slower the reaction; the smaller the E_a is, the faster the reaction.

13.59 Those that collide with each other with energy greater than or equal to E_a and with the proper orientation.

13.61 Incorrect. There is no relationship between E_a and temperature. The energy barrier the reactants have to overcome does not change and E_a remains the same. When the temperature increases, the reaction rate increases because more molecules have sufficiently high energy to overcome that energy barrier (i.e., to collide energetically enough for the reaction to occur). In other words, at higher temperatures there is a better chance of productive collisions. The more the productive collisions in a time period, the faster the reaction.

13.63 The rate generally doubles with every 10 °C increase in temperature.

13.65 Faster, because then all reactants having sufficient energy would be converted to product. Because of the orientation requirement, only a small fraction of energetically favorable molecules are converted to product.

13.67 By providing an alternate pathway that has a lower activation energy E_a.

13.69 Enzymes.

13.71 The shape of an inhibitor mimics the shape of the substrate, so that it can fit into the enzyme's active site, making it unavailable for the substrate. This renders the enzyme ineffective.

13.73 Because they speed up the processes, thereby decreasing production time and making the processes more financially profitable. Catalysts allow biological processes to proceed at the rate necessary to sustain life at normal temperatures. Without catalysts, the reactions would proceed so slowly that death would occur.

13.75 Because decreasing the volume of a gas increases the gas concentration. Increasing the reactant concentration this way increases the number of collisions between reactant molecules, thus increasing the rate of the reaction.

13.77 (a) Increase because k depends partially on the fraction of collisions having energy equal to or greater than E_a. As temperature increases, more collisions have sufficient energy and k increases.

(b) Increase because an addition of a catalyst replaces an "old" reaction mechanism—one with a presumably high E_a, with a "new" mechanism with a lower E_a. Since k depends on the fraction of collisions with energy equal to or greater than E_a, the lower value of E_a for the catalyzed reaction means that a larger fraction of collisions are energetic enough to be productive, which increases the value of k (see Equation 13.4 in your textbook).

13.79 The fraction of collisions having energy equal to or greater than E_a and the fraction of collisions having the proper orientation.

13.81 (a) k. The inherent factors (those not influenced by concentration) in the rate law are fraction of collisions having $E \geq E_a$ and fraction of collisions having proper orientation, and k is by definition the product of these two factors.

(b) Orders.

(c) By adding up all the individual reactant orders.

(d) $x = 2, y = 1$, overall order $= 2 + 1 = 3$.

(e) Its order is 0, and this means the reaction rate does not depend on the concentration of reactant 1.

13.83 By conducting a series of kinetic experiments that determine how reaction rate is affected when the concentrations of the individual reactants are changed one by one. To do so, all concentrations are kept constant except for the concentration of one reactant under current consideration. For the reactant under current consideration, the concentration is typically changed in a simple manner, by being doubled (or tripled). Then the reaction rate is measured and the response of the rate to the concentration change of that reactant is determined. For instance, if the rate is doubled as a response to a concentration of a reactant being doubled, the reaction is first order in that reactant. We repeat this procedure for the remaining reactants. A rule of thumb is that if there are n reactants in a reaction, $(n + 1)$ kinetic experiments are conducted. Experimentally determined reaction orders become the exponents in the experimental rate law equation.

13.85 The rate quadruples. For a reaction where $k = 1$ and the concentration to be doubled is initially 2 M,
Reaction rate $= 1 \times (2\,\text{M})^2 = 4\,\text{M/s}$
Reaction rate $= 1 \times (4\,\text{M})^2 = 16\,\text{M/s}$

13.87 Halving the volume of the balloon doubles the concentration of the reactants. Because rate $= k[\text{A}]$, the reaction is first order with respect to A and zero order with respect to B. Therefore, doubling the concentration of A doubles the rate and doubling the concentration of B has no effect. Therefore, the rate of the reaction doubles. For $k = 1$ and an initial A concentration of 1 M,
Reaction rate $= 1 \times (1\,\text{M}) = 1\,\text{M/s}$
Reaction rate $= 1 \times (2\,\text{M}) = 2\,\text{M/s}$

13.89 Halving the volume of the balloon doubles the concentration of the reactants. Because rate $= k[\text{A}][\text{B}]^2$, the reaction is first order with respect to A and second order with respect to B. Therefore, doubling the concentration of A doubles the rate and doubling the concentration of B quadruples the rate. Doubling both concentrations therefore causes the rate to increase eightfold. For $k = 1$, an initial A concentration of 1 M, and an initial B concentration of 1 M,
Reaction rate $= 1 \times (1\,\text{M}) \times (1\,\text{M})^2 = 1\,\text{M/s}$
Reaction rate $= 1 \times (2\,\text{M}) \times (2\,\text{M})^2 = 8\,\text{M/s}$

13.91 (a) The unit of the rate is M/s.

(b) Units of time (s, min, hr, day, etc.)

13.93 You find the order of each reactant by comparing two experiments in which the concentration of that reactant is changed while the concentrations of all other reactants are held constant. Determine the order with respect to A from experiments 1 and 2. Doubling [A] causes the reaction rate to double from 0.0281 M/s to 0.0562 M/s, making the reaction first order with respect to A:

Reaction rate $= k[\text{A}]^1[\text{B}]^?$

Determine the order with respect to B from experiments 1 and 3. Doubling [B] causes the reaction rate to quadruple from 0.0281 M/s to 0.1124 M/s, making the reaction second order with respect to B:

Reaction rate $= k[\text{A}]^1[\text{B}]^2$

Dropping the exponent 1 by convention, you get the rate law; Reaction rate $= k[\text{A}][\text{B}]^2$

The overall order of the reaction is $1 + 2 = 3$, and this problem confirms that the exponents in the rate law are not necessarily the same as the coefficients in the balanced equation.

13.95 Endothermic in the forward direction because the energy of the products is greater than the energy of the reactants; exothermic in the reverse direction because in this direction the energy of the products is lower than the energy of the reactants:

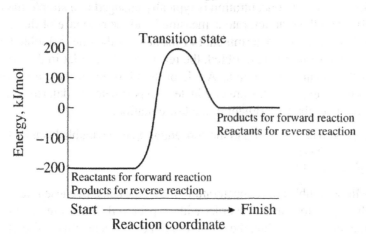

13.97 $E_a = E_{transition\ state} - E_{reactants} = 200\ kJ/mol - (-200\ kJ/mol) = 400\ kJ/mol$.

13.99 Only (c) will slow the reaction down.

13.101 Reaction A: $\Delta E_{rxn} = E_{products} - E_{reactants} = 0\ kJ/mol - (-200\ kJ/mol) = +200\ kJ/mol$

Reaction B: $\Delta E_{rxn} = -200\ kJ/mol - (-100\ kJ/mol) = -100\ kJ/mol$

13.103 The forward rates are the same for the two reactions because the activation energies are the same.

13.105 (a)

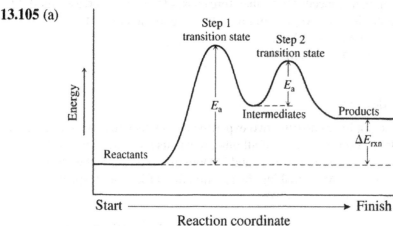

(b) The rate law for the reaction is the rate law for the rate-determining step, which is the slow step. Because this is an elementary step, the exponents in the rate law are the same as the coefficients in the balanced equation:

Reaction rate $= k[A]^1[A]^1 = k[A]^2$

(c) By a factor of 2, which means you double [A]. An order of 2 means that doubling the concentration quadruples the rate. For $k = 1$ and [A] initially 3 M, for instance,

Reaction rate $= 1 \times (3\ M)^2 = 9\ M/s$

Reaction rate $= 1 \times (6\ M)^2 = 36\ M/s$

13.107 A reaction is exothermic whenever ΔE_{rxn} is a negative number and whenever heat is a product of the reaction. This means that (b), (c), (d), and (f) are exothermic.

A reaction is endothermic whenever ΔE_{rxn} is a positive number and whenever heat is a reactant. This means (a) and (e) are endothermic.

13.109 By definition, k = Fraction of collisions with $E \geq E_a$ × Fraction of collisions between molecules having proper orientation. Therefore:

(a) $k = 0.42 \times 0.15 = 0.063$

(b) $k = 0.42 \times 0.30 = 0.13$

(c) $k = 0.84 \times 0.15 = 0.13$

(d) $k = 0.84 \times 0.30 = 0.25$

13.111 The slowest step in a reaction mechanism is called the *rate-determining* step.

13.113 For a reaction mechanism to be valid, the *predicted* rate law must agree with the *experimental* rate law.

13.115 $E_a = E_{\text{transition state}} - E_{\text{reactants}}$. The reaction is exothermic if $\Delta E_{\text{rxn}} = E_{\text{products}} - E_{\text{reactants}}$ is negative and endothermic if ΔE_{rxn} is positive.

(a) $E_a = 150\,\text{kJ} - 100\,\text{kJ} = +50\,\text{kJ}$
$\Delta E_{\text{rxn}} = 130\,\text{kJ} - 100\,\text{kJ} = +30\,\text{kJ}$, endothermic

(b) $E_a = 150\,\text{kJ} - 100\,\text{kJ} = +50\,\text{kJ}$
$\Delta E_{\text{rxn}} = 70\,\text{kJ} - 100\,\text{kJ} = -30\,\text{kJ}$, exothermic

(c) $E_a = 175\,\text{kJ} - 50\,\text{kJ} = +125\,\text{kJ}$
$\Delta E_{\text{rxn}} = 130\,\text{kJ} - 50\,\text{kJ} = +80\,\text{kJ}$, endothermic

(d) $E_a = 40\,\text{kJ} - 20\,\text{kJ} = +20\,\text{kJ}$
$\Delta E_{\text{rxn}} = 10\,\text{kJ} - 20\,\text{kJ} = -10\,\text{kJ}$, exothermic

13.117 False. A catalyst decreases the energy gap between the reactants and the transition state—in other words, it decreases the activation energy. The energy gap between reactants and products remains the same with or without a catalyst present.

13.119 (a) Halving the volume doubles [A]. Because the exponent on [A] is 2, the reaction rate quadruples. For $k = 1$ and initial $[A] = 1\,\text{M}$, $1 \times (1\,\text{M})^2 = 1\,\text{M/s}$; $1 \times (2\,\text{M})^2 = 4\,\text{M/s}$.

(b) The reaction occurs nine times as fast: $1 \times (1\,\text{M})^2 = 1\,\text{M/s}$; $1 \times (3\,\text{M})^2 = 9\,\text{M/s}$.

13.121 When weak bonds are broken (absorbing relatively little energy to do so) and strong bonds are formed (releasing relatively large amounts of energy), the reaction absorbs less energy than it releases and is therefore exothermic.

13.123 The lower the activation energy, the faster the reaction. Therefore, the reaction for which $E_a = 20\,\text{kJ}$ occurs the fastest.

13.125 False. A catalyst is a reactant in the rate-determining step of a reaction. It is not used up because it is regenerated as the product in a subsequent step of the reaction.

13.127 The energy factor is the number of collisions per unit time in which the energy of the colliding particles is equal to or greater than E_a. The orientation factor is the fraction of collisions in which the colliding particles have the proper orientation.

13.129 Catalysts speed up reactions by lowering E_a values. A company can therefore lower its manufacturing costs by consuming less energy in running the chemical reactions it must run to create its products. (This assumes the savings in energy dollars are greater than the cost of using the catalyst, which is usually true.) Additionally, decreasing the time needed to run a reaction by adding a catalyst decreases the cost of the process (because time is money).

13.131 k is defined by two factors describing the collisions between reactant particles. One factor is the fraction of collisions having energy equal to or greater than E_a. When the reaction temperature is increased, the kinetic energy of all the reactant particles is increased. These particles then collide with greater energy, meaning a larger fraction of the collisions have energy equal to or greater than E_a, and therefore k increases.

13.133 (a) Situation (b) represents the reaction being run at the lowest temperature.

      (b) Situation (b) would have the reaction running at the slowest rate, as few if any of the molecules would yield collisions of sufficient energy ($>E_a$).

13.135 The reaction has highest concentration in vessel (b), and this is where the reaction rate is greatest. A higher concentration leads to more collisions per second.

13.137

13.139 The concentration of oxygen appears in the denominator of the rate law, which means that as the concentration of oxygen increases, the rate of the reaction decreases. The rate law could also be written as:

Reaction rate $= k[O_3]^2[O_2]^{-1}$.

13.141 The student is incorrect. The rate of a reaction does not depend on the starting energy of the reactants versus the ending energy of the products. It depends, rather, on the energy decrease between the reactants and the transition state (the activation energy, E_a).

13.143 When Sidney spilled some of his reaction, he decreased the amount of solution but not the concentration of any of its components. Thus, there will be no effect on the rate or expected time of color change. Therefore, the color change will occur in about 10 minutes.

13.145 True, because the slowest elementary step is the rate-determining step, and the balancing coefficients from the rate-determining step are equal to the orders in the rate law.

13.147 If the reaction $A + 2B + C \rightarrow P$ proceeded in one step (a practically impossible occurrence), the equation must represent a rate-determining elementary step (because it is the *only* step), and therefore the equation coefficients can be used as orders to yield:

Reaction rate $= k[A][B]^2[C]$

13.149 Because in any multistep reaction, the rate law for the overall reaction is the same as the rate law for the rate-determining step.

13.151 It cannot be correct because it does not agree with the experimentally determined rate law.

13.153 All you can say is that the postulated mechanism *may possibly* be the correct mechanism. There may be other mechanisms that also agree with the experimental data. Incorrect mechanisms can be ruled out by experimental evidence, but you cannot prove that a given possible mechanism for a reaction is the correct one.

13.155 The overall balanced equation is determined by adding the steps in the mechanism:

$$Cl_2 \rightarrow \cancel{Cl} + Cl$$
$$\cancel{Cl} + CHCl_3 \rightarrow HCl + \cancel{CCl_3}$$
$$\underline{\cancel{Cl} + \cancel{CCl_3} \rightarrow CCl_4}$$
$$Cl_2 + CHCl_3 \rightarrow HCl + CCl_4$$

Chapter 14

14.1 See solution in textbook.

14.3 Because the equilibrium lies so far to the right (products side) that there are essentially no reactants present at equilibrium. Therefore, the reverse reaction can be ignored, and the reaction is considered a one-way reaction in the forward direction.

14.5 $K_{eq} = \dfrac{[HI]^2}{[H_2] \times [I_2]}$

14.7 $K_{eq} = \dfrac{[HI]^2}{[H_2] \times [I_2]} = \dfrac{(0.158 \text{ M})^2}{0.021 \text{ M} \times 0.021 \text{ M}} = 57$

Because K_{eq} is calculated from equilibrium concentrations, the initial concentrations given in the problem statement are extraneous information. All you need are the equilibrium concentrations.

14.9 $K_{eq} = \dfrac{[CS_2] \times [H_2]^4}{[CH_4] \times [H_2S]^2} = \dfrac{6.10 \times 10^{-3} \text{ M} \times (1.17 \times 10^{-3} \text{ M})^4}{2.35 \times 10^{-3} \text{ M} \times (2.93 \times 10^{-3} \text{ M})^2} = 5.67 \times 10^{-7}$

14.11 To the left so that the rate of the reverse reaction, $2SO_2 + O_2 \leftarrow 2SO_3$, increases so as to consume some of the SO_3 you added. Adding product to a reaction at equilibrium shifts the reaction toward reactants.

14.13 See solution in textbook.

14.15 (a) A negative ΔE_{rxn} means the reaction is exothermic. Heat is released to the surroundings and is therefore to be written as a product: $2CO(g) + O_2(g) \rightleftharpoons 2CO_2(g) + \text{Heat}$

(b) To the left because raising the temperature is equivalent to adding something to the right side of the equation. The reaction shifts left to use up the added heat.

14.17 (a) Show no liquids or solids in the K_{eq} expression: $K_{eq} = [O_2]^6 / [CO_2]^6$.

(b) It would shift to the right to use up the added CO_2.

(c) No change in equilibrium because $[H_2O]$ does not appear in the K_{eq} expression. (All liquid concentrations are taken to be constant in equilibrium calculations.)

(d) It would shift to the right because the reaction is endothermic (positive ΔE_{rxn}), making heat a reactant.

(e) Addition of a catalyst has no effect on the position of the equilibrium, only on the time required to reach equilibrium.

14.19 See solution in textbook.

14.21 (a) $Mg(OH)_2(s) \rightleftharpoons Mg^{2+}(aq) + 2OH^-(aq)$

(b) Show only the *(aq)* species, no *(s)* species: $K_{sp} = [Mg^{2+}] \times [OH^-]^2$.

(c) $(1.44 \times 10^{-4} \text{ M Mg}^{2+}) \times \dfrac{2 \text{ OH}^-}{1 \text{ Mg}^{2+}} = 2.88 \times 10^{-4} \text{ M OH}^-$

(d) $K_{sp} = [Mg^{2+}] \times [OH^-]^2 = 1.44 \times 10^{-4} \text{ M} \times (2.88 \times 10^{-4} \text{ M})^2 = 1.19 \times 10^{-11}$

14.23 See solution in textbook.

14.25 $Ag_2S(s) \rightleftharpoons 2Ag^+(aq) + S^{2-}(aq)$

$K_{sp} = [Ag^+]^2 \times [S^{2-}]$

$[S^{2-}] = s = 1.14 \times 10^{-17}\,M$

$[Ag^+] = 2s = 2 \times (1.14 \times 10^{-17}\,M) \times 10^{-17}\,M$

14.27 Reversible reactions. In principle, all reactions can be reversible (as long as the products are *not* allowed to escape!).

14.29 The position of a reaction's equilibrium refers to whether there are more products or more reactants present at equilibrium. When there are more reactants, the equilibrium is said to lie toward the left end of the reaction coordinate; when there are more products, the equilibrium is said to lie toward the right end of the reaction coordinate; when there are significant amounts of both products and reactants, the equilibrium is said to lie toward the middle of the reaction coordinate. In practice, the further to the right the equilibrium lies, the better the conversion of the reactants to products and the better the yield of the reaction, which is typically very desirable when a reaction is conducted with the purpose of making an important or valuable chemical.

14.31 Far to the left. That the reaction appears not to occur means the reaction vessel contains essentially all reactants—in other words, the reaction is still at the starting point. By convention, the left end of the reaction coordinate is taken as the starting point.

14.33 This chemist is likely ignoring the fact that this reaction must be reaching an equilibrium that just does not lie far enough to the right. When such an equilibrium is reached, there are significant amounts of the reactants being left over and contaminating the products. No amount of additional time will remedy this situation.

14.35 Initially there are no forward-reaction products in the vessel. Because these products are the reverse-reaction reactants, the reverse reaction rate is initially zero.

14.37 The reverse reaction because those products present are the reverse-reaction reactants. At the moment after loading, there are no forward-reaction reactants in the vessel, which means the forward reaction cannot occur. Therefore, the reverse reaction must be going faster because the rate of the forward reaction is zero.

14.39 Yes. In the sealed container, the liquid water evaporates until the air in the container is saturated with water vapor. After that point, the rate at which liquid water evaporates is equal to the rate at which water vapor condenses, which means the water level remains constant.

14.41 No, because it is not possible to reach equilibrium if both rates increase, which is what the rate meters show. As the reverse reaction proceeds, the reverse-reaction reactant concentrations decrease, which means $Rate_{reverse} = k_{reverse}[Reactants]^{order}$.

14.43 For simplicity, assume the order of R is 1 for both the forward and the reverse reaction. Then

$Rate_{forward} = k_f[R]$

$Rate_{reverse} = k_r[P]$

At equilibrium, $k_f[R] = k_r[P]$. To solve for k_f/k_r, divide both sides of this expression by k_r:

$$\frac{k_f[R]}{k_r} = \frac{k_r[P]}{k_r}$$

$$\frac{k_f[R]}{k_r} = [P]$$

Then divide both sides by $[R]$:

$$\frac{k_f[\cancel{R}]}{k_r[\cancel{R}]} = \frac{[P]}{[R]}$$

$$\frac{k_f}{k_r} = \frac{[P]}{[R]}$$

14.45 K_{eq}, the equilibrium constant.

14.47 No matter how many reactants you have—1, 2, or 200—all of them go in the denominator because the definition of K_{eq} is in the form:

$$\frac{\text{Concentrations of products}}{\text{Concentrations of reactants}}$$

When there are more than one reactant concentration, you multiply:

$$K_{eq} = \frac{[\text{Product 1}]^a}{[\text{Reactant 1}]^b \times [\text{Reactant 2}]^c \times [\text{Reactant 3}]^d \times \cdots}$$

The same holds true for products:

$$\frac{[\text{Product 1}]^a \times [\text{Product 2}]^b \times [\text{Product 3}]^c \times \cdots}{[\text{Reactant 1}]^d \times [\text{Reactant 2}]^e \times [\text{Reactant 3}]^f \times \cdots}$$

14.49 When a reaction is doubled, all coefficients are doubled:

$$4A(g) + 6B(g) \rightleftharpoons 2C(g) + 2D(g)$$

$$K_{eq,\,doubled} = \frac{[C]^2 \times [D]^2}{[A]^4 \times [B]^6} = \left(\frac{[C] \times [D]}{[A]^2 \times [B]^3}\right)^2 = K_{eq}^2$$

14.51 Put product concentrations in the numerator, reactant concentrations in the denominator. Raise each concentration to a power equal to the substance's stoichiometric coefficient in the balanced equation. Because both species in the reaction are gases, show both of them in the K_{eq} expression:

$$K_{eq} = \frac{[NO]^2}{[N_2O_4]}$$

14.53 You would place some N_2O_4 in a container and allow the reaction to proceed to equilibrium. Then you would determine the concentration of N_2O_4 and NO_2 at equilibrium and substitute those concentration values into the K_{eq} expression shown in the solution to Problem 14.51.

14.55 $K_{eq} = \dfrac{[NO]^2}{[N_2] \times [O_2]} = \dfrac{(0.011 \text{ M})^2}{0.25 \text{ M} \times 1.2 \text{ M}} = 4.0 \times 10^{-4}$

14.57 That the concentration of products at equilibrium is much greater than the concentration of reactants. The general expression shows that the numerator (product concentration) must be much larger than the denominator (reactant concentration) if $K_{eq} > 10^3$.

14.59 That the concentration of products at equilibrium is much less than the concentration of reactants. The general expression shows that the numerator (product concentration) must be substantially smaller than the denominator (reactant concentration) if $K_{eq} < 10^{-3}$.

14.61 $K_{eq} = \dfrac{[CH_4] \times [H_2O]}{[CO] \times [H_2]^3}$

The equilibrium molar concentrations are

$[CO] = \dfrac{0.613 \text{ mol}}{10.0 \text{ L}} = 0.0613 \text{ M}$ $[H_2] = \dfrac{1.839 \text{ mol}}{10.0 \text{ L}} = 0.1839 \text{ M}$

$[CH_4] = \dfrac{0.387 \text{ mol}}{10.0 \text{ L}} = 0.0387 \text{ M}$ $[H_2O] = \dfrac{0.387 \text{ mol}}{10.0 \text{ L}} = 0.0387 \text{ M}$

$K_{eq} = \dfrac{0.0387 \text{ M} \times 0.0387 \text{ M}}{0.0613 \text{ M} \times (0.1839 \text{ M})^3} = 3.93$

The equilibrium lies about in the middle of the reaction coordinate because 3.93 is roughly midway between $K_{eq} = 10^{-3}$, indicating equilibrium essentially all the way to the left (mainly reactant present), and $K_{eq} = 10^3$, indicating equilibrium essentially all the way to the right (mainly product present).

14.63 (a) $K_{eq} = \dfrac{[C]}{[A][B]^2} = \dfrac{1.98 \text{ M}}{0.020 \text{ M} \times (0.040 \text{ M})^2} = 6.2 \times 10^4$.

(b) As long as the temperature at which both reactions are conducted stays the same, the value of the equilibrium constant will remain the same (i.e., 6.2×10^4), regardless of the initial concentrations of the reactants and/or products.

14.65 The textbook notes that any reaction for which K_{eq} is greater than 10^5 goes to completion and any reaction for which K_{eq} is less than 10^{-3} hardly occurs at all. Thus, reaction a $(K_{eq} \approx 10^{81})$ goes essentially to completion, with virtually all of the reactants converted to products at equilibrium, and reaction b $(K_{eq} = 10^{-95})$ does not happen, with almost none of the reactants converted to products at equilibrium.

14.67 All species are gases, and so all appear in the expression for K_{eq}:

$K_{eq} = \dfrac{[HI]^2}{[H_2] \times [I_2]}$

The equilibrium molar concentrations are:

$[HI] = \dfrac{3.00 \text{ mol}}{8.00 \text{ L}} = 0.375 \text{ M}$ $[H_2] = \dfrac{0.650 \text{ mol}}{8.00 \text{ L}} = 0.0812 \text{ M}$ $[I_2] = \dfrac{0.275 \text{ mol}}{8.00 \text{ L}} = 0.0344 \text{ M}$

$K_{eq} = \dfrac{(0.375 \text{ M})^2}{0.0812 \text{ M} \times 0.0344 \text{ M}} = 50.3$

The equilibrium lies near the middle of the reaction coordinate because 50.3 is roughly halfway between $K_{eq} = 10^{-3}$ (indicating equilibrium is essentially all the way to the left and mainly reactants are present) and $K_{eq} = 10^3$ (indicating equilibrium is essentially all the way to the right and mainly products are present).

14.69 The value of K_{eq} does not change because you have not changed the temperature. K_{eq} changes only when the temperature at which the reaction is run changes.

14.71 (a) To the right because Cl_2 appears on the left side of the equation. Equilibrium shifts rightward to consume some of the added Cl_2 via the reaction $PCl_3 + Cl_2 \rightarrow PCl_5$.

(b) To the left so that the reaction $PCl_5 \rightarrow PCl_3 + Cl_2$ produces more Cl_2 to replace some of the removed Cl_2.

(c) To the left so that the reaction $PCl_5 \rightarrow PCl_3 + Cl_2$ uses up some of the added PCl_5.

(d) To the left so that the reaction $PCl_5 \rightarrow PCl_3 + Cl_2$ produces more PCl_3 to replace some of the removed PCl_3.

(e) No shift because H_2 does not take part in the reaction and therefore does not affect the $PCl_3/Cl_2/PCl_5$ equilibrium.

14.73 (a) To the left because the reverse reaction, Products $\rightarrow$ Reactants, speeds up to consume some of the added products and restore equilibrium.

(b) Greater than before the disturbance. Some of the added products are converted to reactants as the new equilibrium is established, but the net result is an increase in the amount of products (and an increase in the amount of reactants, produced when the reaction shifted left). In other words, the equilibrium shift never consumes all of the substance (or substances) added to disturb the equilibrium.

(c) Greater than before the disturbance. See explanation in part (b).

14.75 (a) At a high temperature; this is an endothermic reaction ($\Delta E_{rxn} > 0$), and heat is a "reactant": $2H_2O(g) + \text{Heat} \rightleftharpoons 2H_2(g) + O_2(g)$. Adding a "substance" that appears on the left in the equation (heat in this case) shifts the reaction right, forming greater amounts of H_2 and O_2.

(b) All reactions speed up at higher temperatures and this reaction is not an exception. At higher temperatures a larger fraction of $H_2O(g)$ molecules have energy equal to or greater than E_a. Note that the reverse reaction would also speed up and so the new, more-product rich equilibrium would be reached faster.

14.77 (a) To the left, because K_{eq} is smaller at the higher temperature. The smaller the $K_{eq} = [\text{Products}]/[\text{Reactants}]$ value, the smaller the numerator. That the product concentration decreases tells you the reaction has shifted left.

(b) Exothermic, because increasing the temperature (adding heat) causes the reaction to shift to the left. Heat must therefore be a product in the reaction.

(c) $CO(g) + 2H_2(g) \rightleftharpoons CH_3OH(g) + \text{Heat}$

14.79 (a) That the reaction proceeds in reverse when heated means that is exothermic and heat must be a product, $C_{diamond} \rightleftharpoons C_{graphite} + \text{Heat}$.

(b) That the reaction proceeds in reverse when pressure is increased tells you that diamond has higher density. As the pressure increases, so does the density. In other words, increased pressure favors the side where matter is more condensed. In this case it is the side with $C_{diamond}$ (i.e., the left side of the reaction equation).

14.81 An exothermic reaction means heat is a product: Reactants $\rightleftharpoons$ Products + Heat.

(a) Heating means adding to the right side of the equation, shifting the reaction left. This causes [Products] to decrease and [Reactants] to increase. Therefore, $K_{eq} = [\text{Products}]/[\text{Reactants}]$ decreases.

(b) Cooling means removing from the right side of the equation, shifting the reaction right. This causes [Products] to increase and [Reactants] to decrease. Therefore, K_{eq} = [Products]/[Reactants] increases.

14.83 Decreased. The negative value for ΔE_{rxn} tells you this is an exothermic reaction. Thus, heat is a product, $CO(g) + 2H_2(g) \rightleftharpoons CH_3OH(g)$ + Heat, and adding it to the reaction mixture by raising the temperature shifts the reaction left. [H_2] and [CO] increase, and [CH_3OH] decreases.

14.85 Because a catalyst speeds up the forward and reverse reactions by the same amount, it has no effect on: (a) the position of equilibrium, (b) the value of K_{eq}, and (c) the ratio of k_f/k_r.

14.87 (a) That $K_{eq} = [NO]^2/[N_2] \times [O_2]$ is higher at the higher temperature means that at the higher temperature [NO] must be higher and [N_2] × [O_2] must be lower. Therefore, the reaction shifts to the right.

(b) Endothermic. The fact that heating shifts the reaction to the right means heat must be a "reactant."

14.89 Show only (g) and (aq) species in a K_{eq} expression. Put product molar concentrations in the numerator, reactant molar concentrations in the denominator. Raise each concentration to a power equal to the substance's stoichiometric coefficient in the balanced equation.

(a) $K_{eq} = [HCl]^6/[H_2O]^3$

(b) $K_{eq} = [CO_2]^3/[CO]^3$

(c) $K_{eq} = [Pb^{2+}] \times [SO_4^{2-}]$

(d) $K_{eq} = [H_2CO_3]/[CO_2]$

14.91 The fact that the concentrations of solids and liquids do not change during any heterogeneous reaction.

14.93 (a) Because the sealed vessel keeps the $H_2(g)$ from escaping. In an open container, $H_2(g)$ would escape, continuously shifting the reaction to the left so that no product could form and equilibrium would never be established.

(b) $K_{eq} = \dfrac{1}{[H_2]^2}$

$[H_2]^2 = \dfrac{1}{K_{eq}}$

$\sqrt{[H_2]^2} = \sqrt{\dfrac{1}{K_{eq}}}$

$[H_2] = \dfrac{1}{\sqrt{K_{eq}}}$

(c) The fact that K_{eq} = [Products]/[Reactants] gets smaller tells you the reaction shifts to the left, with [Products] decreasing and [Reactants] increasing.

(d) That adding heat shifts the reaction left means heat must be a product:

$SnO_2(s) + 2H_2(g) \rightleftharpoons Sn(s) + 2H_2O(l)$ + Heat. The reaction therefore is exothermic.

(e) Fired! SnO_2 is a solid and its amount, as long as there is enough of it to produce the products, does not affect the position of equilibrium.

14.95 Table 14.2 of the textbook tells you that for any sparingly soluble salt of the form XY_3, the cation equilibrium concentration is equal to the salt solubility s and the anion equilibrium concentration is equal to $3s$. Therefore,

$$[Al^{3+}] = s = 2.86 \times 10^{-9} \text{ M}, [OH^-] = 3s = 3 \times (2.86 \times 10^{-9} \text{ M}) = 8.58 \times 10^{-9} \text{ M}.$$

14.97 (a) $AgC_2H_3O_2(s) \rightleftharpoons Ag^+(aq) + C_2H_3O_2^-(aq)$

(b) $K_{sp} = [Ag^+] \times [C_2H_3O_2^-]$

(c) To calculate K_{sp}, you need the value of $[Ag^+]$ and $[C_2H_3O_2^-]$. Table 14.2 of the textbook tells you that for any sparingly soluble salt of the form XY, the cation equilibrium concentration and the anion equilibrium concentration are both equal to s. You are given solubility but in grams rather than moles. Because concentrations in the K_{sp} expression are molar concentrations, you must convert:

$$s = \frac{10.6 \text{ g} \cancel{AgC_2H_3O_2}}{L} \times \frac{1 \text{ mol } AgC_2H_3O_2}{166.9 \text{ g} \cancel{AgC_2H_3O_2}} = 0.0635 \text{ mol } AgC_2H_3O_2/L = 0.0635 \text{ M}$$

Therefore, $[Ag^+] = 0.0635 \text{ M}$, $[C_2H_3O_2^-] = 0.0635 \text{ M}$, and

$$K_{sp} = [Ag^+] \times [C_2H_3O_2^-] = (0.0635 \text{ M}) \times (0.0635 \text{ M}) = 4.03 \times 10^{-3}$$

14.99 Lead(II) chloride, because it is the one with the largest K_{sp}.

14.101 (a) Table 14.1 of the textbook gives 3.3×10^{-14} for the 25 °C K_{sp} for $PbCO_3$. Because this is a 1:1 salt,

$$s = \sqrt{K_{sp}} = \sqrt{3.3 \times 10^{-14}} = 1.8 \times 10^{-7} \text{ mol/L}$$

(b) $$\frac{1.8 \times 10^{-7} \text{ mol} \cancel{PbCO_3}}{L} \times \frac{267.2 \text{ g } PbCO_3}{1 \text{ mol} \cancel{PbCO_3}} = \frac{4.8 \times 10^{-5} \text{ g } PbCO_3}{L}$$

(c) You know from Table 14.2 of the textbook that for a 1:1 salt the equilibrium concentration of both cation and anion is equal to s. Therefore, $[Pb^{2+}] = [CO_3^{2-}] = 1.8 \times 10^{-7} \text{ M}$.

(d) $$1.000 \times 10^6 \cancel{\text{gallons}} \times \frac{3.79 \cancel{L}}{1 \cancel{\text{gallon}}} \times \frac{4.8 \times 10^{-5} \text{ g } PbCO_3}{\cancel{L}} = 1.8 \times 10^2 \text{ g } PbCO_3$$

14.103 $PCl_5(g) \rightleftharpoons PCl_3(g) + Cl_2(g); K_{eq} = [PCl_3] \times [Cl_2]/[PCl_5]$

$$[PCl_5] = \frac{[PCl_3] \times [Cl_2]}{K_{eq}} = \frac{5.50 \times 10^{-3} \text{ M} \times 0.125 \text{ M}}{7.50 \times 10^{-2}} = 9.17 \times 10^{-3} \text{ M}$$

14.105 (a) To the right. Because the reverse reaction is so much slower than the forward reaction, there must be a large buildup of product (AB) before the reverse reaction is going fast enough to match the rate of the forward reaction. The large amount of AB means the equilibrium lies to the right.

(b) Because the forward rate is so much faster, k_f must be much larger than k_r. Therefore, the fraction $k_f/k_r = K_{eq}$ must be a large number, which means the equilibrium lies to the right, just as in part (a).

14.107 (a) The reverse reaction is faster because it is the only reaction going on at that moment. The forward reaction is not happening because initially there are no reactants for it in the container.

(b) The balanced equation tells you that if you begin with only NO, the molar amount of O_2 formed must be the same as the molar amount of N_2 formed. Therefore, the O_2 concentration must be equal to the N_2 concentration—both are 0.2159 M.

$$K_{eq} = \frac{[NO]^2}{[N_2] \times [O_2]} = \frac{(0.0683 \text{ M})^2}{0.2159 \text{ M} \times 0.2159 \text{ M}} = 0.100$$

14.109 K_{eq} must change because the temperature has changed. In an endothermic reaction heat is a reactant: Reactants + Heat $\rightleftharpoons$ Products, which means adding heat by increasing the temperature shifts the reaction to the right, creating more product. Therefore, in the fraction: [Products]/[Reactants] = K_{eq}, [Products] increases, [Reactants] decreases, and so K_{eq} increases.

14.111 (a) $K_{eq} = \dfrac{[NH_3]^2}{[N_2] \times [H_2]^3} = \dfrac{(0.090 \text{ M})^2}{0.25 \text{ M} \times (0.15 \text{ M})^3} = 9.6$

(b) To the right so that, by Le Châtelier's principle, some of the added $H_2(g)$ is used up.

(c) Because there is no mention of changing the temperature, you must assume temperature is held constant as the $H_2(g)$ is added. Therefore, K_{eq} does not change because only changes in temperature affect the value of K_{eq}.

(d) The new coefficients mean the expression for K_{eq} changes to

$$K_{eq} = \frac{[NH_3]^4}{[N_2]^2 \times [H_2]^6}$$

To calculate the value of K_{eq}, use the concentrations given at the beginning of the problem, now raised to the new powers:

$$\frac{(0.090 \text{ M})^4}{(0.25 \text{ M})^2 \times (0.15 \text{ M})^6} = 92$$

How can this be? There's been no temperature change, and yet K_{eq} changes from 9.6 in (a) to 92 here just because the equation was multiplied through by 2. The answer is that any numeric value of K_{eq} is constant for a given temperature *but also for a given form of the balanced equation.*

To see why this must be so, think about putting some $N_2(g)$ and $H_2(g)$ together in a reaction vessel held at some constant temperature. (The amounts you use don't matter.) The two gases will react, an equilibrium state will be set up after some time has passed, and at that point the concentrations of N_2, H_2, and NH_3 have attained their (constant) equilibrium values. Plug these values into the two expressions for K_{eq}, and you must get different numbers, as you saw when you calculated your answers to (a) and (d). (Chemists do not usually run into any problem with this interesting phenomenon because published K_{eq} values are by convention taken to be for chemical equations balanced with the smallest whole number coefficients—in this case,

$N_2 + 3H_2 \rightleftharpoons 2NH_3$ rather than $2N_2 + 6H_2 \rightleftharpoons 4NH_3$.)

(e) A shift to the right means either (1) something was added to the left side of the equation $N_2 + 3H_2 \rightleftharpoons 2NH_3$ or (2) something was removed from the right side. Cooling means the removal of heat, and therefore (2) is what happened: heat was removed from the right. Being that the right side is the products side, heat must be a product, which means the reaction is exothermic.

14.113 Because K_{sp} values are calculated with molar concentrations, a conversion is your first step:

$$\frac{7.51 \text{ g AgC}_2\text{H}_3\text{O}_2}{1 \text{ L}} \times \frac{1 \text{ mol AgC}_2\text{H}_3\text{O}_2}{166.9 \text{ g AgC}_2\text{H}_3\text{O}_2} = \frac{4.50 \times 10^{-2} \text{ mol AgC}_2\text{H}_3\text{O}_2}{\text{L}}$$

Because this is a 1:1 salt, this is also the concentration of both Ag^+ and Cl^-. Thus,

$$K_{sp} = [Ag^+] \times [C_2H_3O_2^-] = (4.50 \times 10^{-2} \text{ M}) \times (4.50 \times 10^{-2} \text{ M}) = 2.03 \times 10^{-3}$$

14.115 The given solubility means that, at 25 °C, 1.52×10^{-3} mole of PbI_2 is dissolved in each liter of solution. Therefore,

$$2.50 \times 10^{-6} \text{ gallons} \times \frac{3.79 \text{ L}}{1 \text{ gallon}} \times \frac{1.52 \times 10^{-3} \text{ mol PbI}_2}{1 \text{ L}} \times \frac{461 \text{ g PbI}_2}{1 \text{ mol PbI}_2} = 6.64 \times 10^6 \text{ g PbI}_2$$

14.117 $K_{eq} = \dfrac{[N_2O_5]^2}{[NO_2]^4 \times [O_2]}$

$$[NO_2]^4 = \frac{[N_2O_5]^2}{K_{eq} \times [O_2]}$$

$$[NO_2] = \sqrt[4]{\frac{[N_2O_5]^2}{K_{eq} \times [O_2]}} = \sqrt[4]{\frac{(0.300 \text{ M})^2}{0.150 \times 1.20 \text{ M}}} = \sqrt[4]{0.500}$$

$$[NO_2] = 0.841 \text{ M}.$$

14.119 Greater in water because in water the concentration of I^- is initially zero. In NaI solution, the initial concentration of I^- is greater than zero, with the result that the amount of PbI_2 that can dissolve before the numeric value of $[Pb^{2+}] \times [I^-]^2$ exceeds K_{sp} is reduced.

14.121 (a) K_{sp}, because the reaction is the dissolution in water of a sparingly soluble salt.

(b) $K_{sp} = [Ca^{2+}] \times [F^-]^2$

$$[F^-]^2 = \frac{K_{sp}}{[Ca^{2+}]}$$

$$[F^-] = \sqrt{\frac{K_{sp}}{[Ca^{2+}]}} = \sqrt{\frac{3.9 \times 10^{-11}}{3.3 \times 10^{-4} \text{ M}}} = \sqrt{1.2 \times 10^{-7}} = 3.5 \times 10^{-4} \text{ M}$$

(c) The rate constant k_r for the reverse reaction. Because $K_{eq} < 1$, it must be true that the denominator in the fraction $k_f/k_r = K_{eq}$ is larger than the numerator.

(d) E_a for the forward reaction. The slower the rate of a reaction, the larger the E_a value for the reaction. Because k_f is smaller [see (c)], the forward reaction is slower and therefore has the larger E_a value.

(e) At equilibrium, by definition, the *rates* of the forward and reverse reactions are the same.

(f) $Li_2CO_3(s) \rightleftharpoons 2Li^+(aq) + CO_3^{2-}(aq); K_{eq} = K_{sp} = [Li^+]^2 \times [CO_3^{2-}]$

(g) Lithium carbonate; because these two salts give the same number of moles of ions when they dissolve ($1 \text{ mol Ca}^{2+} + 2 \text{ mol F}^- = 3 \text{ mol ions}; 2 \text{ mol Li}^+ + 1 \text{ mol CO}_3^{2-} = 3 \text{ mol ions}$), their solubilities are directly proportional to their K_{sp} values. Li_2CO_3 has the larger K_{sp} value and therefore is more soluble.

14.123 (a) They are the same. K_{eq} is a constant for a reaction at a given temperature.

(b) The rate meters for the new equilibrium must be wrong. Although they must be equal (after all, we are at equilibrium), they cannot be at the identical setting to the equilibrium before the disturbance. Since we added reactant, the meters should settle at some higher value than before the disturbance.

14.125 Somewhere between 7 and 8 minutes. It is between these limits that the concentrations of product and reactant stop changing.

14.127

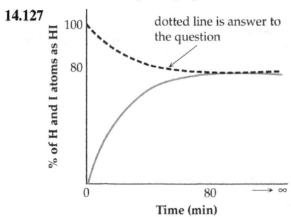

Both scenarios start with 2 moles of gas and thus are capable of achieving identical concentrations of all the reactants and products at any given position from start to finish. Since the position of final equilibrium for a reaction does not depend upon which direction you approach it from, the two scenarios will reach the same final % of HI.

14.129 Decreasing the volume of a gas phase reaction at equilibrium raises the internal pressure, disturbing the equilibrium. The reaction $N_2(g) + 3H_2(g) \rightleftharpoons 2NH_3(g)$ consumes gas as it goes from left (4 total moles of gas) to right (2 total moles of gas). Thus, the reaction will shift to the right to partially lower the pressure (make less gas) and re-establish equilibrium. The reaction $N_2(g) + O_2(g) \rightleftharpoons 2NO(g)$ has equal moles of gas on left and right and thus cannot consume gas and lower the pressure by shifting its position of equilibrium, so it will not shift upon a pressure disturbance.

14.131 Although all reactions in theory (on paper) are reversible, some have an activation energy $E_{a,\,forward}$ or $E_{a,\,reverse}$ so large that the reaction can effectively go in only one direction. Now you might say why not just put in tons of energy, rather than go in the difficult direction, reactants or products will instead just decompose.

14.133 $K_{eq} = \dfrac{[B]}{[A]} = \dfrac{0.050\ M}{0.070\ M} = 0.71$

14.135 For $SO_2(g) + \frac{1}{2}O_2(g) \rightleftharpoons SO_3(g)$, $K_{eq} = \dfrac{[SO_3]}{[SO_2][O_2]^{1/2}} = 0.112$

For $SO_3(g) \rightleftharpoons SO_2(g) + \frac{1}{2}O_2(g)$, $K_{eq\ reverse} = \dfrac{[SO_2][O_2]^{1/2}}{[SO_3]} = ?$

But look, $K_{eq\ reverse}$ is just the inverse of K_{eq}, so:

$K_{eq,\ reverse} = \dfrac{1}{K_{eq}} = \dfrac{1}{0.112} = 8.93$

Chapter 15

15.1 See solution in textbook.

15.3 (a)

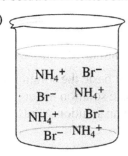

(b) Yes. That NH_4Br dissociates tells you it is an electrolyte.

(c)

15.5 $H_2CO_3 + 2H_2O \rightleftarrows 2H_3O^+ + CO_3^{2-}$

15.7 See solution in textbook.

15.9 It would take 0.05 mole of $Ba(OH)_2$ to neutralize 0.1 mole of hydrochloric acid because each mole of this base dissociates to 2 moles of OH^- ions.

15.11 See solution in textbook.

15.13 Water is behaving as an acid because it donates a proton to NH_2^-.

15.15 As acid: $HCO_3^- + H_2O \rightleftarrows H_3O^+ + CO_3^{2-}$

As base: $HCO_3^- + H_2O \rightleftarrows H_2CO_3 + OH^-$

15.17 See solution in textbook.

15.19 (a) $[H_3O^+] = \dfrac{1.00 \text{ mol } H_3O^+}{0.5000 \text{ L of solution}} = 2.00 \text{ M}$

(b) $[OH^-] = \dfrac{1 \times 10^{-14}}{2.00} = 5.00 \times 10^{-15} \text{ M}$

15.21 See solution in textbook.

15.23 The logarithm is 1 for both 10 and 10^1 because the number 10 written with no exponent is understood to be raised to the first power: $10 = 10^1$.

15.25 See solution in textbook.

15.27 Because the (positive) logarithm of 10^{-7} is -7, the negative logarithm is $-(-7) = +7$.

15.29 The pH of solution B is 10. (Recall that the higher the pH, the more basic the solution.) Because pH is a logarithmic scale, each *one*-unit increase in the pH of a basic solution means a *ten*fold increase in OH^- concentration. Thus, to be ten times more basic than solution A (pH 9), solution B must have a pH of $9 + 1 = 10$.

15.31 See solution in textbook.

15.33 $[H_3O^+] = 10^{-5.55} = 2.82 \times 10^{-6}$ M

$$[OH^-] = \frac{1 \times 10^{-14}}{2.82 \times 10^{-6}} = 3.55 \times 10^{-9} \text{ M}$$

15.35 CO_3^{2-} is the conjugate base of HCO_3^- because any conjugate base has one fewer H^+ ion than its acid.

15.37 See solution in textbook.

15.39 This is called a buffer solution because it contains NH_4^+ (from NH_4Cl) and NH_3. Differing by only one proton, NH_4^+ and NH_3 are a weak acid/base conjugate pair.

15.41 $NH_3 + H_3O^+ \rightarrow NH_4^+ + H_2O$

The added H_3O^+ is converted to the weak acid NH_4^+, resulting in a pH change much smaller than what would occur in an unbuffered solution.

15.43 Substances were classified as bases if they had a bitter taste, turned litmus blue, and formed aqueous solutions that felt slippery to the touch.

15.45 An electrolyte is a substance whose aqueous solution can conduct electricity. A nonelectrolyte is a substance whose aqueous solution cannot conduct electricity. Some examples of electrolytes are NaCl, HCl, and NaOH. Some examples of nonelectrolytes are CH_3OH, $C_6H_{12}OH$, and CH_4. See Table 15.1 of the textbook for additional examples.

15.47 Ethanol does not dissociate into ions when dissolved in water, but magnesium chloride does. In other words, ethanol is a nonelectrolyte, and magnesium chloride is an electrolyte.

15.49 Ionic compounds dissociate into ions when they dissolve in water. The compounds are electrolytes because the ions conduct electricity.

15.51 Ions that are free to move must be present for a solution to conduct electricity. Electrons cannot flow through water, but ions that are free to move do. Electrolyte solutions are well suited for that as, in the process of dissolving, the particles of solute dissociate and form both positive and negative ions. These ions migrate to the oppositely charged wires, closing the circuit and allowing for the electricity to flow.

15.53 (a) Yes, because NaCl does not exist as discrete molecules whose atoms are held together by covalent bonds.

(b) Because it is a water-soluble ionic compound.

15.55 (a) Electrolyte.

  (b) Nonelectrolyte.

  (c) Nonelectrolyte.

  (d) Electrolyte.

  (e) Electrolyte.

  (f) Electrolyte.

  (g) Nonelectrolyte.

15.57 (a) NH_4^+

  (b) NH_4Cl

  (c) Ammonium chloride is a soluble ionic salt. When it dissolves, it also dissociates into ions: NH_4^+ and Cl^-. These ions are free to move and, hence, the solution is an electrolyte and conducts electricity.

15.59 Glucose is a molecular compound that does not dissociate in water.

15.61 The intense glow indicates a strong electric current and thus a large number of ions in the HCl beaker. HCl is a strong electrolyte that dissociates completely in solution. The dim glow indicates a weak current and thus few ions in the HF beaker. HF is a weak electrolyte that dissociates only partially in solution.

15.63 The dissociation equilibrium lies far to the right for a strong electrolyte (we say the reaction *goes to completion*) and to the left for a weak electrolyte.

15.65 (a) They all produce the hydronium ion, H_3O^+.

  (b) All three of these substances are electrolytes because when they dissolve in water, they also dissociate, producing ions in solution.

  (c) $CH_3COOH(aq) + H_2O(l) \rightleftharpoons H_3O^+(aq) + CH_3COO^-(aq)$

15.67 It is a weak electrolyte. The low equilibrium constant of 8.2×10^{-6} means that the numerator in the expression $[Products]/[Reacants]$ for the reaction undissociated molecules $\rightarrow$ ions is much smaller than the denominator. In other words, the product concentration at equilibrium is very low, the equilibrium lies to the left, and most of the molecules are undissociated, which is the definition of a weak electrolyte.

15.69 Numerous answers possible. Examples of two strong molecular electrolytes are hydrobromic acid and hydrochloric acid:

$HBr + H_2O \rightarrow H_3O^+ + Br^-$

$HCl + H_2O \rightarrow H_3O^+ + Cl^-$

Examples of two weak molecular electrolytes are hydrofluoric acid and ammonia:

$HF + H_2O \rightleftharpoons H_3O^+ + F^-$

$NH_3 + H_2O \rightleftharpoons NH_4^+ + OH^-$

15.71 Compounds (a)–(e) are molecular because they do not contain a metal ion or polyatomic ion. Compound (f) is ionic, consisting of ammonium (NH_4^+) and chloride (Cl^-) ions.

15.73 They are all acids because they all produce H_3O^+ ions in solution.

15.75 (a) Weak acids: hydrofluoric, acetic, and the ammonium cation.

  (b) Strong acids: hydrochloric, nitric, and sulfuric.

(c) In solutions of equal concentrations, the strong acids produce more H^+ (or H_3O^+) ions than the weak acids do. Therefore, the 1.0 M solutions of the strong acids are more acidic than the 1.0 M solutions of the weak acids. The acid that is expected to produce the largest concentration of H_3O^+ ions is the sulfuric acid because, in addition to being a strong acid, it is diprotic. Note, H_2SO_4 is not expected to release twice as many protons into solution as the second of the two protons is only partially dissociated.

15.77 Phosphoric acid is triprotic and its three protons dissociate in three stages:

$$H_3PO_4 + H_2O \rightleftharpoons H_3O^+ + H_2PO_4^- \quad K_{eq,1} = 7.5 \times 10^{-3}$$

$$H_2PO_4^- + H_2O \rightleftharpoons H_3O^+ + HPO_4^{2-} \quad K_{eq,2} = 6.2 \times 10^{-8}$$

$$HPO_4^{2-} + H_2O \rightleftharpoons H_3O^+ + PO_4^{3-} \quad K_{eq,3} = 4.2 \times 10^{-13}$$

The decreasing values of each subsequent equilibrium constant mean that each next proton dissociates with more and more difficulty. There are two major reasons why this is so. First, a dissociating proton, which has a +1 charge, must move away from an anion. With each departing proton, the charge of that anion becomes more and more negative. According to Coulomb's law, the strength of electrostatic attraction between oppositely charged ions increases with increased charges, making it more and more difficult to separate charged particles from each other. The second reason goes back to Chapter 5—the amount of energy needed to break bonds. In general, the more electrons are in a bond, the higher the bond order and the stronger the bond. As protons dissociate from the phosphoric acid, each of them leaves one electron behind. These left over electrons add to the number of electrons in the remaining bonds, strengthening it. Therefore, it becomes more and more difficult to break the remaining bonds and for the remaining protons to dissociate.

15.79 (a) After water, H_3O^+ and NO_3^- would be in highest concentration because HNO_3 (nitric acid) is a strong acid and so it dissociates completely in water.

(b) After water, HNO_2 (nitrous acid) would be in highest concentration because it is a weak acid and so it dissociates only to a limited extent in water ($K_{aq} = 4.5 \times 10^{-4}$).

(c) For oxyacids with the general formula HXO_n, the smaller the value of n, the weaker the acid.

15.81 The beaker contains essentially all dissociated acid ions because the acid is strong:

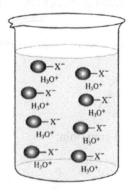

15.83 Section 15.3 of the textbook states that for acetic acid, $K_{aq} = 1.8 \times 10^{-5}$. The small value of K_{aq} tells you that acetic acid dissociates very little in water. In other words, it is a weak acid.

15.85 False. The extent of dissociation is what determines whether an acid is weak or strong, not the number of protons in the chemical formula. A monoprotic acid can be either strong or weak, and a diprotic acid can be either strong or weak. In this case, H_2CO_3 is a weak diprotic acid and HCl is a strong monoprotic acid.

15.87 Acid–base neutralization is the production of water from the reaction between H_3O^+ supplied by an acid and OH^- supplied by a base:

$$H_3O^+ + OH^- \rightarrow 2H_2O$$

15.89 (a) The product of the reaction is the water-soluble salt $CaCl_2$. This neutral compound, like all water-soluble salts, is a strong electrolyte and it completely dissociates into ions in solution. Ca^{2+} and Cl^- ions conduct electricity very well.

(b) $CaCl_2(aq) \rightarrow Ca^{2+}(aq) + 2Cl^-(aq)$

15.91 $0.5 \text{ mol } H_2SO_4 \times \underbrace{\dfrac{2 \text{ mol } H_3O^+}{1 \text{ mol } H_2SO_4}}_{\substack{\text{Monoprotic}}} \times \underbrace{\dfrac{1 \text{ mol } OH^-}{1 \text{ mol } H_3O^+}}_{\substack{H_3O^+ + OH^- \rightarrow 2H_2O \\ \text{Neutralization}}} \times \underbrace{\dfrac{1 \text{ mol } Ba(OH)_2}{2 \text{ mol } OH^-}}_{\substack{\text{From formula} \\ Ba(OH)_2}} = 0.50 \text{ mol } Ba(OH)_2$

The products of the neutralization are $BaSO_4(s)$ and water. Because the $BaSO_4$ is poorly soluble in water (see Table 8.1 of the textbook), there are essentially no ions present to conduct electricity. When an excess of soluble $Ba(OH)_2$ is added, Ba^{2+} and OH^- ions are present to conduct electricity.

15.93 Problem 15.91— $BaSO_4(s)$: $Ba(OH)_2(aq) + H_2SO_4(aq) \rightarrow BaSO_4(s) + 2H_2O(l)$.

Problem 15.92— $Ba(NO_3)_2(aq)$: $Ba(OH)_2(aq) + 2HNO_3(aq) \rightarrow Ba(NO_3)_2(aq) + 2H_2O(l)$.

15.95 $NH_3(g) + H_2O(l) \rightleftharpoons NH_4^-(aq) + OH^-(aq)$

15.97 Because NH_3 contains no OH groups and therefore cannot dissociate to produce OH^- ions in solutions. By definition, an Arrhenius base must produce OH^- ions.

15.99 Water can be thought of as a Brønsted–Lowry acid because it donates a proton to the ammonia:

$$NH_3 + H_2O \rightleftharpoons NH_4^+ + OH^-$$

15.101 (a) The very low value of K_{eq} indicates that the equilibrium lies far to the left, which means there are very few $H_3O^+(aq)$ ions present. The solution is therefore only slightly acidic.

(b) Water is acting as a base because it accepts a proton from NH_4^+.

15.103 (a) $H^- + H_2O \rightarrow H_2 + OH^-$

(b) The Brønsted–Lowry definition says the base H^- is a proton acceptor, so that H_2 gas forms as H^- combined with the proton it accepted from the water molecule: $H^- + H^+ \rightarrow H_2$.

$$H{:}^- + H{:}\ddot{O}{:}H \longrightarrow H{:}H + \bar{:}\ddot{O}{:}H$$

15.105 (a) $2Na^+(aq) + O^{2-}(aq) + H_2O(l) \rightarrow 2Na^+(aq) + 2OH^-(aq)$

(b) Both H^- and O^{2-} have lone pair(s) of electrons.

(c) The lone pairs of electrons allow H^- and O^{2-} to accept protons from water and act as strong bases.

15.107 The C_2H_5OH donates a proton to the H^-.

15.109 (a)

(b) No because there are no more acidic protons in the acetate ion to donate to another substance, such as water.

(c) $CH_3COO^- + H_2O \rightleftharpoons CH_3COOH + OH^-$

(d) By donating a proton to acetate, water acts as a Brønsted–Lowry acid.

15.111 The LiOH solution is more basic. Because lithium is a group 1 metal, LiOH is a strong base that dissociates completely, meaning a 1 M LiOH solution has an OH^- concentration of 1 M. The weak base NH_3 dissociates very little, meaning a 1 M NH_3 solution has an OH^- concentration of much less than 1 M.

15.113 (a) $HPO_4^{2-} + H_2O \rightleftharpoons PO_4^{3-} + H_3O^+$

(b) $HPO_4^{2-} + H_2O \rightleftharpoons H_2PO_4^- + OH^-$

(c) You need the K_{eq} values for both reactions. If K_{eq} for reaction (a) is larger than that for reaction (b), the solution will be weakly acidic. If the opposite is true, the solution will be weakly basic.

15.115 It must produce OH^- ions in water.

15.117 (a) The autoionization of water refers to the reaction in which one water molecule donates a proton to another water molecule:

(b) $H_2O + H_2O \rightleftharpoons H_3O^+ + OH^-$

(c)

Proton donor: Proton acceptor:
acid   base

15.119 The autoionization equilibrium lies far to the left. The extremely small K_w value of 1.0×10^{-14} verifies this.

15.121 No. Due to the autodissociation of water, $2H_2O \rightleftharpoons H_3O^+ + OH^-$, 1 L of pure water contains 1.0×10^{-7} mole of H_3O^+ and 1.0×10^{-7} mole of OH^-.

15.123 Pure water is neutral because $[H_3O^+]$ equals $[OH^-]$.

15.125 (1) An aqueous solution is basic when $[OH^-] > [H_3O^+]$.

(2) An aqueous solution is basic when $[OH^-] > 1.0 \times 10^{-7}$ M.

15.127 (a) The product of $[H_3O^+]$ and $[OH^-]$ must always equal 1.0×10^{-14} in an aqueous solution. If one concentration increases, the other must decrease. Hence, if the concentration of the H_3O^+ ions increases, the concentration of the OH^- ions must decrease.

(b) No, $[H_3O^+] \times [OH^-] = $ constant.

15.129 Solve the K_w expression, $[H_3O^+] \times [OH^-] = 1.0 \times 10^{-14}$, for $[OH^-]$:

$$[OH^-] = \frac{1.0 \times 10^{-14}}{[H_3O^+]} = \frac{1.0 \times 10^{-14}}{1.0 \text{ M}} = 1.0 \times 10^{-14} \text{ M}$$

The solution is acidic because $[H_3O^+] > [OH^-]$.

15.131 LiOH dissociates to produce 1 mole of OH$^-$ for every 1 mole of LiOH. Therefore:

$$[OH^-] = \frac{2.50 \text{ mol OH}^-}{4.00 \text{ L solution}} = 0.625 \text{ M}$$

$$[H_3O^+] = \frac{1.0 \times 10^{-14}}{0.625 \text{ M}} = 1.6 \times 10^{-14} \text{ M}$$

15.133 Molar mass of Ba(OH)$_2$ = 171.342 g/mol.

$$\text{Mol Ba(OH)}_2 = 2.40 \text{ g} \times \frac{1 \text{ mol}}{171.342 \text{ g}} = 0.0140 \text{ mol}$$

Ba(OH)$_2$ dissociates to produce 2 moles of OH$^-$ for every 1 mole of Ba(OH)$_2$, meaning that 0.0140 mole of Ba(OH)$_2$ produces 0.0280 mole of OH$^-$. The concentrations are therefore:

$$[OH^-] = \frac{0.0280 \text{ mol}}{4.00 \text{ L solution}} = 0.00700 \text{ M}$$

$$[H_3O^+] = \frac{1.0 \times 10^{-14}}{0.00700 \text{ M}} = 1.43 \times 10^{-12} \text{ M}$$

15.135 (a) The solution is acidic because the $[OH^-]$ concentration is below 10^{-7} M, meaning that the $[H_3O^+]$ must be above 10^{-7} M.

     (b) Since $[OH^-]$ equals 1.0×10^{-11} M, $[H_3O^+]$ must be 1.0×10^{-3} M. This ensures that the product of the two concentrations, $10^{-11} \times 10^{-3}$, equals $K_w = 10^{-14}$.

15.137 -34

15.139 The logarithm of both 10^0 and 1 is 0 ($10^0 = 1$).

15.141 The logarithm of 10 is 1 ($10^1 = 10$) and the logarithm of 100 is 2 ($10^2 = 100$). Because 60 is between 10 and 100, the logarithm of 60 must be between 1 and 2.

15.143 The logarithm of 0.1 is -1 ($10^{-1} = 0.1$), and the logarithm of 1 is 0 ($10^0 = 1$), by definition. Because 0.73 is between 0.1 and 1, so the logarithm of 0.73 must be between -1 and 0.

15.145 pH $= -\log[H_3O^+]$. Leaving the negative out would result in negative values of pH for all H_3O^+ concentrations that are less than 1 M.

15.147 pH $= -\log[H_3O^+] = -\log 0.0010 = -\log 10^{-3} = 3$. The solution is acidic because the pH < 7.

15.149 pH $= -\log(6.40 \times 10^{-9}) = 8.19$. The solution is basic because the pH > 7.

15.151 If $[OH^-] = 2.0 \times 10^{-3}$ M, $[H_3O^+] = 5.0 \times 10^{-12}$ M (from the K_w equation). The pH is therefore $-\log(5.0 \times 10^{-12}) = 11.30$. The solution is basic because the pH > 7.

15.153 pH $= -\log 10.0 = -1.000$. The solution is acidic because the pH is below 7.

15.155 (a) $[H_3O^+] = 10^{-pH} = 10^{-8}$ M.

     (b) $[OH^-] = K_w/[H_3O^+] = 10^{-14}/10^{-8} = 10^{-6}$ M.

     (c) The solution is basic because the pH of 8 is above 7.

15.157 The pH of the HCl solution is lower because acetic acid is a weak acid and HCl is a strong acid. One mole of the weak acid puts only a few H_3O^+ ions into solution, meaning $[H_3O^+] \ll 1$ M. One mole of the strong acid puts 1 mole of H_3O^+ ions into solution, meaning $[H_3O^+] = 1$ M.

15.159 No, the weak acid always has one more proton than its conjugate base. Because protons have a positive charge, the acid always carries one more positive charge than its conjugate base.

15.161 The $NaNO_3$ solution would be neutral. The NO_3^- ion has no tendency to accept protons because doing so forms the strong acid HNO_3, which dissociates immediately and completely. The conjugate of any strong acid is so weak that it is neither acid nor base.

15.163 Because Cl^- has no tendency to accept protons to form the strong acid HCl. Strong acids have conjugates that are neither acid nor base; they are neutral. A solution of $NaCl$ is therefore neutral.

15.165 No, because it does not contain significant amounts of either a weak acid and its conjugate base or a weak base and its conjugate acid.

15.167 Blood in living organisms is one biological system that is buffered. The buffer is bicarbonate/carbonate, HCO_3^-/CO_3^{2-}.

15.169 Yes, the $NaHCO_3$ dissociates to produce Na^+ and HCO_3^- ions. The solution then contains the weak acid H_2CO_3 and its conjugate base HCO_3^- components of a buffer solution. The HCO_3^- reacts with added acid according to the reaction:

$$HCO_3^- + H_3O^+ \rightarrow H_2CO_3 + H_2O,$$

and the HCO_3^- reacts with added base according to the reaction:

$$H_2CO_3 + OH^- \rightarrow HCO_3^- + H_2O.$$

15.171 The NaOH reacts with half of the acetic acid to produce sodium acetate, according to the neutralization reaction

$$CH_3COOH + NaOH \rightarrow Na^+ + CH_3COO^- + H_2O$$

After the neutralization reaction, the solution contains significant amounts of acetic acid and acetate ion, a conjugate pair and therefore components of a buffer solution.

15.173 Dissolve HClO and NaClO or any other soluble hypochlorite salt in water, so that the solution contains a weak acid, HClO and its conjugate base, ClO^-.

15.175 (a) The sodium acetate dissociates completely, forming 2.0 moles of acetate ions and 2.0 moles of sodium ions. One mole of acetate ions reacts with the 1 mole of HCl according to the reaction:

$$CH_3COO^- + HCl \rightarrow CH_3COOH + Cl^-$$

(b) Na^+, Cl^-, CH_3COO^-, and CH_3COOH. The added 1.0 mole of H_3O^+ reacts with half of the sodium acetate, leaving 1.0 mole of CH_3COO^- 1.0 mole of CH_3COOH, 1.0 mole of Cl^- (from the 1.0 mole of HCl added), and 2.0 moles of Na^+ (from the 2.0 moles of sodium acetate initially dissolved).

(c) Yes, because there is a large amount of both a weak acid, CH_3COOH, and its conjugate base, CH_3COO^- present in solution.

15.177 The acid part of the buffer pair combines with the OH^- ions of the added strong base, locking the OH^- ions up in water molecules. In an acetic acid/acetate buffer, for instance, the mechanism is

$$CH_3COOH + OH^- \rightarrow CH_3COO^- + H_2O$$

The strong base OH^- has been replaced by the weak base CH_3COO^-.

15.179 The weak base in the buffer combines with the added protons to form molecules of a weak acid.

15.181 (a) $HClO_4 + H_2O \rightarrow H_3O^+ + ClO_4^-$.

(b) Because this is a strong acid, it dissociates completely, meaning the solution contains no $HClO_4$, and equal amounts of the two ions from the acid. The relative concentrations are therefore $[H_2O] > [H_3O^+] = [ClO_4^-]$.

(c) Because the acid is monoprotic, each mole of acid yields 1 mole of H^+ and 1 mole of ClO_4^-. The concentrations are therefore: $[H_3O^+] = [ClO_4^-] = 0.100$ M.

(d) $pH = -\log 0.100 = -(-1.000) = 1.000$

15.183 (a) $H_2PO_4^-$; the K_{eq} values are given in Equation 15.2 of the textbook. They are 4.2×10^{-13} for HPO_4^{2-} and 6.2×10^{-8} for $H_2PO_4^-$. The species with the larger K_{eq} is the stronger acid.

(b) H_3O^+

(c) HCO_2H

(d) HI; Section 15.1 of the textbook notes that HI is a strong electrolyte, meaning it dissociates completely and has a very high K_{eq} value. Section 15.2 notes that HF is a weak electrolyte and has a very low K_{eq} value.

15.185 (a) Basic because the number of moles of OH^- is greater than the number of moles of H_3O^+:

$$\frac{0.015 \text{ mol NaOH}}{\cancel{L}} \times 0.0500 \, \cancel{L} = 0.00075 \text{ mol NaOH, yielding } 0.0075 \text{ mol } OH^-$$

$$\frac{0.010 \text{ mol } HNO_3}{\cancel{L}} \times 0.500 \, \cancel{L} = 0.000500 \text{ mol } HNO_3, \text{ yielding } 0.00050 \text{ mol } H_3O^+$$

(b)

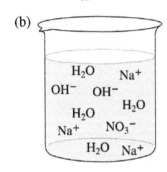

(c) The 0.00050 mole of H_3O^+ ions reacts with 0.00050 mole of OH^- ions, leaving 0.00025 mole of OH^- in the 100 mL of solution. The pH is therefore

$$[OH^-] = \frac{0.00025 \text{ mol}}{0.100 \text{ L}} = 0.0025 \text{ M}$$

$$[H^+] = \frac{1.0 \times 10^{-14}}{0.0025 \text{ M}} = 4.0 \times 10^{-12} \text{ M}$$

$$pH = -\log 4.0 \times 10^{-12} = -(11.40) = 11.40$$

15.187 (a) $HNO_3 + OH^- \rightarrow H_2O + NO_3^-$

(b) $HF + OH^- \rightarrow H_2O + F^-$

(c) $NH_3 + H_2O \rightleftharpoons NH_4^+ + OH^-$

(d) $HCO_3^- + H_2O \rightleftharpoons OH^- + H_2CO_3$

15.189

Substance	Name	Type of electrolyte (strong, weak, or nonelectrolyte)	Type of acid/base (strong or weak)	Reaction(s) in water
NaCl	Sodium chloride	Strong	—	(1) $NaCl \rightarrow Na^+ + Cl^-$
HNO_3	Nitric acid	Strong	Strong acid	(1) $HNO_3 + H_2O \rightarrow H_3O^+ + N$
$Mg(NO_3)_2$	Magnesium nitrate	Strong	—	(1) $Mg(NO_3)_2 \rightarrow Mg^{2+} + 2N$
HF	Hydrofluoric acid	Weak	Weak acid	(1) $HF + H_2O \rightleftarrows H_3O^+ + F^-$
NaF	Sodium fluoride	Strong	Weak base	(1) $NaF \rightarrow Na^+ + F^-$ (2) $F^- + H_2O \rightleftarrows HF + OH^-$
NH_4Cl	Ammonium chloride	Strong	Weak acid	(1) $NH_4Cl \rightarrow NH_4^+ + Cl^-$ (2) $NH_4^+ + H_2O \rightleftarrows H_3O^+ + N$

15.191 A pH of 2.0 means $[H_3O^+]$ is $10^{-2.0} = 0.010$ M. Nitric acid, HNO_3, is monoprotic and strong, meaning each mole yields 1 mole of H_3O^+ ions. To prepare 200.0 mL of a 0.010 M solution, you need:

$$\frac{0.010 \text{ mol}}{\cancel{L}} \times 0.2000 \cancel{L} = 0.0020 \text{ mol of } HNO_3$$

15.193 (a) $HCO_3^- + OH^- \rightarrow H_2O + CO_3^{2-}$

(b) $HCl + F^- \rightarrow HF + Cl^-$

(c) $H_2CO_3 + OH^- \rightarrow H_2O + HCO_3^-$

(d) $HCN + H_2O \rightleftarrows H_3O^+ + CN^-$

15.195 (a) $pH = -\log 1.0 = 0$ 　　　(b) $pH = -\log 0.1 = 1.0$

(c) $pH = -\log 0.001 = 3.0$ 　　(d) $pH = -\log(1.0 \times 10^{-5}) = 5.00$

(e) $pH = -\log(1.10 \times 10^{-7}) = 6.958$

15.197 First determine the molar concentration of OH^-, then use K_w and OH^- concentration to determine H_3O^+ concentration, then convert to pH:

$$\frac{20.0 \text{ g NaOH}}{2.00 \text{ L}} \times \frac{1 \text{ mol NaOH}}{40.0 \text{ g NaOH}} \times \frac{1 \text{ mol } OH^-}{1 \text{ mol NaOH}} = 0.250 \text{ mol } OH^-/L = 0.250 \text{ M}$$

$$[H_3O^+] = \frac{1.00 \times 10^{-14}}{0.250 \text{ M}} = 4.00 \times 10^{-14} \text{ M}$$

$$pH = -\log(4.00 \times 10^{-14}) = -(-13.400) = 13.400$$

15.199 In a 0.010 M NaOH solution, $[OH^-] = 0.0010$ M. Therefore,

$$[H_3O^+] = \frac{1.0 \times 10^{-14}}{0.010 \text{ M}} = 1.0 \times 10^{-12} \text{ M}$$

$$pH = -\log(1.0 \times 10^{-12}) = -(-12.00) = 12.00$$

15.201 KOH and H_2SO_4. The equation is $2KOH + H_2SO_4 \rightarrow 2H_2O + K_2SO_4$.

15.203

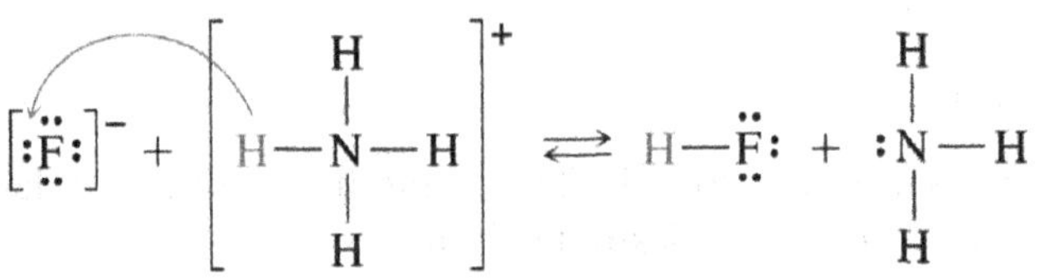

Base   Acid

The fluoride ion is the Brønsted–Lowry base because it accepts a proton. The ammonium ion is the Brønsted–Lowry acid because it donates a proton.

15.205 A solution that has a pH of 4.20 is acidic, and you cannot add the strong base NaOH to pure water and end up with an acidic solution.

15.207

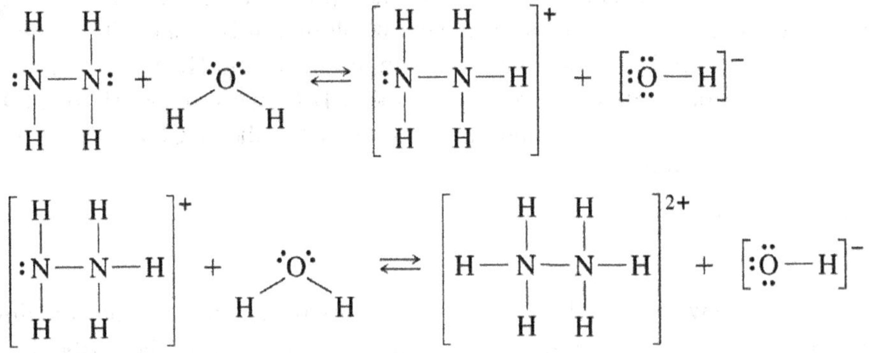

15.209 First figure out what the OH^- ion concentration is, which must also be the LiOH concentration because each LiOH yields one OH^- ion:

$$[OH^-] = \frac{1.0 \times 10^{-14}}{2.30 \times 10^{-13}} = 0.043 \text{ M}$$

Now use the molar mass of LiOH to determine the number of grams in this number of moles of LiOH:

$$\frac{0.04348 \text{ mol LiOH}}{L} \times \frac{23.948 \text{ g LiOH}}{\text{mol LiOH}} \times 0.750 \text{ L} = 0.781 \text{ g LiOH}$$

15.211 One thousand times more acidic means $10 \times 10 \times 10$ times more acidic. Because each one-unit change in pH is a tenfold change in acidity, the more acidic solution must have a pH of $9.20 - 3 = 6.20$. Therefore,

$$[H_3O^+] + 10^{-6.20} = 6.31 \times 10^{-7} \text{ M}$$

$$[OH^-] = \frac{1.0 \times 10^{-14}}{6.31 \times 10^{-7} \text{ M}} = 1.58^{-8} \text{ M}$$

15.213 (a) $LiOH + HI \rightarrow H_2O + LiI$

(b) $CH_3COOH + NaOH \rightarrow H_2O + NaCH_3COO$

(c) $2HBr + Ca(OH)_2 \rightarrow 2H_2O + CaBr_2$

(d) $3KOH + H_3PO_4 \rightarrow 3H_2O + K_3PO_4$

15.215 No. Because HBr is a strong acid, Br⁻ cannot act as a conjugate base by accepting protons. A buffered solution must contain a weak acid.

15.217 $[H_3O^+] = 10^{-2.40} = 3.98 \times 10^{-3} \, \text{mol/L}$

Because each HCl gives one H_3O^+, this is also the molar HCl concentration. Use the molar mass of HCl to get grams per liter and then multiply by the desired volume:

$$\frac{3.98 \times 10^{-3} \, \cancel{\text{mol}}}{\cancel{L}} \times \frac{36.46 \, \text{g}}{\cancel{\text{mol}}} \times 7.50 \, \cancel{L} = 1.09 \, \text{g}$$

15.219 The $NaClO_2$ solution would have the higher pH. Both these sodium salts are soluble in water, which means we must talk about the ions in solution. Because $HClO_4$ is a strong acid, it dissociates completely, with the result that anytime ClO_4^- accepts an H^+ from a water molecule, the $HClO_4$ immediately breaks up. The same is true for Na^+ accepting OH^- from water—the strong base NaOH immediately breaks up. The result is a neutral solution of Na^+ and ClO_4^- ions. The ClO_2^- ions, however, can accept H^+ from water and form the weak acid $HClO_2$. Because the Na^+ in the $NaClO_2$ solution cannot accept OH^- from water, every H_2O that gives an H^+ to ClO_2^- results in a free OH^- and, therefore, a basic solution. Thus, the pH of the $NaClO_4$ solution is 7, and the pH of the $NaClO_2$ solution is above 7.

15.221 $CN^- + H_2O \rightleftharpoons HCN + OH^-$

15.223 $NaOCl + HNO_3 \rightarrow NaNO_3 + HOCl$

Because all sodium salts are water-soluble (Table 8.1 of the textbook), the solution contains $Na^+(aq)$ and $NO_3^-(aq)$ ions. Because HOCl is a weak acid (Table 15.2B of the textbook), the solution also contains HOCl. And because the NaOCl was added in excess, the solution contains OCl^-. This solution is therefore a buffer of the weak acid HOCl and its conjugate base, OCl^-.

15.225

15.227 Weak acid. X^- is the conjugate base of HX. If HX were a strong acid, X^- would not be able to act as a base by accepting a proton from water and thereby causing the solution to be basic:

$$H_2O + X^- \rightarrow HX + OH^-$$

15.229

15.231 Just like the two sides of the mercury column in the u-tube, if you increase the hydroxide concentration, the hydronium concentration will be decreased (or vice versa). And similar to the mercury column, if you multiply (instead of add) the concentration of hydroxide ion by the concentration of hydronium ion, you will get a constant for all cases, but the constant will not equal zero. It will equal 1×10^{-14} (so called K_w for water).

15.233 $[Fe(OH_2)_6]^{3+}$ yields acidity via dissociation of an H^+ ion from one of the bound water molecules.

An H_2O loses on H^+ and becomes on OH^- hydroxide ion.

$$\left[Fe(OH_2)_6\right]^{3+}_{(aq)} \rightleftharpoons \left[Fe(OH_2)_5(OH)\right]^{2+}_{(aq)} + H^+_{(aq)}$$

This occurs because the Fe^{3+} at the center is sufficiently positive to attract electrons out of the Fe–O and O–H bounds to itself, weakening the O–H bounds so an H^+ ion can dissociate. In $[Fe(OH_2)_6]^{2+}$, the central Fe^{2+} ion is not sufficiently positive to do this.

15.235 The lower the pK_a the *stronger* the acid. This is just like pH (low pH means acidic). Indeed, a negative pK_a is a very strong acid.

15.237 It can still provide buffering action. Now, the $H_3{}^+N-$ end is the weak acid that neutralizes the strong

OH$^-$ base, and the carboxylate acid is the weak base that neutralizes strong acid H^+.

15.239 $NH_3(aq) + H_2O(l) \rightleftharpoons NH_4{}^+(aq) + OH^-(aq)$ $K_b = 1.8 \times 10^{-5}$

$$K_b = 1.8 \times 10^{-5} = \frac{[NH_4{}^+][OH^-]}{[NH_3]} \text{ if pH} = 11.50,$$

then $[H^+] = 10^{-11.50} = 3.16 \times 10^{-12} \, M$

Next, since $K_w = 10^{-14} = [H^+][OH^-]$, then

$$[OH^-] = \frac{K_w}{[H^+]} = \frac{10^{-14}}{3.16 \times 10^{-12} \, M} \quad [OH^-] = 0.00316 \, M$$

The $[OH^-]$ and $[NH_4^+]$ concentrations are equal since they are produced in a one-to-one ratio (see balanced reaction), so, finally, plug the values of $[OH^-]$ and $[NH_4^+]$ into the K_b equation and solve for X.

$$x = [NH_3] = \frac{(0.00316)(0.00316)}{1.8 \times 10^{-5}} = 0.55 \text{ M}$$

15.241 The pH is less than 7, so it is certainly an acid. The question is whether it is strong or weak. Let's for the moment assume it is strong. In that case, a 0.0035 M solution of it would yield $[H^+]$ of 0.0035 M, assuming it was monoprotic. This would give a pH or $-\log(0.0035) = 2.4$. The pH is actually 4.76, more than two full pH units higher, so it is a weak acid.

Chapter 16

16.1 See solution in textbook.

16.3 Calculate the mass defect for 1 mole of $_2^4$He:

$$2 \text{ mol protons} \times \frac{1.00730 \text{ g}}{1 \text{ mol protons}} = 2.01460 \text{ g}$$

$$2 \text{ mol neutrons} \times \frac{1.00870 \text{ g}}{1 \text{ mol neutrons}} = 2.01740 \text{ g}$$

$$2 \text{ mol electrons} \times \frac{0.00055 \text{ g}}{1 \text{ mol electrons}} = 0.0011 \text{ g}$$

4.0331 g	Theoretical $_2^4$He molar mass
-4.003 g	Actual $_2^4$He molar mass from periodic table
0.030 g	(mass defect per 1 mol of $_2^4$He)

Because $1 \text{ J} = 1 \text{ kg} \cdot \text{m}^2/\text{s}^2$, you must convert this mass to kilograms for your next calculation:

$$\frac{0.030 \text{ g}}{\text{mol}} \times \frac{1 \text{ kg}}{1000 \text{ g}} = 3.0 \times 10^{-5} \text{ kg/mol}$$

Now use Einstein's energy equation, with the mass defect $3.0 \times 10^{-5} \text{ kg/mol}$ as m:

$$E = mc^2 = \frac{3.0 \times 10^{-5} \text{ kg}}{\text{mol}} \times (3.00 \times 10^8 \text{ m/s})^2$$

$$= 2.7 \times 10^{12} \frac{\text{kg} \cdot \text{m}^2/\text{s}^2}{\text{mol}} = 2.7 \times 10^{12} \frac{\text{J}}{\text{mol}} \times \frac{1 \text{ kJ}}{1000 \text{ J}} = 2.7 \times 10^9 \text{ kJ/mol}$$

16.5 See solution in textbook.

16.7 Beta emission means a neutron in the nucleus splits into a proton plus an electron (beta particle). $_{20}^{40}$Ca has 20 protons, which means the parent must have had 19 protons. The periodic table tells you an atom containing 19 protons is potassium. Because the total number of nucleons does not change in beta emission, the parent nucleus also had 40 nucleons, making the isotope $_{19}^{40}$K. The reaction was $_{19}^{40}$K $\rightarrow$ $_{20}^{40}$Ca $+$ $_{-1}^{0}$e.

16.9 (a) Electron capture means an inner-shell electron and a proton in the nucleus combine to form a neutron. The number of protons therefore decreases by 1. The total number of nucleons does not change because the lost proton is replaced by the newly formed neutron (but the *details* do change, from 18p + 19n = 37 nucleons in the parent to 17p + 20n = 37 nucleons in the daughter). An atom containing 17 protons is chlorine, making the reaction $^{37}_{18}Ar + ^{0}_{-1}e \rightarrow ^{37}_{17}Cl$.

(b) Electron capture increases the n/p ratio, from 19/18 = 1.1 to 20/17 = 1.2. This means the n/p ratio for $^{37}_{18}Ar$ must have been too low, placing this isotope below the dark brown central region of the band of stability.

(c) Chlorine-37

16.11 Positron emission means a proton splits into a neutron plus a positron. Therefore the parent must have contained 12 + 1 = 13 protons, making it aluminum. The gained neutron means the parent must have contained 13 − 1 = 12 neutrons, for a mass number of 13p + 12n = 25. The parent was aluminum-25, $^{25}_{13}Al$. If you compare this result with the one given in the solution to Practice Problem 16.10, you'll see that electron capture and positron emission create the same daughter from a given parent.

16.13 See solution in textbook.

16.15 Alpha emission because, of the decay processes covered in the textbook, only alpha emission creates a daughter having an atomic number 2 less than that of the parent.

16.17 See solution in textbook.

16.19 The sum of $^{206}_{82}Pb$ and $^{238}_{92}U$ atoms present in the rock today is equal to $^{238}_{92}U$ atoms initially present.

$$1.82 \text{ g } ^{238}_{92}U \times \frac{1 \text{ mol } ^{238}_{92}U}{238.029 \text{ g } ^{238}_{92}U} \times \frac{6.022 \times 10^{23} \text{ atoms U}}{1 \text{ mol } ^{238}_{92}U} = 4.60 \times 10^{21} \text{ atoms } ^{238}_{92}U \text{ today}$$

$$4.02 \text{ g } ^{206}_{82}Pb \times \frac{1 \text{ mol } ^{206}_{82}Pb}{205.974 \text{ g } ^{206}_{82}Pb} \times \frac{6.022 \times 10^{23} \text{ atoms Pb}}{1 \text{ mol } ^{206}_{82}Pb} = 1.18 \times 10^{22} \text{ atoms } ^{206}_{82}Pb \text{ today}$$

$$(0.460 \times 10^{22}) + (1.18 \times 10^{22}) = 1.64 \times 10^{22} \text{ atoms } ^{238}_{92}U \text{ initially present}$$

$$\% \ ^{238}_{92}U \text{ remaining in rock} = \frac{4.60 \times 10^{21} \text{ atoms}}{1.64 \times 10^{22} \text{ atoms}} \times 100\% = 28.0\%$$

$$\text{Age} = \frac{-2.303 \times \log\left(\frac{28.0\%}{100}\right) \times 4.46 \times 10^{9} \text{ years}}{0.693} = 8.19 \times 10^{9} \text{ years}$$

16.21 Yes, because the reaction produces more neutrons (three) than it uses up (one).

16.23 The mass of the atom is less than the sum of the masses of the particles of which the atom is composed.

16.25 (a) False, because a nucleon is either a proton or a neutron and the atomic number accounts only for the number of protons.

(b) True, because a nucleon is either a proton or a neutron and mass number is defined as number of protons plus number of neutrons.

(c) False, because the atomic number is already defined as number of protons.

(d) True, because the mass number is defined as number of protons plus number of neutrons and the atomic number equals the number of protons, so the difference between them yields the number of neutrons in an atom.

16.27 In order to transform 1 mole of lead-207 into gold-195 (these isotopes are used as examples), the total number of nucleons that would have to be removed is: 3p + 9n = 12 nucleons. From the plot in Section 16.1, for a nucleus with 207 nucleons, the binding energy is 7.57×10^8 kJ/mol of nucleons. So, the energy needed for this transformation would be:

12 mol of nucleons $\times$ (7.57×10^8 kJ/mol of nucleons) $= 9.08 \times 10^9$ kJ. Clearly, making such changes in a nucleus requires an amount of energy that is millions of times greater than the energy generated in chemical reactions, which—if exothermic—typically produce only from a few hundred to a few thousands of kilojoules.

16.29 The factor c^2. Because c is such a large number, 3.00×10^8 m/s, multiplying the mass defect by c^2 makes E tremendously large.

16.31 Because the solution to Problem 16.30 is in terms of moles of $^{14}_{6}C$ but the graph in the textbook is in terms of moles of nucleons, you must convert:

$$\frac{1.02 \times 10^{10} \text{ kJ}}{1 \text{ mol } ^{14}_{6}C} \times \frac{1 \text{ mol } ^{14}_{6}C}{14 \text{ mol nucleons}} = 7.29 \times 10^8 \text{ kJ/mol nucleons}$$

The textbook graph shows about 7.45×10^8 kJ/mol nucleons for $^{12}_{6}C$ and about 7.25×10^8 kJ/mol nucleons for $^{13}_{6}C$. Thus, $^{14}_{6}C$ is less stable than $^{12}_{6}C$ and just a bit more stable than $^{13}_{6}C$.

16.33 False. All atoms have a mass defect and binding energy, even radioactive ones.

16.35 (a) The ratio increases because the number of neutrons becomes larger than the number of protons.

(b) The more protons in the nucleus, the greater the number of neutrons needed to reduce the proton–proton repulsions.

16.37 $^{16}_{8}O$:8/8 $= 1.00$; $^{17}_{8}O$:9/8 $= 1.12$; $^{19}_{8}O$:11/8 $= 1.38$

16.39 The length of time required for exactly half the atoms in a sample of a radioactive isotope to decay.

16.41 The textbook defines *stable* as meaning both nonradioactive atoms and radioactive atoms that have a measurable half-life. Therefore any atom lying outside the band of stability, both light brown region and dark brown central band, or outside the island of stability is unstable. To answer this question, see where the 60-proton vertical line intersects the bottom and top of the band of stability. (The 60-proton line does not intersect the island of stability, which means you don't have to worry about it here.) Where the 60-proton line intersects the bottom of the light brown region, move your finger leftward to the vertical axis and read about 67 neutrons; where the 60-proton line emerges out the top of the light brown region, move your finger leftward to the vertical axis and read about 92 neutrons. Thus, an atom containing 60 protons is unstable anytime it contains fewer than 67 neutrons or more than 92 neutrons.

16.43 According to the plot in Section 16.1, $Z = 112 - 118$.

16.45 The n/p ratio is too small for these nuclei, making them radioactive and most likely to decay by converting a proton to a neutron, so that n/p gets larger.

16.47 The spontaneous process whereby a radioactive nucleus adjusts its n/p ratio (an effort to become more stable) by capturing or emitting subatomic particles.

16.49 Because it is the oppositely charged version of an electron. Whenever a positron and electron encounter each other, they instantaneously annihilate each other and create a lot of energy as they do so.

16.51 Any nucleus below the central dark brown region of the band of stability, because the n/p ratio of such a nucleus is too low.

16.53 True. Radioactive decay changes the number of protons in the nucleus, thus changing the identity of the element.

16.55 When a neutron in the nucleus converts to a proton, an electron is also created and emitted. This is beta emission: $^1_0n \rightarrow ^1_1p + ^0_{-1}e$. The electron created in this decay is often called a beta particle and given the symbol β^-, so that you may sometimes see the decay written $^1_0n \rightarrow ^1_1p + \beta^-$. The symbols $^0_{-1}e$ and β^- are completely equivalent.

16.57 (a) In positron emission, a proton converts to a neutron plus a positron, which means the nucleus loses a proton, gains a neutron, and ejects a positron.

(b) In electron capture, a proton combines with an inner-shell electron and thereby converts to a neutron, which means that the nucleus loses a proton and gains a neutron.

(c) γ emission does not affect the number of nucleons.

16.59 (a) Conversion of a proton to a neutron increases the n/p ratio, which means the n/p ratio was probably too low to begin with. This puts the isotope below the central dark brown region on the band of stability.

(b) Conversion of a proton to a neutron means the atomic number (number of protons) decreases by 1. The daughter isotope therefore has $74 - 1 = 73$ protons, making it tantalum, Ta. The number of neutrons in the parent $^{162}_{74}W$, as $162 - 74 = 88$, making the mass number of the daughter $73 + 89 = 162$: $^{162}_{73}Ta$.

Two decay processes convert a proton to a neutron, one creating a positron in addition to the neutron, the other capturing an inner-shell electron. The two reactions are therefore $^{162}_{74}W \rightarrow ^0_{+1}e + ^{162}_{73}Ta$ and $^{162}_{74}W + ^0_{-1}e \rightarrow ^{162}_{73}Ta$.

(c) $^{162}_{74}W \rightarrow ^0_{+1}e + ^{162}_{73}Ta$ is positron emission; $^{162}_{74}W + ^0_{-1}e \rightarrow ^{162}_{73}Ta$. is electron capture.

16.61 Because an alpha particle, ^4_2He, consists of two protons and two neutrons, the atomic number (number of protons) decreases by 2 and the mass number (number of protons plus number of neutrons) decreases by 4.

16.63 As kinetic energy (energy of motion) in the emitted particle or as electromagnetic radiation, such as gamma rays.

16.65 By checking superscripts and subscripts. In a balanced reaction, the sum of the superscripts on the left must equal the sum of the superscripts on the right, and the sum of the subscripts on the left must equal the sum of the subscripts on the right.

16.67 Mass number, A, is number of protons plus number of neutrons (n + p) and changes only when this *sum* changes.

(a) Ejection of a beta particle means the decay process is beta emission: $^1_0n \rightarrow ^1_1p + ^0_{-1}e$. One neutron is lost, but one proton is gained, so that the sum n + p = mass number, A, does not change.

One proton is gained, and therefore the atomic number, Z, increases by 1.

(b) Ejection of a positron means the decay process is positron emission: $^1_1p \rightarrow ^1_0n + ^0_{+1}e$.

One proton is lost, but one neutron is gained, so that the sum n + p = mass number, A, does not change.

One proton is lost, and therefore the atomic number, Z, decreases by 1.

(c) Electron capture is $^0_{-1}e + ^1_{+1}p \rightarrow ^1_0n$.

One proton is lost, but one neutron is gained, so that the sum n + p = mass number, A, does not change.

One proton is lost, and therefore the atomic number, Z, decreases by 1.

(d) An alpha particle is a helium nucleus, which consists of two protons and two neutrons.

The ejection of an alpha particle means two neutrons and two protons are lost, so that the sum n + p = mass number, A, decreases by 4.

Also, since two protons are lost, the atomic number, Z, decreases by 2.

16.69 (a) You know alpha decay causes a nucleus to lose two protons (plus two neutrons), and you know number of protons in a nucleus is the same as atomic number. So, moving two elements to the left of the parent in the periodic table gives you the daughter.

(b) You know beta decay creates a proton: $^1_0n \rightarrow ^1_1p + ^0_{-1}e$. You also know that the number of protons in a nucleus is the same as the atomic number. Therefore, moving one element to the right of the parent in the periodic table gives you the daughter.

(c) You know that positron emission destroys a proton: $^1_1p \rightarrow ^1_0n + ^0_{+1}e$. You also know that the number of protons in a nucleus is the same as the atomic number. Therefore, moving one element to the left of the parent in the periodic table gives you the daughter.

(d) You know that electron capture destroys a proton: $^0_{-1}e + ^1_{+1}p \rightarrow ^1_0n$. You also know that number of protons in a nucleus is the same as atomic number. Therefore, moving one element to the left of the parent in the periodic table gives you the daughter.

16.71 Numerous answers possible. One sequence is:

$$^{207}_{82}Pb \rightarrow ^{203}_{80}Hg + ^4_2He$$

$$^{203}_{80}Hg \rightarrow ^{203}_{79}Au + ^0_{+1}e$$

16.73 The superscript sum on the left must equal the superscript sum on the right, and ditto for subscripts. Because the superscript does not change, the missing particle must have a 0 superscript. Because the subscript increases by 1, the missing particle must have a -1 subscript. The particle fitting this bill is the electron, $^0_{-1}e$: $^{47}_{20}Ca \rightarrow ^0_{-1}e + ^{47}_{21}Sc$. The fact that the electron is created coupled with the fact that, in talking about nuclear reactions, *beta particle* is another name for the electron tells you the decay process is beta emission.

16.75 The superscript sum on the left must equal the superscript sum on the right, and ditto for subscripts. The missing particle therefore has superscript 11 and subscript 4. The subscript tells you the number of protons = atomic number; the element having atomic number 4 is beryllium, Be: $^{11}_4Be \rightarrow ^{11}_5B + ^0_{-1}e$. The fact that an electron is given off coupled with the fact that, in nuclear chemistry, *beta particle* is another name for the electron tells you the decay process is beta emission.

16.77 $26.2 \not{h} \times \dfrac{1 \text{ half-life}}{13.1 \not{h}} = 2 \text{ half-lives}$

$39.3 \not{h} \times \dfrac{1 \text{ half-life}}{13.1 \not{h}} = 3 \text{ half-lives}$

After 26.2 h: $100 \text{ g} \times \dfrac{1}{2} \times \dfrac{1}{2} = 25.0 \text{ g}$ or $100 \text{ g} \xrightarrow{13.1 \text{ h}} 50 \text{ g} \xrightarrow{26.2 \text{ h}} 25 \text{ g}$

After 39.3 h: $100 \text{ g} \times \dfrac{1}{2} \times \dfrac{1}{2} \times \dfrac{1}{2} = 12.5 \text{ g}$ or $100 \text{ g} \xrightarrow{13.1 \text{ h}} 50 \text{ g} \xrightarrow{26.2 \text{ h}} 25\text{g} \xrightarrow{39.3 \text{ h}} 12.5\text{g}$

16.79 Because new $^{14}_{6}\text{C}$ is constantly being produced in the upper atmosphere by incoming radiation, always replacing the 50% that disappears every 5715 years.

16.81 (a) That all the $^{206}_{82}\text{Pb}$ came from the decay of $^{238}_{92}\text{U}$.

(b) $14.90 \text{ g} \, {}^{238}_{92}\text{U} \times \dfrac{1 \text{ mol} \, {}^{238}_{92}\text{U}}{238.029 \text{ g} \, {}^{238}_{92}\text{U}} \times \dfrac{6.022 \times 10^{23} \text{ atoms} \, {}^{238}_{92}\text{U}}{1 \text{ mol} \, {}^{238}_{92}\text{U}} = 3.770 \times 10^{22} \text{ atoms} \, {}^{238}_{92}\text{U}$

$26.50 \text{ g} \, {}^{206}_{82}\text{Pb} \times \dfrac{1 \text{ mol} \, {}^{206}_{82}\text{Pb}}{205.974\ 46 \text{ g} \, {}^{206}_{82}\text{Pb}} \times \dfrac{6.022 \times 10^{23} \text{ atoms} \, {}^{206}_{82}\text{Pb}}{1 \text{ mol} \, {}^{206}_{82}\text{Pb}} = 7.748 \times 10^{22} \text{ atoms} \, {}^{206}_{82}\text{Pb}$

(c) $(3.770 \times 10^{22}) + (7.748 \times 10^{22}) = 11.518 \times 10^{22} = 1.1518 \times 10^{23} \text{ atoms} \, {}^{238}_{92}\text{U}$

(d) $\dfrac{3.770 \times 10^{22} \text{ atoms}}{1.1518 \times 10^{23} \text{ atoms}} \times 100\% = 32.73\%$

(e) $\text{Age} = \dfrac{-2.303 \times \log\left(\dfrac{32.73\%}{100\%}\right) \times (4.46 \times 10^{9} \text{ years})}{0.693} = 7.19 \times 10^{9} \text{ years}$

16.83 *Fission* means to break up, and therefore nuclear fission is the breaking up of a large nucleus into two or more smaller ones, while nuclear fusion is the combining of small nuclei into one larger nucleus.

16.85 Because each fission event produces more neutrons than it consumes. The newly formed excess neutrons are available to take part in subsequent fission events.

16.87 The superscript sum on the left must equal the superscript sum on the right, and ditto for subscripts. Therefore the missing particle must have superscript 12 and subscript 6. The subscript is the atomic number, making the missing particle carbon, C: $^{8}_{4}\text{Be} + {}^{4}_{2}\text{He} \rightarrow {}^{12}_{6}\text{C} + \gamma$.

16.89 The mass of fissionable material required to keep most of the neutrons produced in a fission chain reaction in the sample. The result is a runaway reaction and an explosion.

16.91 Numerous answers possible. Some benefits are that electricity generated by fission lessens our dependency on fossil fuels, causes less atmospheric pollution, and may eventually be less expensive to the consumer. Two problems are what to do with nuclear waste and the potential for accidents that may release radioactive materials into the environment.

16.93 Because nuclei, being positively charged, repel one another, and extremely large amounts of thermal energy (in other words, extremely high temperatures) are needed to force the nuclei close enough together to fuse.

16.95 Fusion fuel (heavy isotopes of hydrogen) is cheap and readily available from ocean water; fusion reactions produce no nuclear waste; no danger of an accident releasing toxic substances into the environment. However, fusion requires high temperature (up to 15 million K) so that the colliding nuclei can move fast enough to overcome their mutual repulsions. Because no one has developed a material or confinement method able to withstand such high temperatures, we have no "container" to hold the fusion fuel.

16.97 By ionizing the biological molecules that compose them. This ionization causes bonds in the molecules to weaken and often break, causing irreparable damage.

16.99 *Magnetic resonance imaging* lets doctors see deep inside the body to spot tumors and other trouble spots. *Gamma radiation from cobalt-60* destroys cancer cells. *Radiation imaging with iodine-131* lets doctors check the health of a thyroid gland. In *radioimmunoassays*, radioactive reactants are mixed with a sample of a patient's blood to determine what compounds are present in the blood. *Irradiation of foods* before they reach the supermarket cuts down on cases of food poisoning.

16.101 The graph in Section 16.1 of the textbook tells you that binding energy per nucleon is greatest for ^{56}Fe and begins a steady decrease for all atoms more massive than it. This means that fusing atoms of ^{56}Fe produces a less stable nucleus, making the reaction endothermic. Energy must be put into the reaction to make it go.

16.103 (a) The atomic number of polonium is 84, and an alpha particle is $^4_2$He. The daughter isotope must therefore have superscript $210 - 4 = 206$ and subscript $84 - 2 = 82$. Atomic number 82 is lead, Pb: $^{210}_{84}$Po $\rightarrow$ $^{206}_{82}$Pb $+$ $^4_2$He.

(b) Another way of asking this question is, "How old is the sample when it contains only 5.00% of the initial amount of ^{210}Po?" Therefore, use the age formula:

$$\frac{-2.303 \times \log\left(\dfrac{5.00\,\%}{100\%}\right) \times 138 \text{ days}}{0.693} = 597 \text{ days}$$

16.105 In every nuclear decay, the superscript sum on the left must equal the superscript sum on the right, and ditto for the subscripts.

(a) $^{121}_{51}$Sb $+$ $^4_2$He $\rightarrow$ $^{124}_{52}$Te $+$ $^1_1$H

(b) $^{27}_{13}$AI $+$ $^4_2$He $\rightarrow$ $^{30}_{15}$P $+$ $^1_0$n

(c) $^{238}_{92}$U $+$ $^1_0$n $\rightarrow$ $^{239}_{93}$Np $+$ $^{\ 0}_{-1}$e

16.107 Because beta emission converts a neutron to a proton and lowers the n/p ratio, it is most likely in nuclei that have too high an n/p ratio. Because positron emission converts a proton to a neutron and raises the n/p ratio, it is most likely in nuclei that have too low an n/p ratio. Fluorine nuclei are light enough to be stable with n/p = 1.

$^{17}_{9}$F $\dfrac{\text{n}}{\text{p}} = \dfrac{17 - 9}{9} = 0.89$; too low, positron emission

$^{20}_{9}$F $\dfrac{\text{n}}{\text{p}} = \dfrac{20 - 9}{9} = 1.2$; too high, beta emission

$^{21}_{9}$F $\dfrac{\text{n}}{\text{p}} = \dfrac{21 - 9}{9} = 1.3$; too high, beta emission

16.109 The age of the painting is:

$$\frac{-2.303 \times \log\left(\frac{96.1\%}{100\%}\right) \times 5715 \text{ years}}{0.693} = 328 \text{ years}$$

meaning it was created in the year $2002 - 328 = 1674$, 5 years after Rembrandt's death. However, given the uncertainty in carbon-dating, 1674 is close enough to Rembrandt's time that the painting might indeed have been done by him.

16.111 The atomic number of thorium is 90. Loss of six alpha particles, $_2^4\text{He}$, decreases mass number by 24 and the atomic number by 12. This means daughter 1 has atomic number $90 - 12 = 78$ and mass number $232 - 24 = 208$. Atomic number 78 is for platinum, making daughter 1 $_{78}^{208}\text{Pt}$.

Loss of four beta particles, $_{-1}^0 e$, has no effect on mass number but increases the atomic number by 4. This means daughter 2 has atomic number $78 + 4 = 82$ and mass number 208. Atomic number 82 is lead, making daughter 2 $_{82}^{208}\text{Pb}$. The sequence is:

$$_{90}^{232}\text{Th} \longrightarrow 6\,_2^4\text{He} + _{78}^{208}\text{Pt}$$

$$_{78}^{208}\text{Pt} \longrightarrow 4\,_{-1}^0 e + _{82}^{208}\text{Pb}$$

16.113 (a) The first mechanism is the correct one because it shows the colored O in the water molecule. The arrows indicate which reactant bonds must break:

(b) If ^{18}O were radioactive, you could separate the two products from each other and use a Geiger counter to see which one was radioactive, revealing where the radioactive ^{18}O "tag" went.

16.115 The equation from Problem 16.103, $_{84}^{210}\text{PO} \longrightarrow _{82}^{206}\text{Pb} + _2^4\text{He}$, tells you that each mole of ^{210}Po yields 1 mole of ^4He. Thus, if you know the number of moles of ^{210}Po decaying in 138.4 days, you know how much helium you have. First determine the number of moles of ^{210}Po in the initial sample:

$$1.000 \text{ g PoO}_2 \times \frac{1 \text{ mol PoO}_2}{241.981 \text{ g PoO}_2} \times \frac{1 \text{ mol } ^{210}\text{Po}}{1 \text{ mol PoO}_2} = 4.133 \times 10^{-3} \text{ mol } ^{210}\text{Po}$$

After 1 half-life, half of this amount, 2.066×10^{-3} mol, has decayed, producing 2.066×10^{-3} mole of ^4He. Now use the ideal gas equation to find the volume of this much $_2^4\text{He}$ gas at standard temperature (0.00 °C) and pressure (1.00 atm). Remember that you must use the Kelvin temperature in your calculation.

$$V = \frac{nRT}{P} = \frac{2.066 \times 10^{-3}\text{mol } 0.0821 \text{ L} \cdot \text{atm/K} \cdot \text{mol} \times 273.15 \text{ K}}{1.00 \text{ atm}}$$

$$= 0.0463 \text{ L} \times \frac{1000 \text{ mL}}{1 \text{ L}} = 46.3 \text{ mL}$$

16.117 This is either electron capture or positron emission:

$$_4^7\text{Be} + _{-1}^0 e \longrightarrow _3^7\text{Li} \quad \text{(electron capture)}$$

$$_4^7\text{Be} \longrightarrow _3^7\text{Li} + _{+1}^0 e \quad \text{(positron emission)}$$

16.119 This is alpha decay:

$$_{95}^{241}\text{Am} \longrightarrow _{93}^{237}\text{Np} + _2^4\text{He}$$

16.121 (a) They are positrons; $4\,{}^{1}_{1}\text{H} \longrightarrow {}^{4}_{2}\text{He} + {}^{0}_{1}\text{e} + {}^{0}_{1}\text{e}$

(b) $(2.56 \times 10^{9}\,\text{kJ})\left(\dfrac{\text{mol octane}}{5509.0\,\text{kJ}}\right)\left(\dfrac{114.23\,\text{g octane}}{\text{mol octane}}\right)\left(\dfrac{1\,\text{mL octane}}{0.703\,\text{g octane}}\right)\left(\dfrac{1\,\text{gallon octane}}{3785.408\,\text{mL octane}}\right)$

$= 1.97 \times 10^{4}$ gallons of octane.

That's almost 20,000 gallons of octane that need to be combusted to give the same energy per mole of He formed.

(c) $E = mc^2$, so $m = \dfrac{E}{c^2} = \dfrac{(2.56 \times 10^{9}\,\text{kJ}\,)\left(\dfrac{10^{3}\,\text{J}}{\text{kJ}}\right)\left(\dfrac{\text{kg m}^2/\text{s}^2}{\text{J}}\right)}{(3.00 \times 10^{8}\,\text{m/s})^2}$

$= (2.84 \times 10^{-5}\,\text{kg})\left(\dfrac{10^{3}\,\text{g}}{\text{kg}}\right) = 2.84 \times 10^{2}\,\text{g} = 0.0284\,\text{g lost.}$

16.123 Fission of ${}^{4}\text{He}$ into two ${}^{2}\text{D}$ atoms would absorb energy as we are going from a more stable to less stable atoms on a binding energy per nucleons basis. This means mass would be gained, not lost. From the plot, we are going from an atom with a binding energy per nucleon of 7.0 MeV to two atoms with binding energy per nucleon of 2.8 MeV.

$$\begin{array}{ccccc}
28.0\ \text{MeV} & {}^{4}_{2}\text{He} & \longrightarrow & Z^{2}_{1}\text{D} & 11.2\ \text{MeV}\\
\text{total} & 4\ \text{nucleons} & & 4\ \text{nucleons} & \text{total}\\
\text{binding energy} & \text{at 7.0 MeV/nucleons} & & \text{at 2.8 MeV/nucleons} & \text{binding energy}
\end{array}$$

E (MeV)

$$\begin{aligned}
-11.2\ &Z^{2}_{1}\text{D}\\
&\text{total increase in energy is}\\
&\begin{aligned}28.0\ &\text{MeV}\\ -\ 11.2\ &\text{MeV}\\ \hline 17.0\ &\text{MeV}\end{aligned}\\
-28.0\ &{}^{4}_{2}\text{He}
\end{aligned}$$

Finally, $E = mc^2$, so $m = \dfrac{E}{c^2} = \dfrac{(17.0\,\text{MeV})\left(\dfrac{1.602 \times 10^{-13}\,\text{kg/m}^2\text{s}^2}{\text{MeV}}\right)}{(3.00 \times 10^{8}\,\text{m/s})^2}$

$= 3.03 \times 10^{-29}\,\text{kg} \times \dfrac{10^{3}\,\text{g}}{1\,\text{kg}} = 3.03 \times 10^{-26}\,\text{g}$

16.125 age $= \dfrac{-2.303 \log\left(\dfrac{\%\ \text{isotope remaining}}{100\%}\right) \times \text{half-life}}{0.693}$

So,

$\log\left(\dfrac{\%\ \text{isotope remaining}}{100\%}\right) = \dfrac{(\text{age})\,(0.693)}{(-2.303)\,(\text{half-life})} = -0.821$

and $\dfrac{\%\ \text{isotope remaining}}{100\%} = 10^{-0.821}$

So % isotope remaining $= 100\% \times 10^{-0.821} = 15.1\%$

Finally, $(3.00\,\text{g}\ {}^{132}_{55}\text{Cs})\,(0.151) = 0.453\,\text{g}$ remain after 150 days

Chapter 17

17.1 See solution in textbook.

17.3 The $C \equiv C$ bond is linear, which means you do not show a zigzag in that part of the molecule:

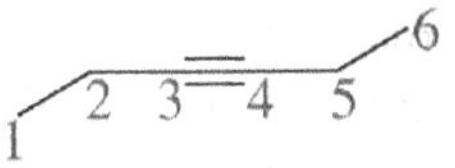

17.5 See solution in textbook.

17.7 The more H atoms in a hydrocarbon, the less unsaturated the molecule. Thus, (c) is the least unsaturated because it has eight hydrogens and the other molecules have six each:

(a) $H_3C - C \equiv C - CH_3$ (b) 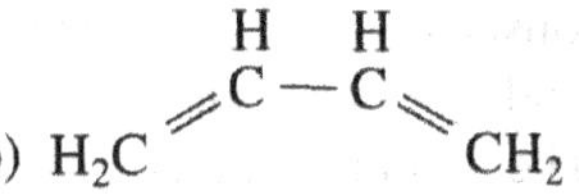 H_2C ... CH_2 (c) 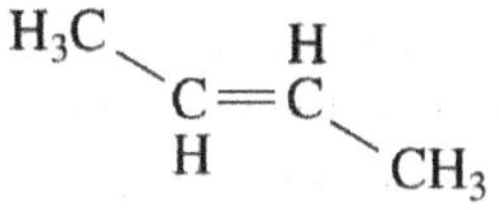 ... CH_3

17.9 *Oct-* means the main chain is eight C long. 2-*octene* means the chain has one double bond and it begins at carbon 2. There is a methyl group, $-CH_3$, on carbon 6 of the chain and a propyl group, $-C_3H_7$, on carbon 4 of the chain:

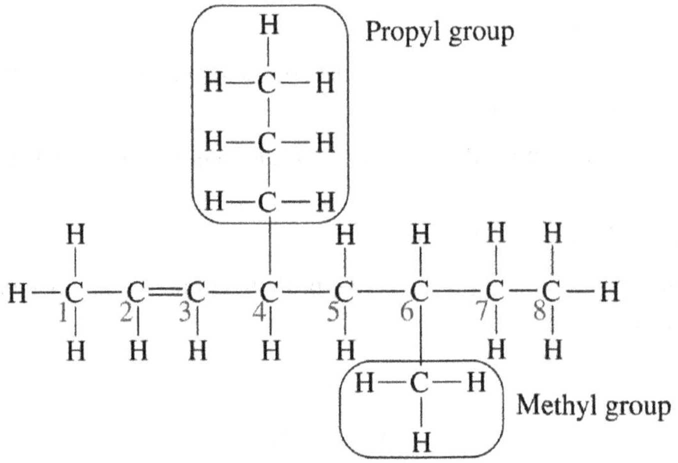

17.11 See solution in textbook.

17.13 2-Aminopropane

17.15 *Penta-*means five carbons in the main chain; 2-pentan*ol* means an alcohol group, $-OH$, on carbon 2; and there is a methyl group, $-CH_3$, on carbon 3:

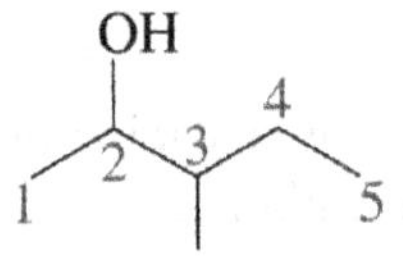

17.17 The ending -*one* signifies a ketone, and *ethan-* means two carbons. Ethanone does not exist because the carbonyl carbon in a ketone must be connected to two other carbons. Hence, the smallest number of carbons that can exist in a ketone is 3.

17.19 (a) The ending for a ketone is -*one*, and four carbons means the root name is butane. Use a number to indicate the position of the carbonyl group, and number the chain so that this number is as low as possible: 2-butanone.

(b) The aldehyde ending is *-al*, and five carbons means the root name is pentane: pentanal. No number is needed because the aldehyde functional group must be at one end of the molecule in order for the carbonyl carbon to be bonded to an H. The lowest-number-possible rule automatically makes the carbonyl carbon C 1.

(c) Carboxylic acids end with *-oic acid,* and a four-carbon main chain makes the parent butane. Use numbers to indicate positions of any branches, and number the main chain to make these numbers as low as possible: 2-methyl-butanoic acid. Numeral 1 before *butanoic* is not needed because any carboxylic acid carbon must be a terminal carbon and the lowest-possible-number rule automatically makes the carboxylic acid carbon C 1.

17.21 Diamond, graphite, and fullerene. Each carbon atom in diamond is tetrahedrally bound to four other carbon atoms. Graphite consists of sheets of hexagons of carbon atoms where each carbon atom is bound to three others. Fullerene consists of a sphere of carbon atoms where each carbon atom is bound to three others. The similarity in the allotropes is that, in all three of them, each carbon atom always has four bonds to it.

17.23

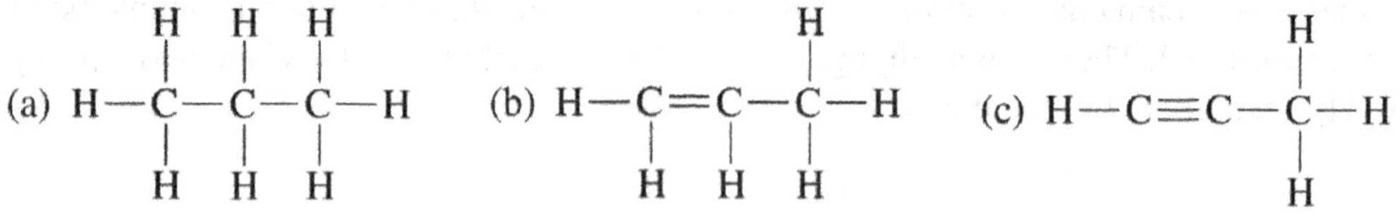

17.25 There is no possibility of double or triple bonds in the molecule because only carbon atoms having all single bonds can have a tetrahedral geometry.

17.27 Organic chemistry, so called because carbon-based molecules are the major molecules found in organisms.

17.29 A compound composed solely of carbon and hydrogen.

17.31 Start with a five-carbon main chain, which means there is one branch and it contains one carbon. The branch can be on carbon 2 or carbon 3:

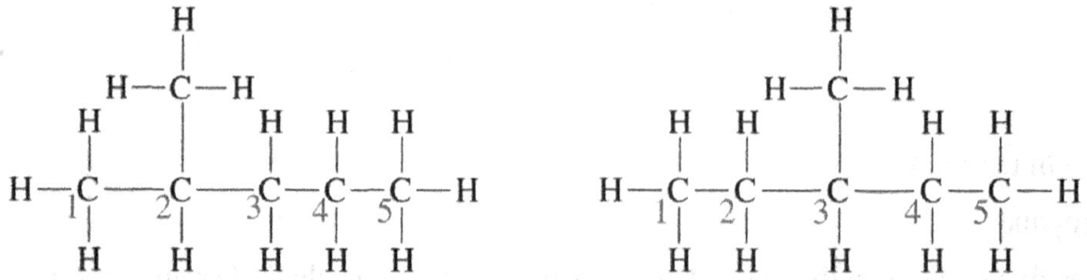

Note that placing the —CH₃ on carbon 4 forms the mirror image of the structure on the left and so is not a third possibility.

Now make a four-carbon main chain, which leaves two carbons for branching, meaning two methyl groups. You may put one on carbon 2 and one on carbon 3 or both on carbon 2:

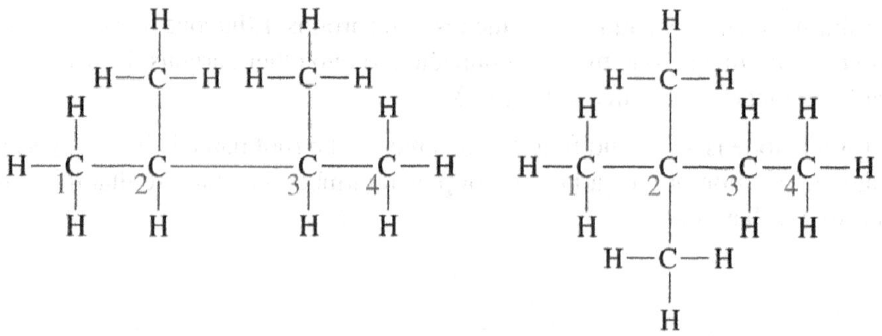

You should understand why forming an ethyl group, —CH_2CH_3, from the two carbons available for branching is not a valid choice. Doing so gives this molecule,

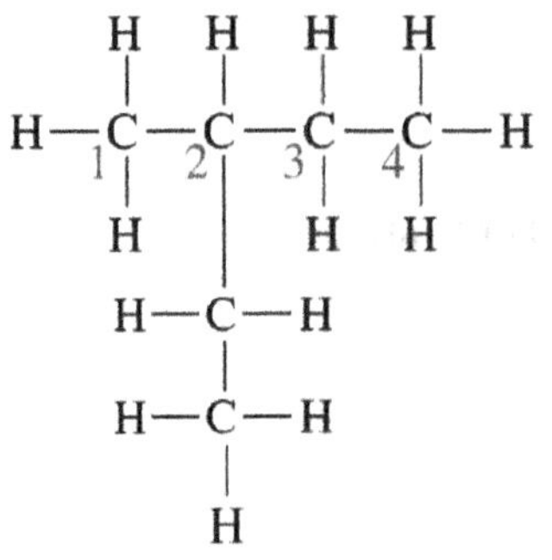

but the numbering is incorrect because the main chain is not the longest. Correctly numbered, this molecule is 3-methylpentane, which you drew earlier.

These are all the possibilities. If you go to a three-carbon main chain, any branching you add will bring you back to one of the four-carbon or five-carbon main chains already shown. Try it.

17.33 The geometry about C—C single bonds is tetrahedral, with bond angles of 109.5°. Thus, a zigzag line drawing represents the true shape of the molecule more accurately than a straight-line drawing does. In a straight-line drawing, the angles appear to be 180°.

17.35 Crude oil (petroleum). Combustion of hydrocarbons leads to the formation of carbon dioxide, CO_2—a green house gas, linked to the global warming. Burning this cheap source of fuel with little restraint both pollutes the environment and depletes resources for future generations.

17.37 (a) Seven carbons:

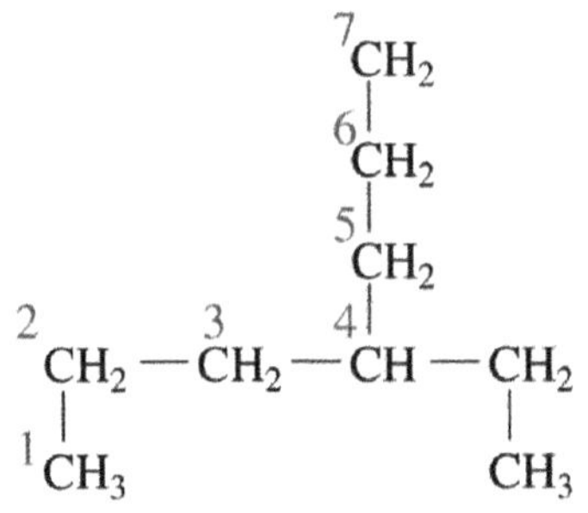

Any other counting sequence gives you a shorter main chain, and in saturated hydrocarbons the main chain is the longest one.

(b) Two carbons.

17.39 (a) Each free end represents one C, and each intersection of two lines represents one C:

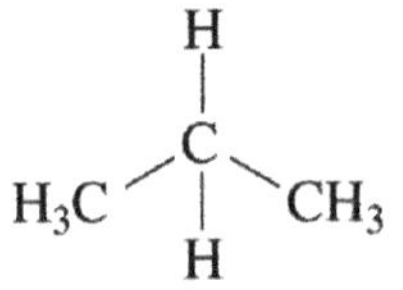

C_3H_8, linear because no C atom is bonded to more than two other C atoms.

(b)

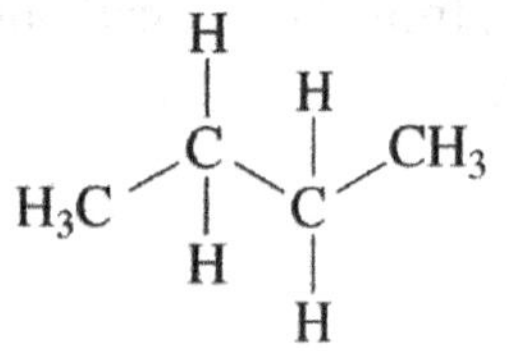

C_4H_{10}, linear because no C atom is bonded to more than two other C atoms.

(c)

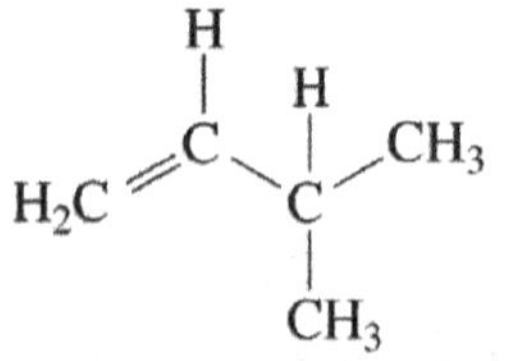

C_5H_{10}, branched because one C atom is bonded to three other C atoms.

17.41

(a)  (b) 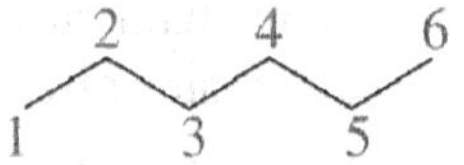 (c)

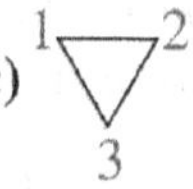

17.43 (c), because it contains a C=C double bond and therefore must contain fewer than the maximum number of H atoms.

17.45 (a) The structural formula corresponding to this line drawing is:

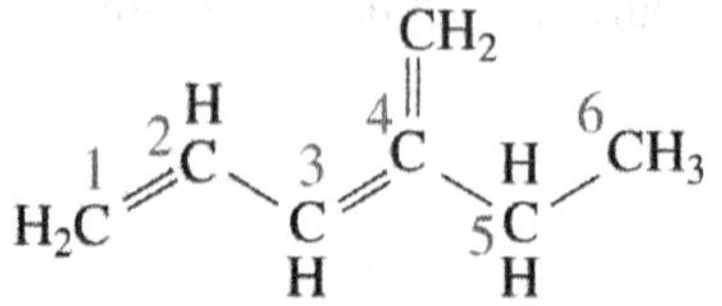

C4 forms five bonds in this drawing, which is not possible.

(b) Remove an H from C2 and an H from C3 to form:

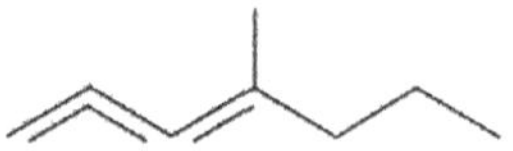

or remove an H from C5 and an H from C6 to form:

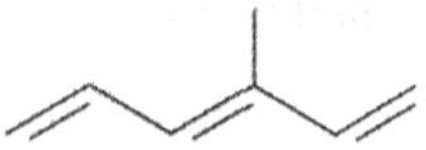

17.47 Unsaturated; the maximum number of hydrogen atoms for a seven-carbon chain is $(2 \times 7) + 2 = 16$. This compound has two fewer than that and is therefore unsaturated.

17.49 Begin with the six-carbon unbranched chain:

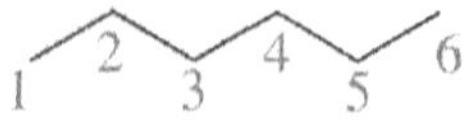

Then draw the five-carbon main chain with the one-carbon branch in all possible locations:

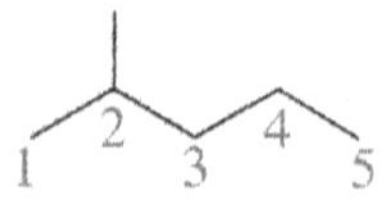

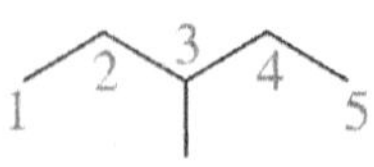

Then go to the four-carbon main chain. Here the branching possibilities are two one-carbon branches on one carbon and two one-carbon branches on different carbons:

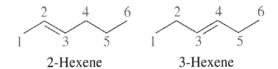

17.51 No, because they do not have the same molecular formula.

17.53 Molecules 1 and 3 because they have the same molecular formula, C_8H_{16}. The molecular formula of molecule 2 is C_7H_{14}, and that of molecule 4 is C_8H_{14}.

17.55 Long-chain hydrocarbons. The word comes from the Greek word *aleiphas*, meaning fat.

17.57 International Union of Pure and Applied Chemistry.

17.59 The general formula for alkanes is C_nH_{2n+2}, and that for alkenes is C_nH_{2n}. This means that, for a given number of carbons, an alkene contains two fewer hydrogens than an alkane.

17.61 *Eth-* means two carbons, *-yne* means a $C\equiv C$ triple bond: $H{:}C{:}{:}{:}C{:}H$ $H-C\equiv C-H$. Ethene is a two-carbon molecule (*eth-*) containing one $C=C$ double bond (*-ene*). Thus, acetylene is more unsaturated because a triple bond is more unsaturated than a double bond.

17.63 Alkane because it fits the general formula C_nH_{2n+2}.

17.65 No; although the three compounds all have three carbons, they belong to different homologous series. Propane belongs to the alkanes, propene to the alkenes, and propyne to the alkynes.

17.67 Each compound has six carbons (*hex-*) and one double bond (*-ene*). The 2 and 3 tell you the position of the double bond is different in the two compounds:

2-Hexene   3-Hexene

Because they have the same molecular formula, C_6H_{12}, but different structural formulas, these two molecules are isomers.

17.69 False. We number unsaturated and branched hydrocarbons from the end that is closer to the multiple bond or branch. One example is:

2-Methylbutane

and two more examples are the molecules drawn in the solution to Problem 17.68.

17.71 Each line end represents one C atom, and each point where two lines intersect represents one C atom.

(a) 3, propane.

(b) , 2-hexyne.

(c) , 2-pentene.

(d) , 1-butene.

17.73 Isomers have the same molecular formula but different structural formulas. Pair 1: (d) and (e); both are linear but the position of the triple bond is different. Pair 2: (c) and (f); both have six carbon atoms and are saturated; (c) is linear, while (f) is branched.

17.75 The numbering is incorrect. The main chain should be numbered starting at the end closer to the double bond because multiple bonds have priority over branches.

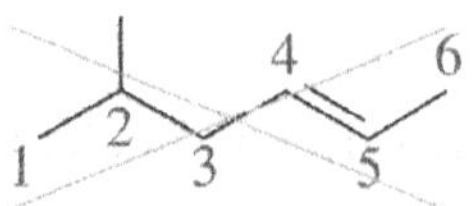

2-Methyl-4-hexene   5-Methyl-2-hexene
WRONG   CORRECT

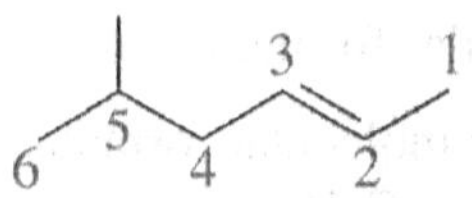

17.77 In each case, get the molecular formula by first drawing the longest chain, indicated by the parent name; then adding side chains as specified; then adding H as needed to have each C form four bonds.

(a) *Pentane* means five carbons in the main chain:

C—C—C—C—C

Add a methyl branch on C3:

C—C—C—C—C
        |
       CH_3

Add H to C as needed:

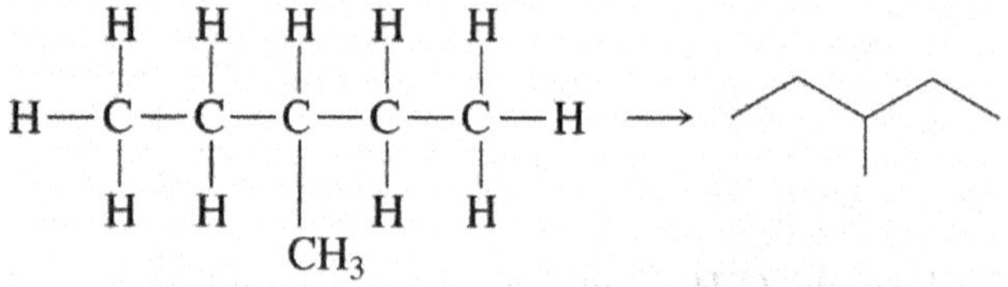

C_6H_{14}

(b) The *2-butene* means four carbons (*but-*) in the main chain and a C=C bond (*-ene*) between C2 and C3. The *2,3-dimethyl* tells you C2 and C3 each carry a methyl group:

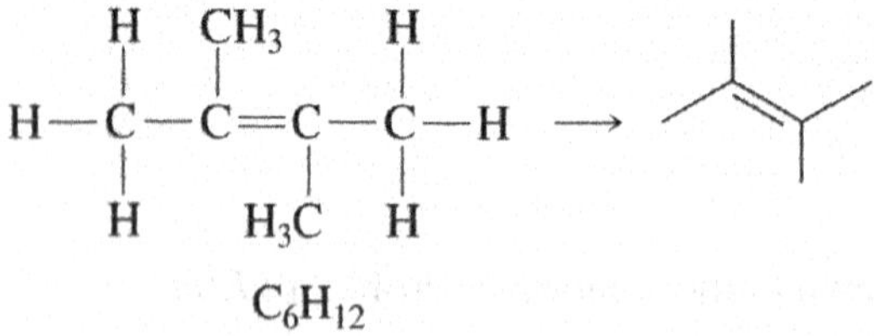

C_6H_{12}

(c) The *1-pentene* means a five-carbon (*pent-*) main chain with a C=C bond (*-ene*) between C1 and C2. The *3-methyl* tells you C3 carries a methyl group:

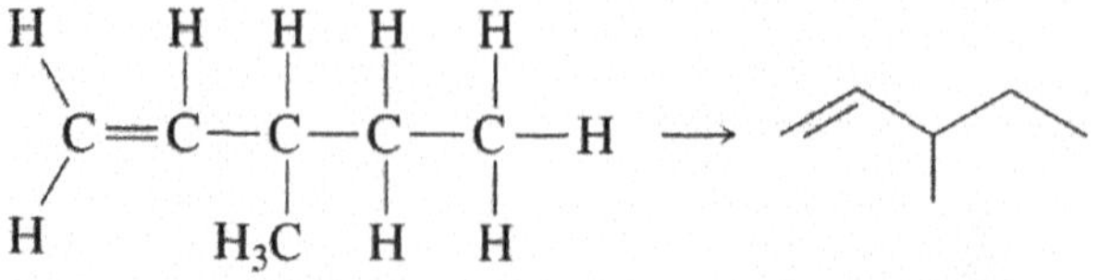

C_6H_{12}

(d) The *1-hexyne* means a six-carbon (*hex-*) main chain with a C≡C bond (*-yne*) between C1 and C2. The *5-methyl* tells you C5 carries a methyl group:

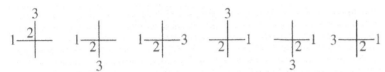

$$C_7H_{12}$$

17.79 To answer this question, you need to identify a main chain:

(a) Any way you number the carbon atoms, the carbon chain is three-carbons long:

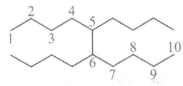

In all cases, there are two branches, both on C2, and both are methyl. The name of the molecule is 2,2-dimethylpropane.

(b) Once again, many ways of numbering the carbon atoms exist and below is just one of them.

There are two branches, both are butyl groups, one on C5 and another one on C6. The name of the molecule is 5,6-dibutyldecane.

17.81 False. If a multiple bond is present, the longest chain that contains the multiple bond is the main chain. This may or may not be the longest chain in the molecule.

17.83 As alkanes get larger, the intermolecular attractive forces get progressively stronger because there can be more and more of them in larger molecules, leading to a steady increase in boiling point.

17.85 The portion of a functionalized hydrocarbon that gives the hydrocarbon its unique properties.

17.87 By putting the prefix *amino-* in front of the name of the parent hydrocarbon and, in molecules containing more than two carbons, using a number to indicate the position of the amino group. For example, $CH_3CH_2NH_2$ is aminoethane and $CH_3CH=CHCH_2CH_2CHCH_2CH_3$ is 6-amino-
                                                                                                    |
                                                                                                   NH_2

2-octene (in numbering carbons, double bond takes precedence over amino group).

17.89 Number each chain to give the functional group as low a number as possible.

(a) 2-chlorohexane

(b) 2-bromopropane

(c) 2,4-dichloro-3-ethylhexane (*not* 3,5-dichloro-4-ethylhexane)

17.91 Number each chain to give the functional group as low a number as possible.

(a) 2-hexanol

(b) 3-aminopentane

(c) 3-hexanol

17.93 Because they destroy the ozone layer when they rise into the upper atmosphere.

17.95 By reacting it with a strong base via the reaction $R-Cl + OH^- \rightarrow R-OH + Cl^-$.

17.97 One H must go with the O to form $-OH$, leaving nine H to bond with C atoms. This many H means no unsaturation in the hydrocarbon chain. Begin with a four-carbon chain and all possible positions for the $-OH$, on C1 and on C2:

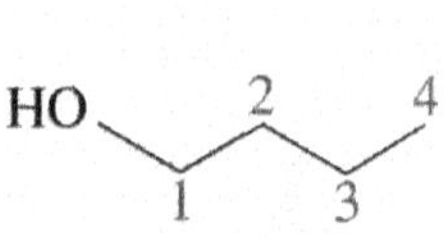

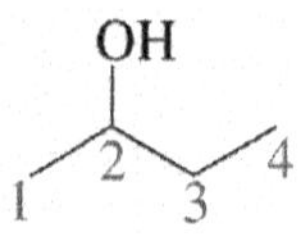

Note that when a functional group is present, the end of a line does *not* represent a C:

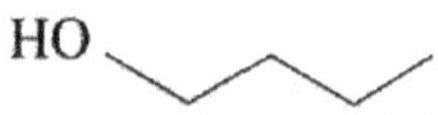

HO is $HO-CH_2CH_2CH_2CH_3$, *not* $HO-CH_2CH_2CH_2CH_2CH_3$.

Next form a three-carbon chain and all possible positions for the $-OH$ on C1 and C2:

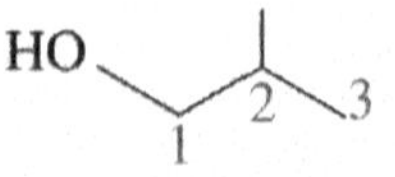

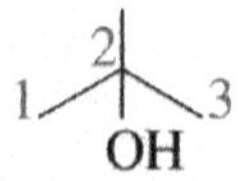

17.99 The electronegative O and Cl atoms in the acid chloride leave the carbonyl C atom partially positive (δ^+). All amines have partially negative (δ^-) N atom due to the large electronegativity of nitrogen. The δ^- N and δ^+ C atoms attract one another. In addition, halide atoms, such as Cl, generally leave easily, making room for the amine.

17.101 Because there is no C at the two line endings adjacent to O, the molecular formula is $C_4H_{10}O$. Because the molecule fits the general formula $R-O-R$, it is an ether–diethyl ether.

17.103

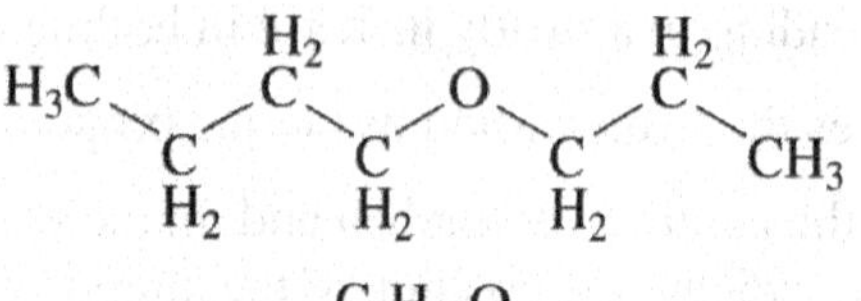

$C_7H_{16}O$

17.105

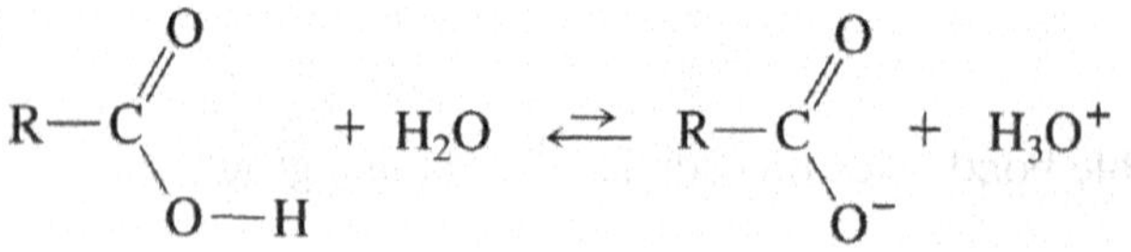

The equilibrium lies to the left, implying that most of the acid is undissociated and that RCOOH molecules are weak acids.

17.107 One of the H₂O hydrogens migrates to the amine and forms a covalent bond with the lone pair of electrons on N:

$$R-\overset{..}{N}-H + HOH \;\rightleftharpoons\; \left[R-\overset{\overset{\displaystyle H}{|}}{\underset{\underset{\displaystyle H}{|}}{N}}-H \right]^{+} + OH^{-}$$

The equilibrium lies to the left, which implies that a primary amine is a weak base in water.

17.109

NH₂

(chain numbered 1 2 3 4 5 6)

This is a primary amine because the N is bonded to two H and one R group (the hexane chain).

17.111

H H H
| | |
Cl—C—C—C—H
| | |
H H H

1-Chloropropane
(halogen-functionalized
hydrocarbon)

H H H
| | |
HO—C—C—C—H
| | |
H H H

1-Propanol
(alcohol)

H H H
| | |
H—C—C—C—NH₂
| | |
H H H

1-Aminopropane
(primary amine)

H O H
| ‖ |
H—C—C—C—H
| |
H H

2-Propanone
(ketone)

H H O
| | ‖
H—C—C—C
| | \
H H OH

Propanoic acid
(carboxylic acid)

H H O
| | ‖
H—C—C—C
| | \
H H H

Propanal
(aldehyde)

17.113 (a)

H H
| |
H—C—C—Ö—H
| ‖
H :O:

(b) The H—O bond is very polar and it gets relatively easily pulled apart by polar water molecules. The products of acetic acid dissociation in water are H₃O⁺ and CH₃COO⁻, both of which are stabilized in water due to the ion–dipole intermolecular interactions. On the other hand, the C—H bonds are negligibly polar and do not break easily, especially in polar environments, such as water.

(c) Acetic acid is found in vinegar. You are likely to encounter it in any kitchen or in a salad.

17.115 Aldehydes, H—$\overset{\overset{\displaystyle O}{\|}}{C}$—R; ketones, R—$\overset{\overset{\displaystyle O}{\|}}{C}$—R; and carboxylic acids, R—$\overset{\overset{\displaystyle O}{\|}}{C}$—OH.

17.117 Methanol dissolves readily in water because it possesses an —OH that forms a hydrogen bond to a water molecule. This helps in the solute-separation step by replacing some of the water-to-water hydrogen bonds that must be broken to create a "hole" in the water for the methanol molecules. Because hydrocarbons lack any functional group that can aid in the same way, hydrocarbons are insoluble in water.

17.119 (a) Because of the —OH group, which can hydrogen-bond with water molecules. Compound (b) is a hydrocarbon, essentially nonpolar and therefore insoluble in water. Compound (c) has an —OH group but also a large, nonpolar hydrocarbon chain. The solvent "hole" needed for this molecule is much larger than for (a) and would require breaking many more solvent–solvent hydrogen bonds.

17.121 If the molecular formula were C_4H_{10}, you would recognize this as a saturated hydrocarbon fitting the general formula C_nH_{2n+2}. The Br replaced one H, which means the molecule is indeed a saturated hydrocarbon. Begin with the longest possible chain and the halogen in every possible position:

1-Bromobutane   2-Bromobutane

Next do the three-carbon chain with the Br in all possible locations, listing the functional group and branch alphabetically in the name:

1-Bromo-2-methylpropane   2-Bromo-2-methylpropane

17.123 (a) Isomers because they both have the molecular formula C_3H_8O.

(b) Identical. Both are the three-carbon alcohol 1-propanol.

(c) Isomers because they both have the molecular formula C_4H_6.

(d) Completely unrelated (C_5H_{10} and C_5H_{12}).

(e) Completely unrelated (C_6H_8 and C_5H_8). Remember that, in the molecule on the left, there is a —CH_3 group at the end of the short line sticking off the ring.

17.125 (a) To make the molecule polar, place both electronegative Cl atoms at the same end. The small number of H tells you there is unsaturation in the molecule. (Assume each Cl replaced an H, meaning original formula must have been $C_4H_8 = C_nH_{2n}$ = alkene.) Some possible polar molecules are:

(b) To make the molecule nonpolar, draw it symmetrically. Place the Cl atoms at opposite ends so that the two δ^- cancel each other, and place the double bond between C2 and C3:

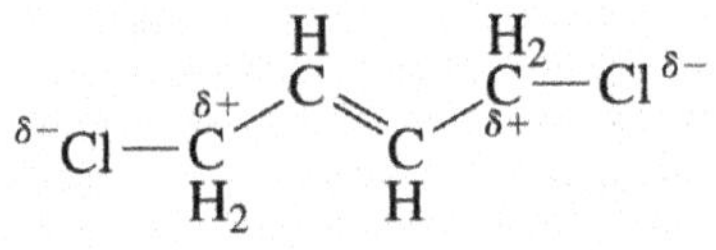

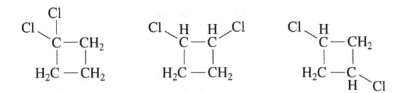

No molecular dipole moment

(c) Only in a ring structure can a molecule with this molecular formula have single C—C bonds alone:

17.127 (a) C_6H_5Cl can have only one isomer because the benzene ring is planar and symmetrical. Therefore all six carbons are equivalent, and it makes no difference which one carries the Cl.

(b) $C_6H_2Cl_2$ can have three isomers:

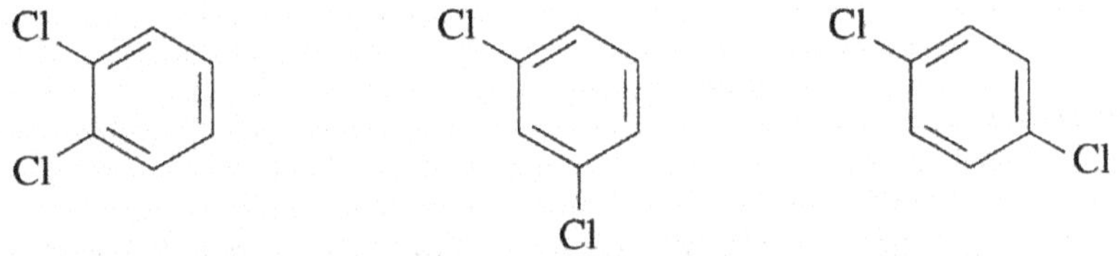

1,2-Dichlorobenzene   1,3-Dichlorobenzene   1,4-Dichlorobenzene

(There is no 1,5 or 1,6 isomer because 1,5 is the same as 1,3 and 1,6 is the same as 1,2.)

(c) $C_6H_2Cl_4$ has three isomers:

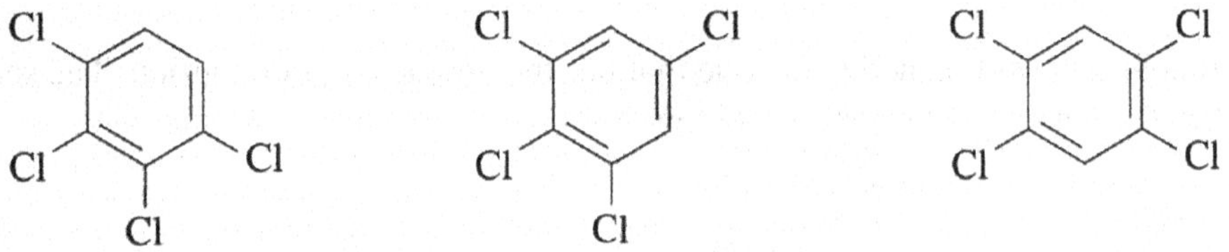

1,2,3,4-Tetrachlorobenzene 1,2,3,5-Tetrachlorobenzene 1,2,3,6-Tetrachlorobenzene

17.129 (a) Glycerol is an alcohol (actually, a tri-alcohol or tri-ol), and it is completely soluble in water due to the fact that its hydrocarbon (nonpolar) portion is small and it has three OH groups, all capable of hydrogen bonding to water.

(b) The product is the sodium salt or the fatty acid and water.
$$RCOOH + NaOH \rightarrow RCOO^-Na^+ + H_2O$$

(c)

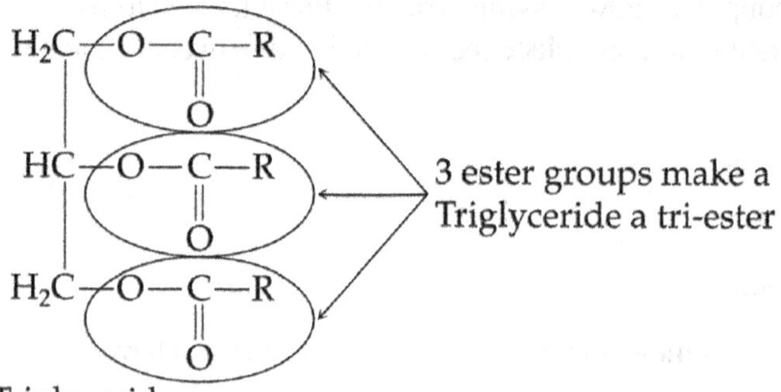

Triglyceride

3 ester groups make a
Triglyceride a tri-ester

17.131 (a) Because it has a sulfur atom attached to the amide portion:

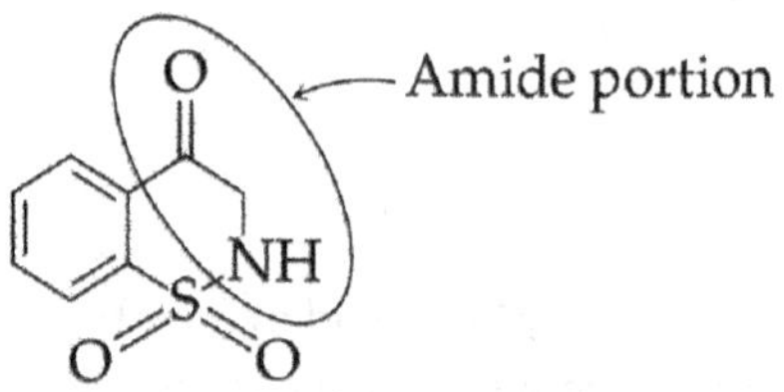

Amide portion

(b) $C_7H_5NO_3S$

Chapter 18

18.1 See solution in textbook.

18.3 (a) $H—C\equiv C—H$

(b) The two electrons in one pair of the $C\equiv C$ triple bond unpair, one electron going to each C
and making each available to form a single $C—C$ bond:

18.5 The first amino acid named is the one that has the unreacted amino group. The second is the one
that has the unreacted carboxylic acid group.

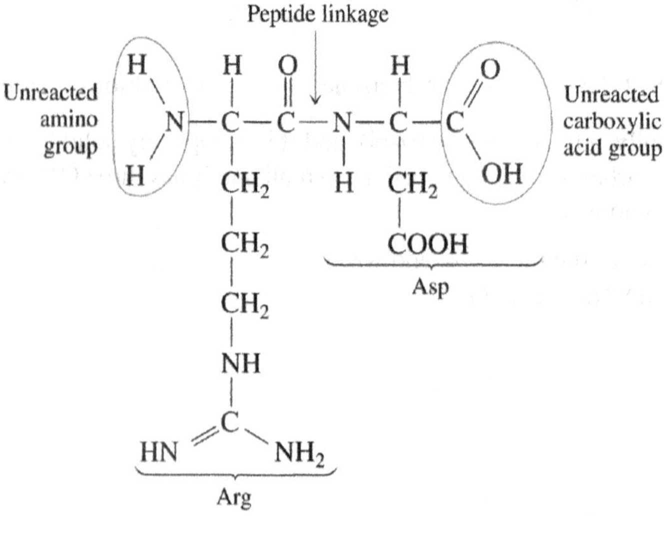

18.7 A polymer is a long-chain molecule consisting of many repeating units of some monomer. A monomer is a relatively small molecule, a "building block," from which a polymer is made.

18.9 A monomer is a small molecule used to produce a polymer. A monomer unit is a monomer once it has become part of a polymer. An example of a monomer is the compound tetrafluoroethene, $F_2C{=}CF_2$. Its monomer unit is the molecule once two electrons of the double bond have unpaired and the unit has been incorporated into a polymer:

18.11 Ethylene is $H_2C{=}CH_2$, and two electrons from its double bond split up:

Each C now having a lone electron, it can form a covalent bond with a lone electron on another monomer:

Because the dimer has a lone electron at each end, the chain can keep adding a monomer at each end.

18.13 (a) Choose an arbitrary position for the opening parenthesis:

Then position the closing parenthesis at the same location on the immediately adjacent monomer unit:

The formula for the monomer unit is C_4H_8.

(b) The carbons in the polymer backbone are the ones that originally had a double bond, which means the monomer must have been:

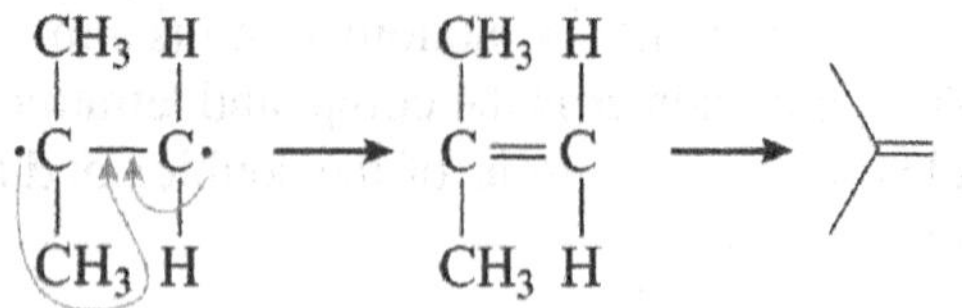

(c) Monomer 1-methylpropene, polymer poly(1-methylpropene).

(d)

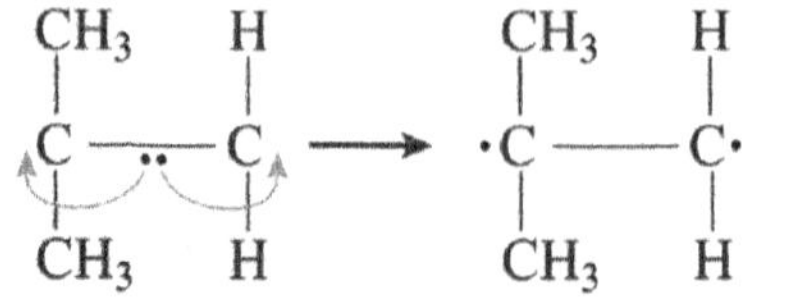

18.15 Teflon is made by the polymerization of tetrafluoroethene:

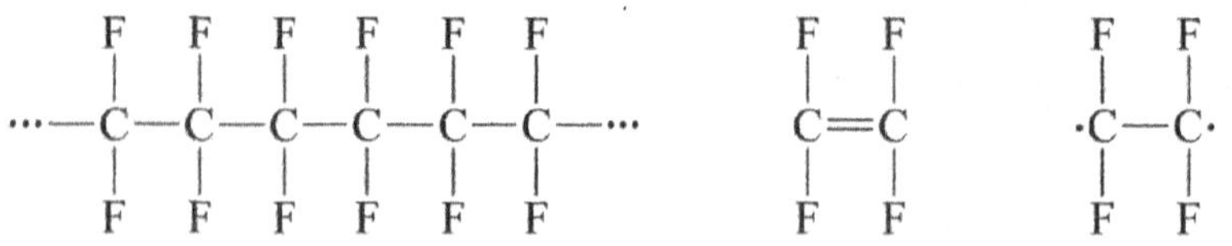

Polymer   Monomer   Free Radical

18.17 Ethylene is a small, nonpolar molecule, and therefore intermolecular attractive forces are very weak; this causes the compound to have a relatively low boiling point. Polyethylene is a huge macromolecule, and even though the molecule is nonpolar, its huge size affords many opportunities for intermolecular attractions to form, resulting in a relatively high boiling point.

18.19 A bond between a carbonyl carbon and an amine nitrogen.

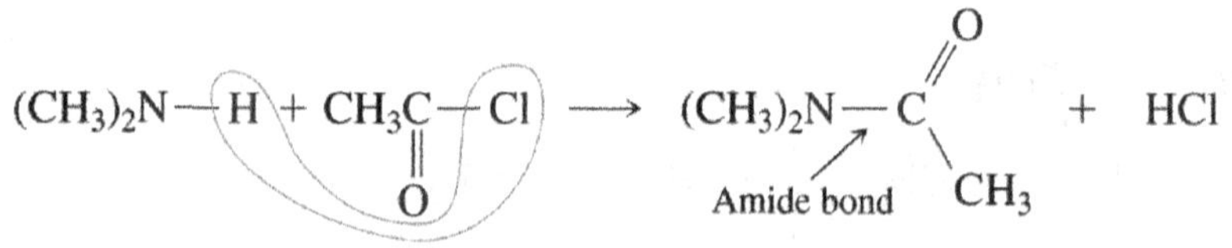

18.21 Hydrogen bonds, as shown below for Nylon 66:

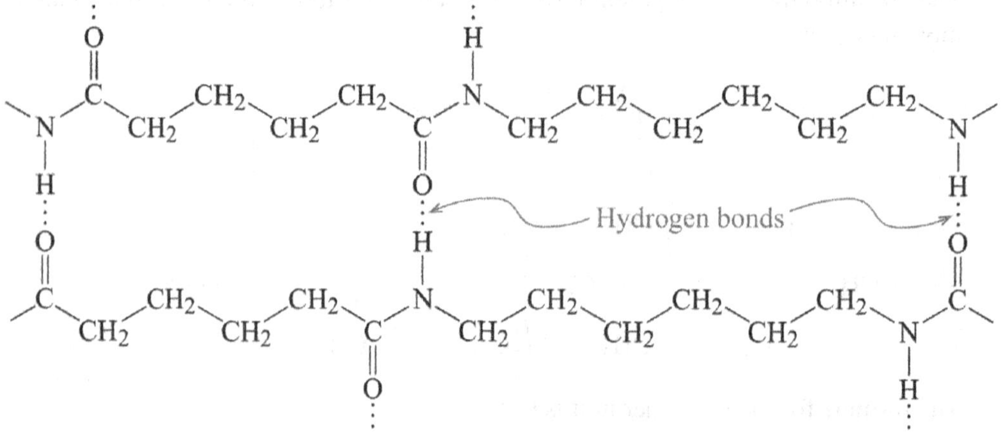

18.23 A monosaccharide is a sugar consisting of only one sugar ring; glucose is an example. A disaccharide is a sugar consisting of two sugar rings bonded together; sucrose is an example. A polysaccharide is a polymer consisting of a large number of sugar rings bonded together; starch and cellulose are examples.

18.25 Cellulose because it forms long linear strands of β-glucose combined into strong hydrogen-bonded fibers. It is also insoluble and indigestible by a vast majority of animals. Starch, on the other hand, is made of shorter, curved chains of α-glucose with little to no hydrogen bonding. Starch does not have the strength needed for cell wall building. It also partially dissolves in water and is digestible.

18.27 They are both polymers of glucose monomer units. They differ in the orientation of the linkage between the glucose units, a difference that makes the three-dimensional shape of cellulose very different from that of starch.

18.29 Glycine, like all the other α-amino acids, has the amino group attached to the α-carbon:

Glycine   β-Alanine

18.31 Any one of the ten amino acids the human body needs but cannot synthesize and therefore must obtain from food. See Table 18.1 of the textbook for their names.

18.33 A short- or medium-length chain of amino acids. Polypeptides and proteins differ only in the length of their amino-acid chains—short- and medium-length chains are called polypeptides; long chains are called proteins.

18.35 The peptide with the unreacted amino group is named first. When the amino group of alanine reacts with the carboxylic acid group of glycine, you get Gly–Ala. When the amino group of glycine reacts with the carboxylic acid group of alanine, you get Ala–Gly.

Gly–Ala   Ala–Gly

18.37 Among other properties, the sequence of amino acids determines the three-dimensional structure of a protein, which in turn determines how the protein functions in the body. So, yes, swapping just two amino acids or replacing one with another can affect its biological functions. Sickle-cell anemia is an example of a disease caused by such a single amino-acid replacement in a protein chain in hemoglobin. Normal hemoglobin molecules contain an amino acid that has a $-CH_2CH_2COOH$ chain, and the carboxylic acid group contributes to hemoglobin's solubility in water. In sickle-cell anemia patients, this chain is replaced by the nonpolar $-CH(CH_3)_2$ group, lowering solubility of the hemoglobin and impeding normal blood flow.

18.39 Alternating sugar molecules, called deoxyriboses, and phosphate groups.

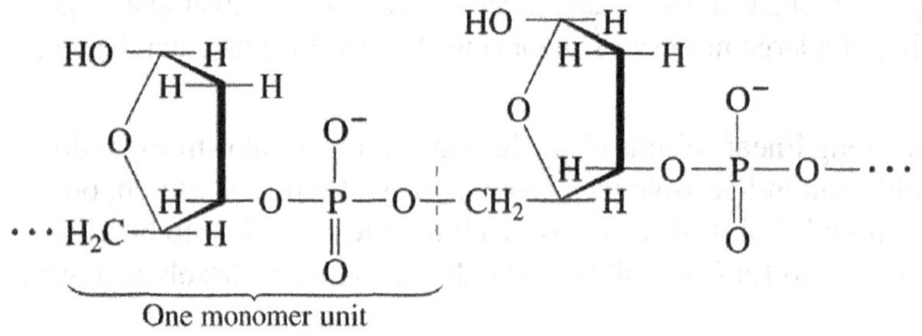

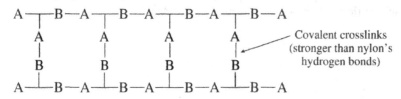

One monomer unit

18.41 Four; adenine (A), guanine (G), cytosine (C), thymine (T).

18.43 A sequence of three adjacent DNA bases; the function is to code for the incorporation of a specific amino acid into a growing protein chain.

18.45 Viruses may allow physicians to incorporate portions of foreign DNA into a cell for the good of the cell. For example, once loaded into a virus cell, the gene for making human insulin might be incorporated into the cells of a diabetic patient.

18.47 In order for the interchain attractions to be stronger than those in nylon, they should be covalent bonds. Therefore, design monomers that can form covalent bonds even after they become monomer units in a polymer. The polymer chains will consist of A—B bonds,

A—B—A—B—A—B

but to get covalent cross-linking, each chain must also have some A and/or B free to cross-link. The two different monomers fitting this bill are:

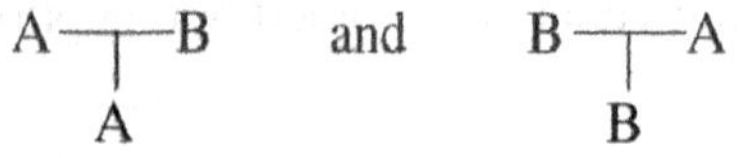

A⊤B   and   B⊤A
 A   B

and the polymer they create is:

```
A⊤B—A⊤B—A⊤B—A⊤B—A
 A     A     A     A         Covalent crosslinks
 |     |     |     |         (stronger than nylon's
 B     B     B     B         hydrogen bonds)
A⊥B—A⊥B—A⊥B—A⊥B—A
```

18.49 (a) Deoxyribonucleic acid (DNA).

        O
        ‖
(b)    ——N is the amide linkage when you are talking about nonbiological polymers and the peptide linkage when you are talking about biological polymers. The shaded boxes represent arbitrary monomer units, either hydrocarbons or amino acids. Therefore, the structure represents both an amide polymer, such as nylon, and a polypeptide.

(c) Polysaccharide, because the ring structures are sugars.

18.51 Aspartic acid and phenylalanine (the carboxylic acid end of which has been modified to $COOCH_3$).

18.53 You have two choices for the first amino acid: Ala − x − x − x, Gly − x − x − x. Because there are two Ala to begin with, you have two choices for the second amino acid: Ala − Ala − x − x, Ala − Gly − x − x, Gly − Ala − x − x, Gly − Gly − x − x. Depending on which amino acid was used for the second one, you have these choices for the third: Ala − Ala − Gly − x, Ala − Gly − Ala − x, Ala − Gly − Gly − x, Gly − Ala − Ala − x, Gly − Ala − Gly − x, Gly − Gly − Ala − x. And the choices for the final amino acid depend on which one amino acid remains in each case: Ala − Ala − Gly − Gly, Ala − Gly − Ala − Gly, Ala − Gly − Gly − Ala, Gly − Ala − Ala − Gly, Gly − Ala − Gly − Ala, Gly − Gly − Ala − Ala. Thus, six tetrapeptides can be made from two of one amino acid and two of another.

18.55 The equation for freezing point depression is given in Section 12.9 of the textbook:

$$\Delta T_f = \frac{K_f \times \text{Moles Solute Particles}}{\text{Kilograms Solvent}}$$

and Table 12.2 of the textbook lists 1.86 °C · kg solvent/mol solute as the value of K_f for water. Proteins do not dissociate into ions when they dissolve in water, which means moles solute particles = moles protein. Thus,

$$\text{Moles Solute Particles} = \text{Moles Protein} = \frac{\Delta T_f \times \text{Kilograms solvent}}{K_f}$$

$$= \frac{3.72 \times 10^{-4}\ °C \times 1\ kg}{1.86\ °C \cdot kg/\text{mol solute}} = 0.000200\ \text{mol solute}$$

$$\text{Molar mass} = \frac{10.0\ g}{2.00 \times 10^{-4}\ \text{mol}} = 5.00 \times 10^4\ g/\text{mol}$$

18.57

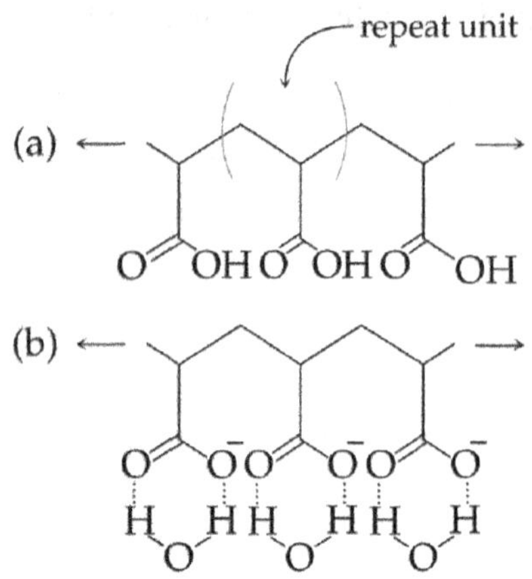

(a) repeat unit

(b)

18.59

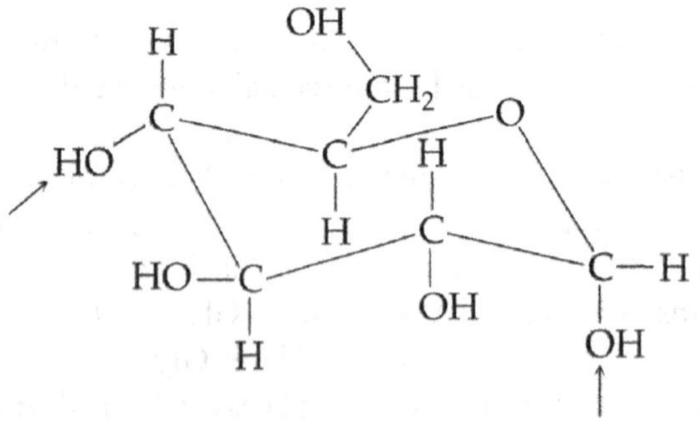

Examine the starch polymer and you will see that the two OH groups used to form glycosidic linkages are the ones indicated with the arrow. Each glycosidic linkage formed produces a molecule of H_2O.

18.61 Adenine and thymine possess adjacent C—H and C—O functions. The H atom in the C—H bond is not capable of hydrogen bonding because it is not bound to a F, O, or N atom.

18.63 First determine the trypsin possibilities. Because there is no arginine (Arg) in either fragment, you care only that trypsin breaks peptide linkages where the COOH part of the bond comes from lysine (Lys). The trypsin fragments therefore could have come from Ala − Gly − Trp − Gly − Lys*Thr − Val − Lys, where the star indicates the broken bond, or from Thr − Val − Lys*Ala − Gly − Trp − Gly − Lys.

Whichever is the original octapeptide, it also reacted with chymotrypsin, which works on tyrosine (Tyr), phenylalanine (Phe), and tryptophan (Trp), with tryptophan being the only one of concern here. If the original peptide was the one beginning with Ala, cleavage by chymotrypsin would give Ala − Gly − Trp and Gly − Lys − Thr − Val − Lys. If the original peptide was the one beginning with Thr, cleavage by chymotrypsin would give Thr − Val − Lys − Ala − Gly − Trp and Gly − Lys. These are indeed the fragments, telling you the octapeptide was the second of the two possibilities you deduced from your trypsin information: Thr − Val − Lys − Ala − Gly − Trp − Gly − Lys.